Mutual Coupling Between Antennas

Mutual Coupling Between Antennas

Edited by

Trevor S. Bird
Distinguished Visiting Professor
University of Technology Sydney & Principal Antengenuity
Sydney, Australia

WILEY

This edition first published 2021

Registered Offices
John Wiley & Sons, Inc., 111 River Street, Hoboken, NJ 07030, USA
John Wiley & Sons Ltd, The Atrium, Southern Gate, Chichester, West Sussex, PO19 8SQ, UK

Editorial Office
The Atrium, Southern Gate, Chichester, West Sussex, PO19 8SQ, UK

For details of our global editorial offices, customer services, and more information about Wiley products, visit us at www.wiley.com.

Wiley also publishes its books in a variety of electronic formats and by print-on-demand. Some content that appears in standard print versions of this book may not be available in other formats.

Library of Congress Cataloging-in-Publication Data Applied

Names: Bird, Trevor S., editor.
Title: Mutual coupling between antennas / edited by Trevor S. Bird, distinguished visiting professor, University of Technology Sydney & principal, Antengenuity, Sydney, Australia.
Description: First edition. | Hoboken, NJ, USA : Wiley, 2021. | Includes bibliographical references and index.
Identifiers: LCCN 2020032113 (print) | LCCN 2020032114 (ebook) | ISBN 9781119564980 (hardback) | ISBN 9781119564881 (Adobe pdf) | ISBN 9781119564973 (epub)
Subjects: LCSH: Antenna arrays. | Electromagnetic interference. | Coupled mode theory.
Classification: LCC TK7871.67.A77 M88 2021 (print) | LCC TK7871.67.A77 (ebook) | DDC 621.382/4–dc23
LC record available at https://lccn.loc.gov/2020032113
LC ebook record available at https://lccn.loc.gov/2020032114

Cover Design: Wiley
Cover Image: Illustration courtesy of Trevor S. Bird; Supersmario/Getty Images; moorsky/Getty Images

Set in 9.5/12.5pt STIXTwoText by Straive, Chennai, India

10 9 8 7 6 5 4 3 2 1

To my parents Norma May Bird (née Beckham) and Roy Murray Bird for their love and teaching, and for always encouraging me in anything worthwhile I attempted.

Contents

Preface

The genesis of this book dates back to the early 1980s. After some early work on the topic of mutual coupling in waveguide arrays, one of my mentors, the late Professor Peter J.B. Clarricoats, FRS of Queen Mary, University of London, inspired me to write a monograph on the topic. The late Dr. Bob (R.C.) Hanson also encouraged me through interactions on this and other topics right up to his passing in 2018. The concept grew to a book proposal in the late 1980s, and drafts of several chapters had been prepared by the early 1990s. Pressures of work and a management role at CSIRO resulted in me shelving the project for over 20 years. Recently, I had the opportunity to review the material I had in hand. It was clear that, over the years, some of the material had dated, and, additionally, I needed to include more recent topics and investigate a wider range of problems than had been originally envisaged. Some of these up-to-date topics have been provided in this volume by a new generation of talented contributors.

My introduction to the topic of mutual coupling in antennas had come much earlier in late 1969 through a short project to design a full-wave dipole backed by a metal plate. This antenna was proposed as an element of a holographic array for microwave link path testing by a local telecommunications provider. Later, at Queen Mary, I worked on a waveguide feed array and this was for illuminating a reflector for creating shaped beams from a geostationary satellite. In the late 1970s, the European Space Agency had proposed a prototype beam for covering Europe and this inspired our research. At Queen Mary, I received expert tuition in Green's functions and mutual coupling from the late George A. Hockham FREng, of optical communications fame. We continued this collaboration at Plessey Radar a few years later, where he was Research Director, when he asked me to work on conformal arrays. All of the topics mentioned above form part of chapters in this book. I continued to work on mutual coupling in feed arrays at CSIRO in Australia and initially for the design of the second-generation Australian domestic satellites. We found that our methods, and especially those relating to mutual coupling, were accurate compared with measurement. Through detailed computer modelling, we demonstrated that an expensive prototype development stage, which was then used throughout the satellite industry, could be eliminated. As a result, satellite development time and costs were reduced. Work on computational techniques for the satellite industry and radio astronomy occupied a significant part of my subsequent research career. Some of the methods developed during that time, and improved upon, are given in this book for the first time including recent work on coupling between metahorns. These topics as well as student collaboration on wire and microstrip antennas have found a place in this book along with topics contributed by well-known international researchers. It is hoped that the topics selected for this book becomes worthwhile reference material for researchers and engineers in the future.

Acknowledgments

I wish to extend my gratitude to several people who enabled this book to see the light of day. First, I sincerely thank the contributors and co-authors of Chapters 8–14, who helped to broaden the range of topics offered – beyond what I could achieve at this time.

In addition, I thank the staff at Wiley – particularly the commissioning editor Sandra Grayson for her encouragement every step of the way, and the project editor Juliet Booker for leading me through the preparation of the final manuscript and Emily Paul, Content Refinement Specialist, for assistance at the proofing stage. I thank John Wiley & Sons for all other aspects of getting this book into the market.

I thank my colleagues and students – both past and present – for their assistance in supporting my research and experimentally verifying some aspects of the chapters reported here. These include the late George A. Hockham from QMC and Plessey Radar days also Robert (Bob) Wills from the latter, and the late Geoffrey T. Poulton, Graeme L. James, Nasiha Nikolic, Stephanie Smith, Stuart G. Hay, David G. Bateman, Geoffrey C. James, Andrew J. Parfitt, John C. Kot, Francis R. Cooray, Doug Hayman, Stephen M. Hanham, Andrew R. Weily. Christophe Granet, Kieran J. Greene, Ken W. Smart, and Dallas R. Rolph from my time at CSIRO.

Finally, grateful thanks are extended to my family (Katie, Nick, Alison, Bee, Elliot, James, and Clare) for their love, assurance, and patience, which has sustained me over the years, especially when this material was first formulated as well as when this book was being prepared. Special thanks to my wife, Val, for her love and unwavering support over all the years we have been together and for helping me focus on the task of completing this book.

List of Contributors

Prof. Eqab Almajali
Department of Electrical Engineering
University of Sharjah
PO Box 27272, Sharjah
United Arab Emirates

Dr Alpha O. Bah
Global Big Data Technology Centre
University of Technology Sydney
PO Box 123 Broadway
New South Wales 2007
Australia

Prof. Trevor S. Bird
Global Big Data Technology Centre
University of Technology Sydney and Principal Antengenuity
PO Box 306, Eastwood
New South Wales 2122
Australia

Dr Ha Bui Van
Louvain School of Engineering
Institute of Information and Communication Technologies
Electronics and Applied Mathematics (ICTM)
Université Catholique de Louvain
Belgium

Dr Jean Cavillot
Louvain School of Engineering
Institute of Information and Communication Technologies
Electronics and Applied Mathematics (ICTM)
Université Catholique de Louvain
Belgium

Prof. Christophe Craeye
Louvain School of Engineering
Institute of Information and Communication Technologies
Electronics and Applied Mathematics (ICTM)
Université Catholique de Louvain
Belgium

Dr Eloy de Lera Acedo
Louvain School of Engineering
Institute of Information and Communication Technologies
Electronics and Applied Mathematics (ICTM)
Université Catholique de Louvain
Belgium

Dr Hoi Shun Antony Lui
Global Big Data Technology Centre
University of Technology Sydney
PO Box 123 Broadway
New South Wales 2007
Australia

Prof. Derek McNamara
School of Electrical Engineering and
Computer Science (EECS)
University of Ottawa
800 King Edward Ave
Ottawa ON K1N 6N5
Canada

Prof. Karl F. Warnick
Department of Electrical and Computer
Engineering
Brigham Young University
450 Engineering Building
Provo UT 84602
USA

Notation

The mathematical notation is consistent throughout the book and is identical with Bird (2016). Equations are numbered consecutively throughout a chapter. For example, Equation 6 in Chapter 4 is Equation (4.6) and is referred to in the text as Eq. (4.6). If the equation is in several parts, each equation is shown as (4.6a), (4.6b), etc. Similarly, for figures by example, the first diagram in Chapter 5 is captioned Figure 5.1 and is referred to in the text in abbreviated format as Fig. 5.1 or Fig. 5.1a and 5.1b if it has two parts.

Vector time-harmonic field quantities are shown in bold type face such as $\boldsymbol{A}(\boldsymbol{r})$, instantaneous vector fields quantities are given in gothic script such as $\mathfrak{A}(\mathbf{r}, t) = \mathrm{Re}\{\mathbf{A}(\mathbf{r}) \exp(j\omega t)\}$, and matrices are given in bold font or equivalently standard typeface in square brackets such as $\boldsymbol{Y} = [Y_{ij}]$. Matrix column and row vectors are made bold font, usually lowercase letters, e.g. $\boldsymbol{b} = \boldsymbol{S}\,\boldsymbol{a}$, where $\boldsymbol{a}$ and $\boldsymbol{b}$ are column vectors and $\boldsymbol{S}$ is a matrix. A product of a matrix column vector in Einstein notation is shown as $b_i = S_{ij}\, a_j$. The product of an operator and a function is shown as $(H(\boldsymbol{\alpha}_j), \boldsymbol{\alpha}_i,)$. The transpose of a matrix is shown as $\boldsymbol{Y}^{\boldsymbol{T}} = [Y_{ij}]^T$, and a Hermitian conjugate or complex transpose is indicated by $\boldsymbol{Y}^{\dagger}$ or $\boldsymbol{Y}^{\boldsymbol{H}}$ as both notations are in common use.

1

Introduction

Trevor S. Bird

Global Big Data Technology Centre, University of Technology Sydney, Ultimo, NSW 2007, Australia
Antengenuity, Eastwood, NSW 2122, Australia

1.1 Aims and Scope

A fundamental problem confronting the designer of antenna arrays is that the excitation of an individual element is not directly proportional to the applied source. This is a result of the array elements interacting with each other through the electromagnetic field, which induces currents on all antennas present in addition to those on the driven element. This interaction is called mutual coupling. Since the radiation characteristics of the elements are different in the array environment, all properties of the array, such as the gain, beam width, and input reflection coefficients, are affected by mutual coupling (Bird 2016). Another effect is that, in arrays of waveguides and horns, modes other than the primary ones may be excited, which can cause array blindness, reflection resonances, and depolarization. Mutual coupling is polarization dependent, and its effect is generally strongest when the inter-element spacing is small. The effect of mutual coupling is not all deleterious, however. Some arrays use, for example, parasitically excited elements to enhance performance. In the last few years, mutual coupling has also been used to improve the performance of arrays by placing the elements closer together so as to create an effective sheet of current. Antennas in arrays or those connected unintentionally have occurred in a wide variety of applications over the years as illustrated in Figure 1.1. The examples shown in Figures 1.1a and 1.1b are early warning radar antennas, the SCR-270 from the 1940s (Stitzer et al. 2007) and the PAVEPAWS array dating from the 1970s (Brookner et al. 2010), respectively. PAVEPAWS is a network of early warning radars. Each face of the array shown in the figure is about 27m in diameter and consists of over 1790 bent, crossed dipoles. The light weight array of horns shown in Figure 1.1c was a standard method of generating shaped beams with a reflector in the 1990s. The feed array in the figure produced a shaped beam for continental USA and Hawaii (Bird and Sroka 1992). Wireless communications in the 2000s have been dominated by the development of multiple input multiple output (MIMO) systems. Figure 1.1d shows a set of dual-band MIMO base station antennas (Guo and Jones 2018) with the cover removed. Platforms and towers with many co-sited antennas such as in Figures 1.1e and 1.1f show clustering which creates potential for coupling in a base station and on a warship, respectively.

The topic of this book is integral to many antenna configurations and is fundamental to understanding their operation. Yet, in the past, before modern computer software, its effect was often neglected when designs were prepared. This is because mutual coupling affects were taken to be of a secondary nature. It is assumed that a design without considering its influence could still provide

Mutual Coupling Between Antennas, First Edition. Edited by Trevor S. Bird.

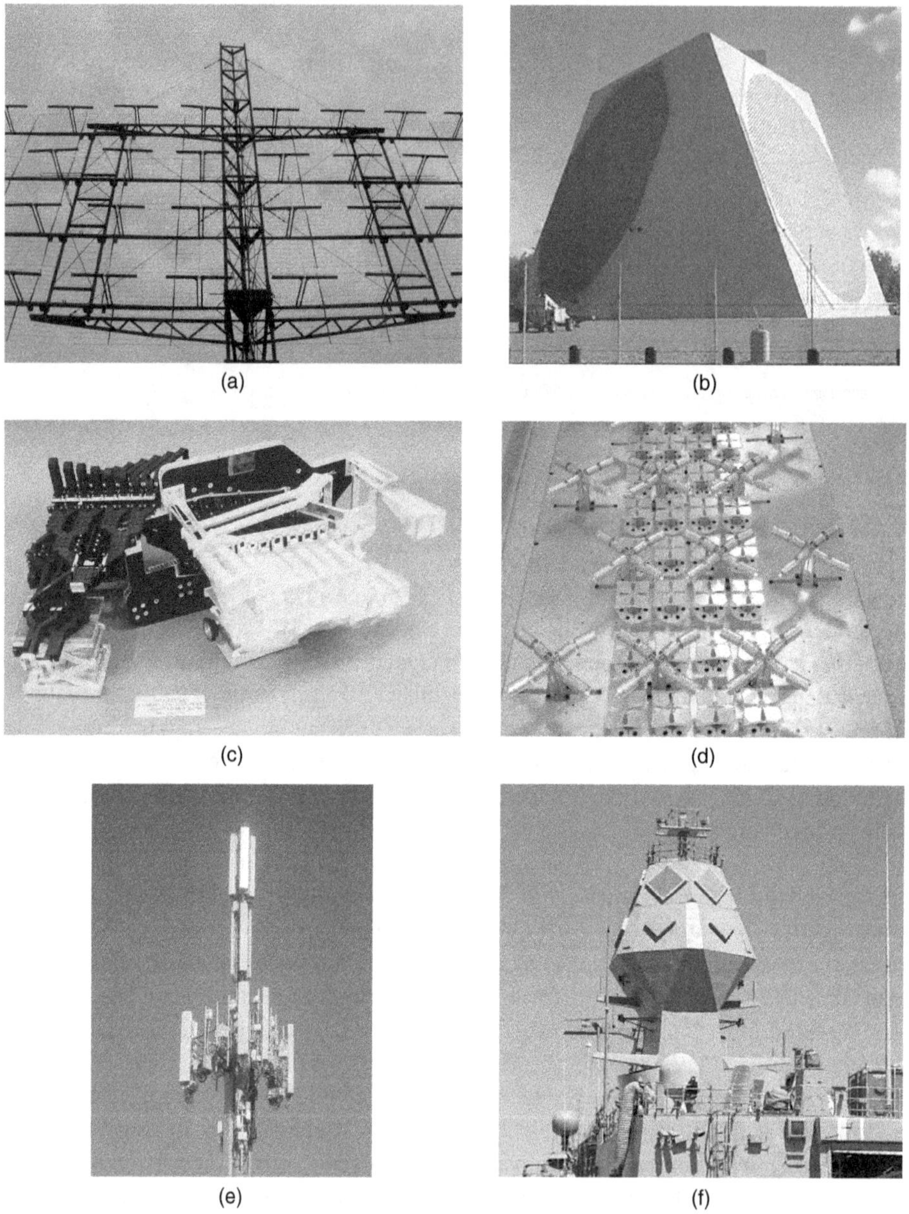

Figure 1.1 Arrays of antennas over the past 80 years in which mutual coupling has a role. (a) One of the first operational early warning radar sets, the SCR-270 deployed in the 1940s, had 32 dipoles in front of a reflecting mesh. Source: brewbooks, https://commons.wikimedia.org/wiki/Category:SCR-270_radar#/media/File:SCR-270_Radar_Antenna2.jpg, CC BY-SA 2.0; (b) the AN/FPS-123 PAVEPAWS phased array radar at Clear Air Base, Alaska (USA) upgraded from earlier models dating from the 1960s (U.S. Army Corps of Engineers). Source: U.S. Army Corps of Engineers Alaska District; (c) a feed array for shaped beam reflector on Galaxy IV satellite in the 1970s. Source: Courtesy of Hughes Space and Communications Group; (d) dual-band multiple input multiple output arrays showing the inter leaved LTE (long-term evolution standard for mobile communications) low- and high – band dipoles. Source: Courtesy of Argus BSA; (e) unintended interaction of antennas on a repeater tower in 2010s. Source: Trevor S. Bird; and (f) co-sited antennas including arrays on frigate HMAS Arunta in the 2000s. Source: Courtesy of Royal Australian Navy.

an approximate expectation of performance. Further, its inclusion in a physical model requires detailed consideration of Maxwell's equations, which, to this day, many engineers try to avoid. Some antennas have a well-established design methodology that allows mutual coupling to be circumnavigated using lumped circuit elements or high-frequency approximations.

The intention of this book is to cover the main techniques that depend clearly on an understanding of the physics through Maxwell's equations. While detailed design is possible through the several accurate computer packages that are readily available, a greater understanding of what may be possible and is achieved through connection with the physics often with simple models. Several approaches will be described, including lumped networks, which should enable both better design and understanding of the physics involved in a more comprehensive manner than hitherto available. The resulting formulations, and several are possible, are applied to a variety of practical antenna configurations and geometries. This includes antennas such as tightly coupled arrays that inherently make use of mutual coupling to increase performance as well as antennas that operate nearby each other, such as those mounted on a mast or ship, where they can interact with a resulting modification in performance. The influence of mutual coupling on stand-alone antennas such as in radar direction finding is considered. In these and other applications, mutual coupling can be undesirable and, as a result, techniques for reducing its effects are important. A range of techniques for doing this is covered here as well. By contrast, some modern-day communications are made possible by means of MIMO systems which require several antennas in close proximity, and this interaction is considered in detail, as well as large, dense arrays of elements that also make inherent use of mutual coupling. The range of array applications is such that the variety of antenna geometries considered here range from arrays of wire antennas to arrays that conform to surfaces such as a cylinder or a sphere. At all times, the intention is to use the basic physics of the problem and Maxwell's equations in its many manifestations. The material in the chapters is well supported by references and appendices.

1.2 Historical Perspective

Mutual coupling was recognized as important by the early pioneers in electromagnetics and antenna theory. Brillouin (1922) was probably one of the earliest to detail a method of analysis, and like many early workers, he was concerned mainly with calculating radiation resistance rather than the complex impedance at the input of array elements. A systematized approach, called the *electromotive force (emf) method*, was applied by Pistolkors (1929) to find the radiation resistance of various dipole array configurations. Another approach adopted at that time was the Poynting vector method, the present-day text book method of calculating radiation resistance from the integrated normal energy flow through a surface surrounding the antenna. This second approach was used, for example, by Bontsch-Bruewitsch (1926) and Knudsen (1952). Both approaches are equivalent and can be converted into the other by means of Gauss's law, as shown by Bechmann (1931). The emf method gained wider acceptance after the work of Carter (1932), who used reciprocity and the emf method to determine expressions for self and mutual impedances for a variety of two-dipole arrangements. This expression for the self-impedance was also derived by Richmond (1961) from the Lorentz reciprocity theorem. Formulae for the mutual impedance of two dipoles in various geometrical arrangements have been derived from assumed currents, and some of these are given by Brown (1937), Barzilai (1948), and Hansen (1966). Storer (1952) developed a variational method to calculate mutual impedances. More sophisticated coupling configurations were subsequently investigated, including the coupling between unequal-size dipoles by Levis and

Tai (1956). Measurement of mutual impedance in dipole arrays was commenced in the 1930s, and the techniques used were described by Carter (1932) and Brown (1937) who detailed a technique for measuring the mutual impedance of elements in the broadcast frequency band.

Carter's paper profoundly influenced much of the subsequent literature on antenna coupling because for the first time the coupling problem was expressed as an equivalent circuit. All the above-mentioned papers were concerned with dipole elements. Slot coupling such as in a waveguide was investigated by Stevenson (1948). However, prior to that, Booker (1946) had identified the complementarity of dipole and slot antennas through Babinet's principle of duality. This paper led to subsequent research on slot antennas, which had been tried during the early 1940s, in the same way Carter did with dipoles. Complementarity was applied by Compton (1950) to slot antennas and the principle was used by Surtees (1950) to obtain the coupling between a slot and a dipole. Expressions for mutual coupling were also derived from duality by Begovich (1950). Other results followed including the derivation by Yee (1961) of the mutual admittance between two magnetic line sources.

The analysis of periodic structures by Floquet's method (e.g. see Collin 1960; Amitay et al. 1972) had been used for decades in solid-state physics (e.g. Merzbacher 1961) before its adoption for antenna arrays in the 1960s even though the significance of periodicity was recognized earlier by Wheeler (1948). The Floquet method found favor at that time for a range of array structures and provided an early means of understanding coupling effects and for analyzing the performance of large arrays. One of the first to apply this approach to a large array was Edelberg and Oliner (1960). An impetus for this was the discovery of unexpected radiation pattern nulls at small scan angles, which stimulated greater investigation of mutual coupling effects. Some of this work is reported by Amitay et al. (1972) and Stark (1974). These pattern nulls were later found to be due to the presence of high-order modes excited in the wave guide apertures by mutual coupling and also the excitation of surface waves on the array aperture. The infinite array solution was also applied to medium-sized arrays where edge and corner effects are important and, therefore, this approximation had varying degrees of success. This led in the 1960s and 1970s to the development of methods to analyze mutual coupling in finite arrays. The methods developed were improved through the 1980s and 1990s as computers and software became more capable. At the same time, the development of numerical methods and computer codes allowed the analysis of more complicated antennas including antennas consisting of several different types of complex elements. The computer codes available today allow mutual coupling to be included as a matter of course. Even so, the physical aspects of mutual coupling needs to be understood to minimize the disadvantages and make full use of the advantages. Indeed, in some applications such as in onboard satellite communications, mutual coupling can provide benefits when even a small gain increase of 0.5–0.6 dB, for example, can result in substantially reduced costs and weight carried into space as well as reduced power consumption.

1.3 Overview of Text

Most antennas are subject to various forms of mutual coupling whether due to internal or external effects. Of the internal kind, antenna arrays are the most important antenna type influenced by it. They are common place in modern systems used for communications, radar, environmental monitoring, and radio astronomy. An antenna array is any group of antennas that are excited together for the purpose of producing a prescribed radiation pattern. Examples of some arrays are pictured in Figure 1.1. Arrays are used because the antenna designer is able, within limits, to tailor the radiation pattern as needs require. The limits are determined by fundamental physics such as the radiation pattern of the elements, the number of elements, element spacing, the physical size of the array, and ohmic losses as well as mutual coupling between the elements.

The interaction of antennas through the electromagnetic field known as mutual coupling is explored through Maxwell's equations, analysis, models, and applications. By highlighting some of its effects, it is intended that the physics of this coupling will become more widely known. It is hoped that with better understanding and modeling, it will be standard practice to include mutual coupling in all design and synthesis of the arrays. There are 14 chapters in this book of which the first 7 are authored by the editor. In Chapter 2, a simplified field model of mutual coupling is introduced. The initial approach is through an incremental form of Maxwell's equations and an inverse square law that applies to a small radiating current element. This enables a simple and direct approach to understanding the presence of mutual coupling in a radiating system. The impact of mutual coupling on the far-field radiation is analyzed and demonstrated through magnetic currents and the dual problem of electric currents. The results obtained in Chapter 2 are applied in later chapters to specific current distributions.

Many of the most important techniques that are used in the analysis and computation of mutual coupling are introduced in Chapter 3. Most of these techniques are used in later chapters. Antenna radiation relies on the reciprocity theorem which links currents and fields. Arising from this theorem is the concept of reaction, which is reviewed. Following the definition of circuit quantities, the chapter specifies these quantities for finite and infinite arrays. The former follows as an extension from two to many elements, while in infinite arrays, periodicity naturally creates a cellular approach within which all details of an element are contained for analysis. Most coupling problems are expressed in the form of integral and differential equations, and a common way to solve these is through the method of weighted residuals, which is described. Within the class of weighted residuals is included many of the common methods such as Galerkin's method and the method of moments. Special techniques for coupling problems are briefly described through the method of characteristic modes and minimum scattering. Finally, some particular numerical methods involved in mutual coupling problems are discussed.

In the following chapter, Chapter 4, dipole arrays are studied through a combination of methods. The background and basic tools for studying wire antennas is described in this chapter to give physical insight into the designs using more sophisticated elements. This description has been supported by practical examples and some measurements. Integral solution techniques have been supported by asymptotic approximations, which are shown to provide useful results when elements are at least a wavelength apart.

An analysis of mutual coupling in finite arrays of different-sized wave guides and horns is described in Chapter 5 using an approach based on a Greens function approach. A range of practical problems are analyzed by this method and expressions obtained for the mutual admittance of all possible combinations of mode coupling. A formula is derived for reducing the order of integration in these expressions, allowing the quadruple integral for mutual admittance to be expressed as the sum of four double integrals in the most general case. The present approach is particularly useful for analysis of large arrays where the formulation can be expressed in efficient code enabling large arrays to be analyzed. The elements range from rectangular, circular, and elliptical as well as to dielectric-and metamaterial-loaded apertures.

Chapter 6 describes the coupling present in microstrip patch antennas arrays. Three approaches are described for coupling between different patches. These were the E-and H-field methods and a full-wave approach. Examples will be given of each approach and compared with measured results.

In Chapter 7, a general formulation for the analysis of mutual coupling on concave surfaces is described. This formulation is based on a blend of canonical asymptotic solutions for the electric and magnetic currents on the surface on a cylinder and a sphere. These canonical solutions are studied separately, and results are obtained for mutual coupling between apertures on surfaces with varying sizes of radius of curvature. Some examples of the use of the general formulation are given for a variety of arrays on a circular cylinder, a sphere, and an ellipsoid.

In Chapter 8, coauthored by Derek McNamara and Eloy de Lera Acedo, describe techniques for analyzing mutual coupling effects of the external kind such as in co-sited antennas and antennas on platforms. The extent of the effects depends on the type of antenna, its isolated directivity, its location on the platform, and the geometry of the platform. Examples are given of antennas mounted on platforms and the assessment of the level of mutual coupling from direct electromagnetic field computations.

A frame work for understanding and computing the effects of mutual coupling in MIMO systems is given in Chapter 9 by contributor Karl Warnick. Mutual coupling impacts MIMO systems in a number of areas, and these are discussed. In the receiving array, strong coupling leads to poor active impedance matching and decreased signal-to-noise ratio (SNR) at the receiver output. Most importantly, a highly coupled array generally has high correlation between the received signals at the array elements and fails to achieve the desired antenna diversity that is at the heart of the increased capacity that MIMO systems are expected to realize. The analytical framework developed in the chapter can be used to model fully all coupling effects from end to end in a MIMO system.

The performance of direction-of-arrival (DOA) estimation in the presence of mutual coupling using linear arrays with different inter element separation and general array apertures is studied in Chapter 10, which is coauthored by Antony Lui and Trevor Bird. Initially, DOA estimation performance is evaluated under an ideal situation using point detectors such that the impact of inter-element separation and array aperture to the accuracies of the estimation is apparent. Mutual coupling is neglected in most antenna synthesis procedures, and this has resulted in suboptimum designs. Maximum gain solutions in the presence of mutual coupling are described for single and multiple beams, although without optimization, there is no guarantee that the solution is better than the ideal situation in the absence of coupling. As an example of optimization, a pattern synthesis is described which includes mutual coupling. In the receive mode, the effect of mutual coupling between elements forms part of the received signals that degrades the estimation performance. The impact of number of array elements within a fixed aperture with monopole antenna elements is studied. It is shown that effective mutual coupling compensation can be achieved using a receiving mutual impedance method.

The following chapter, also co-authored by Antony Lui and Trevor Bird, deals with minimizing mutual coupling in arrays. The desire to achieve the array performance at or near the uncoupled case means that techniques for minimizing mutual coupling will always be in demand. The techniques described include conventional methods as well as some new approaches, for example, a method that uses additional parasitic elements in the array with reactive loading to modify the system scattering matrix and improve performance.

Arrays are increasingly being used for receiving applications, and therefore, the noise performance in the presence of mutual coupling is of importance. In Chapter12, Christophe Craeye et al. show that in densely packed arrays, the traditional noise budget based on noise figure of the amplifier is incomplete. Direction-dependent noise at the output of beamformer, such as thermal noise from the environment, needs to be included. A link is established with the co-polar directivity of the array. In all cases, the accurate computation of the noise budget increasingly relies on the ability to efficiently evaluate standard array quantities such as the array impedance matrix and embedded element patterns.

In Chapter 13, methods for analyzing mutual coupling in large arrays are detailed by Christophe Craeye and HaBui Van. Numerical methods are an essential tool for large array design when mutual coupling is included. Different categories of methods are available, based on Maxwell's equations in differential or integral forms. The latter present an advantage for the study of large arrays made of identical elements, although clever combinations between integral and differential techniques also facilitate the analysis of arrays made of complex elements. Some techniques such as macro basis functions are attractive for the analysis of large arrays. Several techniques are described, and examples are provided.

An important aspect of the study of mutual coupling in arrays is the measurement of its effects. In Chapter 14, techniques used for the measurement of mutual coupling are outlined by Alpha Bah and Trevor Bird. Two basic approaches are adopted, depending on whether the arrays have static beams or are phased so that the beams scan. Accurate methods using vector network analyzers are described for both approaches as well as some basic precautions to achieve greater accuracy. Some radiation methods for determination of coupling effects are described as are measurements of radiation characteristics with an emphasis on mutual coupling effects. The chapters are supported by eight appendices, located at the end of the book.

References

Amitay, N., Galindo, V., and Wu, C.P. (1972). *Theory and Analysis of Phased Array Antennas*. New York, USA: Wiley-Interscience.

Barzilai, G. (1948). Mutual impedance of parallel aerials. *Wireless Engineer* 36: 343–352.

Bechmann, R. (1931). On the calculation of radiation resistance of antennas and antenna combinations. *Proc. IRE* 19 (8): 1471–1480.

Begovich, N.A. (1950). Slot radiators. *Proc. IRE* 38 (7): 803–806.

Bird, T.S. (2016). *Fundamentals of Aperture Antennas and Arrays*. Chichester, UK: Wiley.

Bird, T.S. and Sroka, C. (1992). Design of the Ku-band antennas for the Galaxy HS 601C satellites. *J.Electr. Electron. Eng. Aust.* 12: 267–273.

Bontsch-Bruewitsch, M.A. (1926). Die Strahlung der komplizierten rechtwinkeligen Antennen mit gleichbeschaffenen Vibratoren. *Ann. Phys.* 81: 425–453.

Booker, H.G. (1946). Slot aerials and their relation to complementary wire aerials (Babinet's principle). *J. Inst. Electr. Eng. Pt. IIIA* 93: 620–626.

Brillouin, L. (1922). Sur l'origine de la resistance de rayonnement. *Radio electricite* 3 (4): 147–152.

Brookner, E., Porter, B., Chang, K. et al. (2010). Demonstration of accurate prediction of PAVEPAWS embedded element gain. In: *Proc. IEEE International Symposium on Phased Array Systems and Technology*, 12-15 Oct., Waltham Mass. USA, 417–422.

Brown, G.H. (1937). Directional antennas. *Proc. IRE* 25 (1): 78–145.

Carter, P.S. (1932). Circuit relations in radiating systems and applications to antenna problems. *Proc. IRE* 20 (6): 1004–1041.

Collin, R.E. (1960). *Field Theory of Guided Waves*. NY, USA: McGraw-Hill.

Compton, J.W. (1950). Impedance/frequency characteristics of some slot aerials. *J.IEE* 97: 39–44.

Edelberg, S. and Oliner, A.A. (1960). Mutual coupling effects in large antenna arrays: Part I–slot arrays. *IRE Trans. Antennas Propagat.* 8 (5): 286–297.

Guo, Y.J. and Jones, B. (2018). Base station antennas. In: *Antenna Engineering Handbook*, 5e (ed. J. Volakis), Chap.40. McGraw-Hill.

Hansen, R.C. (1966). *Microwave Scanning Antennas*, New York, Academic, Vol.3, Chap. 2–3.

Knudsen, H.L. (1952). On the calculation of some definite integrals in antenna theory. *Appl. Sci. Res., Sect. B* 3: 51–68.

Levis, C.A. and Tai, C.T. (1956). A method for analyzing coupled antennas of unequal sizes. *IRE Trans. Antennas Propagat.* 4 (2): 128–132.

Merzbacher, E. (1961). *Quantum Mechanics*. New York, USA: Wiley.

Pistolkors, A.A. (1929). The radiation resistance of beam antennas. *Proc. IRE* 17 (3): 562–579.

Richmond, J.H. (1961). A reaction theorem and its application to antenna impedance calculations. *IRE Trans. Antennas Propagat.* AP-9: 515–520.

Stark, L. (1974). Microwave theory of phased-array antennas–a view. *Proc. IEEE* 62: 1661–1701.

Stevenson, A.F. (1948). Theory of slots in rectangular waveguide. *J. Appl. Phys.* 19: 24–38.

Stitzer, S.N., Colton, R.B., and Suffield, F.G. (2007). Reverberations-the SCR-270 radar. *IEEE Microwave Mag.* 8 (3): 88–98.

Storer, J. (1952): 'Variational solution to the problem of symmetrical cylinder antenna', Cruft Lab. Report, TR 101, Cambridge, MA, USA.

Surtees, W.J. (1950). Mutual coupling of a slot with a dipole antenna. *Proc. IRE*: 208–211.

Wheeler, H.A. (1948). The radiation resistance of an antenna in an infinite array or wave guide. *Proc. IRE* 36 (4): 478–487.

Yee, J.S. (1961). Mutual coupling of two thin infinitely-long slots located on a perfectly conducting plane in the presence of a uniform plasma layer. *Proc. IRE* 12 (12): 1837–1845.

2

Basics of Antenna Mutual Coupling

Trevor S. Bird

Global Big Data Technology Centre, University of Technology Sydney, Ultimo, NSW 2007, Australia
Antengenuity, Eastwood, NSW 2122, Australia

2.1 Introduction

The physical phenomenon of mutual coupling occurs through the interaction between fields and sources that are present locally. This interaction is described by Maxwell's equations, which are the physical laws describing them (Jackson 1998). The terminology in the use of these equations often varies depending on the application and whether time or frequency dependence is required. In this chapter, the fundamentals of electromagnetic field equations are reviewed before a model of mutual coupling is introduced.

2.2 Electromagnetic Field Quantities

2.2.1 Definitions

The electromagnetic field can be considered as the superposition of the fields associated with electric (e) and magnetic (m) sources, as follows:

$$\boldsymbol{E} = \boldsymbol{E}_e + \boldsymbol{E}_m \text{ and } \boldsymbol{H} = \boldsymbol{H}_e + \boldsymbol{H}_m \tag{2.1}$$

where $\boldsymbol{E}$ is the electric field intensity in Volts (V)/m, and $\boldsymbol{H}$ is the magnetic field intensity in Ampere (A)/m. When these fields are time harmonic, the differential forms of the physical equations governing the field behavior are the pairs of Maxwell's equations which are listed in Table 2.1.

The other quantities in this table are defined as follows (m.k.s. units given in parentheses):

$\mathbf{J}(\mathbf{r})$ = electric current density (A m^{-2})
$\mathbf{M}(\mathbf{r})$ = magnetic current density (V m^{-2})
$\rho_e(r)$ = electric charge density (C m^{-3})
$\rho_m(r)$ = magnetic charge density (Wb m^{-3})
ε = permittivity (F m^{-1})
μ= permeability (H m^{-1})

where V is the volt, A is the Ampere, m is meter, Wb is Weber, C is the Coulomb, F is the Farad, and H is the Henry. The operator $\nabla = \nabla_t + \hat{z}\,\partial/\partial z$ is the gradient operator, $\nabla\times$is the curl operation, $\nabla\,\cdot$ denotes divergence, and ∇_t is the transverse gradient operator, when z is the direction of

Mutual Coupling Between Antennas, First Edition. Edited by Trevor S. Bird.

Table 2.1 Time-harmonic Maxwell's equations for electric and magnetic sources.

Electric: $\mathbf{J} \neq 0$ $\mathbf{M} = 0$	Magnetic: $\mathbf{J} = 0$ $\mathbf{M} \neq 0$
$\nabla \times \mathbf{E}_e = -j\omega\mu\mathbf{H}_e$	$\nabla \times \mathbf{H}_m = j\omega\varepsilon\mathbf{E}_m$
$\nabla \times \mathbf{H}_e = \mathbf{J} + j\omega\varepsilon\mathbf{E}_e$	$\nabla \times \mathbf{E}_m = \mathbf{M} - j\omega\mu\mathbf{H}_m$
$\nabla \cdot \mathbf{E}_e = \rho_e/\varepsilon$	$\nabla \cdot \mathbf{E}_m = 0$
$\nabla \cdot \mathbf{H}_e = 0$	$\nabla \times \mathbf{H}_m = \rho_m/\mu$

propagation of a wave. In homogeneous materials, the permittivity and permeability are $\varepsilon = \varepsilon_r\varepsilon_o$ and $=\mu_r\mu_o$, respectively, where ε_r is the relative permittivity, μ_r is the relative permeability, $\varepsilon_o = 8.854 \times 10^{-12}$ F m^{-1} and $\mu_o = 4\pi \times 10^{-7}$ H m^{-1} are the permittivity and permeability of free space. The speed of electromagnetic waves in free space is $c = 1/\sqrt{\varepsilon_o\mu_o} = 2.99792458 \times 10^8$ m s^{-1}.

The two partial field pairs, $(\mathbf{E}_e, \mathbf{H}_e)$ and $(\mathbf{E}_m, \mathbf{H}_m)$, satisfy separate sets of Maxwell's equations listed in Table 2.1 that originate from either electric or magnetic sources, which are indicated thus by subscripts e and m, respectively. In general, both sources can be present. The former is due to physical electric currents and charges, while the latter is due to magnetic currents and charges. Although no such latter sources are currently known to exist, field equivalence principles enable them to be created and used to maintain the symmetry of Maxwell's equations (Harrington 1961; Jones 1964). Supplementary to Maxwell's equations is the important relationships between currents and charges (Jackson 1998). The electric charge density and electric current are related through the following equation of continuity:

$$\nabla \cdot \boldsymbol{J} + j\omega\rho_e = 0 \tag{2.2a}$$

Similarly, the magnetic current density and magnetic current are related by

$$\nabla \cdot \boldsymbol{M} + j\omega\rho_m = 0 \tag{2.2b}$$

A consequence of Table 2.1 and the equations of continuity Eqs. (2.2) is that the fields due to electric and magnetic currents are duals of each other. This duality is summarized in Table 2.2.

Table 2.2 Field duality.

Electric dipole		Magnetic dipole
$\mathbf{E}$	$\leftrightarrow$	$\mathbf{H}$
$\mathbf{H}$	$\leftrightarrow$	$-\mathbf{E}$
k	$\leftrightarrow$	k
ε	$\leftrightarrow$	μ
μ	$\leftrightarrow$	ε
η	$\leftrightarrow$	$1/\eta$
$\mathbf{J}$	$\leftrightarrow$	$\mathbf{M}$
ρ_e	$\leftrightarrow$	ρ_m
ω	$\leftrightarrow$	ω

2.2.2 Field Representations in Source-Free Regions

In problems involving sections of uniform structures that guide electromagnetic waves such as waveguides and transmission lines when **J** and **M** are absent, it is sometimes convenient to represent all field components in terms of the field components in the direction of propagation, i.e. in the direction of uniformity. By convention, this direction is usually taken as the z-component in a cylindrical coordinate system with directions denoted by (u, v, z). It is recognized that the field components $E_z(u, v, z)$ and $H_z(u, v, z)$ satisfy Helmholtz wave equations, where (u, v) are transverse co-ordinates. The guiding structures are assumed to exhibit reflection symmetry, and, therefore, it is sufficient to represent the total field as the superposition of forward and reverse traveling wave solutions in the z-direction. For time-harmonic fields of the type defined by Table 2.1, a forward traveling wave (in the +z-direction) has the fields

$$\boldsymbol{E}(u, v, z) = \left[\boldsymbol{E}_t(u, v) + \hat{z}E_z(u, v)\right] exp(-j\gamma z) \tag{2.3a}$$

$$\boldsymbol{H}(u, v, z) = \left[\boldsymbol{H}_t(u, v) + \hat{z}H_z(u, v)\right] exp(-j\gamma z) \tag{2.3b}$$

where $\mathbf{E}_t(u, v)$ and $\mathbf{H}_t(u, v)$ are the transverse electric and magnetic field vectors, and $\gamma = \beta - j\alpha$ is the complex propagation constant. β is the phase shift per unit length, and α is the attenuation constant. For lossless structures, $\alpha = 0$. A representation of the field arising from Eq. (2.3) can be expressed in terms of the field components in the direction of propagation. In this representation, in a source-free region Maxwell's equations give

$$k_z^2\boldsymbol{E}_t(u, v) = j\omega\mu\hat{z} \times \nabla_t H_z - j\gamma\nabla_t E_z$$

$$k_z^2\boldsymbol{H}_t(u, v) = -j\omega\varepsilon\hat{z} \times \nabla_t E_z - j\gamma\nabla_t H_z$$

where $k_z^2 = \omega^2\varepsilon\mu - \gamma^2$ is the axial wave number. The first term on the right-hand side of k_z^2, is the square of $k = \omega\sqrt{\varepsilon\mu}$, which is the propagation constant of a plane wave in the homogeneous material. In free space, $\varepsilon_r = 1$ and $k = 2\pi/\lambda = k_o = \omega\sqrt{\mu_o\varepsilon_o} = \omega/c$. This axial representation of Maxwell's equations is frequently used for describing propagation in uniform homogeneously filled guiding structures such as waveguides (Collin 1960). Modes of such structures are usually denoted as transverse electric (TE) or transverse magnetic (TM) with respect to the propagation (z) direction where either z-component is zero. When the waveguide has anisotropic walls (e.g. corrugated or metamaterial covering) or is inhomogenously-filled such as with dielectric rods, all six field components are present. In like manner to TE and TM modes, these hybrid modes are designated as HE or HM depending on the dominance of the axial component, respectively, the electric or magnetic field (Clarricoats & Olver 1984). All TE, TM, HE, and HM modes have a relationship between the propagation constant and frequency called a characteristic equation of the form $g(\gamma, \omega) = 0$. The mode cut-off frequency is a special result of this equation. In coaxial metallic structures, the first mode does not have a cut-off frequency but can operate down to DC. In homogenously loaded coaxial lines, this mode is the transverse electromagnetic (TEM) mode, and in inhomogeneous structures, the first mode is the hybrid-electromagnetic (HEM) mode. When guiding structures are partially loaded such as for dielectric-loaded rectangular waveguides, it is convenient to represent the field solution in terms of modes that are TE or TM to the interface or parallel to the surface normal. For example, if waveguide is uniform in the z-direction and there is an interface at y = c, where c is a constant bounding the surfaces in the x-direction, the fields can be decomposed as described by Harrington (1961).

For a field that is TM to the x-direction (denoted TMx), i.e. $H_x = 0$ due often to a boundary condition in the region, the field components are given by

$$
\begin{aligned}
E_x &= \frac{1}{j\omega\varepsilon}\left(\frac{\partial^2}{\partial x^2} + k^2\right)\Psi_x \quad H_x = 0 \\
E_y &= \frac{1}{j\omega\varepsilon}\left(\frac{\partial^2 \Psi_x}{\partial x \partial y}\right) \quad H_y = \left(\frac{\partial \Psi_x}{\partial z}\right) \\
E_z &= \frac{1}{j\omega\varepsilon}\left(\frac{\partial^2 \Psi_x}{\partial x \partial z}\right) \quad H_z = -\left(\frac{\partial \Psi_x}{\partial y}\right)
\end{aligned}
\tag{2.4}
$$

where $\left(\nabla_t^2 + k_{ci}^2\right)\Psi_x = 0$ subject to boundary conditions on an x-directed magnetic field in each region bounded by the interface at y = c. In region i $k_{ci}^2 = k^2\varepsilon_i\mu_i - \gamma^2$, ε_i is the permittivity of region i, and μ_i is its permeability. Similarly, for a field that is TE to the x-direction (ie. TEx), i.e. $E_x = 0$:

$$
\begin{aligned}
E_x &= 0 \quad H_x = \frac{1}{j\omega\mu}\left(\frac{\partial^2}{\partial x^2} + k^2\right)\Psi_x \\
E_y &= -\left(\frac{\partial \Psi_x}{\partial z}\right) \quad H_y = \frac{1}{j\omega\mu}\left(\frac{\partial^2 \Psi_x}{\partial x \partial y}\right) \\
E_z &= \left(\frac{\partial \Psi_x}{\partial y}\right) \quad H_z = \frac{1}{j\omega\mu}\left(\frac{\partial^2 \Psi_x}{\partial x \partial z}\right)
\end{aligned}
\tag{2.5}
$$

where $\left(\nabla_t^2 + k_{ci}^2\right)\Psi_x = 0$ is subject to boundary conditions on an x-directed electric field in each region bounded by the interface at y = c where k in all parameters remaining the same. Note that equation sets (2.4) and (2.5) are duals of each other. To convert these equations for interfaces in the other principal directions, interchange all quantities cyclically in the order $x \rightarrow y \rightarrow z$. Another modal notation for uniform waveguide is that an axially propagating mode with no electric field component normal to the interface is called longitudinal-section electric (LSE), while a mode with no magnetic field component normal to the interface is called longitudinal-section magnetic (LSM) (Collin 1960).

Modes in a spherical geometry (r, θ, ϕ) can be defined in a similar way (Harrington 1961). In this case, the fundamental solution to the wave equation is of the form

$$\Psi_{m,n} \sim b_m(kr)\, L_m^n(cos\theta)\, h(n\phi)$$

where $b_m(kr)$ is a spherical Bessel function, $L_m^n(cos\theta)$ is the associated Legendre function, and $h(n\phi)$ is a harmonic function such as cosine or a complex exponential such as exp($jn\phi$).

2.3 Mutual Coupling Between Elementary Sources

2.3.1 Radiation

To study radiation and its interaction with other bodies, consider initially a short electric current element dj_z, which operates harmonically and lies along the z-direction in a spherical co-ordinate system (see Figure 2.1). The field radiated by this current is found by solving the inhomogeneous wave equation

$$\nabla^2 A_z + k^2 A_z = -\mu\, dj_z \tag{2.6}$$

from which the electric and magnetic fields can be obtained using

$$E = -j\omega\left(\hat{r}\,cos\theta - \hat{\theta}\,sin\theta\right)A_z + \frac{1}{j\omega\mu\varepsilon}\nabla\left(\frac{\partial A_z}{\partial z}\right) \tag{2.7a}$$

and

$$\boldsymbol{H} = \frac{1}{\mu}\nabla \times \left(\hat{z}A_z\right) \tag{2.7b}$$

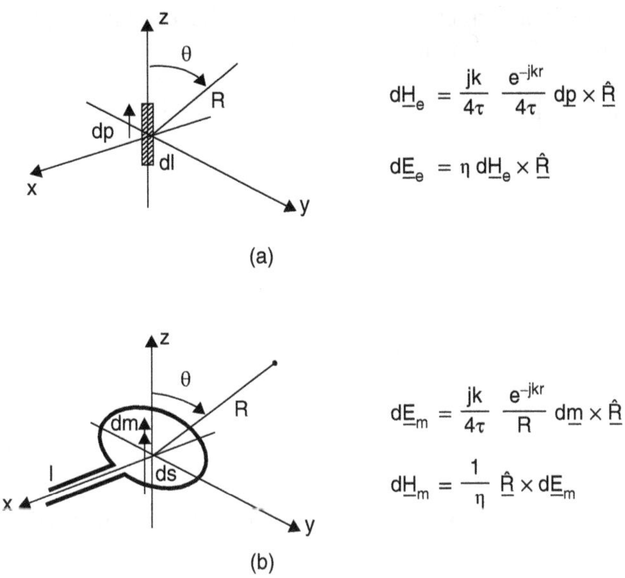

Figure 2.1 Models of electric and magnetic dipole sources. (a) Elementary electric dipole; (b) elementary magnetic dipole.

At a short distance from the current, the right-hand side of Eq. (2.6) is zero. As the current element is small, the geometry is symmetric, so that a solution to the homogenous wave equation is

$$A_z = c_1 \frac{\exp(-jkr)}{r} + c_2 \frac{\exp(jkr)}{r} \tag{2.8}$$

where c_1 and c_2 are constants. For a wave propagating in free space away from the source, the outward traveling wave solution is the only valid one and thus here $c_2 = 0$. To determine the value of c_1, enclose the element in a sphere of radius r and integrate Eq. (2.6) over the volume-enclosed V_o, which gives

$$\iiint_{V_o} dv\left(\nabla^2 A_z + k^2 A_z\right) = -\mu \iiint_{V_o} dv\, dj_z. \tag{2.9}$$

The right side of Eq. (2.9) is constant and equals $-\mu\, I\, d\ell$, where I is the current and $d\ell$ is the length of the element. On the left-hand side, consider the limit as the radius of the sphere shrinks. The second term on the left side of Eq. (2.9) vanishes as $r \to 0$. Taking the limit of the first term, we find

$$\lim_{r\to 0} \iiint_{V_o} dv \nabla \cdot \nabla A_z = \lim_{r\to 0} \iint_{S_o} dS\, \nabla A_z \cdot \hat{r} = \lim_{r\to 0} \left[4\pi r \frac{d}{dr}\left(\frac{c_1 exp(-jkr)}{r}\right)\right] = -4\pi c_1$$

Thus,

$$c_1 = \frac{\mu\, I\, d\ell}{4\pi}$$

Therefore,

$$A_z = \frac{\mu\, I\, d\ell}{4\pi} \frac{\exp(-jkr)}{r} \tag{2.10}$$

Substitution of Eq. (2.10) into Eq. (2.7) yields the fields

$$\boldsymbol{E} = -j\omega A_z \left\{2\hat{r}\, cos\theta \left[\frac{j}{kr} + \frac{1}{(kr)^2}\right] + \hat{\theta}\, sin\theta \left[-1 + \frac{j}{kr} + \frac{1}{(kr)^2}\right]\right\} \tag{2.11a}$$

$$\boldsymbol{H} = j\hat{\phi} \frac{1}{\mu} A_z \left[jk + \frac{1}{r}\right] sin\theta \tag{2.11b}$$

At distances much larger than the length of current element, i.e. $r > d\ell$, the terms in Eq. (2.11) higher than second order can be ignored giving components

$$
\begin{aligned}
E_r &= -\frac{j\omega\mu\, I\, d\ell}{4\pi}\frac{\exp(-jkr)}{r}\cos\theta\left(\frac{2j}{kr}\right)\\
E_\theta &= \frac{j\omega\mu I\, d\ell}{4\pi}\frac{\exp(-jkr)}{r}\sin\theta\left[1-\frac{j}{kr}\right]\\
H_\phi &= \frac{jk\, I\, d\ell}{4\pi}\frac{\exp(-jkr)}{r}\sin\theta\left[1-\frac{j}{kr}\right]
\end{aligned} \tag{2.12}
$$

At greater distances, E_r becomes asymptotically small compared to the other components such that the field radiated by the current element is a spherical transverse electromagnetic wave in the direction of propagation. The ratio of the nonzero transverse field components is

$$\frac{E_\theta}{H_\phi} = \frac{\omega\mu}{k} = \sqrt{\frac{\mu}{\varepsilon}} = \eta \tag{2.13}$$

where η is the wave impedance of the medium into which the current element is radiating. When this medium is free space,

$$\eta_o = \sqrt{\frac{\mu_o}{\varepsilon_o}} = 376.7\, ohms \approx 120\pi$$

is referred to the wave impedance of free space.

2.3.2 Generalized Infinitesimal Current Elements

Consider an infinitesimally small element dS' of a surface A. On this surface element, suppose there are electric and magnetic dipole sources with electric and magnetic dipole moments **dp** and **dm**, respectively. These dipole moments are related to the surface currents on surface A, as follows:

$$\mathbf{dp} = \mathbf{J_s} dS' \tag{2.14a}$$

$$\mathbf{dm} = \mathbf{M_s} dS' \tag{2.14b}$$

Representations of both dipoles are shown in Figure 2.1. A short wire of length $d\ell'$ directed along the z-axis (Figure 2.1a) carrying a uniform current I along its length has an electric dipole moment $\mathbf{dp} = \mathbf{z}\, Id\ell'$. At distance R such that kR ≫ 1 from the wire, the nonzero field components are given by the dominant first-order terms of Eq. (2.12). In vector form, these fields are expressed as

$$\boldsymbol{dH_e} = \frac{jk}{4\pi}\frac{e^{-jkR}}{R}\,\mathbf{dp}\times\hat{R} \tag{2.15a}$$

$$\mathbf{dE_e} = \eta_o\ \mathbf{dH_e}\times\hat{R} \tag{2.15b}$$

where the identity $\hat{z}\times\hat{R} = \hat{\phi}\, \sin\theta$ has been used. For a magnetic dipole, the fields may be obtained from Eq. (2.15) using the principle of field duality (Table 2.2). The fields due to **dm** are, therefore

$$\mathbf{dE_m} = -\frac{jk}{4\pi}\frac{e^{-jkR}}{R}\ \mathbf{dm}\times\hat{R} \tag{2.16a}$$

$$\mathbf{dM_m} = \frac{1}{\eta_o}\ \mathbf{dE_m}\times\hat{R} \tag{2.16b}$$

Equation set (2.16a) and (2.16b) are equivalent to the fields due to a wire loop of cross-section dS' supporting a harmonic current I, as depicted in Figure 2.1b. The loop has a magnetic dipole moment

$$\mathbf{dm} = j\omega\mu_0 I\, dS' \tag{2.17}$$

Thus, a model of the sources at each point P′ on incremental parts of region A are small electric dipoles acting in conjunction with small loops of current. The electric field at P due to both dipole sources is

$$\mathbf{dE} = \mathbf{dE_e} + \mathbf{dE_m} \tag{2.18a}$$

and

$$\mathbf{dH} = \mathbf{dH_e} + \mathbf{dH_m} \tag{2.18b}$$

From Eq. (2.18a), the total electric field due to both sources is

$$\begin{aligned}\mathbf{dE} &= \frac{jk}{4\pi}\frac{e^{-jkR}}{R}\left[-\mathbf{dm}\times\hat{R} + \eta_o\left(\mathbf{dp}\times\hat{R}\right)\times\hat{R}\right] \\ &= \frac{jk}{4\pi}\frac{e^{-jkR}}{R}\left[-\mathbf{M_s}\times\hat{R} + \eta_o\left(\mathbf{J_s}\times\hat{R}\right)\times\hat{R}\right] dS'\end{aligned} \tag{2.19}$$

Adding contributions from all such sources on region A by means of integration results in the electric field

$$\boldsymbol{E} = \frac{jk}{4\pi}\int_A dS' \frac{e^{-jkR}}{R}\left[(\hat{R}\times\boldsymbol{M}_s) + \eta_0(\boldsymbol{J}_s\times\hat{R})\times\hat{R}\right] \tag{2.20a}$$

where $\mathbf{R} = \mathbf{r} - \mathbf{r}'$, as shown in Figure 2.1. The surface integral in Eq. (2.20) is with respect to the primed coordinates, that is, the source coordinates on region A. The magnetic field is obtained similarly and is

$$\boldsymbol{H} = \frac{jk}{4\pi}\int_A dS' \frac{e^{-jkR}}{R}\left[\left(\boldsymbol{J}_s\times\hat{R}\right) + \frac{1}{\eta_0}(\boldsymbol{M}_s\times\hat{R})\times\hat{R}\right] \tag{2.20b}$$

As a consequence, the assumptions leading to Eqs. (2.15, 2.16, 2.20) are valid for all field points P such that $kR \gg 1$ and are applicable at intermediate distances from the source region as well as in the far-field; that is, at distances that are comparable in size to the area itself and beyond (Silver 1946).

2.3.3 Mutual Coupling Between Infinitesimal Current Elements

Two or more current elements interact with each other depending on their orientation. To provide an initial insight into mutual coupling and its effects, consider two infinitesimally short electric dipoles of length $d\ell_1$ and $d\ell_2$ supporting time-harmonic currents of amplitude I_1 and I_2 operating at an angular frequency ω. Dipole 2 is rotated at an angle α in the same plane (z-y plane) relative to dipole 1, as shown in Figure 2.2. The theory of a short electric dipole given in Section 2.3.1 allows the elemental electric and magnetic fields of dipole 1 to be expressed to second order as

$$\boldsymbol{dE}_1 = \frac{\eta_0 I_1 d\ell_1}{4\pi} e^{-jkr_1}\left[\hat{\theta}_1 \sin\theta_1\left(\frac{jk}{r_1} + \frac{1}{r_1^2}\right) + \hat{\boldsymbol{r}}_1\cos\theta_1\frac{2}{r_1^2}\right] \tag{2.21a}$$

$$\boldsymbol{dH}_1 = \frac{1}{\eta_0}\hat{\boldsymbol{r}}_1\times\boldsymbol{dE}_1 \tag{2.21b}$$

where $(r_1, \theta_1, \varphi_1)$ are the spherical polar coordinates at the center of dipole 1. Similar expressions apply to dipole 2 where the subscript 2 replaces the subscript 1 in the above equations. Now, let the

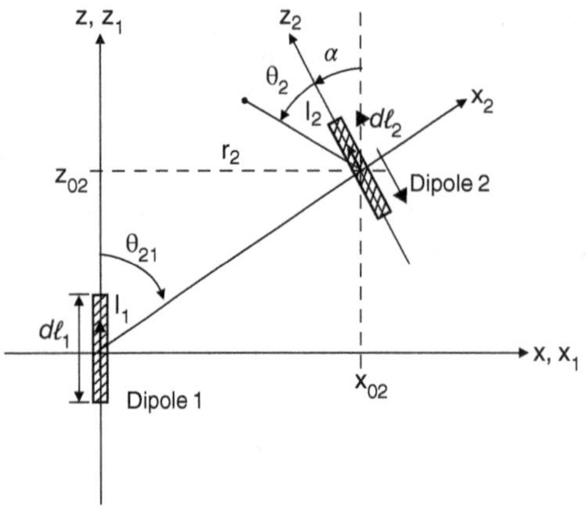

Figure 2.2 Geometry for coupling between two short electric dipoles.

electric field due to dipole 1 on dipole 2, which is of length $d\ell_2$, be given by $\mathbf{dE}_{21}$. Following from an incremental form of Faraday's law, the field due to dipole 1 induces an electromotive force (emf) across terminals at dipole 2 in the form (Schelkunoff and Friis 1952)

$$dV_{21} = -\boldsymbol{dE}_{21} \cdot \hat{\mathbf{z}}_2 d\ell_2 \tag{2.22}$$

Since I_1 is the current producing this emf, the mutual impedance of element 2 due to element 1 is defined as

$$Z_{21} = \frac{dV_{21}}{I_1} = -\frac{\mathbf{dE}_{21} \cdot \hat{\mathbf{z}}_2 d\ell_2}{I_1} \tag{2.23}$$

Similarly, the current I_2 induces an emf on dipole 1 allowing the mutual impedance at dipole 1 due to dipole 2 to be given by

$$Z_{12} = \frac{dV_{12}}{I_2} = -\frac{\mathbf{dE}_{12} \cdot \hat{\mathbf{z}}_1 d\ell_1}{I_2} \tag{2.24}$$

A relationship between these two mutual impedances is found by applying Lorentz's reciprocity theorem (see Section 3.2.2) to the two sets of fields and elementary sources. This theorem results in

$$I_1 dV_{12} = I_2 dV_{21} \tag{2.25}$$

This result along with Eq. (2.22) and its equivalent for antenna 1 requires that

$$Z_{12} = Z_{21} \tag{2.26}$$

A formal expression for the mutual impedance can be obtained from Eqs. (2.21a) and (2.24). Using some vector identities, this mutual impedance of the dipoles is given by

$$Z_{21} = \frac{\eta_o d\ell_1 d\ell_2}{4\pi} e^{-jkr_{21}} \left[\frac{jk}{r_{21}} \sin\theta_{21} \sin\left(\theta_{21} - \alpha\right) \right.$$
$$\left. + \frac{1}{r_{21}^2} \left(\cos\alpha \left(1 - 3\cos^2\theta_{21}\right) + \sin\alpha \sin\theta_{21} \cos\theta_{21} \left(2\cos\theta_{21} - 1\right)\right) \right] \tag{2.27}$$

As shown in Figure 2.2, θ_{21} is the angle subtended at dipole 1 by dipole 2 and r_{21} is the distance between the dipole's centers. In the special case of parallel dipoles ($\alpha = 0$), that is, in a broadside arrangement, ($\theta_{21} = 90°$ and in H-plane), Eq. (2.27) simplifies to

$$Z_{21} = \frac{\eta_o d\ell_1 d\ell_2}{4\pi} e^{-jkr_{21}} \left(\frac{jk}{r_{21}} + \frac{1}{r_{21}^2} \right) \tag{2.28}$$

When the dipoles are in an echelon arrangement, that is, end to end ($\theta_{21} = 0°$ and in E-plane), the mutual impedance is

$$Z_{21} = -\frac{\eta_o\, d\ell_1 d\ell_2}{2\pi\, r_{21}^2}\, e^{-jkr_{21}} \tag{2.29}$$

Therefore, for a broadside configuration (H-plane coupling) of dipoles, the mutual impedance is inversely proportional to the distance between the dipoles, while in the echelon configuration (E-plane coupling), the distance dependence is inverse square (Hansen 1998).

In the same way, the coupling of short magnetic dipoles can be studied. However, the fields due to electric and magnetic dipoles are duals of each other, and, therefore, the corresponding results for the magnetic dipole may be obtained by inspection from the above results. For magnetic dipoles, the dipole moments are, respectively, $\mathbf{dm}_1 = \hat{z}_1\, V_1\, d\ell_1$ and $\mathbf{dm}_2 = \hat{z}_2\, V_2\, d\ell_2$, where V_1 and V_2 are the applied voltages. Now, from the duality summarized in Table 2.1, the mutual admittance of elemental magnetic dipoles is

$$Y_{21} = \frac{dI_{21}}{V_1} = -\frac{\mathbf{dH_{21}} \cdot \hat{\mathbf{z}}_2\, d\ell_2}{V_1} \tag{2.30}$$

Therefore,

$$Y_{21} = \frac{Y_o d\ell_1 d\ell_2}{4\pi} e^{-jkr_{21}} \left[\frac{jk}{r_{12}} sin\theta_{21} sin\left(\theta_{21} - \alpha\right) + \frac{1}{r_{21}^2}\left(cos\alpha\left(1 - 3cos^2\theta_{21}\right) + sin\,\alpha\, sin\theta_{21} cos\theta_{21}\left(2\cos\theta_{21} - 1\right)\right)\right] \tag{2.31}$$

Equation (2.31) shows that when magnetic dipoles are arranged broadside to each other (i.e. E-plane), the mutual admittance varies inversely with the distance between them, while in an echelon arrangement (i.e. H-plane), the dependence is as the square of the distance. This is opposite to what we have shown for electric dipoles and could be expected from duality.

There is a general relationship between the mutual impedance and admittances of electric and magnetic dipoles that finds widespread use. Let the mutual impedance for electric dipoles in free space, Eq. (2.29), be written as Z_{21}^{ef} and the admittance for magnetic dipoles also in free space, given by Eq. (2.30), be Y_{21}^{mf}. The ratio of the first order terms of these quantities is

$$\frac{Z_{21}^{ef}}{Y_{21}^{ef}} = \eta_o^2 \tag{2.32}$$

Equation (2.32) is equivalent to Booker's relation for complementary antennas in free space (Booker 1946). Similar expressions can be found for other arrangements such as for dipoles backed by plane conducting sheets. For example, the mutual admittance of a magnetic dipole located adjacent to a plane electric conductor can be shown to be $Y_{21}^{me} = 2Y_{21}^{ef}$ where the superscript "e" on the right side refers to an electric wall. The Booker's relation between this mutual admittance and the mutual impedance of an electric dipole in free space is, therefore,

$$\frac{Z_{21}^{ef}}{Y_{21}^{me}} = \frac{\eta_o^2}{2} \tag{2.33}$$

The more common form of Booker's formula (Kraus and Carver 1973) is the relation between the input impedance of a half-wave dipole in free space and a half-wave slot in a ground plane, which is its complementary structure. In that instance, the right side of Eq. (2.33) is further divided by 2 to give $Z_{11}^{dipole}/Y_{11}^{slot} = \eta_o^2/4$.

2.4 Network Representation of Mutual Coupling

2.4.1 Extension to Combination of Elements

Up to this point, isolated infinitesimal currents and associated voltage have been defined. When some of these elements are joined, the voltage and currents become integrated sums of equivalent voltages and currents that satisfy Maxwell's equations. The current elements may combine into current lines or sheets, and there will be voltages between their terminals. The impedances become integrated quantities also and dependent on current sheets and integrated radiated fields. The integrated sum of connected elements leads to self-impedances or admittances. The elements may be separated or clustered into different connected groups, as illustrated in Figure 2.3. The interaction between these groups provides continuation of the mutual impedances. The end result of these combinations is a network representation of mutual coupling with corresponding circuit elements that are continuous representations of the incremental quantities that were derived in the previous section.

The terminal voltages are

$$V_i = \int_b^a \boldsymbol{d\ell} \cdot \boldsymbol{E}_{ii} \tag{2.34}$$

where V_i is the sum of a static component V_{stat} and an induced electromotive force, which is the sum of transformer and motion components (Vujević et al. 2011) and here $\boldsymbol{dl}$ (note l is script) is a vector incremental length.

The currents are given by

$$I_i = \iint_{S_i} \boldsymbol{dS} \cdot \boldsymbol{J}_i \tag{2.35}$$

where I_i is the sum of Ampere's circuital current and Maxwell's displacement current and $\boldsymbol{dS} = \hat{n}dS$.

The self-impedance based on eqn. 2.23 and an electric current distribution $\boldsymbol{J}_i$ on i is

$$Z_{ii} = -\frac{1}{I_i I_i} \iint_{S_i} dS\, \boldsymbol{J}_i \cdot \boldsymbol{E}_{ii} \tag{2.36}$$

where $\boldsymbol{E}_{ii}$ is the electric field due to i on i and similarly the mutual impedance between elements i and j is

$$Z_{ij} = -\frac{1}{I_i I_j} \iint_{S_i} dS\, \boldsymbol{J}_i \cdot \boldsymbol{E}_{ij} \tag{2.37}$$

where $\boldsymbol{E}_{ij}$ is the electric field on i due to j.

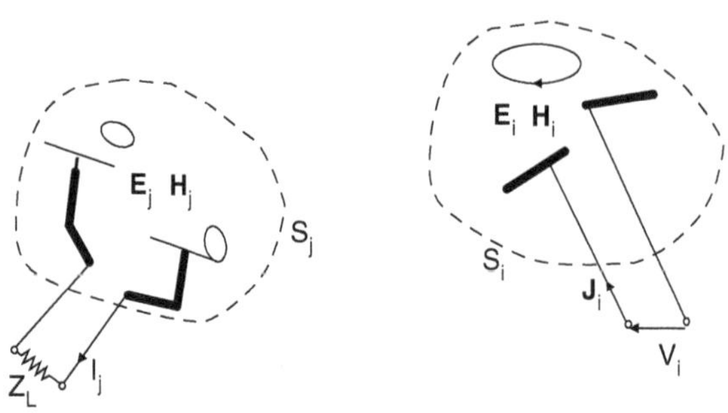

Figure 2.3 Formation of continuous sources.

The relations corresponding to eqns. (2.36) and (2.37) for a magnetic current distribution $\boldsymbol{M}_i$ on i are as follows; the self-admittance is

$$Y_{ii} = -\frac{1}{V_i V_i} \iint_{S_i} dS \, \boldsymbol{M}_i \cdot \boldsymbol{H}_{ii} \tag{2.38}$$

and mutual admittance is

$$Y_{ij} = -\frac{1}{V_i V_j} \iint_{S_i} dS \, \boldsymbol{M}_i \cdot \boldsymbol{H}_{ij} \tag{2.39}$$

where $\boldsymbol{H}_{ij}$ is the magnetic field on i due to j.

2.4.2 Mutual Impedance and Admittance Matrix Formulation

We move now to a larger arrangement of radiators. Consider an N-element array of electric dipole antennas as shown in Figure 2.4. Suppose that the dipoles have external voltages V_p^{in} (p = 1,…, N) applied at their input terminals. Application of the emf method at all dipoles requires that

$$\boldsymbol{V}^{in} = \boldsymbol{ZI} \tag{2.40}$$

where **Z** is the impedance matrix for the mutual coupling in the array, given by

$$\mathbf{Z} = \begin{bmatrix} Z_{11} & Z_{12} & \cdots & Z_{1N} \\ Z_{21} & Z_{22} & \cdots & Z_{2N} \\ \cdots\cdot & \cdots\cdot & \cdots\cdot & \cdots\cdot \\ Z_{N1} & \cdots\cdot & \cdots\cdot & Z_{NN} \end{bmatrix} \tag{2.41}$$

The statement of reciprocity, Eq. (2.26), requires that when there are several elements, the mutual impedance matrix for these elements is symmetric. The currents **I** at the terminals of the dipoles are obtained from the inverse of $\boldsymbol{Z}$, as follows:

$$\mathbf{I} = \mathbf{Z}^{\;1}\mathbf{V}^{\mathbf{in}} \tag{2.42}$$

For arrays of magnetic dipoles, apertures, and slots, the expressions are the dual of those for the electric dipole. In matrix notation, the relation between the externally applied current vector $\mathbf{I}^{\mathbf{in}}$ and the induced voltage is

$$\mathbf{V} = \mathbf{Y}^{-1}\,\mathbf{I}^{\mathbf{in}} \tag{2.43}$$

where **Y** is the admittance matrix.

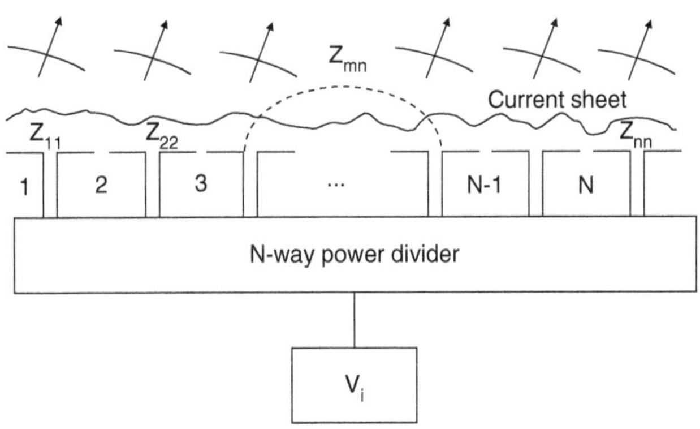

Figure 2.4 An N-element array of dipoles.

2.4.3 Scattering Matrix Representation

When antennas are connected via transmission lines to the applied sources, the currents that are excited are geometry dependent, and, therefore, it is often convenient to use a scattering matrix formulation. To relate this formulation to the impedance and admittance network, representations of voltage and current are specified in terms of the complex amplitudes of forward and backward traveling waves on these lines. Let

a_k = amplitude of a wave traveling toward port k,
b_k = amplitude of a wave traveling away from port k, and
Z_{ok} = characteristic wave impedance at port k.

The associated wave admittance of transmission line k is $Y_{ok} = 1/Z_{ok}$. The voltage and current at port k are defined as follows:

$$V_k = Z_{ok}^{1/2}\left(a_k + b_k\right)$$
$$I_k = Z_{ok}^{-1/2}\left(a_k - b_k\right) \tag{2.44}$$

After substituting Eq. (2.44) into (2.40) and rearranging them, it is found that

$$\boldsymbol{b} = \boldsymbol{Sa} \tag{2.45}$$

where

$$\mathbf{S} = (\mathbf{z} - \mathbf{I})\,/\,(\mathbf{z} + \mathbf{I}) = \mathbf{I} - 2\,(\mathbf{I} + \mathbf{z})^{-1}$$

is the scattering matrix of the array, **a** and **b** are column vectors of forward and backward traveling waves, **I** is the unit matrix, and **z** is a matrix of normalized impedances with elements defined by

$$z_{pq} = Z_{pq}\,Z_{op}^{-1/2}\,Z_{oq}^{-1/2}$$

For an array of only two elements, the scattering matrix is

$$\mathbf{S} = \frac{1}{\Delta}\begin{bmatrix}\left(z_{11} - 1\right)\left(z_{22} + 1\right) - z_{12}z_{21} & 2z_{12} \\ 2z_{21} & \left(z_{11} + 1\right)\left(z_{22} - 1\right) - z_{21}z_{12}\end{bmatrix}$$

where $\Delta = (z_{11}+1)(z_{22}+1) - z_{12}z_{21}$ is the determinant of the matrix $(\boldsymbol{S}+\boldsymbol{I})$. Note that if the elements are identical and their inputs are well matched $z_{11} = z_{22} \approx 1$, then the coupling coefficient between the antennas is $S_{12} = S_{21} \approx z_{12}/2$. That is, as a general rule, in a well-matched array of identical elements, the coupling coefficient is approximately half or about 3 dB below the magnitude of the mutual impedance (or admittance). In the case of magnetic dipoles, Eq. (2.45) remains the same except that the scattering matrix is given by

$$\boldsymbol{S} = (\boldsymbol{I} - \mathbf{y})\,/\,(\boldsymbol{I} + \mathbf{y}) = 2(\boldsymbol{I} + \mathbf{y})^{-1} - \boldsymbol{I} \tag{2.46}$$

where **y** is the normalized admittance matrix with elements

$$y_{pq} = Y_{pq}\,Y_{op}^{-1/2}\,Y_{oq}^{-1/2}$$

A property of the scattering matrix formulation is that the N eigenvalues of $\boldsymbol{S}$ are the reflection coefficients of the ports. Thus,

$$\boldsymbol{b} = \boldsymbol{Sa}$$

and

$$\boldsymbol{Sa} = \lambda\boldsymbol{a}$$

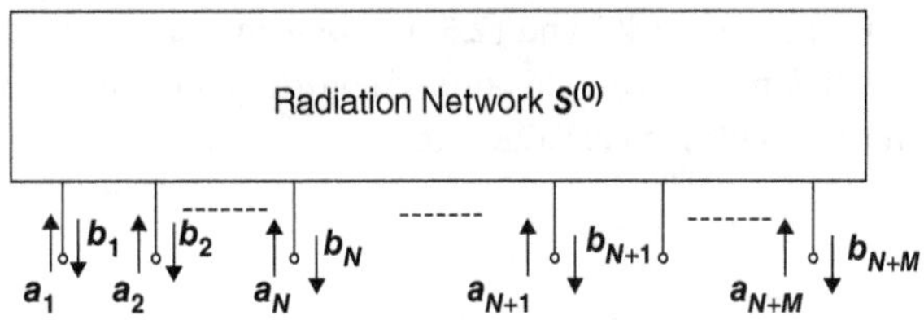

Figure 2.5 Scattering matrix representation of mutual coupling with N driven elements and M parasitics.

where λ is an eigenvalue. Also, for the n-th eigenvalue (n = 1,…,N) ordered from lowest to highest, the reflection coefficient is

$$\Gamma_n = \left(\frac{\boldsymbol{b}}{\boldsymbol{a}}\right)_n = \lambda_n$$

The smallest eigenvalue λ_1 is the port with the smallest reflection coefficient.

A network of radiation by a typical array with mutual coupling is illustrated in Figure 2.5. This shows N driven elements and M parasitics. The parasitics are loaded and are excited by mutual coupling in the array. Initially, it is assumed that there are only driven elements present and $M = 0$. The input traveling waves drive the array, which radiates and couples through the mechanism described by $\boldsymbol{S}^{(0)}$, which is given by Eq. (2.45). This coupling and creation of $\boldsymbol{S}^{(0)}$ is detailed in later chapters. For example, in Section 5.2.2, each array element has another network connected to it, and these are represented by N scattering matrices. The outward traveling waves are related to the driven waves by

$$\boldsymbol{b} = \boldsymbol{S}^{(0)}\boldsymbol{a} \tag{2.47}$$

With M parasitics present, this can be expressed as

$$\begin{bmatrix} \boldsymbol{b}_I \\ \boldsymbol{b}_P \end{bmatrix} = \begin{bmatrix} \boldsymbol{S}_{II}^{(0)} & \boldsymbol{S}_{IP}^{(0)} \\ \boldsymbol{S}_{PI}^{(0)} & \boldsymbol{S}_{PP}^{(0)} \end{bmatrix} \begin{bmatrix} \boldsymbol{a}_I \\ \boldsymbol{a}_P \end{bmatrix} \tag{2.48}$$

where the subscript I refers to the N driven ports and P to the parasitic ports. Suppose the parasitics are loaded with single port networks with a reflection coefficient Γ_i, i = 1,…M. Thus,

$$a_P = \boldsymbol{\Gamma} b_P$$

and, therefore, the second row of Eq. (2.48) gives

$$a_P = \left(\boldsymbol{\Gamma}^{-1} - S_{PP}^{(0)}\right)^{-1} S_{PI}^{(0)}\, a_P$$

As a result,

$$\boldsymbol{b}_I = \boldsymbol{S}_I\, \boldsymbol{a}_I \tag{2.49}$$

where

$$\mathbf{S}_{\mathrm{I}} = \mathbf{S}_{\mathrm{II}}^{(0)} + \mathbf{S}_{\mathrm{IP}}^{(0)}\left(\boldsymbol{\Gamma}^{-1} - \mathbf{S}_{\mathrm{PP}}^{(0)}\right)^{-1}\mathbf{S}_{\mathrm{PI}}^{(0)} \tag{2.50a}$$

or alternatively

$$\mathbf{S}_{\mathrm{I}} = \mathbf{S}_{\mathrm{II}}^{(\mathbf{0})} + \mathbf{S}_{\mathrm{IP}}^{(0)}\boldsymbol{\Gamma}\left(\mathrm{I} - \mathbf{S}_{\mathrm{PP}}^{(0)}\boldsymbol{\Gamma}\right)^{-1}\mathbf{S}_{\mathrm{PI}}^{(0)} \tag{2.50b}$$

which shows $\boldsymbol{b}_I = S_{II}^{(O)}\, a_I$ when $\boldsymbol{\Gamma} = \boldsymbol{0}$.

From Eqs. (2.49) and (2.50), it is seen that the loads can be used to tune the input reflection coefficient of the driven ports by adjusting the value of $\boldsymbol{\Gamma}$. Also, the driven ports have zero reflection at the smallest eigenvalue λ of

$$\boldsymbol{S}^{(I)}\,(\lambda)\,\boldsymbol{X} = 0$$

where the eigenvector $\boldsymbol{X}$ is the input excitation required to achieve this. More generally, the driven ports are reflectionless when

$$\boldsymbol{\Gamma}^{-1} = \boldsymbol{S}_{pp}^{(0)} - \boldsymbol{G}\,\boldsymbol{H}\,\boldsymbol{G} \tag{2.51}$$

where $\boldsymbol{G} = \boldsymbol{S}_{PI}^{(0)}\boldsymbol{S}_{IP}^{(0)}$and $\boldsymbol{H} = \left[\boldsymbol{S}_{PI}^{(0)}\boldsymbol{S}_{II}^{(0)}\boldsymbol{S}_{IP}^{(0)}\right]^{-1}$ both of which are square matrices involving the cross-coupling, and are full rank, indicating they are invertible. If the cross coupling is strong and the parasitics are reasonably well matched

$$\boldsymbol{\Gamma}^{-1} \cong -\boldsymbol{G}\,\boldsymbol{H}\,\boldsymbol{G}.$$

whereas if the coupling is weak,

$$\boldsymbol{\Gamma}^{-1} \cong \boldsymbol{S}_{pp}^{(0)}$$

which gives $\boldsymbol{S}^{(I)} \cong \boldsymbol{S}_{II}^{(0)}$. Substitute Eq. (2.51) into Eq. (2.50a) results in

$$\boldsymbol{b}_I = \left[\boldsymbol{S}_{II}^{(0)} - \boldsymbol{S}_{IP}^{(0)}\mathrm{G\;HG}\,\boldsymbol{S}_{PI}^{(0)}\right]\boldsymbol{a}_I \tag{2.52}$$

In practice Eq. (2.51) is only approximate except when the number of driven and parasitic ports are identical, that is $M = N$ when it is exact. In this identical case, the eigenvalues of $\boldsymbol{S}^{(I)}(\lambda)$ are all zeros. A disadvantage is that the reflection matrix predicted by Eq. (2.51) is that the parasitic loads may not be physically realisable with passive components.

As an example, consider six $L = 0.45\,\lambda$ dipoles in the configurations shown in Figure 2.6. Suppose as well, $s_x = 0.5\,\lambda$, $s_y = 0.3\,\lambda$ and $\alpha_o = 45°$. The methods used to analyse this array will be described in Section 4.3.

The scattering matrix for this configuration is

$$S^{(0)} = \left(\begin{array}{ccc|ccc}
0.991-0.18i & -0.077-0.033i & 0.055+0.078i & 0.09-0.166i & 0.036-0.055i & -2.884\times10^{-3}+0.092i \\
-0.077-0.033i & 1.028-0.094i & -0.072-0.117i & -0.029-3.465i\times10^{-5} & 0.054-0.138i & 0.036-0.055i \\
0.055+0.078i & -0.072-0.117i & 0.965-6.036i\times10^{-3} & -0.036+0.054i & -0.029-3.465i\times10^{-5} & 0.09-0.166i \\
\hline
0.09-0.166i & -0.029-3.465i\times10^{-5} & -0.036+0.054i & 0.965-6.036i\times10^{-3} & -0.072-0.117i & 0.055-0.078i \\
0.036-0.055i & 0.054-0.138i & -0.029-3.465i\times10^{-5} & -0.072-0.117i & 1.028-0.094i & -0.077-0.033i \\
-2.884\times10^{-3}+0.092i & 0.036-0.055i & 0.09-0.166i & 0.055+0.078i & -0.077-0.033i & 0.991-0.18i
\end{array}\right)$$

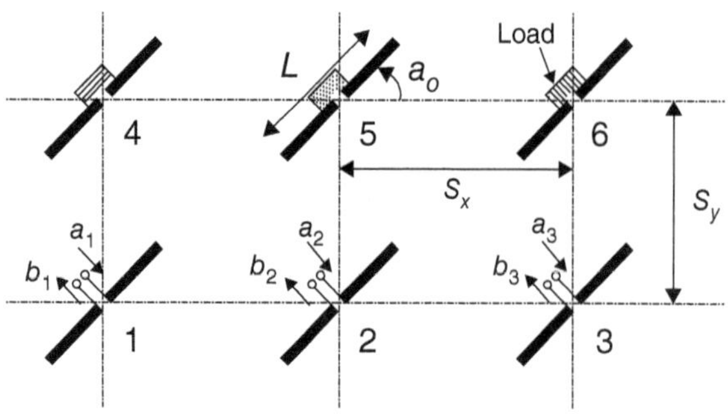

Figure 2.6 Example of reflectionless condition.

When $S^{(0)}$ is partitioned as shown into driven (1 to 3) and loaded (4 to 6) elements as are the matrices for evaluating Eq. (2.51), the result is

$$\boldsymbol{G}\ \mathbf{H}\ \boldsymbol{G} = \begin{pmatrix} -0.023-0.041\mathrm{i} & -3.902\times10^{-3}-2.107\mathrm{i}\times10^{-4} & 0.017+0.025\mathrm{i} \\ -3.902\times10^{-3}-2.107\mathrm{i}\times10^{-4} & -0.014-0.022\mathrm{i} & -4.237\times10^{-3}-0.012\mathrm{i} \\ 0.017+0.025\mathrm{i} & -4.237\times10^{-3}-0.012\mathrm{i} & -0.019-0.037\mathrm{i} \end{pmatrix}$$

Subtracting this matrix from the 3 x 3 matrix in the bottom right corner of $S^{(0)}$ gives $\boldsymbol{\Gamma}^{-1}$. When inverted this gives

$$\boldsymbol{\Gamma} = \begin{pmatrix} 1-0.018\mathrm{i} & 0.058+0.112\mathrm{i} & -0.03-0.047\mathrm{i} \\ 0.058+0.112\mathrm{i} & 0.95+0.083\mathrm{i} & 0.067+0.028\mathrm{i} \\ -0.03-0.047\mathrm{i} & 0.067+0.028\mathrm{i} & 0.972+0.144\mathrm{i} \end{pmatrix}$$

From these matrices, Eq. (2.50) can be shown to give $S^{(I)} = \mathbf{0}$, which implies from Eq. (2.49), the driven dipoles have zero input reflection for all excitations. Further, it is seen that the load matrix $\boldsymbol{\Gamma}$ in this example could be realised with passive components.

2.5 Radiation from Antennas in the Presence of Mutual Coupling

Once the complex amplitudes of the currents have been obtained from a mutual coupling formulation, the radiation and other circuit quantities can be determined as has been described. The conventional approach to calculate radiation is to neglect mutual coupling and to simply superpose the radiation from each element in isolation applied with the excitation in the array. If there is no excitation, there is no contribution from that element. However, in the presence of mutual coupling, although superposition still applies, all elements contribute to the radiation and most other parameters. In this section, the radiation in the far-field of an element is derived, and this is used to evaluate the radiation in the presence of mutual coupling.

2.5.1 Far-Field Radiation

At distances, r, which are large compared with the dimensions of the element when $|\mathbf{r}| >> |\mathbf{r}'|$ (see Figure 2.7), the distance R can be approximated by (e.g. Bird 2016)

$$R = \boldsymbol{r} - \boldsymbol{r}' \approx r - \hat{r}\cdot\boldsymbol{r}' \tag{2.53}$$

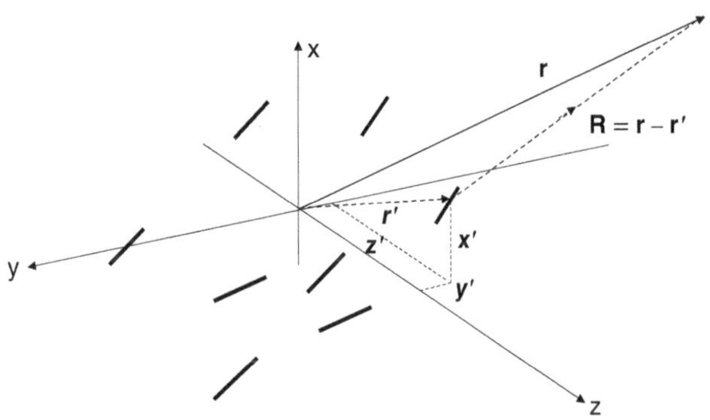

Figure 2.7 Geometry for calculation of far-field radiation from an array.

where $\boldsymbol{r}' = \hat{x}x' + \hat{y}y' + \hat{z}z'$, and the first two terms of a binomial series expansion of the distance are retained. In the far-field, the exponential function in Eq. (2.20) is more sensitive to approximation than the complex amplitude of the integrand. Therefore, to retain the same order of approximation, Eq. (2.53) is used in the exponential but in the amplitude of the integrand

$$\frac{1}{R} \approx \frac{1}{r} \text{ and } \hat{R} \approx \hat{r} \tag{2.54}$$

With these approximations, Eq. (2.20) simplifies to

$$\mathbf{E} \approx \frac{jk}{4\pi}\frac{e^{-jkr}}{r}\hat{r} \times \int_A \left[\mathbf{M_s} - \eta_o \mathbf{J_s} \times \hat{r}\right] \exp\left(jk\hat{r}\cdot\mathbf{r}'\right) \quad dS' \tag{2.55a}$$

and

$$\mathbf{H} = \frac{1}{\eta_o} \mathbf{r} \times \mathbf{E} \tag{2.55b}$$

In the case of an aperture situated in the x-y plane on which there are fields $\boldsymbol{E_a}$ and $\boldsymbol{H_a}$, the surface currents are

$$\boldsymbol{J_s} = \hat{n} \times \boldsymbol{H_a} \text{ and } \boldsymbol{M_s} = -\hat{n} \times \boldsymbol{E_a} \tag{2.56}$$

From Eq. (2.55a), the electric field is

$$\boldsymbol{E} \approx -\frac{jk}{4\pi}\frac{e^{-jkr}}{r}\hat{r} \times \int_A \left[\hat{z} \times \boldsymbol{E_a} + \eta_o\left(\hat{z} \times \boldsymbol{H_a}\right) \times \hat{r}\right] \exp\left(jk\hat{r}\cdot\boldsymbol{r}'\right) dS' \tag{2.57}$$

Now, making use of

$$\hat{x} = \hat{r}\, sin\, \theta\, cos\, \phi + \hat{\theta}\, cos\theta\, cos\phi - \hat{\varphi}\, sin\, \phi \tag{2.58a}$$

$$\hat{y} = \hat{r}\, sin\, \theta\, sin\, \phi + \hat{\theta}\, cos\theta\, sin\phi + \hat{\varphi}\, cos\, \phi \tag{2.58b}$$

$$\hat{z} = \hat{r}\, \cos\theta - \hat{\theta}\, \sin\theta \tag{2.58c}$$

the spherical components of Eq. (2.57) are

$$E_r = 0 \tag{2.59a}$$

$$E_\theta \approx \frac{jk}{4\pi}\frac{e^{-jkr}}{r}\left[\left(N_x cos\phi + N_y sin\phi\right) + \eta_o cos\theta\left(-L_x sin\phi + L_y cos\phi\right)\right] \tag{2.59b}$$

$$E_\phi \approx \frac{jk}{4\pi}\frac{e^{-jkr}}{r}\left[\, cos\theta\left(-N_x sin\phi + N_y cos\phi\right) - \eta_o\left(L_x cos\phi + L_y sin\phi\right)\right] \tag{2.59c}$$

where N_x, N_y and L_x, L_y are rectangular components of the transforms

$$\boldsymbol{N}(u,v) = \int_A \mathbf{E_a}\left(x', y'\right) \exp\left(j2\pi\left(ux' + vy'\right)\right) dx'dy' \tag{2.60a}$$

$$\boldsymbol{L}(u,v) = \int_A \mathbf{H_a}\left(x', y'\right) \exp\left(j2\pi\left(ux' + vy'\right)\right) dx'dy' \tag{2.60b}$$

and

$$u = \frac{1}{\lambda} sin\, \theta\, cos\, \phi \tag{2.61a}$$

$$v = \frac{1}{\lambda} sin\, \theta\, sin\, \phi \tag{2.61b}$$

It is observed from Eq. (2.60) that the components of **N** and **L** are two-dimensional Fourier transforms of the aperture field components. These transforms can be evaluated by numerical quadrature or by means of the fast Fourier transform (Brigham 1974).

2.5.2 Magnetic Current Only

Consider the special case of array elements where $\boldsymbol{J_s} = 0$. A typical example is a waveguide aperture in a ground plane. Suppose the array has M elements in which element p is driven with a complex amplitude a_p. However, because of mutual coupling and reflection, the actual amplitude of the field on element p is $C_p = (a_p + b_p)$. These amplitudes are related via a scattering matrix $\mathbf{S_{\Sigma m}}$ through $\boldsymbol{b} = \boldsymbol{S_{\Sigma m} a}$. The radiation from element p in this array is

$$\boldsymbol{E_p} \approx -\frac{jk}{4\pi}\frac{e^{-jkr}}{r}\left(a_p + b_p\right)\left\{\hat{r}\times\hat{z}\times\boldsymbol{N_p}(u,v)\right\} \tag{2.62}$$

where $(a_p + b_p)$ is the amplitude of aperture field $E_{ap}(x', y')$ and

$$\boldsymbol{N_p}(u,v) = \int_{A_p} \boldsymbol{E_{ap}}\left(x',y'\right)\,\exp\left(j2\pi\left(ux' + vy'\right)\right)dx'dy'$$

in which A_p is aperture p. The positive sign on b_p was chosen to be consistent with Maxwell's equations and Eq. (2.44). The total radiation field is a superposition of all M elements

$$\boldsymbol{E} \approx -\frac{jk}{4\pi}\frac{e^{-jkr}}{r}\hat{r}\times\hat{z}\times\sum_{p=1}^{M}\left(a_p + b_p\right)\boldsymbol{N_p}(u,v) \tag{2.63}$$

Define a column vector of M vector mode functions N_p. Therefore,

$$\boldsymbol{N}(u,v) = \begin{bmatrix}\boldsymbol{N_1}\\ \boldsymbol{N_2}\\ \vdots \\ \boldsymbol{N_M}\end{bmatrix} = \begin{bmatrix}\boldsymbol{N_x}(u,v)\\ \boldsymbol{N_y}(u,v)\end{bmatrix}$$

where $\mathbf{N_x}$ and $\mathbf{N_y}$ are column vectors corresponding to the x- and y-components of the transforms of the mode functions.

The far-fields are now expressed as

$$E_\theta = -\frac{jk}{4\pi}\frac{e^{-jkr}}{r}\left[\boldsymbol{I} + \mathbf{S_{\Sigma m}}\right]\boldsymbol{a}\left(\boldsymbol{N_x}\cos\phi + \boldsymbol{N_y}\sin\phi\right) \tag{2.64a}$$

$$E_\varphi = \frac{jk}{4\pi}\frac{e^{-jkr}}{r}\left[\boldsymbol{I} + \mathbf{S_{\Sigma m}}\right]\boldsymbol{a}\cos\theta\left(\boldsymbol{N_x}\cos\phi - \boldsymbol{N_y}\sin\phi\right) \tag{2.64b}$$

The total fields are related to all the incident modes via all elements of the system scattering matrix and not just the diagonal terms as is assumed in simple representations. To obtain the fields in the near or intermediate field regions, it is preferable to return to Eq. (2.20) and to calculate the magnetic field separately by integrating across the aperture field (magnetic current), as given by Eq. (2.20b), and then obtain the other field components by means of Maxwell's equations. Thus, the transverse magnetic field, $\mathbf{H}_T$, is found this way, while the remaining magnetic field component is obtained from $\nabla\cdot\boldsymbol{H} = 0$. Thereafter, the electric field is calculated from $\boldsymbol{E} = (-j\eta_o/k)\,\nabla\times\boldsymbol{H}$.

2.5.3 Electric Current Only

In the dual case of Section 2.4.2, consider when $\boldsymbol{M_s} = 0$. Again, consider an array of M elements, which is an array of cylindrical metal conductors or several small reflectors. In this case, the electric field due to element p at large distances is given by

$$\boldsymbol{E_p} \approx -\frac{jk\eta_o}{4\pi}\frac{e^{-jkr}}{r}\hat{r}\times\int_{A_p}\left(\boldsymbol{J_{sp}}\times\hat{r}\right)\exp\left(jk\hat{r}\cdot\boldsymbol{r'}\right)dS' \tag{2.65}$$

Suppose the metal conductor is fed some way by a voltage V_p and current on the conductor is described by the vector function $I_p\boldsymbol{g_p}(x, y)$ where I_p is the complex amplitude. In the presence of this and the other conductors, the current produced on this body has a complex amplitude given through the impedance matrix $\boldsymbol{Z}_{\Sigma e}$. The currents on all conductors are obtained by inverting this matrix, and thus, $\boldsymbol{I} = \boldsymbol{Z}_{\Sigma e}{}^{-1}\boldsymbol{V}$. Therefore, the radiation from element p is given by

$$\begin{aligned}\boldsymbol{E}_p &\approx -\frac{jk\eta_o}{4\pi}\frac{e^{-jkr}}{r} I_p \hat{r} \times \int_{A_p} \left(\boldsymbol{j}_p(x,y) \times \hat{r}\right) \exp\left(jk\hat{r}\cdot\boldsymbol{r}'\right) dS' \\ &= -\frac{jk\eta_o}{4\pi}\frac{e^{-jkr}}{r} I_p \left[\hat{r} \times \boldsymbol{j}_p(r,\theta,\phi) \times \hat{r}\right] \end{aligned} \tag{2.66}$$

where $\boldsymbol{j}_p(r,\theta,\phi) = \int_{A_p} \boldsymbol{g}_p(x,y) \exp\left(jk\hat{r}\cdot\boldsymbol{r}'\right) dS'$.

The total field is the summation of all M components. Thus,

$$\boldsymbol{E} \approx -\frac{jk\eta_o}{4\pi}\frac{e^{-jkr}}{r}\left[\hat{r} \times \left(\boldsymbol{Z}_{\Sigma e}{}^{-1}\boldsymbol{V}\right) \boldsymbol{j}(r,\theta,\phi) \times \hat{r}\right] \tag{2.67}$$

If $\boldsymbol{j}(r,\theta,\phi)$ is in the z-direction only,

$$\boldsymbol{E}_\theta = -\frac{jk}{4\pi}\frac{e^{-jkr}}{r} sin\theta \left(\boldsymbol{Z}_{\Sigma e}{}^{-1}\boldsymbol{V}\right) j_z(r,\theta,\phi) \tag{2.68}$$

Once again, the presence of the inverted system impedance matrix ensures that all elements contribute to the total electric field. When both magnetic and electric sources are present, the total electric field is a summation of Eqs. (2.64) and (2.67).

2.6 Conclusion

Fundamental field quantities have been reviewed in this chapter before an elementary model of mutual coupling was introduced. The initial approach is through an incremental form of Maxwell's equations and the inverse square law that applies to an elemental radiator. This enables a simple and direct approach to understanding the presence of mutual coupling in a radiating system. The impact of mutual coupling on the far-field radiation was described and demonstrated through two examples, one involving magnetic currents and a dual problem with electric currents. The results obtained here will be applied in later chapters to specific current distributions.

References

Bird, T.S. (2016). *Fundamentals of Aperture Antennas and Arrays*. Chichester, UK: Wiley.

Booker, H.G. (1946). Slot aerials and their relation to complementary wire aerials (Babinet's principle). *J. Inst. Elect. Eng. Pt IIIA* 93: 620–626.

Brigham, E.O. (1974). *The Fast Fourier Transform*. Eaglewood Cliffs, N.J.: Prentice Hall, Inc.

Clarricoats, P.J.B. and Olver, A.D. (1984). *Corrugated Horns for Microwave Antennas*. London, UK: Peter Peregrinus Ltd.

Collin, R.E. (1960). *Field Theory of Guided Waves*. New York, USA: McGraw-Hill.

Hansen, R.C. (1998). *Phased Array Antennas*. New York, USA: Wiley.

Harrington, R.F. (1961). *Time Harmonic Electromagnetic Fields*. New York, USA: McGraw-Hill.

Jackson, J.D. (1998). *Classical Electrodynamics*, 3e. Hoboken, NJ, USA: Wiley.

Jones, D.S. (1964). *The Theory of Electromagnetism*. London, UK: Pergamon Press Ltd.

Kraus, J.D. and Carver, K.R. (1973). *Electrodynamics*, 2e. Tokyo, Japan: McGraw-Hill Kogakusha.

Schelkunoff, S.A. and Friis, H.T. (1952). *Antennas Theory and Practice*. New York, USA: Wiley.

Silver, S. (1946). *Microwave Antenna Theory and Design'*, first published by. New York: McGraw-Hill Co. Reprint published by Peter Peregrinus Ltd, London, UK, 1984.

Vujević, S., Modrić, T., and Lovrić, D. (2011). The difference between voltage and potential difference. In: *Proc. Joint INDS'11 & ISTET'11*, Klagenfurt, Austria, 25–27 July, 1–7.

3

Methods in the Analysis of Mutual Coupling in Antennas

Trevor S. Bird

Global Big Data Technology Centre, University of Technology Sydney, Ultimo, NSW 2007, Australia
Antengenuity, Eastwood, NSW 2122, Australia

3.1 Introduction

In full-scale antennas and arrays, accurate design is essential in achieving a desired performance. For arrays, interaction between elements modifies the overall radiation pattern as well as the individual element characteristics. If care is not taken, mutual coupling can reduce the overall system performance, for example, by enhancing the generation of grating lobes in arrays (Hansen 1998). However, comprehending the details behind its effect can help to understand mutual coupling and may even be used to improve overall performance. In the previous chapter, coupling between elemental dipoles was introduced from fundamentals. In this chapter, a more rigorous theory is presented to support this description, and techniques for analyzing mutual coupling are provided. The variety of these methods and their relationships are summarized by the tree diagram in Figure 3.1. At the top of the tree, the methods are split between finite and infinite arrays.

Methods for finite arrays generally follow on from those for individual elements. From about the start of the 1950s, individual dipole antennas were analyzed by the finite methods such as the subdomain and collocation methods (Finlayson 1972) that ultimately led to the method of moments (MoM). This method was adopted in electromagnetics with alacrity from the mid-1960s and particularly after the publication of a book on this topic by Harrington (1968). Not long afterwards, wire antenna arrays and more complicated array elements such as multimode waveguide appeared, and finite methods achieved considerable success in practical applications. In parallel with these developments, from the late 1950s period, following on from solid-state physics, particularly connected with lattice structures (Merzbacher 1961), it was recognized that large arrays could be approximated by infinite periodic arrays. In that way, Floquet's method was adopted to simplify the incorporation of practical elements through the building blocks provided by a unit cell approach and for taking mutual coupling effects into account. In practical arrays, there is the issue of the termination at the edges of the array, and various techniques were adopted to include these effects and reduce their influence. There is a substantial body of literature on periodic arrays produced at that time, and also it was convenient because of the limited computing power available.

From the very beginning of wireless, Hertz (1888) used the Poynting vector method to calculate the energy flow through a large sphere surrounding his dipole radiator. This method was adopted for calculating radiation resistance of a general antenna (Balanis 1983, p. 112). Also, Brillouin (1922) used the Poynting vector to obtain the radiation resistance but this time integrating it over the surface of the antenna, which resulted in a power series in frequency. This

Mutual Coupling Between Antennas, First Edition. Edited by Trevor S. Bird.

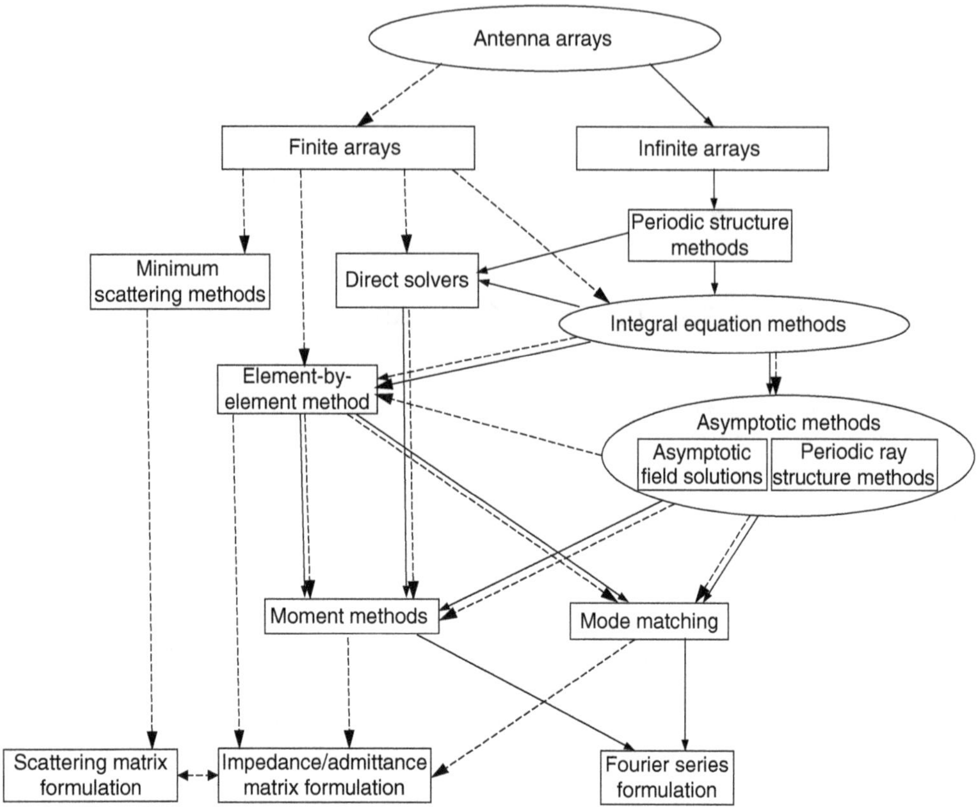

Figure 3.1 Relationships of methods for analyzing mutual coupling. Source: After Bird T. S. (2016): 'Fundamentals of aperture antennas and arrays', John Wiley & Sons, Chichester UK.

surface integration approach, called the electromotive force (emf) method in the literature, was adopted by Pistolkors (1929) to find the radiation resistance of various dipole configurations. Tai (1949) showed the applicability of this method to calculate the input impedance of an antenna. The Poynting vector method was used by Bontsch-Bruewitsch (1926) and also Knudsen (1952a,b) to analyze coupling between dipoles. Both the emf and Poynting methods mentioned above are equivalent and can be converted into the other by means of Gauss's law, as shown by Bechmann (1931). The emf method ultimately gained wide acceptance for calculating the complex mutual impedance after the work of Carter (1932). This method allows the calculation of the susceptance as well as the resistance. Furthermore, Carter used reciprocity with the emf method to determine expressions for self- and mutual impedances for a variety of two-dipole arrangements. Carter's work had a big impact on much of the subsequent literature on antenna coupling because for the first time the coupling problem could be expressed as an equivalent circuit. This development was led by Hallén (1938), followed by King and Harrison (1944) and Tai (1948, 1949). All the above-referenced publications prior to the 1950s were concerned with dipole elements. Coupling to different antennas was studied by Surtees (1952) who obtained expressions for the coupling between a slot and a dipole. His measured results are in good agreement with the theory. An analysis by a variational technique of two different-sized dipoles was reported by Levis and

Tai (1956). Galejs (1969) obtained the mutual impedance of two antennas in the presence of a plasma.

The coupling between apertures such as waveguides and slots was first investigated by Booker (1946), who derived the admittance of a slot from the impedance of its dual, the dipole. The variational nature of the impedance formula was established by Miles (1949) and Storer (1952). General methods of obtaining admittances in ground planes were described by Harrington (1961). An early impetus for analyzing mutual coupling effects in aperture antennas came from the need to counter blind spots in large arrays, which subsequently were found to be due to surface waves on the array excited by coupling and their reflection and transmission at discontinuities. Throughout the 1950s, Carter's procedure was used to determine first the self-impedance of the element and the mutual impedances between neighbors, and then to sum these appropriately to yield the input impedance which correctly accounts for the presence of all the other elements in the array. However, before computers, this process was very tedious, and the computation needed at the time was impractical for a large array. Other more detailed methods arose from the extension of methods for solving integral equations as well as others based on modal methods or periodic structures with Floquet modes. In two-dimensional arrays, Edelberg and Oliner's (1960) periodic structure approach proved highly effective.

The method of periodic structures using the Floquet approach found favor for a range of array structures and provided a powerful means of understanding coupling effects in large arrays. Array edge effects were seen to be important for small- and medium-sized arrays where the infinite array solution is a poor approximation to reality. This plus advances in computers in the 1960s and 1970s led to the development of methods of analyzing mutual coupling in finite arrays. Expressions for self-impedance of rectangular waveguides were developed by Levine and Schwinger (1950), Lewin (1951), and Galejs (1969). One of the earliest to describe a technique for the mutual coupling between rectangular waveguides was Mailloux (1969a,b). This work was complemented by the work of Golden and Stewart (1969), Hockham and Walker (1973), and Harrington and Mautz (1976). Circular apertures were studied by Mishustin (1965) for the self-admittance, although a result for thin-walled waveguide was given earlier by Marcuvitz (1986) and others, Vajnshtejn (1950), and Noble (1958). The mutual admittance between circular waveguides was obtained by Steyskal (1974) and Bailey and Bostian (1974). The former found an asymptotic approximation to the mutual coupling expression, and this was corrected by Bird (1979). The extensions to obtain the mutual admittance of different-sized rectangular and circular waveguide apertures situated in a ground plane were found by Bird (1990, 1996). Throughout the 1980s and 1990s, as computer technology became more capable, new methods were introduced. For example, Fenn et al. (1982) applied the moment method to a rectangular waveguide phased array, and Bird (2016) combined a mode matching (MM) and Greens' function approach to analyze arrays of rectangular horns. At the same time, the development of numerical methods and computer codes allowed the analysis of more complicated antennas such as arrays and their supporting structures. The computer codes available today allow mutual coupling to be included in design as a matter of course. Nevertheless, the physical aspects of mutual coupling need to be understood to minimize its disadvantages and maximize its advantages. Mutual coupling can even provide benefits in some designs. For example, in an array used to feed a reflector antenna used to create a shaped beam, the gain achievable across the beam can be 0.2–0.3 dB higher if mutual coupling is properly incorporated. While this increase appears relatively modest, in some applications, such as in satellite communications, modest gain increases of the order of 0.1 dB are generally achieved at great expense (Bird and Sroka 1992).

As illustrated in Figure 3.1, the two basic approaches for analysing mutual coupling are through finite and infinite arrays. Some methods are common to both, as indicated in the figure. However, the end solution is usually different because the impedance and pattern characteristics of infinite periodic arrays are efficiently obtained by means of Fourier series methods. Methods are described in this chapter for coupling in an infinite array and also by an element-by-element approach for finite arrays. The common links are indicated by a solid line for infinite arrays and dashed for finite arrays. At the next level in Figure 3.1, there is the minimum scattering method, integral equations, and direct solvers. The minimum scattering method is described in a later section of this chapter. Integral equations can be used for both finite and infinite arrays, although there are variations due to periodicity that can simplify the infinite array case. The direct solvers block in Figure 3.1 refers to types of direct numerical methods such as finite element or finite difference where the full finite array is represented. Some parts of this method may use integral equations to simplify the problem such as with special media which can be represented by an appropriate Green's function or with MoM. The range of techniques and their linkage to mutual coupling can be sometimes confusing, and to help keep this in mind, Figure 3.1 will be referred to link the various sections of this chapter together.

The next level down in the schematic hierarchy of Figure 3.1 is the element-by-element approach. This is the approach used in Section 2.3 to fill an impedance or admittance matrix and requires only two elements of the array being considered at a time while the other elements are terminated in such a way that no signal is absorbed or scattered. This is described further in the next section. Asymptotic methods can be used to simplify the expressions for self- and mutual impedances. When used successfully, they can make it easier or faster to compute coupling, especially when integral equations are involved. Later chapters use asymptotic methods to simplify the analysis for several types of elements. To go to the next steps toward calculating actual values of impedance or the scattering parameters, except for direct methods, actual field solutions will still be required. Various methods here can be used, but in coupling problems, two popular approaches are the MoM for general antennas and MM for aperture antennas such as horns. Both methods are described in this chapter. As a result, at the end of these calculations, parameters will be available for additional analysis and calculation of parameters such as radiation patterns and gain and other data for more extensive system assessment.

3.2 Mutual Coupling in Antennas with Continuous Sources

3.2.1 Impedance and Admittance with Continuous Sources

We now extend the incremental results obtained in Section 2.3 to continuous sources, commencing with the general fields and sources illustrated in Figure 3.2. Consider electric and magnetic current pair $(\boldsymbol{J}_a, \boldsymbol{M}_a)$ on an object a that neighbors another body b that are linked through fields $(\boldsymbol{E}_a, \boldsymbol{H}_a)$. In turn, the current pair $(\boldsymbol{J}_b, \boldsymbol{M}_b)$ on body b reacts on body a through fields $(\boldsymbol{E}_b, \boldsymbol{H}_b)$. These currents are continuous; otherwise, electromotive and magnetomotive force potentials would exist due to the gaps and the current would be different. The mutual impedance/admittance is obtained by summing up all the elemental fields from the continuous currents. Following Section 2.4, suppose the elemental electric field at antenna b due to an element of current of antenna a is $\boldsymbol{dE}_b = \overline{\overline{\boldsymbol{G}^e}}(\boldsymbol{b}, \boldsymbol{a}) \cdot \boldsymbol{dJ}_a$, where $\overline{\overline{\boldsymbol{G}^e}}(\boldsymbol{b}, \boldsymbol{a})$ is the dyadic electric field due to a delta function vector source. Thus, from Eq. (2.23), by summing all contributions,

$$Z_{ba} = -\frac{1}{I_a}\iint_{S_b} dS_b \boldsymbol{J_b} \cdot \left[\iint_{S_a} dS_a \overline{\overline{\boldsymbol{G}^e}}(\boldsymbol{b}, \boldsymbol{a}) \cdot \boldsymbol{J_a} \cdot \hat{\boldsymbol{n}}_a\right] \cdot \hat{\boldsymbol{n}}_b \tag{3.1}$$

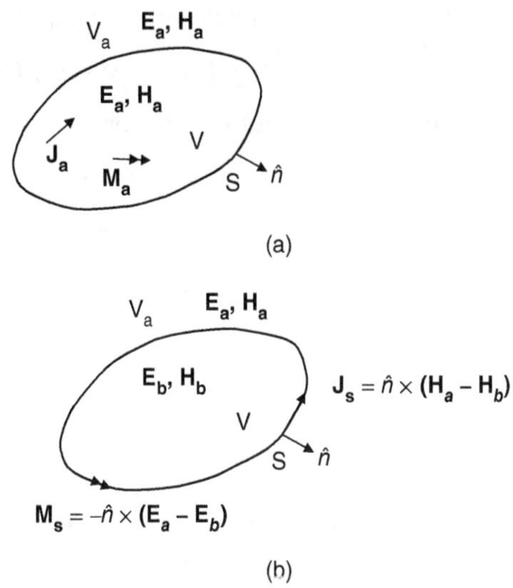

Figure 3.2 Field coupling and reaction of two sources on each other through their fields.

In the same way, for mutual admittance or from duality in Table 2.1, we have

$$Y_{ba} = -\frac{1}{V_a}\iint_{S_b} dS_b \boldsymbol{M}_b \cdot \left[\iint_{S_a} dS_a \overline{\overline{\boldsymbol{G}^h}}(\boldsymbol{b},\boldsymbol{a}) \cdot \boldsymbol{J}_a \cdot \hat{\boldsymbol{n}}_a\right] \cdot \hat{\boldsymbol{n}}_b \tag{3.2}$$

where $\overline{\overline{\boldsymbol{G}^h}}(\boldsymbol{b},\boldsymbol{a})$ is the dyadic magnetic field at point b due to a delta function source at point a.

3.2.2 Reaction

The concept of reaction (Rumsey, 1954) between bodies is related to mutual coupling, as will now be demonstrated. Consider the rotational pair of Maxwell equations in the first row of Table 2.2 due to the source pairs $(\boldsymbol{J}^a, \boldsymbol{M}^a)$ and $(\boldsymbol{J}^b, \boldsymbol{M}^b)$. Thus, for a

$$\nabla \times \boldsymbol{E}^a = -\boldsymbol{M}^a - j\omega\mu\boldsymbol{H}^a \tag{3.3a}$$

$$\nabla \times \boldsymbol{H}^a = \boldsymbol{J}^a + j\omega\varepsilon\boldsymbol{E}^a \tag{3.3b}$$

and, similarly, for pair b. Now, project $\boldsymbol{E}^b$ onto Eq. (3.3b) as

$$\left(\nabla \times \boldsymbol{H}^a\right) \cdot \boldsymbol{E}^b = \boldsymbol{E}^b \cdot \boldsymbol{J}^a + j\omega\varepsilon\boldsymbol{E}^a \cdot \boldsymbol{E}^b \tag{3.4a}$$

and do the same for $\boldsymbol{H}^a$ on the b equivalent of Eq. (3.3a) giving

$$\left(\nabla \times \boldsymbol{E}^b\right) \cdot \boldsymbol{H}^a = -\boldsymbol{M}^b \cdot \boldsymbol{H}^a - j\omega\mu\boldsymbol{H}^b \cdot \boldsymbol{H}^a \tag{3.4b}$$

Subtracting Eq. (3.4b) from Eq. (3.4a) and using the vector identify

$$\nabla \cdot (\mathbf{A} \times \mathbf{B}) = \mathbf{B} \cdot \nabla \times \mathbf{A} - \mathbf{A} \cdot \nabla \times \mathbf{B}$$

which gives

$$-\nabla \cdot \left(\mathbf{E}^b \times \mathbf{H}^a\right) = j\omega\varepsilon\mathbf{E}^a \cdot \mathbf{E}^b + j\omega\mu\mathbf{H}^a \cdot \mathbf{H}^b + \mathbf{E}^b \cdot \mathbf{J}^a + \mathbf{H}^a \cdot \mathbf{M}^b$$

When the sources and fields are swapped and the process repeated, we obtain

$$-\nabla\cdot\left(\mathbf{E}^a\times\mathbf{H}^b\right)=j\omega\varepsilon\mathbf{E}^a\cdot\mathbf{E}^b+j\omega\mu\mathbf{H}^a\cdot\mathbf{H}^b+\mathbf{E}^a\cdot\mathbf{J}^b+\mathbf{H}^b\cdot\mathbf{M}^a$$

Subtraction of these two equations eliminates the energy terms yielding

$$-\nabla\cdot\left(\mathbf{E}^a\times\mathbf{H}^b-\mathbf{E}^b\times\mathbf{H}^a\right)=\mathbf{E}^a\cdot\mathbf{J}^b+\mathbf{H}^b\cdot\mathbf{M}^a-\mathbf{E}^b\cdot\mathbf{J}^a-\mathbf{H}^a\cdot\mathbf{M}^b \tag{3.5}$$

Making use of the divergence theorem, we find that Eq. (3.5) gives

$$-\oint\!\!\!\oint\left(\mathbf{E}^a\times\mathbf{H}^b-\mathbf{E}^b\times\mathbf{H}^a\right)\cdot d\mathbf{S}=\iiint\left(\mathbf{E}^a\cdot\mathbf{J}^b+\mathbf{H}^b\cdot\mathbf{M}^a-\mathbf{E}^b\cdot\mathbf{J}^a-\mathbf{H}^a\cdot\mathbf{M}^b\right)dv \tag{3.6}$$

If the volume integral on the right side contains no sources, we see that Eq. (3.6) becomes

$$\oint\!\!\!\oint\left(\mathbf{E}^a\times\mathbf{H}^b-\mathbf{E}^b\times\mathbf{H}^a\right)\cdot d\mathbf{S}=0 \tag{3.7}$$

which is called the Lorentz reciprocity theorem. This theorem shows that the fields on the bounding surface of a source-free region are reciprocal. It can be shown further that Eq. (3.7) also applies to sources of finite extent (Harrington 1961). In addition, reaction has been defined by Harrington as equivalent to Eqs. (3.6) and (3.7). The reaction of field a on source b is defined as

$$\langle a,b\rangle=\iiint\left(\mathbf{E}^a\cdot\mathbf{J}^b-\mathbf{H}^a\cdot\mathbf{M}^b\right)dv \tag{3.8}$$

and, similarly, for field b on source a. As a result of Eq. (3.7)

$$\langle b,a\rangle=\langle a,b\rangle\,. \tag{3.9}$$

Reaction describes the interaction between sources and field in a given domain. This concept has been likened to the electromagnetic equivalent of Newton's second law (Richmond 1961). The self-reaction $\langle a,a\rangle$ of field a on source a is also possible and is due to reflection of part of the emanating field back to the source.

3.2.3 Definition of Circuit Quantities

Relevant to mutual coupling, reaction can be used to determine network quantities. Consider the reaction of field a on a current source b that is input to a network. From Eq. (3.6), for a current source this is expressed as

$$\langle a,b\rangle=\iiint\mathbf{E}^a\cdot\mathbf{J}^b dv=\int\mathbf{E}^a\mathrm{i}^b\cdot d\boldsymbol{\ell}=\mathrm{i}^b\int\mathbf{E}^a\cdot d\boldsymbol{\ell}=-\mathrm{v}^a\mathrm{i}^b. \tag{3.10}$$

With a voltage source, Eq. (3.6) gives

$$\langle a,b\rangle=-\iiint\mathbf{H}^a\cdot\mathbf{M}^b dv=\oint\mathbf{H}^a\mathrm{v}^b\cdot d\boldsymbol{\ell}=\mathrm{v}^b\oint\mathbf{H}^a\cdot d\boldsymbol{\ell}=\mathrm{v}^b\mathrm{i}^a. \tag{3.11}$$

Suppose all current sources in an $N\times N$ network are open circuited apart from the source at port b ($b=1,\ldots N$). By definition, the mutual impedance at port a ($a=1,\ldots N$) is

$$z_{ab}=\frac{v^a}{i^b}=-\frac{\langle a,b\rangle}{i^a i^b}=-\frac{1}{i^a i^b}\iiint\left(\mathbf{E}^a\cdot\mathbf{J}^b-\mathbf{H}^a\cdot\mathbf{M}^b\right)dv \tag{3.12}$$

The elements of an impedance matrix can be computed from Eq. (3.12).

Another approach also arises from the Lorentz reciprocity theorem, Eq. (3.7). For two antennas a and b surrounded by surfaces S^a and S^b, Eq. (3.6) can be rewritten as

$$\iint_{S^a}\left(\mathbf{E}^a\times\mathbf{H}^b-\mathbf{E}^b\times\mathbf{H}^a\right)\cdot\hat{n}dS=-\iint_{S^b}\left(\mathbf{E}^a\times\mathbf{H}^b-\mathbf{E}^b\times\mathbf{H}^a\right)\cdot\hat{n}dS$$

where $\hat{n}$ is the normal vector to each surface. Each of these integrals is also the reaction quantity; thus,

$$\langle a, b\rangle = \iint_{S^a} (\mathbf{E}^a \times \mathbf{H}^b - \mathbf{E}^b \times \mathbf{H}^a) \cdot \hat{n} dS \quad (3.13)$$

and, therefore, from Eq. (2.37), the impedance can also be obtained from

$$z_{ab} = -\frac{1}{i^a i^b} \iint_{S^a} (\mathbf{E}^a \times \mathbf{H}^b - \mathbf{E}^b \times \mathbf{H}^a) \cdot \hat{n} dS \quad (3.14)$$

Equation (3.14) is a convenient expression for calculation of mutual impedance, where the integration is arbitrary but encloses only one antenna, in this case a. It has some useful approximate forms as well. It can also be used to calculate the coupling between two antennas. The calculation can be simplified if they are in the far-field of each other or computed directly when they are in near-field such as in probe corrected measurement (e.g. Paris et al. 1978).

Suppose that antenna a is transmitting and antenna b is open circuited. By reversing the roles of the antennas such that antenna b transmits fields ($\mathbf{E}^{b\prime}$, $\mathbf{H}^{b\prime}$) in a new environment (Richmond 1961). This new situation does not necessarily include antenna a as illustrated in Figure 3.3. In this case,

$$z_{ab} = -\frac{1}{i^{ba} i^{bb\prime}} \iint_{S^b} (\mathbf{E}^{b\prime} \times \mathbf{H}^a - \mathbf{E}^a \times \mathbf{H}^{b\prime}) \cdot \hat{n} dS \quad (3.15)$$

where the prime indicates the reversal of roles, $i^{bb'}$ is the input current due to b alone, and i^{ba} is the current input to a with b present but with it open circuited. The new environment can be any convenient situation (Richmond 1961) such as the one in which antenna a operates, a perfectly conducting shell on S^b, an extension of the structure of antenna b, and a third antenna or more. Thus, Eq. (3.15) can be used as a basis for filling the impedance matrix of an array.

When S^b is a perfectly conducting shell, the first term in Eq. (3.15) vanishes. If the induced electric current on antenna b is $\mathbf{J}^{b\prime} = -\hat{\mathrm{n}} \times \mathbf{H}^{b\prime}$, Eq. (3.15) simplifies to

$$z_{ab} = \frac{1}{i^{ba} i^{bb\prime}} \iint_{S^b} \mathbf{J}^{b\prime} \cdot \mathbf{E}^a dS \quad (3.16)$$

This equation is convenient for mutual coupling calculations especially when one of the antennas has holes or slots. In that situation, S^b can be chosen on the perfectly conducting surface, and

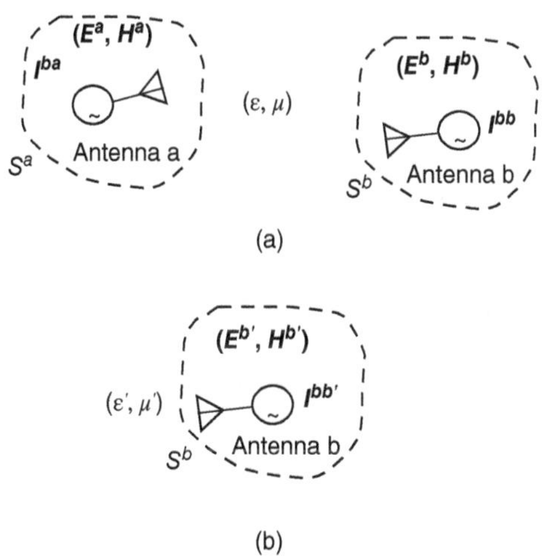

Figure 3.3 Antenna reciprocity. (a) Antenna a transmits and Antenna b receives; (b) Antenna b transmits in a new environment.

therefore, the integral is zero except in the holes, which could be the aperture of a waveguide or horn. Equation (3.15) is not a convenient form for cylindrical wire antennas except when S^b crosses the region between the wires. A more convenient form derived by Richmond (1961) is

$$z_{ab} = -\frac{1}{i^{ba} i^{bb}} \iint_{S^b} \left(\mathbf{E}^b \times \mathbf{H}^{ai} - \mathbf{E}^{ai} \times \mathbf{H}^b\right) \cdot \hat{n} dS \tag{3.17}$$

where now the environment is the same as the one antenna a operates in and the field has been replaced by its incident component. When S^b coincides with the surface of antenna b, the current $\mathbf{J}^b = \hat{\mathrm{n}} \times \mathbf{H}^b$ on the metal surface can be used to conveniently calculate the mutual impedance as follows:

$$z_{ab} = \frac{1}{i^{ba} i^{bb}} \iint_{S^b} \mathbf{J}^b \cdot \mathbf{E}^a dS \tag{3.18}$$

These three cases, Eqs. (3.16)–(3.18), show how adaptable and powerful the reaction concept can be. Applying duality to Eq. (3.18), the mutual admittance between antennas a and b is

$$y_{ab} = \frac{1}{v^{ba} v^{bb}} \iint_{S^b} \mathbf{M}^b \cdot \mathbf{H}^a dS \tag{3.19}$$

where v^{ba} is the voltage applied to antenna a in the presence of antenna b, v^{bb} is the voltage induced across the terminals of antenna b at which there are no reflections, $\mathbf{M}^b$ is the magnetic current induced on b, and $\mathbf{H}^a$ is the magnetic field radiated by antenna a. Eq. (3.17) can also be used to calculate the scattered power back into the illuminating antenna. Such a situation arises in a reflector antenna where the scattered power contributes to the mismatch in the feed antenna (Silver 1984). The reflector corresponds to S^b and the reaction integral $\langle a, b\rangle$ in Eq. (3.17) is the complex peak power scattered into the feed (antenna a). For a metallic scatterer, the only current on b is the electric current $\mathbf{J}^b$. If P^a is the total power radiated by antenna a, the input reflection feed due to the scatterer is

$$\Gamma = -\frac{\langle a, b\rangle}{2\, P^a} \tag{3.20}$$

As an example, suppose a feed antenna radiates an electric field towards a reflector of the form (Bird 2016, chap. 6) $\boldsymbol{E}^a = \frac{\exp(-jkr)}{r}\left[\hat{\theta}\, A(\theta) \cos\phi - \hat{\phi}\, B(\theta) \sin\phi\right]$ wherein the functions $A(\theta)$ and $B(\theta)$ are the E- and H-plane radiation patterns of the feed. This reflector is symmetric about the z-axis with profile $r(\theta)$ and has a normal $\hat{n} = -\hat{r} n_r(r, \theta) + \hat{\theta} n_\theta(r, \theta)$. Therefore,

$$\Gamma = \frac{\pi}{\eta_o P^a} \int_0^{\theta_c} d\theta \exp(-2jkr(\theta)) \left[\frac{(A(\theta)^2 + (B(\theta)^2)}{2}\right] n_r(r(\theta), \theta) \sin\theta \tag{3.21}$$

where η_o is the free-space wave impedance and θ_c is the half-cone angle subtended by the reflector at the feed phase centre. It can be shown that the asymptotic approximation to Eq. (3.21) for a paraboloid of moderate to large diameter is identical to Silver's (1984) formula.

3.3 On Finite and Infinite Arrays

As we have seen in Sec. 3.1, radiation and fields from finite and infinite regular arrays have been a subject of intense research for many years. Among early compilations, the work by Hansen (1966, Chapters 2 and 3) is significant. Clearly, finite arrays have a boundary and, therefore, contain edge elements that are influenced by only part of the array, whereas more central elements are surrounded by other elements. The edge elements are often a major source of investigation in arrays, as they can adversely affect performance especially at wide radiation angles. On the other hand, elements in regular infinite arrays operate under almost constant conditions from the center to far out. Large arrays can often be approximated by infinite arrays as long as care is taken not to assign the properties of central elements to ones near the edge of the array or consider situations where edge elements occur prominently such as wide scan angles. In the subsequent sections, several analysis methods related to finite and infinite arrays will be described. The main distinction between these methods is that infinite array analysis is based on periodic functions, unit cells and Fourier series.

3.3.1 Finite Array Analysis by Element-by-Element Method

The most frequently used way of analyzing arrays with a finite number of elements is by the element-by-element approach. In this approach, an impedance or admittance matrix of an array is filled by considering each element in turn in the presence of another element, with the remaining elements of the array removed either physically or through open or short-circuiting their terminals. A physical justification for this method comes from the Lorentz reciprocity theorem as described in Section 3.2 and by Eq. (3.7), in particular. As illustrated in Figure 3.4a for elements with electric currents, all except two wire antennas being considered are left open circuited, and the mutual impedance is calculated. For the dual situation shown in Figure 3.4b for aperture antennas, the mutual admittance of two elements is found with the remaining elements short-circuited. The special case of an element in isolation yields the diagonal or self-impedance terms of the $\boldsymbol{Z}$ matrix. These conditions just described are what is expected from network theory for calculating elements of the $\boldsymbol{Z}$ and $\boldsymbol{Y}$ matrices. They also come out this way from integral equation solutions for both problems. The methods for calculating mutual impedance or admittance vary widely and often depend on the type of antenna elements and expressions, and several formulae are in common

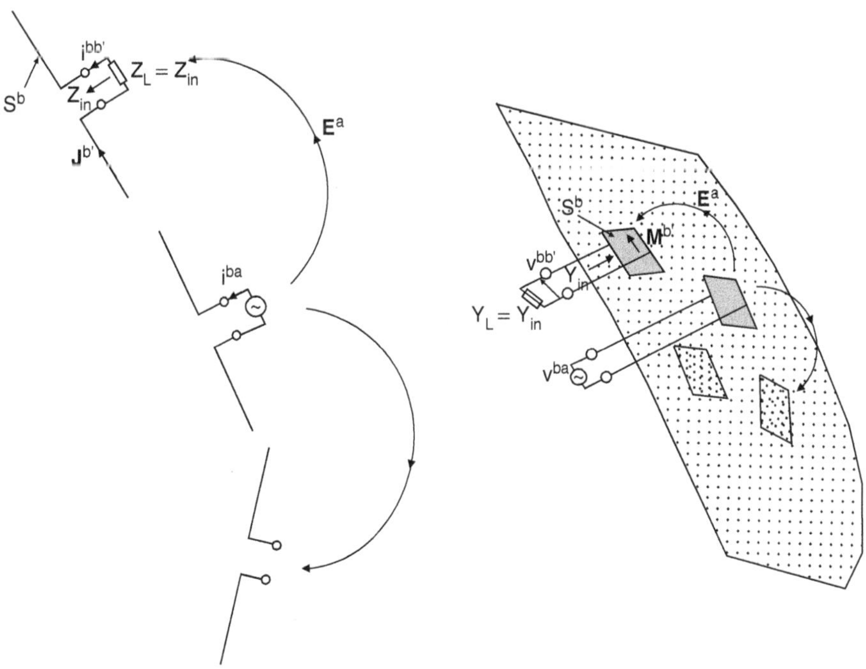

Figure 3.4 Representation of element-by-element method for (a) electric and (b) magnetic currents.

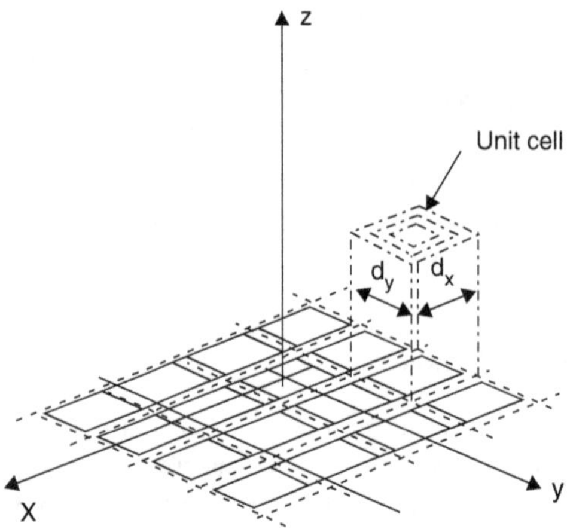

Figure 3.5 Infinite array of apertures and identification of a unit cell.

use. Some expressions that were derived for bodies with electric currents are given in Section 3.2. They will be applied to arrays of wire antennas in Chapter 4. Mutual admittance expressions for aperture antennas excited by magnetic currents can be found from duality. These will be discussed and extended in Chapter 5.

3.3.2 Infinite Periodic Array Analysis

When an array has identical elements and the element spacing is regular (see Figure 3.5), a solution for radiation can be expressed in terms of Floquet modes. Since the array geometry is periodic, the field solution is also periodic, except for a phase function which varies linearly across each cell of the array. This periodic function with phase progression is called a Floquet mode. Floquet's theorem applies to the second-order differential equation $y(t)'' + F(t)y(t) = 0$ where the prime denotes a derivative with respect to t and $F(t)$ is a periodic function that has a period T such that $F(t+T) = F(t)$. This differential equation has two independent solutions $y_{\pm}(t) = \exp(\pm j2\pi t/T)p_{\pm}(t)$ where $p_{+}(t)$ and $p_{-}(t)$ are independent periodic solutions with period T. This property of a periodic second-order differential equation is known as Floquet's theorem.

Periodicity allows considerable simplification of the computing problem because once a solution is obtained for one cell, the solution for other cells is the same except for a progressive phase factor. The field solution in a cell can be obtained in a variety of ways such as numerical methods, integral equations, MM, and Fourier methods. By means of this periodic representation, large arrays have been analyzed with some success. Historically, in the analysis of mutual coupling in antennas, infinite periodic arrays were analyzed first in detail. The literature on infinite arrays is substantial, and for details, the reader should consult the references such as Hansen (1966, 1998), Diamond (1968), Farrell and Kuhn (1968), Amitay et al. (1972), Rudge et al. (1983, Chapters 11 and 12), and Haupt (2010).

3.4 Integral Equation Methods Used in Coupling Analysis

3.4.1 Introduction

Integral equations frequently arise in the analysis of mutual coupling because integrated quantities are matched across boundaries, as illustrated in Figure 3.6. The representation of the fields on either

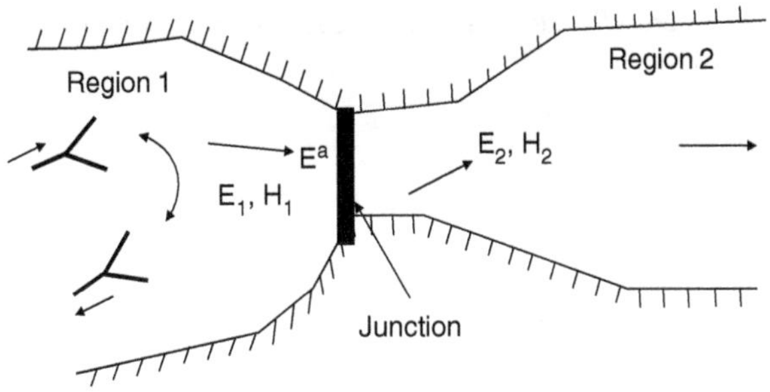

Figure 3.6 Two regions with a common boundary.

side of the boundaries, as well as the solution of the equations, is at the core of the analysis. Their importance is indicated by their position in the second tier of Figure 3.1 and arises in both finite and infinite arrays. A useful way to represent the fields is in terms of their Greens functions, which is discussed in the next section, and these usually lead to integral representations. Most often, there is no closed-form solution to the field matching, and as a result, a solution is usually only possible with numerical methods after suitable reorganization. One important way of doing this is the method of weighted residuals, which is summarized in Section 3.4.3.

In Figure 3.6, a source antenna is located in region 1, and a second antenna is located nearby. Suppose $\boldsymbol{H_o}$ is the field in region 1 in the absence of the junction and the magnetic field in this region with the aperture present is

$$\boldsymbol{H}_1 = \boldsymbol{H}_o + \boldsymbol{L}_1\left(\boldsymbol{E}^a\right)$$

where $\boldsymbol{L}_1$ is an operator related to region 1. Also, in region 2, the magnetic field is related to the aperture electric field via an operator indicated by $\boldsymbol{L}_2$ as

$$\boldsymbol{H}_2 = \boldsymbol{L}_2\left(\boldsymbol{E}^a\right).$$

Continuity of the magnetic field at the junction requires

$$\boldsymbol{H}_o + \boldsymbol{L}_1\left(\boldsymbol{E}^a\right) = \boldsymbol{L}_2\left(\boldsymbol{E}^a\right) \tag{3.22}$$

This resulting equation is an integral equation of the type that arises in many coupling problems. The construction and reduction of equations of this form is the subject of the next section.

3.4.2 Green's Function Methods

The reciprocity expressions given in Eqs. (3.6) and (3.7) describe symmetrical relationships between two field and source pairs. While these relationships are physically based, there are mathematical statements underpinning them. Such statements are called Green's theorems. They can be used to express the field at some external region due to fields or sources in another region. To describe this, suppose we require the fields at r due to a point source located at $\boldsymbol{r}'$ as shown in Figure 3.7. This point source is represented by $\boldsymbol{J} = \boldsymbol{J}_o\delta(\boldsymbol{r}-\boldsymbol{r}')$, where δ is the Dirac delta function, and this is surrounded by an arbitrary surface S'. To find the electric field outside of S', we first consider the wave equation for a dipole in free space (Felsen and Marcuvitz 1973). We commence from the rotational Maxwell equations and initially consider an electric dipole. Vector wave equations can be formed in either electric or magnetic fields, and in the case of an electric dipole source, the electric field is given by

$$\nabla \times \nabla \times \boldsymbol{E}(\boldsymbol{r}) - k^2\boldsymbol{E}(\boldsymbol{r}) = j\omega\mu\boldsymbol{J}(\boldsymbol{r}),$$

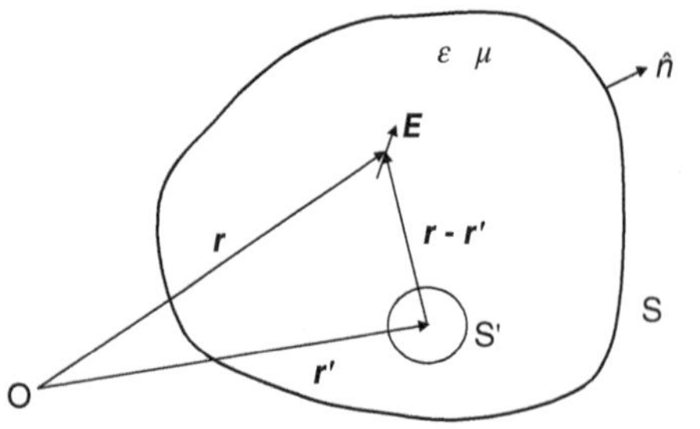

Figure 3.7 Regions for free-space Green's functions.

where

$$\begin{aligned}\nabla \times \nabla \times \boldsymbol{E}(\boldsymbol{r}) &= \nabla(\nabla \cdot \boldsymbol{E}(\boldsymbol{r})) - \nabla^2 \boldsymbol{E}(\boldsymbol{r}) \\ &= \nabla\left(\frac{\rho_e}{\varepsilon}\right) - \nabla^2 \boldsymbol{E}(\boldsymbol{r}) \\ &= \nabla\left(\frac{\nabla \cdot \boldsymbol{J}(\boldsymbol{r})}{-j\omega\varepsilon}\right) - \nabla^2 \boldsymbol{E}(\boldsymbol{r})\end{aligned}$$

in which we have made use of Gauss's law and the continuity equation for electric sources given in Eq. (2.2a). Thus,

$$\nabla^2 \boldsymbol{E}(\boldsymbol{r}) + k^2 \boldsymbol{E}(\boldsymbol{r}) = -j\omega\mu\left(\boldsymbol{I} + \frac{\nabla \cdot \nabla}{k^2}\right) \cdot \boldsymbol{J}(\boldsymbol{r}) \tag{3.23}$$

where $\boldsymbol{I}$ is the unit dyadic.

3.4.2.1 Free-Space Green's Function for Harmonic Sources

To obtain a solution, use is made of the vector Green's theorem and the scalar Green's wave equation. The result is (Felsen and Marcuvitz 1973, Section 1.1)

$$\begin{aligned}\boldsymbol{E}(\boldsymbol{r}) &= -j\omega\mu \iint_{S'} dS' \left(\boldsymbol{I} + \frac{\nabla \cdot \nabla}{k^2}\right) G_o(|\boldsymbol{r} - \boldsymbol{r}'|) \cdot \boldsymbol{J}(r') \\ &= \iint_{S'} dS' \underline{\underline{\boldsymbol{G}}}^{(e)}(\boldsymbol{r}, \boldsymbol{r}') \cdot \boldsymbol{J}(r')\end{aligned} \tag{3.24}$$

where

$$\underline{\underline{\boldsymbol{G}}}^{(e)}(\boldsymbol{r}, \boldsymbol{r}') = -j\omega\mu\left(\boldsymbol{I} + \frac{\nabla \cdot \nabla}{k^2}\right) G_o(|\boldsymbol{r} - \boldsymbol{r}'|) \tag{3.25}$$

is the dyadic free-space Greens function for the electric field and

$$G_o(|\boldsymbol{r} - \boldsymbol{r}'|) = \exp(-jk|\boldsymbol{r} - \boldsymbol{r}'|)/4\pi|\boldsymbol{r} - \boldsymbol{r}'| \tag{3.26}$$

is the scalar free-space Greens function in which $|\boldsymbol{r} - \boldsymbol{r}'| = \sqrt{(x - x')^2 + (y - y')^2 + (z - z')^2}$. From duality, the magnetic field due a magnetic current source $\boldsymbol{M}(r')$ is

$$\boldsymbol{H}(\boldsymbol{r}) = \iint_{S'} dS' \underline{\underline{\boldsymbol{G}}}^{(h)}(\boldsymbol{r}, \boldsymbol{r}') \cdot \boldsymbol{M}(r') \tag{3.27}$$

where

$$\begin{aligned}\underline{\underline{\boldsymbol{G}}}^{(h)}(\boldsymbol{r}, \boldsymbol{r}') &= -j\omega\varepsilon\left(\boldsymbol{I} + \frac{\nabla \cdot \nabla}{k^2}\right) G_o(|\boldsymbol{r} - \boldsymbol{r}'|) \\ &= \frac{\varepsilon}{\mu} \underline{\underline{\boldsymbol{G}}}^{(e)}(\boldsymbol{r}, \boldsymbol{r}') = \frac{1}{\eta^2} \underline{\underline{\boldsymbol{G}}}^{(e)}(\boldsymbol{r}, \boldsymbol{r}')\end{aligned}$$

wherein $\eta = \sqrt{\mu/\varepsilon}$ is the wave impedance of the homogenous medium as illustrated in Figure 3.7. When there is a preferred direction for, say, in stratified media or half-planes, the two-dimensional Greens dyadic can be used. For the magnetic field, this is given by

$$\underline{\underline{G}}^{(h)}\left(\mathbf{r},\mathbf{r}'\right) = \frac{-jk}{2\pi\eta_o}\left(\mathbf{I} + \frac{\nabla_t\cdot\nabla_t}{k^2}\right) g_o\left(\left|\mathbf{r}-\mathbf{r}'\right|\right) \tag{3.28}$$

where ∇_t is the gradient operator in the plane transverse to the preferred direction, say the z-direction, when $\nabla_t = \hat{x}\,\partial/\partial x + \hat{y}\,\partial/\partial y$ and $g_o(R) = \exp(-jkR)/R$.

In some applications, accurate asymptotic solutions based on large separation distance between the elements can give approximate but useful results. Consider two sources on a plane at z and ϕ constant, if the sources and observation points lie on a constant ϕ cut as shown in Figure 3.8a and aperture center spacing R_{ij} is large in terms of wavelengths, the free-space Green's function can be expanded as a Taylor series in R_{ij}. Up to fourth order, this is

$$G(R) \approx G_o\left(R_{ij}\right)\exp\left(-jk\,\widehat{\boldsymbol{R}}_{ij}\cdot\boldsymbol{\Delta}\right)\left[1 - \frac{\widehat{\boldsymbol{R}}_{ij}\cdot\boldsymbol{\Delta}}{R_{ij}} - \frac{jk}{2R_{ij}}\left(|\Delta|^2 - \left|\widehat{\boldsymbol{R}}_{ij}\cdot\boldsymbol{\Delta}\right|^2\right) + \frac{1}{R_{ij}^2}\left(\left|\widehat{\boldsymbol{R}}_{ij}\cdot\boldsymbol{\Delta}\right|^2 \right.\right.$$
$$\left.\left. + \left(jk\widehat{\boldsymbol{R}}_{ij}\cdot\boldsymbol{\Delta} - \frac{1}{2}\right)\left(|\Delta|^2 - \left|\widehat{\boldsymbol{R}}_{ij}\cdot\boldsymbol{\Delta}\right|^2\right) - \frac{k^2}{8}\left(|\Delta|^2 - \left|\widehat{\boldsymbol{R}}_{ij}\cdot\boldsymbol{\Delta}\right|^2\right)^2\right)\right] + O\left(R_{ij}^{-4}\right) \tag{3.29}$$

where $R_{ij} = \sqrt{\left(x_j - x_i\right)^2 + \left(y_j - y_i\right)^2}$ and $G_o(R_{ij}) = exp\,(-jkR_{ij})/4\pi R_{ij}$, $\boldsymbol{\Delta} = \mathbf{r}_j - \mathbf{r}_i$ and $\widehat{\boldsymbol{R}}_{ij}$ is a unit vector. Equation (3.29) can be applied on a ϕ = constant cut to give an asymptotic expansion of the free-space Green's function in spherical coordinates. In terms of the expansion of terms on the right hand side of Eq. (3.29), it can be expressed in terms of local coordinates that are convenient for integration.

$$\begin{aligned}\widehat{\boldsymbol{R}}_{ij}\cdot\boldsymbol{\Delta} = \widehat{\boldsymbol{R}}_{ij}\cdot\mathbf{r}_j - \widehat{\boldsymbol{R}}_{ij}\cdot\mathbf{r}_i &= \mathrm{r}_j\cos\left(\frac{\pi}{2} - \theta_j + \alpha_j - \alpha_{ij}\right) - \mathrm{r}_i\cos\left(\frac{\pi}{2} - \theta_i + \alpha_i - \alpha_{ij}\right)\\ &= \mathrm{r}_j\sin\left(\theta_j - \alpha_j + \alpha_{ij}\right) - \mathrm{r}_i\sin\left(\theta_i - \alpha_i + \alpha_{ij}\right)\end{aligned}$$

and

$$\begin{aligned}|\Delta|^2 - \left|\widehat{\boldsymbol{R}}_{ij}\cdot\boldsymbol{\Delta}\right|^2 &= \mathrm{r}_j^2\cos^2\left(\theta_j - \alpha_j + \alpha_{ij}\right) + \mathrm{r}_j^2\cos^2\left(\theta_i - \alpha_i + \alpha_{ij}\right)\\ &\quad - 2\,\mathrm{r}_i\mathrm{r}_j\left[1 - \sin\left(\theta_j - \alpha_j + \alpha_{ij}\right)\sin\left(\theta_i - \alpha_i + \alpha_{ij}\right)\right]\end{aligned}$$

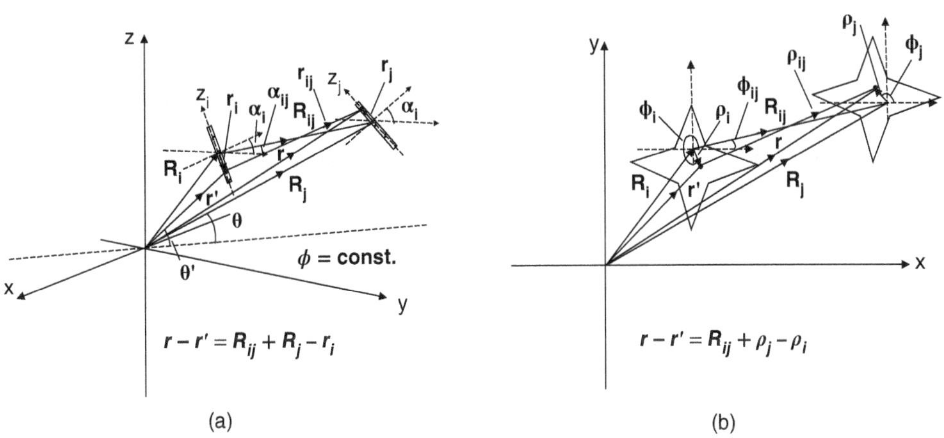

Figure 3.8 Geometry for asymptotic approximation of coupling. (a) A constant ϕ cut in spherical coordinates;(b) the z = 0 plane with polar coordinates.

Another practical case is the application of Eq. (3.26) to the $z = 0$ plane. As shown in Fig. 3.8(b), it can be expressed in polar co-ordinates. To second order in R_{ij}, the Green's function expansion is (Bird, 1979)

$$G(R) \approx G_o(R_{ij}) \exp\left[-jk\rho_i \cos(\phi_i - \phi_{ij}) + jk\rho_j \cos(\theta_j - \phi_{ij})\right]$$
$$\times \left\{1 - \frac{1}{R_{ij}}\left[\rho_i \cos(\phi_i - \phi_{ij}) - \rho_j \cos(\phi_j - \phi_{ij}) + \frac{jk}{4R_{ij}}\right.\right.$$
$$\left[4\rho_i\rho_j \sin(\phi_i - \phi_{ij}) \sin(\phi_j - \phi_{ij}) + \rho_i^2 \cos\left[2(\phi_i - \phi_{ij})\right] + \rho_j^2 \cos\left[2(\phi_j - \phi_{ij})\right]\right.$$
$$\left.\left.\left. - \left(\rho_i^2 + \rho_j^2\right)\right]\right]\right\} \tag{3.30}$$

The polar coordinates in Eq. (3.30) can be easily transformed to other aperture coordinate systems on a plane such as rectangular by replacing the local coordinates $(\rho_i.\ \phi_i)$ and $(\rho_j.\ \phi_j)$ with $(x_i, y_i, 0)$ and $(x_j, y_j, 0)$ by $x_i = \rho_i \cos\phi_i$ and $y_i = \rho_i \sin\phi_i$, and similarly, for the aperture j. Both Eqs. (3.29) and (3.30) will be used in later chapters to obtain asymptotic approximations to mutual impedances.

3.4.2.2 Free-Space Green's Function for Transient Sources

Equation (3.25) is a special case of the general expression of the Green's function in spatial and time domains for harmonic sources. This expression is (Felsen and Marcuvitz 1973, p. 15)

$$\underline{\underline{G}}^{(e)}(\boldsymbol{r}, \boldsymbol{r}' \mid t, t') = \left(\mu_o \frac{\partial}{\partial \tau}\boldsymbol{I}\bigg|_{\tau=t} - \frac{\nabla \cdot \nabla}{\varepsilon_o}\int_{-\infty}^{t} d\tau\right) g_t(\boldsymbol{r}, \boldsymbol{r}' \mid \tau, t') \tag{3.31}$$

where $g_t(\boldsymbol{r}, \boldsymbol{r}' \mid \tau, t') = \frac{s(|\boldsymbol{r}-\boldsymbol{r}'|, \tau - t')}{4\pi|\boldsymbol{r}-\boldsymbol{r}'|}$ in which $s(\boldsymbol{r}-\boldsymbol{r}', \tau - t')$ is the exciting signal. For the two special cases of time-harmonic sources and transient point sources, $s(\boldsymbol{r}-\boldsymbol{r}', \tau - t') = \exp(j\omega\tau - jk|\boldsymbol{r}-\boldsymbol{r}'|)$ and $s(\boldsymbol{r}-\boldsymbol{r}', \tau - t') = \delta(\boldsymbol{r}-\boldsymbol{r}')\delta(\tau - t')$, respectively, where δ is the Dirac delta function. When Eq. (3.31) is evaluated for these time dependencies, Eq. (3.25) is the result in the time-harmonic case, while for transient sources

$$\underline{\underline{G}}^{(e)}(\boldsymbol{r}, \boldsymbol{r}' \mid t, t') = \eta_o \frac{1}{4\pi|\boldsymbol{r}-\boldsymbol{r}'|}\left[(\boldsymbol{I} - \boldsymbol{r}'\boldsymbol{r}')\frac{\delta'\left(t - \frac{|\boldsymbol{r}-\boldsymbol{r}'|}{c}\right)}{c} + (\boldsymbol{I} - 3\boldsymbol{r}'\boldsymbol{r}')\frac{\delta\left(t - \frac{|\boldsymbol{r}-\boldsymbol{r}'|}{c}\right)}{|\boldsymbol{r}-\boldsymbol{r}'|}\right]$$
$$+ \left[(\boldsymbol{I} - 3\boldsymbol{r}'\boldsymbol{r}')\frac{H(t - |\boldsymbol{r}-\boldsymbol{r}'|/c)}{4\pi\varepsilon_o|\boldsymbol{r}-\boldsymbol{r}'|}\right] \tag{3.32}$$

where $\delta'(x) = \partial\delta(x)/\partial x$ and $H(x)$ is the Heaviside step function, which is defined in Appendix A. The first two terms of Eq. (3.32) correspond to the radiation and near fields of the transient source, and the third term is the static electric field that exists only inside the sphere of radius $|\boldsymbol{r}-\boldsymbol{r}'| < ct$ that follows the spherical wavefront.

3.4.2.3 Fields with Sources

Expressed in terms of Green's functions, the fields due to elemental electric and magnetic current sources are

$$d\boldsymbol{E_e}(\boldsymbol{r}) = \boldsymbol{G}^{(e)}(\boldsymbol{r}, \boldsymbol{r}') \cdot d\boldsymbol{J}(\boldsymbol{r}') \text{ and } d\boldsymbol{H_e}(\boldsymbol{r}) = \boldsymbol{G}^{(h)}(\boldsymbol{r}, \boldsymbol{r}') \cdot d\boldsymbol{M}(\boldsymbol{r}')$$

The total field is obtained by integrating over a surface containing these contributions. The associated magnetic and electric fields are found from the relevant Maxwell curl equation as follows:

$$d\boldsymbol{H_e}(\boldsymbol{r}) = \frac{1}{j\omega\mu}\boldsymbol{\nabla}_r \times \boldsymbol{G}^{(e)}(\boldsymbol{r}, \boldsymbol{r}') \cdot d\boldsymbol{J}(\boldsymbol{r}') \text{ and}$$
$$d\boldsymbol{E_h}(\boldsymbol{r}) = -\frac{1}{j\omega\varepsilon}\boldsymbol{\nabla}_r \times \boldsymbol{G}^{(h)}(\boldsymbol{r}, \boldsymbol{r}') \cdot d\boldsymbol{M}(\boldsymbol{r}')$$

where the subscript $\boldsymbol{r}$ on the del operator refers to the observation coordinates. If both sources are present, the total electric and magnetic fields are

$$d\boldsymbol{E}(\boldsymbol{r}) = d\boldsymbol{E}_e(\boldsymbol{r}) + d\boldsymbol{E}_h(\boldsymbol{r}) = \mathbf{G}^{(e)}(\boldsymbol{r},\boldsymbol{r}') \cdot d\boldsymbol{J}(\boldsymbol{r}') + \mathbf{G}^{(eh)}(\boldsymbol{r},\boldsymbol{r}') \cdot d\boldsymbol{M}(\boldsymbol{r}')$$

$$d\boldsymbol{H}(\boldsymbol{r}) = d\boldsymbol{H}_e(\boldsymbol{r}) + d\boldsymbol{H}_h(\boldsymbol{r}) = \mathbf{G}^{(he)}(\boldsymbol{r},\boldsymbol{r}') \cdot d\boldsymbol{J}(\boldsymbol{r}') + \mathbf{G}^{(h)}(\boldsymbol{r},\boldsymbol{r}') \cdot d\boldsymbol{M}(\boldsymbol{r}')$$

where $\mathbf{G}^{(eh)}(\boldsymbol{r},\boldsymbol{r}') = \frac{1}{j\omega\mu}\nabla_r \times \mathbf{G}^{(e)}(\boldsymbol{r},\boldsymbol{r}')$ and $\mathbf{G}^{(he)}(\boldsymbol{r},\boldsymbol{r}') = -\frac{1}{j\omega\varepsilon}\nabla_r \times \mathbf{G}^{(h)}(\boldsymbol{r},\boldsymbol{r}')$.

It is verified that $\mathbf{G}^{(eh)}(\boldsymbol{r}, \boldsymbol{r}') = -\mathbf{G}^{(he)}(\boldsymbol{r}, \boldsymbol{r}')$. This method of creating Green's functions can be generalized, for instance, for other bounding surfaces and materials with dyadic constitutive parameters. A more general approach is described in the references Felsen and Marcuvitz (1973) and Chew (1995).

As an example of a specific solution arising from a magnetic source, the field on the surface of an infinitely long conducting cylinder of radius $\boldsymbol{R_o}$ at $\boldsymbol{r} = (\boldsymbol{R_o}, \phi, z)$ may be found by the methods of Felsen and Marcuvitz (1973) as follows:

$$d\boldsymbol{H}(\boldsymbol{r}) = \underline{\underline{\mathbf{G}}}^{(h)}(\boldsymbol{r},\boldsymbol{r}') \cdot d\boldsymbol{M}(\boldsymbol{r}') \tag{3.33}$$

where

$$\underline{\underline{\mathbf{G}}}^{(h)}(\phi, z \mid \phi', z') = \left(\hat{\phi}\hat{\phi}' G_{\phi\phi} + \hat{\phi}\hat{z}' G_{\phi z} + \hat{z}\hat{\phi}' G_{z\phi} + \hat{z}\hat{z}' G_{zz}\right) \tag{3.34}$$

is the dyadic for the surface magnetic field. A modal solution representation of the dyadic components gives

$$G_{\phi\phi}(\phi, z \mid \phi', z') = -c_h \int_{-\infty}^{\infty} d\xi \sum_{n=-\infty}^{\infty} e^{j(n\bar{\phi} - n\bar{z})} \frac{k_o}{h} \left[\frac{H_n^{(2)\prime}(hR_o)}{H_n^{(2)}(hR_o)} - \left(\frac{n\xi}{k_o h R_o}\right)^2 \frac{H_n^{(2)}(hR_o)}{H_n^{(2)\prime}(hR_o)}\right], \tag{3.35a}$$

$$G_{\phi z}(\phi, z \mid \phi', z') = c_h \int_{-\infty}^{\infty} d\xi \sum_{n=-\infty}^{\infty} e^{j(n\bar{\phi} - n\bar{z})} \left[\left(\frac{n\xi}{k_o h R_o}\right) \frac{H_n^{(2)}(hR_o)}{H_n^{(2)\prime}(hR_o)}\right], \tag{3.35b}$$

$$G_{z\phi}(\phi, z \mid \phi', z') = G_{\phi z}(\phi, z \mid \phi', z'), \tag{3.35c}$$

and

$$G_{zz}(\phi, z \mid \phi', z') = c_h \int_{-\infty}^{\infty} d\xi \sum_{n=-\infty}^{\infty} e^{j(n\bar{\phi} - n\bar{z})} \left[\left(\frac{hR_o}{k_o R_o}\right) \frac{H_n^{(2)}(hR_o)}{H_n^{(2)\prime}(hR_o)}\right] \tag{3.35d}$$

where

$$h = \begin{cases} |h|, |\xi| < k_o \\ -j|h|, |\xi| > k_o \end{cases},$$

$$|h| = \sqrt{|k_o^2 - \xi^2|},$$

$$c_h = \frac{jY_o}{(2\pi)^2 R_o},$$

$$\bar{\phi} = \phi - \phi', \quad \bar{z} = z - z',$$

and

$$k_o = 2\pi/\lambda_o.$$

$H_n^{(2)}$ is the Hankel function of second kind of order n. A prime on this function indicates the first derivative with respect to the argument. Y_o and λ_o are free-space wave admittance and wavelength, respectively. Although Eqs. (3.35a)–(3.35d) may be evaluated numerically either directly or as surface integrals, the computation of the resulting infinite series and infinite integrals is very time consuming even for small cylinders. The three independent components can be reformulated for better numerical calculation as follows:

$$G_{\phi\phi}\left(\phi, z \mid \phi', z'\right) = -\frac{Y_o k^2}{\pi^3} \sum_{n=0}^{\infty} \varepsilon_n \cos n\overline{\phi} \left[\int_0^1 dx\ G_2\left(\overline{z}, K, x\right) + j \int_0^{\infty} dx\ G_2\left(\overline{z}, K, -jx\right) \right] \tag{3.36a}$$

$$G_{\phi z}\left(\phi, z \mid \phi', z'\right) = -2j \frac{Y_o k^2}{\pi^3 R_o} \sum_{n=0}^{\infty} n \sin n\overline{\phi} \left[\int_0^1 dx\, x\ G_3\left(\overline{z}, K, x\right) + j \int_0^{\infty} dx\, x\ G_3\left(\overline{z}, K, -jx\right) \right] \tag{3.36b}$$

$$G_{zz}\left(\phi, z \mid \phi', z'\right) = -\frac{Y_o}{\pi^3 R_o^2} \sum_{n=0}^{\infty} \varepsilon_n \cos n\overline{\phi} \left[\int_0^1 dx\ G_1\left(\overline{z}, K, x\right) + j \int_0^{\infty} dx\ G_1\left(\overline{z}, K, -jx\right) \right] \tag{3.36c}$$

where

$$G_1\left(\overline{z}, K, x\right) = \exp\left(-jkx\,\overline{z}\right) / N_n^2(K\sqrt{1-x^2})$$

$$G_2\left(\overline{z}, K, x\right) = \frac{\exp\left(-jkx\,\overline{z}\right)}{K^2\sqrt{1-x^2})} \left[\frac{1}{M_n^2(K\sqrt{1-x^2})} + \left(\frac{nx}{K\sqrt{1-x^2})} \right)^2 \frac{1}{N_n^2(K\sqrt{1-x^2})} \right]$$

$$G_3\left(\overline{z}, K, x\right) = \exp\left(-jkx\,\overline{z}\right) / \left(K^2\sqrt{1-x^2})N_n^2(K\sqrt{1-x^2})\right)$$

and $K = k\,R_o$, $M_n^2(x) = J_n^2(x) + Y_n^2(x)$, and $N_n^2(x) = J_n'^2(x) + Y_n'^2(x)$, where $J_n(x)$ is the n-th order ordinary Bessel function, and $Y_n(x)$ is the modified Bessel function. Calculation of the surface field due to distributed sources, and hence the mutual admittance of apertures, is similarly affected.

Although the transformed Eqs. (3.36a)–(3.36c) are more suitable for computation, they converge only if the source and field points are not on the same circumferential section of the cylinder. Even with this improvement, computation times are excessive, particularly when many distributed sources are involved such as in array calculations. Furthermore, there are cases (e.g. waveguide self-admittance calculations) where Eqs. (3.35a)–(3.35d) would still need to be used. Typically, for numerical evaluation, the number of terms required to achieve convergence is $N \geq n = |K| + 5$. It is especially important, therefore, to find accurate approximate solutions for cylinders of moderate to large radius.

For large diameter cylinders, an asymptotic solution can be greatly beneficial in computations. To obtain an asymptotic solution for the surface field on the surface of the cylinder due to a source situated also on the surface, we express the dyadic of the type Eq. (3.32) as follows:

$$\underline{\underline{\mathbf{G}}}^{(h)}\left(\phi, z \mid \phi', z'\right) = \sum_{p=0}^{\infty} \left[\underline{\underline{\mathbf{G}}}\left(\overline{\phi} + 2p\pi, \overline{z}\right) + \underline{\underline{\mathbf{G}}}(2\pi - \overline{\phi} + 2p\pi, \overline{z}) \right] \tag{3.37}$$

where $\overline{\phi} = \phi - \phi'$ and $\overline{z} = z - z'$. The dyadic function $\underline{\underline{\mathbf{G}}}$ is the field due to a single surface ray in one direction, while the summation corresponds to creeping waves traveling along helical

Figure 3.9 Geometry for surface field at P due to a magnetic dipole at P′ and the first two surface rays.

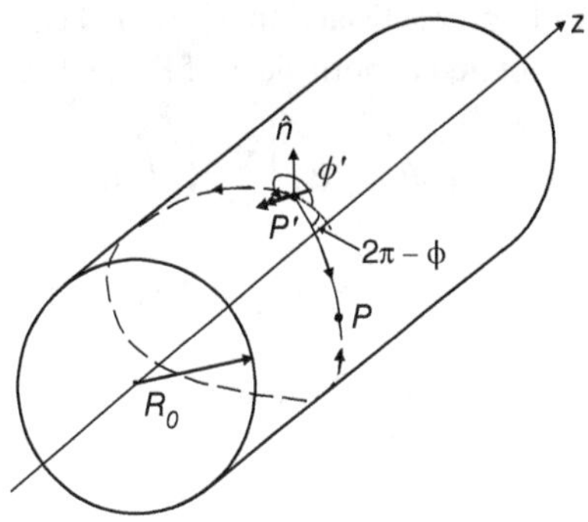

paths between the field and source points. Figure 3.9 illustrates creeping waves on a cylinder. An approximate asymptotic solution can be obtained for each component of the surface field dyadic of a cylinder. Each term of Eq. (3.37) can be interpreted as surface waves and the field on the cylinder as a standing wave produced by contra-rotating surface waves, which propagate from the source to the field point along helical ray paths, as depicted in Figure 3.9. These waves follow the curvature of the cylindrical surface and appear to "creep" from the source to the field point (Watson, 1918). These creeping waves attenuate as they propagate requiring, in most calculations with large cylinders, the consideration of only two contra-rotating creeping waves encircling large cylinders just once, that is,

$$\underline{\underline{G}}^{(h)}\left(\phi, z \mid \phi', z'\right) \sim \underline{\underline{G}}\left(\overline{\phi}, \overline{z}\right) + \underline{\underline{G}}\left(2\pi - \overline{\phi}, \overline{z}\right) \tag{3.38}$$

Several approximate asymptotic solutions have been obtained, and these will be discussed in Chapter 7.

3.4.3 Solution by Weighted Residuals

A class of methods used to solve integral and differential equations, which often arise in formulations involving Green's function, is known as weighted residuals (Finlayson 1972). In general, the integral operator we are attempting to solve is of the form

$$\int_D ds' F\left(s, s'\right) \Psi\left(s'\right) = g\left(s\right) \tag{3.39}$$

where $\Psi(s)$ is the function we require in domain $\mathcal{D}$. The kernel function $F(s, s')$ may be a Green's dyadic or a differential function, which is symmetric in its two arguments. In the method of weighted residuals, the function $\Psi(s')$ is replaced by the expansion

$$\Psi\left(s\right) \cong \sum_{n=1}^{N} \psi_n \alpha_n\left(s\right) \tag{3.40}$$

where ψ_n and $\alpha_n(s)$ are the N coefficients and expansion functions, respectively. The expansion functions should satisfy all boundary conditions unless they are satisfied naturally such as with a variational formation or are enforced through Lagrange multipliers. In this case, it is assumed that the boundary conditions are satisfied and all that is required is the difference between the left and right sides should be as small as possible over domain $\mathcal{D}$ in an average sense. When Eq. (3.40) is substituted into Eq. (3.39) the residual $\mathfrak{R}$ is

$$\sum_{n=1}^{N} \psi_n \left\{ \int_D ds' F\left(s, s'\right) \alpha_n\left(s\prime\right) \right\} - g\left(s\right) = \mathfrak{R}\left(s\right) \tag{3.41}$$

To give satisfaction of original equation in domain D, a weighting function $w_m(s)$ (m = 1,…, M) is applied to both sides of Eq. (3.41), and the result is integrated across this domain to give

$$\int_D ds\, w_m(s) \left[\sum_{n=1}^{N} \left\{ \int_D ds' F(s,s')\, \alpha_n(s')\, \psi_n \right\} - g(s) \right]$$
$$= \int_D ds\, w_m(s) R(s) = 0; \mathrm{m} = 1, \ldots, \mathrm{M} \quad (3.42)$$

It is seen that Eq. (3.42) is the weighted residual of the operator equation (Eq. (3.39)) over the domain D (Finlayson 1972). When expressed in matrix form, Eq. (3.42) becomes

$$[A_{mn}]\,[\psi_n] = [b_m]\,. \quad (3.43)$$

where

$$A_{mn} = \int_D ds\, w_m(s) \int_D ds' F(s,s')\, \alpha_n(s),$$

and

$$b_m = \int_D ds\, w_m(s) g.$$

The matrix $[A_{mn}]$ is, in general, an $M \times N$ matrix and is inverted by multiplying both sides by its transpose before inverting. Thus,

$$[A_{mn}]^T [A_{mn}]\,[\psi_n] = [A_{mn}]^T [b_m]$$

and the unknown coefficients are found by inverting $\boldsymbol{C} = [A_{mn}]^T[A_{mn}]$ so that

$$[\psi_n] = \boldsymbol{C}^{-1} [A_{mn}]^T [b_m] \quad (3.44)$$

The $\psi_n{}'s$ will be approximations to actual values of Ψ at N points or coefficients of the expansion functions depending on the chosen functions.

A variety of numerical methods are possible based on different weighting functions. For example, the Galerkin method corresponds to the weighting function $w_m(s) = \alpha_m(s)$. The least squares method to $w_m(s) = \partial \Re / \partial \psi_m$, where the ψ_m's are the coefficients of Eq. (3.40), and point matching to $w_m(s_m) = \delta(s - s_m)$, where δ is the delta function. For example, when the delta function is applied to the residual

$$\int_D ds\, \delta(s - s_m)\, R(s) = \Re(s_m) = 0; \mathrm{m} = 1, \ldots, \mathrm{N} \quad (3.45)$$

thus satisfying the right side of Eq. (3.42). Another approach is the MoM where one or a succession of weight functions are applied in the integrand. The original MoM due to Yamada (1947, 1950) applied successively higher weighting functions $w_m(s) = 1, s, s^2, \ldots, s^{m-1}$, where s is the distance to some point raised to the m-th power. Each weighted residual is required to be zero. The zero-th moment requires the average residual to be zero over the domain. The first moment is the same as the subdomain method or integral method developed separately by von Karman and Pohlhausen (1921) and is the most commonly used approach. This moment corresponds to a variance in statistics. In Chapter 4, the MoM based on a single weighting function is applied to wire arrays.

The most general approach in electromagnetics (Harrington 1968) is to use different weight and shape functions. The weighting functions are often pulse or delta functions, while the shape functions can vary from overlapping segments to continuous functions. For example, in the Galerkin method to be applied to coupling in aperture arrays, in Chapter 5, the weight and shape functions are often related to waveguide modes of the two-dimensional geometry.

A popular formulation in the MoM in electromagnetics corresponds to a Galerkin representation and is equivalent to Eq. (3.44) with the A matrix replaced by the Z matrix, which is square by construction (as there are as many testing functions as basis functions), and is generally called the impedance matrix as they are closely related. When the unknown source distribution $\Psi(s)$ represents an electrical current distribution, the resulting function we are solving is for the electric field

$$g(s) = E(s) = \int_D ds' G(s, s') \; \Psi(s')$$

that is radiated by that distribution where $G(s, s')$ is an electric dyadic operator. For radiation of electrical currents in free space, the Greens function is given by Eq. (3.25). The mn-th entry of that distribution is given by testing the n-th component of $E(s')$ with a testing function m, as follows:

$$Z_{mn} = \int_D ds\, \overrightarrow{w_m}.\overrightarrow{E_n}(s)$$

The excitation vector is calculated from

$$b_m = -\int_D ds\, \overrightarrow{w_m}.\overrightarrow{E_{inc}}(s)$$

where $\overrightarrow{E_{inc}}(s)$ is the incident field, such as an incoming plane wave from a Dirac delta field distribution at the level of the feedpoint. The treatment of antennas involving dielectric material involves the use of magnetic currents as well; a detailed formulation for that case can be found in Rao and Wilton (1990). It is interesting to note that Galerkin method testing leads to solutions to exactly satisfy reciprocity, which, however, is not a guarantee for accuracy of the solution.

As an example of weighted residuals, consider the one-dimensional version of Laplace's equation

$$\frac{\partial}{\partial x}\left(\varepsilon_r(x)\frac{\partial V(x)}{\partial x}\right) = 0 \quad ; \quad 0 \le x \le s$$

where V is the potential and ε_r is the relative dielectric constant. The aim is to find the potential between two plates at levels V_o and V_s where $V_o = V(0) = 12$ and $V_s = V(s) = 55$. In this example, the dielectric region is a half-space with a relative dielectric constant of $\varepsilon_r(x) = 1 + \Delta\varepsilon\, x$. The exact solution of Laplace is

$$V(x) = (V_s - V_o)\frac{ln(1 + \Delta\varepsilon\, x)}{ln(1 + \Delta\varepsilon\, s)} + V_o \tag{3.46}$$

To evaluate by weighted residuals, take as an approximate solution $V(x) \approx V_o + \sum_{k=1}^{2}\alpha_k x^k$ where the α_k's are constants to be determined. When this trial solution is substituted into the Laplace equation, the residual is

$$\mathfrak{R} = \Delta\varepsilon\, (\alpha_1 + 2\alpha_2 x) + 2\alpha_2\, [1 + \Delta\varepsilon\, (1 + \Delta\varepsilon\, x)]$$

Suppose we require the n-th order residual to be zero,

$$\int_0^s x^n\, \mathfrak{R}\, dx = 0.$$

This results in

$$\alpha_1 = \frac{(V_o - V_o)}{s\left\{1 - \frac{\Delta\varepsilon}{2[1+2(n+1)\Delta\varepsilon s/(n+2)]}\right\}} \text{ and } \alpha_2 = \frac{\alpha_1\Delta\varepsilon}{2\left\{1 + 2\frac{(n+1)}{(n+2)}\Delta\varepsilon s\right\}}.$$

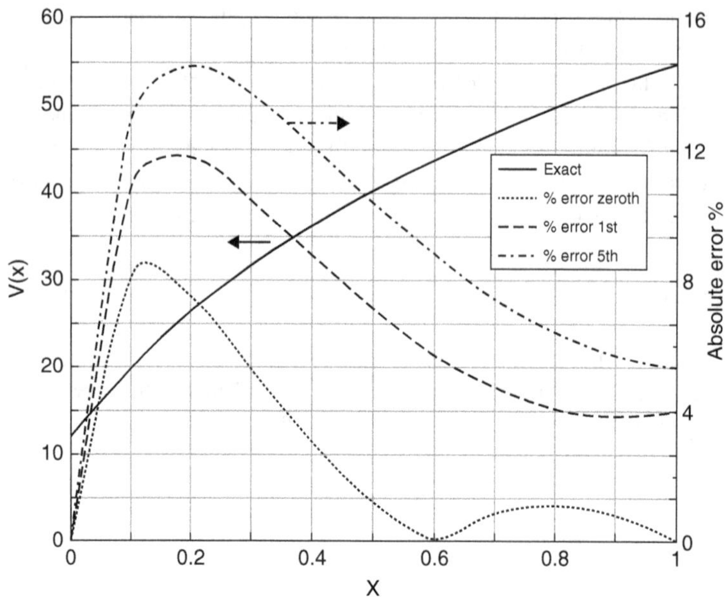

Figure 3.10 Potential in a dielectric with linearly varying dielectric constant as a function of distance. Error in weighted residual approximation with weights of increasing order.

Consider the potential in a layer of thickness $s = 1$ where $\Delta\varepsilon = 3$. The result given by Eq. (3.46) is shown in Figure 3.10. The errors in the approximate solutions obtained by weighted residuals for various moments n=0, 2 and 5 are compared with the exact solution in Figure 3.10. It is seen in this example that the zero-th moment gives the smallest error over the domain.

3.5 Some Other Methods Used in Coupling Analysis

The methods to be discussed in the following section relate to more detailed approaches that are shown in the third tier of Figure 3.1 under the headings of finite and infinite arrays. Most methods outlined could be applied to both types, although, currently, the unit cell analysis is limited to infinite arrays.

3.5.1 Unit Cell Analysis in Periodic Structure Method

The analysis of mutual coupling in a periodic array may be considered automatically in a direct way by using a periodic structure approach. In the following, it is assumed that all elements are identical, the spacing regular, and a uniform phase shift occurs across the array, which is simply a linear phase shift from one element to the next. The application of the periodic structure method is independent of the type of element employed, although it is usually only applied to infinite arrays, as indicated in Figure 3.1. The analysis of the fields in a unit cell as shown in Figure 3.11 can be undertaken in many ways. It may be a structure analyzable by analytic methods or as complicated as the application requires which may require a full-wave code such as HFSS.

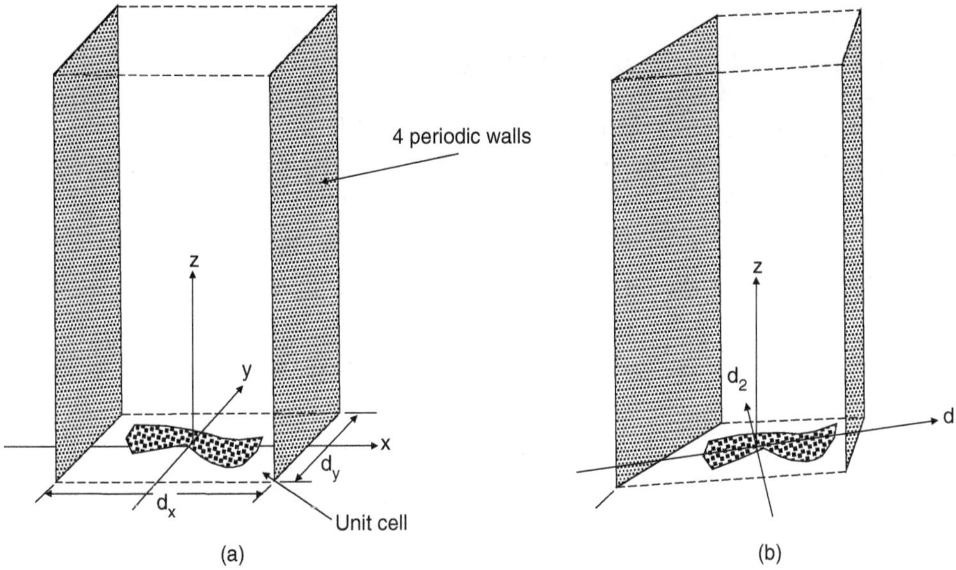

Figure 3.11 Unit cell of a periodic planar array. (a) Orthogonal geometry; (b) no orthogonal geometry.

A formulation is now described that allows mutual coupling to be included in a periodic array given a solution for the unit cell. Consider an infinite array of identical elements as shown in Figure 3.5. A view of the unit cell of a planar array is shown in Figure 3.11. Due to periodicity of the bounding walls of the unit cell, periodic boundary conditions exist there as well as any other conditions that may exist due to the presence of electric or magnetic walls. For a field $\boldsymbol{F}$ in the x-direction in Figure 3.11, this means $\boldsymbol{F}(x) = \boldsymbol{F}(x + d_x)$, where d_x is the unit cell period in that direction. The location of a unit cell in the transverse plane is specified by the vector

$$\rho_{tmn} = m\boldsymbol{d}_1 + n\boldsymbol{d}_2; |m| \leq M \text{ and } |n| \leq N$$

where $\boldsymbol{d}_1$ and $\boldsymbol{d}_2$ are vectors, which define the array lattice. In general, these vectors are not orthogonal, as shown in Figure 3.11b. The electric field in the mn-th unit cell is

$$\boldsymbol{E}_{mn}\left(\rho_{tmn}\right) = \boldsymbol{E}_o\left(\rho - \rho_{tmn}\right) e^{-jk\boldsymbol{w}\cdot\rho_{tmn}}$$

where $\boldsymbol{w} = \hat{x}u + \hat{y}v$ with $u = \sin\theta\cos\varphi$ and $u = \sin\theta\sin\varphi$. The total electric field in the transverse aperture plane $z = 0$ is found by summing these individual contributions, as follows:

$$\boldsymbol{E}_t\left(\rho_{tmn}\right) = \sum_{m=-M}^{M}\sum_{n=-N}^{N} \boldsymbol{E}_o\left(\rho - \rho_{tmn}\right) e^{-jk\boldsymbol{w}\cdot\rho_{tmn}}.$$

In turn, the total radiated field is obtained by integrating over the infinite aperture, thus

$$\boldsymbol{E}(r, u, v) = \frac{1}{4\pi^2}\int_{-\infty}^{\infty}\int_{-\infty}^{\infty}\sum_{m=-M}^{M}\sum_{n=-N}^{N} \boldsymbol{E}_o\left(\rho - \rho_{tmn}\right) e^{-jk\boldsymbol{w}\cdot\rho_{mn}}\, dxdy \tag{3.47}$$

The summations in Eq. (3.47) can be completed in closed form and the Fourier transform of the aperture field of the unit cell is identified. This results in

$$\boldsymbol{E}(r, u, v) = \mathcal{E}_o(r, u, v)\left[\frac{\sin\left(\left(M + \frac{1}{2}\right)\left(\boldsymbol{\kappa}_t \cdot \boldsymbol{d}_1\right)\right)}{\sin\left(\frac{1}{2}\boldsymbol{\kappa}_t \cdot \boldsymbol{d}_1\right)}\right]\left[\frac{\sin\left(\left(N + \frac{1}{2}\right)\left(\boldsymbol{\kappa}_t \cdot \boldsymbol{d}_2\right)\right)}{\sin\left(\frac{1}{2}\boldsymbol{\kappa}_t \cdot \boldsymbol{d}_2\right)}\right] \tag{3.48}$$

where

$$\mathcal{E}_o(r,u,v) = \frac{1}{4\pi^2}\iint_{unitcell} \mathbf{E}_o(x,y)\, e^{-jk(ux+vy)}\, dxdy \tag{3.49}$$

is the transform of the aperture field in the unit cell and $\boldsymbol{\kappa}_t = k\left[\hat{x}(1-u) + \hat{y}(1-v)\right]$.

Consider a field that is transverse electric (TE) to the z-direction in the region above the aperture shown in Figure 3.11. The z-directed magnetic field is a solution of Helmholtz's equation of the form

$$\left[\nabla_t^2 + (k^2 + \gamma^2)\right] h_z(x,y) = 0 \tag{3.50}$$

where $H_z(x,y,z) = h_z(x,y)\,exp(-j\gamma z)$. The array is periodic in the x and y-directions with spacing d_x in the x-direction and d_y in the y-direction. As $h_z(x,y)$ is a periodic function in x and y, a general solution is of the form

$$f\left(x+d_x, y+d_y\right) = h_z(x,y)\,exp\left(-j\left(k_x d_x + k_y d_y\right)\right). \tag{3.51}$$

where k_x and k_y are the propagation constants in the x- and y-directions. Equation (3.48) is a particular form of Floquet's theorem. A separable solution is $h_z(x,y) = X(x)Y(y)$, where the functions X and Y satisfy scalar wave equations with separation constants κ_x^2 and κ_y^2 such that

$$\gamma^2 = k^2 - \kappa_x^2 - \kappa_y^2.$$

As well, assume that the scan phase is progressive on the excitation from cell to cell so that for element mn-th excitation is of the form $V_{mn}\,exp\,(j2\pi(m\psi_x + n\psi_y))$ where steering phases are given by

$$\psi_x = \frac{d_x}{\lambda} sin\theta_d cos\phi_d \text{ and } \psi_y = \frac{d_y}{\lambda} sin\theta_d sin\phi_d \tag{3.52}$$

where θ_d, ϕ_d are the drive angles. For a general displacement Δ in the x-direction, $X(x+\Delta) = X(x)\,exp\,(j2\pi m\psi_x\Delta/d_x)$, and similarly for $Y(y)$ in the y-direction. Furthermore, $X(x)$ can be represented by a Fourier series of the form

$$\begin{aligned} X(x) &= \sum_{m=-\infty}^{\infty} A_m \exp\left(\frac{j2\pi mx}{d_x}\right) \exp\left(j2\pi m\psi_x x/d_x\right) \\ &= \sum_{n=-\infty}^{\infty} A_n \exp\left(j2\pi m\left(1+\psi_x\right)x/d_x\right) \end{aligned} \tag{3.53}$$

Similarly, in the y-direction, it is shown that

$$Y(y) = \sum_{n=-\infty}^{\infty} B_n exp\left(j2\pi n\left(1+\psi_y\right)y/d_y\right).$$

Thus, the axial magnetic field can be expressed as a double summation of plane waves in the form

$$H_z = \sum_{m=-\infty}^{\infty}\sum_{n=-\infty}^{\infty} D_{mn} exp\left(-j\gamma_{mn}z\right) exp\left(j\kappa_{mx}x\right) exp\left(j\kappa_{ny}y\right) \tag{3.54}$$

where $\kappa_{mx} = k_x = 2m\pi(1+\psi_x)/d_x$, $\kappa_{ny} = k_y = 2n\pi(1+\psi_y)/d_y$, and $\gamma_{mn}^2 = k^2 - \kappa_{mx}^2 - \kappa_{ny}^2$. At present, D_{mn} is an unknown coefficient.

The transverse field components can be obtained from Eq. (3.54) and Maxwell's equations in their reduced form for TE fields as given by Eq. (2.4) with $E_z = 0$. In the present instance of a rectangular geometry, $\nabla_t = \hat{x}\partial/\partial x + \hat{y}\partial/\partial y$ and $\kappa_t^2 = \boldsymbol{\kappa}_t \cdot \boldsymbol{\kappa}_t = k_x^2 + k_y^2$, where $\boldsymbol{\kappa}_t = \hat{x}k_x + \hat{y}k_y$. The transverse fields are found to be

$$\boldsymbol{E}_t = -k\eta_o \sum_{m=-\infty}^{\infty} \sum_{n=-\infty}^{\infty} D_{mn} exp\left(-j\gamma_{mn}z\right) \left(\frac{\hat{z} \times \boldsymbol{\kappa}_t}{\kappa_t^2}\right) exp\left(j\kappa_{mx}x\right) exp\left(j\kappa_{ny}y\right) \tag{3.55a}$$

$$\boldsymbol{H}_t = \sum_{m=-\infty}^{\infty} \sum_{n=-\infty}^{\infty} \gamma_{mn} \left\{ D_{mn} \left(\frac{\boldsymbol{\kappa}_t}{\kappa_t^2}\right)\right\} exp\left(-j\gamma_{mn}z\right) exp\left(j\kappa_{mx}x\right) exp\left(j\kappa_{ny}y\right) \tag{3.55b}$$

Suppose in addition the aperture at $z = 0$ is located in a ground plane and the electric field in the aperture is given by $\boldsymbol{E}_t(x, y, 0) = \boldsymbol{E}_A$ over the domain of the unit cell denoted by $\mathfrak{D}$. With this boundary constraint, and after cross-multiplying both sides with $\hat{z}$, Eq. (3.55a) becomes

$$\hat{z} \times \boldsymbol{E}_A = k\eta_o \sum_{m=-\infty}^{\infty} \sum_{n=-\infty}^{\infty} D_{mn} \left(\frac{\boldsymbol{\kappa}_t}{\kappa_t^2}\right) exp\left(j\kappa_{mx}x\right) exp\left(j\kappa_{ny}y\right)$$

Taking the inverse transform of this complex series, then

$$D_{mn}\left(\frac{\boldsymbol{\kappa}_t}{\kappa_t^2}\right) = \frac{1}{k\eta_o d_x d_y} \iint_{\mathfrak{D}'} \left(\hat{z} \times \boldsymbol{E}_A\left(x', y'\right)\right) exp\left(-j\kappa_{mx}x'\right) exp\left(-j\kappa_{ny}y'\right) dx' dy'. \tag{3.56}$$

Let the aperture field components be given by $\boldsymbol{E}_A = \sum_{p=1}^{N} V_p \boldsymbol{e}_p$and $\boldsymbol{H}_A = \boldsymbol{H}_t(x, y, 0) = \sum_{p=1}^{N} I_p\left(\hat{z} \times \boldsymbol{e}_p\right)$, where $\boldsymbol{e}_p(x, y)$ are shape functions that satisfy the boundary conditions as well as orthogonality. These are often chosen to be the modes of the domain $\mathfrak{D}$, and N is the number of shape functions adopted. The term in the braces in Eq. (3.55b) is replaced by Eq. (3.56). The expansions of the transverse fields are substituted into the result. With these substitutions, Eq. (3.55b) now becomes

$$\sum_{p=1}^{N} I_p\left(\hat{z} \times \boldsymbol{e}_p\right) = \frac{1}{k\eta_o d_x d_y} \sum_{m=-\infty}^{\infty} \sum_{n=-\infty}^{\infty} \gamma_{mn} \exp\left(j\kappa_{mx}x\right) \exp\left(j\kappa_{ny}y\right)$$
$$\times \left(\sum_{q=1}^{N} V_q F_q\left(m, n, -\kappa_{mx}, -\kappa_{ny}\right)\right)$$

where

$$F_p\left(m, n, \kappa_{mx}, \kappa_{ny}\right) = \iint_{\mathfrak{D}'} \left(\hat{z} \times \boldsymbol{e}_p\left(x', y'\right)\right) exp\left(j\kappa_{mx}x'\right) exp\left(j\kappa_{ny}y'\right) dx' dy'.$$

The primed coordinates refer to the source radiation. The vector $\left(\hat{z} \times \boldsymbol{e}_p\right)$ $(p = 1, \ldots N)$ is projected onto both sides of this equation and result integrated across the domain $\mathfrak{D}$ of the receiving element. Because of orthogonality, this equation simplifies to

$$I_p N_p = \frac{Y_o}{d_x d_y k} \sum_{q=1}^{N} \gamma_{mn} F_p\left(m, n, \kappa_{mx}, \kappa_{ny}\right) F_q\left(m, n, -\kappa_{mx}, -\kappa_{ny}\right) V_q$$

where $Y_o = 1/\eta_o$ and $N_p = \iint_{\mathfrak{D}'} \boldsymbol{e}_p \cdot \boldsymbol{e}_p dx' dy'$. Therefore, the coefficients of the expansion of $\boldsymbol{H}_A$ are

$$I_p = \sum_{q=1}^{N} Y_{pq}\left(\psi_x, \psi_y\right) V_q \tag{3.57}$$

or, alternatively, in matrix form $\boldsymbol{I} = \boldsymbol{Y}\boldsymbol{V}$, where $\boldsymbol{Y}$ is the admittance matrix with coefficients

$$Y_{pq}\left(\psi_x, \psi_y\right) = \frac{2Y_o}{d_x d_y k N_p} \sum_{m=-\infty}^{\infty} \sum_{n=-\infty}^{\infty} \gamma_{mn} F_p\left(m, n, \kappa_{mx}, \kappa_{ny}\right) F_q\left(m, n, -\kappa_{mx}, -\kappa_{ny}\right) \tag{3.58}$$

for a rectangular periodic array. The series in Eq. (3.58) converges rapidly with increasing indices m and n.

As an example, and to a first approximation, consider an infinite periodic array of rectangular apertures supporting only the TE_{10} mode. When only one mode is used in Eq. (3.58), the result is called a "grating lobe" series (Amitay et al. 1972; Hansen 1998). This series can be a useful approach for understanding coupling problems and the impact of mutual coupling on the grating lobes, although care is required as sometimes it not very accurate. To demonstrate this approach, consider a single shape function with $N = 1$ which is given by $e_1 = \hat{y}\cos\left(\frac{\pi x}{a}\right)$ over the domain defined by $|a/2| \geq x$, $|b/2| \geq y$ and is zero elsewhere. With this approximation, it follows that

$$F_1\left(m, n, \psi_x, \psi_y\right) = -\left(\frac{2ab}{\pi}\right) C\left(\kappa_{mx}a/2\right) S\left(\kappa_{ny}b/2\right)$$

where the functions $C(x) = \cos(x)/(1-(2x/\pi)^2)$ and $S(x) = \sin(x)/x$. Also, the normalization in Eq. (3.57) is $N_1 = (ab)/2$.

Therefore,

$$Y_{11}\left(\psi_x, \psi_y\right) = \frac{Y_o 8ab}{\pi^2 d_x d_y k} \sum_{m=-\infty}^{\infty} \sum_{n=-\infty}^{\infty} \gamma_{mn} C\left(\kappa_{mx}a/2\right) C\left(-\kappa_{mx}a/2\right) S\left(\kappa_{ny}b/2\right) S\left(-\kappa_{ny}b/2\right) \tag{3.59}$$

The series in Eq. (3.59) converges reasonably quickly if the summation in m has limits $M \geq \pm 4(1 + 1.5d_x/a)$ and similarly for n. This equation is similar to the result obtained by Diamond (1968) for the element driving admittance, which can be also written as

$$Y_{11}\left(\psi_x, \psi_y\right) = \sum_{m=-\infty}^{\infty} \sum_{n=-\infty}^{\infty} \overline{Y}(m, n)$$

where

$$\overline{Y}(m, n) = \frac{Y_o 8ab}{\pi^2 d_x d_y k} \sqrt{k^2 - \kappa_{mx}^2 - \kappa_{ny}^2}\, C\left(\kappa_{mx}a/2\right) C\left(-\kappa_x a/2\right) S\left(\kappa_y b/2\right) S\left(-\kappa_y b/2\right).$$

When all sources are turned on and with the specified progressive steering angles, the active reflection coefficient in the presence of only one mode is

$$\Gamma_a\left(\psi_x, \psi_y\right) = \left[\frac{1 - Y_{11}\left(\psi_x, \psi_y\right)}{1 + Y_{11}(\psi_x, \psi_y}\right] \tag{3.60}$$

The coupling coefficients are given by the following complex Fourier series

$$\Gamma_a\left(\psi_x, \psi_y\right) = \sum_{m=-\infty}^{\infty} \sum_{n=-\infty}^{\infty} S_{mn}^0 exp\left(jm\psi_x\right) exp\left(jn\psi_y\right)$$

where S_{mn}^0 is the active coupling coefficient of the mn-th element. When the Fourier series is inverted, it is found that this is

$$S_{mn}^0 = \frac{1}{4\pi^2} \int_{-\pi}^{\pi} \int_{-\pi}^{\pi} \Gamma_a\left(\psi_x, \psi_y\right) exp\left(-jm\psi_x\right) exp\left(-jn\psi_y\right) d\psi_x d\psi_y. \tag{3.61}$$

As an example, the results of single-mode analysis of the active reflection coefficient of a periodic array of square waveguides on a square grid with dimensions $d_x = d_y = 0.5714\ \lambda$ are shown in Figure 3.12. Also shown is the magnitude of a multimode solution (Galindo and Wu 1966). The calculated active reflection coefficient at a scan angle of 20° is found to be $\Gamma_a = 0.30\angle -80°$. This single basis function result compares with the value of $\Gamma_a = 0.32\angle 149.4°$ that is quoted by Amitay et al. (1972) and was obtained with a significantly higher basis of 30 pulse functions.

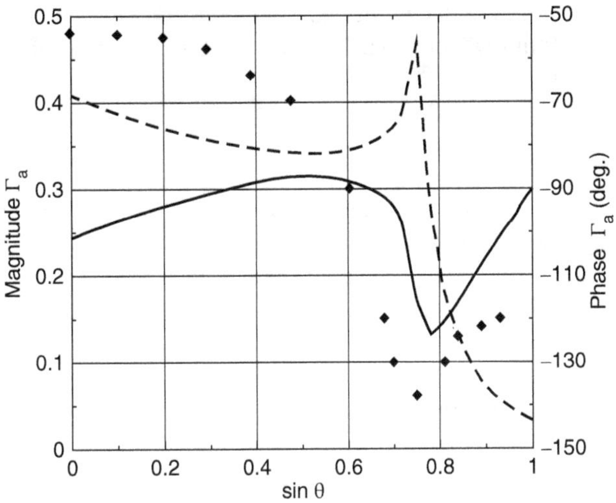

Figure 3.12 Magnitude of input reflection coefficient (Magnitude: solid line, phase: dashed line) in the H-plane scan of a periodic array of square waveguides with dimensions $a = b = 0.5354\lambda$ and $d_x = d_y = 0.5714\lambda$ estimated with a single shape function. This compares with results of Galindo and Wu (1966) using 30 pulse functions (Magnitude: dots). Source: Modified from GALINO, V. and WU C.P. (1966): 'Numerical solutions for infinite phased array of rectangular waveguides with thick walls', IEEE Trans. Antennas Propagat., Vol. 14, No.2, pp. 149-158.

3.5.2 Mode Matching Methods

MM is a classic and powerful method for analyzing spectral and scattering problems. The fundamental idea behind the method is to decompose a domain into "basic" subdomains elements, in which the underlying spectral or scattering field representation can be solved explicitly, and then these analytical solutions are matched at "junctions" between subdomains. This matching leads to functional equations at the junctions and thus reduces the dimensionality of the problem. Although the resulting integral or functional equations are in general more difficult to evaluate than that of the original problem, they are efficient for obtaining analytical estimates and numerical solutions. MM can be applied to both finite arrays or the unit cell of infinite arrays.

A typical example of the use of MM is an L-shaped waveguide shown in Figure 3.13a. It can be decomposed into two rectangular regions. The field in each region can be represented as an infinite sum of products of two sinusoidal functions, one in the x-dimension and the other in the y-direction. The problem is solved by matching the two rectangular regions along their boundary. The resulting two equations can be used to solve for the dispersion of the phase shift with frequency. The other major use of MM is along the direction of wave propagation where there are steps or

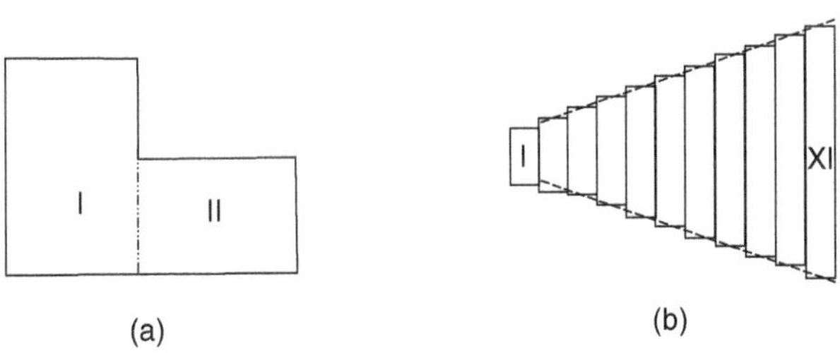

Figure 3.13 Mode matching. (a) L-shaped waveguide; (b) stepped horn.

junctions or the profile of the structure changes as shown in Fig. 3.13b. Both examples of MM mentioned here are used in the analysis of mutual coupling. The latter problem is discussed in Section 5.2.2.

3.5.3 Moment Methods

One of the methods classified under weighted residuals is the moment method (MoM), which was introduced in Section 3.4.3. This technique is widely used in electromagnetics and is used in coupling problems for both wire antennas and waveguides. The application of the moment method to wire antennas is detailed in Chapter 4. For other structures, the moment method is used in the method of characteristic modes (CM), and this will be outlined in Section 3.5.4. Further details of the application of moment methods to coupling problems, are given in the references Fenn et al. (1982), Luzwick and Harrington (1982), Lyon and Sangster (1987), Arndt et al. (1989), and Pereia-Fiho et al. (2018).

3.5.4 Method of Characteristic Modes

In this method, the set of CM is a natural mode set that is based on numerical solutions to the actual antenna geometry (Harrington and Mautz 1971; Angiulli and Di Massa 1996). These solutions can be obtained by a variety of methods, but the most common one is MoM. To begin, consider a uniform plane wave incident upon an array of radiators as illustrated in Figure 3.14. Without loss of generality, assume this field is polarized in the z-direction as follows:

$$\boldsymbol{E}^{inc} = z E_0 exp\left[\text{jk}\rho\cos\left(\phi - \phi o\right)\right].$$

Suppose electric and magnetic currents $\boldsymbol{J(r)}$ and $\boldsymbol{M(r)}$ are setup on the elements of the array. The electric field radiated by the electric current is given by

$$\boldsymbol{L^e}(\boldsymbol{J}) = \frac{\eta k}{4} \oint_{C_1 \cup C_2 \cup \ldots} H_0^2\left(k\left|r - r'\right|\right) \boldsymbol{J}(r)\, dC$$

The associated radiated field by a magnetic current $\boldsymbol{M(r)}$ is obtained from duality. If we satisfy the continuity of the tangential fields on the arrays, it leads to

$$\left[\boldsymbol{L^e}(\boldsymbol{J}) - \boldsymbol{E}^{inc}\right]\Big|_{tan} = 0$$

where

$$\boldsymbol{L^e}(\boldsymbol{J})\big|_{tan} = Z\left(\boldsymbol{J}_n\right) = R\left(\boldsymbol{J}_n\right) + jX\left(\boldsymbol{J}_n\right)$$

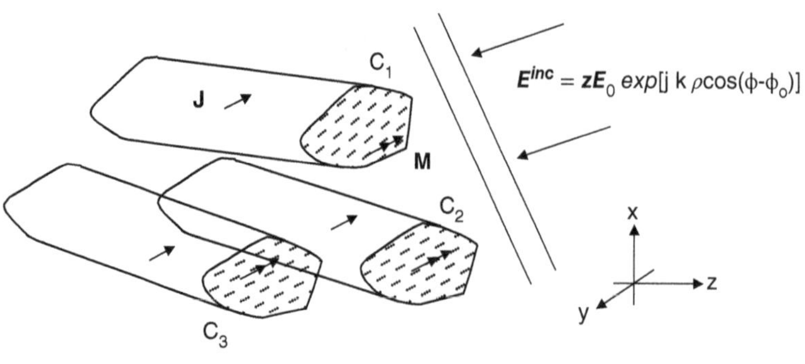

Figure 3.14 Array excited by incident plane wave.

and

$$X\left(\boldsymbol{J}_n\right) = \lambda_n \mathrm{R}\left(\boldsymbol{J}_n\right) \tag{3.62}$$

In Eq. (3.62), R and X are real symmetric operators and λ_n is the n-th eigenvalue of the eigenfunction that diagonalizes Z. The characteristic fields $(E_n.H_n)$ produced by the eigencurrent $\boldsymbol{J}_n$ are then obtained from Maxwell's equations, by solving the matrix equation

$$\left\langle \boldsymbol{J}_n, [\boldsymbol{Z}]\boldsymbol{J}_m \right\rangle = \left(1 + j\lambda_n\right)\delta_{nm}. \tag{3.63}$$

The $\boldsymbol{J}_n$ are the CMs of the conducting body defined on $S = S_1 \cup S_2 \cup \ldots$ and [Z] is impedance matrix obtained from conventional methods such as MoM.

The eigen functions can be used as a basis for a representation of a general current as follows:

$$\boldsymbol{J} = \sum_k \alpha_k \boldsymbol{J_k}$$

where

$$\alpha_k = \frac{V_k}{\left(1 + j\lambda_k\right)}$$

and

$$V_k = \left(\boldsymbol{J}_k, E^{inc}\Big|_{tan}\right)$$

wherein (A, B) is the inner product. In the same way, the characteristic fields $(\boldsymbol{E_n}, \boldsymbol{H_n})$ produced by the n-th eigencurrent $\boldsymbol{J_n}$ are obtained from Maxwell's equations. The total field is found by summing the fields from each of these characteristic fields

$$\boldsymbol{E} = \sum_k \alpha_k \boldsymbol{E_k}$$

$$\boldsymbol{H} = \sum_k \alpha_k \boldsymbol{H}_k$$

As an example of the use of the total fields and characteristic mode analysis (CMA), consider the design of a multiport multiple-input multiple-output (MIMO) antenna. The antenna patterns from CMA are orthogonal, which is required for MIMO. In this application, a critical indicator is the envelope correlation coefficient, which is, given by

$$\rho = \sqrt{\sum_{m,n} \rho_{m,n}^2}$$

where $\rho_{m,n} = \dfrac{\left|\iint \boldsymbol{E}_m \cdot \boldsymbol{E}_n^* d\Omega\right|^2}{\left|\iint |\boldsymbol{E}_m|^2 d\Omega\right| \left|\iint |\boldsymbol{E}_n|^2 d\Omega\right|}$.

The CM set can be used directly to find ρ_{mn} and, therefore. the envelope correlation coefficient. This criterion can be used to design MIMO antennas to minimize antenna correlation and to achieve acceptable values if possible $\rho < 0.25$ (e.g. Kim and Nam 2018).

3.5.5 Minimum Scattering Element Method

A minimum scattering antenna was defined by Montgomery et al. (1948) as a receive antenna for which the scattering is minimum in some sense for each possible eigensolution. In this regime, if $\boldsymbol{a}_k^s$ is the absorbed wave and $\boldsymbol{a}_k^{s\dagger}\boldsymbol{a}_k^s = a_{1k}^2$ is the absorbed power where a_k is the k-th eigenvector and † is the transpose conjugate of the vector, the absorbed power is equal to the scattered power. For this to occur, and with ideal matching, it is necessary that the associated eigenvalue should be $\exp(j\gamma_k)$, where γ_k is real. When the input reflection coefficient is Γ and the local k-th port is terminated in a reactance $\exp(j\xi_k)$, the corresponding result is $\exp(j\gamma_k)(1-\Gamma^*\exp(j\xi_k))/(1-\Gamma\exp(-j\xi_k))$. It can be shown that a terminating reactance can always be found to this eigenvalue and is equal to unity (Rogers 1986), and thus, a reactance can always be found to make the antenna invisible.

Another property of such antennas is that the scattered radiation has the same pattern as the incident radiation. The work on characteristics modes indicates these requirements can be met is an excellent approximation particularly for antennas at or below its first resonance. Should the load be a pure open circuit, the antennas are termed canonical minimum scattering (CMS) antennas, and this property may be employed directly to compute the open-circuit impedance of the array. This method was used in a landmark paper by Wasylkiwskyj and Kahn (1970), which shows that for a special class of antennas, mutual coupling can be determined completely from the radiation of an isolated minimum scattering antenna. This explains why the element-by-element approach gives accurate predictions of mutual coupling in dipole arrays. For example, Anderson et al. (1974) showed that the CMS theory provides a good approximation to the coupling between pairs of cross dipoles and also helical antennas. As a result of invisibility, the mutual impedance between two antennas is easily calculated (Wasylkiwskyj and Kahn 1970). Thus, consider two antennas at a distance d apart, if antenna 2 is open circuited and antenna 1 is excited by a unit amplitude wave, the voltage across the terminals of antenna 2 due to 1 is

$$V_{21} = 2\left(\mathbf{S}_{21}\right)^{INV}\boldsymbol{a}_\beta/\left(1-\Gamma_2\right) \tag{3.64}$$

where $\boldsymbol{a}_\beta$ are is the mode amplitude due to modes incident on all ports other than the driven one and the superscript *INV* on $\mathbf{S}_{21}$ indicates measurement at an invisibility port. The current into antenna 1 is

$$I_1 = \left(1+\Gamma_1\right)/Z_{I1} = \left(1-\Gamma_1\right)$$

where Z_{I1} is the input impedance of antenna 1. As a result, the mutual impedance is given by

$$Z_{21} = \frac{V_{21}}{I_1} = 2\left(\mathbf{S}_{21}\right)^{INV}\boldsymbol{a}_\beta/\left[\left(1-\Gamma_1\right)\left(1-\Gamma_2\right)\right]. \tag{3.65}$$

The vector $\boldsymbol{a}_\beta$ can be found from the field scattered from the other ports as detailed by Wasylkiwskyj and Kahn (1970) and Rogers (1986).

3.6 Practical Aspects of Numerical Methods in Mutual Coupling Analysis

3.6.1 Introduction

An important aspect of any calculation of mutual coupling is the available computing resources, computer packages, and numerical methods used in them. There are several commercial computer packages available that enable the direct calculation of mutual coupling in most antennas such as FEKO (2020), HFSS (2020), and CST Microwave Studio (2020). For large arrays, these packages can use considerable computation time even when care is taken, and alternative approaches may need to be adopted. Techniques for analyzing large arrays are discussed in Chapter 13. To help with preparing data for major computing, packages such as MATLAB (2020) can be of significant help, as described by Wang et al. (2017). Direct calculation using the mathematical equations such as published in this book is possible using software such as MathCad (2000). While there is little control of the methods adopted in software packages, some recommendations can be made for local or personal computation that may assist in preparing data in the future. Another purpose of this section is to alert the reader to the type of methods that are available for mutual coupling calculations. These methods come into play when all the steps shown in Figure 3.1 have been completed. It is not the intention here to discuss the numerical methods used for solving the electromagnetic problem itself but to discuss the main numerical concerns that arise in mutual coupling calculations. These concerns are often numerical integration, and matrix inversion. In all cases, many methods are available and have been dealt with in many numerical texts, but they are discussed here in the context of a mutual coupling calculations which are often repetitive and have singular integrals whilst matrices are often exceptionally large when inversion can be time consuming.

3.6.2 Numerical Quadrature

In computations involving Greens functions, the self-admittance will contain singular integrals. In special cases, this singularity can be removed through change of variable such as a cylindrical or spherical polar transformation. Often, however, the integrals are usually integrated numerically (Fröberg 1974; Preis 1976; Wilton et al. 1984). There are some quadrature methods (e.g. Duffy 1982) and computer packages (e.g. MathCad. 2020) that take account of singularities in the integrand. In direct calculation with some mathematical packages, other alternatives are required. One effective approach is the singularity subtraction method. This entails subtracting out the singularity with an integrable function with the same order of singularity. Thus, if the original integral is

$$I = \int_a^b dx\, F(x)\, G(x)$$

where $F(x)$ is regular function of x, and $G(x)$ has a singularity at $x = x_o$ in $[a, b]$. The subtraction method involves selecting another function $H(x)$ which is also singular at x_o except that over the range $[a, b]$ it has a finite value, say $V(a, b)$. Therefore,

$$\begin{aligned} I &= \int_a^b dx\, F(x)\, G(x) = \int_a^b dx\, F(x)\, [G(x) - H(x)] + \int_a^b dx\, F(x_o)\, H(x) \\ &= \int_a^b dx\, F(x)\, [G(x) - H(x)] + V(a, b) \end{aligned} \tag{3.66}$$

As an example, consider $F(x) = [\cos(3x) - 1]$ and $G(X) = \frac{1}{4\pi}\frac{\exp(-j2\pi X)}{X}$ over the interval $0 \leq x \leq 1$, where $X = x - x_o$, and $x_o = 0.7$. A possible subtractor is the series representation of the Green's function given by

$$H(X) = \frac{1}{4\pi}\sum_{n=0}^{\infty}\frac{(-jk)^n}{n!}X^{n-1}.$$

Therefore, then Eq. (3.66) becomes

$$\begin{aligned} I &= \int_0^1 dx\, [\cos(3x) - 1]\, \frac{1}{4\pi}\frac{\exp(-j2\pi X)}{X} \\ &= \int_0^1 dx\, [\cos(3x) - 1] \left[\frac{1}{4\pi}\frac{\exp(-j2\pi X)}{X} - \frac{1}{4\pi}\sum_{n=0}^{L}\frac{(-j2\pi)^n}{n!}X^{n-1}\right] \\ &+ \frac{1}{4\pi}\sum_{n=0}^{L}\frac{(-j2\pi)^n}{n!}\int_0^1 dx\, [\cos(3x) - 1]\, X^{n-1} \end{aligned}$$

where $X = x - 0.7$. The integrand in square braces on the second line can be made sufficiently smooth for quadrature by choosing L sufficiently large, e.g. $L \geq 3$. The second integral can be expressed in closed form in terms of the exponential-integral function $Ei(x)$ (Gradshteyn and Ryzhik, 1980). The result for L = 0 is $= -0.074 + j0.369$. This can be checked with increasing number of terms. In this example, $L = 3$, 5, and 10 has been evaluated, and although the value of the individual integrals changes, their total remains about the same.

3.6.3 Matrix Inversion

At the center of computing the mutual coupling between antenna elements is the inversion of the impedance/admittance matrix or a matrix derived from it.. That is, the problem is to determine the matrix $\boldsymbol{X}$ which satisfies the relation $\boldsymbol{AX} = \boldsymbol{I}$, where $\boldsymbol{A}$ is the matrix to be inverted. This matrix is usually complex and symmetric, and it is assumed here that $det\ \boldsymbol{A} \neq \boldsymbol{0}$. With exceptionally large arrays, this computation can take a considerable amount of time when following classical elimination methods (Fröberg 1974). There are a variety of alternative approaches that can be adopted in such cases, and some are described here.

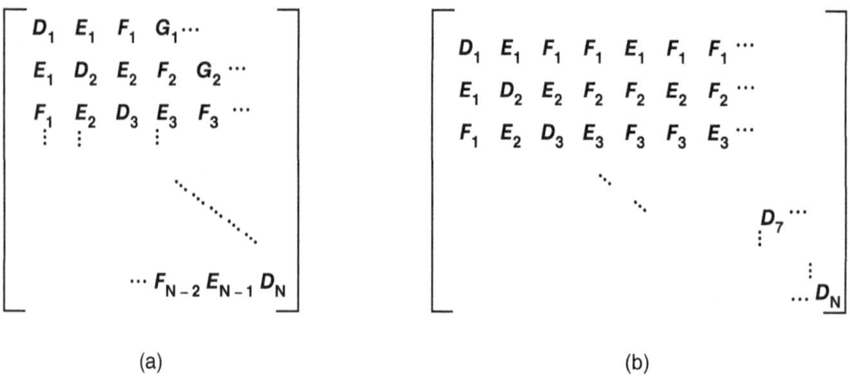

Figure 3.15 Matrix filling. (a) Regular rectangular grid; (b) hexagonal grid.

The structure of matrix $\boldsymbol{A}$ that arises from the geometry and coupling calculations has resulted in several different approaches to inversion. The matrices resulting from an element-by-element analysis of a rectangular uniformly spaced array of elements and a hexagonal array are illustrated in Figure 3.15. The block submatrices $\boldsymbol{D}_1$, $\boldsymbol{E}_1$, $\boldsymbol{F}$, … etc. can be single elements only or several dimensions depending on the approximation in each element. For example, in MM, the order of the submatrix depends on the number of modes chosen in each waveguide, and in MoM, the order corresponds to the number of points in a segment. For the case in Figure 3.15a, for a rectangular array, as pointed out by Sinnott (1973), the symmetry of $\boldsymbol{A}$ is symmetric block-Toeplitz. In the case of a circular arrangement of elements, efficient inversion algorithms can be devised (Vescovo 1996). If there are N equispaced elements each supporting M quantities (e.g. modes), the submatrices are of order $M \times M$.

Let $\boldsymbol{A} = \boldsymbol{A}^{(N)}$ where $\boldsymbol{A}^{(N)}$ denotes that the inverse is obtained recursively on N where $\boldsymbol{A}^{(N)}$ has order $M \cdot (N+1) \times (N+1) \cdot M$. The inverse of Toeplltz matrices can be found in the literature, for example, by Trench (1964). This method was generalized by Sinnott (1973). Thus, if $\boldsymbol{A}$ is given by

$$\boldsymbol{A} = \boldsymbol{A}^{(N)} = \begin{bmatrix} \boldsymbol{A}_0 & \boldsymbol{A}_1 & \cdots & \boldsymbol{A}_N \\ \boldsymbol{A}_1 & \boldsymbol{A}_0 & & \boldsymbol{A}_{N-1} \\ \vdots & \vdots & \cdots & \vdots \\ \boldsymbol{A}_N & \boldsymbol{A}_{N-1} & \cdots & \boldsymbol{A}_0 \end{bmatrix} \tag{3.67}$$

The inverse, $\boldsymbol{X}$, has the same submatrix partitioning and is of the form

$$\boldsymbol{X} = \boldsymbol{X}^{(N)} = \begin{bmatrix} \boldsymbol{X}_{00}^{(N)} & \boldsymbol{X}_{01}^{(N)} & \cdots & \boldsymbol{X}_{0N}^{(N)} \\ \boldsymbol{X}_{10}^{(N)} & \boldsymbol{X}_{11}^{(N)} & & \boldsymbol{X}_{1N}^{(N)} \\ \vdots & \vdots & \cdots & \vdots \\ \boldsymbol{X}_{N0}^{(N)} & \boldsymbol{X}_{N1}^{(N)} & \cdots & \boldsymbol{X}_{NN}^{(N)} \end{bmatrix} \tag{3.68}$$

where the submatrices $\boldsymbol{X}_{ij}^{(N)}$ are of order $M \times M$ and have the symmetry properties $\boldsymbol{X}_{ij}^{(N)} = \boldsymbol{X}_{ij}^{(N)T}$ and $\boldsymbol{X}_{ii}^{(N)} = \boldsymbol{X}_{ii}^{(N)T}$, $(0 \le i, j \le N)$. The block-Toeplitz property vanishes on inversion, although some symmetry remains about the cross diagonal, that is,

$$\boldsymbol{X}_{ij}^{(N)} = \boldsymbol{X}_{N-i,N-j}^{(N)} = \boldsymbol{X}_{N-j,N-i}^{(N)T} \text{ and } \boldsymbol{X}_{i,N-i}^{(N)} = \boldsymbol{X}_{i,N-i}^{(N)T}.$$

The inverse is obtained recursively using the following steps:

$$\boldsymbol{X}_{00}^{(N)} = \boldsymbol{\Delta}^{(N-1)} \tag{3.69a}$$

$$\boldsymbol{X}_{i0}^{(N)} = -\boldsymbol{\psi}_{i-1}^{(N-1)}\boldsymbol{\Delta}^{(N-1)};\ (1 \le i \le N) \tag{3.69b}$$

$$\boldsymbol{X}_{ij}^{(N)} = \boldsymbol{X}_{i-1,j-1}^{(N)} + \boldsymbol{\psi}_{i-1}^{(N-1)}\boldsymbol{\Delta}^{(N-1)}\boldsymbol{\psi}_{j-1}^{(N-1)T} - \boldsymbol{\psi}_{N-i}^{(N-1)}\boldsymbol{\Delta}^{(N-1)}\boldsymbol{\psi}_{N-j}^{(N-1)T};\ (1 \le i,j \le N) \tag{3.69c}$$

commencing with

$$\boldsymbol{\psi}_0^{(0)} = \mathbf{A}_0^{-1}\mathbf{A}_1 \text{ and } \boldsymbol{\Delta}^{(-1)} = \mathbf{D}_0^{-1} \tag{3.69d}$$

$$\boldsymbol{\psi}_m^{(m)} = -\boldsymbol{\Delta}^{(m-1)}\left[\sum_{j=0}^{m-1}\boldsymbol{\psi}_j^{(m-1)T}\boldsymbol{A}_{m-j} - \boldsymbol{A}_{m+1}\right] \tag{3.69e}$$

$$\boldsymbol{\psi}_i^{(m)} = \boldsymbol{\psi}_i^{(m-1)} - \boldsymbol{\psi}_{m-i-1}^{(m)}\boldsymbol{\psi}_m^{(m)};\ (0 \le i \le m-1) \tag{3.69f}$$

$$\boldsymbol{\Delta}^{(m-1)} = \left[\boldsymbol{I}_N - \boldsymbol{\psi}_{m-1}^{(m-1)2}\right]^{-1}\boldsymbol{\Delta}^{(m-2)} \tag{3.69g}$$

Equations (3.69a) and (3.69b) are used to determine the quarter section bounded by the zeroth and two diagonals; symmetry then defines all the other combinations. Using the recurrence algorithm given in Eqs. (3.69a)–(3.69g) to invert $\boldsymbol{A}^{(N)}$, the number of multiplications required is $[(N+1)\cdot M]^2$. Thus, with this algorithm, the time is proportional to the square of the matrix order rather than with the cube of the order as is the case with elimination methods. As a result, Eqs. (3.69a)–(3.69g) give huge savings when the matrix is large.

An approximate inversion method for a uniformly spaced array of identical elements has been described by Gera (1984). It relies on special features of the resulting symmetric block-Toeplitz matrix and having a nilpotent off-diagonal segment (i.e. a square sub-matrix equal to zero), which allows the possibility of a finite series expansion. An approximate inverse matrix can be obtained by truncating the series, which results in a significant reduction in computation time for a small loss of accuracy (typically 6%).

In some applications, the scattering matrix $\boldsymbol{S}$ of the array is the main computing objective, although the impedance/admittance needs to be found first. In terms of the normalized admittance matrix, for example, this matrix can be expressed as

$$\boldsymbol{S} = [\boldsymbol{I} + \boldsymbol{y}]^{-1}[\boldsymbol{I} - \boldsymbol{y}]. \tag{3.70}$$

where $\boldsymbol{I}$ is the unit diagonal matrix. The elements of the admittance decay absolutely with increasing distance from a driven element particularly after about one wavelength spacing. Therefore, it is possible to invert the first term on the right side using a power series expansion, thus

$$[\boldsymbol{I} + \boldsymbol{y}]^{-1} = \sum_{n=0}^{\infty}(-1)^n\boldsymbol{y}^n \tag{3.71}$$

which is valid providing the norm $||\mathbf{y}|| < 1$. When the norm is small, the series will converge rapidly, and only a few terms will be required in the series. Thus,

$$\mathbf{S} = \mathbf{I} - 2\sum_{n=0}^{\infty}(-1)^n \mathbf{y}^n \mathbf{y}.$$

A five term approximation is of the form

$$\mathbf{S} \cong \mathbf{I} - 2\left(\mathbf{I} - \mathbf{y} + \mathbf{y}^2 - \mathbf{y}^3 + \mathbf{y}^4\right)\mathbf{y} \tag{3.72}$$

The elements of the approximate **S**-matrix Eq. (3.72) have an absolute error of <2% when the norm is below unity. The error becomes higher as the norm $||\mathbf{y}||$ increases above unity. In many applications, the admittance matrix is diagonally dominant with element values close to unity. When this occurs, a better approximation is to use the $\mathbf{y1} = (\mathbf{y} - \mathbf{I})/2$ in the expansion of Eq. (3.70). This leads to similar conclusions about accuracy with the resulting alternative approximation to Eq. (3.72)

$$\mathbf{S} \cong -\left(\mathbf{I} - \mathbf{y1} + \mathbf{y1}^2 - \mathbf{y1}^3 + \mathbf{y1}^4\right)\mathbf{y1} \tag{3.73}$$

and to the norm of $||\mathbf{y1}||$. For example, if

$$\mathbf{y} = \begin{bmatrix} 0.9 - j0.1 & -0.5 + j0.3 & 0.08 \\ -0.5 + j0.3 & 0.9 - j0.1 & -0.5 + j0.3 \\ 0.08 & -0.5 + j0.3 & 0.9 - j0.1 \end{bmatrix} \tag{3.74}$$

the exact scattering matrix given by Eq. (3.73) is

$$\mathbf{S} = \begin{bmatrix} 0.09 - j0.033 & 0.28 - j0.192 & 1.9 * 10^{-3} - j0.093 \\ 0.28 - j0.192 & 0.146 - j0.129 & 0.28 - j0.192 \\ 1.9 * 10^{-3} - j0.093 & 0.28 - j0.192 & 0.09 - j0.033 \end{bmatrix}$$

while the fourth-order approximation given by Eq. (3.73) is

$$\mathbf{S} \cong \begin{bmatrix} 0.1 - j0.033 & 0.281 - j0.19 & 4.4 * 10^{-3} - j0.093 \\ 0.281 - j0.19 & 0.151 - j0.128 & 0.281 - j0.19 \\ 4.4 * 10^{-3} - j0.093 & 0.281 - j0.19 & 0.1 - j0.033 \end{bmatrix}$$

The norm of $||\mathbf{y1}||$ in the example is about 0.9, and the error is <1%.

3.7 Conclusion

The intention has been to review some important techniques that are used in the analysis and computation of mutual coupling. Many of these techniques are referred to and used in later chapters. Antenna radiation relies on the reciprocity theorem and through it the concept of reaction which was shown to apply to the interaction of fields and sources in a domain. Reaction leads naturally to circuit quantities such as self- and mutual impedances which are consistent with the derivations from elementary sources. Various forms of circuit quantities are obtained for electric and magnetic sources. Following these derivations, the chapter turned to providing these quantities for finite and infinite arrays. The former follows on as an extension from two to many elements, while in infinite arrays, periodicity is used to naturally create a cellular approach within which all details of an element are contained for analysis. Most coupling problems are expressed in the form of integral and differential equations that cannot be solved in closed form, and methods of solution usually involve reducing the problem to systems of equations. The main way of doing this is the method of weighted residuals. Within this class, many of the common methods are included such as Galerkin's method, method of moments, and point matching. These approaches are outlined, and examples are given. Special techniques for coupling problems are described through the method of characteristic modes and minimum scattering. Finally, in all calculations of mutual coupling, both quadrature and matrix inversion are usually employed. Some techniques for these operations are discussed with an emphasis on fast and accurate computation. In the former, integration in the vicinity of a singularity due to the source occurs, when it is shown that singularity subtraction can be useful in this. For the latter, use of the matrix structure can reduce the number of multiplications required, and for an inversion recurrence algorithm, the time is proportional to the square of the matrix order rather than the cube as occurs with the standard method. With this chapter, the basis of many techniques employed in this book has been provided.

References

Amitay, N., Galindo, V., and Wu, C.P. (1972). *Theory and Analysis of Phased Array Antennas*. New York, USA: Wiley.

Anderson, J.B., Lessow, H.A., and Schjaer-Jacobsen, H. (1974). Coupling between minimum scattering antenna. *IEEE Trans. Antennas Propag.* AP-22 (6): 832–835.

Angiulli, G. and di Massa, G. (1996): Mutual Coupling evaluation by characteristic modes in multiple scattering of electromagnetic field', Digest Electro technical Conference, 1996. MELECON '96., 8th Mediterranean, Contribution in Conference Proceedings, 1996, pp. 580-583, ISBN: 0-7803-3109-5.

Arndt, F., Wolff, K.H., Brunjes, L. et al. (1989). Generalized moment method analysis of planar reactively loaded rectangular waveguide arrays. *IEEE Trans. Antennas Propag.* AP-37: 329–338.

Bailey, M.C. and Bostian, C.W. (1974). Mutual coupling in a finite planar array of circular apertures. *IEEE Trans. Antennas Propag.* AP-22: 178–184.

Balanis, C.A. (1983). *Antenna Theory: Analysis and Design*. New York, USA: Harper & Row.

Bechmann, R. (1931). On the calculation of radiation resistance of antennas and antenna combinations. *Proc. IRE* 19: 1471–1480.

Bird, T.S. (1979). Mode coupling in a planar circular waveguide array. *IEE J. Microwaves Opt. Acoust.* 3: 172–180.

Bird, T.S. (1990). Analysis of mutual coupling in arrays of different-sized rectangular waveguides. *IEEE Trans. Antennas Propag.* AP-38: 166–172.

Bird, T.S. (1996). Improved solution of mode coupling in different sized circular apertures and their application. *IEE Proc. H, Microwave Antennas Propag.* 143: 457–464.

Bird, T.S. (2016). *Fundamentals of Aperture Antennas and Arrays*. Chichester UK: Wiley.

Bird, T.S. and Sroka, C. (1992). Design of the Ku-band antennas for the Galaxy 601C satellites. *J. Electr. Electron. Eng. Aust.* 12: 267–273.

Bontsch-Bruewitsch, M.A. (1926). Die strahlung der kompliziertenrechtwinkeligenantennenmitgleichbescheaffenenvibratoren. *Ann. Phys.* 81: 425–453.

Booker, H.G. (1946). Slot aerials and their relation to complementary wire aerials (Babinet's principle). *J. Inst. Electr. Eng. Pt IIIA* 93: 620–626.

Brillouin, L. (1922). Sur l'origine de la resistance de rayonnement. *Radioélectricité* 3 (4): 147–152.

Carter, P.S. (1932). Circuit relations in radiating systems and applications to antenna problems. *Proc. IRE* 20 (6): 1004–1041.

Chew, W.C. (1995). *Waves and Fields in Inhomogeneous Media*. New York, USA: IEEE Press Inc.

CST MWS (2020): https://au.mathworks.com/products/connections/product_detail/cst-microwave-studio.html, CST MICROWAVE STUDIO® software package for designing, analyzing and optimizing electromagnetic (EM) components and systems.

Diamond, B.L. (1968). A generalized approach to the analysis of infinite planar antenna. *Proc. IEEE* 56: 1837–1851.

Duffy, M.G. (1982). Quadrature over a pyramid or cube of integrands with a singularity at a vertex. *SIAM J. Numer. Anal.* 19: 1260–1262.

Edelberg, S. and Oliner, A.A. (1960). Mutual coupling effects in large arrays I: slot arrays. *IRE Trans. Antennas Propag.* AP-8 (3): 286–297.

Farrell, G.F. and Kuhn, D.H. (1968). Mutual coupling in infinite planar arrays of rectangular waveguide horns. *IEEE Trans. Antennas Propag.* AP-16 (4): 405–414.

FEKO (2020): https://altairhyperworks.com/product/FEKO, computational electromagnetics software offers several frequency and time domain EM solvers under a single license.

Felsen, L.B. and Marcuvitz, N. (1973). *Radiation and Scattering of Waves*. Upper Saddle River, NJ, USA: Prentice-Hall.

Fenn, A.J., Thiele, A.J., and Munk, B.A. (1982). Moment method analysis of finite rectangular waveguide phased arrays. *IEEE Trans. Antennas Propag.* AP-30: 554–564.

Finlayson, B.A. (1972). *The Method of Weighted Residuals and Variational Principles*. New York, USA: Academic Press.

Fröberg, C.-E. (1974). *Introduction to Numerical Methods*, 2e. Addison-Wesley Publishing Company. World Series.

Galejs, J. (1969). *Antennas in Inhomogeneous Media*. Oxford, UK: Pergamon Press.

Galindo, V. and Wu, C.P. (1966). Numerical solutions for infinite phased array of rectangular waveguides with thick walls. *IEEE Trans. Antennas Propag.* 14 (2): 149–158.

Gera, A.E. (1984). A simplified computational procedure for the analysis of planar arrays. *IEEE Trans. Antennas Propag.* AP-32 (6): 647–651.

Golden, K.E. and Stewart, G.E. (1969). Self and mutual admittances of rectangular slot antennas in the presence ofan inhomogeneous plasma layer. *IEEE Trans. Antennas Propag.* AP-17: 763–771.

Gradshteyn, I.S. and Ryzhik, I.M. (1980). *Table of Integrals, Series and Products*, 5e (ed. A. Jeffrey). London, UK: Academic Press.

Hallén, E. (1938). *Theoretical Investigation into the Transmitting and Receiving Qualities of Antennas*, Nova Acta Regiae Soc. Sci Ups., Ser VI, vol. 11, 1–44.

Hansen, R.C. (1966). *Microwave Scanning Antennas*, vol. 3. New York: Academic Press.

Hansen, R.C. (1998). *Phased Array Antennas*. New York, USA: Wiley.

Harrington, R.F. (1961). *Time Harmonic Electromagnetic Fields*. New York, USA: McGraw-Hill.

Harrington, R.F. (1968). *Field Computation by Moment Methods*. New York, USA: The Macmillan Company.

Harrington, R.F. and Mautz, J.R. (1971). Theory of characteristic modes for conducting bodies. *IEEE Trans. Antennas Propag.* AP-19 (5): 622–628.

Harrington, R.F. and Mautz, J.R. (1976). A generalized network formulation for aperture problems. *IEEE Trans. Antennas Propag.* AP-24: 870–873.

Haupt, R.L. (2010). *Antenna Arrays: A Computational Approach*. Hoboken, NJ, USA: IEEE Press, Wiley.

Hertz, H. (1888). Untersuchingenüber die ausbreitung der elektrischen kraft. *Wied. Ann.* 36: 1.

HFSS (2020): https://www.ansys.com/products/electronics/ansys-hfss, Ansys HFSS 3D electromagnetic field simulator for RF and wireless design.

Hockham, G.A. and Walker, G.H. (1973): Study of a finite phased array antenna', Proc. European Microwave Conf., 4-7 Sept. Brussels, Belgium, paper C.1.3.

Kim, D.-W. and Nam, S. (2018). Systematic design of a multiport MIMO antenna with bilateral symmetry based on characteristic mode analysis. *IEEE Trans. Antennas Propag.* AP-66 (3): 1076–1085.

King, R.W.P. and Harrison, C.W. (1944). Mutual and self-impedance for coupled antennas. *J. Appl. Phys.* 15: 481–495.

Knudsen, H.L. (1952a). Mutual radiation resistance of aerials and arrays. *Wireless Eng.* 29: 301–305.

Knudsen, H.L. (1952b). On the calculation of some definite integrals in antenna theory. *Appl. Sci. Res., Sect. B* 3: 51–68.

Levine, H. and Schwinger, J. (1950). On the theory of electromagnetic wave diffraction by an aperture in an infinite plane conducting screen. In: *The Theory of Electromagnetic Waves* (ed. M. Kline), 355–391. New York, USA: Dover Publications Inc.

Levis, C.A. and Tai, C.T. (1956). A method of analyzing coupled antennas of unequal sizes. *IRE Trans. Antennas Propag.* AP-4: 128–132.

Lewin, L. (1951). *Advanced Theory of Waveguides*. London, UK: Iliffe & Sons, Ltd.

Luzwick, J. and Harrington, R.F. (1982). Mutual coupling analysis in a finite planar rectangular waveguide antenna array. *Electromagnetics* 2: 25–42.

Lyon, R.W. and Sangster, A.J. (1987). Efficient moment analysis of radiating slot in a thick-walled rectangular waveguide. *IEE Proc.H* 128 (4): 197–205.

Mailloux, R.J. (1969a). Radiation and near-field coupling between two collinear open-ended waveguides. *IEEE Trans. Antennas Propag.* AP-17: 49–55.

Mailloux, R.J. (1969b). First-order solutions for mutual coupling between waveguides which propagate twoorthogonal modes. *IEEE Trans. Antennas Propag.* AP-17: 740–746.

Marcuvitz, N. (1986). *Waveguide Handbook*. London, UK: Peter Peregrinus.

MathCad (2020): https://www.mathcad.com/en PTC MathCad Prime 6.0 Overview, live editing of typeset mathematical notation, combined with automatic computation.

MATLAB (2020): https://au.mathworks.com/products/matlab.html?s_tid=hp_products_matlab, programming language for mathematics & graphing

Merzbacher, E. (1961). *Quantum Mechanics*. New York, USA: Wiley.

Miles, J.W. (1949). On the diffraction of an electromagnetic wave through a plane screen. *J. Appl. Phys.* 20: 760–768.

Mishustin, B.A. (1965). Radiation from the aperture of a circular waveguide with an infinite flange. *Sov. Rad. Phys.* 8: 852–858.

Montgomery, C.G., Dicke, R.H., and Purcell, E.M. (1948). *Principles of Microwave Circuits*. New York: Dover Publications Inc.

Noble, B. (1958). *Methods Based on the Weiner-Hopf Technique*. New York, USA: Chelsea Publishing Company.

Paris, D.T., Leach, W.M., and Joy, E.B. (1978). Basic Theory of Probe-Compensated Near-Field Measurement. *IEEE Trans. Antennas & Propagat.* 26 (3): 373–379.

Pereia-Fiho, O.M.C., Barbosa, H.B., Diniz, C.A., and Tinoco, A.F. (2018). Method of moments analysis of arrays of cylindrical microstrip antennas with superstrate. *IET Microwaves Antennas Propag.* 12 (4): 561–568.

Pistolkors, A.A. (1929). The radiation resistance of beam antennas. *Proc. IRE* 17: 562–579.

Pohlhausen, K. (1921). The approximate integration of the differential equation for the laminar boundary layer. *Z. Angew. Math. Mech.* 1: 252–268.

Preis, D.H. (1976). A comparison of methods to evaluate potential integrals. *IEEE Trans. Antennas Propag.* AP-24 (2): 223–229.

Rao, S.M. and Wilton, D.R. (1990). 'E-field, H-field, and combined field solution for arbitrary shaped three dimensional dielectric objects'. *Electromagnetics* 10: 407–421.

Richmond, J.H. (1961). A reaction theorem and its application to antenna impedance calculations. *IRE Trans. Antennas Propag.* AP-9: 515–520.

Rogers, P.G. (1986). Application of minimum scattering antenna theory to mismatched antennas. *IEEE Trans. Antennas Propag.* AP-34 (10): 1223–1228.

Rudge, A.W., Milne, K., Olver, A.D., and Knight, P. (1983). *The Handbook of Antenna Design*, vol. 2. London, UK: Peter Peregrinus.

Rumsey, V.H. (1954). Reaction concept in electromagnetic theory. *Phys. Rev.* 94 (6): 1483–1491.

Silver, S. (ed.) (1984). *Microwave Antenna Theory and Design*, 155–158. London, UK: Peter Peregrinus.

Sinnott, D.H. (1973). Matrix analysis of linear arrays of equally spaced elements. *IEEE Trans. Antennas Propag.* AP-21 (3): 385–386.

Steyskal, H. (1974). Mutual coupling analysis of a finite planar waveguide array. *IEEE Trans. Antennas Propag.* AP-22: 594–597.

Storer, J. (1952): 'Variational solution to the problem of the symmetrical cylindrical antenna', Cruft Lab Report TR 101, Cambridge, MA, USA.

Surtees, W.J. (1952). Mutual coupling of a slot with a dipole antenna. *Proc. IRE* 40 (2): 208–211.

Tai, C.T. (1948). Coupled antennas. *Proc. IRE* 36: 487–500.

Tai, C.T. (1949). A study of the e.m.f method. *J. Appl. Phys.* 20: 717–723.

Trench, W.F. (1964). An algorithm for the inversion of finite Toeplitz matrices. *J. Soc. Ind. Appl. Math.* 12 (3): 515–522.

Vajnshtejn, L.A. (1950). Radiation of asymmetric electromagnetic waves from the open end of circular waveguide. *Dokl. Akad. Nauk* 74: 485–488.

Vescovo, R. (1996). Inversion of block-circulant matrices and circular array approach. *IEEE Trans. Antennas Propag.* AP-45 (10): 1565–1567.

Wang, Z., Simeoni, M., and Lager, I.E. (2017). A complete tool for analysing mutual couplings in nonuniform arrays of rectangular aperture radiators. *IEEE Antennas Propag. Lett.* 16: 3192–3195.

Wasylkiwskyj, W. and Kahn, W.K. (1970). Theory of mutual coupling among minimum-scattering antennas. *IEEE Trans. Antennas Propag.* AP-18 (2): 204–216.

Watson, G.N. (1918). The diffraction of electric waves by the earth. *Proc. Royal Soc. A* 95: 83–99.

Wilton, D.R., Rao, S.M., Glisson, A.W. et al. (1984). Potential integrals for uniform and linear source distributions on polygonal and polyhedral domains. *IEEE Trans. Antennas Propag.* AP-32 (3): 276–281.

Yamada, H. (1947). A method of approximate integration of the laminar boundary-layer equation. *Rept. Res. Inst. Fluid Engng., Kyushu Univ.* 2: 29.

Yamada, H. (1950). A method of approximate integration of the laminar boundary-layer equations. *Appl. Mech. Rev.* 3 (762).

4

Mutual Coupling in Arrays of Wire Antennas

Trevor S. Bird[1,2]

[1] *Glocal Big Data technology Centre, University of Technology Sydney, Ultimo, NSW 2007, Australia*
[2] *Antengenuity, Eastwood, NSW 2122, Australia*

4.1 Introduction

Wire antennas were the first type of antenna to be analyzed in any detail and for mutual coupling in particular. Some elementary aspects of mutual coupling of wire antennas have been mentioned in Chapter 2. The early work of Carter (1932), Brown and King (1934), King and Harrison (1944), Tai (1948), and others were especially important for giving a complete picture of wire antennas and for developing numerical tools for analyzing more complex aspects of coupling in these structures. In the intervening years, these tools have been extended to antennas with greater complexity. One of these tools is the moment method (MoM) formulation of Harrington (1967, 1968), who made this well-known method more accessible to antenna engineers. This formulation has become very popular in the antennas area since the 1970s and has continued to this day. The moment method is now widely used for solving many types of electromagnetic problems including coupling in other types of antennas. As well as describing coupling in wire antennas in this chapter, more details of the moment method will be provided also.

4.2 Formulation of the Problem

A fundamental wire antenna in the form of a dipole is illustrated in Figure 4.1a. It has a radius a and length $2L$. Initially, this dipole is assumed to be excited symmetrically from terminals at its center. Other driving points that lead to asymmetrical excitation can be advantageous in some applications. However, for the present, we assume that the driving point is at the center of the dipole (King et al., 1965). The basic symmetrical antenna is the half-wave dipole, and the monopole driven from its base above an imperfect earth is a basic asymmetrical fed dipole. However, when the monopole is driven above a perfect ground plane, by image theory, this antenna is equivalent to the dipole. Figure 4.1b illustrates several types of dipoles and the ideal excitation current. A circuit model of the dipole is shown in Figure 4.1c. A more wideband circuit model that represents the antenna and the coupling to other antennas as a two-port network can be expressed in impedance or admittance matrix form or as a scattering matrix, both of which relate directly to the analysis of mutual coupling. When a dipole is driven symmetrically, the currents that are excited travel away from and toward the feedpoint. If the frequency is reasonably high, these currents flow close to the

Mutual Coupling Between Antennas, First Edition. Edited by Trevor S. Bird.

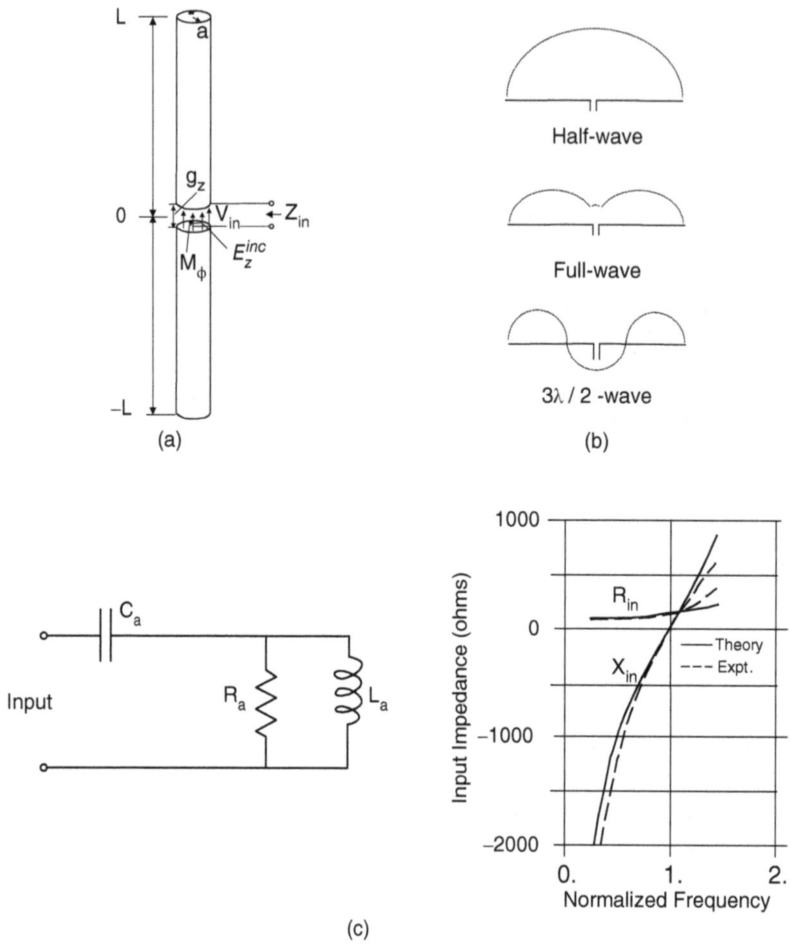

Figure 4.1 Basic dipole antenna: (a) geometry; (b) some types; and (c) equivalent circuit.

surface of the dipole due to the skin effect, and, usually, it can be assumed to be negligibly thin, which simplifies the representation of the current.

The impedance of the dipole can be obtained by classical analytical methods. The real part is derived by integrating the average Poynting vector over a far-field sphere. On the other hand, the reactance is found by the induced electromotive force (emf) method over the conductor surface on which the total tangential electric field must vanish. When this is carried out (Balanis 1982), the impedance obtained is

$$R_{in} = \frac{\eta_o}{2\pi \sin^2(k_o L)} \left\{ \gamma_e + \ln(2k_o L) - Ci(2k_o L) + \frac{1}{2}\sin(2k_o L)\left[Si(4k_o L) - 2\,Si(2k_o L)\right] \right.$$
$$\left. + \frac{1}{2}\cos(2k_o L)\left[\gamma_e + \ln(k_o L) + Ci(4k_o L) - 2\,Ci(2k_o L)\right] \right\} \tag{4.1a}$$

$$X_{in} = \frac{\eta_o}{2\pi \sin^2(k_o L)} \left\{ Si(2k_o L) + \frac{1}{2}\sin(2k_o L)\left[Ci(4k_o L) - 2\,Ci(2k_o L) + Ci\left(\frac{k_o a^2}{L}\right)\right] \right.$$
$$\left. + \frac{1}{2}\cos(2k_o L)\left[2\,Si(2k_o L) - Si(4k_o L)\right] \right\} \tag{4.1b}$$

where $2L$ is the total length, a is the wire radius, $\gamma_e = 0.5772156649$ is Euler's constant, and $Si(z)$ and $Ci(z)$ are the sine and cosine integrals, which are defined in Appendix A. The length-to-radius parameter of the dipole is $\Omega = 2\ln(2L/a)$ (King et al. 1965). This approximate impedance is reasonably accurate and can be used in the diagonal terms of an impedance matrix representation of an array of dipole antennas.

An antenna with a finite radius can be analyzed by satisfying the boundary conditions for the vanishing total tangential electric field everywhere on its surface. For a dipole directed along the z-direction, the currents flow on the surface and follow this direction. The surface current $\boldsymbol{J}_s$ can be found by commencing from Eq. (3.24). The electric field just above the surface of the dipole is

$$\boldsymbol{E}(a, a, 0+) = -j\omega\mu \iint_{S'} dS' \left(\boldsymbol{I} + \frac{\nabla \cdot \nabla}{k^2}\right) G_o(R) \cdot \boldsymbol{J}(r') \tag{4.2}$$

where S refers to the surface of the dipole, $G_0(R) = \exp(-jkR)/R$ and $R = \sqrt{a^2 + (z - z')^2}$ is the distance from the source point on the dipole to a field point also just above the dipole surface (denoted by 0+), which has radius a. The wire lies along the z-axis, and therefore, the current and electric field are z-directed,

$$\begin{aligned} E_z(0) &= -\frac{j}{\omega\varepsilon}\int_A dS' J_z(z')\left(\frac{\partial^2}{\partial z^2} + k^2\right)\frac{\exp(-jkR)}{R} \\ &\cong -\frac{ja}{\omega\varepsilon}\int_0^{2\pi} d\varphi \int_{-L}^{L} dz' J_z(z')\left(\frac{\partial^2}{\partial z^2} + k^2\right)\frac{\exp(-jkR)}{R} \end{aligned} \tag{4.3}$$

where E_z is the field radiated by the current density J_z. As the current is thin, $J_z = I_z(z')/2\pi a$. This current is assumed to follow the surface of the dipole, and to avoid field singularities, the observation point can be taken just inside the surface along the z-axis. Another form of Eq. (4.3) known as Polkington's equation is given by

$$\int_{-L}^{L} dz' I_z(z')\left(\frac{\partial^2}{\partial z'^2} + k^2\right)\frac{\exp(-jkR)}{R} = j\omega\varepsilon E_z(0) \tag{4.4}$$

For an axisymmetric dipole, Eq. (4.4) can be expressed more conveniently for calculation, as follows (Richmond 1965):

$$\int_{-L}^{L} dz' I_z(z')\frac{\exp(-jkR)}{R^3}\left[(1 + jkR)\left(2 - 3\left(\frac{a}{R}\right)^2\right) + (ka)^2\right] = j\frac{k}{\eta_0}E_z(0) \tag{4.5}$$

The boundary condition of zero tangential electric field at the surface is satisfied when $E_z^{in} + E_z(0) = 0$. As is shown in Figure 4.1a, the gap width g_z is small, and the field in this region is uniform and is symmetric approximating a magnetic ring current $M_\varphi = -\hat{\rho} \times \hat{z} E_z^{in}$. The voltage applied between the terminals V_{in} creates an incident field given by $E_z^{in} = \frac{V_{in}}{g_z}$, from which the surface boundary condition sets

$$E_z(0) = -\frac{V_{in}}{g_z}.$$

Thus, Eq. (4.5) becomes

$$\int_{-L}^{L} dz' I_z(z')\frac{\exp(-jkR)}{R^3}\left[(1 + jkR)\left(2 - 3\left(\frac{a}{R}\right)^2\right) + (ka)^2\right] = -j\frac{k}{\eta_0}\frac{V_{in}}{g_z} \tag{4.6}$$

All parameters are now fully specified allowing a solution for I_z to be found by solving the integral equation, Eq. (4.6), usually by numerical methods such as the moment method as to be described in Section 4.2.1. Once the current at the drive point has been obtained, the self-impedance can be found from

$$Z_{in} = \frac{V_{in}}{I_z(0)} \tag{4.7}$$

where $I_z(0)$ is the current at the midpoint of the solution. In the numerical solution, this may be an average of several discrete values at input points.

4.2.1 Moment Method

The input impedance can be accurately predicted using the moment method by using more and more functions approximating the unknown current. As we have seen in Chapter 3, the moment method is one approach in the class of weighted residuals (Finlayson 1972). It is one way of solving the integral equation Eq. (4.6). Expressed more generally, this equation is of the form

$$\int_D ds' F(s, s') \Psi(s') = g(s) \tag{4.8}$$

where $\Psi(s)$ is the function over the domain D. The kernel function $F(s, s')$ is symmetric in its two arguments and may be a Green's dyadic or a differential function. The function $\Psi(s)$ in Eq. (4.8) can be represented by a finite expansion

$$\Psi(s) \cong \sum_{n=1}^{N} \psi_n \alpha_n(s) \tag{4.9}$$

It was shown in Section 3.4.3 that when Eq. (4.9) is substituted in Eq. (4.8) and a weighting function $w_m(s)$, m = 1,…, M is applied to both sides

$$\int_D ds\, w_m(s) \left[\sum_{n=1}^{N} \left\{ \int_D ds' F(s, s') \alpha_n(s') \psi_n \right\} - g(s) \right] = 0; \text{m} = 1, \ldots, \text{M} \tag{4.10}$$

In matrix format, Eq. (4.10) is expressed as

$$[A_{mn}] [\psi_n] = [b_m]$$

where

$$A_{mn} = \int_D ds\, w_m(s) \int_D ds' F(s, s') \alpha_n(s),$$

and

$$b_m = \int_D ds\, w_m(s)\, g(s)$$

A wide variety of shape and weighting functions are possible which belong to two broad categories, subdomain and entire domain functions.

The subdomain methods are typified by the delta function and the pulse function

$$\alpha_m(s) = \begin{cases} 1 & s_{m-1} \le s \le s_m \\ 0 & elsewhere \end{cases} \tag{4.11}$$

which are illustrated in Figure 4.2. For the pulse function, when a solution to ψ_n is obtained, the final result is shaped like a staircase and hence the function goes by that name. The delta function in Figure 4.2a corresponds to Eq. (4.11) when $s_m - s_{m-1} \to 0$. The point matching method is a single order moment method that uses delta functions as a shape function.

Figure 4.2 Shape and weighting functions. (a) Delta; (b) pulse; (c) triangular; and (d) entire domain function.

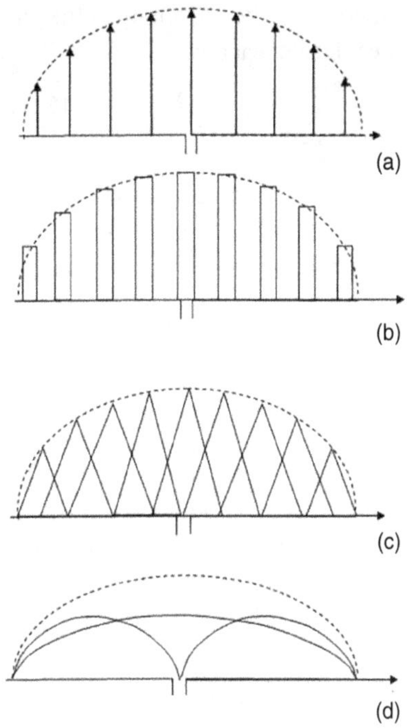

Another popular subdomain function is the triangle or piecewise linear function defined by

$$\alpha_m(s) = \begin{cases} (s - s_{m-1}) / (s_{m+1} - s_{m-1}) & s_{m-1} \leq s \leq s_m \\ (s_{m+1} - s) / (s_{m+1} - s_{m-1}) & s_m \leq s \leq s_{m+1} \\ 0 & \textit{elsewhere} \end{cases} \tag{4.12}$$

As shown in Figure 4.2c, these functions overlap two segments and adjacent functions.

Of the variety of entire domain functions that can be selected, some are related to the original problem, such as in Figure 4.2d. These may be trigonometric, Bessel, or Hermite functions. Sometimes, polynomials can be chosen to satisfy the boundary conditions. Thus, for example, on the wire antenna in Figure 4.1a, the current at the endpoints $z = \pm L$ should be zero. A suitable entire domain function for symmetrical excitation at z = 0 is

$$\Psi(z) \cong \sum_{n=1}^{N} \psi_n (L - |z|)^n; -L \leq z \leq L \tag{4.13}$$

Another possible entire function is

$$\Psi(z) \cong \sum_{n=1}^{N} \psi_n \cos\left[\frac{(2n-1)\pi z}{2L}\right]; -L \leq z \leq L \tag{4.14}$$

The advantage of entire domain functions is that the assumed function can be chosen to follow the physical result a priori.

4.2.2 Moment Method Solution for the Dipole

A moment method solution is now obtained for the integral equation in Eq. (4.6). In this approach, pulse functions will be used as shape functions for the current $I(z)$ and delta functions as weighting functions. If the full length of the dipole is considered and this is subdivided into N equal segments, the width of each is $\Delta = 2L/N$. When Eq. (4.11) is used in Eq. (4.6), we get

$$\sum_{n=1}^{N} \psi_n \int_{-L+(n-1)\Delta}^{-L+n\Delta} dz' \frac{\exp(-j\,k_o R)}{R^3} \left[(1 + j\,k_o R) \left(2 - 3 \left(\frac{a}{R} \right)^2 \right) + (k_o a)^2 \right] = -\frac{j\,k_o}{\eta_o} \frac{V_{in}}{\Delta} \tag{4.15}$$

where in this case the gap width is chosen to be $g_z = \Delta$. A voltage is applied across the gap $V_{in} = \Delta E_z^{in}$ and, in this example, there are no other voltage sources on the dipole. When the set of delta functions $\delta(z - z_m)$; $m = 1, \ldots, M$ are applied as weighting functions, Eq. (4.15) becomes

$$\sum_{n=1}^{N} \psi_n \frac{j\,\eta_o\,\Delta}{k_o} \int_{-L+(n-1)\Delta}^{-L+n\Delta} dz' \frac{\exp(-j\,k_o R_m)}{R_m^3} \left[(1 + j\,k_o R_m) \left(2 - 3 \left(\frac{a}{R_m} \right)^2 \right) + (k_o a)^2 \right] = V_{in}\,\delta(m - m_o) \tag{4.16}$$

where $R_m = \sqrt{a^2 + (z_m - z')^2}$ and m_o is the segment number that contains the feed gap, which is usually near the center of the range. In that case, $m_o = M/2$ if M is even or $m_o = (M+1)/2$ if it is odd. In matrix form, Eq. (4.16) becomes

$$[Z_{mn}]\,[\psi_n] = [V_m] \tag{4.17}$$

where

$$Z_{mn} = \frac{j\eta_o\Delta}{k_o} \int_{-L+(n-1)\Delta}^{-L+n\Delta} dz' \frac{\exp(-jkR_m)}{R_m^3} \left[(1 + jk_o R_m) \left(2 - 3 \left(\frac{a}{R_m} \right)^2 \right) + (ka)^2 \right] \tag{4.18}$$

is the impedance matrix and

$$[V_m] = \begin{bmatrix} 0 \\ \vdots \\ V_{in} \\ 0 \\ \vdots \end{bmatrix}$$

is the excitation vector. The currents on N segments can be found by solving

$$[\psi_n] = \left([Z_{mn}]^T [Z_{mn}] \right)^{-1} \left([Z_{mn}]^T [V_m] \right)$$

as M is generally not equal to N.

As an example of the results calculated with Eq. (4.17) for a dipole of finite diameter and varying length. Figure 4.3 shows a comparison of results obtained from a moment method using $N = 51$ pulse shape functions and $M = 51$ delta functions as weights and the formula given in Eqs. (4.1a) and (4.1b). Plotted are real and imaginary parts of the input admittance $Y_{in} = \psi_{(N+1)/2}/V_{in}$ with increasing length of a dipole with a radius of a = 0.025 λ. We see there are differences which are probably due to the finite diameter, the gap as well as the approximations made in both approaches. For the integral equation solution, the field in the gap is assumed to be uniform.

4.3 Mutual Impedance

The theory given in Section 2.3 is readily extended to antennas with distributed currents as in the representation given in Section 3.2.1. The extension requires taking account of the phase delay between the radiation from elements of antenna 2, and the current elements on the surface, S_2; the

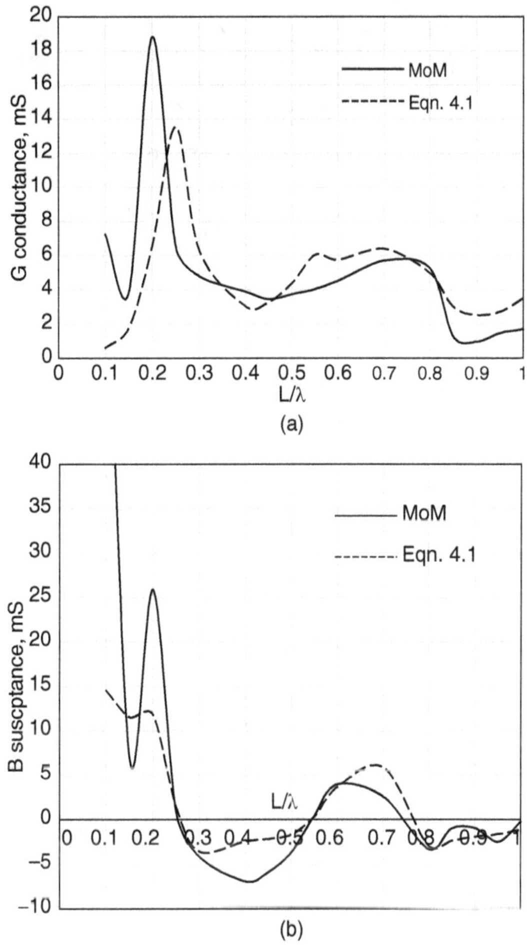

Figure 4.3 Comparison of input admittance of a dipole of half-length L and constant radius $a = 0.025\,\lambda$ calculated by a method of moments (MoM) analysis (51 pulse shape functions and 51 delta function weights) and by Eqs. (4.1a) and (4.1b) (Balanis 1982). (a) Conductance G; (b) susceptance B.

phase delay on antenna 1 is accounted for in the same way. Consider an infinitesimal element of antenna 2, which is illuminated by an incident electric field from antenna 1 given by $\mathbf{E}_{21}$. Suppose an electric current, $\mathbf{J}_2\,dS$, is induced on the element of antenna 2. At the terminals of antenna 2, an emf induced at the terminals is dV_{21}. For continuity of complex power, it is required that

$$I_2 dV_{21} = -\mathbf{J_2}\cdot\mathbf{E_{21}}\,(\mathbf{R})\;dS \tag{4.19}$$

where I_2 is the current at terminals of antenna 2, and $\mathbf{R}$ is the vector from antenna 1 to antenna 2. Now, summing all such current elements over the surface of antenna 2 results in an expression for mutual impedance:

$$Z_{21} = \frac{V_2}{I_1} = -\frac{1}{I_1 I_2}\iint_{S_2} \boldsymbol{J_2}\cdot\mathbf{E_{21}}\,(\mathbf{R})\,dS. \tag{4.20}$$

where

$$\mathbf{E_{21}}\,(\mathbf{R}) = \iint_{S_1} dS'\,\overline{\overline{\mathbf{G^e}}}\left(\mathbf{R},\mathbf{R}'\right)\cdot \boldsymbol{J_1}$$

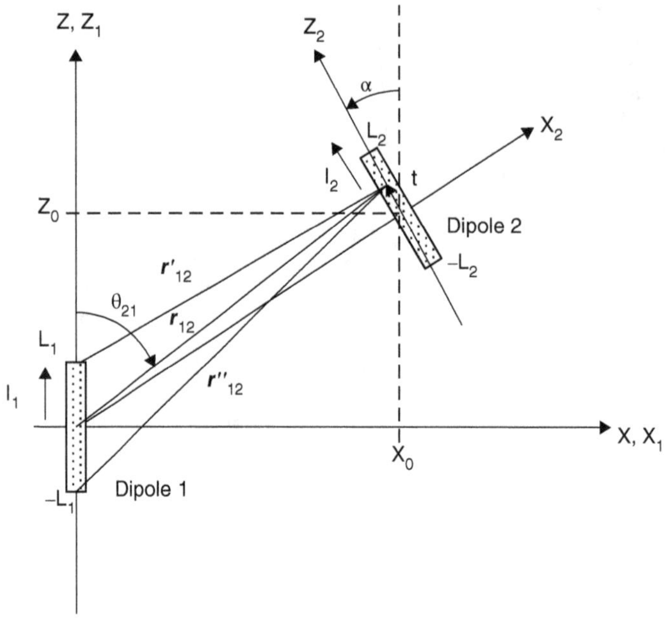

Figure 4.4 Geometry of two thin dipoles coupling in x-z plane.

in which $\overline{\overline{\boldsymbol{G}^e}}$ is the Green's function for the electric field. For thin wire antennas, Eq. (4.20) can be expressed as

$$Z_{21} = \frac{V_2}{I_1} = -\frac{1}{I_1 I_2} \int_0^{L_2} dl\, I_2 \int_0^{L_1} dl' G\left(r - r'\right) I_1 \tag{4.21}$$

4.3.1 Closed Form Expressions for Mutual Impedance

Consider two dipoles in arbitrary dispositions as shown in Figure 4.4. Expressions for the mutual impedance can be derived from Eq. (4.21) for assumed current distributions and dipole arrangements, and some of these are given by Brown (1937), Barzilai (1948), and Hansen (1966, 1998). For the case of two thin dipoles with lengths $2L_1$ and $2L_2$, shown in Figure 4.4, that support sinusoidal currents of the form $I_o\, sin\,(k(L_{1,2} - |z|))$, the mutual impedance is given by

$$Z_{21} = -\int_{-L_2}^{L_2} dt \frac{I_2(t)}{I_{in2}} E_{21z}(t) \tag{4.22}$$

where

$$E_{21z}(t) = \int_{-L_1}^{L_1} ds \frac{\exp\left(-jk\left|r_{12} - \hat{z}_1 s\right|\right)}{\left|r_{12} - \hat{z}_1 s\right|} \frac{I_1(s)}{I_{in1}} \tag{4.23}$$

and

$$\boldsymbol{r}_{12} = \hat{x}_1\left(x_0 - t\,sin\alpha\right) + \hat{z}_1\left(z_0 + t\,cos\alpha\right).$$

I_{in1} and I_{in2} are the currents at the input terminals at the center of the elements, and for a sinusoidal current, $I_{ini} = I_o \sin k_o L_i$, where $i = 1{,}2$. A minor difficulty occurs with the approximation when $L_i = \frac{\lambda}{2}$. In those cases, in numerical integration in the vicinity of $z = 0$, Hospital's rule is used to obtain the ratio of the antenna current to the input current in the form $I_i(x)/I_{ini} \approx \cos k_o x|_{x \to 0}$.

An approximation of Eq. (4.22), which has a single integral only, is given by Balanis (1982, p.298), as follows

$$Z_{21} = \frac{j\eta_o}{4\pi sin\left(kL_1/2\right) sin\left(kL_2/2\right)} \int_0^{L_2} \left(\frac{e^{-jkr'_{21}}}{r'_{21}} + \frac{e^{-jkr''_{21}}}{r''_{21}} - 2\cos\left(kL_1\right) \frac{e^{-jkr_{21}}}{r_{21}} \right) sin(k\left(L_2 - |t|\right) dt \tag{4.24}$$

where $\boldsymbol{r}'_{12} = \boldsymbol{r}_{12} - \hat{z}_1 L_1, \boldsymbol{r}''_{12} = \boldsymbol{r}_{12} + \hat{z}_1 L_1$, and $\boldsymbol{r}_{12} = \hat{x}_1\left(x_0 - t\, sin\alpha\right) + \hat{z}_1\left(z_0 + t\, cos\alpha\right)$. Equation (4.24) is valid for thin dipoles of arbitrary length. The integral can be evaluated in terms of sine and cosine integrals in the case of a parallel H-plane arrangement ($\alpha = 0$, $z_o = 0$, $x_o = s$) as shown by Cox (1947) and King (1957)

$$Z_{21} = \frac{R^a_{21} + jX^a_{21}}{\sin\left(kL_1/2\right) sin\left(kL_2/2\right)}$$

where

$$\begin{aligned} R^a_{21} = \frac{\eta_o}{4\pi} & \left[\cos\left(k\left(L_2 - L_1\right)\right)\left\{Ci\left(k\left(s_{21} - L_2 + L_1\right)\right) + Ci\left(k\left(s_{21} + L_2 - L_1\right)\right)\right.\right. \\ & - Ci\left(k\left(s_1 - L_1\right)\right) - Ci\left(k\left(s_1 + L_1\right)\right) - Ci\left(k\left(s_2 - L_2\right)\right) - Ci\left(k\left(s_2 + L_2\right)\right) + 2Ci\left(ks\right)\} \\ & + \cos\left(k\left(L_1 + L_2\right)\right)\left\{Ci\left(k\left(s_{22} - L_2 - L_1\right)\right) + Ci\left(k\left(s_{22} + L_1 + L_2\right)\right) - Ci\left(k\left(s_1 - L_1\right)\right)\right. \\ & - Ci\left(k\left(s_1 + L_1\right)\right) - Ci\left(k\left(s_2 - L_2\right)\right) - Ci\left(k\left(s_2 + L_2\right)\right) + 2Ci\left(ks\right)\} \\ & + \sin\left(k\left(L_2 - L_1\right)\right)\left\{-Si\left(k\left(s_{21} - L_2 + L_1\right)\right) + Si\left(k\left(s_{21} + L_2 - L_1\right)\right) - Si\left(k\left(s_1 - L_1\right)\right)\right. \\ & + Si\left(k\left(s_1 + L_1\right)\right) + Si\left(k\left(s_2 + L_2\right)\right) - Si\left(k\left(s_2 - L_2\right)\right)\} \\ & + \sin\left(k\left(L_1 + L_2\right)\right)\left\{-Si\left(k\left(s_{22} - L_1 - L_2\right)\right) + Si\left(k\left(s_{22} + L_1 + L_2\right)\right)\right. \\ & \left.\left. - Si\left(k\left(s_1 - L_1\right)\right) + Si\left(k\left(s_1 + L_1\right)\right) + Si\left(k\left(s_2 + L_2\right)\right) - Si\left(k\left(s_2 - L_2\right)\right)\right\}\right] \end{aligned} \tag{4.25a}$$

$$\begin{aligned} X^a_{21} = \frac{\eta_o}{4\pi} & \left[\cos\left(k\left(L_2 - L_1\right)\right)\left\{-Si\left(k\left(s_{21} - L_1 + L_2\right)\right) - Si\left(k\left(s_{21} + L_1 - L_2\right)\right)\right.\right. \\ & + Si\left(k\left(s_1 - L_1\right)\right) + Si\left(k\left(s_1 + L_1\right)\right) + Si\left(k\left(s_2 - L_2\right)\right) + Si\left(k\left(s_2 + L_2\right)\right) - 2Si\left(ks\right)\} \\ & + \cos\left(k\left(L_1 + L_2\right)\right)\left\{-Si\left(k\left(s_{22} - L_1 - L_2\right)\right) - Si\left(k\left(s_{22} + L_1 + L_2\right)\right) + Si\left(k\left(s_1 - L_1\right)\right)\right. \\ & + Si\left(k\left(s_1 + L_1\right)\right) + Si\left(k\left(s_2 - L_2\right)\right) + Si\left(k\left(s_2 + L_2\right)\right) - 2CSi\left(ks\right)\} \\ & + \sin\left(k\left(L_2 - L_1\right)\right)\left\{-Ci\left(k\left(s_{21} - L_1 + L_2\right)\right) + Ci\left(k\left(s_{21} + L_1 - L_2\right)\right) + Ci\left(k\left(s_1 - L_1\right)\right)\right. \\ & - Ci\left(k\left(s_1 + L_1\right)\right) - Ci\left(k\left(s_2 + L_2\right)\right) + Ci\left(k\left(s_2 - L_2\right)\right)\} \\ & + \sin\left(k\left(L_1 + L_2\right)\right)\left\{-Ci\left(k\left(s_{22} - L_1 - L_2\right)\right) + Ci\left(k\left(s_{22} + L_1 + L_2\right)\right)\right. \\ & \left.\left. + Ci\left(k\left(s_1 - L_1\right)\right) - Ci\left(k\left(s_1 + L_1\right)\right) - Ci\left(k\left(s_2 + L_2\right)\right) + Ci\left(k\left(s_2 - L_2\right)\right)\right\}\right] \end{aligned} \tag{4.25b}$$

where $Si(z)$ and $Ci(z)$ are the sine and cosine integrals (see Appendix A), $s_1 = \sqrt{s^2 + L_1^2}$, $s_2 = \sqrt{s^2 + L_2^2}$, $s_{21} = \sqrt{s^2 + (L_1 - L_2)^2}$, and $s_{22} = \sqrt{s^2 + (L_1 + L_2)^2}$, Eq. (4.25) reduces to the formulae given by Schelkunoff and Friis (1952, p. 409) when the lengths of the dipoles are identical. In that case, the term in curly brackets with $\sin(k(L_2 - L_1)) = 0$ as a multiplier does not contribute to the impedance. Another expression for coupling between different length dipoles is given by Barzilai (1948). The mutual impedance in the case of a identical half-wave dipoles ($L_1 = L_2 = \lambda/4$) is listed in the first row of Table 4.1. In this expression, if the dipole spacing $s \rightarrow 0$, the input impedance approaches

$$Z_{in} = \frac{\eta_o}{4\pi}\left[\ln\left(2\pi\gamma_e\right) - Ci\left(2\pi\right) + jSi\left(2\pi\right)\right] = 73.13 + j\,42.54\text{ohms} \tag{4.26}$$

Table 4.1 Mutual impedance of dipole configurations as illustrated. (a) Parallel in-line; (b) collinear; (c) general parallel displaced. $2L = n\,\lambda/2;\ n = 1, 2$.

(a) Parallel (H-plane)

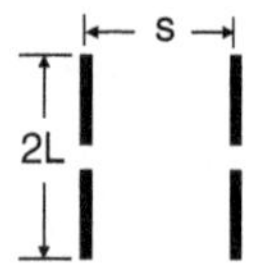

$$R_{21} = \frac{\eta_o}{4\pi}\left[2\,Ci\,(k_o s) + Si\,(k_o\,(s_2 + 2L)) + Ci\,(k_o\,(s_2 - 2L))\right]$$

$$X_{21} = \frac{\eta_o}{4\pi}\left[Si\,(k_o\,(s_2 + 2L)) + Si\,(k_o\,(s_2 - 2L)) - 2Si\,(k_o s)\right]$$

$$s_2 = \sqrt{s^2 + (2L)^2}$$

(b) Collinear (E-plane)

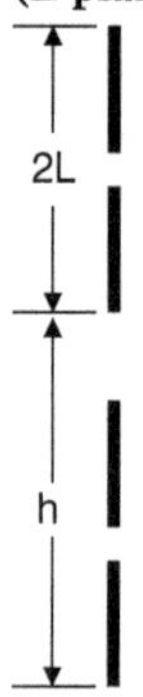

$$R_{21} = -\frac{\eta_o}{8\pi}\cos\,(k_o h)\,[-2Ci\,(2k_o h) + Ci\,(u_1) + Ci\,(u_2) - \ln\,(u_3) + \frac{\eta_o}{8\pi}\sin\,(k_o h)\;[2Si\,(2k_o h) - Si\,(u_1) - Si\,(u_2)]$$

$$X_{21} = -\frac{\eta_o}{8\pi}\cos\,(k_o h)\,[2Si\,(2k_o h) - Si\,(u_1) - Si\,(u_2))] + \frac{\eta_o}{8\pi}\sin\,(k_o h)\;[2\,Ci\,(2k_o h) + Ci\,(u_1) + Ci\,(u_2) - \ln\,(u_3)]$$

$$u_1 = 2k_o\,(h + 2L) \quad u_2 = 2k_o\,(h - 2L) \quad u_3 = 1 - (2L/h)^2$$

(c) General parallel displaced

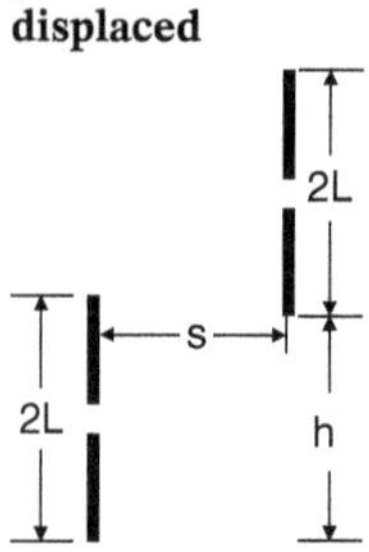

$$R_{21} = -\frac{\eta_o}{8\pi}\cos\,(v_1')\,[-2Ci\,(v_1 + v_1') - 2Ci\,(v_1 - v_1') + Ci\,(v_2 + v_2') + Ci\,(v_2 - v_2') + Ci\,(v_3 + v_3') + Ci\,(v_3 - v_3')] + \frac{\eta_o}{8\pi}\sin\,(v_1')\,[2Si\,(v_1 + v_1') - 2Si\,(v_1 - v_1') - Si(v_2 + v_2') + Si\,(v_2 - v_2') - Si\,(v_3 + v_3') + Si\,(v_3 - v_3')]$$

$$X_{21} = -\frac{\eta_o}{8\pi}\cos\,(v_1')\,[2Si\,(v_1 + v_1') + 2Si\,(v_1 - v_1') - Si\,(v_2 + v_2') - Si\,(v_2 - v_2') - Si\,(v_3 + v_3') - Si\,(v_3 - v_3'))] + \frac{\eta_o}{8\pi}\sin\,(v_1')\,[2Ci\,(v_1 + v_1') - 2Ci\,(v_1 - v_1') + Ci(v_2 + v_2') + Ci\,(v_2 - v_2') - Ci\,(v_3 + v_3') + Ci\,(v_3 - v_3')]$$

$$v_1 = k_o\sqrt{s^2 + h^2} \qquad v_2 = k_o\sqrt{s^2 + (h - 2L)^2} \qquad v_3 = k_o\sqrt{s^2 + (h + 2L)^2}$$

$$v_1' = k_o h \qquad v_2' = k_o\,(h - 2L) \qquad v_3' = k_o\,(h + 2L)$$

Source: Balanis (1982).

which is the expected result for a thin half-wave dipole (Balanis 1982). For the full-wave dipole with a small but finite radius a and $s \to 0$, the input impedance is $Z_{inf} = R_f + jX_f$ where

$$R_f \cong \frac{\eta_o}{4\pi}\left[6\,Ci\left(2\pi\frac{a}{\lambda}\right) - 3\,Ci\left[\frac{\pi}{2}\left(\frac{a}{\lambda}\right)^2\right] - 4Ci\,(2\pi) + Ci\,(4\pi)\right]$$

$$X_f \cong -\frac{\eta_o}{4\pi}\left[6\,Si\left(2\pi\frac{a}{\lambda}\right) - 3\,Si\left[\frac{\pi}{2}\left(\frac{a}{\lambda}\right)^2\right] - 4Si\,(2\pi) + Si\,(4\pi)\right] \tag{4.27}$$

If the radius is allowed to become infinitely thin, Eq. (4.27) gives $R_f \to \infty$ and $X_f \to \frac{\eta_o}{4\pi}\,[4Si\,(2\pi) - Si\,(4\pi)] = 125.317$ ohms.

Closed-form expressions for the mutual impedance in other parallel arrangements of identical thin dipoles have been summarized by Balanis (1982) and these are listed in Table 4.1. For completeness, the first entry of this table is the special case of Eqs. 4.25 for multiple half-wavelength dipoles. The assumption in these equations is that the dipoles support a predominantly symmetric sinusoidal current. Note, the expressions listed in Table 4.1 are for dipoles in free space when $k = k_o$. The most general parallel arrangement shown is the general parallel displaced of dipoles seen in the third case of Table 4.1. The mutual impedances of the other two configurations are special cases of this displaced case and, therefore, they can be derived from the equations in the third row. Historically, the parallel arrangement was first considered for the half-wave dipole by Bontsch-Bruewitsch (1926), Pistolkors (1929), followed by Bechmann (1931) who was primarily concerned with the radiation resistance of two antennas above a perfectly conducting ground. By allowing the spacing $s \to 0$, he was able to reproduce the radiation resistance of a loaded conductor previously obtained by Van der Pol in 1918.

In common with the single dipole, there are direct integral equation solutions for the current and the mutual impedances. An approach based on Polkington's integral equation, Eq. (4.4), but extended for $p = 1, 2, \ldots NE$ dipoles is now described. Thus,

$$\sum_{p=1}^{NE} \int_{-L_p}^{L_p} dz' I_{zp}(z') \left(\frac{\partial^2}{\partial z'^2} + k^2 \right) \frac{\exp(-jkR)}{R} = \frac{jk}{\eta_0} E_z^{in} \tag{4.28}$$

where L_p is the half-length of element p and $R = \sqrt{(x - x')^2 + (y - y')^2 + (z - z')^2}$. The kernel can be simplified using Richmond's expansion as follows:

$$\left(\frac{\partial^2}{\partial z'^2} + k^2 \right) \frac{\exp(-j\,kR)}{R} = \frac{\exp\left(-j\,kR_{mp}\right)}{4\pi R_{mp}^3} \left[\left(1 + j\,kR_{mp}\right) \left(2 - 3\left(\frac{a}{R_{mp}}\right)^2\right) + \left(ka_p\right)^2 \right] \tag{4.29}$$

where $R_{mp} = \sqrt{(x - x')^2 + (y - y')^2 + (z - z')^2 + a_p^2}$ in which a_p is the radius of dipole p. The integral equation, Eq. (4.28), can be solved as described in Section 4.2.1. It has been found that an efficient representation of the current is with an entire domain function for element p such as (Thiele 1969)

$$I_{zp}(z) = \sum_{n=1}^{N} I_{np} \cos\left[(2n - 1) \frac{\pi z}{2L_p} \right]. \tag{4.30}$$

4.3.2 Asymptotic Approximations to Mutual Impedance

In array applications, the spacing between elements quickly increases. In this calculation, the impedances of non-nearest elements take the longest time to compute. Therefore, mutual impedance calculations can take up considerable computing resources. This computing time can be reduced for a small cost in overall accuracy by the use of asymptotic expressions for the mutual impedances of elements with large spacings. This has been done by several workers with good results. Different asymptotic approaches can be adopted such as in the Green's function for MoM (Susuga et al. 2019) or through approximate sinusoidal currents as was done by Mahony (1989) using an approach based on a classic mutual impedance formula such as the one of Balanis (1982, pp. 298–299).

An approximation to the mutual impedance of dipoles operating in the same azimuth plane can be found by substituting the asymptotic approximation of the free-space Green's function into the expression for mutual impedance Eq. (4.20) as discussed in Section 3.4.2.1. Sinusoidal approximations to the current can be used to determine an approximate expression for the mutual impedance to any order of approximation. In the case of the free-space Green's function in Section 3.4.2.1, this has been expanded to third order in Eq. (3.29). The integrals in the resulting impedance can be evaluated in closed form. They consist of sums of permutations of Fourier transforms of powers of

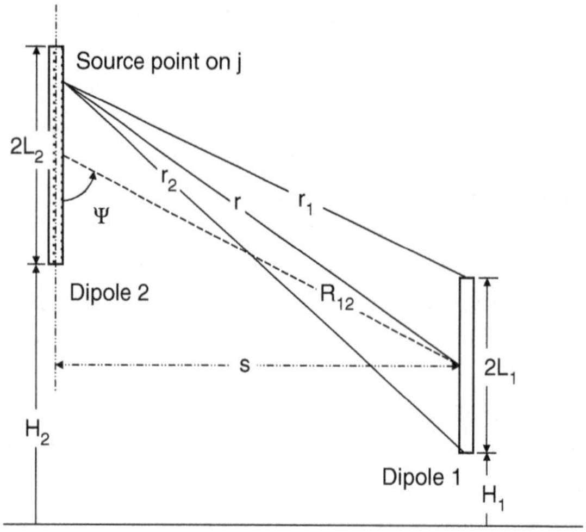

Figure 4.5 Geometry of dipole coupling of dipole 2 to radiation from dipole 1.

the axial direction and sinusoidal functions. These transforms are recursive functions and can be evaluated very rapidly from a base function.

An alternative approach is to make approximations within an existing mutual impedance formulation (Balanis 1982; Mahony 1989) such as for the simplified configuration shown in Figure 4.5. The result is

$$Z_{21} \cong j\frac{\eta_o}{\pi k}g_o(R_{12})\left\langle\left[\frac{\cos\left(\frac{\pi}{2}\cos(\Psi)\right)}{\sin(\Psi)}\right]^2\left\{1+(kL_1+kL_2-\pi)\left[\frac{1-\cos(\Psi)\sin\left(\frac{\pi}{2}\cos(\Psi)\right)}{\cos\left(\frac{\pi}{2}\cos(\Psi)\right)}\right]\right\}\right.$$
$$-\frac{1}{2jkR_{ij}}\left\{\left[\frac{\cos\left(\frac{\pi}{2}\cos(\Psi)\right)}{sin^2(\Psi)}\right]^2\left(\frac{\pi^2}{2}sin^4(\Psi)+2sin^2(\Psi)-4\right)\right.$$
$$\left.\left.+\frac{2\pi\cos(\Psi)\cos\left(\frac{\pi}{2}\cos(\Psi)\right)\sin\left(\frac{\pi}{2}\cos(\Psi)\right)}{sin^2(\Psi)}-\frac{\pi^2}{2}sin^2\left(\frac{\pi}{2}\cos(\Psi)\right)\right\}\right\rangle \tag{4.31}$$

where $g_o(r) = \exp(-jkr)/r$, $\sin(\Psi) = s/R_{12}$, $R_{12} = \sqrt{(H2 - H1)^2 + s^2}$, in which s is the separation distance as shown in Figure 4.5. Equation (4.31) is accurate to second order with respect to separation and first order in variations in dipole lengths $2L_1$ and $2L_2$. Results obtained with Eq. 4.31 for the case of two one-half-wavelength dipoles in parallel is shown in Figure 4.6 compared with direct integration and closed-form solution in Eqs. (4.25a) and (4.25b).

Important special cases of Eq. (4.31) can be obtained for half-wave dipoles arranged in parallel (i.e. H-plane coupling), as shown in Table 4.1a, which have been verified by others (e.g. Gera 1988). For this case, the asymptotic mutual impedance is

$$Z_{21} \approx \frac{j\eta_o}{\pi}\frac{e^{-jks}}{ks}\left[1-\frac{\left(\frac{\pi^2}{4}-1\right)}{(jks)}\right] \tag{4.32}$$

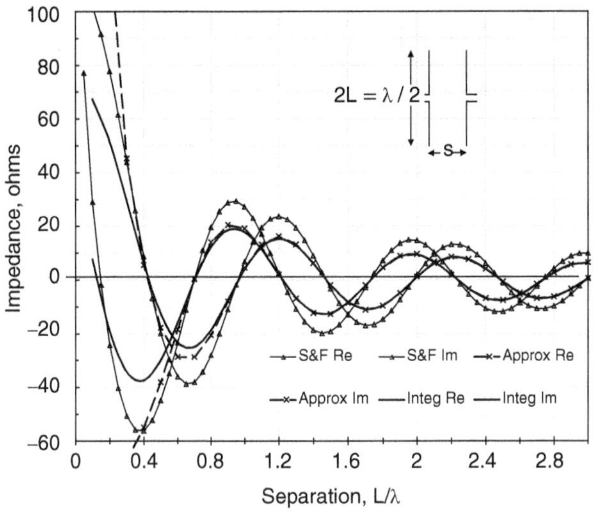

Figure 4.6 Mutual impedance for two parallel half-wave dipoles as a function of spacing. Key: solid line – direct integration Eq. (4.28); dashed line with x – asymptotic approximation Eq. (4.32); and dark line with △–. Source: SCHELKUNOFF, S.A. and FRIIS, H.T. (1952).

which is accurate for s > 1 λ. For the full-wave dipole, the corresponding result for H-plane coupling is

$$Z_{21} \approx \frac{j\eta_o}{\pi} \frac{e^{-jks}}{ks} \left[(1+\pi) - \frac{\left(\frac{\pi^2}{4} - 1\right)}{(jks)} \right] \tag{4.33}$$

Another important case is this one for two collinear half-wave dipoles (i.e. E-plane coupling) as shown in Table 4.1b. An asymptotic approximation for this case is

$$Z_{21} \approx -\frac{\eta_o \pi}{8} \frac{e^{-jks}}{(ks)^2} \tag{4.34}$$

This is also accurate for s > λ. We note from Eqs. (4.32–4.34) that at large separations the mutual impedance decays as the inverse of the distance (s/λ) in the H-plane and inverse of the square of the distance $(s/\lambda)^2$in the E-plane. This compares with the dual problem of slots where the reverse occurs.

Examples of some of the results for the coupling between co-linear half-wave dipoles are shown in Figure 4.7. This figure compares direct integration, Eq. (4.21), the closed-form expression given in Table 4.1, and the asymptotic approximation given by Eq. (4.34). In another comparison, results for the full-wave dipole are shown in Figure 4.8 for H-plane coupling. It may be verified from these figures that the approximate formula Eq. 4.33 is also reasonably accurate for s> λ, which has error typically <5%, although there are some discrepancies in Eq. (4.33) due to a phase error of about $(\pi s/20\lambda)$ and the coefficient $(1+\pi)$ in Eq. (4.33) seems about 15% too high, both of which are probably due to the approximations employed.

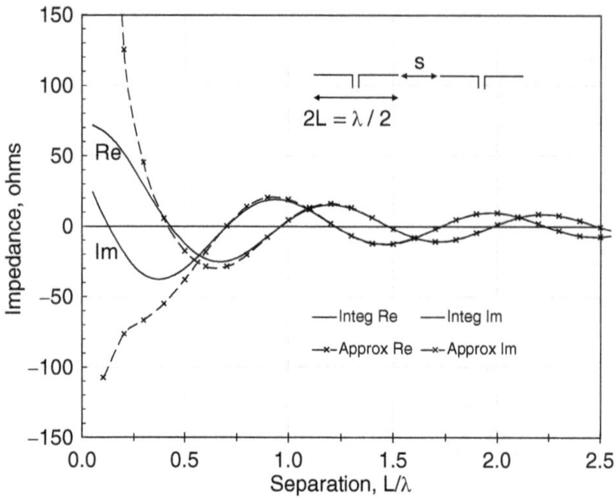

Figure 4.7 E-plane coupling impedance for two colinear half-wave dipoles with sinusoidal current approximation as a function of spacing. Key: Solid line – direct integration Eq. (4.28); dashed line with x – asymptotic approximation Eq. (4.34).

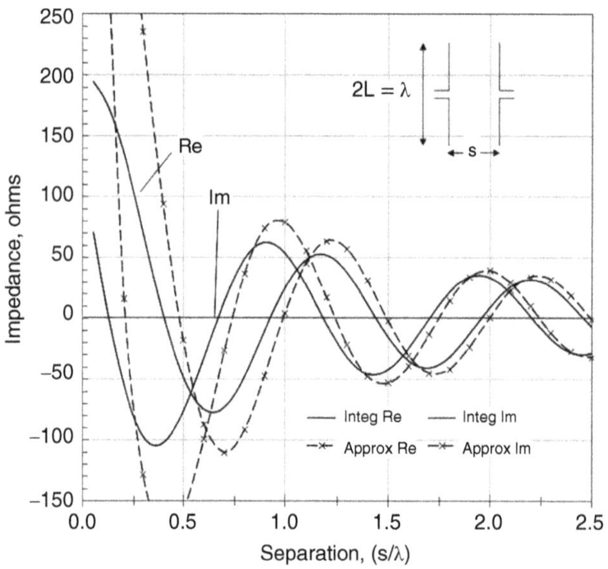

Figure 4.8 Mutual impedance for two parallel full-wave dipoles as a function of spacing. Key: solid line – direct integration Eq. (4.28); and dashed line with x – asymptotic approximation Eq. (4.33).

4.4 Arrays of Wire Antennas

An array of wire antennas can be modeled using several techniques such as direct numerical calculation using the MoM or asymptotically to approximately calculate the Z-matrix. The performance of each element depends on its nearest neighbors particularly but also on other elements of the array. Those in the central part of a uniformly spaced array will have similar behavior as in a periodic array, and those on the boundaries will be affected differently depending on the currents on

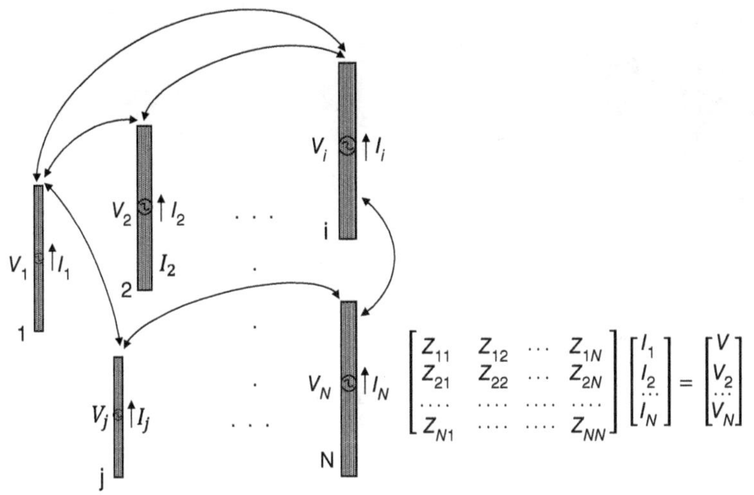

Figure 4.9 Model of array of N wire antennas.

neighboring elements, nearby structures, and the boundedness of the array. A finite array of N dipoles is illustrated in Figure 4.9. Suppose element i of this array is driven by a voltage V_i as shown. As well, some arrays may have parasitic elements such as in the Yagi–Uda antenna in which the element may be shorted or terminated in an impedance. Whether elements are driven directly or by parasitics with loads or short circuits, the array can be analyzed through its Z-matrix. As illustrated in Figure 4.9, this matrix relates the currents at the driven ports to the excitation voltages through

$$\boldsymbol{ZI} = \boldsymbol{V}. \tag{4.35}$$

The currents on the elements, at the inputs or in loads, are obtained by inverting the Z-matrix; thus,

$$\boldsymbol{I} = \boldsymbol{Z}^{-1}\boldsymbol{V}. \tag{4.36}$$

as was described in Section 2.4.2.

Arrays with many elements are described in Chapter 13 and also in the literature and will not be covered here except for the special case of centre-fed dipole arrays. In this section, we will show how the Z-matrix can be filled and how it can be used to accommodate mutual coupling in arrays with parasitic elements for two examples. When there is more than one driven element, the procedures are essentially the same and will not be described here. In a final example, the aperture field of a tightly coupled dipole array is examined.

4.4.1 Full-Wave Dipole Above a Perfect Ground

When a thin dipole lies above an infinite ground plane, the ground can be replaced by an image antenna. If the ground plane is finite, diffraction occurs and these effects should be included as well. If this plane is comparable in size to the dipole, image theory is only approximate. Nevertheless, the antenna shown in Figure 4.10 can be visualized as in Figure 4.11. The potential application of the antenna in Figure 4.10 was for a holographic array, which used $3\lambda/\pi$ dipoles in front of a ground plane as scattering elements. Almost full-wave dipoles were chosen because these elements have the greatest radar scattering cross-section (Hu 1958). The dipole was driven by a low-frequency

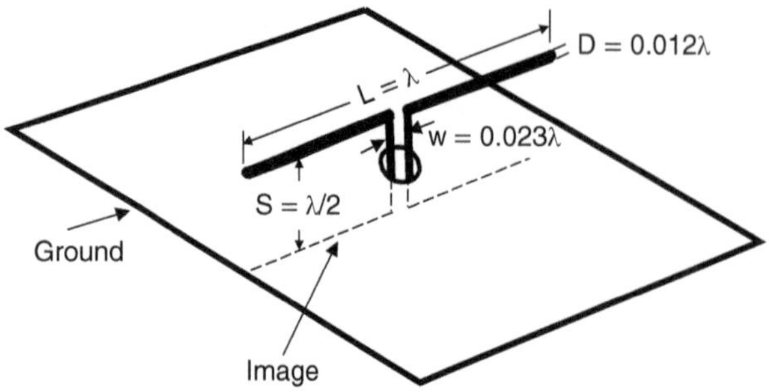

Figure 4.10 Full-wave dipole backed by a perfect conductor.

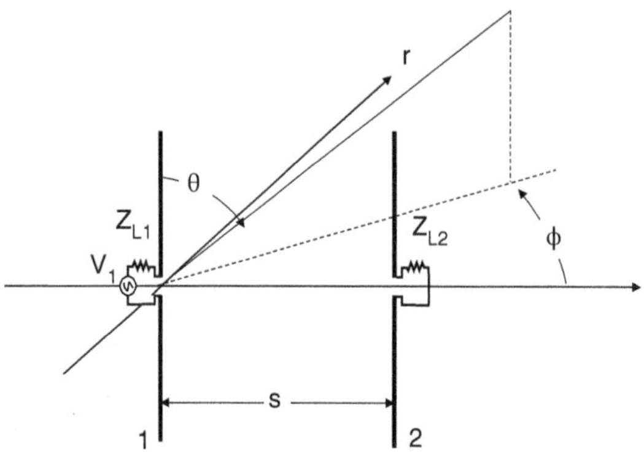

Figure 4.11 Center driven element (1) with a parasitic (2).

square-wave oscillator attached to its terminals, which modulated the radar cross-section, and any scattered signal could be readily detected (Bird 1970; Ja 1975).

The $3\lambda/\pi$ dipole is fed from a parallel wire line of characteristic impedance Z_{of}. The line is driven at the back of the ground plane through a balun transformer, which gave an input impedance close to 50 Ω. The aim here is to find the input impedance of this element. The antenna can be modeled with a Z-matrix. Above a large ground plane, the image (port 2) gives $V_2 = -V_1$ and $I_2 = -I_1$. The impedance at the dipole is

$$Z_1 = \frac{V_1}{I_1} = Z_{11} - Z_{12} \tag{4.37}$$

This impedance is transformed to the input over a distance of s. The input impedance is given by

$$Z_{in} = Z_{of}\frac{Z_1 + jZ_{of}\tan(k_o s)}{Z_{of} + Z_1 \tan(k_o s)}.$$

When $s = \frac{\lambda}{4}$, the impedance is

$$Z_{in} = \frac{Z_{0f}^2}{Z_1} = \frac{Z_{0f}^2}{Z_{11} - Z_{12}}. \tag{4.38}$$

For a scatterer designed to operate at 3.94 GHz, the two-wire feeder has a spacing of 1.75 mm and wire diameter of 0.92 mm. The characteristic impedance of the feeder is

$$Z_{of} = 120\cosh^{-1}\left(\frac{w}{2D}\right)$$

This gives $Z_{of} = 151\ \Omega$. The input impedance of the element is estimated to be $Z_{in} = 49.1 - j\,26.1$, which shows that antenna could be matched with a 1 nH inductance provided by a short wire. A prototype element was built for the operation at 3.94 GHz (Bird 1970). The element length was $L = 7.3$ cm, and this was backed by a finite ground plane $10 \times 10\,\text{cm}^2$, the size of which was tuned to maximize the gain (Schelkunoff and Friis 1952, p. 571). When matched, the input VSWR was 1.1, the maximum gain was 4.8 dB above a full-wave dipole (3.8 dBi), and the radiation pattern had a front-to-back ratio of >20 dB as shown in Figure 4.12.

The performance of the $3\lambda/\pi$ dipole in Figure 4.10 is compared to a conventional full-wave dipole with a parasitic reflector as illustrated in Figure 4.11. Once again, the antenna is modeled with a Z-matrix. A single parasitic element is now considered in conjunction with a driven element (Figure 4.11), where this time element #2 is shorted out. For two coupled dipoles as shown can be represented as (King et al., 1965)

$$\begin{aligned} V_1 &= I_1 Z_{11} + I_2 Z_{12} \\ 0 &= I_1 Z_{21} + I_2 Z_{22} \end{aligned} \tag{4.39}$$

where $Z_{11} = Z_{L1} + Z_{s1}$, and $Z_{22} = Z_{L2} + Z_{s2}$, wherein Z_{s1} and Z_{s2} are the input impedances of the antenna elements, Z_{L1} and Z_{L2} are load impedances, and Z_{12} and Z_{21} are the mutual impedances. The current at the center of the parasitic element 2 is

$$I_2 = -I_1 \frac{Z_{21}}{Z_{22}} \tag{4.40}$$

As a result,

$$V_1 = \left(Z_{11} - \frac{Z_{21} Z_{12}}{Z_{22}} \right) I_1 \tag{4.41}$$

and

$$Z_{in} = \frac{V_1}{I_1} = Z_{11} - \frac{Z_{21} Z_{12}}{Z_{22}}$$

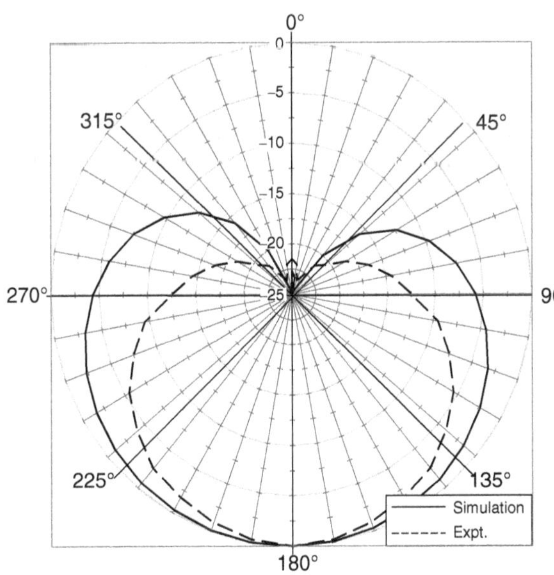

Figure 4.12 Radiation pattern of full-wave dipole in the azimuth plane. Key: solid – theory full-wave dipole and parasitic; dashed line – experiment 0.95 λ dipole spaced λ/4 in front of 1.2λ × 1.2 λ backing reflector. Source: BIRD, T.S. (1970), 'Antenna element for a holographic array of modulated scatters', unpublished report, Australian Post Office Research Laboratories.

The far-field of the two antennas at (r, θ, Φ) is

$$E_\theta = j\frac{\eta_o}{2\pi r}e^{-jk_o r}F(\theta)\left(I_1 + I_2 e^{jk_o\, s\, \sin\theta\cos\Phi}\right)$$
$$= j\frac{\eta_o}{2\pi r}e^{-jk_o r}F(\theta)\left(1 - \frac{Z_{21}}{Z_{22}}e^{jk_o\, s\, \sin\theta\cos\Phi}\right)I_1 \tag{4.42}$$

where s is the element spacing, $F(\theta) = \frac{\cos^2\left(\frac{\pi}{2}\cos\theta\right)}{\sin\theta}$ is the radiation pattern of the full-wave dipole, θ is elevation angle in a spherical coordinates taken along the axis of the dipole, and Φ is the azimuth angle. The horizontal pattern ($\theta = \pi/2$) is determined by the full-wave dipole pattern and an angular dependence Φ in which $\Phi = 0$ or π is the peak radiation. The ratio Z_{21}/Z_{22} and the spacing can be adjusted to maximize the field in the front or back directions such that the parasitic can act as a director or a reflector, respectively. For thin dipoles, maximum gain occurs with a parasitic spacing of $0.15-0.25\,\lambda$ from the driven element in front of the parasitic. The radiation pattern of a full-wave dipole and a parasitic separated by $0.15\,\lambda$ is illustrated by the full-line in Figure 4.12. This antenna has a computed gain of 6.7 dBi and an input VSWR of 1.15. This compares with a measured gain of 8.6 dBi and a VSWR of 1.1 with the tuned reflector. The improvement in gain justifies the use of a backing reflector over a parasitic element in the intended application.

4.4.2 The Yagi–Uda Array

The principles described in Section 4.4.1 for a driven element with parasitic elements are used to full effect with the Yagi–Uda antenna illustrated in Figure 4.13. Elements that are slightly longer than the driven elements act as reflectors, and shorter elements in the forward direction act as directors. The Yagi antenna can have any number of directors but usually only one reflector. Here, we will consider a five-element Yagi as an example. The driven element in most Yagi's is slightly less than half-wavelengths (usually $0.45-0.49\ \lambda$), the directors are still shorter (typically $0.4-0.45\ \lambda$), and the reflectors are larger ($0.5-0.52\ \lambda$). As the directors's dimensions are slightly below resonance, they will be capacitive and their currents lead the phase of the driven element, while the reflector is longer, will be inductive, and will provide a phase lag.

Although there is usually a variation between different designs and the way Yagi–Uda antennas are constructed, it is possible to estimate the gain against the number of elements in the design, which is a very useful design guide. These approximate values are listed in Table 4.2 and are given here as a design aid. As an additional guide, above about 4 or 5 directors, each additional director adds about an extra 1 dB of gain for up to about 15 directors. After that, the gain increment falls

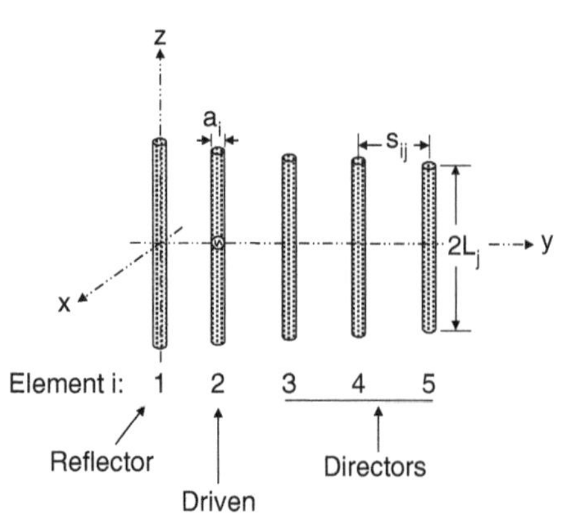

Figure 4.13 Five-element Yagi–Uda array geometry.

Table 4.2 Approximate gain of Yagi–Uda array with number of elements (one reflector).

No. Elements	Approx. Gain dBi
2	7.2
3	8.5
4	9.8
5	11.1
6	12.2
7	13.3
8	14.4
9	15.5

Table 4.3 Geometry of a five-element Yagi–Uda array. All dimensions in wavelengths.

Element no. (i)	1	2	3	4	5
pair ij	11	12	13	14	15
s_{ij}		0.38	0.2	0.2	0.2
L_i	0.253	0.248	0.215	0.215	0.215
a_i	0.005	0.005	0.005	0.005	0.005

with the increasing number of directors. Note that there is negligible increase in performance by adding more reflectors greater than one.

As an example in the use of the Z-matrix in array calculations, for the five-element array with dimensions listed in Table 4.3, the Z-matrix is filled element by element, by assuming element i operates in the presence of another element in the array, say j, when all other elements are open circuited. When considering element 2, for example, we do that for elements $j = 1, 3, 4, 5$ in this case. When the elements are open circuited, they appear virtually invisible so that the self-impedance and mutual impedance can be calculated as described in Section 4.2. This is done in turn to fill the Z-matrix. For the array shown in Figure 4.13, the Z-matrix is a 5×5 symmetric matrix and is given by

$$Z = \begin{bmatrix} 75.775 - j5.069 & 10.615 - j37.725 & -21.372 - j18.59 - & 19.814 + j9.908 & 1.786 + j17.999 \\ 10.615 - j37.725 & 71.329 - j18.233 & 50.266 - j19.011 & 6.115 - j36.637 & -22.762 - j15.561 \\ -21.372 - j18.59 & 50.266 - j19.011 & 45.794 - j91.899 & 50.266 - j19.011 & 6.115 - j36.637 \\ -19.814 + j9.908 & 6.115 - j36.637 & 50.266 - j19.011 & 45.794 - j91.899 & 50.266 - j19.011 \\ 1.786 + j17.999 & -22.762 - j15.561 & 6.115 - j36.637 & 50.266 - j19.011 & 45.794 - j91.899 \end{bmatrix} \tag{4.43}$$

The self- and mutual impedances were calculated using the expressions given in Eqs. (4.1a), (4.1b), and (4.24). All elements apart from the second one is shorted, and as a result, the drive voltage vector is

$$V_d = \begin{bmatrix} 0 \\ 1. \\ 0 \\ 0 \\ 0 \end{bmatrix}$$

The current at the center of the elements, I_d, is found from

$$ZV_d = I_d. \tag{4.44}$$

The currents at the center of each elements are found by inverting the Z-matrix

$$I_d = \begin{bmatrix} -3.836 * 10\wedge - 3 + J2.486 * 10\wedge - 3 \\ 0.015 + j8.335 * 10\wedge - 3 \\ -2.315 * 10\wedge 3 - j0.01 \\ -6.452 * 10\wedge - 3 - J8.824 * 10\wedge - 4 \\ 9.745 * 10\wedge - 4 + j0.01 \end{bmatrix} \tag{4.45}$$

from which the input impedance, radiation pattern, and gain can be found by standard methods (Bird 2016). The input impedance is

$$Z_{in} = \frac{V_1}{I_{d1}} = \frac{1.}{0.015 + j8.335x10\wedge - 3} = 51.614 - j29.419 \text{ ohms.}$$

Without matching, this gives a reflection coefficient into a 50 Ω feeder cable of −11.8 dB. The computed gain of this five-element Yagi is 11.85 dBi, and the radiation pattern is shown in Figure 4.14. It has a front-to-back ratio of 22.6 dB, and a peak sidelobe is 12.3 dB below the main beam. The computed pattern is shown in the figure compared with the radiation pattern of a similar five-element design (DX-antennas, 2020). Further optimization of the geometry is possible

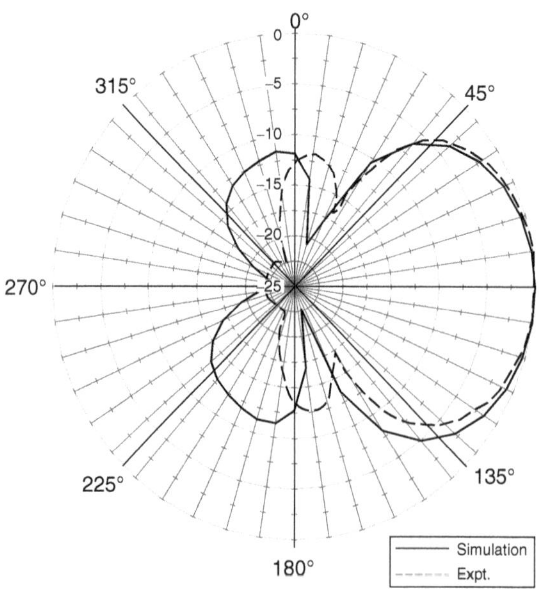

Figure 4.14 Polar pattern of five-element Yagi–Uda array simulation (here) compared with experimental data (dashed line) for a similar five-element design (*DX-antennas*, 2020).

to achieve a higher isolation especially in the backward direction, but this is usually achieved at the expense of lower gain.

4.4.3 7 x 7 array of closely packed elements

There has been increasing interest in large arrays of closely-packed elements to make use of the effects of mutual coupling to achieve a current sheet as suggested by Wheeler (1965). To achieve this, the spacing between elements is typically less than a quarter of a wavelength, which means the element size is usually smaller than this. The purpose of the present section, is to show the influence of mutual coupling in a small array of closely-packed dipoles of length 0.15λ. Dipoles of this size normally have high input impedance but it is assumed in the example that the geometry has been modified and suitable matching (eg. Doane et al., 2013) is applied to each element in a 7 x 7 array. The elements are arranged on a rectangular grid of spacing 0.2λ in both orthogonal-directions. The impedance matrix has been filled assuming mutual impedances and self-impedances are given by Eqs 4.25 and 4.27, respectively. A uniform voltage was applied to each element and the radiation field was computed from Eq. 4.23 assuming a sinusoidal current. The resulting aperture field in the plane of the array is shown in Fig. 4.15(b). This is compared to the case when there is no mutual coupling, which is shown in Fig. 4.15(a).

The presence of the individual elements in the no coupling case (Fig 4.15(a)) is evident and less so in Fig. 14.15(b). In both plots the outer ring of 24 edge elements can be clearly seen, and their prominence makes the aperture field less uniform. Even in this small sized array, the impact of close proximity is that the aperture field of the central 5 x 5 array is filled-in through coupling and is less 'lumpy' than with no coupling. It is expected that in a bigger array of N x N elements, coupling will make the aperture field more uniform over the central (N-2) x (N-2) elements leaving the outer ring of 4(N-1) edge elements to be excited or terminated in a suitable manner.

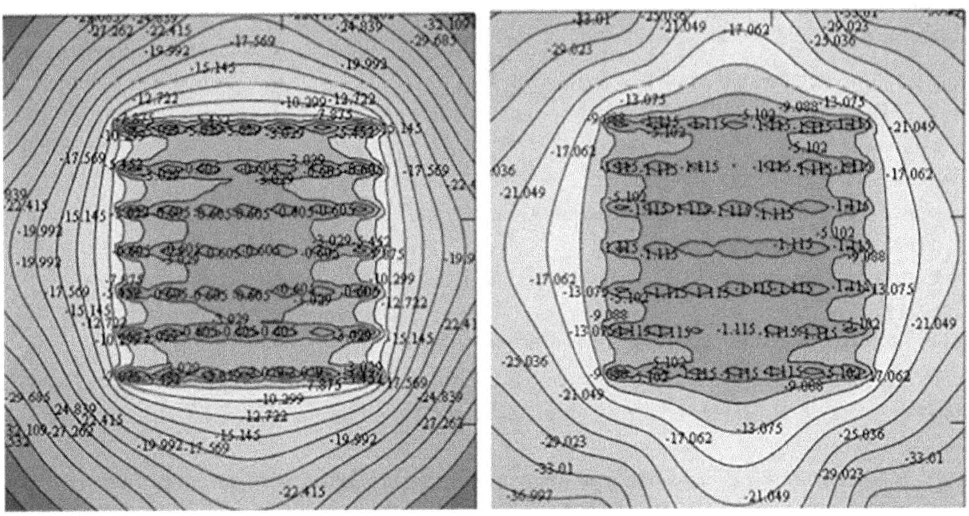

(a) (b)

Figure 4.15 Aperture field of 7 x 7 array of 0.15λ dipoles on a regular square grid with spacing 0.2λ. (a) No coupling; and (b) with dipole coupling.

4.5 Concluding Remarks

The study of mutual coupling in wire antennas has continued from the early 1930s to the present day with increasing insights into what is achievable with both small and large arrays. For instance, in recent years, arrays of closely coupled dipoles have been used to facilitate greater mutual coupling to improve bandwidth and overall performance. This has been aided by increasingly sophisticated computer software, such as those using a combination of discrete methods and the method of moments, which are able to handle closely packed dipoles that may be printed instead of wires. The background and basic tools for studying wire antennas has been described in this chapter to give physical insight into the designs using more sophisticated elements. This description has been supported by practical examples and some measurements for a Yagi–Uda array and a candidate full-wave scatterer. Integral solution techniques have been supported by asymptotic approximations which allow increased understanding of these methods and of mutual coupling itself. For evaluating mutual impedances of elements spaced greater than a wavelength apart, asymptotic approximations can be used to significantly reduce computing time for a small cost in overall accuracy. Another important conclusion is that for wire elements, at very large distances, the coupling decays as the inverse distance in the H-plane, while, in the E-plane, it decays as the square of the distance.

References

Balanis, C.A. (1982). *Antenna Theory: Analysis and Design*. New York: Harper & Row.

Barzilai, G. (1948). Mutual impedance of parallel aerials. *Wireless Engineer* 36: 343–352.

Bechmann, R. (1931). On the calculation of radiation resistance of antennas and antenna combinations. *Proc. IRE* 19 (8): 1471–1480.

Bird, T.S. (1970), 'Antenna element for a holographic array of modulated scatterers', unpublished report, Australian Post Office Research Laboratories.

Bird, T.S. (2016). *Fundamentals of Aperture Antennas*. Chichester, UK: Wiley.

Bontsch-Bruewitsch, M.A. (1926). Die strahlung der komplizierten rechtwinkeligen antennen mit gleichbeschaffenen vibratoren. *Ann. Phys.* 81: 425–453.

Brown, G.H. and King, R.W.P. (1934). High-frequency models in antenna investigations. *Proc.IRE* 22 (4): 457–480.

Brown, G.H. (1937). Directional antennas. *Proc. IRE* 25 (1): 78–145.

Carter, P.S. (1932). Circuit relations in radiating systems and applications to antenna problems. *Proc.IRE* 20 (6): 1004–1041.

Cox, C.R. (1947). Mutual impedance between vertical antennas of unequal heights. *Proc.IRE* 35 (11): 1367–1370.

Doane, J.P., Sertel, K. and Volakis, J.L. (2013): 'A wideband, wide scanning tightly coupled dipole array with integrated balun (TCDA-IB)', IEEE Trans. Antennas & Propagat., Vol. 61, No. 9, pp. 4538–4548.

Finlayson, B.A. (1972). *The Method of Weighted Residuals and Variational Principles*. New York, USA: Academic Press.

Gera, A.E. (1988). Simple expressions for mutual impedances. *IEE Proc. H, Microwaves Antennas Propag.* 135: 395–399.

Harrington, R.F. (1967). Matrix methods for field problems. *Proc. IEEE* 55 (2): 136–149.

Harrington, R.F. (1968). *Field Computation by Moment Methods*. New York, NY: Macmillan Company.

DX-antennas, http://dx-antennas.com/Yagi5.htm, accessed Aug. 2020.

Hu, Y.-Y. (1958). Back-scattering cross-section of a center-loaded cylindrical antenna. *IRE Trans. Antennas Propag.* AP-8: 140.

Ja, Y.H. (1975). Measurement of angles of arrival of waves by microwave holographic techniques. *IEEE Trans. Antennas Propag.* AP-23 (5): 720–722.

King, H.E. (1957). Mutual Impedance of Unequal Length Antennas in Echelon. *IRE Trans. Antennas Propag.* AP-5 (3): 306–313.

King, R. and Harrison, C.W. (1944). Mutual and self-impedance for coupled antennas. *J. Appl. Phys.* 15 (6): 481–495.

King, R.W.P., Mimno, H.R., and Wing, A.H. (1965). *Transmission Lines, Antennas and Waveguides*. New York: Dover.

Mahony, J.D. (1989). Planar array design: approximation to mutual impedances between two remotely spaced "half-wavelength" dipoles. *Electron. Lett.* 25: 1711–1712.

Pistolkors, A.A. (1929). The radiation resistance of beam antennas. *Proc.IRE* 17 (3): 562–579.

Schelkunoff, S.A. and Friis, H.T. (1952). *Antennas: Theory and Practice*. London: Wiley.

Susuga, T., Konno, K. , and Chen, Q. (2019): 'Far-field approximation of a dyadic Green's function in a free-space and its application to method of moments', IEICE Technical Report A-P2019-96 (2019-10), pp. 85-88.

Tai, C.T. (1948). Coupled antennas. *Proc.IRE* 46 (4): 487–500.

Thiele, G.A. (1969). Analysis of Yagi-Uda-type antennas. *IEEE Trans. Antenna Propag.*. AP-17 (1): 24–31.

Hansen, R.C. (1966): '*Microwave Scanning Antennas*', New York, Academic, Vol. 3, Chap. 2-3

Hansen, R.C. (1998). Phased array antennas. In: *Wiley-Interscience*. John Wiley & Sons, Inc., New York: USA.

Richmond, J.H. (1965). Digital computer solutions of the rigorous equations for scattering problems. *Proc. IEEE* 53 (8): 796–804.

Wheeler, H.A. (1965): 'Simple relations derived from a phased-array antenna made of an infinite current sheet', IEEE Trans. Antenna & Propagat., Vol. AP-13, No.4, pp. 506–514.

5

Arrays of Planar Aperture Antennas

Trevor S. Bird[1,2]

[1] *Global Big Data Technology Centre, University of Technology Sydney, Ultimo, NSW 2007, Australia*
[2] *Antengenuity, Eastwood, NSW 2122, Australia*

5.1 Introduction

Aperture antennas such as waveguides and horns are commonly used as elements for directly radiating arrays and in array feeds for reflectors. The array radiation pattern, polarization properties, and element driving impedance are all influenced by mutual coupling between elements. In the case of aperture antennas, mutual coupling alters the mode content of the aperture distribution by causing the amplitude of modes to differ from that of isolated elements, as well as sometimes generating other modes. For accurate prediction of array performance, this coupling should be included in any design procedure.

Analysis of radiation from an open-ended waveguide involves consideration of both the field radiated external to the waveguide and the field inside the waveguide. When the waveguide terminates in a large ground plane, as illustrated in Figure 5.1, the problem can be solved by means of an integral equation and Green's function approach (Levine and Schwinger 1965), which is described later in this chapter. This approach is applicable to apertures of arbitrary shape, as will be discussed. A variety of apertures have been analyzed with this technique over the years with excellent results.

One of the most powerful approaches for finite arrays is Galerkin's method in which expansion functions satisfy the wave equation and some of the boundary conditions. One approach (Hockham and Walker 1973; Luzwick and Harrington 1982) uses linearly polarized expansion functions that are cosinusoidal over rectangular apertures, while over circular apertures, Bessel functions (Steyskal 1974; Bird 1979) are used. In a further approach, Fenn et al. (1982) use two orthogonally polarized sets of overlapping piecewise sinusoidal-uniform surface patches as expansion and weighting functions. This leads to mutual coupling calculations involving only a single integration. Arrays of rectangular waveguides have been studied (Fenn et al. 1982), during which the method of images was used to determine the generalized admittance for the internal waveguide region. Another formulation (Arndt et al. 1989) adopts a combination of pyramidal and triangular functions. An alternative but similar approach to this aperture problem through an integral equation is detailed in this chapter.

For apertures of waveguides or horns, the integral equation is usually solved by replacing the fields in the apertures with finite series of waveguide modes and determining the coefficients of the series by Galerkin's method. For example, rectangular waveguide arrays of identical elements have been analyzed by several authors. Of these, Mailloux (1969a,b) and also Hockham and Walker

Mutual Coupling Between Antennas, First Edition. Edited by Trevor S. Bird.

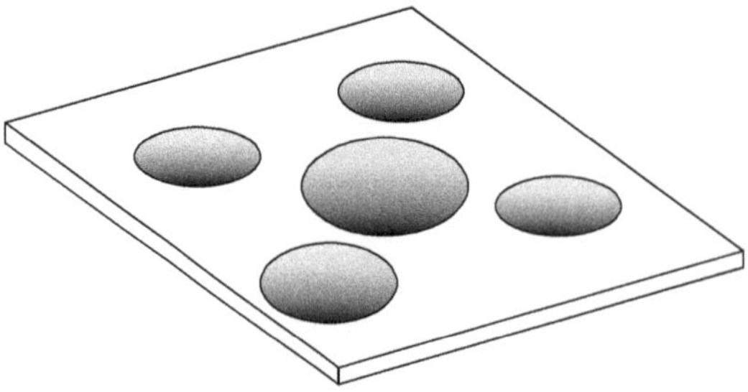

Figure 5.1 Mutual coupling in a waveguide array terminating in a large ground plane.

(1973) used a series of longitudinal-section electric (LSE) modes, while Luzwick and Harrington (1982) and Bird (1987) employed transverse electric (TE) and transverse magnetic (TM) modes. Experimental results have also been presented, and, in particular, Mailloux (1969a, b) demonstrated excellent agreement with theory for the coupling of two open-ended waveguides. Mutual coupling in circular aperture waveguides and horns have been studied in some detail as well and applied to a variety of array applications. A significant part of the integral solution can be reduced to standard functions and expressed partly in closed form, although usually some two-dimensional integration is required. Nonetheless, a significant amount of computation time and effort has been removed. In arrays of identical apertures, the computation can be reduced further, although different-sized or different elements are worth persevering with to provide design flexibility in beam shaping and sidelobe level control. For example, arrays of different-sized horns are sometimes used as feeds for reflector antennas to create shaped beams from geostationary satellites (Bird and Sroka 1992, Newell and Bird 1994). Another application is to beam steering, where guard elements, normally smaller in size to the main array elements, are sometimes used to control the element's driving impedance (Hockham 1974).

Other aperture shapes are important too. For example, elliptical apertures find application as sources of circular polarization and as feeds for reflectors giving a shaped beam from satellites (Vokurka 1979). Radiation from elliptical corrugated horns has been studied in the past by Müller (1960) and Jansen and Jeuken (1973) who described solutions for the radiation characteristics of elliptical waveguides with anisotropic wall lining. Experience with smooth-wall horn radiators of rectangular and circular cross-section has shown that significant reflection can occur from apertures of moderate size, resulting in mismatch at the horn input. It was only later that reflection in open-ended elliptical waveguide was formulated by Chew and Habashy (1986) and mode coupling presented by Bird (1990b) for the fundamental mode. A formulation for even more general cross-section shapes has been described by Kuehne and Marquardt (2001). In this chapter, a method is described for analyzing coupling in arbitrary-shaped apertures and is demonstrated by providing computationally efficient solutions and detailed results for the most commonly used types of rectangular, circular, coaxial, and elliptical apertures. By similar means, mutual coupling in dielectric and metamaterial-loaded rectangular apertures is also described. An asymptotic approximation for mutual admittance in apertures of arbitrary shape is derived, and when applied to waveguides with rectangular and coaxial cross-sections, good agreement is achieved compared with the exact expressions.

5.2 Mutual Coupling in Waveguide and Horn Arrays

5.2.1 Integral Equation Formulation

Arrays of apertures in metallic surfaces occur in a variety of applications in communications and radar. The analysis of these antennas is therefore important, especially where accuracy, efficiency,

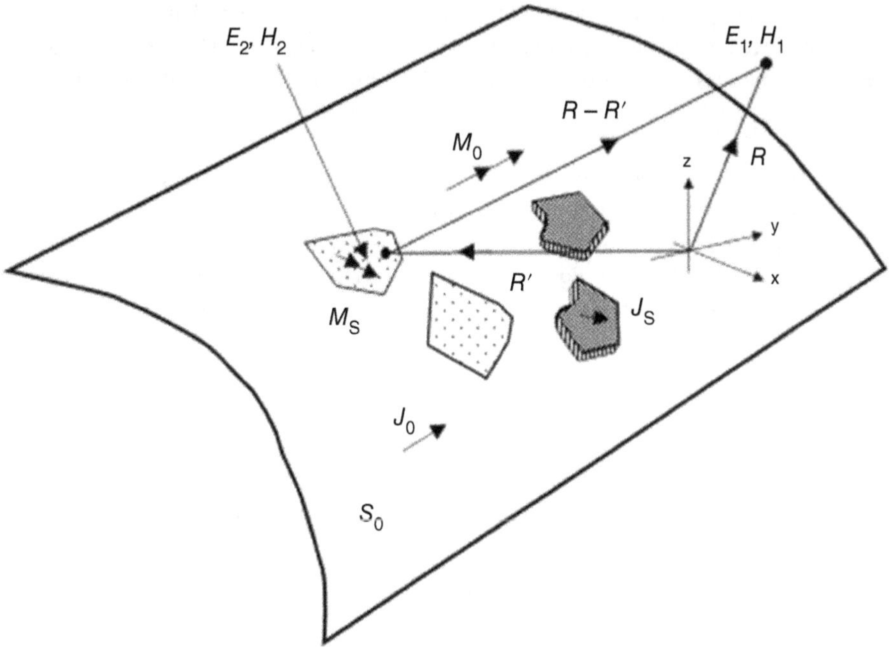

Figure 5.2 Sources and fields in the vicinity of an arbitrary surface $\boldsymbol{S_o}$.

and low sidelobes are required. The apertures may be physical openings, as in Figure 5.1, or two-dimensional current patches (see Chapter 6). Suppose the array is located on an arbitrary surface $\boldsymbol{S_o}$, as in Figure 5.2. When the array is operating, let the fields radiated external to this surface be $\boldsymbol{E}_1, \boldsymbol{H}_1$ and the fields inside the surface be $\boldsymbol{E}_2, \boldsymbol{H}_2$. It is convenient to use the field equivalence principle to replace the field inside $\boldsymbol{S_o}$ by a null field. To do this and maintain continuity of the field, electric and magnetic currents $\boldsymbol{J_o}, \boldsymbol{M_o}$ are introduced on the surface. In this situation, two types of array of practical interest can be identified which permit further simplification: (i) the array which comprises separate isolated electric conductors separated by homogeneous materials and (ii) the complementary situation when $\boldsymbol{S_o}$ is an electric conductor and the array comprises apertures in the conductor. Both of these antennas can be analyzed rigorously by the integral equation method. The solution for type (i) has been discussed in Chapter 4, and in this section, we consider the solution of (ii).

In a basic Green's function approach to Figure 5.1, the aperture field and ground plane are replaced by a magnetic current distribution, which is coincident with the original aperture and situated in free space. The field in the external region remains the same but is now considered to be produced by the equivalent source. At the same time, the field inside the waveguide, and hence the source distribution, is represented as a sum of waveguide modes with unknown coefficients. The problem is solved by satisfying boundary conditions at the aperture and determining the unknowns, allowing the mode reflection and conversion coefficients as well as the radiated field to be determined.

Returning to the more general situation in Figure 5.2, the equivalent sources are an electric current $\boldsymbol{J_o}$, which exists only on patches on the surface, and a magnetic current, $\boldsymbol{M_o}$, which is nonzero in the apertures only. To remove the need to determine both sources, it is convenient to introduce an electric conductor into the null field region inside $\boldsymbol{S_o}$ and place it adjacent to the surface so that it shorts out $\boldsymbol{J_o}$ completely. The same could be done with a magnetic conductor to short out all sources of magnetic current and to follow through with a dual of the previous description. In the former, the result is the equivalent problem of an array of patches of magnetic current that are backed by an electric conductor. Since the tangential component of the electric field should be continuous across the aperture, the magnetic current inside $\boldsymbol{S_o}$ must be $-\mathbf{M}_o$, thus, (Harrington and Mautz 1976)

$$\boldsymbol{H}_1^{inc} + \boldsymbol{H}_t^{rev}\left(-\boldsymbol{M}_o\right) = \boldsymbol{H}_t^{ext}\left(\boldsymbol{M}_o\right). \tag{5.1}$$

where each field quantity may be expressed in a variety of forms such as series or transforms expressed in terms of elemental sources. For the purpose of driving the array, the terms on the left-hand side of Eq. (5.1) are usually given in terms of the modes representing the driver as an outcome of the analysis. To satisfy Eq. (5.1) and solve for the magnetic current, express the magnetic current in terms of N_e general expansion functions, here denoted by $\{\boldsymbol{\alpha}_n \mid n = 1, \ldots, N_e\}$, with unknown coefficients V_n, as follows:

$$\boldsymbol{M}_o = \sum_{n=1}^{N_e} V_n\, \boldsymbol{\alpha}_n \tag{5.2}$$

This approximation is now substituted in Eq. (5.1) and use is made of the linearity of the fields. However, since it cannot be assumed that continuity is satisfied exactly, the residual of Eq. (5.1) is expressed as follows:

$$\mathcal{R} = \sum_{n=1}^{N_e} \left(\boldsymbol{H}_t^{rev}\left(\boldsymbol{\alpha}_i\right) + \boldsymbol{H}_t^{ext}\left(\boldsymbol{\alpha}_i\right)\right) V_i - \boldsymbol{H}_t^{inc} \tag{5.3}$$

A set of weighting functions $\{\boldsymbol{\beta}_n \mid n = 1, \ldots, N_e\}$ that are valid over $\boldsymbol{S_o}$ is chosen. As described in Section 3.4.3 and in Finlayson (1972), the inner product (i.e. the integral over S_o) of the weighting functions and the residual is set zero, thus

$$\left(\boldsymbol{\beta}_i, \mathcal{R}\right) = \iint_{S_o} dS\, \boldsymbol{\beta}_i \cdot \mathfrak{R} = 0. \tag{5.4}$$

As described in Section 3.4.3, the simplest set of weighting functions are constants, and in that case, Eq. (5.4) requires that the average residual should be zero. Another simple set of weighting functions is delta functions and use of these results in a point-matching solution. In the Galerkin method, identical weighting and expansion functions, i.e. $\{\boldsymbol{\alpha}_n = \boldsymbol{\beta}_n \mid n = 1, \ldots, N_e\}$, are used, thus

$$\begin{aligned}\left(\boldsymbol{\beta}_i, \mathcal{R}\right) &= \iint_{S_o} dS\, \boldsymbol{\alpha}_i \cdot \mathcal{R} \\ &= \left[\left(\left(\boldsymbol{\beta}_i, \boldsymbol{H}_t^{rev}\left(\alpha_j\right)\right) + \left(\left(\boldsymbol{\beta}_i, \boldsymbol{H}_t^{ext}\left(\alpha_j\right)\right)\right)\right] \boldsymbol{V}_j - \left(\boldsymbol{\beta}_i, \boldsymbol{H}_t^{inc}\right) = 0.\end{aligned} \tag{5.5}$$

The first two terms in Eq. 5.5 correspond to admittance matrices for the reflected field and the exterior region which are defined by

$$\boldsymbol{Y}^{rev} = \left[Y_{ij}^{rev}\right] = \left(\boldsymbol{\beta}_i, \boldsymbol{H}_t^{rev}\left(\boldsymbol{\alpha}_j\right)\right)$$

and

$$\boldsymbol{Y}^{ext} = \left[Y_{ij}^{ext}\right] = \left(\boldsymbol{\beta}_i, \boldsymbol{H}_t^{ext}\left(\boldsymbol{\alpha}_j\right)\right).$$

In addition, define a current source vector

$$\boldsymbol{I}^{inc} = \left[\left(\boldsymbol{\beta}_i, \boldsymbol{H}_t^{inc}\left(\boldsymbol{\alpha}_j\right)\right)\right].$$

Therefore, the method of weighted residuals results in the matrix equation

$$\left(\boldsymbol{Y}^{rev} + \boldsymbol{Y}^{ext}\right) \boldsymbol{V} = \boldsymbol{I}^{inc}. \tag{5.6}$$

The physical picture provided by Eq. (5.6) is of two general admittances in parallel with a current source. By inverting the sum of the admittance matrices, the coefficients of the magnetic current expansion can be found from

$$\boldsymbol{V} = \left(\boldsymbol{Y}^{rev} + \boldsymbol{Y}^{ext}\right)^{-1}\boldsymbol{I}^{inc}. \tag{5.7}$$

Once $\boldsymbol{V}$ has been found, the complex amplitudes of the reverse traveling fields in the waveguides can be obtained.

Continuing with the magnetic current-based solution, an approach will be adopted that has proved particularly powerful across a wide range of applications. In the region external to $\boldsymbol{S_o}$, let the tangential component of the magnetic field be expressed as

$$\boldsymbol{H}^{ext} = \iint\limits_{S_o} d\boldsymbol{S}\, \mathbf{G}^{(h)}\left(\boldsymbol{R}|\boldsymbol{R}'\right) \cdot \boldsymbol{M_o} \tag{5.8}$$

where $\boldsymbol{G}^{(h)}$ is the dyadic Green's function for the magnetic field, and $\boldsymbol{R}$ and $\boldsymbol{R}'$ are vectors from the origin to the field and source points as shown in Figure 5.2. To solve this problem, one needs to know the Green's function for a magnetic dipole radiating in the presence of an electric conductor specified by $\boldsymbol{S_o}$. Closed-form solutions are available for some elementary surfaces, e.g. the plane, the cylinder, and the sphere (Felsen and Marcuvitz 1973), as discussed in Section 3.4. One may also consider the external region to comprise several homogeneous layers or shells, but here, for simplicity's sake, we will consider a homogeneous external region only. When the surfaces have large curvatures, asymptotic solutions may be usefully employed. Further, by following the general principles of the geometrical theory of diffraction, asymptotic solutions for elementary surfaces may be generalized to approximate fields on arbitrary surfaces. For the special case when $\boldsymbol{S_o}$ is a plane, the appropriate Green's dyadic is best obtained by the method of images. Since the remaining patches of magnetic current are backed by a plane, electric conductor image theory says the field due to this arrangement is identical to a magnetic current of twice the strength that operates in free space. The Green's dyadic for a magnetic dipole in a plane is given by Eq. (3.28) and is repeated below for convenience:

$$\mathbf{G}^{(h)}\left(\boldsymbol{R}|\boldsymbol{R}'\right) = -\frac{jk_o}{2\pi\,\eta_o}\left(\underline{\boldsymbol{I}} + \frac{1}{k_o^2}\nabla_t\nabla_t\right) g_o\left(|\boldsymbol{R} - \boldsymbol{R}'|\right) \tag{5.9}$$

where

$$|\boldsymbol{R} - \boldsymbol{R}'| = \sqrt{(x-x')^2 + (y-y')^2 + (z-z)^2},$$

$$g_o(R) = \exp\left(-jk_oR\right)/R,$$

$\underline{\boldsymbol{I}}$ is the unit dyadic and ∇_t is the gradient operator in the transverse x-y plane.

5.2.2 Modal Representation

If the inputs to the array elements are waveguides or coaxial transmission lines, the fields inside the antenna apertures may be approximated by a summation of traveling modes. Modal field solutions may be obtained by analytical methods for waveguides with circular, elliptical, and rectangular cross-section, while structures with general cross-sections can be analyzed by numerical means, such as the finite element method. Whatever method is used, a set of known modes with transverse fields is available in the form $(\boldsymbol{e}_{pi}, \boldsymbol{h}_{pi})$ $i = 1, \ldots, N$; $p = 1, \ldots, M(i)$, where N is the number of waveguides and $M(i)$ is the number of modes in waveguide i. The transverse field $\left(\boldsymbol{E}_t^{(i)}, \boldsymbol{H}_t^{(i)}\right)$ in waveguide i consists of both forward and backward traveling terms, as follows:

$$\boldsymbol{E}_t^{(i)} = \sum_{p=1}^{M(i)} \left(a_{pi}e^{-j\gamma_{pi}z} + b_{pi}e^{+j\gamma_{pi}z}\right) \boldsymbol{e}_{pi}(x,y)\, Y_{pi}^{-1/2} \tag{5.10a}$$

$$\boldsymbol{H}_t^{(i)} = \sum_{p=1}^{M(i)} \left(a_{pi}e^{-j\gamma_{pi}z} - b_{pi}e^{+j\gamma_{pi}z}\right) \boldsymbol{h}_{pi}(x,y)\, Y_{pi}^{+1/2}, \tag{5.10b}$$

where

$$\gamma_{pi} = \beta_{pi} - j\alpha_{pi} = \text{propagation constant of mode } p \text{ in region } i,$$

$$\boldsymbol{h}_{pi} = \hat{z} \times \boldsymbol{e}_{pi} \tag{5.11}$$

and

$$\int_{S_i} dS \left(\boldsymbol{e}_{pi} \times \boldsymbol{h}_{qi} \cdot \hat{z}\right) = 2\delta_{pq} \tag{5.12}$$

where δ_{pq} is the Kronecker delta and Y_{pi} is the mode admittance of mode p in waveguide i, given by

$$Y_{pi} = \frac{1}{\eta_0} \begin{cases} \gamma_{pi}/k & \textit{TE modes} \\ \varepsilon_r^{(i)}\, k/\gamma_{pi} & \textit{TM modes} \end{cases} \tag{5.13}$$

where $\varepsilon_r^{(i)}$ is the relative permittivity of the filling in aperture i. In the following, it will be convenient from time to time to replace the double subscript pi with a single subscript m in which the mode of each aperture is numbered sequentially.

At this stage, we can make use of a magnetic current on the surface through $\boldsymbol{M_o} = \mathbf{E}_t \times \hat{n}$, where $\hat{\mathrm{n}}$ is the unit outward normal to the surface $\boldsymbol{S_o}$ be $\hat{\mathrm{n}}$. When Eqs. (5.10) is substituted into Eq. (5.1), the residual of Eq. (5.4) at the i-th aperture at $z = 0$ is

$$\boldsymbol{R_i} = \sum_{p=1}^{M(i)} \left(a_{pi} - b_{pi}\right) \boldsymbol{h}_{pi}(x,y)\, Y_{pi}^{1/2} - \sum_{j=1}^{N}\sum_{q=1}^{M(i)} \left(a_{qj} + b_{qj}\right) Y_{qj}^{-\frac{1}{2}} \iint_{\boldsymbol{S_j}} \boldsymbol{dS}\, \mathbf{G}^{(\boldsymbol{h})}\left(\mathbf{R}|\mathbf{R}'\right) \cdot \left(\boldsymbol{e}_{qj}(x,y) \times \hat{n}\right). \tag{5.14}$$

Using $\boldsymbol{h}_{pi}$ as testing functions, take the inner product over $\boldsymbol{S_i}$ of both sides of Eq. (5.8) to give

$$\begin{aligned} &\iint_{\boldsymbol{S_i}} \boldsymbol{dS}\, \boldsymbol{h}_{pi} \cdot \boldsymbol{R_i} = 0 \\ &= \sum_{p=1}^{M(i)} \left(a_{pi} - b_{pi}\right) 2\, Y_{pi}^{1/2} - \sum_{j=1}^{N}\sum_{q=1}^{M(i)} \left(a_{qj} + b_{qj}\right) Y_{qj}^{-1/2} \\ &\quad \times \iint_{\boldsymbol{S_i}} \boldsymbol{dS}\, \boldsymbol{h}_{pi} \cdot \iint_{\boldsymbol{S_j}} \boldsymbol{dS}\, \mathbf{G}^{(\boldsymbol{h})}\left(\mathbf{R}|\mathbf{R}'\right) \cdot \left(\boldsymbol{e}_{qj}(x,y) \times \hat{n}\right) \\ &= \sum_{p=1}^{M(i)} 2\, I_{pi} Y_{pi}^{1/2} - \sum_{j=1}^{N}\sum_{q=1}^{M(i)} 2\, V_{pi}\, Y_{pi}^{1/2}\, y_{ij}(p|q), \end{aligned} \tag{5.15}$$

where use has been made of mode orthogonality given by Eq. (5.12) and $V_{pi} = (a_{pi} + b_{pi})$ and $I_{pi} = (a_{pi} - b_{pi})$. $y_{ij}(p\,|\,q)$ is the normalized mutual admittance between modes p and q in apertures i and j which is given by

$$y_{ij}(p|q) = \frac{1}{2\sqrt{Y_{pi}Y_{pj}}}\int_{S_i} dS\, \boldsymbol{h}_{pi}(u_1, u_2) \cdot \int_{S_j} dS'\, \underline{\underline{\boldsymbol{G}}}^{(h)}(\boldsymbol{R} - \boldsymbol{R}') \cdot \left[\boldsymbol{e}_{qj}(u_1', u_2') \times \hat{n}\right], \tag{5.16}$$

where $(u_1,\ u_2)$ are coordinates on aperture i, and, similarly, the primed coordinates refer to aperture j. It is seen that the mutual admittance Eq. (5.16) is symmetric with respect to interchange of modes and apertures; thus,

$$y_{ji}(q|p) = y_{ij}(p|q). \tag{5.17}$$

When the weighted residual in Eq. (5.15) is set to zero, we obtain

$$\boldsymbol{I} = \boldsymbol{y}\,\boldsymbol{V}.$$

In terms of scattering mode amplitudes, this relation is

$$\boldsymbol{a} - \boldsymbol{b} = \boldsymbol{y}(\boldsymbol{a} + \boldsymbol{b}) \tag{5.18}$$

and

$$\boldsymbol{b} = (1 + \boldsymbol{y})^{-1}(1 - \boldsymbol{y})\,\boldsymbol{a} = \boldsymbol{S}\boldsymbol{a} \tag{5.19}$$

where

$$\boldsymbol{S} = (\boldsymbol{I} + \boldsymbol{y})^{-1}(\boldsymbol{I} - \boldsymbol{y}) \tag{5.20}$$

is the mode scattering matrix of the apertures. The scattering matrix enables the amplitude of the reverse traveling mode at each port to be obtained. In the special case of apertures in a conducting ground plane, Eq. (5.6) may be simplified using the fact that the mode functions satisfy the boundary conditions on the waveguide walls. It is found that the elements of the normalized admittance matrix can be expressed (Bird 1979) as

$$y_{ij}(p|q) = \frac{jk}{4\pi\eta_o\sqrt{Y_{pi}Y_{qj}}}\int_{S_i} dS\, \boldsymbol{\Psi}_{pi}(u_1, u_2) \cdot \int_{S_j} dS'\, \boldsymbol{\Psi}_{qj}(u_1', u_2')\, G(|\boldsymbol{R} - \boldsymbol{R}'|), \tag{5.21}$$

where the vector mode function for the admittance is defined as

$$\begin{aligned}\boldsymbol{\Psi}_{pi}(x, y) &= \boldsymbol{h}_{pi} + \hat{z}\frac{1}{jk_o}\nabla_t \cdot \boldsymbol{h}_{pi}\\ &= \left(\boldsymbol{h}_{pi} + \hat{z}\frac{\gamma_{pi}}{k_o}h_{z,pi}\right)\end{aligned} \tag{5.22}$$

in which γ_{pi} is defined in Eq. (5.11) and Y_{pi} is given by Eq. (5.13). It is observed that Eq. (5.21) is valid for arbitrarily shaped apertures, which are homogeneously filled, provided that the fields are solutions of Maxwell's equations and that they satisfy the boundary conditions for the given geometry. Eq. (5.21) is a consequence of a modal representation of the fields in the aperture.

A significant computational issue for mode matching concerns the number of modes needed for an accurate representation of the aperture fields, and this depends on the operating frequency. If the waveguide or horn operates in the fundamental mode and all other modes are well below cut-off, a good estimate of reflection is obtained by including the fundamental mode in the scattering matrix and the next coupled mode in the aperture. Use of several high-order modes is recommended, however, for accurate predictions. Satisfaction of the edge condition is not critical except when there is

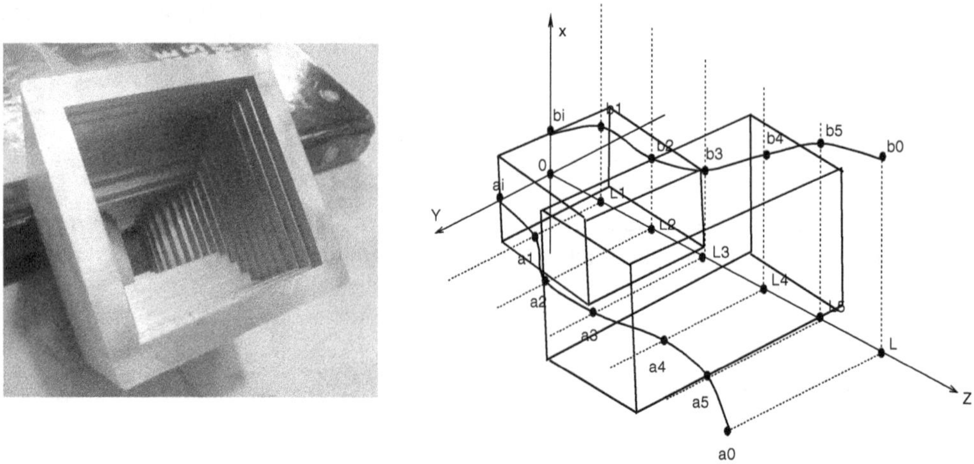

(a) (b)

Figure 5.3 Profiled rectangular electromagnetic horn. (a) Front view and Source: Trevor S. Bird (b) representation of profiles in two principal planes and by uniform waveguide sections.

a thin iris near the aperture, although Hockham (1975) showed that, for rectangular waveguides, inclusion of TE_{m0} ($m = 3,5,\ldots$) and TE_{0n} ($n = 2,4,\ldots$) modes improves solution convergence.

Several different waveguide cross-sections will be analyzed in later sections using Eq. (5.16). However, before proceeding, the mode matching approach to modeling a waveguide with an arbitrary cross-section is summarized in Section 5.2.3.

5.2.3 Modeling of Profiled Horns and Mode Matching

The representation of the field in stepped or more generally profiled structures can be handled through the method of mode matching, which was outlined in Sec. 3.5.2. In this method, the structure is modeled by a series of connected stepped sections in which the fields are expressed as a sum of modes. This method has been detailed in many publications, and particularly relevant are those of Olver et al. (1994) and the author (Bird 2016). The modal representation in the apertures of an array is fundamental to a complete mode matching approach.

Consider the profiled electromagnetic horn in Figure 5.3 which radiates into a semi-infinite region. Internal to the horn, transition sections may be employed for matching purposes, and in the aperture, plane choke rings (not shown in Figure 5.3) are used to control the radiation. There are essentially two types of approach needed to handle the transverse discontinuities in the closed waveguide or semi-infinite regions such as dielectrics, and each approach requires knowledge of the modes in that region. A prototype section for mode matching in a closed waveguide is shown in Figure 5.4. Modal field solutions may be obtained by solving the wave equation applicable to the geometry.

For some structures with circular, elliptical, and rectangular cross-section, analytical methods can be used to obtain closed-form solutions. For other waveguides, with more general cross-sections, numerical methods, such as the finite elements, must be used to find the modal solutions.

It will be assumed that a set of known modes with transverse fields is available in the form $(\boldsymbol{e}_{pi}, \boldsymbol{h}_{pi})$ where the subscript p refers to the mode and i is the section number. In section i of the

Figure 5.4 Prototype section for mode matching in waveguides.

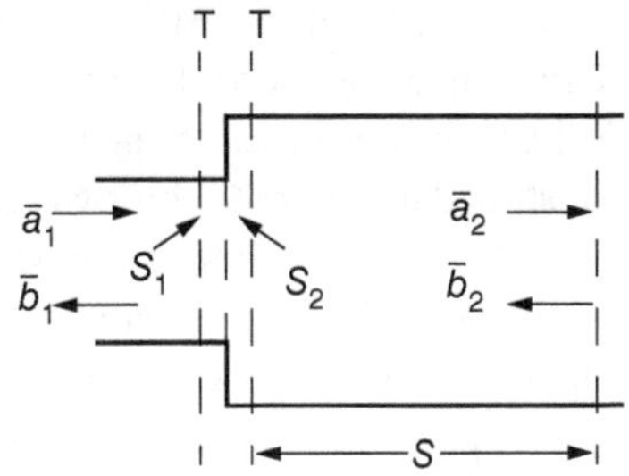

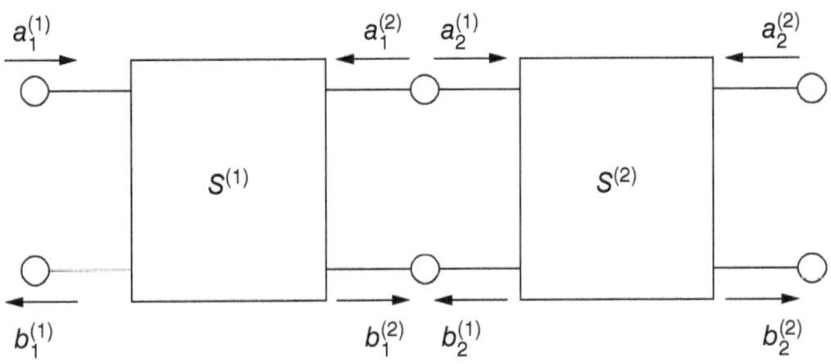

Figure 5.5 Concatenation of two scattering matrices.

horn, the transverse fields $\left(\boldsymbol{E}_t^{(i)}, \boldsymbol{H}_t^{(i)}\right)$ are approximated as a finite sum of $M(i)$ modes as given in Eqs. (5.10)–(5.13). At the junction of two sections at $z = 0$, continuity of the fields is enforced. As a result, two equations are formed, namely $\boldsymbol{E}_t^{(1)} = \boldsymbol{E}_t^{(2)}$ and $\boldsymbol{H}_t^{(1)} = \boldsymbol{H}_t^{(2)}$. The first of these is vector post-multiplied by $\boldsymbol{h}_{q1}$ and the latter post-multiplied by $\boldsymbol{e}_{q2}$. Both equations are then integrated across the junction in a weighted residuals solution. By taking Eq. (5.12) into account, the following is obtained:

$$\begin{bmatrix} \bar{\boldsymbol{b}}_1 \\ \bar{\boldsymbol{a}}_2 \end{bmatrix} = \begin{bmatrix} \boldsymbol{S}_{11} & \boldsymbol{S}_{12} \\ \boldsymbol{S}_{21} & \boldsymbol{S}_{22} \end{bmatrix} \begin{bmatrix} \bar{\boldsymbol{a}}_1 \\ \bar{\boldsymbol{b}}_2 \end{bmatrix} \tag{5.23}$$

where $\boldsymbol{S}_{11}$, $\boldsymbol{S}_{12}$, etc. are scattering parameters of the junction given by

$$\begin{aligned} \boldsymbol{S}_{11} &= -(\boldsymbol{I} + \boldsymbol{XY})^{-1} (\boldsymbol{I} - \boldsymbol{XY}), \\ \boldsymbol{S}_{12} &= 2(\boldsymbol{I} + \boldsymbol{XY})^{-1} \boldsymbol{X}, \\ \boldsymbol{S}_{21} &= 2(\boldsymbol{I} + \boldsymbol{YX})^{-1} \boldsymbol{Y}, \text{ and} \\ \boldsymbol{S}_{22} &= (\boldsymbol{I} + \boldsymbol{YX})^{-1} (\boldsymbol{I} - \boldsymbol{YX}) \end{aligned} \tag{5.24}$$

Further, $\boldsymbol{I}$ is the unit matrix, $\boldsymbol{X} = \boldsymbol{D}_1 \boldsymbol{C}^T \boldsymbol{D}_2^{-1}$ and $\boldsymbol{Y} = \boldsymbol{D}_2^{-1} \boldsymbol{C} \boldsymbol{D}_1 = \boldsymbol{X}^T$ where

$$D_{1,2} = \begin{bmatrix} Y_{11,12}^{1/2} & 0 & 0 & \cdots \\ 0 & Y_{21,22}^{1/2} & 0 & \cdots \\ 0 & 0 & Y_{31,32}^{1/2} & \cdots \\ \cdots & \cdots & \cdots & \cdots \end{bmatrix}.$$

and elements of the matrix $\boldsymbol{C}$ are given by

$$C_{pq} = \frac{1}{2} \iint_{S_1} dS \quad \boldsymbol{e}_{p2} \times \boldsymbol{h}_{q1} \cdot \hat{z}. \tag{5.25}$$

Two additional operations are used to analyze more than a single step in cross-section. The first is a uniform waveguide interconnection, which is treated by a reference plane extension. The second is the cascading of scattering matrices by concatenating them section-by-section. This cascading is illustrated in Figure 5.5 for two sections in scattering matrix notation.

Let $M(i)$ be the number of modes present at port i. To find the equivalent scattering matrix for a cascade of two two-port networks that have known scattering matrices $S^{(1)}$ and $S^{(2)}$, let the scattering matrix for network i be partitioned as follows:

$$\begin{bmatrix} \boldsymbol{b}_1^{(i)} \\ \boldsymbol{b}_2^{(i)} \end{bmatrix} = \begin{bmatrix} \boldsymbol{S}_{11}^{(i)} & \boldsymbol{S}_{12}^{(i)} \\ \boldsymbol{S}_{21}^{(i)} & \boldsymbol{S}_{22}^{(i)} \end{bmatrix} \begin{bmatrix} \boldsymbol{a}_1^{(i)} \\ \boldsymbol{a}_2^{(i)} \end{bmatrix} \tag{5.26}$$

When continuity is enforced between the networks

$$\boldsymbol{a}_2^{(1)} = \boldsymbol{b}_1^{(2)} \text{ and } \boldsymbol{a}_1^{(2)} = \boldsymbol{b}_2^{(1)},$$

the following is obtained for the cascade combination:

$$\boldsymbol{b}_1^{(1)} = \left(\boldsymbol{S}_{11}^{(1)} + \boldsymbol{S}_{12}^{(1)}\boldsymbol{\Delta}\boldsymbol{S}_{11}^{(2)}\boldsymbol{S}_{21}^{(1)}\right)\boldsymbol{a}_1^{(1)} + \boldsymbol{S}_{12}^{(1)}\boldsymbol{\Delta}\boldsymbol{S}_{12}^{(2)}\boldsymbol{a}_2^{(2)}$$

$$\boldsymbol{b}_2^{(2)} = \boldsymbol{S}_{21}^{(2)}\left(\boldsymbol{S}_{21}^{(1)} + \boldsymbol{S}_{22}^{(1)}\boldsymbol{\Delta}\boldsymbol{S}_{11}^{(2)}\boldsymbol{S}_{21}^{(1)}\right)\boldsymbol{a}_1^{(1)} + \left(\boldsymbol{S}_{22}^{(2)} + \boldsymbol{S}_{21}^{(2)}\boldsymbol{S}_{22}^{(1)}\boldsymbol{\Delta}\boldsymbol{S}_{12}^{(2)}\right)\boldsymbol{a}_2^{(2)}$$

where $\boldsymbol{\Delta} = \left(\boldsymbol{I} - \boldsymbol{S}_{11}^{(2)}\boldsymbol{S}_{22}^{(1)}\right)^{-1}$. As a result

$$\boldsymbol{S} = \begin{bmatrix} \boldsymbol{S}_{11}^{(1)} + \boldsymbol{S}_{12}^{(1)}\boldsymbol{\Delta}\boldsymbol{S}_{11}^{(2)}\boldsymbol{S}_{21}^{(1)} & \boldsymbol{S}_{12}^{(1)}\boldsymbol{\Delta}\boldsymbol{S}_{12}^{(2)} \\ \boldsymbol{S}_{21}^{(2)}\left(\boldsymbol{U} + \boldsymbol{S}_{22}^{(1)}\boldsymbol{\Delta}\boldsymbol{S}_{11}^{(2)}\right)\boldsymbol{S}_{21}^{(1)} & \boldsymbol{S}_{22}^{(2)} + \boldsymbol{S}_{21}^{(2)}\boldsymbol{S}_{22}^{(1)}\boldsymbol{\Delta}\boldsymbol{S}_{12}^{(2)} \end{bmatrix}. \tag{5.27}$$

Eq. (5.27) is the scattering matrix of concatenated Sections 1 and 2.

The representation of the external region of a radiating aperture has been described in Section 5.2.1. This is applicable to modeling of horns, of course, and is easily integrated with

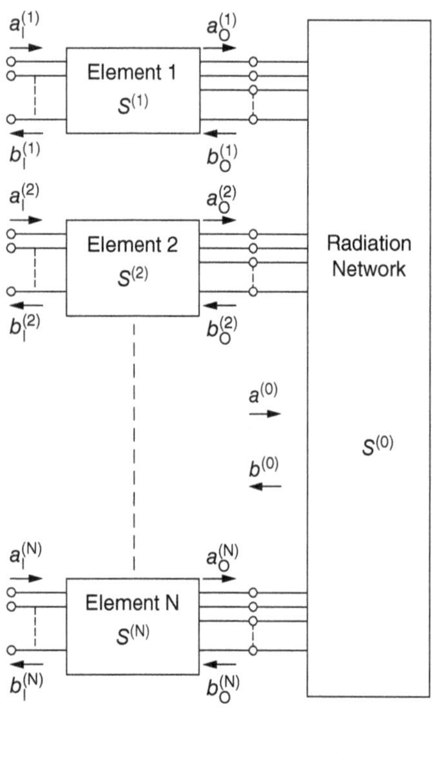

Figure 5.6 Network representation of a horn array.

the mode matching model for the internal fields using a mode scattering matrix in Eq. (5.20) designated here as $\boldsymbol{S}^{(0)}$, as will now be shown. A network model of an array of horns is illustrated in Figure 5.6. In this network model, the forward and reverse amplitudes of modes present in the aperture field as given in Eq. (5.20) and also in (2.47) are expressed here through

$$\boldsymbol{b}^{(0)} = \boldsymbol{S}^{(0)}\boldsymbol{a}^{(0)}. \tag{5.28}$$

where $\boldsymbol{a}^{(o)}$ and $\boldsymbol{b}^{(o)}$ are the amplitudes of incident and reverse travelling modes at the apertures.

A scattering matrix representing all the scattering matrices for elements 1 to N that are attached to the radiating aperture is given by

$$\begin{bmatrix} \boldsymbol{b}_I \\ \boldsymbol{a}_O \end{bmatrix} = \begin{bmatrix} \boldsymbol{S}_{11} & \boldsymbol{S}_{12} \\ \boldsymbol{S}_{21} & \boldsymbol{S}_{22} \end{bmatrix} \begin{bmatrix} \boldsymbol{a}_I \\ \boldsymbol{b}_O \end{bmatrix}, \tag{5.29}$$

where $\boldsymbol{a}_O^{(p)}$ and $\boldsymbol{b}_O^{(p)}$ are the amplitudes of incident and reflected modes at the aperture, which are identical to $\boldsymbol{a}^{(0)}$ and $\boldsymbol{b}^{(o)}$. Combining Eqs. (5.23) and (5.29), the reflection coefficient at the input and the complex amplitudes of the modes in the aperture is expressed in terms of the input excitation as follows:

$$\boldsymbol{b}_I = \left[\boldsymbol{S}_{11} + \boldsymbol{S}_{12}\boldsymbol{S}^{(0)}\left(\boldsymbol{I} - \boldsymbol{S}_{22}\boldsymbol{S}^{(0)}\right)^{-1}\boldsymbol{S}_{21}\right]\boldsymbol{a}_I \tag{5.30a}$$

$$\boldsymbol{a}^{(0)} = \left[\left(\boldsymbol{I} - \boldsymbol{S}_{22}\boldsymbol{S}^{(0)}\right)^{-1}\boldsymbol{S}_{21}\right]\boldsymbol{a}_I \tag{5.30b}$$

$$\boldsymbol{b}^{(0)} = \left[\boldsymbol{S}^{(0)}\left(\boldsymbol{I} - \boldsymbol{S}_{22}\boldsymbol{S}^{(0)}\right)^{-1}\boldsymbol{S}_{21}\right]\boldsymbol{a}_I \tag{5.30c}$$

For a horn array system, the reflected wave amplitudes are related to the complex incident wave amplitudes, as shown in Eq. (5.30a). Clearly, all coefficients in Eq. (5.30) depend on the matrix $(I - S_{22}S^{(0)})^{-1}$, which is almost unity for most arrays of reasonably well-matched horns. This matrix can have a significant influence on the mode amplitudes if this coupling is strong.

In a similar fashion to the radiation network in Section 2.4.3, this extended network is reflectionless when $\boldsymbol{b}_I = 0$ in Eq. (5.30a). This occurs at the smallest eigenvalue λ of

$$\left[S_{12}S^{(0)}\left(I - S_{22}S^{(0)}\right)^{-1}S_{21} + \lambda\, S_{11}\right]X_I = 0$$

when the input excitation is X_I.

5.2.4 Asymptotic Approximation of Mutual Admittance

In large arrays of apertures, the spacing between elements quickly increases. A rapid but accurate way of finding these admittances for widely spaced elements that will reduce computer time but also allow larger arrays to be analyzed is an advantage in many applications. As mentioned in Section 3.4.2.1, the free-space Green's function can be replaced by its Taylor series expansion, which allows an asymptotic approximation to the mutual admittance to be obtained. It is convenient to relate each local aperture coordinate system such as (x_i, y_i) shown in Figure 5.7 to the global array system (X, Y). Commencing with Eq. (5.21), the free-space Green's function is replaced by an asymptotic expansion to second order given by Eq. (3.29), allowing the mutual admittance to be expressed as

$$\begin{aligned} y_{ij}(p|q) \cong {} & \frac{jk\, G_o(R_{ij})}{\eta_0\sqrt{Y_{pi}Y_{qj}}}\int_{S_i} dS\, \Psi_{pi}(x_j, y_j) \cdot \int_{S_j} dS'\Psi_{qj}(x_i, y_i)\, exp(-jk\, v) \\ & \times \left[1 - \frac{1}{R_{ij}}\left(v + \frac{jk}{2}\Theta\right) + \frac{1}{R_{ij}^2}\left(v^2 + \left(jkv - \frac{1}{2}\right)\Theta - \frac{k^2}{8}\Theta^2\right)\right] \end{aligned} \tag{5.31}$$

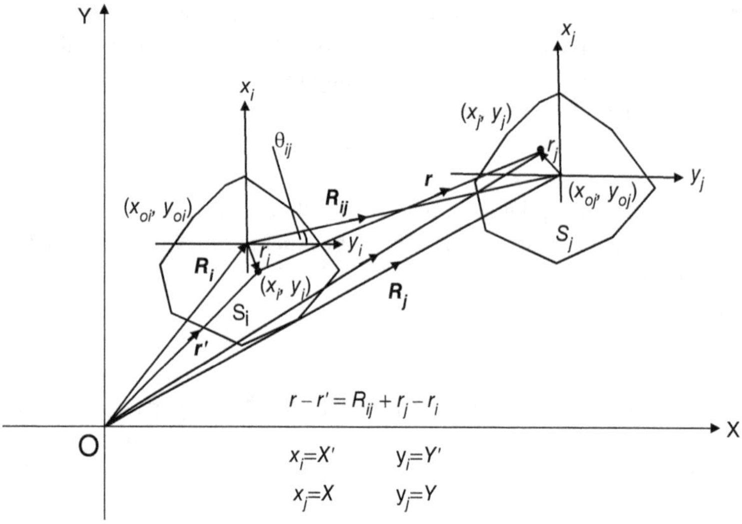

Figure 5.7 Geometry for asymptotic approximation in the X-Y plane.

where (x_i, y_i) are the local coordinates of aperture i, $R_{ij} = \sqrt{(x_{oj} - x_{oi})^2 + (y_{oj} - y_{oi})^2}$, $G(R_{ij}) = \exp(-jkR_{ij})/4\pi R_{ij}$, $\mathbf{\Delta} = \mathbf{r}_j - \mathbf{r}_i$, $v = \hat{\mathbf{R}}_{ij} \cdot \mathbf{\Delta}$, $\Theta = |\Delta|^2 - |\hat{\mathbf{R}}_{ij} \cdot \Delta|^2$, and $\hat{\mathbf{R}}_{ij} = [\hat{\mathbf{x}}(x_{oj} - x_{oi}) + \hat{\mathbf{y}}(y_{oj} - y_{oi})]/R_{ij} = \mathbf{x}\cos\theta_{ij} + \hat{\mathbf{y}}\sin\theta_{ij}$. With these definitions, $\mathbf{\Delta} = \hat{\mathbf{x}}(x_j - x_i) + \hat{\mathbf{y}}(y_j - y_i) = \hat{\mathbf{x}}(X - X') + \hat{\mathbf{y}}(Y - Y')$ and $v = \cos\theta_{ij}(X - X') + \sin\theta_{ij}(Y - Y')$, where (X, Y) is the field point and (X', Y') is the source point in the global coordinates. When $\boldsymbol{\Psi}_{pi}$ is expressed in terms of its rectangular components, then Eq. (5.31) becomes

$$y_{ij}(p|q) \cong \frac{jk\,G(R_{ij})}{\eta_o\sqrt{Y_{pi}Y_{qj}}}\left[I_{xx}(p,q) + I_{yy}(p,q) + \left(\frac{\gamma_{pi}}{k_o}\right)\left(\frac{\gamma_{qj}}{k_o}\right)I_{zz}(p,q)\right] \tag{5.32}$$

where

$$\begin{aligned} I_{\mu\nu}(p,q) = {} & u_{00}(p,q) - \frac{1}{R_{ij}}\left\{\cos\theta_{ij}\,u_{10}(p,q) + \sin\theta_{ij}\,u_{01}(p,q) + \frac{jk}{2}\left[\sin^2\theta_{ij}\,u_{20}(p,q)\right.\right. \\ & \left.\left. + \cos^2\theta_{ij}\,u_{02}(p,q) - \sin 2\theta_{ij}\,u_{11}(p,q)\right]\right\} + \frac{1}{R_{ij}^2}\left\{\left(\frac{3}{2}\cos^2\theta_{ij} - \frac{1}{2}\right)u_{20}(p,q)\right. \\ & + \left(\frac{3}{2}\sin^2\theta_{ij} - \frac{1}{2}\right)u_{02}(p,q) + \frac{3}{2}\sin 2\theta_{ij}\,u_{11}(p,q) + jk\left[\cos\theta_{ij}\sin^2\theta_{ij}\,u_{30}(p,q)\right. \\ & \left.\left. + \sin\theta_{ij}\cos^2\theta_{ij}\,u_{03}(p,q) + \cos\theta_{ij}\left(3\cos^2\theta_{ij} - 2\right)u_{12}(p,q)\right]\right\}, \end{aligned} \tag{5.33a}$$

and

$$\begin{aligned} u_{mn}(p,q) = {} & \iint_{S_j} dx_j dy_j \iint_{S_i} dx_i\,dy_i\,\exp\left[-jk\left(\cos\theta_{ij}(X - X') + \sin\theta_{ij}(Y - Y')\right)\right] \\ & \times (X - X')^m (Y - Y')^n \begin{Bmatrix} h_{xpi}h_{xqj} \\ h_{ypi}h_{yqj} \\ h_{zpi}h_{zqj} \end{Bmatrix} \end{aligned} \tag{5.33b}$$

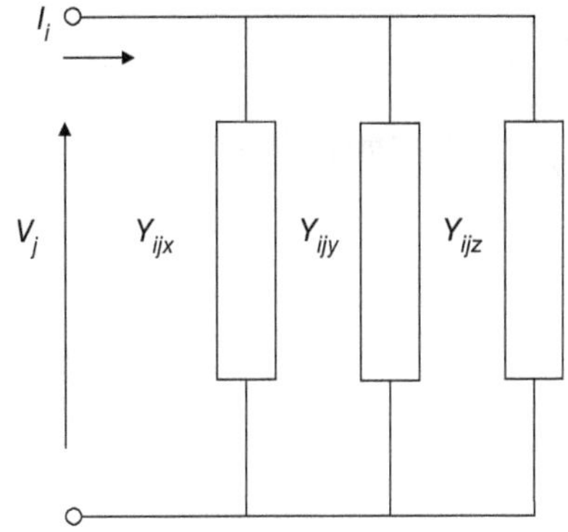

Figure 5.8 Network equivalent model of coupling between two rectangular apertures *i* and *j*.

Equation (5.33b) can be adapted for different geometries. One such geometry is for rectangular apertures, which will be given in Section 5.3, and coaxial apertures in Section 5.4.3. However, the form of Eqs. (5.32) and (5.33) prompts a general comment. The admittance consists of transform contributions from all three magnetic field components if they are nonzero. These give weighted moment contributions as shown by Eq. (5.33b) which increase with the order of the Taylor approximation. Apart from a scale factor, Eq. (5.32) has a network equivalent circuit shown in Figure 5.8 where the three admittances are given by $Y_{ijx} = I_{xx}$, $Y_{ijy} = I_{yy}$, and $Y_{ijz} = (\gamma_{pi}/k_o)(\gamma_{pj}/k_o)I_{zz}$.

5.3 Coupling in Rectangular Waveguides and Horns

In a direct approach to coupling in rectangular apertures, the expression for mutual admittance, Eq. (5.21), is evaluated using the vector mode function for the TE_{mn} modes. In an aperture with side lengths a_i and b_i, this vector mode function is

$$\Psi_{pi}(x,y) = \Psi_{opi}\left[\hat{x}\left(\frac{m\pi}{a_i}\right)\sin\left(\frac{m\pi}{a_i}x\right)\cos\left(\frac{n\pi}{b_i}y\right) + \hat{y}\left(\frac{n\pi}{b_i}\right)\cos\left(\frac{m\pi}{b_i}x\right)\sin\left(\frac{n\pi}{b_i}y\right)\right.$$
$$\left. - \hat{z}\left(\frac{j\,k_{cmn}^2}{k_o}\right)\cos\left(\frac{m\pi}{a_i}x\right)\cos\left(\frac{n\pi}{b_i}y\right)\right], \tag{5.34}$$

where $k_{cmn}^2 = \left(\frac{m\pi}{a_i}\right)^2 + \left(\frac{n\pi}{b_i}\right)^2$ and Ψ_{opi} is determined by the requirement that modal power is normalized through Eq. (5.12). This results in

$$\Psi_{opi} = \sqrt{\frac{2\,\epsilon_{om}\,\epsilon_{on}}{a_i b_i\,k_{c,mn}^2}}.$$

For TM_{mn} modes, the mode function expression is obtained from Eq. (5.34) by omitting the z-component and replacing the coefficient on the x-component by $\frac{na_i}{mb_i}$ and the y-component by $-\frac{mb_i}{na_i}$, where $m, n > 0$. The normalization factor Ψ_{opi} in the $\mathrm{TM_m}$ mode remains the same.

When these vector mode functions are substituted into Eq. (5.16), the result can be expressed as

$$y_{ij}\left(mn, m'n'\right) = \frac{jk^2\,\Psi_{opi}^2}{4\pi\,\gamma_{10}}\left[Y_{ijx}\left(mn, m'n'\right) + Y_{ijy}\left(mn, m'n'\right) + Y_{ijz}\left(mn, m'n'\right)\right] \tag{5.35}$$

where

$$
\begin{aligned}
Y_{ijx}(mn, m'n') &= \left(\frac{n\pi}{b_i}\right)\left(\frac{n'\pi}{b_j}\right)\int_0^{b_i} dy \sin\left(\frac{n\pi}{b_i}y\right)\int_0^{b_j} dy' \sin\left(\frac{n'\pi}{b_j}y'\right) \\
&\times \int_0^{a_i} dx \cos\left(\frac{m\pi}{a_i}x\right)\int_0^{a_j} dx' \cos\left(\frac{m'\pi}{a_i}x'\right) G\left(|r-r'|\right) \\
Y_{ijy}(mn, m'n') &= \left(\frac{m\pi}{a_i}\right)\left(\frac{m'\pi}{a_j}\right)\int_0^{b_i} dy \cos\left(\frac{n\pi}{b_i}y\right)\int_0^{b_j} dy' \cos\left(\frac{n'\pi}{b_j}y'\right) \\
&\times \int_0^{a_i} dx \sin\left(\frac{m\pi}{a_i}x\right)\int_0^{a_j} dx' \sin\left(\frac{m'\pi}{a_i}x'\right) G\left(|r-r'|\right) \\
Y_{ijz}(mn, m'n') &= -\left(\frac{k_{cmn}^2}{k_o}\right)\left(\frac{k_{cm'n'}^2}{k_o}\right)\int_0^{b_i} dy \cos\left(\frac{n\pi}{b_i}y\right)\int_0^{b_j} dy' \cos\left(\frac{n'\pi}{b_j}y'\right) \\
&\times \int_0^{a_i} dx \cos\left(\frac{m\pi}{a_i}x\right)\int_0^{a_j} dx' \cos\left(\frac{m'\pi}{a_i}x'\right) G\left(|r-r'|\right). \qquad (5.36)
\end{aligned}
$$

This format of the admittance Eq. (5.35) once again suggests a parallel circuit model for element i, j as shown in Figure 5.8. Each mode contributes to the x, y, and z admittances. In apertures where the mode set is transverse to any of the principal axes (eg. x), the relevant admittance is zero.

For the self-admittance, there are special convergence difficulties in the integrations due to the singularity in the scalar Green's function as the field point approaches the source point. This is overcome in the rectangular geometry, and the result is expressed in closed form using a transformation of variables based on a method originally due to Lewin (1951). This method can be extended to handle mode coupling in separate and different-sized rectangular apertures. In the latter case, integration of the four-dimensional integrals in (x, y) and (x′, y′) corresponding to the field and the source regions, respectively, is required, but this can be altered to an integration over two separate domains. These are the (x, x′) and the (y, y′) domains, which results in two surface integrals of the same form. Considering the (x, x′) domain, the integral that has to be evaluated is of the general form

$$
I_{xx'} = \int_0^{a_i} dx \int_0^{a_j} dx' \, B\left(x-x'\right) \begin{matrix}\cos\\ \sin\end{matrix}\left(\frac{m\pi x}{a_i}\right) \begin{matrix}\cos\\ \sin\end{matrix}\left(\frac{m'\pi x'}{a_j}\right) \qquad (5.37)
$$

where $B(x-x')$ is a function with integrable singularities due to the scalar Green's function, and a_i is the height of aperture i. Consider the new domains illustrated in Figure 5.9. Initially, let $\sigma = x - x'$ and $\lambda = y - y'$. As two other variables are required for this four-dimensional integration, let $\nu = x + x' - a_i$ and $\mu = y + y' - b_i$. Note that $dxdx' = d\sigma d\nu/2$. The original and new regions of integration are shown in Figure 5.9. Now, introduce the change of variable into the admittance formula and in Eq. (5.37). With these substitutions, the integral $I_{xx'}$ becomes

$$
\begin{aligned}
I_{xx'} = \pm\frac{1}{4}\Bigg\{ \int_0^{a_i} d\sigma &\begin{bmatrix} B(-\sigma)M_{\pm}\left(-\sigma, \sigma - a_i, 2a_j - a_i - \sigma, s_m, d_m\right) + \\ B\left(\sigma + a_i - a_j\right)M_{\pm}\left(\sigma + a_i - a_j, \sigma - a_j, a_j - \sigma, s_m, d_m\right) \end{bmatrix} \\
- \int_0^{a_i - a_j} &d\sigma \, B(\sigma)M_{\pm}(\sigma, \sigma - a_i, 2a_j - a_i + \sigma, s_m, d_m)\Bigg\} \qquad (5.38)
\end{aligned}
$$

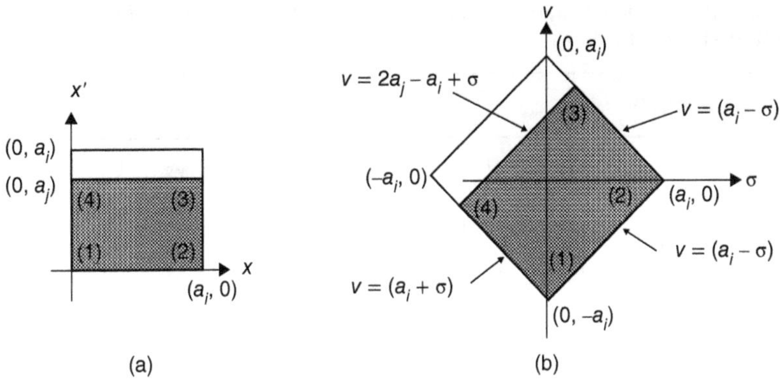

Figure 5.9 Regions of integration in analysis of mutual coupling in different-sized apertures. (a) Original domain; (b) domain after substitution of variables.

where $s_m, d_m = \frac{1}{2}\left(m \pm m'\frac{a_i}{a_j}\right)$ and

$$M_{\pm}\left(\sigma, v_1, v_2, s, d\right) = \frac{2a_i}{\pi}\left\{\frac{1}{s}\sin\left[\frac{s\pi}{2a_i}\left(v_1 - v_2\right)\right]\cdot\cos\left[s\pi\left(1 + \frac{\left(v_1 + v_2\right)}{2a_i}\right) + \frac{\sigma d\pi}{a_i}\right]\right.$$
$$\left.\pm\frac{1}{d}\sin\left[\frac{d\pi}{2a_i}\left(v_1 - v_2\right)\right]\cdot\cos\left[d\pi\left(1 + \frac{\left(v_1 + v_2\right)}{2a_i}\right) + \frac{\sigma s\pi}{a_i}\right]\right\}$$

The above integral reduction formula, Eq. (5.38), is also applicable to the (y, y′) domain with relevant substitutions, e.g. $a_i \to b_i$. With both integrals specified this way, the mutual admittance in Eq. (5.21) can be expressed as

$$y_{ij}\left(m, n, m', n'\right) = -\frac{jk\pi Y_o}{64}\alpha_{mn}\alpha_{m'n'}\left[\int_0^{a_j} d\sigma\left\{\int_0^{b_j} d\lambda[T(-\sigma, \sigma_1, \sigma_2| - \lambda, \lambda_1, \lambda_2) + \right.\right.$$
$$T\left(-\sigma, \sigma_1, \sigma_2|\lambda_3, \lambda_4, -\lambda_4\right) + T\left(\sigma_3, \sigma_4, -\sigma_4| - \lambda, \lambda_1, \lambda_2\right) + T\left(\sigma_3, \sigma_4, -\sigma_4|\lambda_3, \lambda_4, -\lambda_4\right)] +$$
$$\left.\int_0^{b_i - b_j} d\lambda[T\left(-\sigma, \sigma_1, \sigma_2|\lambda, \lambda_1, \lambda_2\right) + T(\sigma_3, \sigma_4, -\sigma_4|\lambda, \lambda_1, \lambda_2)]\right\} +$$
$$\int_0^{a_i - a_j} d\sigma\left\{\int_0^{b_j} d\lambda\left[T\left(\sigma, \sigma_1, \sigma_2| - \lambda, \lambda_1, \lambda_2\right) + T\left(-\sigma, \sigma_1, \sigma_2|\lambda_3, \lambda_4, -\lambda_4\right)\right] + \right.$$
$$\left.\left.\int_0^{b_i - b_j} d\lambda T\left(\sigma, \sigma_1, \sigma_2|\lambda, \lambda_1, \lambda_2\right)\right\}\right] \tag{5.39}$$

where

$$\sigma_1 = \sigma - a_i \qquad \sigma_2 = 2a_j - a_i - \sigma$$
$$\sigma_3 = \sigma + a_i - a_j \qquad \sigma_4 = \sigma - a_j$$
$$\lambda_1 = \lambda - b_i \qquad \lambda_2 = 2b_j - b_i - \lambda$$
$$\lambda_3 = \lambda + b_i - b_j \qquad \lambda_4 = \lambda - b_j$$

and

$$T\left(x, v_1, v_2|y, \mu_1, \mu_2\right) = G\left(X_{0i} - X_{0j} + x, Y_{0i} - Y_{0j} + y\right)$$
$$\times[c_x M_-\left(x, v_1, v_2, s_m, d_m\right) M_+\left(y, \mu_1, \mu_2, s_n, d_n\right) + c_y M_+\left(x, v_1, v_2, s_m, d_m\right) M_-\left(y, \mu_1, \mu_2, s_n, d_n\right)$$
$$+ c_z M_+\left(x, v_1, v_2, s_m, d_m\right) M_+\left(y, \mu_1, \mu_2, s_n, d_n\right)]$$

Table 5.1 Function $F(\sigma, \lambda, m, n, m', n')$ in integrand of Eq. (5.39) for the mutual admittance of modes in identical rectangular apertures (after Bird 1987).

Mode coupling	$F(\sigma, \lambda, m, n, m', n')$
$TE_{mn} \leftrightarrow TE_{m'n'}$	$\frac{mm'}{a^2}L_-(\sigma,a,s_m,d_m)L_+(\lambda,b,s_n,d_n) + \frac{nn'}{b^2}L_+(\sigma,a,s_m,d_m)L_-(\lambda,b,s_n,d_n) + \left(\frac{k_{c,mn}k_{c,m'n'}}{\pi k}\right)^2 L_+(\sigma,a,s_m,d_m)L_+(\lambda,b,s_n,d_n)$
$TM_{mn} \leftrightarrow TM_{m'n'}$	$\frac{nn'}{b^2}L_-(\sigma,a,s_m,d_m)L_+(\lambda,b,s_n,d_n) + \frac{mm'}{a^2}L_+(\sigma,a,s_m,d_m)L_-(\lambda,b,s_n,d_n)$
$TE_{mn} \leftrightarrow TM_{m'n'}$	$\frac{nm'}{ab}L_+(\sigma,a,s_m,d_m)L_-(\lambda,b,s_n,d_n) - \frac{mn'}{ab}L_-(\sigma,a,s_m,d_m)L_+(\lambda,b,s_n,d_n)$
$TM_{mn} \leftrightarrow TE_{m'n'}$	$\frac{mn'}{ab}L_+(\sigma,a,s_m,d_m)L_-(\lambda,b,s_n,d_n) - \frac{nm'}{ab}L_-(\sigma,a,s_m,d_m)L_+(\lambda,b,s_n,d_n)$
$L_\pm(x,a,p,q) = \cos\left[\pi\left(p+\frac{qx}{a}\right)\right]K(-x,a,p) \pm \cos\left[\pi\left(q+\frac{px}{a}\right)\right]K(-x,a,q)$	
$K(-x,a,p) = (a-x)\,S\left[p\left(1-\frac{x}{a}\right)\right];\ s_m = (m+m')/2\quad d_m = (m-m')/2\quad k_{cmn} = \sqrt{((m\pi)/a)^2 + (n\pi/b)^2}$	

In the special case of identical apertures, i.e. $a_i = a_j = a$ and $b_i = b_j = b$, the integrand simplifies to a single double integral, which is due to the first integral on the rightside of Eq. (5.39) as there are no contributions from the other integrals. The mutual admittance in that case is expressed as

$$y_{ij}(m,n|m'n') = \frac{jk\pi Y_o}{8}\alpha_{mn}\alpha_{m'n'}\left\{\left[\int_0^a d\sigma\left\{\int_0^b d\lambda\, F(\sigma,\lambda,m,n,m',n') \times [G(\sigma,\lambda)\right.\right.\right.$$
$$+ \cos m\pi \cos m'\pi G(-\sigma,\lambda) + \cos n\pi \cos n'\pi G(\sigma,-\lambda)$$
$$\left.\left.\left. + \cos m\pi \cos m'\pi \cos n\pi \cos n'\pi G(-\sigma,-\lambda)\right]\right\}\right\}. \tag{5.40}$$

The function $F(\sigma, \lambda, m, n, m', n')$ in the integrand of Eq. 5.40 is summarized in Table 5.1 for each mode combination (Bird 1987). Further simplification is possible for modes coupling in the same aperture, as will now be described.

5.3.1 Self-Admittance of TE_{10} Mode

The self-admittance of the TE_{10} mode at the aperture is required in all calculations of mutual coupling in open rectangular waveguide and horns. When the waveguide is flared into a horn, the admittance of the uniform guide is still applicable. The aperture impedance of horns has been discussed by Silvestro and Collin (1989). In some cases, the TE_{10} self-admittance can be a good first approximation to the practical value as other higher modes are below cut-off. Now return to Eq. 5.40 and consider the fundamental mode in the same aperture by letting $m = m' = 1$ and $n = n' = 0$. As a result of the simplifications obtained for same-sized apertures in Section 5.3, the self-admittance of the fundamental mode is

$$y_{11}(1,0) = \left(\frac{2j\pi k^2}{a^3 b\beta_{10} k_{c,10}^2}\right)\left[\int_0^a d\sigma \int_0^b d\lambda\, G(\sigma,\lambda) F(\sigma,\ \lambda, 1, 0, 1, 0),\right. \tag{5.41}$$

where

$$F(\sigma,\lambda,1,0,1,0) = (b-\lambda)\left[\left(1+\left(\frac{\pi}{ka}\right)^2\right)\left(\frac{a}{\pi}\right)\sin\left(\frac{\pi\sigma}{a}\right) + \left(1-\left(\frac{\pi}{ka}\right)^2(a-\sigma)\cos\left(\frac{\pi\sigma}{a}\right)\right)\right]$$

$G(\sigma,\lambda) = \exp\left(-jk\sqrt{\sigma^2+\lambda^2}\right)/\sqrt{\sigma^2+\lambda^2}$, $k_{c,10} = \pi/a$, and β_{10} is the TE_{10} propagation constant. The reflection coefficient is approximately obtained from Eq. (5.41) in the usual way from $\Gamma \approx (1-y_{11}(1,0))/(1+y_{11}(1,0))$. Equation (5.41) agrees with the result obtained by Lewin (1951, p. 126) for a waveguide with no dielectric filling. There is a singularity at the origin due to $G(\sigma,\lambda)$, but this can be eliminated by the use of polar coordinates as follows. Let $\sigma = t\cos\alpha$ and $\lambda = t\sin\alpha$, where $d\sigma d\lambda = tdtd\alpha$. Through this change of variable, Eq. (5.38) is expressed alternatively as

$$y_{11}(1,0) = \left(\frac{2j\pi k^2}{a^3 b\beta_{10} k_{c,10}^2}\right)\left\{\left(\int_0^{\alpha_o} d\alpha \int_0^{a\,sec\alpha} tdt + \int_{\alpha_o}^{\frac{\pi}{2}} d\alpha \int_0^{b\,cosec\alpha} t\,dt\right)\right.$$
$$\left.\times\ exp(-jkt)\,F(t\,cos\alpha, t\,sin\alpha, 1, 0, 1, 0)\right\}, \tag{5.42}$$

where $tan\,\alpha_o = b/a$. The inner integral with respect to the variable t can be integrated out in terms of elementary functions, but it is usually simpler to integrate Eq. (5.42) numerically.

The overall computation time is significant for most large arrays. Whatever the array configuration, by far, the greatest computational effort and time goes into evaluating the elements of the admittance matrix. Matrix inversion also takes significant computing time for large arrays with several modes in each element (see Section 3.6.1). In rectangular apertures, the admittance of modes coupling in the same waveguide can be obtained using one-dimensional numerical integration. For all other mode admittances, integration in two dimensions is needed. Numerical integration techniques have been described in Section 3.6.2. Others that work well on single and double integrals is a Simpson's rule algorithm with interval bisection until the error in the approximation of the integral is $<10^{-3}$ say. The method used for reducing the order of integration, i.e. Lewin's method, has wide application for removing singularities in rectangular regions. In the present context, with minor modification, it can be used in coupling problems involving different-sized rectangular subregions that occur, for example, in single or multiple apertures partially loaded with dielectric or anisotropic regions.

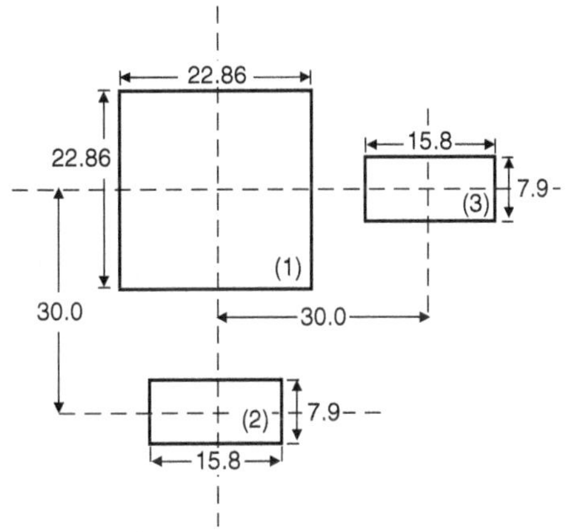

Figure 5.10 Three rectangular waveguides terminating in a ground plane. Waveguides (2) and (3) are WR62 waveguides. All dimensions are in millimeters.

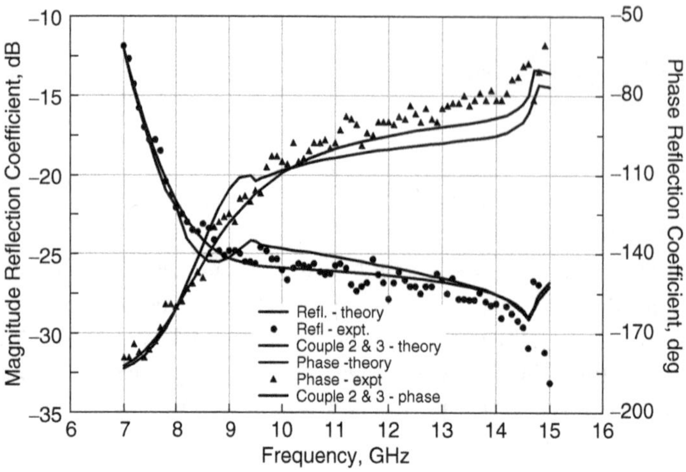

Figure 5.11 Computed reflection coefficient of TE_{10} mode in a single open-ended square waveguide with dimensions a = b = 2.286 cm (solid curves). Also shown (dashed curves) is the computed and measured reflection coefficient with neighboring WR16 waveguides #2 and #3 shown in Figure 5.10. Computed results are curves; experimental results are symbols.

As an example of the use of Eqs. (5.40) and (5.41), the reflection coefficient of an open-ended square waveguide, which is mounted in the center of a ground plane (see Figure 5.10), was computed with 10 modes in the aperture giving the results shown in Figure 5.11. The dimensions of the square waveguide are a = b = 2.286 cm, in which the fundamental mode has a cut-off frequency of 6.557 GHz. This waveguide corresponds to waveguide no. 1 in Figure 5.10 with the smaller nearby apertures shorted out. When these shorts were removed, the reflection coefficient in waveguide no. 1 was repeated to see the effect of the adjacent apertures. Both situations were simulated with 10 modes in each aperture, and these results are shown in Figure 5.11. The reflection coefficient was measured with an experimental model and with aperture shorts in place, and this is also shown in Figure 5.11. Both measurements were made with an early computer-controlled vector analyzer, which had a noise floor at about −40 dB down, and as a result, the results show greater variation at low levels of reflection or transmission. Calibration of the apertures was obtained using sliding loads, a short and offset short, which gave the measurement plane for both reflection and transmission measurements at the ground plane surface.

As can be seen from Figure 5.11, the coupling to the nearby waveguides has an impact of the reflection coefficient of the central waveguide, especially near the onset of the TE_{10} mode of the smaller WR62 waveguides near 9.47 GHz and also the onset of TE_{11} and TM_{11} modes in square waveguide near 9.27 GHz, which is not evident when the adjacent waveguides are absent. At the upper end of the frequency range near 14.63 GHz, the effect of the cut-off of the TE_{12} and TM_{12} modes in square waveguide can also be seen.

5.3.2 Example of Mutual Coupling Between Different-Sized Waveguides

The mutual coupling analysis has been verified for different-sized waveguides, and some results are given by Bird (2016). The techniques described earlier in this section were applied to the three-element array in Figure 5.10, and they were compared with experimental data. Results for the reflection coefficient of the central square element have already been described in Subsection 5.3.1. Mutual coupling results are now given for coupling from the square waveguide to the smaller WR62 waveguides (2) and (3) and between the two WR62 waveguides in the presence of a matched square waveguide. The computed and experimental results are shown in Figure 5.12.

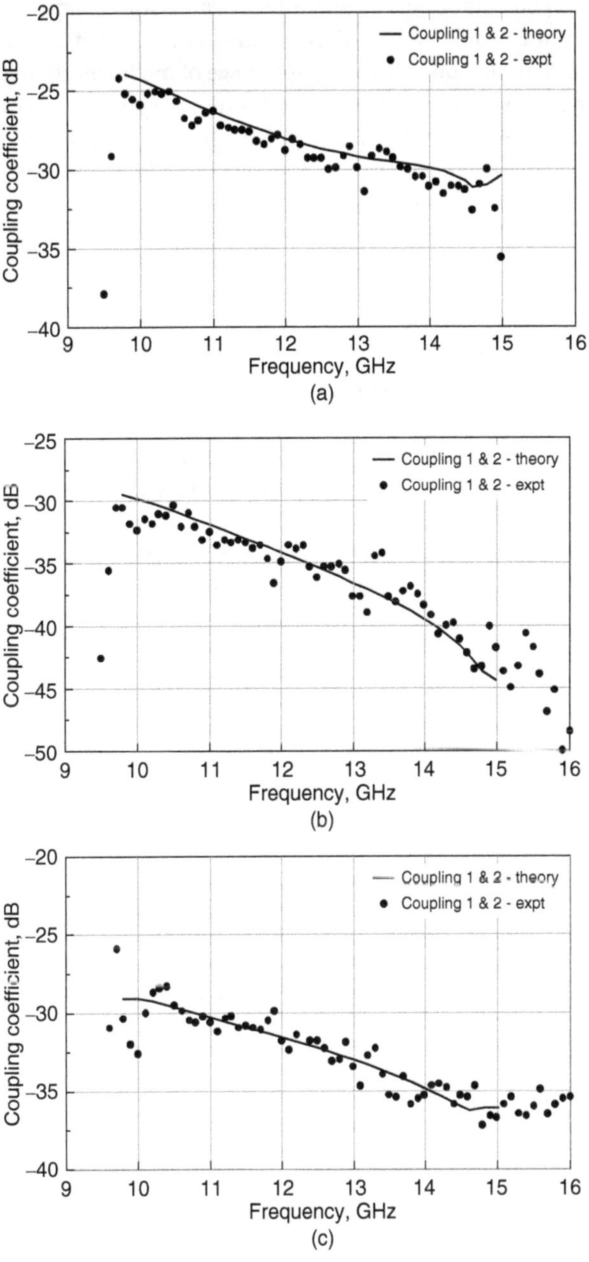

Figure 5.12 Coupling coefficient of the TE_{10} modes in different-sized rectangular waveguides as shown in Figure 5.9. Waveguide (a) (1)–(2) E-plane case; (b) (1)–(3) H-plane case; and (c) (2)–(3). Solid line is theory, and dots are experimental values Source: Modified from BIRD, T.S. (1990a), 'Analysis of mutual coupling in arrays of different-sized rectangular waveguides', IEEE Trans. Antennas Propagat., vol. AP-38, pp. 166–172. [Erratum AP-38, 1727].

It is seen from Figure 5.12 that there is good agreement between theory and experiment. The impact of the onset of the TE_{12} in square waveguide at 14.63 GHz mentioned in Section 5.3.1 is present in the coupling coefficient as well. This demonstrates the importance of mode selection in the field representation, and this depends on symmetry as well as mode order. Some modes have only second-order effects. However, if the incident mode in one waveguide is the TE_{mn} mode where m is even and n is odd, the selected modes should include the first few higher modes with m even and n odd as well as the first few modes in order of increasing cut-off frequency.

5.3.3 Application to Horns

The theory is now applied to arrays of rectangular horns. In a first example, results are obtained for a pair of rectangular horns rotated relative to each other, as shown in Figure 5.13. In this case, the reduction outlined at the beginning of this section cannot be used on this occasion. Therefore, we return to the original formulation where four-dimensional integration is required for the mutual coupling calculations between elements that are rotated relative to each other. One pair of integrals represents the transmit radiation, and the second pair corresponds to the aperture of the receiving horn.

If rectangular apertures are located on a plane surface but on coordinate systems that are rotated at angles α_i and α_j about centers (X_{ci}, Y_{ci}) and (X_{cj}, Y_{cj}) in the global rectangular coordinate system (X,Y) as shown in Figure 5.13, the mutual admittance can be expressed as

$$\begin{aligned} y_{ij}\left(m, n|m', n'\right) = \frac{jk_o Y_o}{4} A_{mn} A_{m'n'} \left[c_{zi}c_{zj}I_{zz} + \left(c_{xi}c_{xj}I_{xx} + c_{yi}c_{yj}I_{yy}\right)\cos\left(\alpha_i - \alpha_j\right)\right. \\ \left. + \left(c_{xi}c_{yj}I_{xy} - c_{yi}c_{xj}I_{yx}\right)\sin\left(\alpha_i - \alpha_j\right)\right], \end{aligned} \tag{5.43}$$

where

$$A_{mn} = \sqrt{\frac{2\,\epsilon_{om}\epsilon_{on}}{a_i\, b_i Y_{mn} k^2_{c,mn}}}$$

and

$$\left\{\begin{array}{c} I_{xx} \\ I_{yy} \\ I_{zz} \\ I_{xy} \\ I_{yx} \end{array}\right\} = \int_{D_i} dS \int_{D_j} dS'\, g_o(s) \begin{array}{c} \sin \\ \cos \\ \cos \\ \sin \\ \cos \end{array} \left(\frac{m\pi x}{a_i}\right) \begin{array}{c} \cos \\ \sin \\ \cos \\ \cos \\ \sin \end{array} \left(\frac{n\pi y}{b_i}\right) \begin{array}{c} \sin \\ \cos \\ \cos \\ \cos \\ \sin \end{array} \left(\frac{m\pi x'}{a_j}\right) \begin{array}{c} \cos \\ \sin \\ \cos \\ \sin \\ \cos \end{array} \left(\frac{n\pi y'}{b_j}\right) \tag{5.44}$$

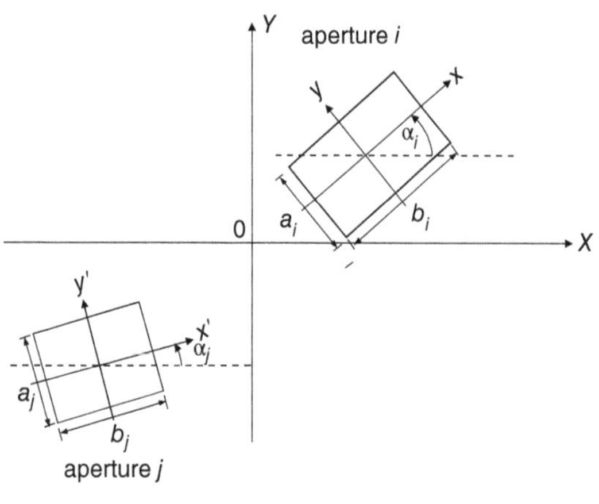

Figure 5.13 Geometry of rotated apertures in the (X, Y) plane.

where

$$\varepsilon_{0p} = \begin{cases} 1; p = 0 \\ 2; p > 0 \end{cases},$$

$$g_o(s) = \exp(-jks)/s$$

$$s = \sqrt{(X - X')^2 + (Y - Y')^2}$$

$$X = \left(x - a_i/2\right)\, cos\alpha_i - \left(y - b_i/2\right)\, sin\alpha_i + X_{ci}$$

and

$$Y = \left(x - a_i/2\right)\, sin\alpha_i + \left(y - b_i/2\right)\, cos\,\alpha_i + Y_{ci}$$

The coefficients c_{xi}, c_{yi}, etc. are listed in Table 5.2.

Modeling of rotated arrays of rectangular horns should be treated carefully since the use of four-dimensional integration in the mutual coupling calculations can result in a large increase in computer time. In the array software, called RECAR (Bird and Bateman 1992), Eq. 5.43 was used for coupling in different-sized rectangular horns when the rotation angles of i and j did not satisfy $\left|\alpha_i - \alpha_j\right| < \frac{\lambda\pi}{100\,\max(a_i, b_i)}$, where function max(...,...) is the maximum value of its arguments, otherwise Eq. 5.35 was used. However, arrays containing only a few rotated elements can still be analyzed reasonably quickly.

The two-element array in Figure 5.14 was analyzed using a range of relative rotation $|\alpha_i - \alpha_j|$ angles. Coupling between the TE_{10} mode in aperture 1 and the TE_{10} and TE_{01} modes in aperture 2 is given in Figure 5.15 for E- and H-plane coupling. Results computed from mode matching in the horn and mode coupling in the aperture given by Eq. (5.43) were used in conjunction with Eq. (5.30). These were compared with the measured results of Lyon et al. (1966) as shown in the figure. As expected, physically, the orthogonal modes are equally excited when the rotation angle is 45°, and at an angle of 90°, the TE_{01} mode only is coupled. A more detailed description of the methods used and some other practical rotated horn examples are given in Bird and Bateman (1994). The importance of including modes other than the fundamental TE_{10} mode in the calculations has been noted previously, particularly when these modes are above cut-off. Calculations for reflection coefficient, E-plane coupling coefficients, and cross-polar radiated fields are especially sensitive (Figure 5.15).

As a further example of an application of mutual coupling analysis to horns, the radiation pattern of a five-element array of stepped rectangular horns was computed, and the results were compared with measurement. The horns have approximately equal length but have different-sized apertures and step lengths. The geometry of these horns is defined in Figure 5.16 and Table 5.3. All horns

Table 5.2 Coefficients for the mutual admittance in Eq. (5.43) for rotated aperture *i* in a ground plane.

Mode	c_{xi}	c_{yi}	c_{zi}
TE_{mn}	$\left(\frac{m}{a_i}\right)$	$\left(\frac{n}{b_i}\right)$	$\left(\frac{k_{c,mn}}{\pi k_o}\right)$
TM_{mn}	$-\left(\frac{n}{b_i}\right)$	$\left(\frac{m}{a_i}\right)$	0

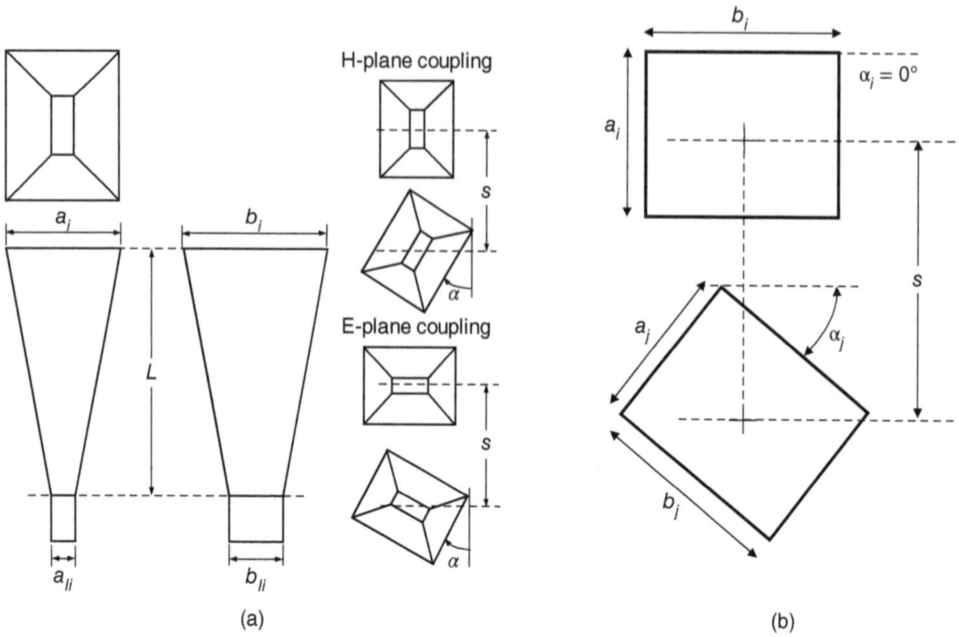

Figure 5.14 Two rotated pyramidal horns. (a) Dimensions are: $a_i = 22.8$ mm, $b_i = 60.4$ mm, $a_{li} = 10.1$ mm, $b_{li} = 47.7$ mm, $L = 101$ mm, and $s = 366$ mm and (b) geometry $a_j = a_i$ and $b_j = b_i$.

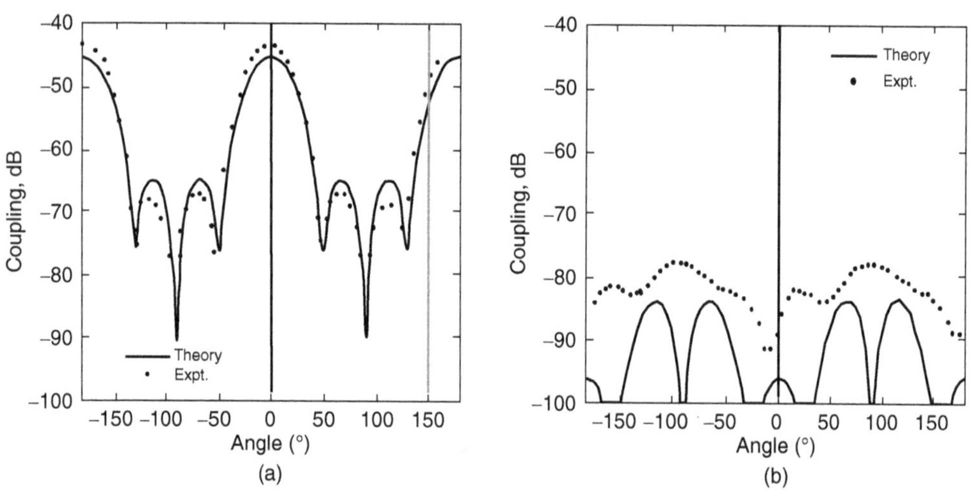

Figure 5.15 Coupling coefficient as a function of rotation angle for two pyramidal horns shown in Figure 5.14 that operate at 9.03 GHz. (a) E-plane; (b) H-plane. Solid line – theory; dot – experiment Source: BIRD, T.S. & BATEMAN, D.G. (1994): Mutual coupling between rotated horns in a ground plane, IEEE Trans. Antennas Propagat., vol. AP-42, pp. 1000–1006.

except the driven element (no. 5) have a 6.0 mm input section connecting the input to the start of the horn. In the array, the horns were mounted in a 500 mm × 500 mm ground plane and configured as illustrated in Figure 5.16a.

The analysis was performed by mode matching using 26 modes in each horn using the documented program RECAR (Bird and Bateman 1992). In this analysis, the first waveguide section

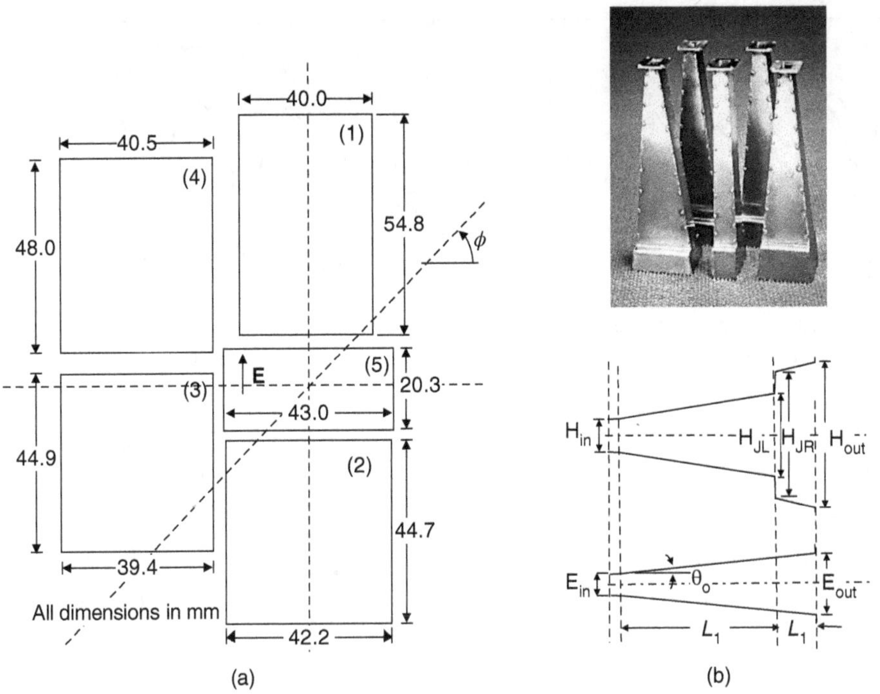

Figure 5.16 Geometry of five-element array of H-plane stepped horns. (a) Front view dimension in mm; (b) horn parameters. Source: Trevor S. Bird

Table 5.3 Dimensions of horns in millimeters in the five-element array in Figure 5.16.

Horn no.	Excitation	H_{out}	E_{out}	H_{JL}	H_{JR}	L_1	L_2	X_c	Y_c
1	0	40.00	54.82	20.65	40.00	158.60	19.20	0.00	43.00
2	0	42.20	44.66	24.33	40.69	153.00	25.50	0.00	−38.00
3	0	39.43	44.90	22.99	39.20	156.70	21.60	−47.00	19.00
4	0	40.50	48.00	26.80	40.50	148.30	30.60	−47.00	33.00
5	1	43.00	20.30	24.49	40.98	157.60	26.82	0.00	0.00

Input waveguide: $H_{in} = 19.0$, $E_{in} = 9.45$.
Modes: TE10, TE30, TE12, M12, TE11, TM11, TE01, TE50, TE32, TM32, TE14, TM14, TE70, TE52, TM52, TE34, TM34, TE16, TM16, TE90, TE02, TE20, TE21, TM21, TE22, TM22.
All horns except no. 5 have 6 mm input section connecting to the input before profile.

is the input of the horn, the final one is the aperture, and the ground plane is assumed infinite. Before combining the horns into an array, the reflection coefficients of the isolated horns were measured, and the results are plotted in Figure 5.17. There is mostly good agreement between the results, although there are some discrepancies for horn 5, which, with the smallest aperture, seems dependent on conditions at the aperture plane and size of the flange as shown in Figure 5.17e.

For the array radiation test, only element 5 in Figure 5.16 is excited while the remaining elements are terminated in matched loads. The co- and cross-polar radiation patterns in the E- and H-plane of the five-element array at 11.7 GHz are shown in Figure 5.18. Element 5 is located in

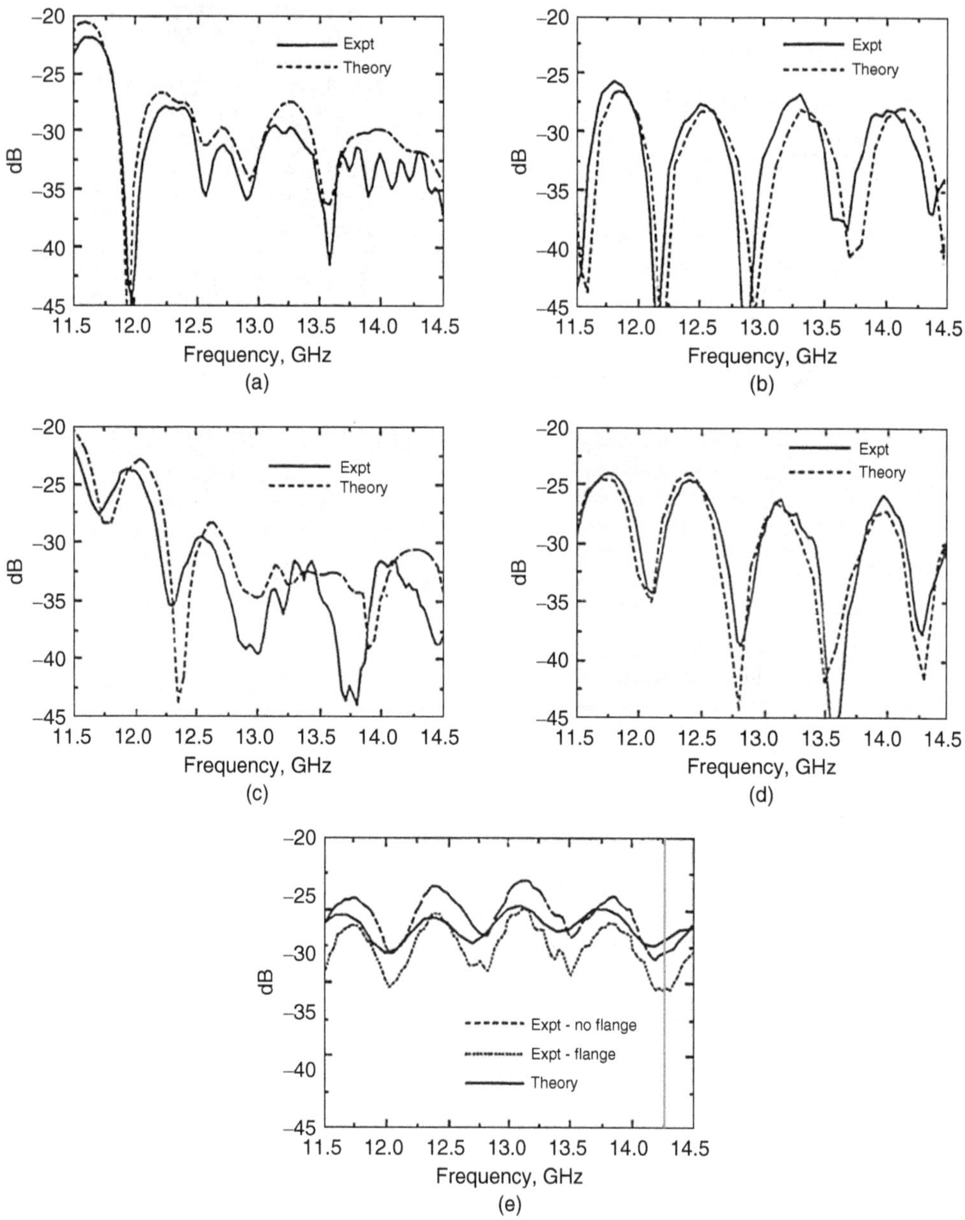

Figure 5.17 Reflection coefficients of the isolated horns of the five-element array, theory, and experiment. (a) Horn 1; (b) horn 2; (c) horn 3; (d) horn 4; and (e) horn 5.

the center of the ground plane, and ideally, the isolated horn would have zero cross-polarization in the principal planes. The results are in general agreement for the copolar patterns except in the E-plane where diffraction from the ground plane is significant. Figure 5.18 also shows significant cross-polarization in these planes, and this is attributed to high-order modes due to the steps and mutual coupling between the adjacent elements as well as some diffraction from the edges of the ground plane. It was found that that higher-order modes excited by the array, particularly the $TE_{11}$and TM_{11} modes, are important for predicting the cross-polarized field, although their effect

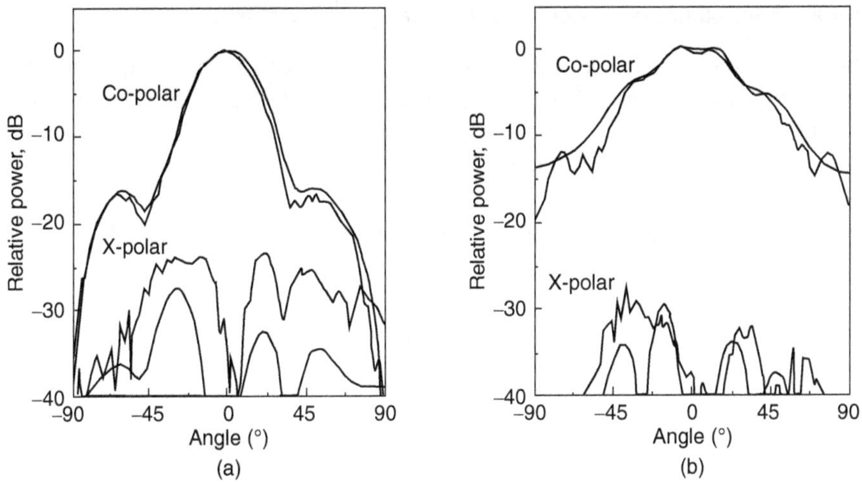

Figure 5.18 Principal plane radiation patterns of element (no. five excited) in the five-element array (Figure 5.16) at 11.7 GHz. (a) E-plane; (b) H-plane. Measured: solid curve; computed: dashed curve.

on the copolar field is relatively small over the main beam. Schennum et al. (1978) showed that by exciting higher-order modes in the aperture of a rectangular horn, the E- and H-plane patterns can be equalized, and the mutual coupling was reduced. In the present case, there is good agreement between theory and experiment for the basic structure of all patterns, although the measured cross-polar levels are generally slightly greater than the computed ones, due possibly to the horn fabrication, which have minor irregularities between the input and output faces that may have caused additional cross-polarization.

5.3.4 Waveguide-Fed Slot Arrays

The theory outlined above has been used to analyze mutual coupling in waveguide-fed slot arrays (Bird 2016). These arrays are generally of two types: standing-wave fed, where adjacent slots are one-half guide wavelength apart, and traveling-wave fed. Whatever the feeding mechanism, the analysis for these sideways-fed apertures is similar and tends to have greater complexity than the analysis given above for apertures fed directly by the waveguide, because coupling must be considered between the interior waveguide and the exterior region via a ground plane of finite thickness. A way of doing this with the current formulation has been described by the author (Bird 2016). Single waveguide-fed slots have been analyzed by several workers including Stevenson (1948), Oliner (1957), Vu Khac and Carson (1973), Quintez and Dudley (1976), Butler and Yung (1979), Lyon and Sangster (1981), and Stern and Elliott (1985). Vu Khac and Carson (1973) describe a moment method for solving the integral equations governing the coupling between the internal and exterior regions and determined the characteristics of a longitudinal slot in a rectangular waveguide. Lyon and Sangster (1981) extended this method to thick-walled rectangular waveguide. The case of an array of slots is considered by Stern and Elliott (1985) where the problem is treated by modeling the slots as center-fed dipoles. The moment method is used to calculate the coupling between two slots in the broad wall by Shan-Wei et al. (1985). They applied Galerkin's method to the equations of continuity derived from the tangential components of the magnetic fields, and sinusoidal basis functions were used for the magnetic currents in the slots. Excellent agreement with experiment was obtained with this approach.

5.3.5 Asymptotic Approximation of Coupling in Rectangular Apertures

The asymptotic formula given for general apertures in Section 5.2.4 can be readily applied to the rectangular shapes. In particular, the TE and TM mode functions in Eq. (5.34) can be substituted directly into Eqs. (5.32) and (5.33). As a result, Eq. (5.44) is approximated by the $u_{\mu\nu}(p, q)$ function, Eq. (5.33b), for these mode functions as follows:

$$
\begin{aligned}
u_{\mu\nu}(p,q) = \iint_{S_j} dx_j dy_j \iint_{S_i} dx_i\, dy_i\, exp\left[-jk\left(cos\theta_{ij}\left(X-X'\right)+sin\theta_{ij}\left(Y-Y'\right)\right)\right] \\
\times \left(X-X'\right)^{\mu}\left(Y-Y'\right)^{\nu}\left\{\begin{matrix} sin \\ cos \\ cos \\ sin \\ cos \end{matrix}\left[\frac{m\pi}{a_i}\left(x_i+\frac{a_i}{2}\right)\right] \begin{matrix} cos \\ sin \\ cos \\ cos \\ sin \end{matrix}\left[\frac{n\pi}{b_i}\left(y_i+\frac{b_i}{2}\right)\right]\right\} \\
\times \left\{\begin{matrix} sin \\ cos \\ cos \\ cos \\ sin \end{matrix}\left[\frac{m'\pi}{a_j}\left(x_i+\frac{a_j}{2}\right)\right] \begin{matrix} cos \\ sin \\ cos \\ cos \\ sin \end{matrix}\left[\frac{n'\pi}{b_j}\left(y_i+\frac{b_j}{2}\right)\right]\right\}
\end{aligned}
\tag{5.45}
$$

In aperture i, mode p has indices mn, while in aperture j, mode q has indices $m'n''$. An advantage of Eq. (5.45) is that integrals in it can be evaluated in closed form. The result consists of sums of permutations of functions that are Fourier transforms of powers of coordinates and trigonometric functions. These transforms are recursive functions and, therefore, can be calculated very quickly.

Commencing with some initial definitions, let

$$
\begin{aligned}
l_x = k\cos\left(\theta_{ij}-\alpha_i\right), l_y = k\sin\left(\theta_{ij}-\alpha_i\right), l'_x = k\cos\left(\theta_{ij}-\alpha_j\right), l'_y = k\sin\left(\theta_{ij}-\alpha_j\right), \\
k_x = m\pi/\alpha_i, k_y = n\pi/b_i, k'_x = m'\pi/\alpha_j, \text{and } k'_y = n'\pi/b_j.
\end{aligned}
$$

Next, define the quadruple product

$$
M(s,t,m,n) = L_{\alpha}^{(s)}\left(-l_x,k_x\right).L_{\beta}^{(t)}\left(-l_y,k_y\right).L_{\alpha'}^{(m)}\left(l'_x,k'_x\right).L_{\beta'}^{(n)}\left(l'_y,k'_y\right). \tag{5.46}
$$

Here, the subscripts α, β, α', and β' correspond to + or − arguments depending on whether the function being integrated is cosine or sine, respectively, and

$$
\begin{aligned}
L_{\pm}^{(n)}(l,k) &= \int_{-a/2}^{a/2} dx\, x^n\, exp\left[jl\left(x+\frac{a}{2}\right)\right] \times \left\{\begin{matrix} cos \\ sin \end{matrix} kx\right. \\
&= \frac{1}{2}\left[exp\left(\frac{jla}{2}\right) J^{(n)}(l,k) \pm exp\left(-\frac{jla}{2}\right) J^{(n)}(l,-k)\right]\left\{\begin{matrix} 1 \\ -j \end{matrix}\right.,
\end{aligned}
\tag{5.47}
$$

where

$$
\begin{aligned}
J^{(n)}(l,k) &= \frac{1}{2}\int_{-a/2}^{a/2} dx\, x^n \exp\left[j(l+k)x\right] \\
&= \frac{1}{j(l+k)}\left\{\begin{matrix} \left(\frac{a}{2}\right)^n cos\left[(l+k)a/2\right] - n\,J^{(n-1)}(l,k);\, n\ odd \\ j\left(\frac{a}{2}\right)^n sin\left[(l+k)a/2\right] - n\,J^{(n-1)}(l,k);\, n\ even \end{matrix}\right.,
\end{aligned}
\tag{5.48}
$$

which has a base function

$$J^{(0)}(l,k) = \frac{sin\left[(l+k)\,a/2\right]}{(l+k)}.$$

It can be shown that for a given pq mode pair,

$$u_{\nu\mu}\left(sin\alpha_i, -cos\alpha_i, sin\alpha_j, -cos\alpha_j\right) = u_{\mu\nu}\left(cos\alpha_i, sin\alpha_i, cos\alpha_j, sin\alpha_j\right). \tag{5.49}$$

The zero-th function of the sequence of transform functions given in Eq. (5.46) is

$$u_{00} = M(0,0,0,0). \tag{5.50a}$$

The remaining functions contain the rotation angles of the apertures α_i and α_j and represent the transformation from the local aperture coordinate system to the array. The next five functions in the sequence in increasing order are

$$u_{10} = M(1,0,0,0)\ cos\alpha_i - M(0,1,0,0)\ sin\alpha_i - M(0,0,1,0)\ cos\alpha_j + M(0,0,0,1)\ sin\alpha_j \tag{5.50b}$$

$$u_{01} = M(1,0,0,0)\ sin\alpha_i + M(0,1,0,0)\ cos\alpha_i - M(0,0,1,0)\ sin\alpha_j - M(0,0,0,1)\ cos\alpha_j \tag{5.50c}$$

$$\begin{aligned} u_{11} = {} & [M(2,0,0,0) - M(0,2,0,0)]\ cos\alpha_i sin\alpha_i + [M(0,0,2,0) - M(0,0,0,2)]\ cos\alpha_j sin\alpha_j \\ & + M(1,1,0,0)\ cos2\alpha_i + M(0,0,1,1)\ cos2\alpha_j - [M(1,0,0,1) + M(0,1,1,0)] \\ & \times cos\left(\alpha_i + \alpha_j\right) - [M(1,0,1,0) - M(0,1,0,1)]\ \sin\left(\alpha_i + \alpha_j\right) \end{aligned} \tag{5.50d}$$

$$\begin{aligned} u_{20} = {} & M(2,0,0,0)\ cos^2\alpha_i + M(0,2,0,0)\ sin^2\alpha_i + M(0,0,2,0)\ cos^2\alpha_j + M(0,0,0,2)\ sin^2\alpha_i \\ & - M(1,1,0,0)\ sin2\alpha_i - M(1,0,1,0)\ 2\,cos\alpha_i cos\alpha_j + M(0,1,1,0)\ 2\,sin\alpha_i cos\alpha_j \\ & - M(0,1,0,1)\ 2\,sin\alpha_i sin\alpha_j - M(0,0,1,1)\ sin2\alpha_j \end{aligned} \tag{5.50e}$$

$$\begin{aligned} u_{02} = {} & M(2,0,0,0)\ sin^2\alpha_i + M(0,2,0,0)\ cos^2\alpha_i + M(0,0,2,0)\ sin^2\alpha_j + M(0,0,0,2)\ cos\alpha_i \\ & + M(1,1,0,0)\ sin2\alpha_i - M(1,0,1,0)\ 2\,sin\alpha_i sin\alpha_j - M(0,1,1,0)\ 2\,cos\alpha_i sin\alpha_j \\ & - M(0,1,0,1)\ 2\,cos\alpha_i cos\alpha_j + M(0,0,1,1)\ sin2\alpha_j \end{aligned} \tag{5.50f}$$

Higher-order functions than Eq. (5.50) can be generated using Eqs. (5.46)–(5.49). In summary, the functions listed in Eq. (5.50) enable a second-order asymptotic approximation to the original admittance Eq. (5.32) to be made through Eq. (5.33b) and Eq. (5.45).

Table 5.4 Coupling coefficients of TE_{10} and TE_{01} modes in two 0.6λ square waveguides centered one wavelength apart as a function of rotation angle of waveguide #2.

α deg.	$(TE_{10})_1 \leftrightarrow (TE_{10})_2$		$(TE_{01})_1 \leftrightarrow (TE_{01})_2$	
	Exact dB/deg.	Approx. dB/deg	Exact dB/deg	Approx. dB/deg
0°	−29.10/−6.12°	−29.2/−5.98°	0.	0.
10°	−29.39/−5.26°	−29.48/−5.56°	−46.26/−139.06°	−47.83/−115.4°
20°	−30.25/−2.52°	−30.34/−3.91°	−40.63/−143.91°	−42.46/−125.1°
30°	−31.60/2.41°	−31.86/0.04°	−37.53/−151.64°	−39.42/−140.95°
40°	−33.31/9.74°	−34.04/7.99°	−35.27/−161.08°	−36.69/−158.53°
45°	−34.27/14.17°	−35.33/14.02°	−34.27/−165.83°	−35.33/−165.98°

Initially, the waveguides are aligned along their E-planes ($\alpha = 0$).

As an example of the accuracy that is possible, Table 5.4 shows the computed coupling coefficients of TE_{10} and TE_{01} modes in two 0.6λ square waveguides aligned initially in their E-planes as waveguide #2 is rotated by an angle α. The center-to-center spacing is one wavelength. At 45°, the modes become mirror images even with a 180° phase change. There is excellent agreement between the asymptotic approximation and direct numerical integration (exact).

5.3.6 Coupling in Horns Approximated with Quadratic Phase

Flaring into pyramidal horns is often approximated by a quadratic phase factor instead of following through with a full mode matching analysis as was done in Sec. 5.3.3. This approximation works remarkably well for radiation providing the flare angles are not too great (Bird 2016). In this approach, a quadratic phase is included in both principal directions of the E and H planes of the horn, which have flare angles θ_E and θ_H respectively. An exponential factor is inserted in the mode functions to approximate the quadratic phase distribution across the aperture as follows (Bird, 2016, p. 76):

$$exp\left[\frac{-jk}{2}\left(\frac{x^2}{\rho'_E}+\frac{y^2}{\rho'_H}\right)\right]$$

where $\rho'_E = a^2/2\theta_E$ and $\rho'_H = b^2/2\theta_E$ are approximate radii of curvature in the principal planes. This exponential factor or a similar one for conical horns can be included in the expressions for mutual admittance such as Eq. (5.43), for example. In turn, the exponential appears under the quadruple integrals in the functions defined in Eq. (5.44), which becomes part of the usual numerical integration. This approximation works best for flare angles less than 10 degrees.

5.4 Coupling in Arrays of Coaxial Waveguides and Horns

Arrays of waveguides and horns with circular and coaxial apertures are used in many special applications such as satellite communications and radio astronomy. In this section, mutual coupling between such elements is considered in detail and follows the analysis of Bird (2004). The geometry for the coupling between two coaxial apertures is shown in Figure 5.19. The inner and outer radii of elements i and j are $a_{i,j}$ and $b_{i,j}$. The circular aperture corresponds to the limiting case of a single-connected waveguide when $b_{i,j} \to 0$. Commencing with the mutual admittance expression Eq. (5.21), the mode functions for the three types of coaxial modes are listed below:

$(TE)_m \equiv TE_{pq}$ Modes:

$$\begin{aligned}
\Psi_{mix} &= \frac{\chi_{mi}}{2}\left[-Z_{p-1}\left(k_{mi}\rho,\beta_{mi}\right)\cos\left[(p-1)\varphi-\psi_{mi}\right]+Z_{p+1}(k_{mi}\rho,\beta_{mi})\cos\left[(p+1)\varphi-\psi_{mi}\right]\right]\\
\Psi_{miy} &= \frac{\chi_{mi}}{2}\left[Z_{p-1}\left(k_{mi}\rho,\beta_{mi}\right)\sin\left[(p-1)\varphi-\psi_{mi}\right]+Z_{p+1}(k_{mi}\rho,\beta_{mi})\sin\left[(p+1)\varphi-\psi_{mi}\right]\right]\\
\Psi_{miz} &= -j\,\frac{\chi_{mi}k_{mi}}{k_o}\,Z_p\left(k_{mi}\rho,\beta_{mi}\right)\cos\left[p\varphi-\psi_{mi}\right]
\end{aligned} \tag{5.51a}$$

$(TM)_m \equiv TM_{pq}$ Modes:

$$\begin{aligned}
\Psi_{mix} &= -\frac{\chi_{mi}}{2}\left[\Lambda_{p-1}\left(k_{mi}\rho,\beta_{mi}\right)\sin\left[(p-1)\varphi-\psi_{mi}\right]+\Lambda_{p+1}(k_{mi}\rho,\beta_{mi})\sin\left[(p+1)\varphi-\psi_{mi}\right]\right]\\
\Psi_{miy} &= \frac{\chi_{mi}}{2}\left[-\Lambda_{p-1}\left(k_{mi}\rho,\beta_{mi}\right)\cos\left[(p-1)\varphi-\psi_{mi}\right]+\Lambda_{p+1}(k_{mi}\rho,\beta_{mi})\cos\left[(p+1)\varphi-\psi_{mi}\right]\right]
\end{aligned} \tag{5.51b}$$

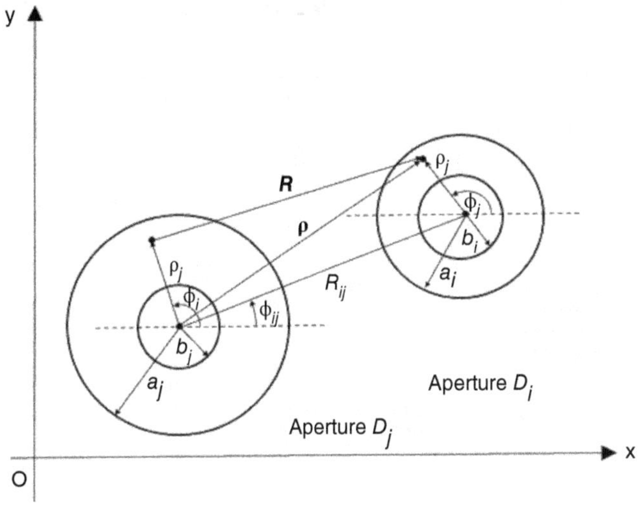

Figure 5.19 Coupling between two different coaxial waveguides.

$TEM \equiv TM_{00}$ Mode:

$$\Psi_{mix} = -\frac{\chi_{mi}}{\rho}\sin\varphi,\ \Psi_{miy} = \frac{\chi_{mi}}{\rho}\cos\varphi \tag{5.51c}$$

where, by definition, $\Psi_{miz} = 0$ for the TM and transverse-electromagnetic (TEM) modes. The modes can be numbered in an order of increasing cut-off frequency from $m = 0$ for the TEM mode. In applications not involving the TEM mode, m commences from $m = 1$. The functions Z_p and Λ_p are combinations of ordinary Bessel functions of order p and are defined as

$$Z_n(x,y) = J_n(x) - \frac{J'_n(y)}{Y'_n(y)}Y_n(x) \text{ and } \Lambda_n(x,y) = J_n(x) - \frac{J_n(y)}{Y_n(y)}Y_n(x),$$

k_{mi} is the cut-off wavenumber of mode m in aperture i, $\alpha_{mi} = k_{mi}a_i$, $\beta_{mi} = k_{mi}b_i$, a_i and b_i are the outer and inner conductor radii of aperture i, ψ_{mi} is the polarization angle of mode m, and χ_{mi} is a scale factor for each mode case as indicated

$$\chi_{mi} = \sqrt{\frac{2\varepsilon_{0p}}{\pi}}\left\{\begin{array}{ll} \left[a_i^2 Z_p^2(\alpha_{mi},\beta_{mi})\left(1 - p^2/\alpha_{mi}^2\right) - b_i^2 Z_p^2(\beta_{mi},\beta_{mi})\left(1 - p^2/\beta_{mi}^2\right)\right]^{-1/2} & TE \\ \left[a_i^2 \Lambda_p'^2(\alpha_{mi},\beta_{mi}) - b_i^2 \Lambda_p'^2(\beta_{mi},\beta_{mi})\right]^{-1/2} & TM \\ \left[2\ln(a_i/b_i)\right]^{-1/2} & TEM \end{array}\right\} \tag{5.52}$$

where $\varepsilon_{0p} = \begin{cases} 1; p = 0 \\ 2; p > 0 \end{cases}$ is Neumann's number, a prime on Z_p or Λ_p indicates the first derivative with respect to the first argument of these functions. As $b_i \to 0$, the functions Z_p and Λ_p limit to ordinary Bessel functions.

The vector mode functions listed above are substituted into Eq. (5.21). In this case, the scalar Green's function in the $z = 0$ plane can be expressed in polar coordinates in terms of a contour integral in the complex w-plane, as follows:

$$G(|\mathbf{R} - \mathbf{R}'|) = G\left(\left|\rho - \rho_j\right|\right) = -jk_o\sum_{s=0}^{\infty}\epsilon_{os}\cos\left[s\left(\phi - \phi_j\right)\right]\int_0^{\infty} dw\,\frac{wJ_s(k_o\rho w)\,J_s(k_o\rho_j w)}{\sqrt{1 - w^2}} \tag{5.53}$$

Table 5.5 Mutual admittance $y_{ij}(m, n)$ of modes $m \equiv (ps)$ and $n \equiv (qt)$ in separate coaxial apertures i and j. Self-admittance is recovered when $R_{ij} \to 0$.

	$(TE_{qt})_j$	$(TM_{qt})_j$	$(TEM)_j$
$(TE_{ps})_i$	$\kappa_{mn}\left[\frac{G^x_{mn}}{k^2 a_i a_j} + \frac{pq}{(\alpha_{mi}\alpha_{nj})^2} F^x_{mn}\right]$	$\kappa_{mn} p N^x_{mn}$	$\kappa_{mn}(-1)^p p C^{(1)}_{p,0,p}\left(d_{mi}, 1; 0, 0; R_{ij}\right) \times \sin\left(p\varphi_{ij} - \psi_{mi}\right)$
$(TM_{ps})_i$	$\kappa_{mn} q M^x_{mn}$	$\kappa_{mn} L^x_{mn}$	$\kappa_{mn}(-1)^p C^{(1)}_{p,0,p}\left(c_{mi}, 1; u_m, 0; R_{ij}\right) \times \cos\left(p\varphi_{ij} - \psi_{mi}\right)$
$(TEM)_i$	$\kappa_{mn}(-1)^q q C^{(1)}_{0,q,q}\left(1, d_{nj}; 0, 0; R_{ij}\right) \times \sin\left(q\varphi_{ij} - \psi_{nj}\right)$	$\kappa_{mn}(-1)^q C^{(1)}_{0,q,q}\left(1, c_{nj}; 0, u_n; R_{ij}\right) \times \cos\left(q\varphi_{ij} - \psi_{nj}\right)$	$\kappa_{mn} C^{(1)}_{0,0,0}\left(1, 1; 0, 0; R_{ij}\right)$

The integrals over the apertures in the azimuth variables can be integrated in closed form, and the radial integrals can be expressed in terms of Hankel transforms, most of which can be evaluated in terms of Bessel functions (see Appendices B and C). The analysis is tedious but straightforward. The final results can be conveniently displayed, as shown in Table 5.5.

The functions used in Table 5.5 are defined as

$$F^x_{mn} = -S^{(1)}_{p,q} C^{(1)}_{p,q,p+q}\left(d_{mi}, d_{nj}; 0, 0; R_{ij}\right) + D^{(1)}_{p,q} C^{(1)}_{p,q,q-p}\left(d_{mi}, d_{nj}; 0, 0; R_{ij}\right) \tag{5.54a}$$

$$G^x_{mn} = S^{(1)}_{p,q} C^{(2)}_{p,q,p+q}\left(d'_{mi}, d'_{nj}; u_m, u_n; R_{ij}\right) + D^{(1)}_{p,q} C^{(2)}_{p,q,q-p}\left(d'_{mi}, d'_{nj}; u_m, u_n; R_{ij}\right) \tag{5.54b}$$

$$L^x_{mn} = S^{(1)}_{p,q} C^{(1)}_{p,q,p+q}\left(c_{mi}, c_{nj}; u_m, u_n; R_{ij}\right) + D^{(1)}_{p,q} C^{(1)}_{p,q,q-p}\left(c_{mi}, c_{nj}; u_m, u_n; R_{ij}\right) \tag{5.54c}$$

$$M^x_{mn} = S^{(2)}_{p,q} C^{(1)}_{p,q,p+q}\left(c_{mi}, d_{nj}; u_m, 0; R_{ij}\right) - D^{(2)}_{p,q} C^{(1)}_{p,q,q-p}\left(c_{mi}, d_{nj}; u_m, 0; R_{ij}\right) \tag{5.54d}$$

$$N^x_{mn} = S^{(2)}_{p,q} C^{(1)}_{p,q,p+q}\left(d_{mi}, c_{nj}; 0, u_n; R_{ij}\right) + D^{(2)}_{p,q} C^{(1)}_{p,q,q-p}\left(d_{mi}, c_{nj}; 0, u_n; R_{ij}\right) \tag{5.54e}$$

where

$$S^{(1,2)}_{p,q} = (-1)^p \begin{cases} \cos\left[(p+q)\varphi_{ij} - \psi_{mi} - \psi_{nj}\right] \\ \sin\left[(p+q)\varphi_{ij} - \psi_{mi} - \psi_{nj}\right] \end{cases},$$

$$D^{(1,2)}_{p,q} = \begin{cases} \cos\left[(p-q)\varphi_{ij} - \psi_{mi} + \psi_{nj}\right] \\ \sin\left[(p-q)\varphi_{ij} - \psi_{mi} + \psi_{nj}\right] \end{cases},$$

$$c_{mi} = \frac{b_i \Lambda'_p\left(\beta_{mi}, \beta_{mi}\right)}{a_i \Lambda'_p\left(\alpha_{mi}, \beta_{mi}\right)}, \quad d_{mi} = \frac{Z_p\left(\beta_{mi}, \beta_{mi}\right)}{Z_p\left(\alpha_{mi}, \beta_{mi}\right)}, \quad d'_{mi} = \frac{b_i Z_p\left(\beta_{mi}, \beta_{mi}\right)}{a_i Z_p\left(\alpha_{mi}, \beta_{mi}\right)}.$$

Table 5.6 Scale factors κ_{mn} for mutual admittances of coaxial waveguide modes.

	$(TE_{qt})_j$	$(TM_{qt})_j$	$(TEM)_j$
$(TE_{ps})_i$	$\dfrac{\pi a_i a_j y_m y_n \alpha_{mi}\alpha_{nj}\chi_{mi}\chi_{nj}Y_o}{2\sqrt{Y_{mi}Y_{nj}}}$	$\dfrac{-\pi a_i a_j y_m y'_n \chi_{mi}\chi_{nj}Y_o}{2\alpha_{mi}\sqrt{Y_{mi}Y_{nj}}}$	$\dfrac{\pi a_i y_m \chi_{mi}\chi_{nj}Y_o}{\alpha_{mi}\sqrt{Y_{mi}Y_{nj}}}$
$(TM_{ps})_i$	$\dfrac{-\pi a_i a_j y'_m y_n \chi_{mi}\chi_{nj}Y_o}{2\alpha_{nj}\sqrt{Y_{mi}Y_{nj}}}$	$\dfrac{\pi a_i a_j y'_m y'_n \chi_{mi}\chi_{nj}Y_o}{2\sqrt{Y_{mi}Y_{nj}}}$	$\dfrac{-\pi a_i y'_m \chi_{mi}\chi_{nj}Y_o}{\sqrt{Y_{mi}Y_{nj}}}$
$(TEM)_i$	$\dfrac{\pi a_j y_n \chi_{mi}\chi_{nj}Y_o}{\alpha_{nj}\sqrt{Y_{mi}Y_{nj}}}$	$\dfrac{-\pi a_j y'_n \chi_{mi}\chi_{nj}Y_o}{\sqrt{Y_{mi}Y_{nj}}}$	$\left(\sqrt{\varepsilon_{ri}}ln\dfrac{a_i}{b_i}.\sqrt{\varepsilon_{rj}}ln\dfrac{a_j}{b_j}\right)^{-1}$

The remaining functions in Table 5.5 are Hankel transforms involving products of Bessel functions

$$C^{(1)}_{p,q,\nu}\left(\alpha,\beta;u_m,u_n;s\right) = \int_0^\infty dw \frac{w^3}{\sqrt{1-w^2}} \frac{\left[J_p\left(ka_iw\right) - \alpha J_p\left(kb_iw\right)\right]}{\left(w^2-u_m^2\right)} \times \frac{\left[J_q\left(ka_jw\right) - \beta J_q\left(kb_jw\right)\right]}{\left(w^2-u_n^2\right)} J_\nu\left(ksw\right) \tag{5.55a}$$

and

$$C^{(2)}_{p,q,\nu}\left(\alpha,\beta;u_m,u_n;s\right) = \int_0^\infty dw\, w\sqrt{1-w^2} \frac{\left[J'_p\left(ka_iw\right) - \alpha J'_p\left(kb_iw\right)\right]}{\left(w^2-u_m^2\right)} \times \frac{\left[J'_q\left(ka_jw\right) - \beta J'_q\left(kb_jw\right)\right]}{\left(w^2-u_n^2\right)} J_\nu\left(ksw\right) \tag{5.55b}$$

where $u_m = \alpha_{mi}/ka_i$, $y_m = Z_p(\alpha_{mi}, \beta_{mi})$, $y'_m = \Lambda'_p\left(\alpha_{mi}, \beta_{mi}\right)$, $\alpha_{mi} = k_{mi}a_i$, and the scale factor of the admittance, κ_{mn}, is listed in Table 5.6. Some properties of the Hankel transforms in Eq. (5.55) are described in more detail in Appendix C. The scale factors in Table 5.6, make use of the following:

$$Y_{mi} = Y_o \begin{cases} \gamma_{mi}/k_o;\ TE \\ k_o\,\varepsilon_r/\gamma_{mi}\ :\ TM \end{cases} \quad \text{is the mode admittance of waveguide } i,$$

Y_o is the wave admittance of free space, ε_r is the dielectric constant of the waveguide filling, and γ_{mi} is the propagation constant of mode m.

The mutual admittances listed in Table 5.5 are applicable for self-admittance as well in which case $R_{ij} \to 0$ when the functions limit to (Bird 1996b)

$$F^x_{mn} \Rightarrow F_{mn} = C^{(1)}_{p,q,0}\left(d_{mi}, d_{nj}; 0, 0; 0\right) \cos\left(\psi_{nj} - \psi_{nj}\right) \tag{5.56a}$$

$$G^x_{mn} \Rightarrow 2G_{mn}/\varepsilon_{om} = 2C^{(2)}_{p,q,0}\left(d'_{mi}, d'_{nj}; u_m, u_n; 0\right) \cos\left(\psi_{nj} - \psi_{nj}\right)/\varepsilon_{0m} \tag{5.56b}$$

$$L^x_{mn} \Rightarrow 2L_{mn}/\varepsilon_{0m} = 2C^{(1)}_{p,q,0}\left(c_{mi}, c_{nj}; u_m, u_n; 0\right) \cos\left(\psi_{nj} - \psi_{nj}\right)/\varepsilon_{0m} \tag{5.56c}$$

$$M^x_{mn} \Rightarrow M_{mn} = C^{(1)}_{p,q,0}\left(c_{mi}, d_{nj}; u_m, 0; 0\right) \sin\left(\psi_{nj} - \psi_{nj}\right) \tag{5.56d}$$

$$N^x_{mn} \Rightarrow N_{mn} = C^{(1)}_{p,q,0}\left(d_{mi}, c_{nj}; 0, u_n; 0\right) \sin\left(\psi_{nj} - \psi_{nj}\right). \tag{5.56e}$$

Table 5.7 Mutual admittance $y_{ij}(m, n)$ of modes $m \equiv (ps)$ and $n \equiv (qt)$ in separate circular apertures i and j.

	$(TE_{qt})_j$	$(TM_{qt})_j$
$(TE_{ps})_i$	$\kappa_{mn}\left[\frac{G^x_{mn}}{k^2 a_i a_j} + \frac{pq}{\left(\alpha_{mi}\alpha_{nj}\right)^2} F^x_{mn}\right]$	$\kappa_{mn} p N^x_{mn}$
$(TM_{ps})_i$	$\kappa_{mn} q M^x_{mn}$	$\kappa_{mn} L^x_{mn}$

In another special case when the radii of the inner conductors shrink to zero, i.e. $b_{i,j} \to 0$, the mode functions for TE and TM modes reduce to those of circular waveguide where also $c_{mi} \to 0$, $d_{mi} \to 0$, and $d'_{mi} \to 0$. Thus, the mode coupling formulae for the TE and TM modes coupling in circular waveguides are given as in Table 5.7.

In circular apertures i and j when both $b_{i,j} \to 0$, the functions in Table 5.7 become

$$F^x_{mn} = -S^{(1)}_{p,q} C^{(1)}_{p,q,p+q}\left(0,0;0,0;R_{ij}\right) + D^{(1)}_{p,q} C^{(1)}_{p,q,q-p}\left(0,0,0;R_{ij}\right) \tag{5.57a}$$

$$G^x_{mn} = S^{(1)}_{p,q} C^{(2)}_{p,q,p+q}\left(0,0;u_m,u_n;R_{ij}\right) + D^{(1)}_{p,q} C^{(2)}_{p,q,q-p}\left(0,0;u_m,u_n;R_{ij}\right) \tag{5.57b}$$

$$L^x_{mn} = S^{(1)}_{p,q} C^{(1)}_{p,q,p+q}\left(0,0;u_m,u_n;R_{ij}\right) + D^{(1)}_{p,q} C^{(1)}_{p,q,q-p}\left(0,0;u_m,u_n;R_{ij}\right) \tag{5.57c}$$

$$M^x_{mn} = S^{(2)}_{p,q} C^{(1)}_{p,q,p+q}\left(0,0;u_m,0;R_{ij}\right) - D^{(2)}_{p,q} C^{(1)}_{p,q,q-p}\left(0,0;u_m,0;R_{ij}\right) \tag{5.57d}$$

$$N^x_{mn} = S^{(2)}_{p,q} C^{(1)}_{p,q,p+q}\left(0,0;0,u_n;R_{ij}\right) + D^{(2)}_{p,q} C^{(1)}_{p,q,q-p}\left(0,0;0,u_n;R_{ij}\right) \tag{5.57e}$$

wherein

$$C^{(1)}_{p,q,\nu}\left(0,0;u_m,u_n;s\right) = \int_0^\infty dw \frac{w^3}{\sqrt{1-w^2}} \frac{J_p\left(ka_i w\right)}{\left(w^2-u_m^2\right)} \frac{\left[J_q\left(ka_j w\right)\right]}{\left(w^2-u_n^2\right)} J_\nu(ksw) \tag{5.58a}$$

and

$$C^{(2)}_{p,q,\nu}\left(0,0;u_m,u_n;s\right) = \int_0^\infty dw\, w\sqrt{1-w^2}\, \frac{J'_p\left(ka_i w\right)}{\left(w^2-u_m^2\right)} \frac{J'_q\left(ka_j w\right)}{\left(w^2-u_n^2\right)} J_\nu(ksw) \tag{5.58b}$$

The Hankel transforms in Eq. (5.58) should be integrated with care in the vicinity of the poles and around the branch cut located at $w = 1$. In some cases, the transforms can be represented by series. Appendix C should be consulted for further details of these types of Hankel transforms.

5.4.1 Self-Admittance of TE_{11} Mode in Coaxial Waveguide

The self-admittance of the TE_{11} mode in the aperture of an open-ended circular coaxial waveguide is an important practical result both for single apertures and also for mutual coupling calculations. For the TE_{11} mode $m = 1 = n$, the self-admittance of the TE_{11} mode is given by the top-left element of Table 5.5. Thus,

$$y_{11}(1,1) = \frac{2\frac{k}{\gamma}\left[\frac{G^x_{mn}\,\alpha_{11}^2}{k^2\,a^2} + \frac{1}{\alpha_{11}^2}F^x_{mn}\right]}{\left[\left(1-\frac{1}{\alpha_{11}^2}\right) - \left(\frac{\beta_{11}}{\alpha_{11}}\right)\frac{Z_1\left(\beta_{11},\beta_{11}\right)^2}{Z_1\left(\alpha_{11},\beta_{11}\right)^2}\left(1-\frac{1}{\beta_{11}^2}\right)\right]} \tag{5.59}$$

For circular apertures with $\beta_{11} \to 0$, G^x_{mn} and F^x_{mn} are replaced by the functions G_{mn} and F_{mn}, as described in Section 5.4. As a result, the self-admittance of the TE_{11} mode in a circular aperture is

$$y_{11}(1,1) = \frac{2k/\gamma}{\left(1 - 1/x_1^2\right)} \left(\frac{G_{11}x_1^2}{(ka)^2} + \frac{F_{11}}{x_1^2} \right)$$

where $x_1 = 1.841184$ is the first zero of $J_1'(x_1) = 0$ corresponding to the cut-off wavenumber of the TE_{11} mode, γ is the propagation constant, and

$$F_{11} = C^{(1)}_{1,1,0}(0,0;0,0;0) = \int_0^\infty dw \frac{\left(J_1(kaw)\right)^2}{w\sqrt{1-w^2}}$$

$$= \int_0^1 dw \frac{\left(J_1(kaw)\right)^2}{w\sqrt{1-w^2}} + j\int_1^\infty dw \frac{\left(J_1(kaw)\right)^2}{w\sqrt{w^2-1}}$$

$$= \left(\frac{1}{2} - \frac{J_1(2ka)}{2ka} \right) + j\frac{\widetilde{H}_1(2ka)}{(2ka)}, \tag{5.60}$$

where $u_1 = x_1/ka$ and $\widetilde{H}_1$ is a Struve function of order 1. The latter function can be expressed as a power series (Abramowitz and Stegun 1965)

$$\widetilde{H}_1(x) = \frac{2}{\pi}\left(1 + \sum_{q=0}^{\infty} \frac{\varepsilon_{0n} J_{2q}(x)}{\left(4q^2 - 1\right)} \right).$$

The remaining function in the self-admittance of the TE_{11} mode is

$$G_{11} = C^{(2)}_{1,1,0}\left(0,0;u_1,u_1;0\right) = \int_0^\infty dw\, w\sqrt{1-w^2}\, \frac{\left[J_1'(kaw)\right]^2}{\left(w^2 - u_1^2\right)^2} \tag{5.61}$$

This function and the more general one, $C^{(1,2)}_{p,q,v}$, can be integrated directly with some care. The contours of both integrals run along the real positive w-axis, which can lead to oscillatory integral values and slow convergence. The integration path can be deformed into the complex plane, and this improves convergence. For all integrals of this type, one of the Bessel functions in the integrand can be substituted by its equivalent in terms of Hankel functions of the first and second kinds. For instance, in G_{11}, one of the Bessel functions in the product is replaced with $J_p'(x) = (H_p^{(1)'} + H_p^{(2)'})/2$. Thus,

$$G_{11} = \frac{1}{2}\int_0^\infty dw\, w\sqrt{1-w^2}\, \frac{J_1'(kaw)\left(H_1^{(1)'} + H_1^{(2)'}\right)}{\left(w^2 - u_1^2\right)^2} = G_1 + G_2 \tag{5.62}$$

The contour for G_1 involving $H_1^{(1)'}$ can now be deformed into the $w = u + jv$ plane along the positive jv axis, and this enables the replacement of the Hankel functions by their modified Bessel equivalents (see Appendix B), which helps to achieve better convergence. Similarly, the contour for G_2 is deformed around the branch cut and along the $-jv$ axis. The details of this approach are given by Bird (1979). The final result is

$$G_1(ka) = -\frac{j}{\pi}\int_0^\infty dv \frac{v\sqrt{1+v^2}\, I_1'(kav)\, K_1'(kav)}{\left(v^2 + u_1^2\right)^2} \tag{5.63a}$$

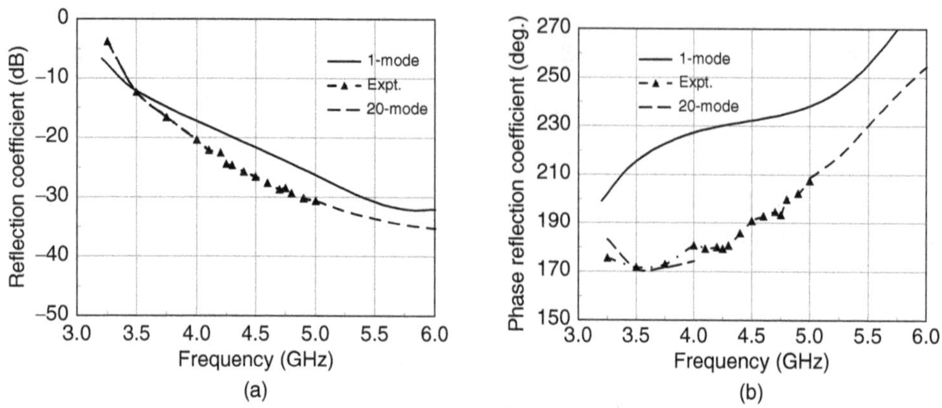

Figure 5.20 Reflection coefficient of circular waveguide with diameter 57 mm. (a) Magnitude in dB; (b) phase. Solid curve: 1-mode (theory); dashed 20-mode (theory); dashed with triangle marker: experiment.

and

$$G_2(ka) = \int_0^1 du \frac{u\sqrt{1-u^2}\, J_1'(kau)\, H_1^{(2)'}(kau)}{\left(u^2-u_1^2\right)^2} - \frac{j}{\pi}\int_0^\infty dv \frac{v\sqrt{1+v^2}\, I_1'(kav)\, K_1'(kav)}{\left(v^2+u_1^2\right)^2}$$
$$+ \frac{\pi ka}{4u_1}\sqrt{1-u_1^2}\, J_1''\left(kau_1\right) Y_1'\left(kau_1\right)\, H\left(u_1-1\right), \tag{5.63b}$$

where I_1' and K_1' are derivatives of the modified Bessel function (Appendix B), and $H(..)$ is the Heaviside step function. The third term on the right side of Eq. (5.62b) is a contribution from the pole on the real u-axis when the mode is operating above its cut-off (i.e. $u_1 > 1$).

As an example, consider an isolated circular waveguide in a ground plane. The waveguide has a diameter of 57 mm, and its reflection coefficient was measured from 3.3 to 5.0 GHz. The measured results are shown in Figure 5.20 along with the results calculated using 20 modes. The latter result closely approximates the measured data, which becomes noisy above 4.5 GHz as the calibration of the network analyzer became more difficult due to overmoding in the transition to the antenna under test. For comparison, the result given by the TE_{11} mode self-admittance (1-mode) in the top-left column of Table 5.5 is also shown in Figure 5.20.

5.4.2 TEM Mode Coupling in Coaxial Waveguide

The TEM mode is present in coaxial waveguide and cable because the cross-section is doubly connected in a complex contour sense. This means a TEM mode does not exist in waveguide without a center conductor. Singly connected structures have modes that cut-off at specific wavenumbers, which are given as roots of a characteristic equation. In coaxial waveguide, both TEM and waveguide modes are present, and these have cut-off wavenumbers that are roots of Bessel function combinations. The TEM mode has a zero wavenumber and, therefore, conducts DC by a return path.

The explicit expression for TEM mode coupling can be found from Table 5.5 and is (Bird 1996a)

$$y_{i,j}(0,0) = \frac{\pi\, \chi_{0i}\chi_{0i}}{\left(\varepsilon_{ri}\varepsilon_{rj}\right)^{1/4}}\sqrt{\frac{\varepsilon_r}{\mu_r}}\int_0^\infty \frac{dw}{w\sqrt{1-w^2}}\left(J_0\left(k_ia_iw\right) - J_0\left(k_ib_iw\right)\right)\left(J_0\left(k_ja_jw\right) - J_0\left(k_jb_jw\right)\right)$$
$$= G_{ij} + jB_{ij} \tag{5.64}$$

where $k_i = k_o\sqrt{\varepsilon_{ri}\mu_{ri}}$, $k_o = \frac{\omega}{c}$, and ε_{ri} and μ_{ri} are the relative permeability and permittivity of the coaxial cable filling. This expression applies to apertures radiating into a lossless half-space.

Equation (5.63) needs to be modified if the cable radiates into a lossy medium. In this instance, such as when the aperture is used for material characterization where the measured reflection and mutual coupling coefficients are used to determine the complex permittivity and permeability of the medium (e.g. Bird and Hayman 1997), then it is convenient to define a complex wavenumber of the external medium such as

$$k_e = k\sqrt{\Pi_e}$$

where $k = k_o\sqrt{\varepsilon_r'\mu_r'}$, and $\Pi_e = \left(1 - j\varepsilon_r''/\varepsilon_r'\right)\left(1 - j\mu_r''/\mu_r'\right)$. The complex relative permittivity and permeability are $\varepsilon_r = \varepsilon_r' - j\varepsilon_r''$ and $\mu_r = \mu_r' - j\mu_r''$. In this case, Eq. (5.64) now becomes

$$\begin{aligned} y_{i,j}(0,0) = {} & \frac{\pi\,\chi_{0i}\chi_{0i}}{\left(\varepsilon_{ri}\varepsilon_{rj}\right)^{1/4}}\sqrt{\frac{\Pi_e\varepsilon_r'}{\mu_r'}}\int_0^\infty \frac{dw}{w\sqrt{\Pi_e - w^2}}\left(J_0\left(k_i a_i w\right) - J_0\left(k_i b_i w\right)\right) \\ & \times \left(J_0\left(k_j a_j w\right) - J_0\left(k_j b_i w\right)\right) \end{aligned} \tag{5.65}$$

For separate apertures, this equation can be expressed more conveniently for computation by deforming the contour in the w-plane as described by Bird (1979) giving

$$\begin{aligned} y_{i,j}(0,0) = {} & \frac{\pi\,\chi_{0i}\chi_{0i}}{\left(\varepsilon_{ri}\varepsilon_{rj}\right)^{1/4}}\sqrt{\frac{\Pi_e\varepsilon_r'}{\mu_r'}}\left\{\int_0^1 \frac{du}{u\sqrt{\Pi_e - u^2}}\left(J_0\left(k_i a_i u\right) - J_0\left(k_i b_i u\right)\right)\right. \\ & \times \left(J_0\left(k_j a_j u\right) - J_0\left(k_j b_i u\right)\right) H_0^{(2)}\left(k_e R_{ij} u\right) \\ & - \frac{2j}{\pi}\int_0^\infty \frac{dv}{u\sqrt{\Pi_e - v^2}}\left(I_0\left(k_i a_i v\right) - I_0\left(k_i b_i v\right)\right) \\ & \times \left(I_0\left(k_j a_j v\right) - I_0\left(k_j b_i v\right)\right) K_0\left(k_e R_{ij} v\right) \end{aligned} \tag{5.66}$$

A series solution exists to Eq. (5.66), which was found by replacing the triple product of Bessel functions by power series (see Appendix B). These series depend on the conductor radii, as shown in Bird (1996a). The final solution is

$$G_{ij} = \frac{\pi\,\chi_{0i}\chi_{0i}}{\left(\varepsilon_{ri}\varepsilon_{rj}\right)^{1/4}}\sqrt{\frac{\varepsilon_r}{\mu_r}}\sum_{m=0}^{\infty} P_m\left(a_i, b_i, a_j, b_j\right)\sum_{\mu=0}^{\infty}\alpha(m,\mu)\left(\frac{kR_{ij}}{2}\right)^{2\mu},$$

and

$$B_{ij} = \frac{\pi\,\chi_{0i}\chi_{0i}}{\left(\varepsilon_{ri}\varepsilon_{rj}\right)^{1/4}}\sqrt{\frac{\varepsilon_r}{\mu_r}}\sum_{m=0}^{\infty} P_m\left(a_i, b_i, a_j, b_j\right)\sum_{\mu=0}^{\infty}\beta(m,\mu)\left(\frac{kR_{ij}}{2}\right)^{2(\mu-m)+1} \tag{5.67}$$

where

$$P_m\left(a_i, b_i, a_j, b_j\right) = (-1)^m\left(F_m\left(a_i.a_j\right) - F_m\left(a_i.b_j\right)\right) - \left(F_m\left(a_j.b_i\right) - F_m\left(b_i.b_j\right)\right),$$

$$\alpha(m,\mu) = \frac{\sqrt{\pi}}{2}\frac{(-1)^\mu}{(\mu!)^2}\frac{\Gamma(m+\mu)}{\Gamma(m+\mu+1/2)},$$

$$\beta(m,\mu) = \frac{\sqrt{\pi}}{2}\frac{(-1)^\mu}{\mu!}\frac{\Gamma(m-\mu-1/2)}{\Gamma(\mu-m+3/2)\Gamma(1/2-\mu)}$$

and

$$F_m(a,b) = \left(\frac{a}{2}\right)^{2m}\sum_{n=0}^{m}\left[\frac{(b/a)^n}{(n!)^2(m-n)!}\right]^2. \tag{5.68}$$

where $\Gamma(..)$ is the gamma function. In the special case when the arguments of Eq. (5.68) are identical, then

$$F_m(a,a) = \left(\frac{a}{2}\right)^{2m} \frac{\Gamma(2m+1)}{(m!)^4}$$

The first two terms of the series in Eq. (5.67) dominate at low frequencies. In that case,

$$G_{ij} \cong \frac{\pi\, \chi_{0i}\chi_{0i}}{24(\varepsilon_{ri}\varepsilon_{rj})^{1/4}} \sqrt{\frac{\varepsilon_r}{\mu_r}} k^4 \left(1 - \frac{1}{5}(kR_{ij})^2\right)(a_i^2 - b_i^2)\left(a_j^2 - b_j^2\right)$$

and

$$B_{ij} \cong -\frac{\pi\, \chi_{0i}\chi_{0i}}{16(\varepsilon_{ri}\varepsilon_{rj})^{1/4}} \sqrt{\frac{\varepsilon_r}{\mu_r}} \frac{k}{R_{ij}^3} \left(1 - \frac{1}{2}(kR_{ij})^2\right)(a_i^2 - b_i^2)\left(a_j^2 - b_j^2\right) \tag{5.69}$$

As an example of the use of Eq. (5.65), the loss tangent of ultrapure water was measured. Figure 5.21 shows the result of such a measurement over a bandwidth from near 100 MHz to 25 GHz using a dual coaxial probe, which is illustrated in Figure 5.22. This probe was constructed by mounting two Storm 421–336, semirigid, coaxial cables ($a_i = 2.67$ mm, $b_i = 0.94$ mm, $\varepsilon_{ri} = 1.56$) in a ground plane with a center spacing of $R_{ij} = 8$ mm. The dual probe was placed in contact with the material under test, and the two-port scattering parameters are measured using a Hewlett Packard HP8510C vector network analyzer. The cut-off frequency of the first high-order mode (TE_{11}) in the cables is 21.7 GHz, although the connectors and the onset of the higher-order mode probably sets the useful upper frequency of the measurement to <20 GHz.

A multimode analysis of the mutual coupling between coaxial cables radiating into a half-space was developed, and the solution is used to calculate the complex permittivity. To determine the complex permittivity (a unity permeability is assumed), computed and measured data are matched by adjusting the real and imaginary parts of the relative permittivity of the sample ε_{re}. The data are matched at each frequency sample by determining the minimum of a function by optimization. This is a more flexible approach than the alternative root-finding technique as it is easily modified and extended for matching mixed data, such as S_{11}, Y_{11}, or S_{12} ; μ_{ri} could be included for magnetic

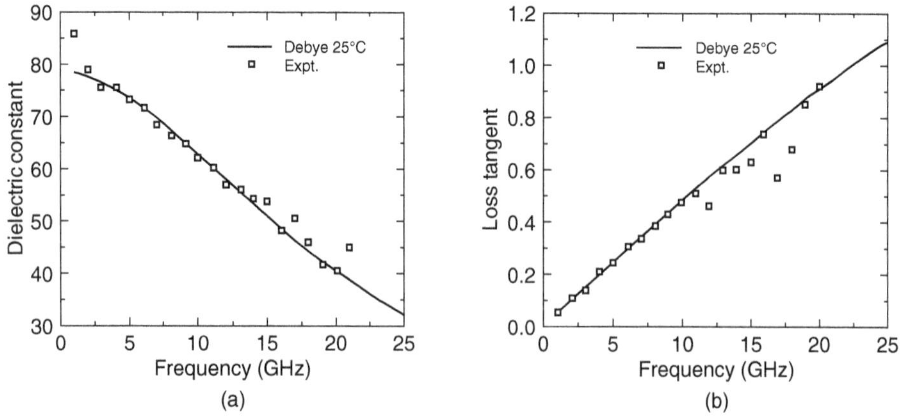

Figure 5.21 Complex permittivity of ultrapure water measured using a dual coaxial probe.

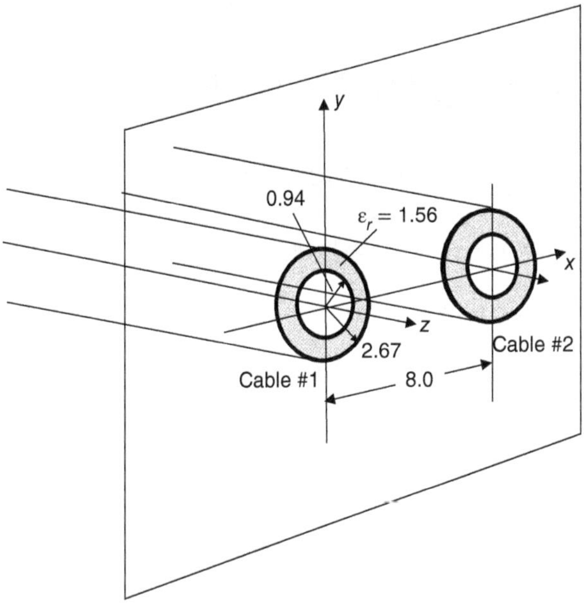

Figure 5.22 Geometry of two semirigid coaxial cables terminating in a $300 \times 300\,\text{mm}^2$ ground plane.

materials if needed. A performance index that proved effective for minimization with a gradient search method is

$$I\left(\varepsilon_{re}\right) = log\left\{ \left(20\log\left(\left|S_{11}^{exp}/S_{11}^{theory}\left(\varepsilon_{re}\right)\right| + \left|M_{11}\right|\right)\right)\left(20\log\left(\left|S_{12}^{exp}/S_{12}^{theory}\left(\varepsilon_{re}\right)\right| + \left|M_{12}\right|\right)\right)\right.$$

$$\left. \times\left(\left(arg\left(S_{11}^{exp}/S_{11}^{theory}\left(\varepsilon_{re}\right)\right)\right)^2 + \left|P_{11}^2\right|\right)\right)\left(\left(arg\left(S_{12}^{exp}/S_{12}^{theory}\left(\varepsilon_{re}\right)\right)\right)^2 + \left|P_{12}^2\right|\right)\right)\right\}$$

where S_{11} and S_{12} are the TEM-mode scattering parameters, the superscripts "expt" and "theory" refer to measured and computed data, and M_{11}, P_{11}, M_{12}, and P_{12} are the uncertainties in the measured magnitude and phase. The measured scattering parameters are often slightly asymmetric due to measurement error, and to enforce symmetry, the geometric mean of the measured data is used, i.e. $S_{11}^{exp} = \sqrt{S_{11meas}S_{22meas}}$ and $S_{12}^{exp} = \sqrt{S_{12meas}S_{21meas}}$.

The measured results are compared with the Debye spectral function at 25 °C in Figure 5.21. Excellent agreement is shown between the experimental results and the Debye function over the frequency range of the probe. The slight discrepancy in the dielectric constant at 1 GHz is probably due to reflections from the bottom of the water vessel.

5.4.3 Asymptotic Approximation of Coupling in Coaxial Waveguide Apertures

With separations between coaxial apertures greater than one wavelength, an asymptotic approximation to coupling becomes an alternative. To do this, the Taylor series expansion in polar coordinates of the free-space Green's function given by Eq. (3.30) is substituted into the formula for mutual admittance Eq. (5.21). The result is

$$y_{ij}(m|n) \cong \frac{jk\, g_o(R_{ij})}{4\pi\eta_o\sqrt{Y_{mi}Y_{nj}}}\int_{S_i} dS\, \boldsymbol{\Psi}_{mi}(u_1, u_2)\cdot\int_{S_j} dS'\boldsymbol{\Psi}_{nj}(u_1', u_2')\, exp\left[-jk\rho_i cos\left(\phi_i - \psi_{ij}\,\phi_{ij}\right)\right.$$

$$\left.+jk\rho_j cos\left(\theta_j - \phi_{ij}\right)\right] \times \left\{1 - \frac{1}{R_{ij}}\left[\rho_i cos\left(\phi_i - \phi_{ij}\right) - \rho_j cos(\phi_j - \phi_{ij}\right)\right.$$

$$+\frac{jk}{4R_{ij}}\left[\rho_i\rho_j sin\left(\phi_i - \phi_{ij}\right) sin\left(\phi_j - \phi_{ij}\right) + \rho_i^2 cos\left[2\left(\phi_i - \phi_{ij}\right)\right]\right.$$

$$\left.\left.+\rho_j^2 cos\left[2\left(\phi_j - \phi_{ij}\right)\right] - \left(\rho_i^2 + \rho_j^2\right)\right]\right\} \tag{5.70}$$

where $\boldsymbol{\Psi}_{mi}$ and $\boldsymbol{\Psi}_{nj}$ are given by Eq. (5.51). The following circular transforms are now defined to simplify Eq. (5.70) further

$$\boldsymbol{S}_{mi}(w) = 2j\frac{k_{mi}}{k}\int_0^{2\pi} d\phi_i \int_{a_i}^{b_i} d\rho_i\, exp\left[jw\rho_i\cos\left(\phi_i - \phi_{ij}\right)\right]\boldsymbol{\Psi}_{mi}$$
$$= 2\pi j^{p-1}\boldsymbol{V}_{mi}(w) \tag{5.71a}$$

$$\boldsymbol{T}_{mi}(w) = 2j\frac{k_{mi}}{k}\int_0^{2\pi} d\phi_i \sin\left(\phi_i - \phi_{ij}\right)\int_{a_i}^{b_i} d\rho_i\, \rho_i exp\left[jw\rho_i\cos\left(\phi_i - \phi_{ij}\right)\right]\boldsymbol{\Psi}_{mi}$$
$$= 2\pi j^{p-2}\frac{1}{w}\boldsymbol{U}_{mi}(w) \tag{5.71b}$$

$$\boldsymbol{X}_{mi}(w) = \boldsymbol{A}_{mi}(w) + \frac{jw}{4}\left(\boldsymbol{B}_{mi}(w) - \boldsymbol{C}_{mi}(w)\right)$$

in which k_{mi} is the cut-off wavenumber of mode m in waveguide i, and

$$\boldsymbol{A}_{mi}(w) = \int_0^{2\pi} d\phi_i\, \cos\left(\phi_i - \phi_{ij}\right)\int_{a_i}^{b_i} d\rho_i\, \rho_i\, exp\left[jw\rho_i\cos\left(\phi_i - \phi_{ij}\right)\right]\boldsymbol{\Psi}_{mi}$$

$$\boldsymbol{B}_{mi}(w) = \int_0^{2\pi} d\phi_i\, \cos\left[2\left(\phi_i - \phi_{ij}\right)\right]\int_{a_i}^{b_i} d\rho_i\, \rho_i^2\, exp\left[jw\rho_i\cos\left(\phi_i - \phi_{ij}\right)\right]\boldsymbol{\Psi}_{mi}$$

$$\boldsymbol{C}_{mi}(w) = \int_0^{2\pi} d\phi_i\, \cos\left(\phi_i - \phi_{ij}\right)\int_{a_i}^{b_i} d\rho_i\, \rho_i^2\, exp\left[jw\rho_i\cos\left(\phi_i - \phi_{ij}\right)\right]\boldsymbol{\Psi}_{mi}$$

so that

$$\boldsymbol{X}_{mi}(w) = 2\pi j^{p-2}\frac{1}{w}\boldsymbol{W}_{mi}(w) \tag{5.71c}$$

After some manipulations, three main vectors can be identified in Eq. 5.70, namely $\mathbf{V}_{\mathrm{m}}$, $\mathbf{U}_{\mathrm{m}}$ and $\mathbf{W}_{\mathrm{m}}$, which have components:

$$V_{mix}(w) = -\left(Q_{p-1}(w)\ cos\Delta_{p-1} + Q_{p+1}(w)\ cos\Delta_{p+1}\right)$$
$$V_{miy}(w) = Q_{p-1}(w)\ sin\Delta_{p-1} - Q_{p+1}(w)\ sin\Delta_{p+1}$$
$$V_{miz}(w) = -2\frac{k_{mi}}{k}Q_p(w)\ cos\Delta_p \tag{5.72}$$

where $\Delta_p = p\Delta_p - \psi_{mi}$ and

$$Q_p(w) = \int_{b_i}^{a_i} d\rho_i\, \rho_i\, Z_p\left(k_{mi}\rho_i,\ \ \alpha_{mi}\right) J_p\left(w\rho_i\right) \tag{5.73}$$

In addition,

$$U_{mix}(w) = (p-1)\,Q_{p-1}(w)\; sin\Delta_{p-1} + (p+1)\,Q_{p+1}(w)\; sin\Delta_{p+1}$$
$$U_{miy}(w) = (p-1)\;Q_{p-1}(w)\; cos\Delta_{p-1} - (p+1)\;Q_{p+1}(w)\; cos\Delta_{p+1}$$
$$U_{miz}(w) = 2\frac{k_{mi}}{k} p\,Q_p(w)\; sin\Delta_p \tag{5.74}$$

and

$$W_{mix}(w) = -\left(\Omega_{p-1}(w)\; cos\Delta_{p-1} + \Omega_{p+1}(w)\; cos\Delta_{p+1}\right)$$
$$W_{miy}(w) = \Omega_{p-1}(w)\; sin\Delta_{p-1} - \Omega_{p+1}(w)\; sin\Delta_{p+1}$$
$$W_{miz}(w) = -2\frac{k_{mi}}{k}\Omega_p(w)\; cos\Delta_p \tag{5.75}$$

where

$$\Omega_p = \frac{1}{2}\left(w\,\mathrm{O}_p(w) + p^2 Q_p(w)\right) \tag{5.76}$$

in which

$$\mathrm{O}_p(w) = \int_{b_i}^{a_i} d\rho_i\, \rho_i^2\, Z_p\left(k_{mi}\rho_i,\ \ \alpha_{mi}\right)\, J_p'\left(w\rho_i\right)$$

The three vectors defined in Eq. (5.72), namely $\boldsymbol{V}_{mi}, \boldsymbol{U}_{mi}$, and $\boldsymbol{W}_{mi}$, apply to the TE modes only. For the TM modes, they are found through the following translations

$$\left(\boldsymbol{V}_{mi}\right)^{TM} <=> \hat{\boldsymbol{z}}\times\left(\boldsymbol{V}_{mi}\right)^{TE},\ \ \left(\boldsymbol{U}_{mi}\right)^{TM} <=> \hat{\boldsymbol{z}}\times\left(\boldsymbol{U}_{mi}\right)^{TE},\ \ \left(\boldsymbol{W}_{mi}\right)^{TM} <=> \hat{\boldsymbol{z}}\times\left(\boldsymbol{W}_{mi}\right)^{TE}$$

where $<=>$ means translate relevant parameters for TE to TM such as the wavenumber and Z_p to Λ_p. In that way, the equations above apply except that in Q_p and O_p, Z_p is replaced by Λ_p. The integrals $\mathrm{Q_p}$ and $\mathrm{O_p}$ are Lommel-type integrals. Some properties of these integrals are summarized in Appendix B.

Making the various substitutions in Eq. (5.70) results finally in

$$y_{ij}(m|n) \cong \frac{k\,\pi\,\chi_{mi}\chi_{ni}}{\varepsilon_{0p}\varepsilon_{0q}\eta_0\sqrt{Y_{mi}Y_{nj}}} j^{p+q-1} g_o\left(R_{ij}\right)\left\{\boldsymbol{V}_{mi}(-k)\cdot\boldsymbol{V}_{nj}(k) + \frac{j}{k\,R_{ij}}\left[\boldsymbol{U}_{mi}(-k)\cdot\boldsymbol{U}_{nj}(k)\right.\right.$$
$$\left.\left. -\boldsymbol{V}_{mi}(-k)\cdot\boldsymbol{W}_{nj}(k) - \boldsymbol{W}_{mi}(-k)\cdot\boldsymbol{V}_{nj}(k)\right]\right\} \tag{5.77}$$

where ε_{0p} is the Neumann number. Equation (5.77) can be computed very quickly and gives results accurate to second order in R_{ij}.

As an example, consider the case of TEM modes for which $m = 1 = n$ and $p = 0 = q$. In this case,

$$\boldsymbol{V}_{1i}(w) = \hat{\phi}\frac{J_0\left(wa_i\right) - J_0\left(wb_i\right)}{w}$$
$$\boldsymbol{U}_{1i}(w) = -\hat{\rho}\,\frac{J_0\left(wa_i\right) - J_0\left(wb_i\right)}{w}$$
$$\boldsymbol{V}_{1i}(w) = \hat{\phi}\frac{a_i J_1\left(wa_i\right) - b_i J_1\left(wb_i\right)}{w}$$

As a first-order approximation, taking only the first term in the curly braces in Eq. (5.77) for two coaxial apertures supporting TEM modes results in

$$y_{ij}(1|1) \cong \frac{j\,\pi\,\chi_{1i}\chi_{1i}}{k\left(\varepsilon_{0p}\varepsilon_{0q}\right)^{1/4}} g_o\left(R_{ij}\right)\left(J_0\left(ka_i\right) - J_0\left(kb_i\right)\right)\left(J_0\left(ka_j\right) - J_0\left(kb_j\right)\right) \tag{5.78}$$

It is interesting to compare this result with the exact solution given in Eq. (5.64). Most coaxial cables are operated well below the cut-off of the first high-order mode and $|ka_i|$, $|kb_i| \ll 1$. In that case, the zero-order Bessel functions can be approximated by the first three terms of their series expansions to give

$$y_{ij}(1|1) \cong \frac{jk^3 \pi \chi_{mi}\chi_{ni}}{16\sqrt{\varepsilon_r}} \left(a_i^2 - b_i^2\right)\left(a_j^2 - b_j^2\right)\left[1 - \frac{k^2}{16}\left(a_i^2 + b_i^2\right)\right]\left[1 - \frac{k^2}{16}\left(a_j^2 + b_j^2\right)\right] g_o\left(R_{ij}\right). \tag{5.79}$$

Now, suppose two coaxial cables terminate in a ground so that they can radiate and couple. The coupling coefficients for this case is

$$S_{21}(1|1) = \frac{-2y_{ij}(1|1)}{\left[1 + y_{ii}(1|1)\right]\left[1 + y_{jj}(1|1)\right] - y_{ij}(1|1)\,y_{ji}(1|1)} \tag{5.80}$$

At medium frequencies above DC for the TEM modes $|y_{ii}(1\,|\,1)|$, $y_{ij}(1\,|\,1) \ll 1$

$$S_{21}(1|1) \cong -2y_{ij}(1|1) \tag{5.81}$$

Any of the approximations in Eqs. (5.70)–(5.79) could be used although with decreasing accuracy.

To demonstrate the typical accuracy of these expressions from the integral to asymptotic approximations, the results obtained have been compared with experiment. This consisted of two identical semirigid cables with inner and outer radii of 0.94 and 2.67 mm, respectively, filled with a dielectric with relative permittivity $\varepsilon_r = 1.56$ mounted in a ground plane, as shown in Figure 5.22. One of the cables was excited in the TEM mode, and the two-port scattering parameters were measured in the frequency range 1–40 GHz using a network analyzer. A plot of the computed cross-coupling between the two cables over the frequency band is shown in Figure 5.23, and for comparison, the experimental data are also shown. Results given by the TEM mode only using the exact formula Eq. (5.40) and the asymptotic approximation Eq. (5.77) are in excellent agreement with the measured data, even at frequencies above the cut-off of the TE_{11} mode (21.2 GHz), although the five-mode solution gives slightly better agreement at high frequencies. Equation (5.77) does not predict the phase variation at low frequencies very well, but in most other aspects, it works well.

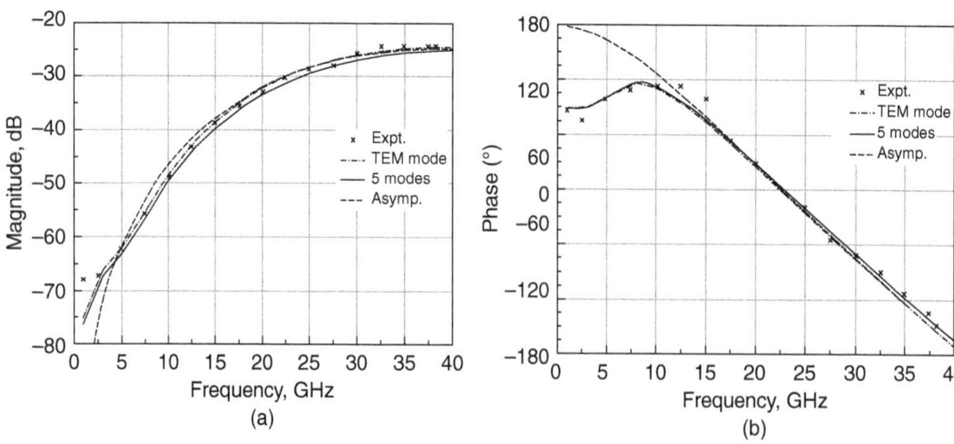

Figure 5.23 Coupling coefficient of coaxial cables terminated in a ground plane as shown in Figure 5.22. (a) Magnitude; (b) phase. Key: + – measured; dash-dot line – theory (TEM only); solid line – theory (five modes); and dash – asymptotic approximation Eq. (5.77).

Another example of the use of Eq. (5.77) will be demonstrated for coupling between coaxial waveguide modes in Section 5.4.4.

5.4.4 Coaxial and Circular Aperture Array Examples

An array example is illustrated in Figure 5.24. This array consists of a combination of a coaxial waveguide and two circular waveguides located nearby. All apertures terminate in a large metal ground plane. The 5.4-cm diameter coaxial waveguide (1) is located at the center of a large ground plane, and two circular apertures (2) and (3) have a diameter of 4.08 cm.

Measurements were made over a 3.25–6 GHz frequency range for the two polarizations $\boldsymbol{E}_1$ and $\boldsymbol{E}_2$, as defined in Figure 5.24. The results are plotted in Figures 5.25–5.27. Computations were made with nine modes in each waveguide, namely TE_{1n} and TM_{1n} ($n = 1,2$), TE_{2n} and TM_{2n} ($n = 1,2$), and TM_{01}. The results obtained are shown in the figures and are compared with the experimental reflection and coupling coefficients for the TE_{11} mode in each waveguide. Figure 5.27 shows the result of including mutual coupling in the radiation pattern calculations. In all cases, excellent agreement is obtained between experiment and theory.

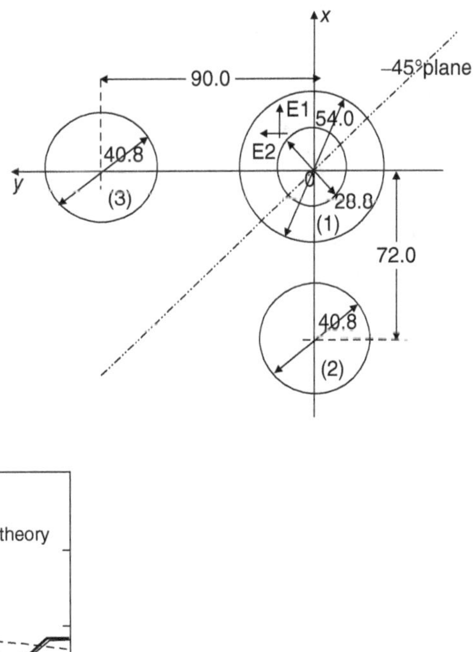

Figure 5.24 Array consisting of a coaxial waveguide and two circular waveguides opening into an infinite ground plane. All dimensions are in mm.

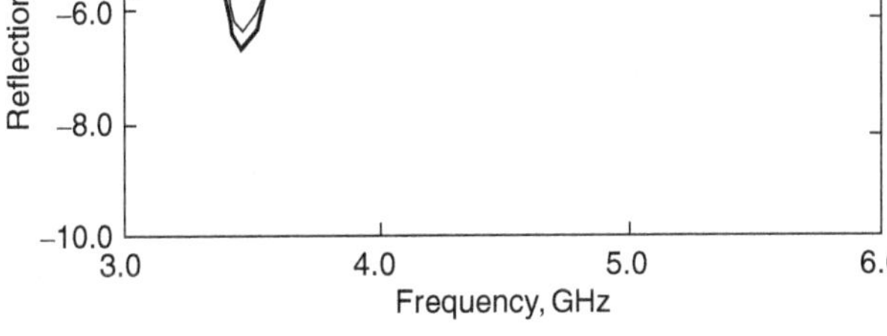

Figure 5.25 Reflection coefficient in the coaxial waveguide of the presence of two circular waveguides with incident TE_{11} mode polarized in the E1 and E2 directions shown in Figure 5.24.

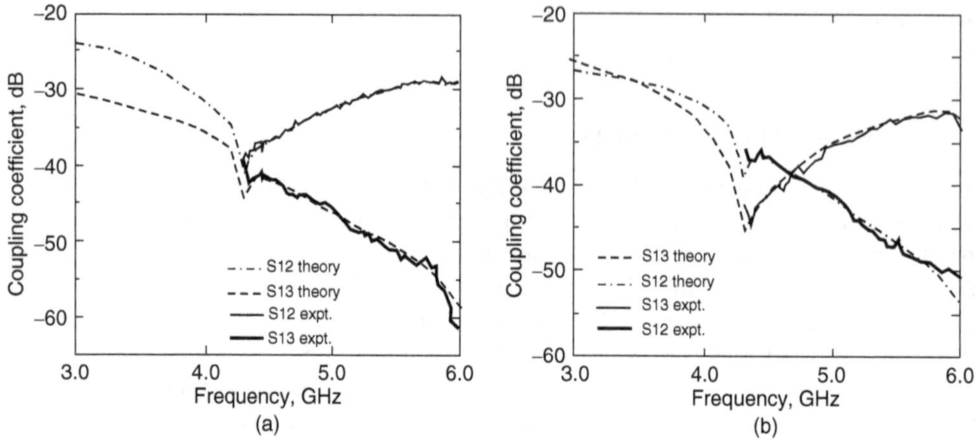

Figure 5.26 Coupling coefficients for the three-element array of circular waveguides for elements 1 and 2 (S_{12}) and elements 1 and 3 (S_{13}) corresponding to polarizations $\boldsymbol{E}_1$ and $\boldsymbol{E}_2$ (Figure. 5.24). Solid curve: theory; Dash curve: experiment. (a) $\boldsymbol{E}_1$ polarization; (b) $\boldsymbol{E}_2$ polarization.

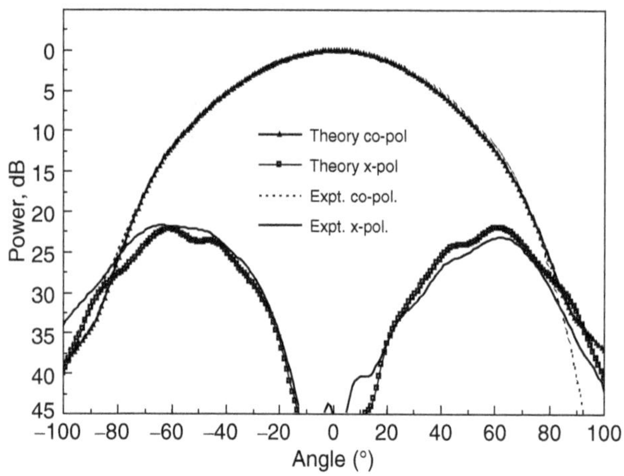

Figure 5.27 Radiation pattern of coaxial waveguide (1) at 4.5 GHz in the three-element array Figure 5.22 when excited in polarization E1.

An example of circular aperture coupling using the expressions for admittance listed in Table 5.7 is given in Bird (2016). In this case, a seven-element hexagonal array of identical conical horns with aperture diameters of 38.1 mm is described. Both computed and measured mutual coupling is shown for the central horn excited with either vertical or horizontal polarization. Once again, the simulated results are in excellent agreement with measured data.

As a final example, Figure 5.28 shows the coupling between coaxial waveguide apertures with a spacing $R_{ij} = 0.405a$ at a normalized frequency $ka = 1.861$ where a is the outer conductor radius as the center conductor radius (b) is increased. Four modes were included in the calculations, namely TE_{11}, TM_{11}, TE_{12}, and TM_{01}, using the exact and asymptotic expressions. The asymptotic solution with four modes is in good agreement with the exact approach. Figure 5.28a shows that in both planes cross-coupling peaks at about $b/a = 0.2$ and falls thereafter as the inner conductor gets larger.

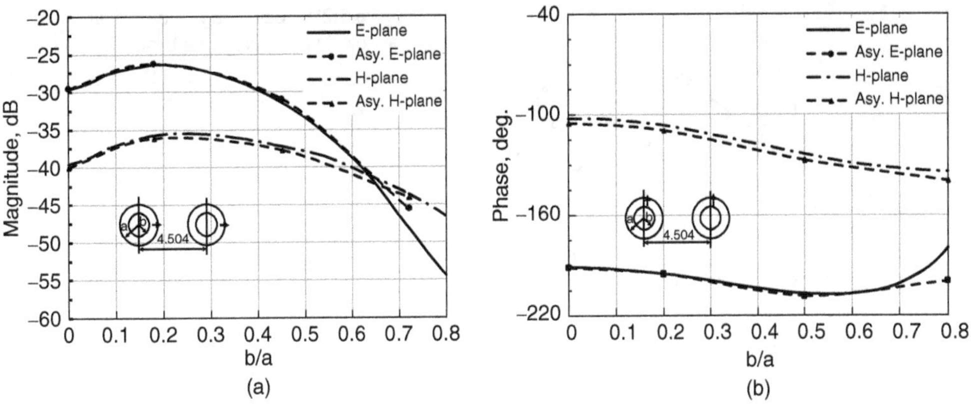

Figure 5.28 Coupling coefficient as a function of inner to-outer conductor radius ratio (b/a) for two coaxial waveguides in a ground plane. (a) Magnitude; (b) phase. Key: solid line – E-plane; dash and square – asymptotic E-plane; dash-dot line – H-plane; dash and triangle – asymptotic H-plane.

5.5 Mutual Coupling Between Apertures of General Cross-Section

Mutual coupling in waveguides and horns with general cross-sections can be analyzed using the same admittance expression given by Eq. (5.21). In separable coordinate systems, mode functions can be readily obtained in closed form, and in some instances, the associated integrals can similarly be evaluated in closed form. This depends on the availability of transform functions in those domains and whether the apertures are aligned. Otherwise, as we have seen for rectangular horns, fourfold integrals must be evaluated for general cross-sections. For some separable coordinate systems, the number of integrals may be reduced by replacing the free-space Green's function, G, with an equivalent transform representation as was done in Section 5.4 for circular/coaxial apertures. Such a representation exists for elliptical coordinates, for example, but how useful this is to the remainder of the analysis depends on the tractability of the resulting integrals,

5.5.1 Elliptical Apertures

Horns with an elliptical cross-section (see Figure 5.29) find application as sources of circular polarization and as feeds for reflectors. Radiation from elliptical corrugated horns (Vokurka 1979) has been studied in the past, but relatively little attention has been paid to ones with a smooth wall up to the aperture and the aperture reflection, which is needed in all horn designs. Mention is made of the early work of Müller (1960) who described solutions for the radiation characteristics of smooth-wall elliptical waveguides and gave results for radiation by the fundamental TE_{c11} mode. In elliptical waveguide, there are infinite sets of TE and TM modes, designated as TE_{cmn}, TE_{smn}, TM_{cmn}, and TM_{smn}, where the subscripts c and s refer to the even and odd types of Mathieu functions (McLachlan 1947) involved in each mode set. Experience with smooth-wall horn radiators of rectangular and circular cross-section has shown that significant reflection can occur from apertures of moderate size, resulting in mismatch at the horn input. A complete solution to the elliptical aperture problem has been given by the author (Bird 1990b), in which both theoretical and experimental results were given. The analysis uses the free-space Green's function for elliptical apertures in a ground plane. The aperture field and ground plane are replaced by a magnetic current distribution as described earlier in this chapter. The four basic modes sets in the solution

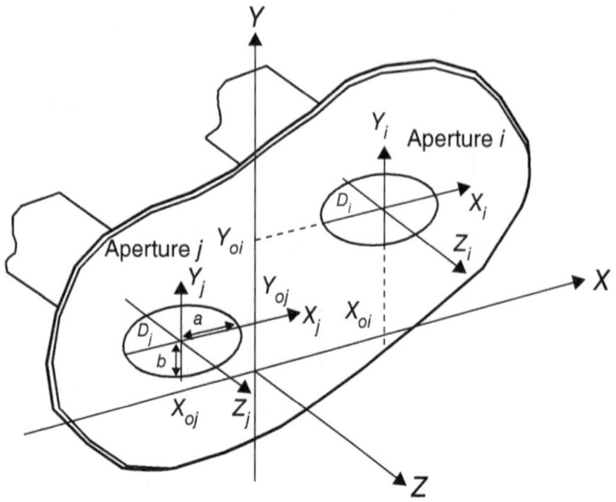

Figure 5.29 Geometry of elliptical waveguide apertures in a ground plane.

of Mathieu's equation in elliptical regions have the electric field parallel to the minor axis (y-axis in Figure 5.29), the 'c' modes, and the 's' modes which have the electric field parallel to the major axis.

The fundamental mode of elliptical waveguide TE_{cmn} limits to the TE_{11} mode in circular waveguide as $b \to a$. The mutual admittance of the TE_{cmn} and $TE_{cm'n'}$ modes in elliptical waveguide apertures D and D′ is given by

$$y_{ij}\left(mn|m'n'\right) = \frac{jkY_o hh' D_{mn} D_{m'n'}}{\sqrt{Y_{mn} Y_{m'n'}}} \int_0^{\xi_o} d\xi \int_0^{2\pi} d\eta \int_0^{\xi_o} d\xi' \int_0^{2\pi} d\eta' \, G\left(X - X|Y - Y'\right)$$
$$\times \left[f_x(\xi, \eta, q) f_x(\xi', \eta', q') + f_y(\xi, \eta, q) f_y(\xi', \eta', q') - \left(\frac{k^2_{c,mn} h k^2_{c,m'n'} h'}{k^2} \right) \right.$$
$$\left. \times f_z(\xi, \eta, q) f_z(\xi', \eta', q') \right] \tag{5.82}$$

where in Eq. (5.82) $G(x|y) = \exp\left(-jk\sqrt{x^2+y^2}\right)/4\pi\sqrt{x^2+y^2}$ and the functions in the integrand are as follows:

TE_{cmn} modes

$$f_{xm}(\xi, \eta, q) = Ce'_m(\xi, q)\, ce_m(\eta, q)\, v - Ce_m(\xi, q)\, ce'_m(\eta, q)\, u$$
$$f_{ym}(\xi, \eta, q) = Ce'_m(\xi, q)\, ce_m(\eta, q)\, u + Ce_m(\xi, q)\, ce'_m(\eta, q)\, v$$
$$f_{zm}(\xi, \eta, q) = \left(l^2/h^2\right) Ce_m(\xi, q)\, ce_m(\eta, q) \tag{5.83}$$

in which (ξ, η, z) are elliptical cylinder coordinates, $u = \cosh\xi \sin\eta$, $v = \sinh\xi \cos\eta$, and $l^2 = h^2(cosh^2\xi - cosh^2\eta)$,

$$h = \sqrt{a^2 - b^2}, \text{and} \quad \cosh\xi_o = a/b$$

and q is the separation variable (McLachlan, 1947). The Mathieu functions $ce_m(\eta, q)$ and $se_m(\eta, q)$ are, respectively, even and odd angular functions of order m, and $Ce_m(\xi, q)Se_m(\xi, q)$ are the even and odd modified radial functions where the orthogonality and normalizations are given by

$$\int_0^{2\pi} d\eta\, ce_m(\eta, q)\, ce_n(\eta, q) = \int_0^{2\pi} d\eta\, se_m(\eta, q)\, se_n(\eta, q) = \begin{cases} \pi \text{ if} & m = n \\ 0 \text{ if} & m \neq n \end{cases}, \tag{5.84a}$$

$$\int_0^{2\pi} d\eta\, ce_m(\eta, q)\, se(\eta, q) = 0, \tag{5.84b}$$

$$M_e^2(\eta) = \int_0^{2\pi} d\phi [ce_m(\eta, cos\phi)]^2 \text{ and } M_o^2(\eta) = \int_0^{2\pi} d\phi [se_m(\eta, cos\phi)]^2. \tag{5.84c}$$

The multiplier D_{mn} in Eq. (5.82) is a normalization constant, which is chosen to give unit power transfer in each mode and a symmetric scattering matrix, and is

$$D_{mn} = \sqrt{\frac{2}{\pi}} \left[\int_0^{\xi_o} d\xi \begin{Bmatrix} \left(Ce_m'^2(\xi, q) + \Theta_m Ce_m^2(\xi, q)\right) \\ \left(Se_m'^2(\xi, q) + \Phi_m Se_m^2(\xi, q)\right) \end{Bmatrix} \right]^{-1/2} \quad \begin{matrix} TE_{c,mn} \\ TE_{s,mn} \end{matrix} \tag{5.85}$$

$$\Theta_m = \begin{cases} A_0^{(m)} A_2^{(m)} + \sum_{r=0}^{\infty} A_{2r}^{(m)} A_{2r+2}^{(m)}; \; m \text{ even} \\ \frac{1}{2}\left(A_1^{(m)}\right)^2 + \sum_{r=0}^{\infty} A_{2r+1}^{(m)} A_{2r+3}^{(m)}; \; m \text{ odd} \end{cases} \tag{5.86}$$

where $A_n^{(m)}$ are the coefficients of the series defining the ce_m Mathieu function that are expressed as

$$ce_m(\eta, q) = \sum_{n=0}^{\infty} \begin{cases} A_{2n}^{(m)} \cos(2n\eta); \; m \text{ even} \\ A_{2n+1}^{(m)} \cos[(2n+1)\eta]; \; m \text{ odd} \end{cases}$$

Similarly, Φ_m are defined in terms of $B_n^{(m)}$ coefficients of the se_m function

$$se_m(\eta, q) = \sum_{n=0}^{\infty} \begin{cases} A_{2n+2}^{(m)} \sin[(2n+2)\eta]; \; m \text{ even} \\ A_{2n+1}^{(m)} \sin[(2n+1)\eta]; \; m \text{ odd} \end{cases}$$

For the remaining modes, the mode functions are obtained from Eq. (5.83) as described below:

TE_{smn} modes:

Same as for Eq. (5.83) except the even functions are replaced with odd functions, as follows: $Ce \to Se$ and $ce \to se$.

TM_{cmn} modes:
Similar to Eq. (5.83) with the replacements $(f_y)_{TE} \Rightarrow (f_x)_{TM}$, $(f_x)_{TE} \Rightarrow (-f_y)_{TM}$, and $(f_z)_{TM} = 0$.
TM_{smn} modes:
Same as for the TM_{cmn} modes as described above except with replacements $Ce \to Se$ and $ce \to se$.

In addition, the mode cut-off wavenumber is $k_{c,mn} = 2h\sqrt{q_{mn}}$, in which q_{mn} is the n-th zero of the following identities:

$$\text{TE Modes: } \begin{Bmatrix} Even; \; Ce_m'(\xi_o, q) \\ Odd; \; Se_m'(\xi_o, q) \end{Bmatrix} = 0$$

$$\text{TM Modes: } \begin{Bmatrix} Even; \; Ce_m(\xi_o, q) \\ Odd; \; Se_m(\xi_o, q) \end{Bmatrix} = 0.$$

The computation time required for the fourfold integration in Eq. (5.82) can be significantly reduced by utilizing the symmetry of the elliptical cross-section and integrating only over a quadrant of the ellipse, e.g. $\xi = [0, \xi_o]$ and $\eta = [0, \pi/2]$. Quadrant reduction of the integrals leads to

$$y_{12}\left(mn|m'n'\right) = \frac{jkY_o hh' D_{mn} D_{m'n'}}{\sqrt{Y_{mn} Y_{m'n'}}} \int_0^{\xi_o} d\xi \int_0^{\pi/2} d\eta \int_0^{\xi_o} d\xi' \int_0^{\pi/2} d\eta' \left[f_{xm} f'_{xm'}\, \sigma_{xm}\, \overline{G}\, \sigma_{xm'}{}^T + f_{ym} f'_{ym'}\, \sigma_{ym}\, \overline{G}\, \sigma_{ym'}{}^T - \frac{k_{c,mn}^2\, h\, k_{c,m'n'}^2\, h'}{k^2} f_{zm} f'_{zm'}\, \sigma_{zm}\, \overline{G}\, \sigma_{zm'}{}^T \right] \tag{5.87}$$

where

$$\sigma_{xm} = \begin{cases} \left[1\ (-1)^m\ (-1)^m\ \ 1\right] & ; TE_{cmn}/TM_{smn} \\ \left[1\ \ (-1)^{m+1}\ \ \ (-1)^m\ \ -1\ \right] & ; TE_{smn}/TM_{cmn} \end{cases}$$

$$\sigma_{ym} = \begin{cases} \left[1\ \ (-1)^{m+1}\ \ \ (-1)^m\ \ -1\ \right] & ; TE_{cmn}/TM_{smn} \\ \left[1\ \ 1\ \ 1\ \ 1\ \right] & ; TE_{smn}/TM_{cmn} \end{cases}$$

$$\sigma_{zm} = \begin{cases} \left[1\ \ (-1)^{m+1}\ \ \ (-1)^m\ \ 1\ \right] & ; TE_{cmn} \\ \left[1\ \ 1\ \ -1\ \ -1\ \right] & ; TE_{smn} \end{cases}$$

and

$$\overline{G} = \begin{bmatrix} G(X-X'|Y-Y') & G(X+X'|Y-Y') & G(X+X'|Y+Y') & G(X-X'|Y+Y') \\ G(X+X'|Y-Y') & G(X-X'|Y-Y') & G(X-X'|Y+Y') & G(X+X'|Y+Y') \\ G(X+X'|Y+Y') & G(X-X'|Y+Y') & G(X-X'|Y-Y') & G(X+X'|Y-Y') \\ G(X-X'|Y+Y') & G(X+X'|Y+Y') & G(X+X'|Y-Y') & G(X-X'|Y-Y') \end{bmatrix}$$

which is Toeplitz due to physical symmetry.

An integral transform of the scalar Green's function expressed in elliptical coordinates is (Morse and Feshbach 1953)

$$G\left(x-x'|y-y'\right) = 2\int_0^{\infty} d\zeta \frac{\zeta}{\sqrt{k^2-\zeta^2}} \sum_{r=0}^{\infty} \left[ce_r(\eta,w)\, ce_r\left(\eta',w\right) Ce_r(\xi,w)\, Ce_r\left(\xi',w\right)/p_r^2 + se_r(\eta,w)\, cs\left(\eta',w\right) Se_r(\xi,w)\, Se_r(\xi',w)/s_r^2\right] \tag{5.88}$$

where $w = (h\,\zeta/2)^2$, $x = h\cosh\xi\cos\eta$, and $y = h\sinh\xi\sin\eta$,

$$p_r = ce_r(0,q) \begin{cases} -ce'_r\left(\frac{\pi}{2},q\right)/kA_1^{(r)}; & r\ odd \\ ce_r\left(\frac{\pi}{2},q\right)/A_0^{(r)} & ; r\ even \end{cases}$$

and

$$s_r = se'_r(0,q) \begin{cases} se_r\left(\frac{\pi}{2},q\right)/kB_1^{(r)}; & r\ odd \\ se'_r\left(\frac{\pi}{2},q\right)/B_2^{(r)} & ; r\ even \end{cases}$$

Equation (5.88) could be used in Eq. (5.87), but this appears to offer few advantages for numerical evaluation due to the paucity of closed form solutions to Mathieu function integrals compared with Bessel functions. Therefore, in this case, the fourfold integration in Eq. (5.82) has been carried out directly. The integral over an aperture quadrant, Eq. (5.87), can be evaluated effectively using numerical quadrature such as Simpson or Gauss (Fröberg 1974). However, the transform representation Eq. (5.88) is useful for identifying modes which couple and those that do not. As shown in Figure 5.30, the dominant TE_{c11} mode couples only to the TE_{cmn} and TM_{smn} modes, where m is an

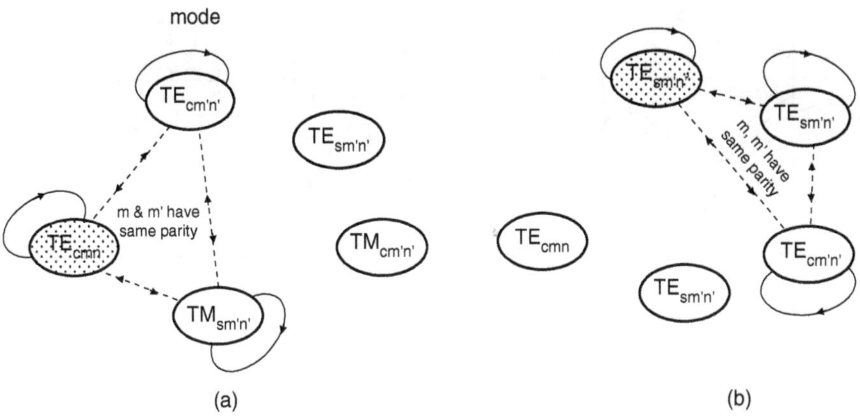

Figure 5.30 Mode coupling in elliptical apertures. Solid line indicates self-coupling, and dashed line shows mode coupling between apertures. Excited mode shown shaded. (a) Even; (b) odd excitation.

odd integer. Thus, the TE_{c11} mode and its odd mode counterpart, the TE_{s11} mode, are not coupled at the open end (McLachlan 1947, pp. 368–369). This is not unexpected as the dominant field polarizations are orthogonal (see Figure 5.30). The TE_{cmn} mode does not couple at all to the $TE_{sm'n'}$ and $TM_{cm'n'}$ modes.

The reflection coefficient of the TE_{c11} mode in a single aperture can be calculated from Eq. (5.87) by letting $m = 1 = m'$ and $n = 1 = n'$. Convergence can be slow due to the singularity of the Green's function, but this singularity can be removed by subtraction (see Section 3.6.2). In this approach, an integral having the same type of singularity with a known integral value is added to the right-hand side outside the integral and also is subtracted under the integral sign. A suitable integral for this purpose is found from the static fields, namely

$$I(x, y) = \iint_{D'} \frac{dx'\,dy'}{\sqrt{(x - x')^2 + (y - y')^2}}. \tag{5.89}$$

When this singularity is subtracted, stable solutions result. For instance, Bird (1990b) obtained the admittance of elliptical waveguide as a function of normalized frequency ka with waveguide ellipticity as a parameter. Some computed results are compared with experiment in Figure 5.31. In

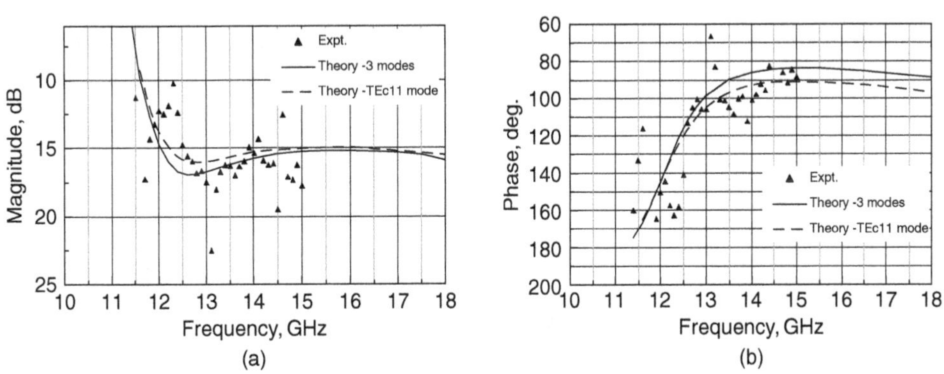

Figure 5.31 Reflection coefficient of elliptical waveguide aperture in a ground plane a/b = 2 where a = 7.9 mm, experiment and theory. (a) Magnitude; (b) phase.

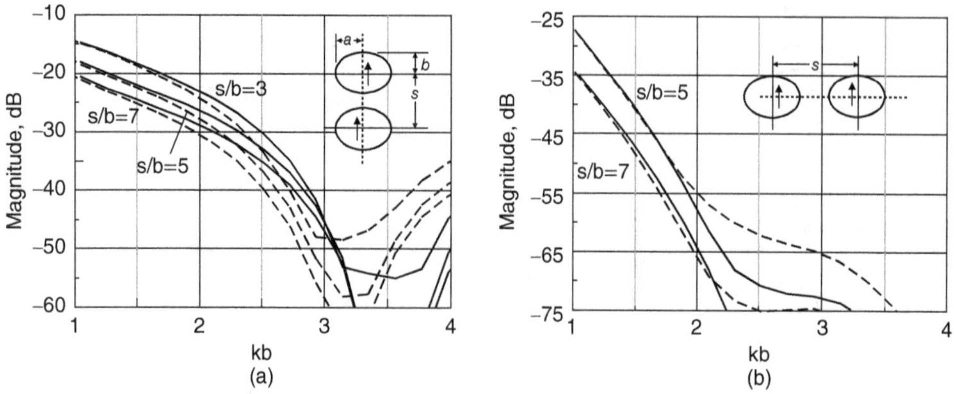

Figure 5.32 Magnitude of coupling coefficient S_{12} for identical (solid) elliptical and (dashed) rectangular waveguides with the same aspect ratio a/b = 2 as a function of separation distance (s). (a) E-plane; (b) H-plane. Source: After Bird, T.S. (1990b). 'Behaviour of multiple elliptical waveguides opening into a ground plane', IEE Proc. H, vol. 137, pp. 121-126.

these calculations, the approximation that coupling to other modes in the aperture was ignored. The single mode computation results correlate reasonably well with experiment. The reflection coefficient in elliptical waveguide increases as a/b increases in much the same way as the reflection level does for the TE_{10} mode of rectangular waveguide. Well above cut-off, the level of reflection of the TE_{c11} mode is approximately the same as for the TE_{10} mode in open-ended rectangular waveguide with the same aspect ratio, a/b of the ellipse (Bird 1990b).

Mutual coupling has been calculated between two elliptical apertures operating in the fundamental TE_{c11} mode only, which is a good approximation to the actual coupling. E- and H-plane coupling computed from Eq. (5.82) is shown in Figure 5.32. Also shown is the coupling for rectangular waveguides with the same aspect ratio and operating in the TE_{11} mode. It is seen that the coupling coefficient in both types of aperture decreases comparable until close to the onset of the first high-order mode.

5.5.2 General Apertures

More general aperture cross-sections can be analyzed with Eq. (5.21) by expressing the mode functions as series of basic functions such as Bessel functions as has been done by Kuehne and Marquardt (2001), for example, as

$$F_{zm}(\rho,\phi,z) = \sum_{p=0}^{\infty} C_{pm} J_p\left(k_{cm}\rho\right) \cos\left(p\phi - \psi_m\right) exp\left(-jk_{zm}z\right) \tag{5.90}$$

where k_{cm} and k_{zm} are the wavenumbers of m mode in the transverse and axial directions, which satisfy $k_{cm}^2 + k_{zm}^2 = k^2$, ψ_m is the phase reference, and C_{pm} is the p-th mode amplitude. More general functions such as polynomials can also be used as well as discrete regions such as triangular subelements (Bird 2016). As an example of the results obtained, Figure 5.33 shows the coupling coefficient S_{21} between two rectangular waveguides with rounded corners (Kuehne and Marquardt 2001) compared with experiment, which are in excellent agreement.

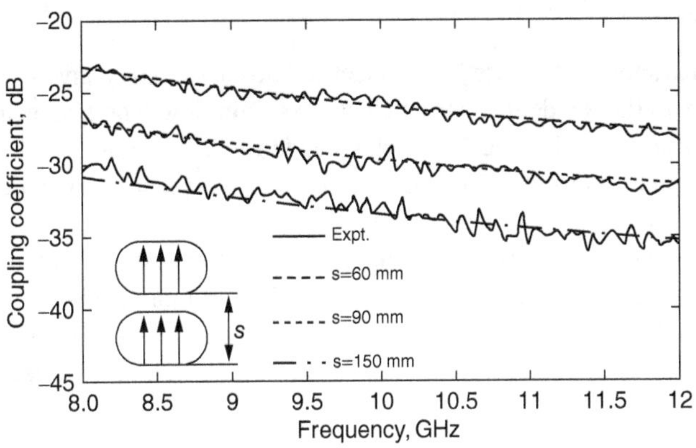

Figure 5.33 Coupling coefficient S_{21} for E-plane coupling between to rectangular waveguides with rounded corners. (Source: After Kuehne and Marquardt 2001).

5.6 Coupling in Apertures Loaded with Dielectrics and Metamaterials

The element patterns provided by common horns such as conical or pyramidal ones often do not provide sufficient flexibility or performance for some applications such as requiring high gain, low sidelobes, or shaped beams. Ways of providing this flexibility as well as improved bandwidth include loading the wall or the interior with dielectric or metamaterials. Dielectric loading of horns has been employed for many years, but with the improvement in understanding and design of metamaterials, they have also been adopted in horn technology (e.g. Wu et al. 2013a,b). Dielectric loading of horns is a simple method of controlling the radiation pattern. For instance, in rectangular horns, dielectric lining only on the walls parallel to the E-plane (E-plane slab load) is used to increase gain and to make the aperture field in the H-plane pattern almost uniform much like the E-plane (see Figure 5.34). This is achieved when the wavenumber in the dielectric slab (#1) is zero corresponding to a dielectric thickness on walls d given by:

$$d = \frac{\lambda}{4\sqrt{\varepsilon_{r1} - \varepsilon_{r2}}}, \tag{5.91}$$

where ε_{r1}, ε_{r2} are the dielectric constants of the dielectric slab and the horn filling, respectively. This has been used to improve the illumination of a reflector in shaped beam from satellites (Bird 2016). By contrast, dielectric lining on the walls parallel to the H-plane only, i.e. H-plane slab loading can be used to reduce the sidelobes in the E-plane. Horn array elements can be designed to be low or high gain depending on the demands of the application. Low gain horns are widely used as feeds for reflector antennas due to their wide beamwidth and low side lobes, which results in low reflector spillover loss and high overall antenna efficiency. High gain horns on the other hand are characterized by nearly uniform aperture amplitude and phase distribution with corresponding narrow beamwidth and high side lobes. They are widely used as array elements for multifeed reflector satellite antennas and as efficient antenna elements for direct radiating arrays for radar and satellites.

5.6.1 Dielectric-Loaded Apertures

In the E-plane slab horns and apertures shown in Figure 5.34, dielectric slabs of thickness given by Eq. (5.91) line the inside walls parallel to the incident electric field. Once again, it will be assumed that the apertures are located in a large ground plane. The fields inside the waveguides can be represented as a sum of longitudinal-section electric (*LSE*) and of longitudinal-section magnetic (*LSM*) modes. A consequence of the geometry is that these modes are TE and TM to the dielectric interface, i.e. x so that the formulation in Section 5.2.2 has to be modified. This means the dominant mode in this structure is the LSE_{mn} or TE_{xmn} mode where m is odd. Both notations will be used interchangeably. Following the approach given in Section 5.2.2, the mutual admittance of modes p and q in apertures i and j is given by Eq. (5.16). A structure that is TE_x or TM_x, the composite field vector in this case becomes

$$\Psi_{pi} = \hat{z} \times \boldsymbol{e_{pi}} + \hat{z}\,\eta_o h_{z,pi}, \tag{5.92}$$

where, as before, $\boldsymbol{e_{pi}}$ and $h_{z,\,pi}$ are the transverse electric field and longitudinal magnetic field, respectively, of the TE_{xmn} or TM_{xmn} mode p (or mn) in waveguide i. The transverse fields are normalized over the cross-section D, as follows:

$$\frac{1}{2} \iint\limits_{D} dS\, \boldsymbol{e_{pi}} \times \boldsymbol{h_{pi}} \cdot \hat{z} = 1. \tag{5.93}$$

The resulting admittance expression in terms of *LSE* and *LSM* modes can be applied to conventional homogeneously filled waveguide, which yields identical expressions such as Eq. (5.21). It is used to advantage when the domain is separated by an interface such as in Figure 5.34b. In this instance, the x-domain is split between regions (1) and (2). The fundamental mode $\mathrm{TE}_{x10}/\mathrm{LSE}_{10}$ of this structure has nonzero field components given by (Woodward 1947)

$$e_y^{(1)} = E_o\, c_o \sin\left[k_1\left(\frac{a}{2} - |x|\right)\right];\ (a/2 - d) \le |x| \le a/2 \tag{5.94a}$$

$$e_y^{(2)} = E_o \cos\left[k_2 x\right]\ ;\ |x| \le (a/2 - d) \tag{5.94b}$$

$$h_x^{(1)} = -\left(\frac{\gamma}{j\omega\mu}\right) E_o\, c_o \sin\left[k_1\left(\frac{a}{2} - |x|\right)\right];\ (a/2 - d) \le |x| \le a/2 \tag{5.94c}$$

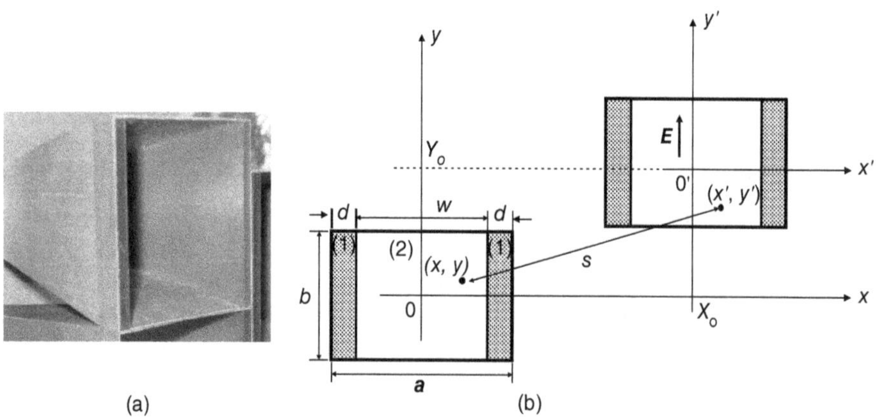

Figure 5.34 Aperture with E-plane slab dielectric loading. (a) Horn Source: Trevor S. Bird; (b) geometry of coupled apertures.

$$h_x^{(2)} = -\left(\frac{\gamma}{j\omega\mu}\right) E_o \cos\left[k_2 x\right];\ |x| \leq (a/2-d) \tag{5.94d}$$

$$h_z^{(1)} = \left(\frac{k_1}{j\omega\mu}\right) E_o c_o \cos\left[k_1\left(\frac{a}{2}-|x|\right)\right];\ (a/2-d) \leq |x| \leq a/2 \tag{5.94e}$$

$$h_z^{(2)} = \left(\frac{k_2}{j\omega\mu}\right) E_o \sin\left[k_2 x\right];\ |x| \leq (a/2-d) \tag{5.94f}$$

where E_o is a normalization factor, which is found by satisfying Eq. (5.93), the superscripts refer to the region numbers shown in Figure 5.34b, $c_o = \cos\left(k_2 \frac{w}{2}\right)/\sin(k_1 d)$, $w = a - 2d$, and $k_i = \sqrt{k^2\varepsilon_{ri} - \gamma_{10}^2}$; $(i = 1, 2)$ are wavenumbers in regions (1) and (2). When continuity is satisfied at the dielectric boundary, it is found that:

$$k_2 tan\left(\frac{k_2 w}{2}\right) = k_1 \cot\left(k_i d\right), \tag{5.95}$$

which is the characteristic equation of the structure. This has an infinite number of solutions corresponding to the TE_{x1n} $n = 0,1,\ldots$ modes. A lower bound on the approximate propagation coefficient γ based in the perturbation method (Bird 2016) for the LSE_{mn} mode is

$$\frac{\gamma}{k} \cong \sqrt{\begin{array}{c} 1 + (\varepsilon_{r1}-1)\frac{2d}{a}\left(1 - \frac{a}{2m\pi d}\sin\left(\frac{2m\pi d}{a}\right)\right) + (\varepsilon_{r2}-1)\frac{2d}{a}\left(1-\frac{2d}{a}\right)\times \\ \left(\frac{a}{d} - 2 + \frac{a}{2m\pi d}\sin\left(\frac{2m\pi d}{a}\right)\right) - \frac{k_{o(0)}^2}{k^2} \end{array}} \tag{5.96}$$

where $k_{o(0)}^2 = \left(\frac{m\pi}{a}\right)^2 + \left(\frac{n\pi}{a}\right)^2$. Equation (5.96) can be used as an initial value for finding the roots of Eq. (5.95) by iteration.

With the field quantities listed in Eq. (5.95), the composite field vector for the $TE_{x10}$is

$$\boldsymbol{\Psi}_{1i} = -\hat{x}\, e_y^{(i)} + \hat{z}\,\eta_o h_z^{(i)} \tag{5.97}$$

The mutual admittance of this mode in waveguides *i* and *j* is therefore:

$$y_{ii}(10|10) = \frac{jk}{4\pi\eta_o\sqrt{Y_{pi}Y_{qj}}}\int_{S_i} dS\,\Psi_{1i}(x,y)\cdot\int_{S_j} dS'\Psi_{1j}(x',y')\,g_o(|r-r'|) \tag{5.98}$$

where the inner integral over S_j is:

$$\begin{aligned}\int_{S_j} dS'\Psi_{1j}(x',y')\,g_o(|r-r'|) = -\hat{x}\,b&\left[\int_{-a/2}^{-a/2+d} dx'\,e_y^{(1)} g_o(|r-r'|) + \int_{-a/2+d}^{a/2-d} dx'\,e_y^{(2)} g_o(|r-r'|)\right.\\ &\left.+\int_{a/2-d}^{a/2} dx' e_y^{(1)} g_o(|r-r'|)\right] + \hat{z}b\,\eta_o\left[\int_{-a/2}^{-a/2/+d} dx'\,h_z^{(1)} g_o(|r-r'|)\right.\\ &\left.+\int_{-a/2+d}^{a/2-d} dx'\,h_z^{(2)} g_o(|r-r'|) + \int_{-a/2+d}^{a/2} dx' h_y^{(1)} g_o(|r-r'|)\right],\end{aligned}$$

in which $g_o(R) = \frac{\exp(-jkR)}{R}$ and $|\boldsymbol{r}-\boldsymbol{r}'| = \sqrt{(x-x')^2+(y-y')^2}$. As the Ψ_{1i} and Ψ_{1j} are symmetric functions, this can be used to reduce the integration to one-half of each aperture. In the case of the self-admittance of the TE_{x10} mode,

$$y_{11}(10|10) = \frac{jk\,Y_o\,E_o}{\pi}\int_0^b d\lambda\,(b-\lambda)\left[2\,c_o^2 I_1 + 2\,c_o^2 I_3 + 4\,c_o I_2 + I_4\right],$$

where I_1, I_2, I_3, etc., are integrals which can be simplified further by means of Lewin's transformation. The final result is

$$y_{11}\left(10|10\right) = \frac{2jk^2}{\pi\,\gamma_{10} ab\, N_{10}} \int_0^b d\lambda\,(b-\lambda)\left\{\int_0^w d\sigma\left[g_1(\lambda) + g_2(\lambda) + g_{12}(\lambda)\right]\right\} \tag{5.99}$$

in which

$$\begin{aligned} g_1(\lambda) = \int_0^w d\sigma & \Bigg[g_o(\sigma,\lambda)\left(A_+\left(\sin(k_2(w-\sigma)\right) + A_- k_2(w-\sigma)\cos\left(k_2\sigma\right)\right. \\ & - B_+\cos\left(k_1 d + k_2\left(\sigma - \frac{w}{2}\right)\right) - B_-\cos\left(k_1 d + k_2\left(\sigma - \frac{w}{2}\right)\right)\Big) \\ & + g_o(\sigma - a + d, \lambda)\frac{4k_1k_2}{\left(k_1+k_2\right)\left(k_1-k_2\right)} A_-\cos\left(k_2(w-\sigma)\right)\Bigg] \end{aligned}$$

$$\begin{aligned} g_2(\lambda) = \int_0^d d\sigma & \Bigg[g_o(\sigma,\lambda)\Bigg(-D_+\left(\sin(k_1(2d-\sigma)\right) - \sin\left(k_1\sigma\right) + 2D_- k_1(d-\sigma)\cos\left(k_1\sigma\right) \\ & + B_+\cos\left(k_1(d-\sigma) - \frac{k_2 w}{2}\right) + B_-\cos\left(k_1(d-\sigma) + \frac{k_2 w}{2}\right)\Bigg)\Bigg] \end{aligned}$$

$$\begin{aligned} g_{12}(\lambda) = \int_0^d & d\sigma\left[g_o(\sigma - a + d,\lambda)\left(D_+\sin\left(k_1(d-\sigma)\right)\right) - D_- k_1(d-\sigma)\cos\left(k_1(\sigma+d)\right)\right. \\ & - B_+\cos\left(k_1\sigma + k_2 w\right) - B_-\cos\left(k_1\sigma - k_2 w\right)\big)\big) + g_o(\sigma - a,\lambda)\left(D_+\sin\left(k_1\sigma\right)\right. \\ & \left. -D_- k_1\sigma\cos\left(k_1\sigma\right)\right)\big) + g_o(\sigma - a + d,\lambda)\left(D_+\sin\left(k_1\sigma\right) - D_- k_1\sigma\cos\left(k_1\sigma\right)\right)\big)\big]. \end{aligned}$$

Other parameters are $A_\pm = \frac{1}{k_2}\left(1 \pm \frac{k_2^2}{k^2}\right)$, $B_\pm = \frac{2c_o}{(k_1 \pm k_2)}\left(1 \pm \frac{k_1k_2}{k^2}\right)$, $D_\pm = \frac{c_o}{k_1}\left(1 \pm \frac{k_1^2}{k^2}\right)$, and

$$N_{10} = \left[\left(1 - \frac{2d}{a}\right) + \frac{w}{a}S\left(k_2 w\right) + \frac{2d}{a}c_o^2\left(1 + S\left(2k_1 d\right)\right)\right],$$

in which $S(..)$ is the sinc function (Appendix A). It can be shown that Eq. (5.99) simplifies to Eq. (5.41) when the waveguide is fully filled with air or a dielectric, i.e. $d \to a/2$ or $d \to 0$. In addition, the mutual admittance can be generalized for TE_{xmn} modes by modifying the integral over λ from a transformation in the y-y′ domain by a straightforward application of Lewin's transformation.

The reflection coefficient of the aperture can now be obtained from the admittance in Eq. (5.98) and extended to high-order modes. As an example of the results given by Eq. (5.99), Figure 5.35 shows the reflection coefficient at 12.85 GHz of the fundamental LSE mode at a 22.86 mm square horn loaded with a dielectric material having a permittivity $\varepsilon_{r1} = 3.84$ and thickness d satisfying Eq. (5.91). The remainder of the horn is air-filled, and the dielectric is tapered from the throat to the aperture These results were computed assuming eight modes in the aperture. The modes included were LSE_{10}, LSM_{12}, LSE_{12}, LSE_{30}, LSM_{32}, LSM_{14}, LSE_{14}, and LSE_{10}. As well, results predicted by the single-mode theory Eq. 5.98 are plotted also to demonstrate the importance of high-order modes in this structure. Mode selection is important to the satisfaction of the edge condition at the aperture as was observed by Hockham (1975).

As an example of this, an E-plane slab-loaded waveguide with a = b = 0.6λ, $d = 0.0908\,\lambda$ and $\varepsilon_{r1} = 4$ was computed and measured. The computed reflection coefficient was found to yield a stable input value of −27.12 dB/_−73.58 deg (Bird and Hay 1990) with 19 selected modes. It was noted that the amplitudes of the modes in region (2) in ascending cut-off frequency decayed with increasing order. In particular, the LSE_{m0} (m odd) and LSE_{1n} (n even) modes that contribute to the

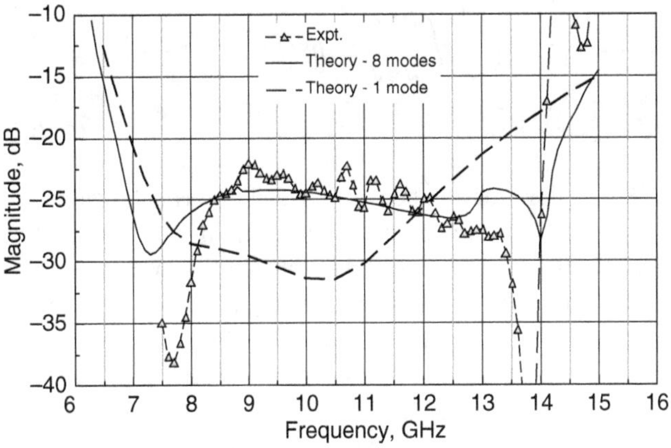

Figure 5.35 Reflection coefficient in E-plane slab-loaded waveguide with a = 22.86 mm = b, lined with dielectric of thickness $d = 3.45$ mm, $\varepsilon_{r1} = 3.84$, and filling with $\varepsilon_{r2} = 1$. Solid line – theory (8 modes); dash - theory (1 mode); dash with Δ marker – experiment.

edge condition at the flange decayed in proportional to $m^{-3.44}$ and $n^{-1.78}$, respectively. By increasing the number of modes to 41, the proportionality changed to $m^{-2.32}$ and $n^{-1.754}$. This compares with a 90° metal edge condition of $m^{-2.667}$ and $n^{-1.667}$, which correlates quite well with modal decay and indicates convergence (Hockham 1975). The measured results shown in Figure 5.34 have some ripple most probably due to minor mismatches at the throat, aperture, and the horn taper at a level of about −35 dB. Even so, there is quite good agreement between theory and experiment although the minima near 7.6 and 14 GHz are far deeper than predicted.

The mutual admittance between two identical apertures can be determined in the same way. This calculation is tedious as the integration must be undertaken over three regions of each of the two domains. It is more direct to carry out the fourfold integral in Eq. (5.21) numerically. The mutual admittance is obtained between the fundamental LSE_{10} modes in each waveguide as given in Eqs. 5.94. Once again, we can commence from Eq.5.98, where the argument of the scalar Green's function $g_o(..)$ is now $|\mathbf{r} - \mathbf{r}'| = \sqrt{(x - x' - X_o)^2 + (y - y' - Y_o)^2}$. As shown in Figure 5.34b, (X_o, Y_o) are the coordinates of the origin of the primed coordinate system relative to the global system. Hence,

$$y_{ij}(10,10) = \frac{jk^2 E_o}{\pi \gamma_{10} ab}\left\{\int_0^b dy \int_0^b dy' dy' \left(\int_{-a/2}^{-a/2+d} dx\, e_y^{(1)}(x)\, U_y(x,y,y')\right.\right.$$
$$+ \int_{-a/2+d}^{-a/2+d} dx\, e_y^{(2)}(x)\, U_y(x,y,y') + \int_{-a/2+d}^{a/2} dx\, e_y^{(1)}(x)\, U_y(x,y,y')$$
$$+ \eta_0^2 \left[\int_{-a/2}^{-a/2+d} dx\, h_z^{(1)}(x)\, U_z(x,y,y') + \int_{-a/2+d}^{a/2-d} dx\, h_z^{(2)}(x)\, U_z(x,y,y')\right.$$
$$\left.\left. + \int_{a/2-d}^{a/2} dx\, h_z^{(1)}(x)\ U_z(x,y,y')\right\} \quad (5.100)$$

where

$$U_y(x,y,y') = \int_{-a/2}^{-a/2+d} dx' e_y^{(1)}(x')\, g_o(|r - r'|) + \int_{-a/2+d}^{a/2-d} dx'\, e_y^{(2)}(x')\, g_o(|r - r'|)$$
$$+ \int_{a/2-d}^{a/2} dx' e_y^{(1)}(x')\ g_o(|r - r'|) \quad (5.101a)$$

$$U_z(x, y, y') = \int_{-a/2}^{-a/2+d} dx'\, h_z^{(1)}(x')\, g_o(|r - r'|) + \int_{-a/2+d}^{a/2-d} dx'\, h_z^{(2)}(x')\, g_o(|r - r'|)$$
$$+ \int_{a/2-d}^{a/2} dx' h_y^{(1)}(x')\, g_o(|r - r'|)\Big]. \tag{5.101b}$$

It can be seen from the form of Eq. (5.100) that, for an aperture divided into N subregions, the overall admittance is obtained from N×N contributions of subregion couplings. If the aperture geometries have a symmetry about one plane, the number of different contributions is N×(N + 1)/2. In a coupling analysis through a subregion approach such as with triangular finite elements as described in Bird (2016), there are many contributions, and, therefore, it is important to simplify the cross-coupling calculation as much as possible to reduce computation time. In the present case of three separate regions (Figure 5.34), there are nine contributions to Eq. (5.100). It should be possible to simplify this expression using Lewin's transformation, but this requires considerable tedious algebra, and to date, Eq. (5.100) and its extension to other modes have been computed directly. As an example, Figure 5.36 shows the E- and H-plane coupling between two identical square waveguides described previously with a = b = 0.6λ, d = 0.0908 λ, ε_{r1} = 4, and reflection coefficient of −27.123 dB/_−73.581°. As a result of the uniform field in the H-plane and also in the E-plane, the coupling coefficients decay similarly with spacing s. This compares with the H-plane coupling coefficient in the empty (no slabs) waveguide, which decays as $1/s^2$ with distance (see Figure 5.36).

The dependence of the mutual coupling on spacing and dielectric constant is demonstrated in Figure 5.37 for apertures that satisfy condition Eq. (5.91) when $\varepsilon_{r1} = 4$. The coupling coefficient for $\varepsilon_{r1} = 8$ is comparable to $\varepsilon_{r1} = 0.4$ and is not shown on the graph. Highest coupling occurs when the dielectric constant is significantly different from $\varepsilon_{r1} = 4$ either larger, i.e. $\varepsilon_{r1} = 8$ or smaller, i.e. $\varepsilon_{r1} = 0.4$. The rate of decay of the coupling depends on the dielectric loading. It varies from about $1/s$ to about $1/s^{1.9}$ when $\varepsilon_{r1} = 2$. Plots of the coupling coefficient in the E-plane as function of dielectric loading for the same case are shown in Figure 5.38. The decay in the E-plane varies between about $1/s^{1.2}$ to $\sim 1/s^2$ depending on the loading from uniform when $\varepsilon_{r1} = 4$ and is most rapid when $\varepsilon_{r1} = 0.4$. The initial level shifts too with dielectric constant depending on the proximity to $\varepsilon_{r1} = 4$. Finally, the self-admittance of the slab-loaded aperture illustrated in Figure 5.34 with various dielectric constants was calculated from Eq. (5.99), and the values obtained are listed in Table 5.8.

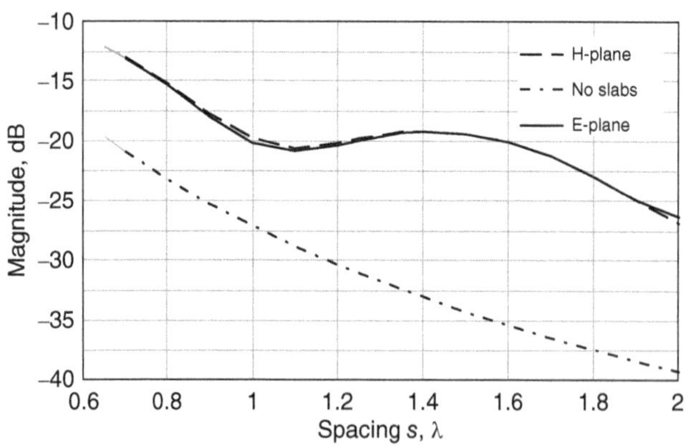

Figure 5.36 Fundamental mode coupling coefficient for identical E-plane slab-loaded waveguides as a function of spacing and disposition. Aperture dimensions are a = 0.6 λ = b. E-wall lined with dielectric of thickness d = 0.0908 λ, ε_{r1} = 4, and ε_{r2} = 1. The E-plane is parallel to the dielectric slabs. solid line - E-plane; dash line - H-plane; dash dot line - H-plane for H-plane coupling for unfilled (no slabs) waveguides.

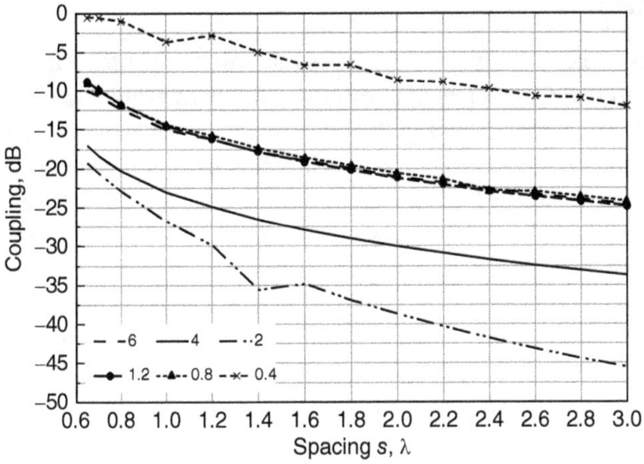

Figure 5.37 Coupling coefficient in the H-plane of the fundamental mode in identical E-plane slab-loaded waveguides as a function of spacing and dielectric constant ε_{r1} of the loading in region 1. Aperture dimensions are a = 0.6 λ = b. E-wall lined with dielectric of thickness d = 0.1443 λ, and ε_{r2} = 1.

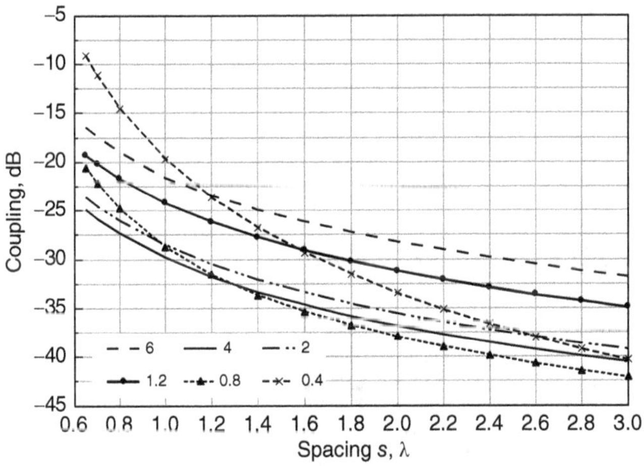

Figure 5.38 E-plane coupling coefficient of the fundamental mode in identical E-plane slab-loaded waveguides for the same range of aperture dimensions as given in Figure 5.37.

Table 5.8 Self-admittance of square aperture with dimensions a = 0.6 λ = b with E-plane slab wall thickness d =0.1443 λ in region 1 (see Figure 5.34) and dielectric constant ε_{r1}.

Dielectric constant ε_{r1}	y_{ii}(10, 10) (Re + j Im)	Reflection coefficient S_{11} (mag, phase) dB	deg.
0.4	0.119–j0.431	−1.741	47.123
0.8	0.777–j0.288	−15.004	52.995
1.	1.236 + j0.047	−19.366	−169,91
1.2	0.187–j0.443	−2.722	30.055
2	0.616–j0.171	−11.733	30.055
4	0.740 + j0.176	−14.923	−39.949
6	0.786 + j0.474	−11.006	−80.538
8	0.801 + j0.678	−8.696	−94.225

5.6.2 Metamaterial-Loaded Apertures

Metamaterial structures have composite electromagnetic properties that mimic the properties of natural materials with effective constitutive parameters of permittivity (ε), permeability (μ), and conductivity (σ). Often, the constitutive properties differ significantly from natural materials, such as ε_r, μr < 1 or zero as illustrated in Figure 5.39. The structures can vary in size from nanoperiodic that are a fraction of a wavelength in size or larger fractions of a wavelength such as lightweight wire mesh. Some examples are shown in Figure 5.40. In this way, a full range of permittivities or permeabilities can be created in many different ways with a variety of bandwidth properties (Lier et al. 2011; Scarborough et al. 2013; Scarborough et al. 2017). Metamaterial-loaded horns, often called metahorns, support hybrid modes with similar properties to modes in conventional corrugated horns (Lier et al. 2019). Such metahorns can support hybrid modes with either soft or hard wall characteristics. They can also consist of conventional metal horns containing

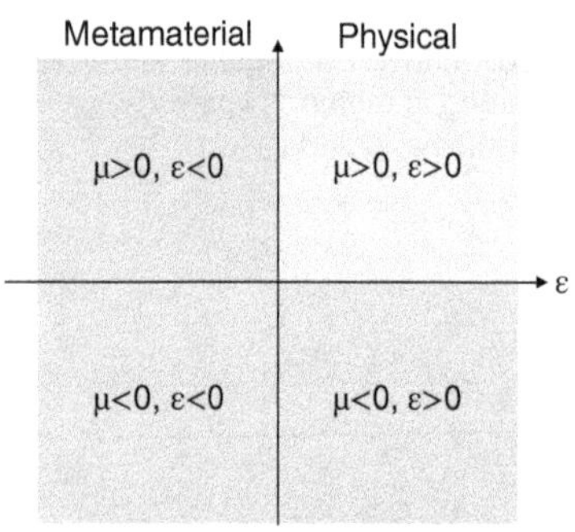

Figure 5.39 The quadrants of materials in the real plane.

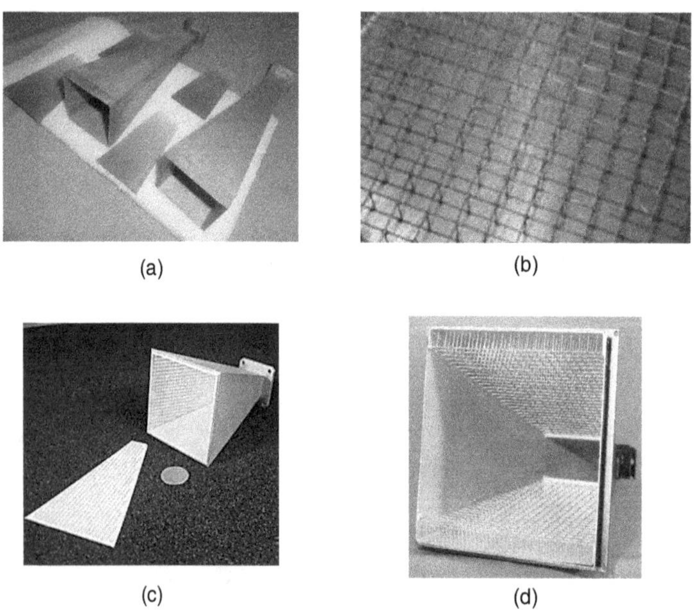

Figure 5.40 Metamaterial loading of apertures. (a) Dielectric tapers for loading metal horns Credit: Trevor S. Bird; (b) wire mesh metamaterial Source: Lockheed Martin; Used with permission; (c) horn with lining implemented on printed circuit board Source: Lockheed Martin; Used with permission; (d) C-band horn with wire mesh lining on the E-plane walls Source: Lockheed Martin; Used with permission.

meta-lenses. Metahorns are potentially lighter in weight than metal horns, and their properties can be tailored to suit the application. For example, low cross-polarization is possible with both soft and hard metahorns over a wide bandwidth. This increased bandwidth occurs because the meta surface permittivity versus frequency can have Drude-like dispersion on the trailing end of a Drude resonance. This simple approach assumes a classical mechanical model of the free electron in the permittivity through the equation:

$$\varepsilon_r(\omega) = \varepsilon_o \left[1 - \omega_p^2 / \left(\omega^2 - j\omega\,\omega_d\right)\right]$$

where ω_p is the plasma resonance frequency and ω_d is the damping frequency. Due to their lightweight, metamaterials are attractive for applications where weight is important such as in space-borne antennas. In this and other applications, arrays of metahorns are of interest. It is possible to synthesize liners for metahorns to possess the dispersive properties of a suitable mode over a required bandwidth. Powerful synthesis procedures are used, which are typically based on global stochastic optimizers such as the genetic algorithm. The cost functions utilized in the genetic algorithm to synthesize broadband soft, hard, or balanced conditions for hybrid modes (Lier et al. 2019) are typically based on the TE and TM surface reactances, X^{TE} and X^{TM}, respectively. For example, for a balanced hybrid mode, a typical cost function is

$$Cost_{balanced} = \sum_{frequencies} \left(X^{TE} X^{TM} + \eta_0^2\right)$$

Figure 5.41 shows the various options for metamaterial lining of rectangular horns. The horns mainly used so far in applications have metamaterial linings of the type shown in Figures 5.41b and 5.41c. The E-plane slab-loading case offers improved efficiency, a reduced beamwidth in the horizontal plane, and higher overall mutual coupling as the H-plane side lobes are of comparable level as the E-plane ones. The H-plane slab loading case (c) potentially offers overall lower mutual coupling, improved beam symmetry, and has lower sidelobes in the E-plane. To demonstrate the lower sidelobes with E-plane loading and, consequently, mutual coupling between such horns, consider a rectangular horn with the H-plane walls lined with a metamaterial of thickness t as illustrated in Figure 5.41c. The aperture distribution with a loading chosen to give a suitable aperture taper is shown in Figure 5.42 along with the resulting radiation pattern in the E-plane of a horn. The horn

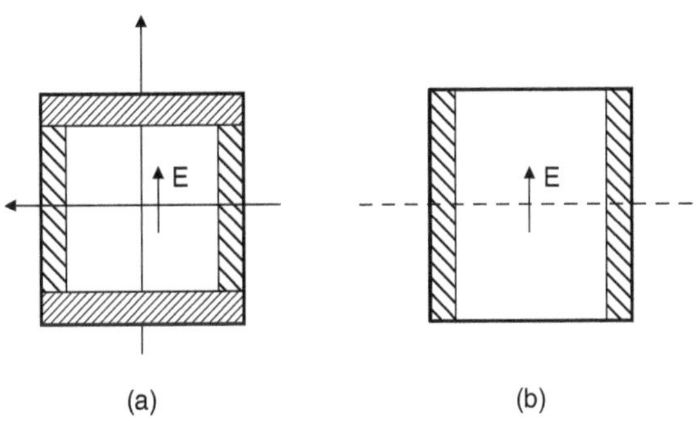

Figure 5.41 Lined rectangular horns. (a) Trifurcated horn modified from CHAN, K.K., HUANG, C.C. and RAAB, A.R. (1980), 'Dielectric-loaded trifurcated horn for H-plane stacked reflector feed array', IEE Proc. Microw., Optics and Antennas, vol 127, no.1, pp. 61–64.; (b) E-plane slab-loading modified from TSANDOULASAND, G.N. and FITZGERALD, W.D. (1972), 'Aperture efficiency enhancement in dielectric loaded horns', IEEE Trans. Antennas Propagat., vol. AP-20, pp. 69–74.; and (c) H-plane slab-loading Modified from LIER, E. and SHAW, R. (2008): 'Design and simulation of metamaterial-based hybrid mode horn antennas', Electron. Lett., vol.44, no. 25, pp. 1444–1445.

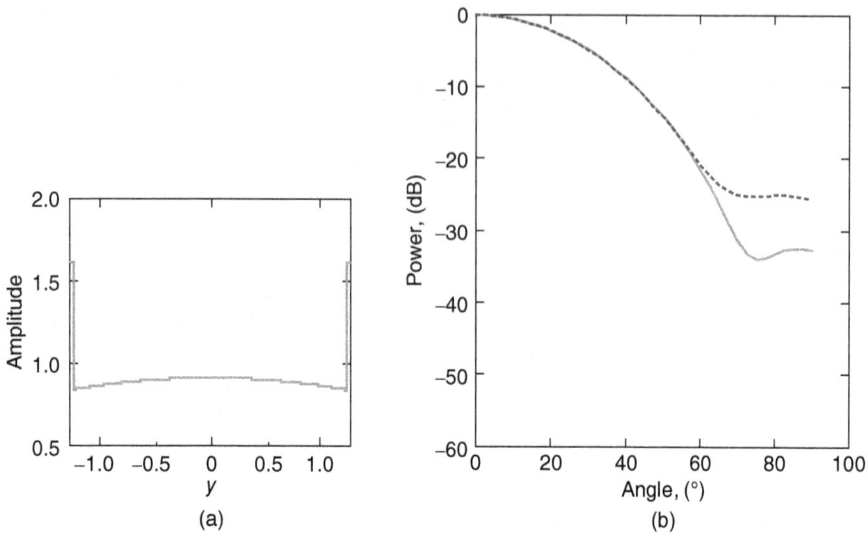

Figure 5.42 Tapering the aperture with E-plane dielectric loading to reduce sidelobe level. (a) Aperture amplitude distribution along the H-plane (y); and far-field radiation pattern – resulting radiation patterns (solid line) and original without loading (dashed).

has aperture dimensions a = 61 mm = b, an inclination length from apex to aperture in the E-plane 331 mm (i.e. half-cone angle 5.26°), wall lining thickness d = 1 mm, and dielectric fillings $\varepsilon_{r1} = 0.45$, $\varepsilon_{r2} = 1$, and $\mu_{r1,2} = 1$ and $t = 1$ mm. For this aperture, based on a simple rule of thumb mentioned earlier, the mutual coupling in the E-plane of the metahorn is <−30 dB compared with −25 dB with a fully air-filled one. A metahorn could also have a metamaterial lens mounted internally. In that case, mutual coupling between metahorns should be analyzed from the fields transformed to the aperture.

We now consider the case of H-plane slab loading as shown in Figures 5.41c and 5.43. The fundamental mode is a TMx_{01} hybrid mode. The central filling (region (2)) and the slabs (regions 1 and 3) of thickness t have a relative permeability and permittivity of $(\mu_{ri}\varepsilon_{ri})$. The fields are given by

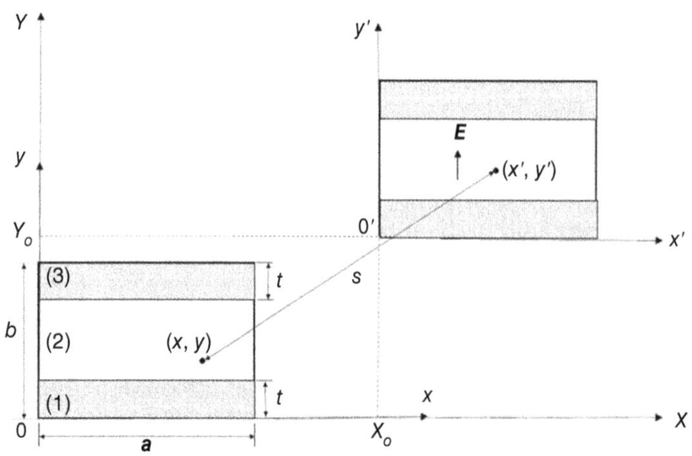

Figure 5.43 Geometry of coupling between two H-plane slab-loaded waveguides.

(Harrington 1961)

$$
\begin{aligned}
e_x^{(1)} &= -\left(\frac{m\pi}{a}\right) k_1 \frac{E_o\eta_o}{jk\varepsilon_{r1}} \sin\left[k_1 y\right] \cos\left(\frac{m\pi x}{a}\right) = \overline{e}_x^{(1)}(\mathrm{y}) \cos\left(\frac{m\pi x}{a}\right); 0 \le y \le t \\
e_x^{(2)} &= -\left(\frac{m\pi}{a}\right) k_2\, C_{21} \frac{E_o\eta_o}{jk\varepsilon_{r2}} \sin\left[k_2 y\right] \cos\left(\frac{m\pi x}{a}\right) = \overline{e}_x^{(2)}(\mathrm{y}) \cos\left(\frac{2n\pi x}{a}\right); t \le y \le b-t \\
e_x^{(3)} &= \left(\frac{m\pi}{a}\right) k_1\, C_{32} \frac{E_o\eta_o}{jk\varepsilon_{r1}} \sin\left[k_1 (b-y)\right] \cos\left(\frac{m\pi x}{a}\right) = \overline{e}_x^{(3)}(\mathrm{y}) \cos\left(\frac{2n\pi x}{a}\right); b-t \le y \le b \\
e_y^{(1)} &= \left(\gamma^2 + \left(\frac{m\pi}{a}\right)^2\right) \frac{E_o\eta_o}{jk\varepsilon_{r1}} \cos\left[k_1 y\right] \sin\left(\frac{m\pi x}{a}\right) = \overline{e}_y^{(1)}(\mathrm{y}) \sin\left(\frac{m\pi x}{a}\right); 0 \le y \le t \\
e_y^{(2)} &= \left(\gamma^2 + \left(\frac{m\pi}{a}\right)^2\right) C_{21} \frac{E_o\eta_o}{jk\varepsilon_{r2}} \cos\left[k_2 y\right] \sin\left(\frac{m\pi x}{a}\right) = \overline{e}_y^{(2)}(\mathrm{y}) \sin\left(\frac{m\pi x}{a}\right); t \le y \le b-t \\
e_y^{(3)} &= \left(\gamma^2 + \left(\frac{m\pi}{a}\right)^2\right) C_{32} \frac{E_o\eta_o}{jk\varepsilon_{r1}} \cos\left[k_1 (b-y)\right] \sin\left(\frac{m\pi x}{a}\right) = \overline{e}_y^{(3)}(\mathrm{y}) \sin\left(\frac{m\pi x}{a}\right); b-t \le y \le b \\
e_z^{(1)} &= (j\gamma)\, k_1 \frac{E_o\eta_o}{jk\varepsilon_{r1}} \sin\left[k_1 y\right] \sin\left(\frac{m\pi x}{a}\right); 0 \le y \le t \\
e_z^{(2)} &= (j\gamma)\, k_2 C_{21} \frac{E_o\eta_o}{jk\varepsilon_{r2}} \sin\left[k_2 y\right] \sin\left(\frac{m\pi x}{a}\right); t \le y \le b-t \\
e_z^{(3)} &= (-j\gamma)\, k_1 C_{32} \frac{E_o\eta_o}{jk\varepsilon_{r1}} \sin\left[k_1 (b-y)\right] \sin\left(\frac{m\pi x}{a}\right); b-t \le y \le b \\
h_x^{(1)} &= (j\gamma)\, E_o \cos\left[k_1 y\right] \sin\left(\frac{m\pi x}{a}\right); 0 \le y \le t \\
h_x^{(2)} &= (j\gamma)\, C_{21} E_o \cos\left[k_2 y\right] \sin\left(\frac{m\pi x}{a}\right); t \le y \le b-t \\
h_x^{(3)} &= (j\gamma)\, C_{32} E_o \cos\left[k_1 (b-y)\right] \sin\left(\frac{m\pi x}{a}\right); b-t \le y \le b \\
h_z^{(1)} &= \left(\frac{m\pi}{a}\right) E_o \cos\left[k_1 y\right] \cos\left(\frac{m\pi x}{a}\right) = \overline{h}_z^{(1)}(\mathrm{y}) \cos\left(\frac{m\pi x}{a}\right); 0 \le y \le t \\
h_z^{(2)} &= \left(\frac{m\pi}{a}\right) C_{21} E_o \cos\left[k_2 y\right] \cos\left(\frac{m\pi x}{a}\right) = \overline{h}_z^{(2)}(\mathrm{y}) \cos\left(\frac{m\pi x}{a}\right); t \le y \le b-t \\
h_z^{(3)} &= \left(\frac{m\pi}{a}\right) C_{32} E_o \cos\left[k_1 (b-y)\right] \cos\left(\frac{m\pi x}{a}\right) = \overline{h}_z^{(3)}(\mathrm{y}) \cos\left(\frac{m\pi x}{a}\right); b-t \le y \le b
\end{aligned} \tag{5.102}
$$

where $k_i = \sqrt{k^2\mu_{ri}\varepsilon_{ri} - \left(\frac{m\pi}{a}\right)^2 - \gamma^2}$, $C_{21} = \frac{\cos[k_1 t]}{\cos[k_2 t)]}$, $C_{32} = C_{21}\frac{\cos[k_2(b-t)]}{\cos[k_1 t]}$ and

$$
E_o = \frac{-j}{\sqrt{\eta_o \frac{\gamma}{k}\left(\gamma^2 + \left(\frac{m\pi}{a}\right)^2\right) \frac{a}{4} \left[\begin{array}{l} \frac{t}{\varepsilon_{r1}}\left(1 + sinc\left(2k_1 t\right)\right) + \left(\frac{C_{21}^2}{\varepsilon_{r2}}\right) \times \\ \left(b - 2t + (b-t)\, sinc(2k_2 (b-t)) - t\, sinc\left(2k_1 t\right)\right) + \left(\frac{C_{32}^2\, C_{12}^2}{\varepsilon_{r1}}\right) t\left(1 + sinc\left(2k_1 t\right)\right) \end{array}\right]}}
$$

The propagation constant is a solution of the transcendental equation

$$
\frac{k_1}{\varepsilon_{r1}} \tan\left[k_1 t\right] = -\frac{k_2}{\varepsilon_{r2}} \tan\left[k_2 (b-t)\right] \tag{5.103}
$$

which was obtained from the continuity of the x-directed electric field and the y-directed electric displacement at $y = b - t$. An estimate of the fundamental LSM_{01} mode propagation constant is

$$
\gamma \cong \sqrt{k^2\left[1 + (2t/b)\left(\varepsilon_{r1}\mu_{r1} - 1\right) + (1 - 2t/b)\left(\varepsilon_{r2}\mu_{r2} - 1\right)\right] - (\pi/a)^2} \tag{5.104}
$$

This approximation gives the exact results in the extremes of when $t = 0$ and $t = b/2$.

The relative electric field amplitude of the fundamental TM mode along the y-axis and axis of symmetry was studied and shown to undergo a discontinuity at the interfaces, but as expected, the displacement field has uniform amplitude for all permittivity values.

The mutual admittance between two identical H-plane slab-loaded waveguide apertures illustrated in Figure 5.43 can be shown to be

$$y_{ij}(01,01) = \frac{jk^2 E_o}{4\pi\,\gamma_{10}ab}\left[Y_{ijx}(01,01) + Y_{ijy}(01,01) + Y_{ijz}(01,01)\right] \tag{5.105}$$

where

$$\begin{aligned} Y_{ijx}(mn,m'n') = &\int_0^a dx\cos\left(\frac{m\pi x}{a}\right)\int_0^a dx'\cos\left(\frac{m'\pi x'}{a}\right)\int_0^t dy\,\overline{e}_x^{(1)}(y)\,T_x(x,x',y) \\ &+ \int_t^{b-t} dy\,\overline{e}_x^{(2)}(y)\,T_x(x,x',y) + \int_{b-t}^{b} dy\,\overline{e}_x^{(3)}(y)\,T_x(x,x',y) \end{aligned} \tag{5.106a}$$

$$\begin{aligned} Y_{ijy}(mn,m'n') = &\int_0^a dx\sin\left(\frac{m\pi x}{a}\right)\int_0^a dx'\sin\left(\frac{m'\pi x'}{a}\right)\int_0^t dy\,\overline{e}_y^{(1)}(y)\,T_y(x,x',y) \\ &+ \int_t^{b-t} dy\,\overline{e}_y^{(2)}(y)\,T_y(x,x',y) + \int_{b-t}^{b} dy\,\overline{e}_y^{(3)}(y)\,T_y(x,x',y) \end{aligned} \tag{5.106b}$$

$$\begin{aligned} Y_{ijz}(mn,m'n') = \eta_o^2 &\int_0^a dx\cos\left(\frac{m\pi x}{a}\right)\int_0^a dx'\cos\left(\frac{m'\pi x'}{a}\right)\int_0^t dy\,\overline{h}_z^{(1)}(y)\,T_z(x,x',y) \\ &+ \int_t^{b-t} dy\,\overline{h}_z^{(2)}(y)\,T_z(x,x',y) + \int_{b-t}^{b} dy\,\overline{h}_z^{(3)}(y)\,T_z(x,x',y) \end{aligned} \tag{5.106c}$$

in which

$$\begin{aligned} T_x(x,x',y) = &\left[\int_0^t dy'\,\overline{e}_x^{(1)}(y')\,g_o(|r-r'|) + \int_t^{b-t} dy'\,\overline{e}_x^{(2)}(y')\,g_o(|r-r'|)\right. \\ &\left.+ \int_{b-t}^{b} dy'\,\overline{e}_x^{(3)}(y')\,g_o(|r-r'|)\right] \end{aligned} \tag{5.107a}$$

$$\begin{aligned} T_y(x,x',y) = &\left[\int_0^t dy'\,\overline{e}_y^{(1)}(y')\,g_o(|r-r'|) + \int_t^{b-t} dy'\,\overline{e}_y^{(2)}(y')\,g_o(|r-r'|)\right. \\ &\left.+ \int_{b-t}^{b} dy'\,\overline{e}_y^{(3)}(y')\,g_o(|r-r'|)\right] \end{aligned} \tag{5.107b}$$

$$\begin{aligned} T_z(x,x',y) = &\left[\int_0^t dy'\,\overline{h}_z^{(1)}(y')\,g_o(|r-r'|) + \int_t^{b-t} dy'\,\overline{h}_z^{(2)}(y')\,g_o(|r-r'|)\right. \\ &\left.+ \int_{b-t}^{b} dy'\,\overline{h}_z^{(3)}(y')\,g_o(|r-r'|)\right]. \end{aligned} \tag{5.107c}$$

The field functions, such as $\overline{e}_x^{(1)}(y)$, are the y-dependencies of Eq. (5.102). Note that the circuit model of mutual admittance shown in Figure 5.8 is applicable to Eq. (5.105) as well.

The mutual coupling between metamaterial-loaded waveguides and horns has only been studied to a limited extent, but using methods described in this chapter, some basic conclusions can be drawn. For example, Figure 5.44 shows the E-plane coupling coefficient of the fundamental mode in H-plane slab-loaded waveguides as a function of spacing and slab permittivity. The waveguides

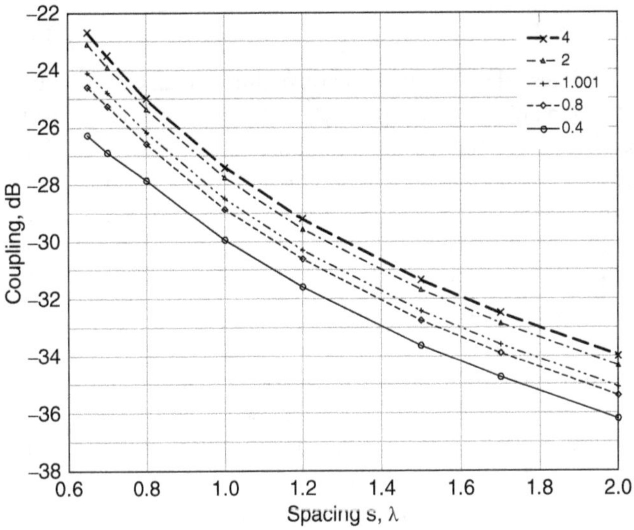

Figure 5.44 H-plane slab-loaded waveguides fundamental mode E-plane coupling coefficient as a function of spacing and slab permittivity. Aperture dimensions are a = 0.6 λ = b. E-wall lined with dielectrics of thickness $t = 0.025\lambda$ and $\varepsilon_{r2} = 1$ and various relative permittivities as shown.

and range of permittivities used are similar to those for the E-plane slab-loaded case shown in Figure 5.38. The dominant contributions to the admittance in Eq. (5.105) arise from Y_{ijy} and Y_{ijz}, and these partially cancel with distance s. At large separations, the coupling decays as $1/s$. The difference is the levels shift downward with decreasing permittivity. The H-plane coupling follows a similar trend except the decay with distance is about $1/s^{1.8}$. This is illustrated in Figure 5.45 where the E- and H-plane coupling coefficients of H-plane-lined square waveguides with a = 0.6 λ = b and dielectric of thickness $t = 0.025\lambda$, $\varepsilon_{r1} = 0.45$, and $\varepsilon_{r2} = 1$ are displayed.

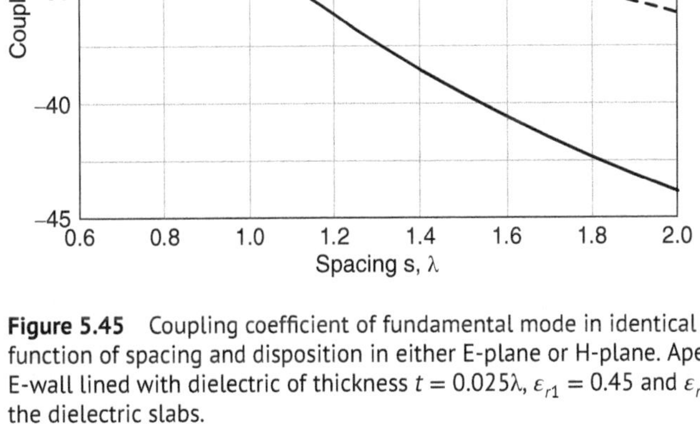

Figure 5.45 Coupling coefficient of fundamental mode in identical H-plane slab-loaded waveguides as a function of spacing and disposition in either E-plane or H-plane. Aperture dimensions are a = 0.6 λ = b. E-wall lined with dielectric of thickness $t = 0.025\lambda$, $\varepsilon_{r1} = 0.45$ and $\varepsilon_{r2} = 1$. Note the H-plane is parallel to the dielectric slabs.

5.7 Concluding Remarks

An analysis of mutual coupling has been described for finite arrays of different-sized waveguides and horns using an approach based on a Green's function approach. A range of practical problems can be analyzed by this method. Expressions for the mutual admittance of all possible combinations of mode coupling were obtained. A formula was derived for reducing the order of integration in these expressions, allowing the quadruple integral for mutual admittance to be expressed as four double integrals in the most general case. The efficacy of the theory was demonstrated by comparing computations with published results and results obtained from experiment or numerical methods. Good agreement has generally been obtained. The present approach is particularly useful for analysis of large arrays where the formulation can be expressed in efficient computer code so that large arrays can be analyzed. The importance of including modes other than the fundamental TE_{10} mode in the calculations was observed, particularly when the higher-order modes are close to or just above cut-off. Calculations of reflection coefficient, E-plane coupling coefficients, and cross-polar radiated fields are especially sensitive in this respect.

Lewin's formula for reducing the order of integration has wider application than originally intended. With minor modification, it has been used in coupling between apertures loaded with dielectric and metamaterials with an effective permittivity and permeability. A general asymptotic formulation was described and applied to several practical cases with excellent agreement with exact results. This formulation has potentially wider application than used up to now.

References

Abramowitz, M. and Stegun, I.A. (1965). *Handbook of Mathematical Functions*. New York: Dover Inc.

Arndt, F., Wolff, K.-H., Brunjes, L. et al. (1989). Generalized moment method analysis of planar reactively loaded rectangular waveguide arrays. *IEEE Trans. Anten. Propagat.* AP-37: 329–338.

Bird, T.S. (1979). Mode coupling in a planar circular waveguide array. *IEE J. Microw., Opt. Acoustics* 3: 172–180.

Bird, T.S. (1987). Mutual coupling in finite coplanar rectangular waveguide arrays. *Electron. Lett.* 23: 1199–1201.

Bird, T.S. (1990a). Analysis of mutual coupling in arrays of different-sized rectangular waveguides. *IEEE Trans. Anten. Propagat.* AP-38: 166–172. [Erratum AP-38, 1727].

Bird, T.S. (1990b). Behaviour of multiple elliptical waveguides opening into a ground plane. *IEE Proc. H* 137: 121–126.

Bird, T.S. (1996a). Cross-coupling between open-ended coaxial radiators. *IEE Proc. Microw. Anten. Propagat.* 143: 265–271.

Bird, T.S. (1996b). Improved solution for mode coupling in different-sized circular apertures and its application. *IEE Proc. Microw. Anten. Propagat.* 143: 457–464.

Bird, T.S. (2004). Mutual coupling in arrays of coaxial waveguides and horns. *IEEE Trans. Anten. Propagat.* AP-52: 821–829.

Bird, T.S. (2016). *Fundamentals of Aperture Antennas and Arrays*. Chichester, UK: Wiley.

Bird, T.S. and Bateman, D.G. (1992). Software for modelling finite arrays of rectangular waveguides or horns with mutual coupling. In: *IEEE Antennas & Propagation Symposium*, Chicago, Illinois, 20-25 July, 633–636.

Bird, T.S. and Bateman, D.G. (1994). Mutual coupling between rotated horns in a ground plane. *IEEE Trans. Anten. Propagat.* AP-42: 1000–1006.

Bird, T.S. and Hay, S.G. (1990). Mismatch in a dielectric-loaded rectangular waveguide antenna. *Electron. Lett.* 26: 59–61.

Bird, T.S. and Hayman, D.B. (1997). Complex permittivity measurement from coaxial cable mutual coupling. In: *IEEE AP-S/URSI Symposium*, Montréal, Canada, 13-18 July, 742.

Bird, T.S. and Sroka, C. (1992). Design of the Ku-band antennas for the Galaxy HS601C satellites. *J. Electr. Electron. Eng. Aust.* 12: 267–273.

Butler, C.M. and Yung, E.K. (1979). Analysis of a terminated parallel-plate waveguide with a slot in its upper plate. *Ann. Telecommun.* 34: 451–458.

Chan, K.K., Huang, C.C., and Raab, A.R. (1980). Dielectric-loaded trifurcated horn for H-plane stacked reflector feed array. *IEE Proc. Microw. Opt. Anten.* 127 (1): 61–64.

Chew, W.C. and Habashy, T.M. (1986). The use of vector transforms in solving some electromagnetic scattering problems. *IEEE Trans. Anten. Propagat.* AP-34: 871–879.

Felsen, L.B. and Marcuvitz, N. (1973). *Radiation and Scattering of Waves*. New Jersey, USA: Prentice-Hall.

Fenn, A.J., Thiele, G.A., and Munk, B.A. (1982). Moment method analysis of finite rectangular waveguide phased arrays. *IEEE Trans. Anten. Propagat.* AP-30: 554–564.

Finlayson, B.A. (1972). *The Method of Weighted Residuals and Variational Principles*. New York: Academic Press.

Fröberg, C.-F. (1974). *Introduction to Numerical Analysis*. London, UK: Addison-Wesley.

Harrington, R.F. (1961). *Time-Harmonic Electromagnetic Fields*. New York: McGraw-Hill.

Harrington, R.F. (1968). *Field Computation by Moment Methods*. New York: Macmillan.

Harrington, R.F. and Mautz, J.R. (1976). A generalized network formulation for aperture problems. *IEEE Trans. Anten. Propagat.* AP-24: 870–873.

Hockham, G.A. (1974). TEM waveguide coupling in presence of short-circuited slot. *Electron. Lett.* 10: 62–63.

Hockham, G.A. (1975). Use of the 'edge condition' in the numerical solution of waveguide antenna problems. *Electron. Lett.* 11: 418–419.

Hockham, G.A. and Walker, G. H. (1973): 'Study of a finite phased array antenna', Proc. European Microwave Conf., 4–7 Sept., Brussels, Belgium, paper C. 1.3.

Jansen, J.K.M. and Jeuken, M.E.J. (1973):'Propagation and radiation properties of elliptical waveguides with anisotropic boundary', Research Report, Eindhoven University of technology, Dept. of Electrical Eng., 20 October.

Kuehne, R. and Marquardt, J. (2001). Modal analysis of waveguide antennas with arbitrary cross sections. *IEEE Trans. Microw. Theory Tech.* MTT-49 (11): 2152–2156.

Levine, H. and Schwinger, J. (1965). On the theory of electromagnetic wave diffraction by an aperture in a infinite plane conducting sheet. In: *The Theory of Electromagnetic Waves* (ed. M. Kline), S1–S37. New York: Dover.

Lewin, L. (1951). *Advanced Theory of Waveguides*. London: Iliffe and Sons Ltd, Chap. 6.

Lier, E. and Shaw, R. (2008). Design and simulation of metamaterial-based hybrid mode horn antennas. *Electron. Lett.* 44 (25): 1444–1445.

Lier, E., Werner, D.H., Scarborough, C.P. et al. (2011). An octave bandwidth, negligible loss radiofrequency Metamaterial. *Nat. Mater.* 10 (3): 216–222.

Lier, E., Werner, D.H., and Bird, T.S. (2019). The evolution from metal horns to metahorns. *IEEE Anten. Propagat. Mag.* 61 (4): 6–18.

Luzwick, J. and Harrington, R.F. (1982). Mutual coupling analysis in a finite planar rectangular waveguide antenna array. *Electromagnetics* 2: 25–42.

Lyon, R.W. and Sangster, A.J. (1981). Efficient moment method analysis of radiating slots in a thick-walled rectangular waveguide. *IEE Proc. H* 128: 197–205.

Lyon, J.A.M., Kalafus, R.M., Kwon, Y-K., Digenis, C.J., Ibrahim, M.A.H., and Chen, C-C, (1966): 'Derivation of aerospace antenna coupling-factor interference prediction techniques', Radiation Lab. Uni. Of Michigan, Tech. Rep. AFAL-TR-66-57, AD-483061, April.

Mailloux, R.J. (1969a). Radiation and near-field coupling between two collinear open-ended waveguides. *IEEE Trans. Anten. Propagat.* AP-17: 49–55.

Mailloux, R.J. (1969b). First-order solutions for mutual coupling between waveguides which propagate two orthogonal modes. *IEEE Trans.* AP-17: 740–746.

McLachlan, N.W. (1947). *Theory and Application of Mathieu Functions*. London, UK: Oxford University Press.

Morse, P.M. and Feshbach, H. (1953). *Methods of Theoretical Physics*. New York USA: McGraw-Hill Book Company, Chapters 6 and 11.

Müller, K.E. (1960). Untersuchung des Strahlungsverhaltens elliptisher Hohleiter sowie Möglichkeit zur eines zircular polarisierten Strahlungsfeldes. *Hochfrequencztechn. U Elektroak.* 69: 140–151.

Newell, P. and Bird, T.S. (1994). Effects of mutual coupling in the design of high performance multifeed satellite antennas. *Radio Sci.* 29: 145–152.

Oliner, A.A. (1957). The impedance properties of narrow radiating slots in the broad face of rectangular waveguide: part – theory. *IRE Trans. Anten. Propagat.* AP-5: 4–11.

Olver, A.D., Claricoats, P.J.B., Kishk, A.A., and Shafai, L. (1994). *Microwave Horns and Feeds*. Stevenage, UK: IEEE Press.

Qintez, J.P. and Dudley, D.G. (1976). Slots in a parallel plate waveguide. *Radio Sci.* 11 (8–9): 713–724.

Scarborough, C.P., Wu, Q., Werner, D.H. et al. (2013). Demonstration of an octave bandwidth negligible loss metamaterial horn antenna for satellite applications. *IEEE Trans. Anten. Propagat.* AP-61 (3): 1081–1088.

Scarborough, C.P., Wu, Q., Werner, D.H. et al. (2017). Broadband low-loss Metamaterial-enabled horn antennas. In: *Broadband Metamaterials in Electromagnetics: Technology and Applications* (ed. D.H. Werner), 45–80. Pan Stanford Publishing.

Schennum, G.H., Han, C.C., and Gould, H.J. (1978). *Reduction of Mutual Coupling in a Waveguide Array*, 412–415. Washington, DC: IEEE APS.

Shan-wei, L., Ji-Shi, R., and Yin, F. (1985). Study of coupling properties of slotted waveguide in rectangular waveguide. *Acta Electron. Sin.* 13: 92–101.

Silvestro, J.W. and Collin, R.E. (1989). Aperture impedance of flared horns. *IEE Proc. H* 136: 235–240.

Stern, G.J. and Elliott, R.S. (1985). Resonant length of longitudinal slots and validity of circuit representation: theory and experiment. *IEEE Trans. Anten. Propagat.* AP-33: 1264–1271.

Stevenson, A.F. (1948). Theory of slots in rectangular waveguides. *J. Appl. Phys.* 19: 24–38.

Steyskal, H. (1974). Mutual coupling analysis of a finite planar waveguide array. *IEEE Trans. Anten. Propagat.* AP-22: 594–597.

Tsandoulas, G.N. and Fitzgerald, W.D. (1972). Aperture efficiency enhancement in dielectric loaded horns. *IEEE Trans. Anten. Propagat.* AP-20: 69–74.

Vokurka, V.J. (1979). Elliptical corrugated horn for broadcasting satellite antennas. *Electron. Lett.* 15: 652–654.

Vu Khac, T. and Carson, C.T. (1973). Impedance properties of a longitudinal slot in the broad face of rectangular waveguide. *IEEE Trans. Anten. Propagat.* AP-21: 708–710.

Woodward, A.M. (1947). Transmission in waveguides – cross section partly of solid dielectric. *Wireless Eng.* 24 (July): 192–196.

Wu, Q., Scarborough, C.P., Martin, B.G. et al. (2013a). Ku-band dual polarization hybrid mode horn antenna enabled by printed circuit board metasurfaces. *IEEE Trans. Anten. Propagat.* AP-61 (3): 1089–1098.

Wu, Q., Scarborough, C.P., Werner, D.H. et al. (2013b). Inhomogeneous metasurfaces with engineered dispersion for broadband hybrid mode horn antennas. *IEEE Trans. Anten. Propagat.* AP-61 (10): 4947–4956.

6

Arrays of Microstrip Patch Antennas

Trevor S. Bird[1,2]

[1] *Global Big Data Technology Centre, University of Technology Sydney, Ultimo, NSW 2007, Australia*
[2] *Antengenuity, Eastwood, NSW 2122, Australia*

6.1 Introduction

Since the 1970s, microstrip patch arrays have become increasing popular due to factors such as low cost, lighter weight than waveguide options, planar geometry, and excellent performance. Some patch antennas that are used in arrays are illustrated in Figure 6.1. The standard rectangular patch (Figure 6.1a) gives reasonable performance through careful design although better performance is possible using other shapes such as E-shaped patches or Jerusalem crosses as in Figure 6.1c or by stacking patches as in Figure 6.1b. The basic properties of microstrip have been studied extensively (e.g. Van de Capelle and Luypaert 1973). The fundamental mode of microstrip is a quasi-transverse electromagnetic (TEM) mode with wave propagation possible down to DC. It is called quasi-TEM as both longitudinal components are present although small and are often designated as EH_m (m = 0, 1, …) with the quasi-TEM mode as the EH_0. High-order modes can be present at discontinuities, and these have cut-off frequencies as in closed waveguide. Modes with even symmetry have m as an even number, while modes with odd symmetry have m as an odd number. The dielectric constant of the substrate is usually different from that in free space so that the wave is traveling in an inhomogeneous medium. As a result, the propagation velocity of the quasi-TEM mode is partway between the wave speed in the substrate and the speed of waves in air. This is described by an effective dielectric constant of the microstrip ε_{eff} and is the dielectric constant of an equivalent homogeneous medium. As the line is dispersive, the effective dielectric constant increases toward that of the substrate as the operating frequency increases because the wave is more confined in the dielectric and the phase velocity gradually decreases. For a strip of width w and thickness t, on a substate of thickness h, the effective dielectric constant is approximately given by (Getsinger 1973; Hammerstad and Jensen 1980)

$$\varepsilon_{eff}(f) = \varepsilon_r - \frac{\varepsilon_r - \varepsilon_{DC}}{(1 + G\,(f/f_c)^2)}, \tag{6.1}$$

where ε_r is the dielectric constant of the substrate relative to the upper-medium,

$$\varepsilon_{DC} \cong \frac{\varepsilon_r + 1}{2} + \frac{\varepsilon_r - 1}{2}\left[1 + \frac{12h}{w}\right]^{-1/2} \tag{6.2}$$

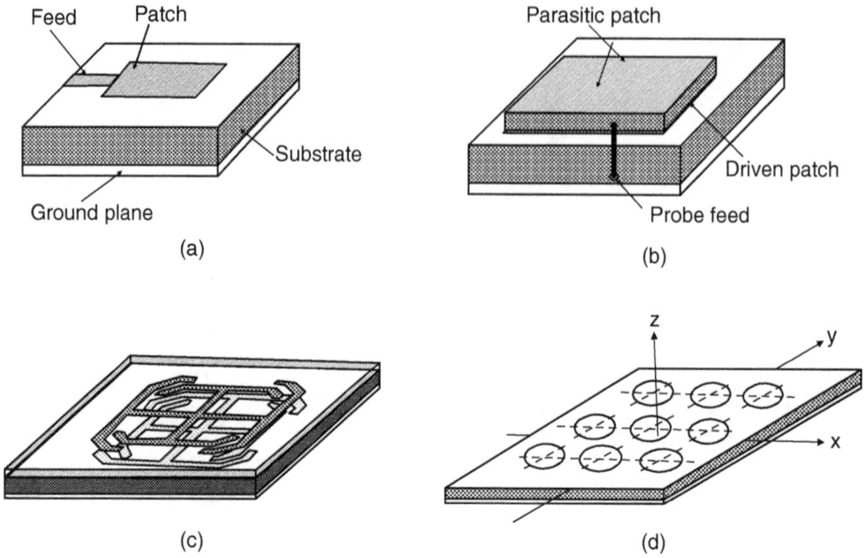

Figure 6.1 Various microstrip patch antennas. (a) Rectangular patch; (b) stacked rectangular patches; (c) double-sided Jerusalem crosses; and (d) array of circular patches.

is the effective dielectric constant at low frequencies, f_c is related to the cut-off frequency of the first high-order EH_1 mode and is approximated by $f_c = \frac{Z_o c}{2\eta_o h}$, where

$$Z_o(w, t, \varepsilon_r) = \frac{Z_{01}\left(\frac{w}{h}\right)}{\sqrt{\varepsilon_{DC}}} \tag{6.3}$$

is the characteristic impedance of the line in which $Z_{01}(x) = \frac{\eta_o}{2\pi} \ln\left[\frac{f_a(x)}{x}\sqrt{1+\left(\frac{2}{x}\right)^2}\right]$ and $f_a(x) = 6 + (2\pi - 6)\exp[-(30.666/x)^{0.7528}]$. Finally, in Eq. (6.1) $G = \frac{\pi^2}{12}\frac{\varepsilon_r - 1}{\varepsilon_{DC}}\sqrt{\frac{2\pi Z_o}{\eta_o}}$ is a geometric factor.

A quasi-TEM mode is the mode most commonly used in patch antennas. The properties of the higher-order modes can be used to improve the radiation performance of patch antennas and also to make the antenna more compact. The fundamental mode uniform microstrip line does not radiate at low frequencies because the electric field is strongly bound to the microstrip line and the ground plane. Some higher-order modes of microstrip lines radiate as leaky waves. The first higher-order mode (EH_1) has a phase reversal and an electric field null at the center, while the distributions of the second higher-order mode (EH_2) have two phase reversals and two nulls that are approximately $w/4$ away from the edges, where w is the width of the microstrip. A simple yet effective technique to excite microstrip line in the first higher-order mode is to create a conducting wall with metal pins to approximate a septum along the center of the microstrip so as to mimic an electric field null at the center. This suppresses the fundamental mode and excites the first higher-order mode. Likewise, two arrays of vias can be employed to suppress the first two modes and to excite the EH_2 mode. This approach can be used to tailor the pattern of a patch element in an array.

In this chapter, techniques are described for modeling planar arrays of microstrip patches and, in particular, the effects of mutual coupling. Along the way, the self-impedance is discussed, and the characteristics of mutual coupling between two patches are examined in detail. The chapter concludes with two applications of microstrip arrays. The first example is a switched array for beam steering, while the second describes a compact array of Jerusalem patches of the type shown in Figure 6.1c to create a wide-angle incidence material used to improve the performance of an array.

6.2 Representation of Mutual Coupling Between Patch Antennas

An isolated microstrip patch which is excited by a feed or a probe resonates at a central frequency that is determined by the patch geometry. The simplest microstrip patch antenna shown in Figure 6.1a is fed by a microstrip transmission line. The patch antenna resonates between the input face of width w and end face over a length ℓ. It can be seen from Eq. 6.3 that the width of the patch controls the input impedance and the radiating faces. Larger widths decrease the bandwidth and also reduce the patch impedance. Typically, the height h of the patch above the ground plane is usually much smaller than the wavelength of operation so that $h > 0.025\ \lambda$ or the antenna efficiency will be degraded. An equivalent circuit of a single rectangular patch antenna has been developed based on a transmission line model shown in the single upper circuit in Figure 6.2 (Sainati 1996) but does not include the transformer inductors L_m. Thus, two resonant end-slot radiators with dimensions w x h are assumed to be connected by a transmission line of length ℓ that resonate at the frequency is $f_r = \frac{c}{2\sqrt{\varepsilon_r}}\left(\frac{1}{L_e}\right)$, where $L_e = \ell + 2\ \Delta\ell$ and $\Delta\ell$ is the effective increase in length due to the fringing capacitance at one end. This is obtained from $C_e = tan(\beta\ell_{eff})/(\omega Z_{op})$. For short lengths, $\Delta\ell \cong C_e Z_{op} c/\sqrt{\varepsilon_r}$, where C_e can be calculated using Laplace's equation for a given geometry; typically, $\Delta\ell \sim 0.1 - 0.15\ \lambda$. A better approximation is

$$\Delta\ell \cong 0.412\, h \frac{(\varepsilon_{eff}(f) + 0.3)}{(\varepsilon_{eff}(f) - 0.258)} \frac{\left(\frac{W}{h} + 0.264\right)}{\left(\frac{W}{h} + 0.8\right)}. \tag{6.4}$$

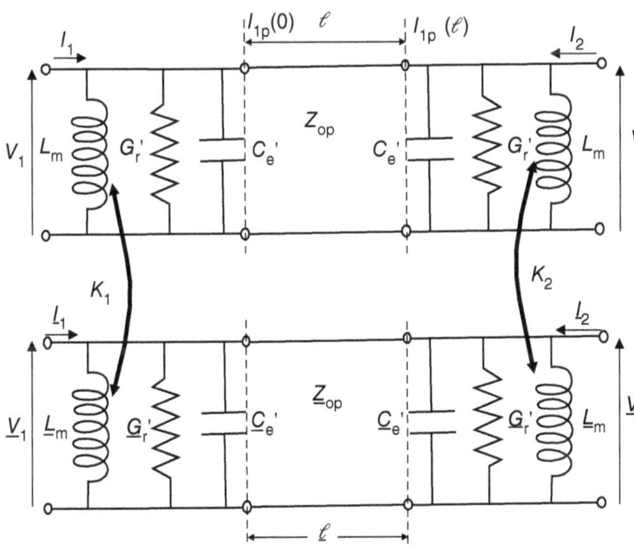

Figure 6.2 Equivalent circuit representation of mutual coupling between two identical microstrip patches. The inductances L_m are absent in an isolated patch.

In the same way, an effective width can be defined as $W_e = W + 2\,\Delta w$, where

$$\Delta w = \left(\frac{ln4}{\pi}\right) h.$$

Returning to the single patch model, the end slots are represented by a conductance G_r in parallel with a capacitance C_e, representing, respectively, radiation and energy storage. The characteristic impedance of the parallel plate transmission line of the patch is given by $Z_{0p} \approx \eta_o h/w\sqrt{\varepsilon_r}$. An approximate value for the conductance can be found from $G_r = P_r/|E_o h|^2$, where P_r is the total radiated power by a single slot and $E_o h$ is the peak voltage across it. The result is

$$G_r = \frac{4\sqrt{\varepsilon_r}}{\eta_o}\left(\frac{w}{\lambda}\right)^2 \int_0^{\pi} d\theta\,(M_0(\theta) - sin^2\theta M_2(\theta)), \tag{6.5}$$

where $M_p(\theta) = \int_0^{2\pi} d\phi\, S^2(\pi w\, sin\,\theta\, sin\,\phi)\, S^2(\pi 2h\, sin\,\theta\, cos\,\phi) sin^p\,\phi$.

After evaluating M_p, an approximation to the integral in Eq. (6.5) can be found using its Maclaurin series from which the first two terms give (Pues and de Van 1984)

$$G_r \cong \frac{1}{\pi^2\eta_o}[W\,Si(W) + S(W) + \cos(W) - 2]\left(1 - \frac{s^2}{24}\right) + (s^2/12)\left[\frac{1}{3} + \frac{1}{W^2}(\cos(W) - S(W))\right],$$

where $s = k_o\Delta\ell$, $W = k_{ow}$, where $k_o = 2\,\pi/\lambda_o$ is the wavenumber of free-space, and $Si()$ is the sine integral function (see Appendix A). In the single patch model, the two end-slot radiators couple as the radiation of one end impinges on the other and vice versa. The mutual admittance is found from

$$Y_{p12} = \frac{1}{|V_1||V_2|}\int_\Sigma \mathbf{dS}\cdot \boldsymbol{E_1} \times \boldsymbol{H_2} = G_{p12} + j\,B_{p12},$$

where $\boldsymbol{E_1}$ is the electric field at one end and $\boldsymbol{H_2}$ is the radiated magnetic field at the second and Σ is a hemisphere surrounding the upper surface of the patch. The real and imaginary parts have been obtained (Pues and de Van 1984) and are given by

$$G_{p12} = \left[J_0(k_o\ell) + \frac{s^2}{24 - s^2}J_2(k_o\ell)\right] G_r, \tag{6.6a}$$

$$B_{p12} = \frac{\pi}{2}\frac{\left[Y_0(k_o\ell) + \frac{s^2}{24-s^2}Y_2(k_o\ell)\right]}{\left[ln\left(\frac{s}{2}\right) - 1.5 + \gamma_e\right] + \left[\frac{s^2}{24-s^2}\right]}[1 - \exp(-0.21w)]\,\omega C_e, \tag{6.6b}$$

where $\gamma_e = 0.577216\ldots$ is Euler's constant and J_0, Y_0, etc. are Bessel functions (see Appendix B).

When a second patch is placed nearby on the substrate, coupling between the patches at the input end slot is represented by the transformer inductances L_m and the coefficient of coupling K_1 indicated by the double headed arrow in Figure 6.2. Similarly, there is now a transformer with a coefficient of coupling K_2 on the output end slot of the transmission line. Although the two patches shown in Fig. 6.2 are the same, generally, the second patch will have a different geometry, and hence, the circuit parameters are different. The mutual impedance between the two circuits can be calculated in terms of the circuit parameters. This will only be approximate, and it is usually preferable to calculate the mutual impedance directly and from this determine the inductance and coupling coefficient from the Z-matrix of the network. The calculation of mutual impedance and Z-matrix is described in the remainder of this section.

Another important parameter in the design of a patch antenna and, arrays of them, is the self-impedance. This has been studied by several workers over the years using different techniques, namely modal (Carver 1980), cavity (Lee and Chen 1997), and transmission line

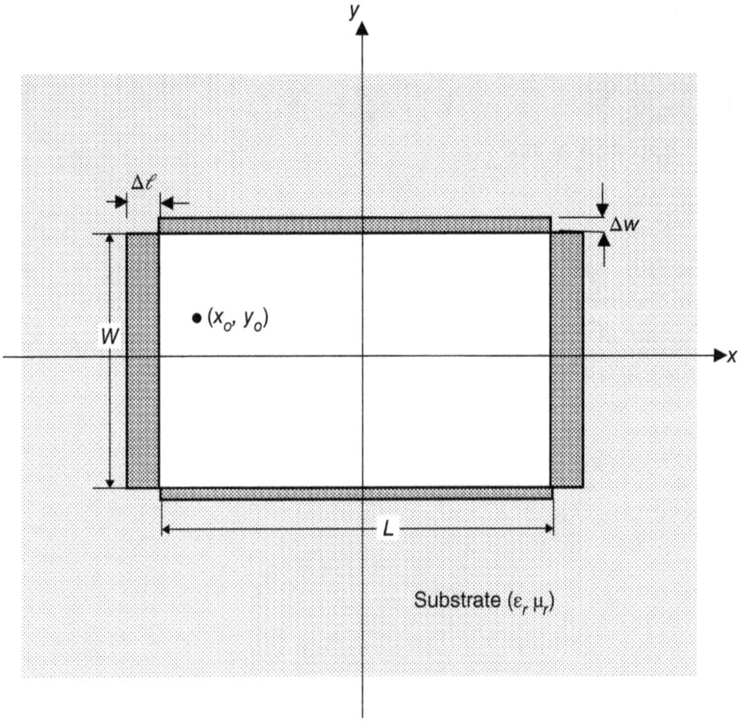

Figure 6.3 Geometry of rectangular patch antenna and probe location.

(Sainati 1996) representations. The results given by the first two representations show impressive agreement with full-wave numerical methods and, therefore, the expression for only one of them will be described here.

To derive a self-impedance, consider the patch shown in Figure 6.3. For a patch with effective length and width $L_e = L + 2\Delta l$ and $W_e = W + 2\Delta w$, respectively, located on a lossless substrate of thickness h, relative permittivity and permeability ε_r and μ_r and a feed that has dimensions d_x in lengthwise and d_y parallel to the width of the strip, which is positioned at (x_o, y_o), a cavity is created in the substrate between the ends and sides. With a probe feed, $d_x = d_y$ and the cross-sectional area $d_x d_y$ is set equal to the area of the probe. In that case, if a is the radius of the probe, $d_x = d_y = a\sqrt{\pi}$. The field in the cavity can be represented as a sum of resonant modes (Carver 1980). The ohmic loss in the substrate and the conductors, the radiation loss into the space wave, and power converted into a surface wave propagating in the substrate can be included by representing them as a loss factor in an effective wavenumber. Using this representation, Carver (1980) expressed the self-impedance as

$$Z_{in} = -jk_o h\, \eta_o \mu_r \sum_{m=0}^{M} \sum_{n=0}^{N} \frac{\psi_{mn}^2(x_{oe},\ y_{oe})}{k_e^2 - \left(\frac{m\pi}{L_e}\right)^2 - \left(\frac{n\pi}{W_e}\right)^2} S\left(\frac{m\pi d_x}{2L}\right) S\left(\frac{n\pi d_y}{2W}\right), \tag{6.7}$$

where

$$\psi_{mn}(x_{oe}, y_{oe}) = \frac{2}{\sqrt{\epsilon_{m0}\epsilon_{n0} L_e W_e}} \cos\left(\frac{m\pi x_{oe}}{L_e}\right) \cos\left(\frac{n\pi y_{oe}}{W_e}\right),$$

$$M = 4L/d_x\, N = 4W/d_y\, k_e^2 = k_o^2 \varepsilon_r \mu_r (1 - j\, l_e),$$

$$x_{oe} = x_o + \Delta\ell, y_{oe} = y_o + \Delta w,$$

$$\epsilon_{n0} = \begin{cases} 2 & if\, n = 0 \\ 1 & otherwise \end{cases},$$

the function $S(\cdot\cdot)$ is defined in Appendix A, and

$$l_e = tan\delta + \frac{1}{Q_c} + \frac{1}{Q_{sp}} + \frac{1}{Q_{sw}} \tag{6.8}$$

is the effective total loss. In Eq. (6.8), $tan\delta$ is the loss tangent of the dielectric substrate, Q_c is due to conductor loss, Q_{sp} is due to the radiation loss in the space wave, and Q_{sw} is due to the loss to the surface wave propagating in the substrate. These Q values have been defined elsewhere (Sainati 1996; Lee and Chen 1997) and are given by

$$Q_c = h\,\mu_r\sqrt{\pi f \mu_o \sigma},$$

$$Q_{sp} = \omega\frac{U_{sp}}{P_{sp}},$$

where

$$U_{sp} = \frac{h}{4}\mu_o\mu_r L_e W_e,$$

and

$$P_{sp} = \left(\frac{2}{\pi}L_e W_e\right)^2 \frac{\eta_o k^2}{3\pi}(k_o h)^2\left(1 - \frac{1}{\varepsilon_r} + \frac{2}{5\varepsilon_r^2}\right)$$

is the power that is radiated in the space wave due to a horizontal electric dipole attributed the patch. Finally,

$$Q_{sw} = Q_{sp}\left(\frac{\xi_{hor}}{1 - \xi_{hor}}\right)$$

where ξ_{hor} is the radiation conversion efficiency of the horizontal electric dipole, which is given by

$$\xi_{hor} = \frac{P_{sp}}{P_{sp} + P_{sw}},$$

where P_{sw} is the power produced as a TM_0 surface wave by a horizonal electric dipole located at (0,0, d). A second-order approximation to this has been given by Pozar (1990)

$$P_{sw} = \frac{\eta_o k_o^2}{8}\frac{\varepsilon_r(\tau_o^2 - 1)}{\varepsilon_r[(\tau_o^2 - 1)^{-1/2} + (\tau_o^2 - 1)^{1/2}/(\varepsilon_r - \tau_o^2)] + (k_o h)[1 + \varepsilon_r^2(\tau_o^2 - 1)/(\varepsilon_r - \tau_o^2)]},$$

where

$$\tau_o = 1 + \frac{\alpha_o\alpha_1 - \varepsilon_r^2 + \varepsilon_r\sqrt{\varepsilon_r^2 - 2\,\alpha_o\alpha_1 + \alpha_o^2}}{\varepsilon_r^2 - \alpha_1^2},$$

$$\alpha_o = s\,\tan(k_o hs), s = \sqrt{\varepsilon_r - 1}, \alpha_1 = -\frac{1}{s}[\tan(k_o hs) + (k_o hs)\,sec^2(k_o hs)].$$

Note that $k_o\tau_0$ is an approximation to the TM_0 propagation constant.

As an example of the self-impedances calculated from Eq. (6.7), Figure 6.4 shows the resistance and reactance obtained for a 30 mm × 20 mm patch antenna on a substrate with relative permittivity of 2.2 and height of 1.524 mm and a probe of radius 0.5 mm located at x_o = 15 mm and

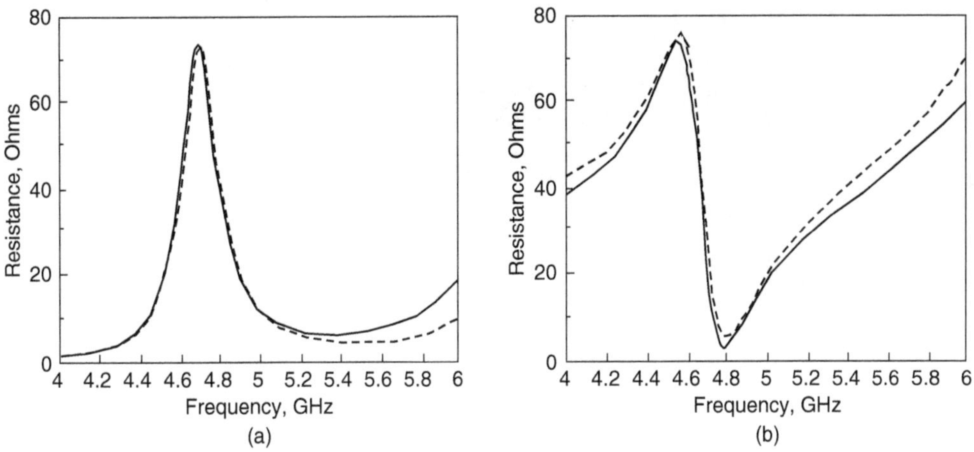

Figure 6.4 The self-admittance of a 30 mm × 20 mm microstrip antenna on a substrate of thickness 1.524 mm and relative permittivity 2.2. (a) Real part; (b) imaginary part. Solid curve – Eq. (6.7); dashed curve – Ensemble. Source: After SMITH, S.L. (2000): 'Investigation of switched parasitic antenna arrays with application to electronic beam steering for mobile communications systems', PhD thesis, Griffith University, Brisbane, Australia.

y_o = 5 mm. Typically, the probe is located just off the center of the patch to achieve good excitation and input match. These results are compared with simulations using the MoM-based computer software package Ensemble® (Smith 2000). The two results are in close agreement, as can be seen in Figure 6.4. Further tuning could be accomplished by adjusting the probe location and dimensions.

6.2.1 E-Current Model of Coupling

In the analysis of mutual coupling between microstrip antennas (see Figure 6.1e), two approaches are usually adopted in the literature (Pozar 1982; Newman et al. 1983; Mohammadian et al. 1989). The first approach (Pozar 1982; Newman et al. 1983) represents the patches as electric current sources, and the fields are given in terms of the Green's function for the elementary electric currents on a grounded dielectric slab. This method is similar to the one described in Chapter 5 for apertures. Once the current on the patches is determined, the mutual impedance is calculated from Eq. (3.1).

Consider two patches on a grounded dielectric slab of thickness h as shown in Figure 6.5. Patch i has width W_i and resonant length L_i causing an electric current density $\boldsymbol{J_i}$ to be excited on it. The slab has permeability and permittivity ε_2 and μ_2 , while the free-space medium above it has properties ε_1 and μ_1. The patches are oriented as shown in Figure 6.5. Patch i radiates an electric field $\boldsymbol{E}_i(\boldsymbol{r_j})$ at $\boldsymbol{r_j}$ on patch j. In this notation, the mutual impedance between these patches is given by

$$Z_{ji}(\boldsymbol{r_j}) = -\iint_{S_j} dS' \boldsymbol{E}_i(\boldsymbol{r_j}) \cdot \boldsymbol{J}_j(\boldsymbol{r_j}). \tag{6.9}$$

To determine the field due to patch i, suppose the current on patch i is separable and of the form

$$\boldsymbol{J_i}(x,y) = \hat{x}J_i(x,y) = \hat{x}\, X(x)Y(y)\,, \tag{6.10}$$

where $X(x)$ and $Y(x)$ are even functions, which satisfy the following transform relations

$$\overline{X}_i(\beta) = 2\int_0^\infty dx\, X_i(x)\cos(\beta x), \overline{Y}_i(\xi) = 2\int_0^\infty dy\, Y_i(y)\cos(\xi y). \tag{6.11}$$

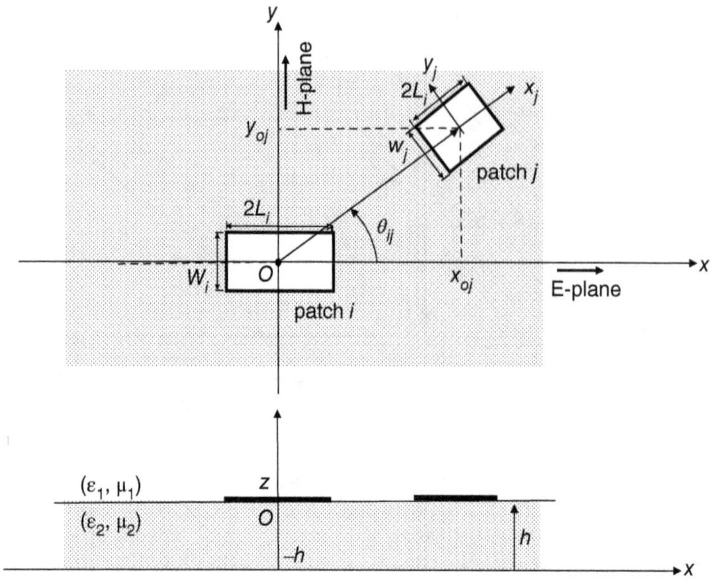

Figure 6.5 Geometry for coupling between patches using electric current model. Source: Based on NEWMAN, E.H., RICHMOND, J.H. and KWAN, B.W. (1983): 'Mutual impedance computation between microstrip antennas', IEEE Trans. Microw. Theory Tech., Vol. MTT-31, pp. 941-945.

Assume that the current on the patch is sinusoidal in the x-direction and uniform in the y-direction, so that

$$X_i(x) = sin\left[K\left(\frac{L_i}{2} - |x|\right)\right] / \sin(KL_i/2) \text{ and } Y_i(y) = 1/W_i, \tag{6.12}$$

where K can be chosen for best fit but typically $K = k_o$ *or* $K = k_o\sqrt{\mu\,\varepsilon_{eff}}$, where ε_{eff} is given by Eq. (6.1). From Eq. (6.11), we find

$$\overline{X}_i(\beta) = \frac{2K[\cos(\beta L_i/2) - \cos(KL_i/2)]}{sin[KL_i\,(K^2 - \beta^2)/2]} \tag{6.13a}$$

$$\overline{Y}_i(\xi) = S\left(\xi\frac{W_i}{2}\right). \tag{6.13b}$$

The field at the surface of the dielectric slab can be found by expressing the fields in both regions in terms of TE and TM modes with respect to the z-direction. Spectral functions are found by satisfying the boundary conditions at $z = h$. The resulting field components are

$$E_{xi}(x, y, t) = \int_0^\infty dk \int_0^{\pi/2} d\phi\, F_x(x, y)\, cos\beta x\, cos\xi y \tag{6.14a}$$

and

$$E_{yi}(x, y, t) = \int_0^\infty dk \int_0^{\pi/2} d\phi\, F_y(x, y)\, sin\beta x\, sin\xi y \tag{6.14b}$$

where

$$F_x(x, y, \zeta) = C_o(\zeta)(\alpha_1(\zeta)\xi^2 + \alpha_2(\zeta)\beta^2)\overline{X}_i(\beta)\overline{Y}_i(\xi)$$

and

$$F_y(x, y, \zeta) = C_o(\zeta)(\alpha_1(\zeta) + \alpha_2(\zeta))\beta\overline{X}_i(\beta)\,\xi\overline{Y}_i(\xi)$$

$$\beta = \zeta\, \cos(\alpha), \xi = \zeta\, \sin(\alpha), \zeta^2 = \beta^2 + \xi^2,$$

in which

$$C_o(\zeta) = -\frac{\sinh(gh)}{j\omega\varepsilon_2\pi^2\zeta},\ a_1(\zeta) = \frac{\gamma_2^2}{\mu_r\gamma\sinh(gh) + g\cosh(gh)}, a_2(\zeta) = \frac{\varepsilon_r\gamma g}{\varepsilon_r\gamma\cosh(gh) + g\sinh(gh)},$$

$$\varepsilon_r = \varepsilon_2/\varepsilon_1, \mu_r = \mu_2/\mu_1, \gamma = \sqrt{\zeta^2 - \omega^2\mu_1\varepsilon_1}, \gamma_2 = j\omega\sqrt{\mu_2\varepsilon_2},\ g^2 = \zeta^2 + \gamma_2^2.$$

Using these relations, Eq. (6.9) can now be re-expressed as

$$Z_{ji}(\mathbf{r_j}) = -\int_{-W_j/2}^{W_j/2} dy \int_{-L_{j/2}}^{L_j/2} dx\,\mathbf{E}_i(\mathbf{r_j})\cdot\mathbf{J}_j(\mathbf{r_j}) = -\int_{-\infty}^{\infty} dy\int_{-\infty}^{\infty} dx\,\mathbf{E}_i(\mathbf{r_j})\cdot\mathbf{J}_j(\mathbf{r_j})$$

where, in the latter, it has been assumed the current is zero outside patch j. On patch j, the current $\mathbf{J}_j$ is assumed to take the same form as given in Eq. (6.10) with the index i replaced by j. In that case,

$$Z_{ji}(\mathbf{r_j}) = -\int_0^{\infty} d\zeta\int_0^{\frac{\pi}{2}} d\alpha\left\{\cos\theta_{ij}\frac{F_x(\beta\cos(\alpha),\xi\sin(\alpha),\zeta)}{4}\sum_{m=-1}^{1}\overline{\sigma}_0(m)\times\right.$$
$$\sum_{n=-1}^{1}\overline{\sigma}_0(n)\,e^{j(m\beta x_{oji}+n\xi y_{oj})}\overline{X}_j(\beta')\overline{Y}_j(\xi') + \sin(\theta_{ij})\frac{F_y(\beta\cos(\alpha),\xi\sin(\alpha),\zeta)}{4}\times$$
$$\left.\sum_{m=-1}^{1}\overline{\sigma}_0(m)\,m\sum_{n=-1}^{1}\overline{\sigma}_0(n)\,n\,e^{j(m\beta x_{oj}+n\xi y_{oj})}\overline{X}_j(\beta')\overline{Y}_j(\xi')\right\} \tag{6.15}$$

where
$\beta' = m\beta\cos(\theta_{ij}) + n\,\xi\sin(\theta_{ij})$, and $\xi' = -m\beta\sin(\theta_{ij}) + n\xi\cos(\theta_{ij})$, (x_{oj}, y_{oj}) are the center coordinates of patch j, and

$$\overline{\sigma}_0(i) = \begin{cases} 0 & if\ i = 0 \\ 1 & otherwise \end{cases}.$$

Special cases can be identified when $\theta_{ij} = 0$, Eq. (6.15) simplifies to

$$Z_{ji}(\mathbf{r_j}) = -\int_0^{\infty} d\zeta\int_0^{\frac{\pi}{2}} d\alpha\ F_x(\beta\cos(\alpha),\xi\sin(\alpha),\zeta)\overline{X}_j(\beta)\,\overline{Y}_j(\xi)\cos(\beta x_{oij})\cos(\xi y_{oij}) \tag{6.16a}$$

since $\overline{X}_j(\gamma) = \overline{X}_j(-\gamma)$ and $\overline{Y}_j(\gamma) = \overline{Y}_j(-\gamma)$. When $\theta_{ij} = \pi/2$, the mutual impedance is

$$Z_{ji}(\mathbf{r_j}) = \int_0^{\infty} d\zeta\int_0^{\frac{\pi}{2}} d\alpha\ F_y(\beta\cos(\alpha),\xi\sin(\alpha),\zeta)\overline{X}_j(\xi)\,\overline{Y}_j(\beta)\sin(\beta x_{oij})\sin(\xi y_{oij}) \quad . \tag{6.16b}$$

Equation (6.16a) corresponds to E-plane coupling when $y_{oij} = 0$ and H-plane coupling when $x_{oij} = 0$.

As an example, the coupling in the two principal planes versus spacing between adjacent edges was measured by Jedlicka et al. (1981) and is compared with results obtained by Newman et al. (1983) using a formulation similar to Eq. (6.15). The geometry consists of two identical microstrip patches of width 6.55 cm and resonant length 10.57 cm mounted on a substrate of thickness 1.575 mm with dielectric constant 2.5. The measurements were made at 1.417 GHz. Similar results were obtained here with Eq. (6.15) providing care was taken with the integration in the vicinity of the poles of a_1 and a_2 by introducing a small amount loss to dampen their effect. The results obtained are shown in Figure 6.6.

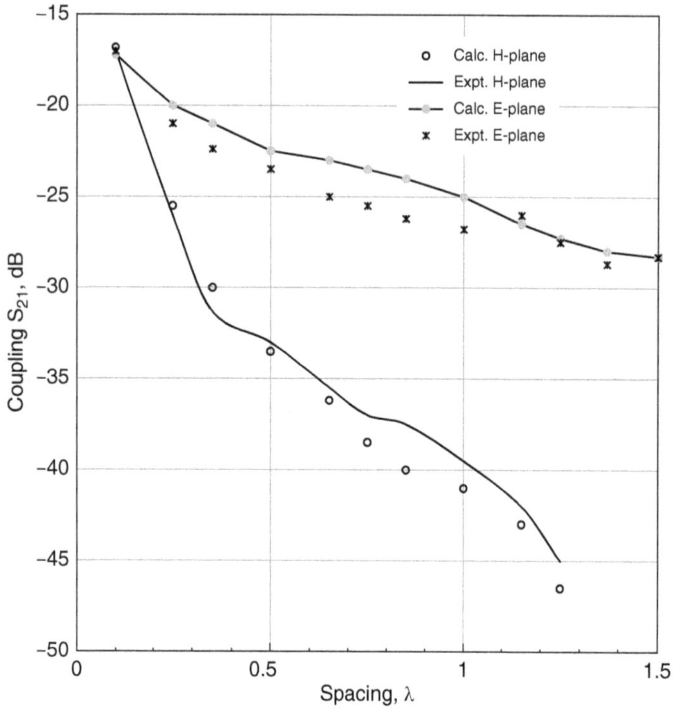

Figure 6.6 Principal plane coupling between two identical patch antennas versus spacing between adjacent edges calculated using the E-field model. The patches have width $W_i = 6.55$ cm and length $L_i = 10.57$ cm and are mounted on a substrate with thickness $h = 1.575$ mm and $\varepsilon_r = 2.5$. The operating frequency is 1.417 GHz. Legend: line is calculated (Newman et al. 1983) – solid E-plane, dash H-plane, marker is experiment (Jedlicka et al. 1981) – x E-plane, o H-plane. Source: NEWMAN, E.H., RICHMOND, J.H. and KWAN, B.W. (1983): 'Mutual impedance computation between microstrip antennas', IEEE Trans. Microw. Theory Tech., Vol. MTT-31, pp. 941–945.

6.2.2 Cavity Model (H-Model) of Coupling

A second approach to mutual coupling (Mohammadian et al. 1989) is based on a cavity model for microstrip. In this approach, the microstrip patch is modeled as a grounded dielectric slab with the magnetic current distributions located at the open walls formed between the edge of the patch and the ground plane. Since the electric field is zero everywhere on the surface of the cavity except on the end slots, field equivalence may be used to replace the fields on the surface by an equivalent magnetic current in front of a perfect electric conductor. Mutual coupling occurs between these sources, and the magnetic field due to these equivalent sources is needed to calculate the effect.

Central to this is determining a dyadic Green's function of the magnetic type for the grounded dielectric slab. From this Green's dyadic, the magnetic field due to the magnetic currents on the open boundary may be calculated and the mutual impedance obtained from

$$Z_{21} = \frac{1}{I_1 I_2} \iint_S dS\, \mathbf{H_{21}}(S) \cdot \mathbf{M}_2, \tag{6.17}$$

where the input currents to the two patches shown in Figure 6.7 are I_1 and I_2, $\mathbf{M}_2$ is the magnetic current on the second patch, and $\mathbf{H}_{21}$ is the magnetic field at patch 2 due to an elementary source on patch 1. The magnetic current on patch 2 is given by $\boldsymbol{M_2} = E_{2z}\,\hat{z} \times \hat{n}$, where $\hat{n}$ is the normal to the open circuit walls on patch 2. The TM_{01} mode is assumed to be generated on each patch. E_{2z} is

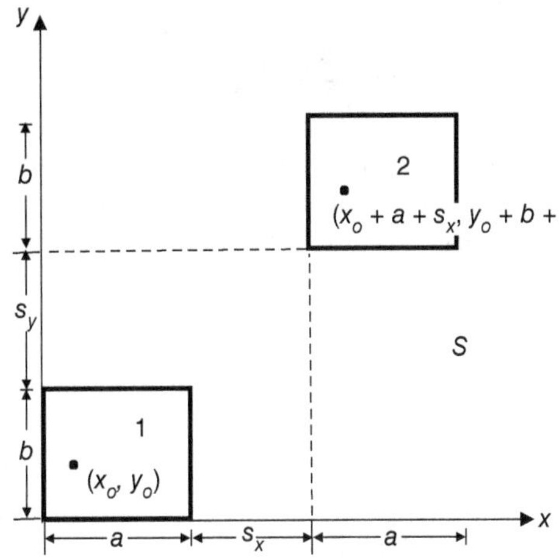

Figure 6.7 Geometry of mutual coupling between two identical microstrip patches showing the probe locations.

derived from the modal expansion of the cavity model. The magnetic field on patch 2 due to patch 1 can be obtained in closed form as can the magnetic field in the dielectric slab. When these are combined, the mutual impedance can be expressed as

$$\begin{aligned} Z_{21} = {} & \frac{-32jk_o\eta_o}{\pi^2} K_f \int_{-\infty}^{\infty} d\kappa_1 \int_{-\infty}^{\infty} d\kappa_2 \frac{1}{(\kappa_1^2 + \kappa_2^2)} sin(\tau_2 h) \\ & \times \{(\alpha\kappa_2^2 + \beta\kappa_1^2)S_1^2 + (\alpha\kappa_1^2 + \beta\kappa_2^2)C_1^2 + 2\kappa_1\kappa_2(\alpha - \beta)S_1C_1\} \\ & \times exp[j\kappa_2(a + s_x) + \kappa_2(h + s_y)], \end{aligned} \tag{6.18}$$

where

$$K_f = \left[\frac{k_2/ab}{(k_2^2 - k_{10}^2)} S\left(\frac{\pi y_p}{2W}\right) \cos\left(\frac{\pi y_o}{W}\right) \right]^2,$$

$$\alpha = \frac{\mu_{r1}\mu_{r2}}{(\tau_2\mu_{r1} - \tau_1\mu_{r2})e^{-j\tau_2 h} + (\tau_2\mu_{r1} + \tau_1\mu_{r2})e^{j\tau_2 h}},$$

$$\beta = \frac{\mu_{r1}\mu_{r2}h_1h_2}{(k_2^2\mu_{r1}\tau_1 - k_1^2\mu_{r2}\tau_2)e^{-jh} + (k_2^2\mu_{r1}\tau_1 + k_1^2\mu_{r2}\tau_2)e^{j\tau_2 h}},$$

$$S_1 = \frac{a}{2} S\left(\frac{\kappa_1 L}{2}\right) sin\left(\frac{\kappa_2 W}{2}\right), \text{ and}$$

$$C_1 = -\kappa_2 \left(\frac{b}{\pi}\right)^2 \cos\left(\frac{\kappa_1 L}{2}\right) C\left(\frac{\kappa_2 W}{2}\right).$$

In Eq. (6.18), h is the substrate thickness, μ_{r1}, μ_{r2}, ε_{r1}, and ε_{r2} are the relative permeabilities and permittivities of the surrounding medium (region 1) and of the substrate (region 2), $\tau_i = \sqrt{k_i^2 - \kappa_1^2 - \kappa_2^2}$, where $k_i = k_o\sqrt{\mu_{ri}\varepsilon_{ri}}$ $(i = 1, 2)$. The functions $S(\cdot\cdot)$ and $C(\cdot\cdot)$ are defined in Appendix A. The spacing between the patches in the x- and y-directions is s_x and s_y as shown in Figure 6.7. The functions α and β contain the transverse electric and transverse magnetic surface wave poles, which have singularities in the integration domain. These poles are found by finding the roots of the denominators of α and β given below Eq. (6.18). The multiplying factor K_f is a term contributed from the input probe where y_p is its diameter and (x_o, y_o) are the coordinates of its connection on patch 1 as shown in Figure 6.7. Typically, the probe is slightly off from the center of the patch, and in the above equation, $y_o \approx 0.47b$. The eigenvalue k_{10} is a root of the transcendental equations

$$tan(k_{10}b) = \frac{2k_{10}\alpha_{10}}{k_{10}^2 - \alpha_{10}^2},$$

where $\alpha_{10} = jk_o(h/L)Y_{rw}\eta_o/\sqrt{\varepsilon_r}$. The real part of α_{10} is approximately $0.98\pi/W$. Y_{rw} is the admittance of the radiating walls at y = 0 and y = W, which are the transitions from a parallel plate waveguide filled with the substrate of thickness h. An approximate result is (Carver and Mink 1981)

$$Y_{rw} \approx \left(0.5 + j\left(\frac{\Delta\ell}{h}\right)\varepsilon_{eff}\right)\left(\frac{kL}{\eta_o}\right), \tag{6.19}$$

where ε_{eff} is the effective dielectric constant of the patch given by Eq. (6.1) and $\Delta\ell$ is given by Eq. (6.4).

The double integrals in Eq. (6.18) can be integrated numerically or by means of a fast Fourier transform (Brigham 1974). Before proceeding with the former approach, it is convenient to change the variables of integration. To do this, let $\kappa_1 = \rho\cos\xi$ and $\kappa_2 = \rho\sin\xi$ so that $d\kappa_1 d\kappa_2 = \rho d\xi d\rho$. As a result, Eq. (6.18) becomes

$$Z_{21} = \frac{-32jk_o\eta_o}{\pi^2}K_f\int_0^\infty d\rho\,\rho\int_0^{2\pi} d\xi\bar{Z}(\rho,\xi)exp(-j\rho(cos\xi(L+s_x)+sin\xi(W+s_y))) \tag{6.20}$$

where

$$\bar{Z}(\rho,\xi) = sin(h_2t)\left\{(\alpha(\rho)sin^2\xi + \beta(\rho)cos^2\xi)S_1^2 + (\alpha(\rho)cos^2\xi + \beta(\rho)sin^2\xi)C_1^2 + sin2\xi(\alpha(\rho) - \beta(\rho))S_1C_1\right\}$$

Care should be taken with the poles in the integrand when evaluating the transform integral of Z_{21}. One approach is described by Mohammadian et al. (1989) in which the poles are subtracted out and their contributions added separately much as described in Section 3.6.2. Typical results obtained from Eq. (6.18) are shown in Figure 6.8 where good agreement is demonstrated with measured results (Mohammadian et al. 1989) for E- and H-plane coupling. These cases correspond in Figure 6.7 when $s_x = -L$ and $s_y = -W$, respectively. When the patches are well matched, the coupling coefficient is approximately given by $S_{21} \approx Z_{21}/2\eta_o$.

One of the complicating features of the analysis of the microstrip patch is that the probe feed to the patch is an integral part of the antenna. Therefore, for accurate results, an accurate model for the probe is required. In practice, the coaxial transmission line fed probe is relatively easy to implement, although, in theory, it is quite difficult to handle. The most common approach is to use an idealized feed model comprising a uniform electric current filament. Although this model gives reasonable predictions for the radiation and cross-coupling properties, it gives inaccurate results for self-impedance because the boundary conditions are not satisfied where the probe and patch connect. Aberle and Pozar (1989) describe an improved method where continuity is ensured at the connection and a moment method is used to calculate the current on the probe and patch. This has been applied successfully to both finite and infinite arrays of patches. It is noted that the mutual impedances calculated from Eq. (6.18) can be used to obtain values of the circuit elements due to mutual coupling in the equivalent circuit model shown in Figure 6.2. A potential problem is that there is no rigorous solution available for the wall admittance of the patch, which affects the resonant frequency, but this can be accommodated intuitively in the model. Fortunately, this sensitivity to resonance has little effect on the calculation of mutual coupling, as results obtained are in good agreement with experiment.

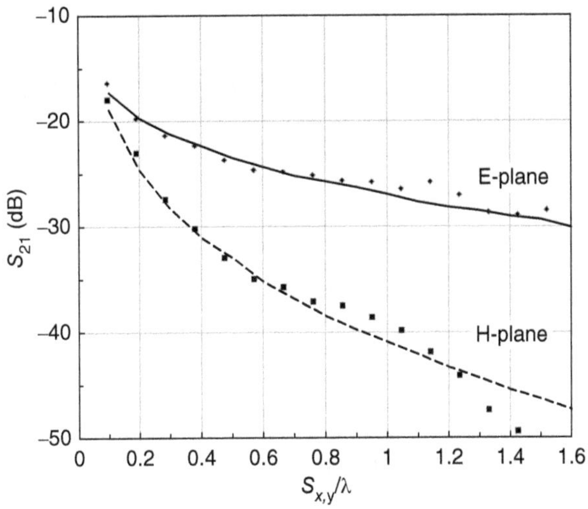

Figure 6.8 Coupling coefficients in the (a) E-plane ($s_x = -L$) and (b) H-plane ($s_y = -W$) of two microstrip patches at 1.405 GHz. Solid and dashed curves: theory (Mohammadian et al. 1989); points: measured (Carver and Mink 1981). Patch dimensions a = 10.57 cm, b = 6.55 cm. Substrate thickness $t = 0.32$ cm and dielectric constant $\varepsilon_r = 2.5$. Probe position $x_o = 1.7$ cm and $y_o = 3.28$ cm.

6.2.3 Full-Wave Solution

Although the expressions given above for microstrip are rigorous, they are however approximate. Some useful design information for microstrip arrays is provided by Sainati (1996). These will enable the designer to get a useful result, but there will be limitations. Using approximate formulas in the analysis of microstrip antennas can lead to inaccuracies; therefore, a full-wave solution is necessary to model the antenna structure accurately. The moment method has been applied with great success to microstrip arrays (Skrivervik and Mosig 1993). One MoM approach is to use the Rao–Wilton–Glisson (RWG) triangular basis functions (Rao et al. 1982), the free-space dyadic Green's function, and Galerkin's method to solve in a mixed potential integral equation, which has the advantage of being capable of representing arrays of patches of various shapes such as the E-shaped patch (e.g. Ghannay and Samet 2009).

The mixed potential integral equation relates an incident electric field upon the patches to the vector and scalar potential generated by yet-to-be-determined surface currents and charge densities, as follows (Mosiq 1988):

$$\boldsymbol{E}_i^{tan} = -(j\omega \boldsymbol{A}_i + \nabla\phi_i)|_{tan} \tag{6.21}$$

where

$$\boldsymbol{A}_i(\boldsymbol{R}) = \mu \iint_{S_i} dS' \boldsymbol{G}^{(e)}(\boldsymbol{R}, \boldsymbol{R}') \cdot \boldsymbol{J}_i(\boldsymbol{R}') \tag{6.22a}$$

$$\phi_i(\boldsymbol{R}) = -\frac{1}{j\omega\varepsilon} \iint_{S_i} dS' \boldsymbol{G}^{(e)}(\boldsymbol{R}, \boldsymbol{R}') \cdot \hat{\tau}\, [\nabla' \cdot \boldsymbol{J}_i(\boldsymbol{R}')] \tag{6.22b}$$

The problems consist of determining the surface currents $\boldsymbol{J}_i$ (i = 1,…, N) on the N patches and the impedance matrix when patch i is driven. The currents are approximated in terms of a set of

RWG vector basis functions $\{\boldsymbol{f}_k\}$ as

$$\boldsymbol{J}_i(\boldsymbol{r}) \cong \sum_{k=1}^{N_R} \alpha_{ik}\boldsymbol{f}_k \tag{6.23}$$

where α_{ik} is the coefficient to be determined. The incident field and the excitation at the probe satisfy the boundary condition

$$\boldsymbol{E}_i^{tan} + \boldsymbol{E}_{prob}^{inc} = \boldsymbol{0}$$

where $\boldsymbol{E}_{prob}^{inc}$ is the electric field due to the input probe. Following the Galerkin method through for the solution of the equations, the usual matrix system is obtained

$$[Z_{mn}][\alpha_n] = [V_m] \tag{6.24}$$

where $[Z_{mn}]$ is the impedance matrix, $[\alpha_n]$ is the vector of surface currents, and $[V_m]$ is the source vector, which contains integrals of the probe field and testing functions or values related to known loads, e.g. open circuit. The impedance matrix elements are given by

$$Z_{mn} = \langle (j\omega\boldsymbol{A_n} + \boldsymbol{\nabla}\phi_n), \boldsymbol{f_m}\rangle \tag{6.25}$$

where

$$\langle \boldsymbol{\nabla}\phi_n, \boldsymbol{f_m}\rangle = \oint_{\ell} d\ell\ \phi_n \boldsymbol{f_m} \cdot \hat{n} - \iint_{S_i} dS\, dS\ \phi_n \boldsymbol{\nabla} \cdot \boldsymbol{f_m}$$

In this equation, ℓ is the edge of the RWG triangles. The integration over the testing triangle surface and the line can be avoided through using an approximate Galerkin method, in which testing functions are implemented at the centroid of the triangle and line. The surface integral over the source triangle is calculated using standard quadrature rules or an approximation to them that is distance dependent. The elements of the impedance matrix are quadruple integrals and extremely computationally intensive to evaluate. There are several ways around this. One way is to approximate the double integral associated with the inner product by a low-order Gaussian quadrature, perhaps with only a single point in the quadrature. This approach resembles point matching. When the trial and shape functions overlap, Z_{mn} has a singularity (Rao et al. 1982). These singularities can be handled with special quadrature methods, such as singularity subtraction or closed-form approximations. To ensure convergence with standard quadrature methods, several terms of the Taylor series expansion of the scalar Green's function may be required to ensure continuity. For patches separated by more than half a wavelength, an asymptotic approximation such as that described by Bateman et al. (1998) or another by Gera (1988) can greatly speed up computation with little loss of overall accuracy (<0.1%).

A number of computer packages are available specifically for patch antennas, but general electromagnetic software such as CST-Microwave Studio and HFSS can also be used for a full-wave design. In such packages, the excitation, coupling between layers, surface wave excitation, and dielectric losses are considered. Rather than describing the use of one package or another, the author refers the reader to the literature and to the website of suitable packages referenced here (CST (2020); HFSS (2020); FEKO (2020); IE3D (2020)). Each have their own advantages such as computational speed, ease of data entry, and visualization of results.

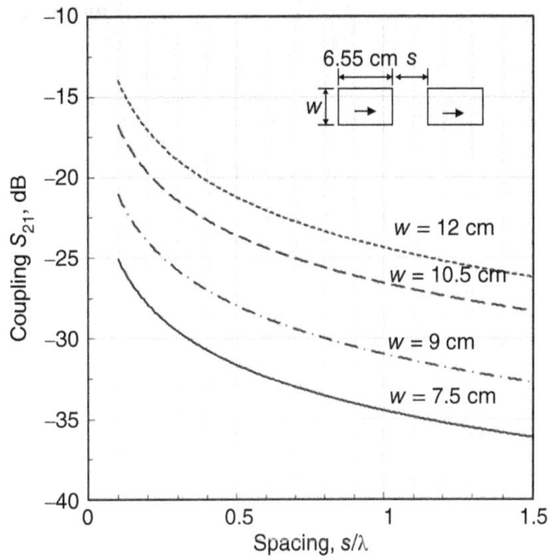

Figure 6.9 Coupling coefficient in the E-plane of identical microstrip patches versus spacing for various patch widths. Substrate has thickness 1.575 mm and dielectric constant 2.5.

6.3 Applications of Microstrip Arrays

6.3.1 Mutual Coupling Between Microstrip Patches

The variation of mutual coupling with size of elements can be used to control the radiation from arrays. As a first step, the variation in coupling with patch size and spacing should be investigated. For example, Figure 6.9 shows the variation in E-plane coupling of identical patches with width w and spacing s on a substrate with thickness 1.575 mm and dielectric constant 2.5 that is predicted by Eq. (6.16a). It is seen that coupling increases with width and decreases by about 3 dB as the spacing is doubled. The coupling increases as the width of the radiators increases due to greater coupling cross-section. The H-plane coupling with varying patch width remains much the same as shown in Figure 6.8 as the coupling length remains the same in each case.

6.3.2 Steering by Switching Parasitic Elements

Steerable antenna arrays are widely used for radar and satellite communication. Electronic steering is often preferred because of lower maintenance costs and speed. Phased arrays are obviously an option but are prohibitive for low-cost systems. A possible alternative uses a central driven element with surrounding parasitic elements operating near resonance. In one approach, an RF switch is used to switch the source between active elements while leaving the parasitic elements at or near resonance to control the beam shape. In the second approach, the parasitic elements are switched in and out of operation with switches such as PIN diodes. The number of possible directions is determined by the number of parasitic elements. The advantage of this parasitic switching approach is that only one element is active at a time leading to a lower insertion loss and a simpler control system (Preston et al. 1997).

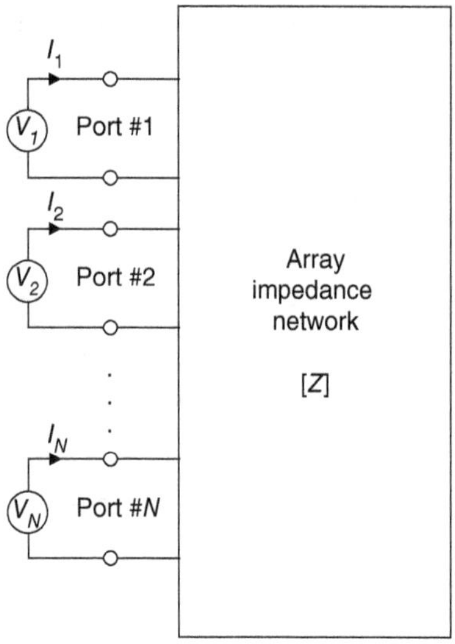

Figure 6.10 Network model of an N-element array.

When a parasitic element is located near an active element, a current is induced in the element through mutual coupling, and as a consequence, the radiation and input characteristics are modified over the operating band. For patch antennas, the current distribution of the parasitic element can be altered by shorting a point located in the maximum current position on the appropriately positioned parasitic element. When the point of maximum current is shorted, the parasitic element acts as a reflector, and when it is not shorted, it operates as a director element of the array.

Consider the network model of the N-element array shown in Figure 6.10. Suppose element #1 is the only active element and all the remaining ports 2, …, M are short-circuited and the other $M+1, \ldots, N$ are open circuited. Then,

$$\begin{bmatrix} V_1 \\ 0 \\ \vdots \\ 0 \\ V_{M+1} \\ \vdots \\ V_N \end{bmatrix} = \begin{bmatrix} Z_{11} & Z_{12} & \cdots & \cdots & \cdots & \cdots & Z_{1N} \\ Z_{21} & Z_{22} & \cdots & \cdots & \cdots & \cdots & Z_{2N} \\ \vdots & \vdots & \vdots & \vdots & \vdots & \vdots & \vdots \\ \vdots & \vdots & \vdots & \vdots & \vdots & \vdots & \vdots \\ \vdots & \vdots & \vdots & \vdots & \vdots & \vdots & \vdots \\ \vdots & \vdots & \vdots & \vdots & \vdots & \vdots & \vdots \\ Z_{N1} & Z_{N2} & \cdots & \cdots & \cdots & \cdots & Z_{NN} \end{bmatrix} \begin{bmatrix} I_1 \\ I_2 \\ \vdots \\ I_M \\ 0 \\ \vdots \\ 0 \end{bmatrix}, \tag{6.26}$$

Thus, there are $N-M$ open circuits, and these are shown in Eq. (6.26) by zero currents. The elements of the impedance matrix in Eq. (6.26) can be calculated using the equations given in Section 6.3.1. The M rows of shorted voltages in Eq. (6.26) give M equations in $N-M$ unknowns, and these can be used to express all short-circuited currents in terms of I_1. The input impedance and short-circuited currents can thus be calculated for a given driving current I_1 through $Z_{in} = V_1/I_1$.

As an example, consider a five-element array consisting of one driven element, the second and third are shorted and the fourth and fifth are open circuited. In that case, Eq. (6.26) becomes

$$\begin{bmatrix} V_1 \\ 0 \\ 0 \\ V_4 \\ V_5 \end{bmatrix} = \begin{bmatrix} Z_{11} & Z_{12} & Z_{13} & Z_{14} & Z_{15} \\ Z_{21} & Z_{22} & Z_{23} & Z_{24} & Z_{25} \\ Z_{31} & Z_{32} & Z_{33} & Z_{34} & Z_{35} \\ Z_{41} & Z_{42} & Z_{43} & Z_{44} & Z_{45} \\ Z_{51} & Z_{52} & Z_{53} & Z_{45} & Z_{55} \end{bmatrix} \begin{bmatrix} I_1 \\ I_2 \\ I_3 \\ 0 \\ 0 \end{bmatrix}. \tag{6.27}$$

The second and third rows of Eq. (6.27) give the short-circuited currents in terms of the driven current, as follows:

$$I_2 = -\left(\frac{Z_{23}Z_{31} - Z_{21}Z_{23}}{Z_{23}Z_{32} - Z_{22}Z_{33}}\right) I_1$$

and

$$I_3 = -\left(\frac{Z_{22}Z_{31} - Z_{21}Z_{32}}{Z_{22}Z_{33} - Z_{23}Z_{32}}\right) I_1,$$

where the factors in the brackets are ratios of minors of the $\boldsymbol{Z}$ matrix. The open circuit voltages at the inputs are

$$V_4 = Z_{41}I_1 + Z_{42}I_2 + Z_{43}I_3$$

and

$$V_5 = Z_{51}I_1 + Z_{52}I_2 + Z_{53}I_3$$

and the input impedance is

$$Z_{in} = \left[Z_{11} - \frac{Z_{12}(Z_{23}Z_{31} - Z_{21}Z_{33}) + Z_{13}(Z_{22}Z_{31} - Z_{21}Z_{32})}{Z_{23}Z_{32} - Z_{22}Z_{33}}\right]. \tag{6.28}$$

This approach can be easily extended to more or fewer elements with short- or open-circuited inputs. By doing this, the currents can be switched to scan the main beam.

As an example of a practical design of a switched parasitic array, Figure 6.11 shows a simple five-element patch array that has been designed to operate near 1.5 GHz (Preston et al. 1997). The substrate has a thickness of 1.6 mm, and it has a complex relative permittivity of 4.35 – j0.08. Two parasitic elements are placed on either side of a central square patch of 45.7 mm. The first parasitic is 45.5 mm square and can be switched, and a second outer patch is 45 mm square and is purely parasitic. A probe feed was placed in the larger central patch at an offset position of 12.5 mm from the center to give a 50 Ω match. The five element array was found to have an increased gain above a single patch. As only two parasitic patches are shorted, this small array can switch only in two directions, forward and backward. These beams at 1.55 GHz are shown in Figure 6.12, for both computed and measured radiation patterns. The measured and computed input reflection coefficient is shown in Figure 6.13.

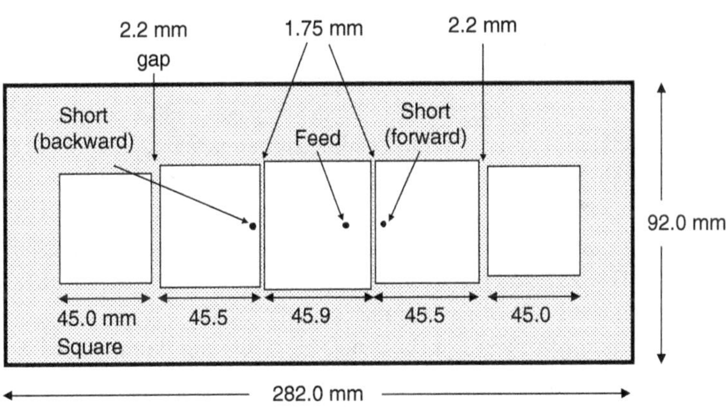

Figure 6.11 Geometry of a five-element switched parasitic array of patches. All dimensions in mm. Source: after Preston et al. 1997.

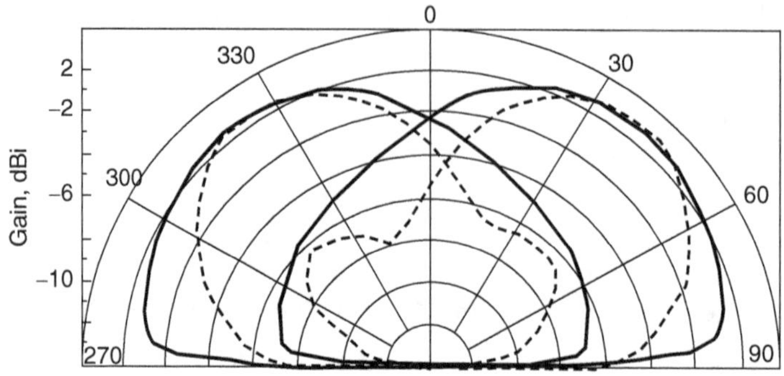

Figure 6.12 E-plane radiation patterns measured for excitation in forward (~320°), and backward (~40°) directions. Solid curve – computed; dash curve – measured. Source: after Preston et al. 1997.

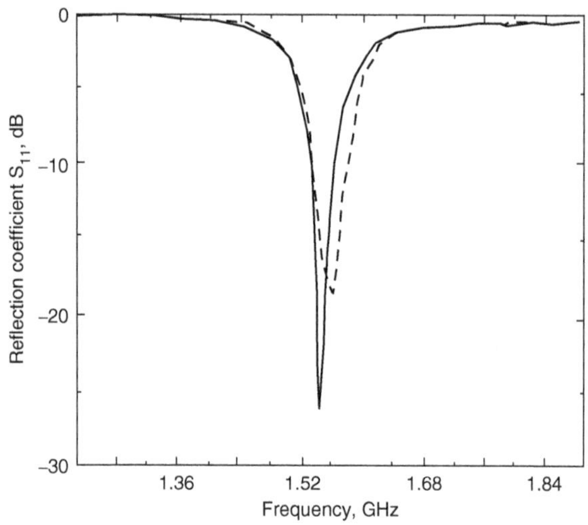

Figure 6.13 Input reflection coefficient for excitation in forward and backward directions of switched array. Source: after Preston et al. 1997. Solid curve – computed; dashed measured.

6.3.3 A Metasurface from Microstrip Patches

Recent advances in radar, communications, and sensing, combined with the shrinking size of antenna platforms, have highlighted the need for wideband array antennas. These wideband arrays can produce multiple scanned beams in two polarizations, and operate over wide frequency bands. There have been numerous investigations into increasing antenna array bandwidth using a wide variety of techniques. One such approach is to use a tightly coupled antenna array, an element of which is illustrated in Figure 6.14 (Bah 2019; Bah et al. 2019). The radiator is a printed dipole, and to improve the scan range of an array of these dipoles, a wide-angle impedance material (WAIM) covers the array. This WAIM uses an array of microstrip patches to produce a thin metasurface (MS) with high effective permittivity and zero permeability. The design of a realistic wideband feed network for the array is summarized here with an emphasis on the WAIM.

The array unit cell shown in Figure 6.14 consists of the dipole element with double-sided MS for wide-angle impedance matching. The E-plane is in the y-z direction, and the H-plane is along the x-z direction. In this figure, the dipole is oriented vertically along the y-z plane with its arms of length $L_{el} = 12$ mm printed on opposite sides of a substrate with a thickness of $t = 1.016$ mm and relative dielectric constant $\varepsilon_r = 10.2$. The period of the unit cell is $d_x = d_y = 22$ mm. The dipole in

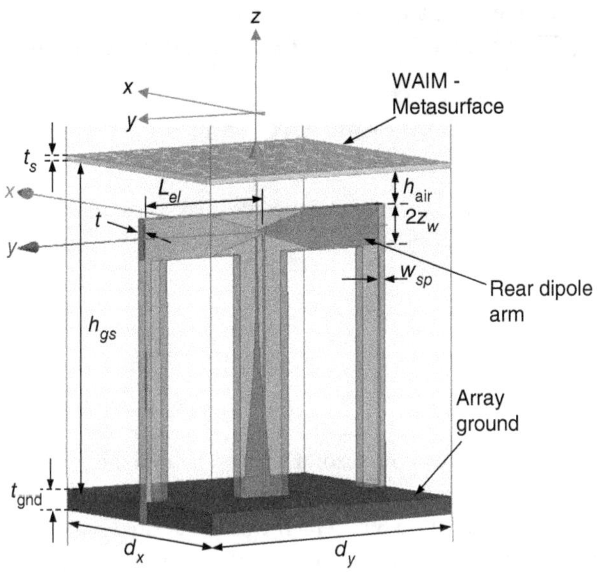

Figure 6.14 A perspective view of the unit cell of a tightly coupled array with a WAIM - metasurface superstrate, and a perforated ground plane. The dipole is mounted on a Rogers RT/Duroid™ 6010 substrate with a thickness, $t = 1.016$ mm and relative dielectric constant, $\varepsilon_r = 10.2$. Source: BAH, A.O., QIN, P-Y, ZIOLKOWSKI, R.W., GUO, Y.J. and BIRD, T.S., 'A wideband low-profile tightly coupled antenna array with a very high figure of merit', IEEE Trans. Antenna. Propagat., Vol. AP-67, no. 4, pp. 2332-2343. © 2019 IEEE.

the cell is covered with a 5×5 array of the WAIM at a height of $h_{air} = 6.76$ mm. The overall height of the array from the ground plane to the top of the MS is $h_{gs} = 37.26$ mm.

In a compact array, a change in the scan angle or the polarization causes a change in the mutual coupling between elements in an array. This effect is greatest for wide-angle scanning that is free of grating lobes (Bah, 2019). To design a WAIM to achieve wide-angle scanning with a compact array, the approach described by Magill and Wheeler (1966) was adopted. In this approach, the superstrate is represented as a material with physical or effective permeability and permittivity. A periodic surface was used to achieve these parameters over the desired bandwidth. The extraction of the constituent parameters of this surface is integral to the design. In the present design, an array of tightly coupled double-sided Jerusalem crosses as shown in Figure 6.1c was used. The crosses are designed in the usual way to create a periodic surface. The variation of the effective value of the susceptance of the surface varies with the polarization and this is used to compensate for reflection in the array. The susceptance varies inversely with increasing scan angle in both principal planes. An analysis of reflection and transmission at a sheet of material was undertaken by considering plane waves incident on a thin sheet of material with parameters ε_{eff} and μ_{eff}. Some advantage can be gained if $\mu_{eff} = 0$ as the angle of refraction from this type of material is zero degrees irrespective of the angle of incidence within the material.

For simplicity in the following description, a material with an effective permittivity ε_{eff} (which is different from Eq. (6.1)) will be described (Collin 1960). For a dielectric with an effective permittivity, the transmission through a thin sheet of thickness h, the effective susceptance $B_{eff}(\theta)$ for the surface is given by (Magill and Wheeler 1966)

$$\frac{B_{eff}(\theta)}{Y_o} = (\varepsilon_{eff} - \cos^2(\theta))\frac{2\pi h}{\lambda_o}, \tag{6.29}$$

where Y_o and λ_o are the wave conductance of free space and wavelength there, respectively. Within the material, the wavelength is $\lambda = \lambda_o/\sqrt{\varepsilon_{eff}}$. At broadside, when $\theta = 0$, the substrate thickness required is

$$h = \frac{\lambda_o}{2\pi(\varepsilon_{eff} - 1)} \frac{B_{eff}(0)}{Y_o}. \tag{6.30}$$

Using the broadside susceptance as a reference, the variation of the effective susceptance with scan angle is

$$\frac{B_{eff}(\theta)}{B_{eff}(0)} = \frac{1}{\cos(\theta)} \quad \text{H-plane} \tag{6.31a}$$

$$\frac{B_{eff}(\theta)}{B_{eff}(0)} = \cos(\theta) - \frac{1}{\varepsilon_{eff}} \tan(\theta)\sec(\theta) \quad \text{E-plane.} \tag{6.31b}$$

By increasing the effective permittivity, the influence of the second term of the E-plane variation in Eq. (6.31b) decreases, and consequently, the variation in the two planes is approximately reciprocal. For $\varepsilon_{eff} \sim 3$, both planes have similar behavior out to 60°, which was the desired maximum scan angle. For a thin sheet, compensation of susceptance out to 60° can be achieved providing

$$\frac{B_{eff}(\theta) - B_{eff}(0)}{B_{eff}(0)} < \frac{\pi}{2} \frac{\varepsilon_{eff} - 1}{\sqrt{\varepsilon_{eff}}} \quad ; \quad 0 < \theta < \frac{7\pi}{18}. \tag{6.32}$$

Clearly for wide scan angles, ε_{eff} should be sufficiently high enough to satisfy Eqs. (6.31a) and (6.31b).

A closer view of the unit cell of the MS is shown in Figure 6.15 (Bah et al. 2019). It consists of unequal-arm Jerusalem cross elements (Figure 6.1c) printed on both sides of a dielectric substrate.

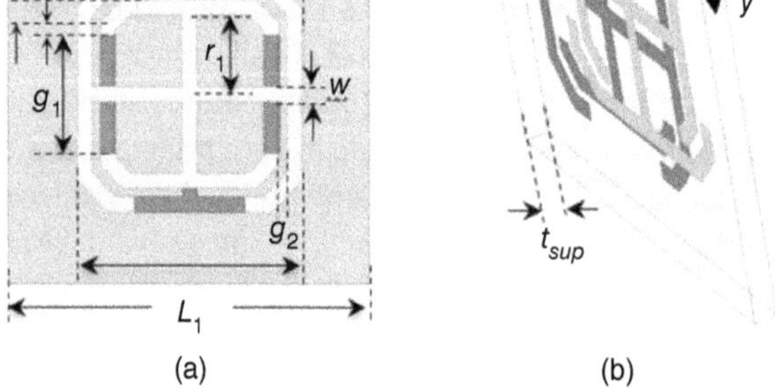

Figure 6.15 Unit cell of double-sided metasurface. The direction of propagation of the incident wave is along the z-axis. (a) Top view; (b) Perspective view. The metasurface unit cell dimensions in mm are $w = 0.2$, $t_{sup} = 0.254$, $g_1 = 1.5$, $g_2 = 0.1$, $L_1 = 4.8$, $L_2 = 3.0$, $L_3 = 0.35$, $L_4 = 0.4$, $L_5 = 0.15$, $L_6 = 0.3$, and $r_1 = 1.0$. The substrate is Rogers RT/Duroid 5880 with $\varepsilon_r = 2.2$ and tanδ = 0.0009. Source: BAH, A.O., QIN, P-Y, ZIOLKOWSKI, R.W., GUO, YJ. and BIRD, T.S., 'A wideband low-profile tightly coupled antenna array with a very high figure of merit', IEEE Trans. Antenna. Propagat., Vol. AP-67, no. 4, pp. 2332-2343. © 2019 IEEE.

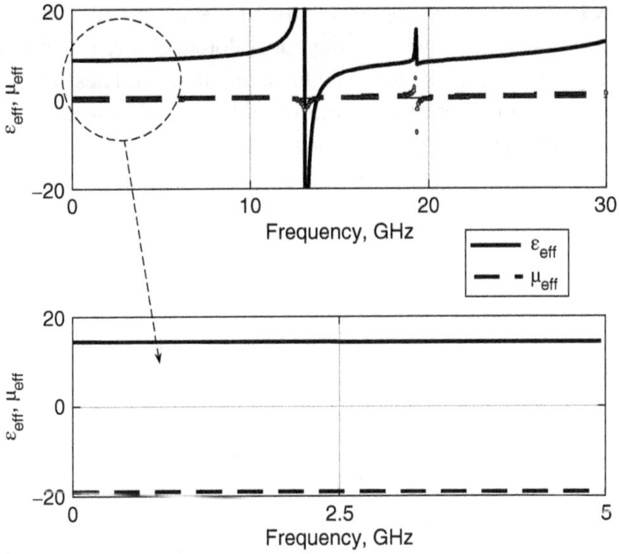

Figure 6.16 Extracted effective permittivity and permeability for the double-sided Jerusalem cross metasurface (after Bah, 2019). Solid curve – ε_{eff}; dashed curve – μ_{eff}.

Figure 6.17 Computed active VSWR of an infinite dipole array at broadside. Key: solid curve – the final array with tuned feed and WAIM superstrate; dotted curve is the array with no superstrate; and dashed curve – array with the MS with unbalanced currents. Source: BAH, A.O., QIN, P-Y, ZIOLKOWSKI, R.W., GUO, Y.J. and BIRD, T.S., 'A wideband low-profile tightly coupled antenna array with a very high figure of merit', IEEE Trans. Antenna. Propagat., Vol. AP-67, no. 4, pp. 2332-2343. © 2019 IEEE.

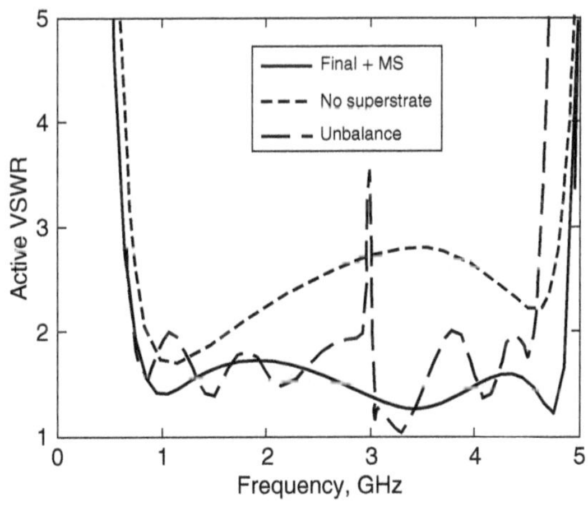

The extracted effective medium parameters of the MS-WAIM based on a full-wave solution are shown in Figure 6.16. It is seen that μ_{eff} is approximately zero, while ε_{eff} has a constant value of about 8.6 within the frequency range of interest (0–5.0 GHz). The former means that according to Snell's law, the angle of refraction is zero for all angles of incidence. As a result, the double-sided MS can transform the spherical waves from the array into plane waves with minimal phase variation to enable wide-angle scanning. This waveform transformation can also help to lower the array profile since all the energy is concentrated in the forward direction. This phenomenon can be used to improve the active VSWR even at broadside as is demonstrated in Figure 6.17. The solid curve in Figure 6.17 shows the effectiveness of matched surface WAIM as a wideband impedance matching layer. Without the WAIM, the array is highly unmatched across most of the band. When the matched surface structure is integrated with the tightly coupled array below, it is usual to adjust the WAIM dimensions to compensate for the near-field interactions.

An impedance bandwidth from 0.79 to 4.92 GHz is shown in Figure 6.17 (solid curve). The performance of the array without the superstrate is also shown as the upper dotted curve. The strong mutual coupling between adjacent array elements causes the feed to be unbalanced. These unbalanced currents led to an induced common-mode resonance around mid-band as shown by the dashed curve in Figure 6.17. To eliminate these currents, the array unit cell size was adjusted, and a shorting pin to ground was also introduced on the excited arm of the dipole shown in Figure 6.14.

6.4 Concluding Remarks

This chapter has examined the two types of coupling present in microstrip patch antenna arrays. One type is the coupling between the ends of a patch, and the other is the coupling between different patches. Three approaches were described for coupling between different patches. These were the E-field method, the H-field method, and the full-wave approaches. Examples were given of each approach and compared with measured results. As an application of some of the techniques, firstly results were described for a switched array of patch elements used for beam steering, and secondly the design of a wide-angle impedance material composed of double sided Jerusalem cross elements was used to improve the scanning performance of a closely coupled array of dipoles.

References

Aberle, J.T. and Pozar, D.M. (1989). Analysis of infinite arrays of probe-fed rectangular microstrip patches using a rigorous feed model. *IEEE Proc. H, Microw., Antenna. and Propag.* 136 (2): 110–119.

Bah, A.O. (2019): Metasurface enhanced ultra-wideband multifunctional antenna arrays and Fabry-Perot antennas, PhD thesis, School of Electrical and Data Engineering, Faculty of Engineering and Information Technology, University of Technology Sydney, Australia.

Bah, A.O., Qin, P.-Y., Ziolkowski, R.W. et al. (2019). A wideband low-profile tightly coupled antenna array with a very high figure of merit. *IEEE Trans. Antennas Propagat.* AP-67 (4): 2332–2343.

Bateman, D.G., Bird, T.S., and Mohan, A.S. (1998). Asymptotic approximation of some integrals involved in EFIE solutions using the method of moments. *IEE Proc. Microw. Antenna. Propag.* 145 (4): 361–365.

Brigham, E.O. (1974). *The fast Fourier transform*. Eaglewood Cliffs, N.J.: Prentice Hall, Inc.

Carver, K.R. (1980). Input impedance of probe fed microstrip antennas. *IEEE Int. Symp. Antenna. Propagat.*: 617–620.

Carver, K.R. and Mink, J.W. (1981). Microstrip antenna technology. *IEEE Trans. Antennas Propagat.* AP-29: 2–24.

Collin, R.E. (1960). *Field Theory of Guided Waves*, 87–96. McGraw-Hill.

CST (2020). https://www.mathworks.com/products/connections/product_detail/cst-microwave-studio.html.

FEKO (2020). https://www.altair.com/resource/feko-microstrip-antennas-and-circuits.

Gera, A.E. (1988). Simple expressions for mutual impedances (dipoles and microstrip elements). *IEE Proc. H.* 135: 395–399.

Getsinger, W.J. (1973). Microstrip dispersion model. *IEEE Trans. Microw. Theory Tech.* MTT-21 (1): 34–39.

Ghannay, N. and Samet, A. (2009): 'E-shaped patch antenna modeling with MoM and RWG basis functions', 16th IEEE Int. Conf. Electron. Circuits & Syst. – (ICECS 2009), pp. 199–201.

Hammerstad, E. and Jensen, O. (1980). Accurate models for microstrip computer aided design. *IEEE MTT-S Int. Microw. Symp. Digest*: 407–409.

HFSS (2020) https://www.ansys.com/products/electronics/ansys-hfss.

IE3D (2020) https://www.rfglobalnet.com/doc/full-wave-3-d-em-simulator-for-both-planar-an-0001.

Jedlicka, R.P., Poe, M.T., and Carver, K.R. (1981). Measured mutual coupling between microstrip antennas. *IEEE Trans Antennas Propagat.* AP-29 (1): 147–149.

Lee, K.F. and Chen, W. (1997). *Advances in Microstrip and Printed Antennas*. USA: Wiley.

Magill, E.G. and Wheeler, H.A. (1966). Wide-angle impedance matching of a planar array antenna by a dielectric sheet. *IEEE Trans. Antennas. Propagat.* AP-14 (1): 49–53.

Mohammadian, A.H., Martin, N.W., and Griffin, D.W. (1989). A theoretical and experimental study of mutual coupling in microstrip antenna arrays. *IEEE Trans. Antenna. Propagat.* AP-37: 1217–1223.

Mosiq, J.R. (1988). Arbitrarily shaped microstrip structures and their analysis with a mixed potential integral equation. *IEEE Trans. Microw. Tech.* MTT-36 (2): 314–323.

Newman, E.H., Richmond, J.H., and Kwan, B.W. (1983). Mutual impedance computation between microstrip antennas. *IEEE Trans. Microw. Theory Tech.* MTT-31: 941–945.

Pozar, D.M. (1982). Input impedance and mutual coupling computation between microstrip antennas. *IEEE Trans. Antennas Propagat.* AP30 (11): 1191–1196.

Pozar, D.M. (1990). Rigorous closed-form expressions for the surface wave loss of printed antennas. *Electron. Lett.* 26 (13): 954–955.

Preston, S.L., Thiel, D.V., Lu, J.W. et al. (1997). Electronic beam steering using switched parasitic patch elements. *Electron. Lett.* 33 (1): 7–8.

Pues, H. and de Van, C. (1984). Accurate transmission line model for the rectangular microstrip antenna. *IEE Proc. H, Microw., Optics & Acoustics* 131 (6): 334–340.

Rao, S.M., Wilton, D.R., and Glisson, A.W. (1982). Electromagnetic scattering by surfaces of arbitrary shape. *IEEE Trans. Antennas Propagat.* 30 (3): 409–418.

Sainati, R.A. (1996). *CAD of microstrip antennas for wireless applications*. Norwood, MA, USA: Artech House.

Skrivervik, A.K. and Mosig, J.R. (1993). Analysis of finite phase arrays of microstrip patches. *IEEE Trans. Antennas Propagat.* 41 (8): 1105–1114.

Smith, S.L. (2000): 'Investigation of switched parasitic antenna arrays with application to electronic beam steering for mobile communications systems', PhD thesis, Griffith University, Brisbane, Australia.

Van de Capelle, A.R. and Luypaert, P.J. (1973). Fundamental and higher-order modes in open microstrip lines. *Electron. Lett.* 9 (15): 345–1346.

7

Mutual Coupling Between Antennas on Conformal Surfaces

Trevor S. Bird[1,2]

[1]*Global Big Data Technology Centre, University of Technology Sydney, Ultimo, NSW 2007, Australia*
[2]*Antengenuity, Eastwood, NSW 2122, Australia*

7.1 Introduction

A well-known limitation of the planar phased array is that, as the beam is scanned from broadside, the beam broadens and the pattern deteriorates. On the other hand, an array that is conformal to a convex surface, such as a circular cylinder or a cone, allows the beam to be scanned in discrete steps through an arc while maintaining a constant pattern. Although there are some well-documented examples and an established theory of conformal arrays (Hansen 1981; Josefson and Persson 2006), their use is often not widespread mainly because of the complexity of the feed network. Recent developments in microwave and optical componentry have simplified the design of feed networks, thereby making conformal arrays an attractive alternative for applications such as wide-angle scanned beam antennas with ultralow sidelobes (e.g. Singh et al. 2017). Of importance in the design of low sidelobe antenna arrays is predicting the effect of mutual coupling between the array elements.

In this chapter, mutual coupling in finite antenna arrays that are conformal to curved surfaces is investigated. As a background interest to this investigation are uniform asymptotic representations of basic solutions for the fields due to point sources on a curved surface that have moderate to large radii of curvature. Although there is an exact modal solution to some canonical problems such as an infinitely long cylinder, this is not useful for practical array analysis except for cylinders of small to moderate curvature.

The asymptotic representations to be described here are uniform in the sense that they provide a smooth transition across the shadow boundary from a region of direct illumination by the source to the shadow region where there is a diffracted field only. In common with the geometrical theory of diffraction (GTD) (James 1976), this representation admits a ray-based interpretation of the field on a cylinder in terms of "creeping waves." Also, following the general principles of GTD, we consider the cylinder problem as well as one for the sphere as canonical problems, and, from their asymptotic solutions, a generalized solution can be devised for any smooth conducting convex surface.

A Green's function representation of two quadric surfaces, the cylinder and sphere, is described in the next section, Section 7.2, and these are placed in a ray-based coordinate system. Section 7.3 gives a brief review of the literature on the application of asymptotic techniques to the analysis of mutual coupling in apertures in convex surfaces. The reference list is representative of the material published in the open literature, but it does not pretend to be exhaustive. An analysis of mutual

Mutual Coupling Between Antennas, First Edition. Edited by Trevor S. Bird.

coupling between apertures using canonical solutions for the cylinder and sphere is described in Section 7.4 and, in Section 7.5, these solutions are extended to develop a general formulation for coupling in slowly varying convex surfaces. Some examples of the use of this general formulation is given in Section 7.6 for three cases of an array of rectangular waveguides on a circular cylinder, two-short monopoles on a circular cylinder, and rectangular apertures in an ellipsoid. Lastly, in Section 7.7, some conclusions are provided.

7.2 Mutual Admittance of Apertures on Slowly Curving Surfaces

7.2.1 Green's Function Formulation for Curved Surfaces

The method for creating Green's functions that was outlined in Chapter 3 can be generalized for other bounding surfaces as well as materials with dyadic constitutive parameters. A general approach to this is given in the references (Felsen and Marcuvitz 1973; Chew 1995). The basic formulation depends on whether the source is magnetic or electric or a combination of both as well as the type of surface such as a conductor or substrate on which the two sources are situated. To separate out the two principal components, the sources are resolved tangential and normal to the surface.

In common with Section 5.2.1 for a planar surface, consider here an array of apertures on a slowly curving surface S_0. An array of dipoles can be considered similarly commencing initially from duality. The apertures can be replaced by areas of equivalent magnetic current, $\boldsymbol{M}_0$, which are adjacent to the curved surface there and a null electric field on the inside S_0. As given in Eq. (5.1), the total magnetic field on such a surface is

$$\boldsymbol{H}_1^{inc} + \boldsymbol{H}_t^{rev}(-\boldsymbol{M}_0) = \boldsymbol{H}_t^{ext}(\boldsymbol{M}_0).$$

The magnetic current is expanded in terms of N_e general expansion functions, $\{\boldsymbol{\alpha}_n \mid n = 1, \ldots, N_e\}$, with unknown coefficients V_n as in Eq. (5.2). Near the curved surface, the fields excited by this source may be found by the methods of Felsen and Marcuvitz (1973). As described in Section 3.4.2.3, the magnetic field at the surface due to an element of magnetic current $d\boldsymbol{M}(\boldsymbol{r}')$ is

$$d\boldsymbol{H}(\boldsymbol{r}) = \underline{\underline{\mathbf{G}}}^{(h)}(\boldsymbol{r}, \boldsymbol{r}') \cdot d\boldsymbol{M}(\boldsymbol{r}'),$$

where $\underline{\underline{\mathbf{G}}}^{(h)}(\boldsymbol{r}, \boldsymbol{r}')$ is the magnetic Green's dyadic of the surface, and it describes the field at a point on S_0 due to an element of magnetic current. As a result, the total external field to the surface is

$$\boldsymbol{H}_t^{ext}(\boldsymbol{M}_0) = \iint_{S_0} \underline{\underline{\mathbf{G}}}^{(h)}(\boldsymbol{r}, \boldsymbol{r}') \cdot \boldsymbol{M}_0 \, dS', \tag{7.1}$$

where $\underline{\underline{\mathbf{G}}}^{(h)}(\boldsymbol{R} \mid \boldsymbol{R}') = \begin{bmatrix} G_{tt} & G_{tb} \\ G_{bt} & G_{bb} \end{bmatrix}$ is the dyadic for the surface magnetic field expressed in matrix form, where $\hat{t}$ and $\hat{b}$ are orthogonal vectors on the surface (see Figure 7.1) such that $\hat{t} \times \hat{b} = \hat{n}$ where $\hat{n}$ is the normal. This dyadic for the surface field is also expressed as

$$\underline{\underline{\mathbf{G}}}^{(h)}(t, b \mid t', b') = (\hat{t}\hat{t}' G_{tt} + \hat{t}\hat{b}' \; G_{tb} + \hat{b}\hat{t}' G_{bt} + \hat{b}\hat{b}' G_{bb}).$$

There are four quadric surfaces in separable coordinates for which there are closed-form solutions, namely the plane, cylinder, sphere, and ellipsoid (Bowman et al. 1969). The solution for the degenerate quadric surface, the plane, has been discussed in detail in Chapter 5. In this section, the closed form solutions for the surface field on two quadric surfaces, the cylinder and sphere, are described and placed in the generalized coordinate system.

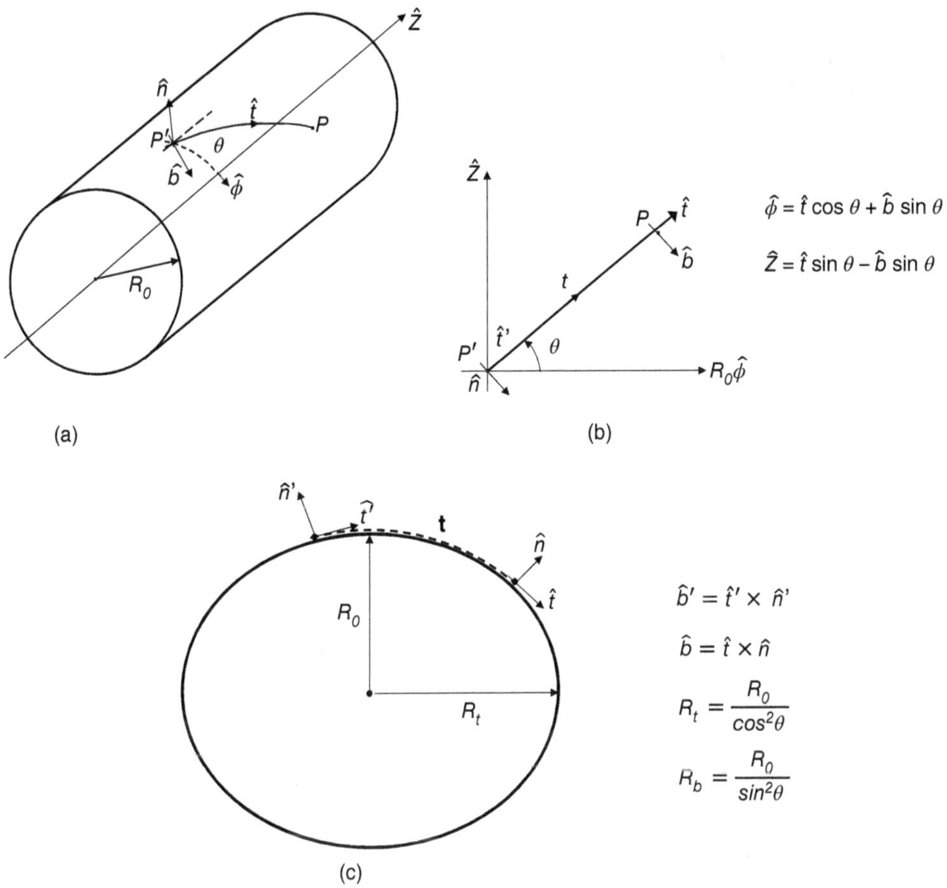

Figure 7.1 Surface ray from source point P1 to field point P on a cylinder of radius R_o. (a) Normal view; (b) developed cylinder; and (c) cut in θ direction.

7.2.2 The Cylinder

With reference to the geometry of the cylinder in Figure 7.1, the dyadic components are listed in Section 3.4.2.3 but are repeated here for convenience

$$G_{\phi\phi}(\phi, z \mid \phi', z') = -c_h \int_{-\infty}^{\infty} d\xi \sum_{n=-\infty}^{\infty} e^{j(n\bar{\phi}-\xi\,\bar{z})} \frac{k_o}{h} \left[\frac{H_n^{(2)\prime}(hR_o)}{H_n^{(2)}(hR_o)} - \left(\frac{n\xi}{k_o h R_o} \right)^2 \frac{H_n^{(2)}(hR_o)}{H_n^{(2)\prime}(hR_o)} \right], \tag{7.2a}$$

$$G_{\phi z}(\phi, z \mid \phi', z') = c_h \int_{-\infty}^{\infty} d\xi \sum_{n=-\infty}^{\infty} e^{j(n\bar{\phi}-\xi\,\bar{z})} \left[\left(\frac{n\xi}{k_o h R_o} \right) \frac{H_n^{(2)}(hR_o)}{H_n^{(2)\prime}(hR_o)} \right], \tag{7.2b}$$

$$G_{z\phi}(\phi, z \mid \phi', z') = G_{\phi z}(\phi, z \mid \phi', z'), \tag{7.2c}$$

and

$$G_{zz}(\phi, z \mid \phi', z') = c_h \int_{-\infty}^{\infty} d\xi \sum_{n=-\infty}^{\infty} e^{j(n\bar{\phi}-\xi\,\bar{z})} \left[\left(\frac{hR_o}{k_o R_o} \right) \frac{H_n^{(2)}(hR_o)}{H_n^{(2)\prime}(hR_o)} \right], \tag{7.2d}$$

where

$$h = \begin{cases} |h|, & |\xi| < k_o \\ -j|h|, & |\xi| > k_o \end{cases},$$

$$|h| = \sqrt{|k_o^2 - \xi^2|},$$

$$c_h = \frac{jY_o}{(2\pi)^2 R_o},$$

$$\bar{\phi} = \phi - \phi', \ \bar{z} = z - z',$$

and $k_o = 2\pi/\lambda_o$. $H_n^{(2)}$ is the Hankel function of second kind of order n. A prime on this function indicates the first derivative to the argument. Y_o and λ_o are free-space wave admittance and wavelength, respectively. It is assumed here that the media external to the cylinder is free space. If it is not, the wavenumber k_o should be changed in the usual way.

As Eqs. (7.2) and a transformed version, Eqs. (3.36), that is suitable for computation are very computer intensive even for cylinders of moderate diameter in array calculations, it is therefore important to find accurate approximate solutions for cylinders of moderate to large radius. One such approach is to make use of the Watson transformation to convert the summations in Eq. (7.2) into contour integrals. From this transformation, it is often possible to easily evaluate these integrals at the poles of the integrands, and the result is a summation in terms of the residues of the integrals.

To demonstrate this approach, consider the general infinite series

$$S(\phi) = \sum_{n=-\infty}^{\infty} f(n) e^{jn\phi}, \tag{7.3}$$

where the complex function $f(n)$ is an even function of n. We can consider n to indicate the n-th pole of an arbitrary function, and this series is a sum of residues. Suppose this function is $1/\sin \nu\pi$. Therefore, through Cauchy's integral theorem, $S(\phi)$ can be rewritten as

$$S(\phi) = \frac{j}{2} \oint_C d\nu\, f(\nu) \frac{\exp j\nu(\phi - \pi)}{\sin \nu\pi}, \tag{7.4}$$

where C is a clockwise contour around the real axis enclosing the poles of $\sin \nu\pi$. In addition, as $f(\nu)$ is an even function of ν, $S(\phi)$ can also be expressed as

$$S(\phi) = -j \oint_{C_\nu} d\nu\, f(\nu) \frac{\cos[\nu(\phi - \pi)]}{\sin \nu\pi}, \tag{7.5}$$

where C_ν in Eq. (7.5) runs just below the real axis from left to right. This form of the transformation was first used by Debye (Watson 1918).

If $f(\nu)$ has single-order complex poles ν_p ($p = 1, 2, \ldots, \infty$) in the bottom half of the complex ν-plane, the integral can be expressed as a residue series of these poles. Suppose $f(\nu) = h(\nu)/g(\nu)$ where $g(\nu_p) = 0$. Application of Cauchy's residue theorem to Eq. (7.5) results in

$$S(\phi) = -2\pi \sum_{p=1}^{\infty} \frac{\cos[\nu_p(\phi - \pi)]}{\sin \nu_p\pi} \frac{h(\nu_p)}{\left.\frac{\partial g}{\partial \nu}\right|_{\nu=\nu_p}}, \tag{7.6}$$

This representation of $S(\phi)$ is known as the Watson transformation (Watson 1918). In addition Eq. (7.5) can be represented alternatively as creeping waves. Noting that

$$\frac{1}{\sin \nu\pi} = 2j \sum_{q=0}^{\infty} \exp[-j\nu(2q+1)\pi] \ ; \ Im\, \nu < 0$$

so that

$$\frac{\cos[\nu(\phi-\pi)]}{\sin \nu\pi} = j\sum_{q=0}^{\infty} e^{-j\nu 2q\pi}\ [e^{-j\nu\phi} \ + e^{-j\nu(2\pi-\phi\ )}].$$

Therefore

$$S(\phi) = \sum_{q=0}^{\infty}[Q(\phi+2q\pi) + Q(2\pi-\phi+2q\pi)], \tag{7.7}$$

where

$$Q(\phi) = -\int_{C_\nu} d\nu\, f(\nu)\exp(-j\nu\phi). \tag{7.8}$$

Equation (7.8) indicates there are two basic contributions to the original series; one occurs in a clockwise direction, and the other is anticlockwise. By the same arguments used to derive Eq. (7.6), Eq. (7.8) can also be expressed as a residue series; thus,

$$Q(\phi) = -2\pi j\sum_{p=1}^{\infty} \frac{h(\nu_p)}{\left.\frac{\partial g}{\partial \nu}\right|_{\nu=\nu_p}} \exp(-j\nu_p\phi). \tag{7.9}$$

Equations (7.4)–(7.9) are alternative ways of representing the same function given by series Eq. (7.3). Depending on the functions involved, there will be instances where one representation is preferable to the others. For example, for certain functions and parameters, the residue series may converge very rapidly, so an accurate solution is obtained from only a few terms, whereas the original series may involve considerably more computational effort for the same accuracy.

As an example, when Watson's transformation is applied to the integrand of Eq. (7.2d), we have initially

$$S(\bar{\phi}) = \frac{j}{2}\oint_C d\nu \left[\left(\frac{hR_o}{k_oR_o}\right)\frac{H_n^{(2)}(hR_o)}{H_n^{(2)'}(hR_o)}\right]\frac{exp[j\nu(\bar{\phi}-\pi)]}{sin(\nu\pi)}, \tag{7.10}$$

where C is a clockwise contour around the real axis enclosing the zeros of $sin(\nu\pi)$. As the function in the square brackets in Eq. (7.10) is an even function, the summation can also be expressed as

$$S(\bar{\phi}) = -j\int_{C_\nu} d\nu \left[\left(\frac{hR_o}{k_oR_o}\right)\frac{H_\nu^{(2)}(hR_o)}{H_\nu^{(2)'}(hR_o)}\right]\frac{cos[j\nu(\bar{\phi}-\pi)]}{sin(\nu\pi)}. \tag{7.11}$$

Once again, C_ν runs just below the real axis from left to right. As $H_\nu^{(2)'}(hR_o)$ has single-order complex zeros ν_p' $(p = 1, 2, \cdots \infty)$ in the bottom half of the complex ν-plane, the integral over ν can be expressed as a residue series of these poles. Application of the Cauchy residue theorem to Eq. (7.5) results in

$$S(\bar{\phi}) = -2\pi\sum_{p=1}^{\infty}\frac{cos[j\nu_p(\bar{\phi}-\pi)]}{sin(\nu_p\pi)}\left[\left(\frac{hR_o}{k_oR_o}\right)\frac{H_{\nu_p}^{(2)}(hR_o)}{\left.\frac{\partial H_\nu^{(2)'}(hR_o)}{\partial \nu}\right|_{\nu=\nu_{p'}}}\right]. \tag{7.12}$$

In like manner, Watson's transformation can be applied to the other components of the dyadic listed in Eq. 7.2.

In many applications involving large cylinders, only the first two terms of the series, corresponding to the two dominant creeping waves at $q = 0$, provide sufficient accuracy in Eq. (7.7) such that

$$S(\bar{\phi}) \cong Q(\bar{\phi}) + Q(2\pi - \bar{\phi}).$$

Continuing with Eq. 7.11, suppose in $Q(\bar{\phi})$ the argument kR_o of the Hankel functions is large. The complex zeros of the Hankel function occur when $\nu \sim kR_o \gg 1$. As well, the zeros of the derivative $H_\nu^{(2)\prime}(hR_o)$ are approximately the zeros of $w_2'(t)$, the derivative of the Airy integral $w_2(t)$ (see Appendix D), where $t = (\nu - kR_o)/M$ and $M = (kR_o/2)^{1/3}$. Let t_p' $(p = 1, 2, \cdots, \infty)$ be the zeros of w_2' and $\nu_p' \cong kR_o + M\, t_p'$.

When the order and argument of the Hankel function is large, this function and its derivative can be represented by uniform asymptotic expansions of Airy functions as follows

$$H_\nu^{(2)}(hR_o) \sim \left(\frac{j}{M\sqrt{\pi}}\right)\left[w_2(t) - \left(\frac{1}{60}M^2\right)(4t\; w_2(t) + t^2\; w_2'(t))\right], \tag{7.13a}$$

$$H_\nu^{(2)\prime}(hR_o) \sim \left(-\frac{j}{M^2\sqrt{\pi}}\right)\left[w_2'(t) + \left(\frac{1}{60}M^2\right)((6 - t^3)\, w_2(t) + 4t\, w_2'(t))\right]. \tag{7.13b}$$

Therefore, from Eq. (7.13b)

$$\partial H_\nu^{(2)\prime}(hR_o)/\partial\nu\Big|_{\nu=\nu_p'} \cong (-j/M^3\sqrt{\pi})\,[t_p'\, w_2(t_p')],$$

and in terms of tabulated functions

$$\partial H_\nu^{(2)\prime}(hR_o)/\partial\nu\Big|_{\nu=\nu_p'} \cong \left(-\frac{2}{M^3}\right)\,[\alpha_p'\, Ai(-\alpha_p')], \tag{7.14}$$

where α_p' is the p-th zero of the derivative of the Airy function $Ai'(t)$, which is defined in Appendix D. The Airy function zeros and associated functions values are tabulated by Abramowitz and Stegun (1965, Chapter 10) and approximations are given in Appendix D. Introducing Eq. (7.14) into Eq. (7.11), we now have

$$Q(\bar{\phi}) = (hR_o)^2\left(\frac{j\pi}{2\,k_oR_o}\right)\sum_{p=1}^{\infty}\left[\frac{e^{-jMt_p'\,(\bar{\phi}-\pi/2)}}{\alpha_p'\,Ai(-\alpha_p')}\right] \tag{7.15}$$

The terms of this series are called creeping waves (Watson 1918; Chan et al. 1977). Since α_p' is complex with a negative imaginary part, creeping waves decay with angular distance. The angles $(\bar{\phi} - \pi/2)$ and $(3\pi/2 - \bar{\phi})$ in Eq. (7.11) are the angles subtended by the creeping wave paths. Figure 7.2 illustrates creeping waves on a cylinder. An approximate asymptotic solution can be obtained for each component of the surface field dyadic of a cylinder. Each term of Eq. (7.2) can be interpreted as surface waves and the field on the cylinder as a standing wave produced by the contrarotating surface waves, which propagate from the source to the field point along helical ray paths, as depicted in Figure 7.2. These waves follow the curvature of the cylindrical surface and appear to "creep" from the source to the field point. These creeping waves attenuate as they propagate requiring, in any practical calculation, the consideration of only two contrarotating creeping waves encircling the cylinder just once as indicated above. Close to the source, the series converges very slowly, and in computations, it is usually replaced by direct integration or an equivalent hard acoustic Fock function, so-called after V.A. Fock, an early investigator of diffraction by curved surfaces (Fock 1965). These functions are described in Appendix D.

7.2.3 The Sphere

The corresponding result of the dyadic for the surface field for a sphere with uniform surface impedance Z_s, which is excited by a magnetic current element $\boldsymbol{dm}$ located at (a, 0, 0) arranged at an angle ϕ_o to the x-axis as shown in Figure 7.3, is (Alisan and Ertürk 2010)

$$\underline{\underline{\mathbf{G}}}(\boldsymbol{R} \mid \boldsymbol{R}') = \begin{bmatrix} T_{\theta\theta} & 0 \\ 0 & T_{\phi\phi} \end{bmatrix}$$

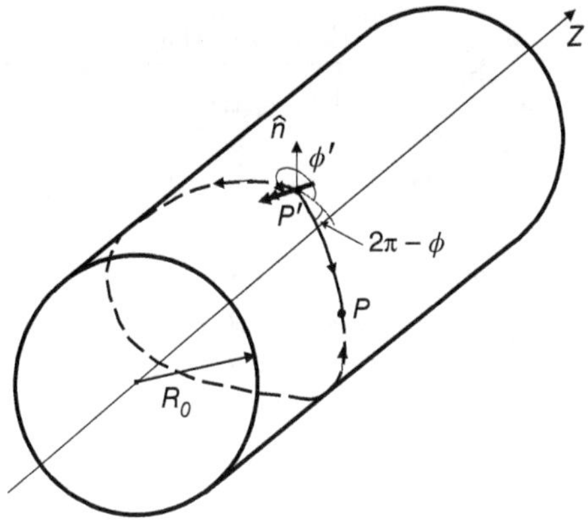

Figure 7.2 Creeping waves launched on the surface of a cylinder by a source.

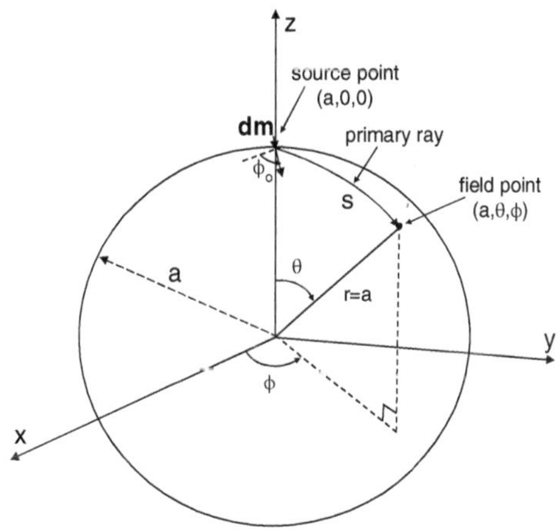

Figure 7.3 Geometry for surface field on an impedance sphere of radius a due to a magnetic surface ***dm*** located at (a, 0, 0).

where

$$G_{\theta\theta}(a,\theta,\phi) = \frac{1}{a}\cos(\phi-\phi_o)\frac{k\,dm}{4\pi j}\left(\frac{1}{ka}\right)\sum_{n=1}^{\infty}\frac{2n+1}{n(n+1)}\left[-F_{n\theta}(a,\theta)+\frac{1}{jk\ sin\theta}\,A_{nr}(a,\theta)\right], \tag{7.16a}$$

$$G_{\phi\phi}(a,\theta,\phi) = \frac{1}{a}\sin(\phi-\phi_o)\frac{k\,dm}{4\pi j}\left(\frac{1}{ka}\right)\sum_{n=1}^{\infty}\frac{2n+1}{n(n+1)}\left[\frac{1}{sin\theta}F_n(a,\theta)-\frac{1}{jk}A_{n\theta}(a,\theta)\right], \tag{7.16b}$$

in which

$$F_n(a,\theta) = (\hat{J}_n(ka) - C_n\hat{H}_n^{(2)}(ka))\hat{H}_n^{(2)\prime}(ka)\ P_n^1(cos\theta)$$

$$F_{n\theta}(a,\theta) = (\hat{J}_n(ka) - C_n\hat{H}_n^{(2)}(ka))\hat{H}_n^{(2)\prime}(ka)\ P_n^2(\cos\theta)$$

$$A_{nr}(a,\theta) = (\hat{J}_n^{\prime}(ka) - D_n\hat{H}_n^{(2)\prime}(ka))\hat{H}_n^{(2)}(ka)\ P_n^1(\cos\theta)$$

$$A_{n\theta}(a,\theta) = (\hat{J}_n^{\prime}(ka) - D_n\hat{H}_n^{(2)\prime}(ka))\hat{H}_n^{(2)}(ka)\ P_n^2(\cos\theta)$$

$$C_n = \frac{\hat{J}_n(ka) + j\Lambda\hat{J}'_n(ka)}{\hat{H}_n^{(2)}(ka) + j\Lambda\hat{H}_n^{(2)\prime}(ka)} \quad D_n = \frac{\hat{J}_n(ka) + j\Lambda^{-1}\hat{J}'_n(ka)}{\hat{H}_n^{(2)}(ka) + j\Lambda^{-1}\hat{H}_n^{(2)\prime}(ka)}$$

$$\hat{H}_n^{(2)}(z) = zh_n^{(2)}(z); \hat{J}_n(ka) = zj_n(z); P_n^s(\cos\theta) = \frac{\partial P_n(\cos\theta)^s}{\partial\theta^s}; \text{and } \Lambda = \frac{Z_s}{\eta_o}$$

where Z_s is the surface impedance in which $Re\{Z_s\}>0$, the functions $j_n(z)$, $h_n^{(2)}(z)$ and $P_n^s(\cos\theta)$ are the spherical Bessel and Hankel functions, and Legendre functions, respectively (Abramowitz and Stegun, 1965). When $Z_s = 0$, the solution reduces to the surface current on a perfectly conducting cylinder. Once again, by means of Watson's transformation, the relatively slowly varying infinite series for the dyadic components can also be converted to contour integrals in the complex plane. The result is

$$G_{\theta\theta}(a,\theta,\phi) = \cos(\phi - \phi_o)\frac{dm}{4\pi ja^2}\frac{1}{2\pi j}\int_{C_v} dv\, \frac{2v+1}{v(v+1)}\frac{(-1)^v\pi}{sin(\pi v)}\left[-F_{v\theta}(a,\theta) + \frac{1}{jk\ sin\theta}A_{vr}(a,\theta)\right],$$

$$G_{\phi\phi}(a,\theta,\phi) = \sin(\phi - \phi_o)\frac{dm}{4\pi ja^2}\frac{1}{2\pi j}\int_{C_v} dv\, \frac{2v+1}{v(v+1)}\frac{(-1)^v\pi}{sin(\pi v)}$$

$$\left[\frac{1}{sin\theta}F_v(a,\theta) - \frac{1}{jk}A_{vr}(a,\theta)\right]. \quad (7.17)$$

These can be further simplified as described by Alisan and Ertürk (2010).

7.3 Asymptotic Solution for Fields Near Convex Surfaces

7.3.1 Review of Literature for Convex Surfaces

The problem of diffraction by a smooth convex surface on which, or near to, a source is located has been of considerable interest since the early days of radio. A review of the literature concerning this problem is given by Wait (1959), Kouyoumjian (1965), and Mittra and Safavi-Naini (1979). Additional relevant material can be found in Logan and Lee (1962), Jones (1962), Fock (1965), Bowman et al. (1969), and James (1976). It is apparent from these references that most of the early work involved asymptotic analysis of the radiation field, and, with a few notable exceptions, asymptotic analysis of the surface field held little practical interest.

The formal solution of the fields due to a source on a circular cylinder was first obtained by Silver and Saunders (1950), although earlier publications, e.g. Pistolkors (1947), described special solutions, including Pappas and King (1949) who discussed the surface current excited by an axial slot. As the series for the surface fields converge slowly, even for cylinders of moderate radius, it was not until much later that the surface field solution was studied in any depth when Hasserjian and Ishimaru (1962a, b) obtained asymptotic solutions for the surface field due to axial and circumferential slots. This work of Hasserjian and Ishimaru (1962b) was motivated by the need to determine mutual coupling between slots on a cylinder. Asymptotic solutions for the currents on a sphere were obtained earlier by Franz (1954), Wait (1956), and Fock (1965). Another significant feature of the paper by Hasserjian and Ishimaru (1962b) was that they obtained a small argument approximation that is valid near the shadow boundary by expressing the contour integral for the field as an inverse Laplace transform (Bromwich integral) of a power series which is easily evaluated. This technique, apparently originally due to Bremmer, had been used earlier by Wait (1956). It is now a standard tool for determining the small argument approximation to functions which arise in the asymptotic analysis of the current on large conducting bodies, the so-called surface Fock functions.

The asymptotic solution obtained by Hasserjian and Ishimaru (1962a, b) for the surface field on a cylinder mentioned earlier retains only the lowest-order terms in $q = (jk_o t)$ and 1/M (the size). In these terms, as shown in Figure 7.1, t is the path length between the source and field points on the cylinder and $M = (kR_t/2)^{1/3}$, where R_t is the radius of curvature of the surface in the ray direction. This solution predicts, for example, zero field along the axis for an axial magnetic point source and hence zero H-plane coupling for TE_{10} mode excited circumferentially polarized waveguide arrays. Only by including q^2 and higher-order terms does one accurately represent the field parallel to the magnetic point source. Chang et al. (1976a,b) and Chan et al. (1977) of the Polytechnic Institute of New York (PINY) give solutions for the cylinder and the cone, respectively, which include all terms up to q^2. Both solutions are based on an asymptotic evaluation of the rigorous Green's function solution for each geometry. By contrast, Lee and Safavi-Naini (1976, 1978) and Lee (1978), from the University of Illinois (UI), obtained a solution containing terms up to q^3 and $1/M^2$, using a GTD-style approach. They modified the asymptotic solution for a source on a conducting sphere such that it provided accurate results for the cylinder as well. The contribution from Pathak and Wang (1978) and Pathak et al. (1981) from Ohio State University (OSU) described an asymptotic solution for surface field on electrically large conducting convex surfaces. This solution was devised from asymptotic solutions for the cylinder and the sphere and, therefore, recovers these results as limiting cases. It contains all terms up to q^3 and $1/M$. All three solutions, those of PINY, UI, and OSU, recover the solution for a magnetic source on a plane in the limit that the radii of curvature become infinitely large. However, each solution predicts a different limiting value for the field in the vicinity of the axis (paraxial region) of a cylinder excited by a circumferential magnetic point source. This discrepancy occurs because the solution is dependent on some $1/M^2$ terms in the region of the axis. At about the same time, Boersma and Lee (1978) developed an asymptotic surface field solution for the circular cylinder that retains terms up to q^3 and $1/M^2$. They demonstrated that the UI solution gives the correct limiting value.

Another high-order solution was derived by Bird (1985) that contains terms up to q^2 and $1/M^3$. To this solution was added approximate q^3 terms of Boersma and Lee (1978) to improve accuracy near the source. The combined solution agrees well with the exact solution for the surface field. Further, in the limit that the cylinder radius is infinite, the solution becomes identical to the exact solution for a point source on a plane conductor. This solution is described in detail in the next section. The development of accurate asymptotic representations of the surface field due to a magnetic point source on a cylinder can simplify the computational task of analyzing mutual coupling in aperture arrays, not only for a cylinder but also on general surfaces with slowly varying curvature. When used for waveguide mutual admittance calculations, the approximate solution gives results in excellent agreement with values obtained from the exact solution. Additional calculations have shown that accuracy improves with cylinders of larger radius.

As mentioned above, some asymptotic solutions for the surface fields on a source excited on perfect electric conductor (PEC) circular cylinder were modified by the addition of some correction terms via an alternative technique (Boersma and Lee 1978), which led to better accuracy within the paraxial region where the conventional asymptotic methods lose accuracy. In a somewhat related study, an alternative representation was also obtained for describing an edge wave within the paraxial region of a PEC wedge for large axial separations (Pearson 1986). Thus, the surface field behavior and mutual coupling phenomenon are well understood for PEC convex surfaces and wedges. However, the analysis of the surface field on a source excited circular cylinder with an impedance boundary condition (IBC) remains a challenging problem, especially for paraxial propagation. An early asymptotic solution proposed a numerical scheme for the evaluation of the vector potentials associated with the nonazimuthal propagation as well as the azimuthal propagation on a circular cylinder with an anisotropic IBC (Pogorzelski 1996). An alternative approximate asymptotic closed-form solution was given by Togöz et al. (2005) for the accurate representation of the tangential surface magnetic field within the paraxial region of a tangential magnetic current excited circular cylinder with an IBC. The approach was similar for the PEC case and was shown to reduce to the PEC solution when the surface impedance approaches zero.

7.3.2 Asymptotic Solution for the Surface Fields

When the radius of curvature of surfaces is large, asymptotic solutions can be constructed that are accurate and therefore greatly beneficial in computations. An asymptotic solution for arbitrary surfaces can be devised from solutions to canonical surfaces such as a cylinder or sphere. The cylinder has the advantage over a sphere in that there are two different radii of curvature although one of them is infinite. To obtain an asymptotic solution for the surface field on a cylinder due to a source situated also on the surface using Eqs. (3.37), express the dyadic in the following form:

$$\underline{\underline{G}}^{(h)}(\phi, z \mid \phi', z') = \sum_{p=0}^{\infty} [\underline{\underline{G}}(\bar{\phi} + 2p\pi, \bar{z}) + \underline{\underline{G}}(2\pi - \bar{\phi} + 2p\pi, \bar{z})]. \tag{7.18}$$

The dyadic function $\underline{\underline{G}}$ is the field due to a single surface ray in one direction, while the summation corresponds to creeping waves traveling along helical paths between the field and source points. For cylinders of large radius, it is usually sufficient to approximate Eq. (7.18) by the two dominate creeping waves namely

$$\underline{\underline{G}}^{(h)}(\phi, z \mid \phi', z') \sim \underline{\underline{G}}(\bar{\phi}, \bar{z}) + \underline{\underline{G}}(2\pi - \bar{\phi}, \bar{z}). \tag{7.19}$$

In addition to the creeping wave representation, closer to the source, several approximate asymptotic solutions have been obtained as described in Section 7.3.1. Five of them were compared with the exact solution for a range of parameters (Bird 1984). All these asymptotic solutions become identical to the exact solution for a point source on a plane conductor when the radius of the cylinder approaches infinity. The one adopted here (Bird 1985) was shown to have best overall performance.

Let t and θ be the coordinates relative to the source on the projected surface of the cylinder, as shown in Figure 7.1, where $t = \sqrt{\bar{z}^2 + (R_o\bar{\phi})^2}$ and $\theta = \text{atan}\left(\frac{\bar{z}}{R_o\bar{\phi}}\right)$. An accurate solution for the three nonzero components of the dyadic has been derived (Bird 1985). Let an asymptotic approximation be

$$\underline{\underline{T}}(\phi, z \mid \phi', z') \sim \underline{\underline{G}}^{(h)}(\phi, z \mid \phi', z')$$

to distinguish an approximation from a direct asymptotic solution. Further, it is convenient to express the asymptotic solution in terms of t and θ. For a cylinder, the components of the dyadic can be shown to be (Bird 1985)

$$T_{\phi\phi}(t, \theta) \sim v(\beta)G_{ho}(t)[sin^2\theta + q(2 - 3sin^2\theta)] + C^{(1)}_{\phi\phi}(t, \theta), \tag{7.20a}$$

where

$$C^{(1)}_{\phi\phi}(t, \theta) = G_{ho}(t)q\left[sec^2\theta[u(\beta) - v(\beta)] + v(\beta)\left(\frac{8}{9}tan^2\theta - \frac{31}{72}sin^2\theta\right)\right.$$
$$\left.+v_1(\beta)\left(\frac{17}{36}sin^2\theta - \frac{43}{45}tan^2\theta\right) + v_2(\beta)\left(\frac{1}{15}tan^2\theta - \frac{1}{24}sin^2\theta\right) + \frac{j\beta}{5}v_0^{(1)}(\beta)tan^2\theta\right],$$

$$T_{\phi z}(t, \theta) \sim v(\beta)G_{ho}(t)sin\theta\cos\theta(1 - 3q) + C^{(1)}_{\phi z}(t, \theta) \tag{7.20b}$$

where

$$C^{(1)}_{\phi z}(t, \theta) = -G_{ho}(t)q\ sin\theta\ cos\theta\left[v(\beta)\left(\frac{5}{9}sec^2\theta - \frac{31}{72}\right) + v_1(\beta)\left(\frac{17}{36} - \frac{28}{45}sec^2\theta\right)\right.$$

$$+v_2(\beta)\left(\frac{1}{15}sec^2\theta - \frac{1}{24}\right) + \frac{j\beta}{5}v_0^{(1)}(\beta)sec^2\theta\Bigg]$$

$$T_{z\phi}(t,\theta) = T_{\theta z}(t,\theta) \tag{7.20c}$$

and

$$T_{zz}(t,\theta) \sim v(\beta)G_{ho}(t)[cos^2\theta + q(2 - 3cos^2\theta)] + C_{zz}^{(1)}(t,\theta) \tag{7.20d}$$

where $C_{zz}^{(1)}(t,\theta) = G_{ho}(t)q\left[v(\beta)\left(\frac{31}{72}sin^2\theta - \frac{5}{24}\right) + v_1(\beta)\left(\frac{11}{60} - \frac{17}{36}sin^2\theta\right) + v_2(\beta)\left(\frac{1}{24}sin^2\theta + \frac{1}{40}\right) + \frac{j\beta}{5}v_0^{(1)}(\beta)\right]$.

In Eq. (7.20), $\beta = \frac{kt}{2M^2} = kt\left(\frac{cos^2\theta}{\sqrt{2kR_o}}\right)^{2/3}$ is the distance parameter, $G_{co}(t) = \frac{kY_o}{2\pi j}\frac{exp(-jkt)}{t}$ where $q = \frac{j}{kt}$, and $v(\beta)$,$v_1(\beta)$,$v_2(\beta)$, and $v_0^{(1)}(\beta)$ are the surface Fock functions, which are detailed in Appendix D. A more accurate asymptotic solution is possible by including higher-order terms such as in the vicinity of the source when terms up to third order in q have been shown to improve accuracy (Bird 1984).

An asymptotic approximation for the surface current on a large sphere has also been derived (Pathak and Wang 1978; Alisan and Ertürk 2010). The nonzero components of the surface dyadic are

$$T_{\theta\theta}(a,\theta,\phi) = dm\,cos(\phi - \phi_o)\left[\left(\frac{j}{ks}\right)\breve{U}_z(\xi) - \frac{D^2}{2}\left(\frac{j}{ks}\right)^2 cos\theta\, U_z(\xi) + D^2\left(\frac{j}{ks\pm}\right)V_z(\xi)\right]D\,G_o(s),$$

$$T_{\phi\phi}(a,\theta,\phi) = dm\,\sin(\phi - \phi_o)\left[\breve{V}_z(\xi) - \frac{D^2}{2}\left(\frac{j}{ks}\right)cos\theta\, V_z(\xi) + D^2\left(\frac{j}{ks}\right)^2 U_z(\xi)\right]D\,G_o(s), \tag{7.21}$$

where

$$D = \sqrt{\frac{\theta}{sin\theta}},\ \xi^{\pm} = M\begin{cases}\theta\\(2\pi - \theta)\end{cases},\ s^{\pm} = a\begin{cases}\theta\\(2\pi - \theta)\end{cases},\ \text{and } M = \left(\frac{ka}{2}\right)^{1/3}$$

is the size parameter for the sphere. The asymptotic surface current for the sphere is expressed in terms of different Fock-type functions (Felsen and Marcuvitz 1973) from the last section and these are defined

$$U_z(\xi) = e^{j3\pi/4}\xi^{3/2}\frac{1}{\sqrt{\pi}}\int_{-\infty}^{\infty}d\tau\, e^{-j\xi\tau}\frac{-R_w(\tau)\,q_m}{(R_w(\tau) - q_m)};$$

$$V_z(\xi) = \sqrt{\frac{j\xi}{4\pi}}\int_{-\infty}^{\infty}d\tau\, e^{-j\xi\tau}\frac{1}{(R_w(\tau) - q_e)},$$

$$\breve{U}_z(\xi) = e^{j3\pi/4}\xi^{3/2}\frac{1}{\sqrt{\pi}}\int_{-\infty}^{\infty}d\tau\, e^{-j\xi\tau}\frac{-R_w(\tau)\,q_m}{(R_w(\tau) - q_m)}\left(1 + \frac{\tau}{2M^2}\right),\ \text{and}$$

$$\breve{V}_z(\xi) = \sqrt{\frac{j\xi}{4\pi}}\int_{-\infty}^{\infty}d\tau\, e^{-j\xi\tau}\frac{1}{(R_w(\tau) - q_e)}\left(1 + \frac{\tau}{2M^2}\right). \tag{7.22}$$

in which $R_w(\tau) = \frac{w_2'(\tau)}{w_2(\tau)}$, $w_2(\tau)$ is the Airy integral (Appendix D), and $q_m = -jM\Lambda^{-1}$ and $q_e = -jM\Lambda$. The path s can be chosen in a positive direction (+) along a geodesic with path length $s^+ = a\theta$ or in the opposite (−) direction with path length $s^- = a(2\pi - \theta)$ depending on the situation.

7.4 Mutual Coupling of Apertures in Quadric Surfaces

Suppose two apertures lie on the surface of a cylinder or a sphere. Following Section 3.2, the magnetic field at a point on the surface due to a magnetic source on the surface is

$\boldsymbol{dH}(\boldsymbol{r}) = \overline{\overline{\boldsymbol{G}}}^h(\boldsymbol{r}, \boldsymbol{r}') \cdot \boldsymbol{M}_1(\boldsymbol{r}')$. The mutual admittance at $\boldsymbol{r}$ on a source located there $\boldsymbol{M}_2(\boldsymbol{r})$ is

$$y_{21} = -\boldsymbol{M}_2(\boldsymbol{r}) \cdot \frac{\boldsymbol{dH}(\boldsymbol{r})}{V_1} \tag{7.23}$$

when a voltage $V_1$is applied at point 1. When all sources across the two apertures are summed, Eq. (7.23) becomes

$$y_{21} = -\frac{1}{V_1}\iint_{S_2} dS\,\boldsymbol{M}_2(\boldsymbol{r}) \cdot \iint_{S_1} dS' \;\; \overline{\overline{\boldsymbol{G}}}^h(\boldsymbol{r}, \boldsymbol{r}') \cdot \boldsymbol{M}_1(\boldsymbol{r}'). \tag{7.24}$$

The Green's function or its asymptotic forms can be used to determine the mutual admittance of apertures in the surface. Similarly, we can obtain the dual result for the coupling between electric sources located on this surface. The difference is in this dual impedance calculation is the electric field dyadic and this needs to be known in the vicinity of the cylinder, i.e. $\boldsymbol{r}$ is usually slightly above the cylinder say at $\boldsymbol{d}$ so that $\boldsymbol{r} = \boldsymbol{d} + \boldsymbol{r_s}$, where $\boldsymbol{r_s}$ is distance along the surface between the two sources. In this section, the canonical solutions for the cylinder and sphere will be used to calculate the mutual coupling between waveguides and microstrip patches on a cylinder and between circular waveguides in a sphere.

7.4.1 Closed-Form Expressions for Mutual Coupling Between Rectangular Waveguides in a Cylinder

Consider an array of identical rectangular waveguides of width a and height b that terminate in a cylindrical surface. It is assumed that each waveguide supports only the dominant TE_{10} mode. Starting from Eq. (7.24) and introducing Eq. (7.2) for a cylinder, the normalized mutual admittance of these modes (i.e. mode 1) in the surface is

$$y_{ij}(1,1) = -\frac{1}{A_i A_j}\iint_{D_i} \frac{1}{Y_j}(\mathbf{E_j} \times \hat{p}) \cdot \boldsymbol{H}_{ij}(R) dS_j, \tag{7.25}$$

where

$$\mathbf{E}_i = A_i \hat{p}\sqrt{\frac{2}{ab}}\cos\left(\frac{\pi x_i}{a}\right), \tag{7.26}$$

with $\hat{p} = (\hat{\phi} cos\omega_i + \hat{z} sin\omega_i)$ being the polarization of the electric field, $Y_i = Y_{10} = Y_o\beta_{10}/k$ the TE_{10} mode wave admittance, β_{10} its propagation constant, and A_i its complex excitation amplitude all in aperture i. The aperture is rotated at an angle ω_i, which is the angle relative to the axis of the cylinder as shown in Figure 7.4. Some arrangements of two apertures are illustrated in Figure 7.5. The magnetic field at j due to the field point in the aperture on the surface is given by Eq. (7.1) and the dyadic $\underline{\mathbf{G}}^{(\boldsymbol{h})}(\boldsymbol{R} \mid \boldsymbol{R}')$, and the magnetic current on aperture i is $\mathbf{M_i} = \mathbf{E_i} \times \hat{p}$. Therefore, Eq. (7.25) becomes

$$\begin{aligned} y_{ij}(1,1) = &-\frac{2k}{abY_o\beta_{10}}\iint_{D_i} dS_i \cos\left(\frac{\pi x_i}{a}\right) \cdot \iint_{D_j} dS_j \cos\left(\frac{\pi x_j}{a}\right) \times [\, G^{(h)}_{\phi\phi}\, cos\omega_i cos\omega_j \\ &+ G^{(h)}_{zz}\, sin\omega_i sin\omega_j + G^{(h)}_{\phi z}\, sin(\omega_i + \omega_j)\,]. \end{aligned} \tag{7.27}$$

The three independent components of the dyadic function for the cylinder, namely $G^{(h)}_{\phi\phi}$, etc. corresponding to the magnetic field on the surface of a metallic surface, are given in Eq. (7.2).

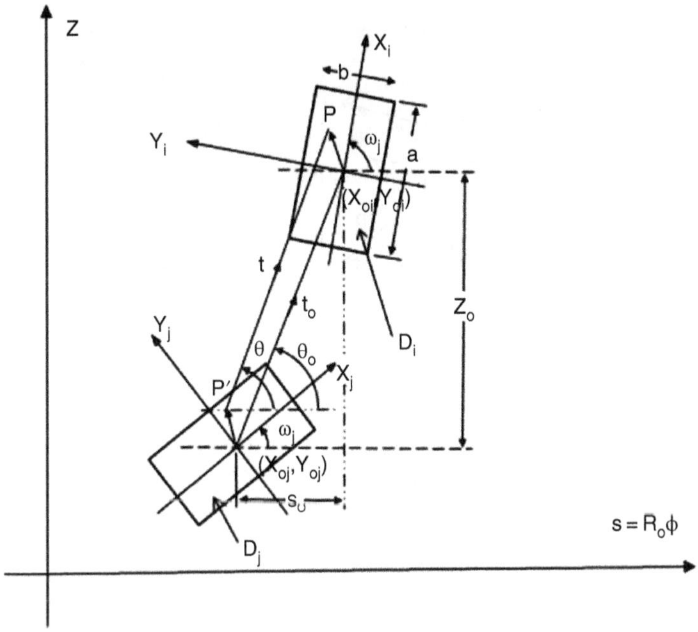

Figure 7.4 Geometry of rectangular waveguides on the surface of a cylinder.

The dyadic components can be substituted into Eq. (7.27), and a closed-form result can be obtained for the mutual admittance on a cylindrical surface. The result is given by Bird (1983) as follows:

$$y_{ij}(1,1) = -\frac{4j\,abk}{\pi^4 R_o \beta_{10}} \int_{-\infty}^{\infty} d\xi e^{-j\xi\overline{Z}} \sum_{n=-\infty}^{\infty} e^{jn\overline{\Phi}}\, P_{ij}(n,\xi)[\, G^c(n,\xi)\cos\omega_i \cos\omega_j + G^a(n,\xi)\sin\omega_i \sin\omega_j$$
$$+\, G^x(n,\xi)\sin(\omega_i+\omega_j)\,], \tag{7.28}$$

where

$$P_{ij}(n,\xi) = C(A_i)C(A_j)S(B_i)S(B_j),$$

$C(x)$ and $S(x)$ are the cosine-shape and sinc functions (see Appendix A),

$$A_i(n,\xi) = \frac{a}{2}\left(\frac{n}{R_o}\cos\omega_i - \xi\sin\omega_i\right),\ B_i(n,\xi) = \frac{b}{2}\left(\frac{n}{R_o}\sin\omega_i + \xi\cos\omega_i\right),$$

$$G^c(n,\xi) = -\frac{k}{h}\left[\frac{H_n^{(2)\prime}(D)}{H_n^{(2)}(D)} - \left(\frac{n\xi}{kD}\right)\frac{H_n^{(2)}(D)}{H_n^{(2)\prime}(D)}\right],$$

$$G^a(n,\xi) = \frac{h}{k}\frac{H_n^{(2)}(D)}{H_n^{(2)\prime}(D)},\ G^x(n,\,\xi) = \frac{n\xi}{kD}\frac{H_n^{(2)}(D)}{H_n^{(2)\prime}(D)},$$

$$D = hR_o,\ h = \begin{cases} \sqrt{k^2-\xi^2} & ;\xi < k \\ -j\sqrt{k^2-\xi^2} & ;\xi > k \end{cases}$$

$\overline{\Phi} = \phi_i - \phi_j$ and $\overline{Z} = z_i - z_j$.

It is obvious from Eq. (7.28) that the function $G^c(n,\xi)$ determines the coupling of circumferential slots, i.e. $\omega_i = 0 = \omega_j$ axially polarized apertures. The expression for coupling in circumferential apertures was first derived by Golden et al. (1974). Similarly, $G^a(n,\xi)$ gives the coupling of axially

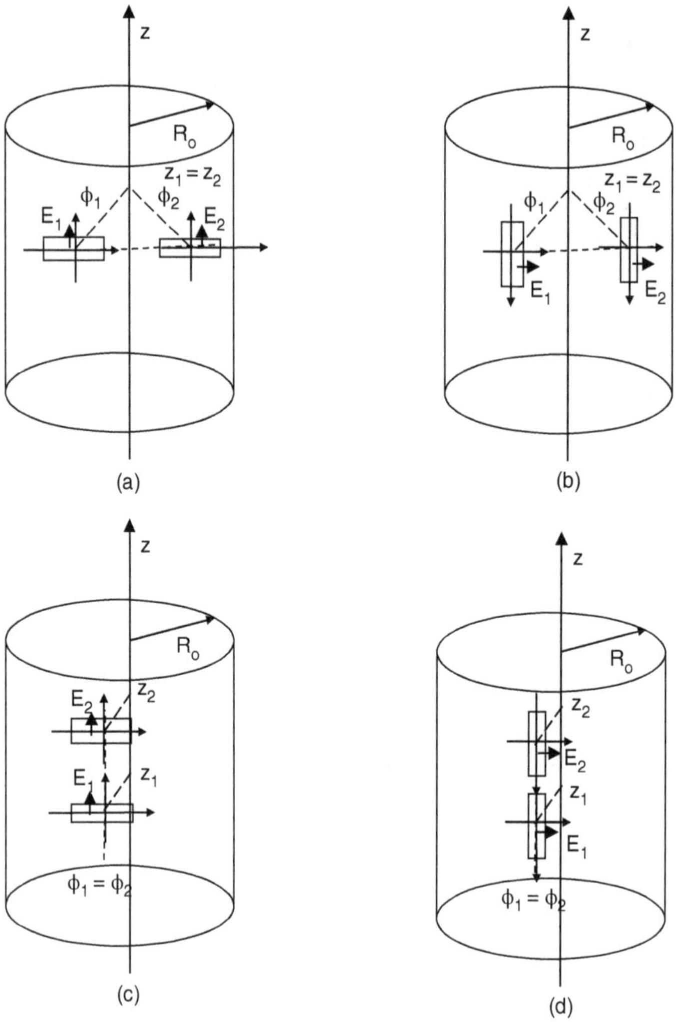

Figure 7.5 Typical arrangements of waveguides in the surface of a cylinder. (a) Axial polarization arranged in H-plane; (b) circumferential polarization in E-plane; (c) axial polarization in E-plane; and (d) circumferential polarization in H-plane.

oriented slots, i.e. $\omega_i = \frac{\pi}{2} = \omega_j$ (Stewart and Golden 1971), and $G^x(n, \xi)$ gives the coupling of angled slots. The latter is required in cases when the waveguide axes are not parallel to the axis of the cylinder and/or when the waveguide axes are not parallel. In its present form, Eq. (7.28) is not very convenient for numerical evaluation. It is possible to express it in a more useful form as described by Bird (1983) and Lee and Mittra (1977), by first splitting the integrand into real and imaginary parts. Then following Duncan (1962), the integral for the imaginary part of the admittance is transformed into another integral, which decays more quickly. Another approach is to make use of the various symmetries of the functions in the integrand, for instance,

$$P_{ij}(n, \xi) = P_{ij}(-n, -\xi), \qquad P_{ij}(-n, \xi) = P_{ij}(n, -\xi)$$

to make the summation and integration one-sided from 0 to ∞. Also, when $\xi > k$, the Hankel function can be replaced by a Bessel function of the second kind (see Appendix B).

In the special case of $j = i$, Eq, (7.28) can also be used to calculate self-admittance by setting $\overline{\Phi}$ and $\overline{Z}$ to zero. In that case, a preferred form for computation has been found to be

$$y_{ii}(1,1) = \frac{2j\,abk}{\pi^4 R_o \beta_{10}} \int_0^1 \frac{dw}{\Lambda(w)} \sum_{n=0}^{\infty} \varepsilon_n \left\{ (P_{ii}(n,w) + P_{ii}(-n,w))\cos^2\omega_i \frac{H_n^{(2)'}(\Omega)}{H_n^{(2)}(\Omega)} \right.$$
$$- \left[P_{ii}(n,w)Q(n,w) + P_{ii}(-n,w)Q(-n,w) \frac{H_n^{(2)}(\Omega)}{H_n^{(2)'}(\Omega)} \right] - \int_1^{\infty} \frac{dw}{\Lambda(w)} \sum_{n=0}^{\infty} \varepsilon_n$$
$$\left\{ (P_{ii}(n,w) + P_{ii}(-n,w))\cos^2\omega_i \frac{K_n'(\Omega)}{K_n(\Omega)} \right.$$
$$- \left[P_{ii}(n,w)Q(n,w) + P_{ii}(-n,w)Q(-n,w) \frac{K_n(\Omega)}{K_n'(\Omega)} \right], \tag{7.29}$$

where

$$\Lambda(w) = \sqrt{|1 - w^2|},\ \Omega(w) = kR_o\Lambda(w)$$
$$Q(n,w) = \left(\frac{n\,w}{kD}\cos\omega_i + \frac{h}{k}\sin\omega_i \right)^2,\ \varepsilon_n = \begin{cases} 1 & ; n = 0 \\ 2 & ; n > 0 \end{cases},$$

and K_n is the modified Bessel function of order n (Appendix B). The equations for axial and circumferential slots have been coded by Lee and Mittra (1979), and numerical results have been obtained for a wide variety of structures. Earlier theoretical and experimental results were obtained by Stewart and Golden (1971), Golden et al. (1974), and Bird (1988) for self-admittance. However, to date, apparently no results for general angled slots have been published in the open literature.

Results can be time-consuming to compute numerically for cylinders of moderate to large radius (i.e. $k\,R_o > 10$). However, in that latter case, accurate approximations of the dyadic components can be obtained from the asymptotic solutions. From these, it is then relatively straightforward to compute elements of the admittance matrix from Eq. (7.27) by numerical integration. A similar approach is possible for apertures on a conducting sphere or an arbitrary surface (Hessel et al. 1979; Wills 1986). As a check on the accuracy of results obtained from Eq. (7.28), calculations obtained have been compared with previous results. For example, Figure 7.6 compares the mutual admittance obtained from Eq. (7.28), and the results obtained from Stewart and Golden's (1971) measured results of isolation ($|S_{21}|$) for the case of axial-aligned WG-16 (a = 2.286 cm, b = 1.016 cm) waveguides ($\omega_i = \pi/2 = \omega_j$) in a cylinder of radius 5.057 cm at a frequency of 9 GHz. As illustrated in Figure 7.6, a second axial slot is displaced along the axis at a distance of $\overline{Z} = 3.81$ cm which it is allowed to follow a circumferential path away from the source waveguide. At various azimuth angles, the isolation has been measured. The magnitude of the mutual admittance is estimated from $|y_{21}| \cong |S_{21}(1 + y_{11})^2/2|$, where y_{11} is the normalized input admittance, which is assumed to be close to the value quoted in the reference Stewart and Golden (1971). It is seen that theory and measurement are in excellent agreement, which is typical for the other configurations shown in Fig. 7.5.

As a further example, the mutual admittance of two WG-16 rectangular waveguides in the same cylinder and frequency as in the previous example has been calculated using Eq. (7.28) as a function of distance along the axis for TE_{10} mode polarization alignment parallel ($\omega_i = 0$, E-plane) and circumferential ($\omega_i = \pi/2$, H-plane) as shown in Figures 7.5c,d, respectively. These results are plotted in Figure 7.7a. In a further example, Figure 7.7b shows the E- ($\omega_i = \pi/2$) and H-plane ($\omega_i = 0$) coupling when these waveguides are arranged as shown in Figure 7.5a,b as their spacing is altered along a circumferential direction. For convenience, the mutual admittance is normalized to the TE_{10} mode admittance, and the magnitude is given in dB. As expected, the E-plane coupling in

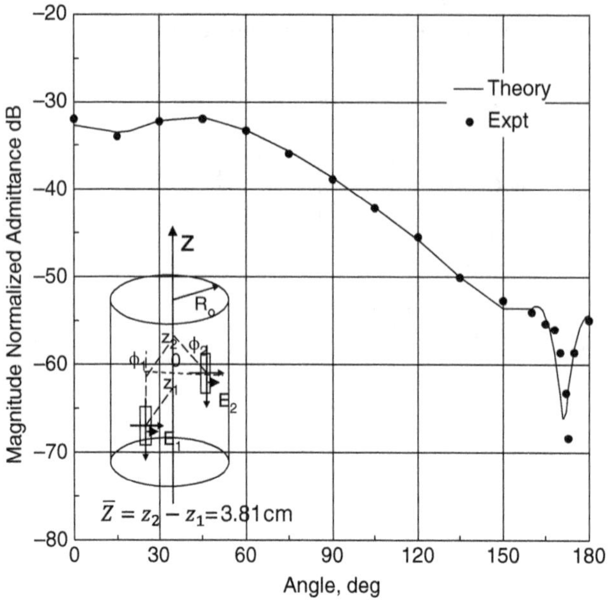

Figure 7.6 Theory and experiment for two axial WG-16 waveguides (ie. $\omega_i = \pi/2 = \omega_j$) on a cylinder of radius $R_o = 5.057$ cm and axial displacement $\overline{Z} = 3.81$ cm. Solid line: theory; dot: experiment Source: Based on STEWART, G.E. and GOLDEN, K.E. (1971), 'Mutual admittance for axial rectangular slots in a large conducting cylinder', IEEE Trans. Antennas & Propagat., AP-19, pp. 120-122.

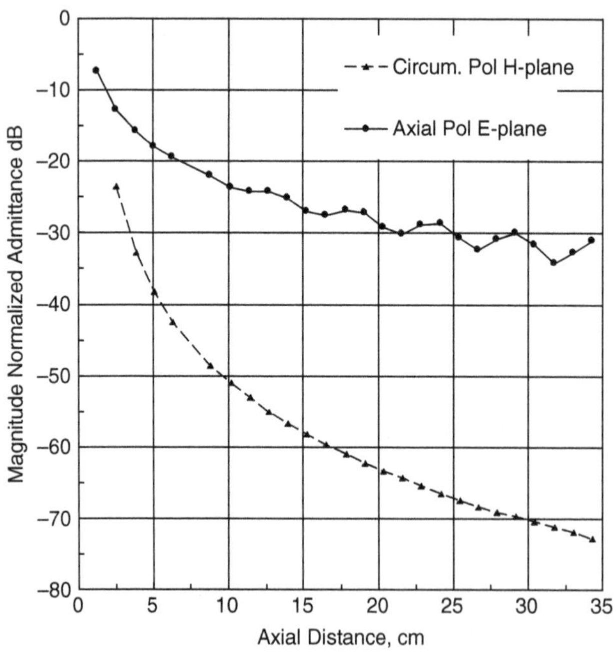

(a)

Figure 7.7 Mutual admittance between two WG-16 waveguides (a = 2.286 cm, b = 1.016 cm) in a cylinder of radius 5.057 cm at a frequency of 9 GHz, which is normalized to the TE_{10} mode admittance for the cases: (a) E- ($\omega_i = 0 = \omega_j$) and H-($\omega_i = \pi/2 = \omega_j$) plane coupling along the axis of the cylinder for axial and circumferential polarization, respectively, (Figure 7.5c,d); (b) E-($\omega_i = \pi/2 = \omega_j$) and H-($\omega_i = 0 = \omega_j$) plane coupling around the circumference for axial and circumferential polarization (Figures 7.5a and 5b respectively).

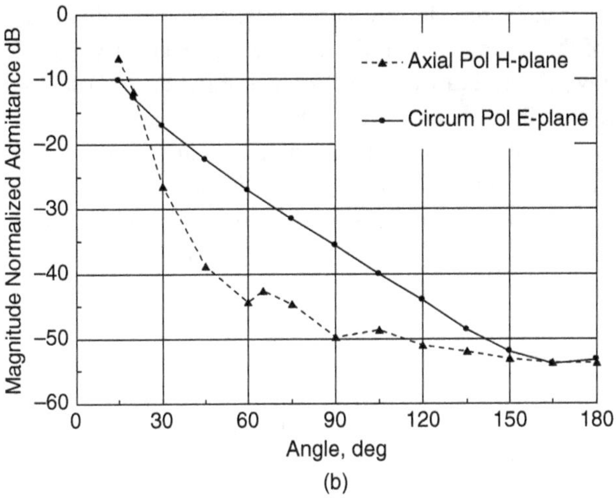

Figure 7.7 *(Continued)*

both instances is considerably stronger than in the H-plane. It is noted that E-plane coupling is higher and H-plane coupling is lower for waveguides aligned axially (Figure 7.7a).

Finally, coupling is calculated for WG-16 waveguides rotated relative to the axis of cylinder. One waveguide, the source, is rotated at 45° (ie. $\omega_i = \pi/4$) and the second is rotated at angles $\omega_j = 0°$, 45° and 90° as illustrated in Figure 7.8. The spacing of this second waveguide follows a helical path on the surface away from the source resulting in the computed results shown in Figure 7.8. This path is described by the equations (Struik 1961) $x = R_o cos\overline{\Phi}, y = R_o sin\overline{\Phi}$, and $z = b_o\ \overline{\Phi}$ where $2b_o\pi$

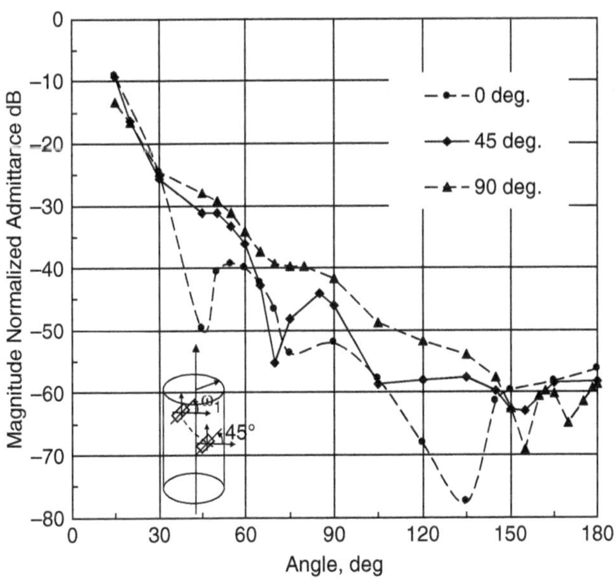

Figure 7.8 Mutual admittance between two WG-16 waveguides (a = 2.286 cm, b = 1.016 cm) in a cylinder of radius 5.057 cm at a frequency of 9 GHz. As illustrated, the source waveguide is rotated by $\omega_i = 45°$ while the second waveguide, which rotated at various angles of $\omega_j = 0°$, 45°, and 90°, is moved along helical paths between them.

is the pitch of the helix. In this example, $b_o = 0.809$ cm. The mutual admittances have similar trends and levels for all three rotation angles of the second waveguide, although the geometry with the strongest coupling is the circumferentially polarized case ($\omega_1 = 90°$) when G^c in Eq. (7.28) does not contribute to the integrand and the only contributions are from $G^a$and G^x.

The coupling coefficients are the elements of the scattering matrix for an array on the cylinder or sphere, and this matrix is obtained in much the same way as for a planar array. The element-by-element approach is used to calculate admittances between two apertures at a time. The self-admittances can be calculated directly from Eq. (7.16) for the sphere or Eq. (7.29) for the cylinder. Alternatively, if the surface has a moderate to large radius of curvature, the admittance of a single aperture in a ground plane is a good approximation. Once the admittance matrix $\mathbf{y}$ has been found, the scattering matrix can be computed in the same way as for the planar case using $\boldsymbol{S} = 2(\boldsymbol{I}+\boldsymbol{y})^{-1} - \boldsymbol{I}$, where $\boldsymbol{I}$ is the unit matrix. The complex amplitudes of the electric and magnetic fields are $\boldsymbol{A} = (\boldsymbol{a}+\boldsymbol{b})$ and $\mathbf{B} = (\mathbf{a} - \mathbf{b})$, respectively, and the amplitudes of the incident $\boldsymbol{a}$ and outgoing $\boldsymbol{b}$ waves are related through $\mathbf{b} = \mathbf{Sa}$.

7.4.2 Expressions for Mutual Coupling Between Circular Waveguides in a Sphere

Radiation from arrays in a spherical surface has the advantage that wide angle coverage can be achieved in all planar cuts unlike an array in a cylindrical surface. To design such arrays, knowledge of the coupling coefficients of elements in a sphere is needed. An example is the evaluation of the mutual admittances of two identical circular waveguides terminating in a sphere. This has been described by Hessel et al. (1979) for circular waveguides supporting only a TE_{11} mode. The field in the region external to the sphere is expanded in transverse electric (TE) and transverse magnetic (TM) spherical vector modes. The nature of the TE_{11} mode field distribution and the symmetry of the sphere result in the mutual admittance depending on five parameters only. These are the radii of the sphere and waveguides, the separation between the apertures, and the two angles between the polarization direction relative to the great circle-path. Without loss of generality, the center of one waveguide is placed at the north pole as shown in Figure 7.9. Referring to this figure, when the radius of the sphere $R_s > 5\lambda$, the tangential components of the aperture field of the TE_{11} mode on the cap are approximately the same as on a terminal plane, and for waveguide 1, the aperture field

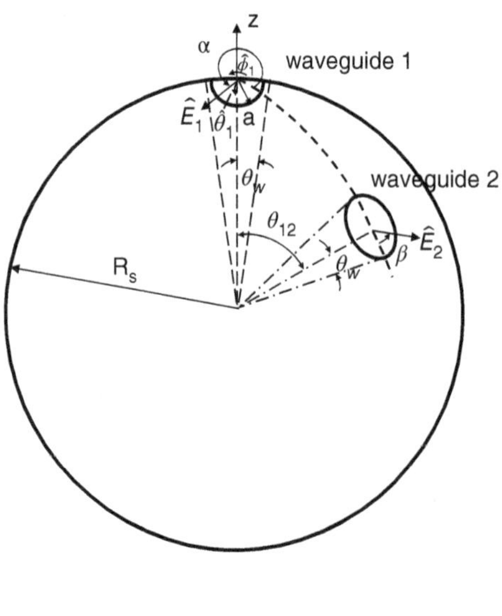

Figure 7.9 Circular waveguides terminating in a sphere.

is

$$\boldsymbol{E}_t = V_1 A(u) \left[\hat{\theta}_1 \frac{J_1(u\theta/\theta_w)}{(u\theta/\theta_w)} \cos(\phi - \alpha) - \hat{\phi}_1 J_1'(u\theta/\theta_w) \sin(\phi - \alpha) \right], \tag{7.30}$$

where

$$A(u) = \sqrt{\frac{2}{\pi(u^2 - 1)}} \frac{u}{a J_1(u)},$$

in which a is the waveguide radius, J_1 is the ordinary Bessel function of order unity on which a prime denotes its first derivative, $u = 1.84118$, $J_1(u) = 0.58187$, $\hat{\theta}_1$ and $\hat{\phi}_1$ are unit vectors in the elevation and azimuthal directions in waveguide 1, and $\theta_w = a/R_s$. Since the mutual admittance is independent of the excitation, let $V_1 = 1$. The mutual admittance of the two waveguides in the sphere can be shown to be (Hessel et al. 1979)

$$Y_{12} = cos\alpha\ cos\beta\ Y^{\|}(R_s, \theta_{12}) + sin\alpha\ sin\beta\ Y^{pp}(R_s, \theta_{12}), \tag{7.31}$$

where

$$Y^{\|}(R_s, \theta_{12}) = \sum_{n=1}^{\infty} \left[\frac{(V_n'(R_s))^2 B_n(\theta_{12})}{Z_n'(kR_s)} + \frac{(V_n''(R_s))^2 C_n(\theta_{12})}{Z_n''(kR_s)} \right] \tag{7.32a}$$

and

$$Y^{pp}(R_s, \theta_{12}) = \sum_{n=1}^{\infty} \left[\frac{(V_n'(R_s))^2 C_n(\theta_{12})}{Z_n'(kR_s)} + \frac{(V_n''(R_s))^2 B_n(\theta_{12})}{Z_n''(kR_s)} \right] \tag{7.32b}$$

The function $Y^{\|}$ is the mutual admittance of apertures polarized parallel to the great circle (i.e. $\alpha = 0 = \beta$), and Y^{pp} corresponds to the mutual admittance when both waveguides are polarized perpendicular to the great circle (i.e. $\alpha = \pi/2 = \beta$). The coefficients are

$$B_n(\theta) = \left[\frac{2}{n(n-1) sin\theta_{12}} \right] [(n-1) P_{n-1}^1(\cos\theta) - n \cos\theta\, P_n^1(\cos\theta)] \tag{7.32c}$$

and

$$C_n(\theta) = \left[\frac{2}{n(n-1)\, sin\theta_{12}} \right] P_n^1(\cos\theta) \tag{7.32d}$$

where

$$P_k^m(cos\theta_{12}) = \frac{(-1)^k \Gamma(1 + m + k)}{k! \Gamma(1 + m)} {}_2F_1(k + m + 1, -k; m + 1; cos(\theta_{12}/2)^2)$$

is the Jacobi polynomial expressed in terms of ${}_2F_1$ the hypergeometric function (Appendix A).

The TM and TE mode voltages of the electric field due to a source on the cylinder are, respectively,

$$V_n'(r) = -\left(\frac{A(u) R_s \pi \theta_w}{N_n u} \right) P_n^1(cos\theta_w) \left[\frac{h_n^{(2)\prime}(kr)}{h_n^{(2)\prime}(kR_s)} \right], r \geq R_s \tag{7.33a}$$

and

$$V_n''(r) = -\left(\frac{A(u) R_s \pi}{N_n} \right) P_n^1(cos\theta_w) \left[\frac{h_n^{(2)}(kr)}{h_n^{(2)}(kR_s)} \right] \int_0^{\theta_w} d\theta\ (u\theta/\theta_w) P_n^1(cos\theta) \tag{7.33b}$$

where $N_n = n(n+1)\left[\frac{2\pi}{2n+1}\right]^{1/2}$ is a normalization constant and $h_n^{(2)}$ is the spherical Hankel function of the second kind. The characteristic impedances of the modes can be expressed as follows:

$$\frac{Z_n'(kR_s)}{\eta_o} = \frac{n}{jkR_s} + \cfrac{1}{\cfrac{2n-1}{jkR_s} + \cfrac{1}{\cfrac{2n-3}{jkR_s} + \cfrac{\ddots}{\cfrac{1}{\cfrac{3}{jkR_s} + \cfrac{1}{\cfrac{1}{jkR_s} + 1}}}}} \tag{7.34a}$$

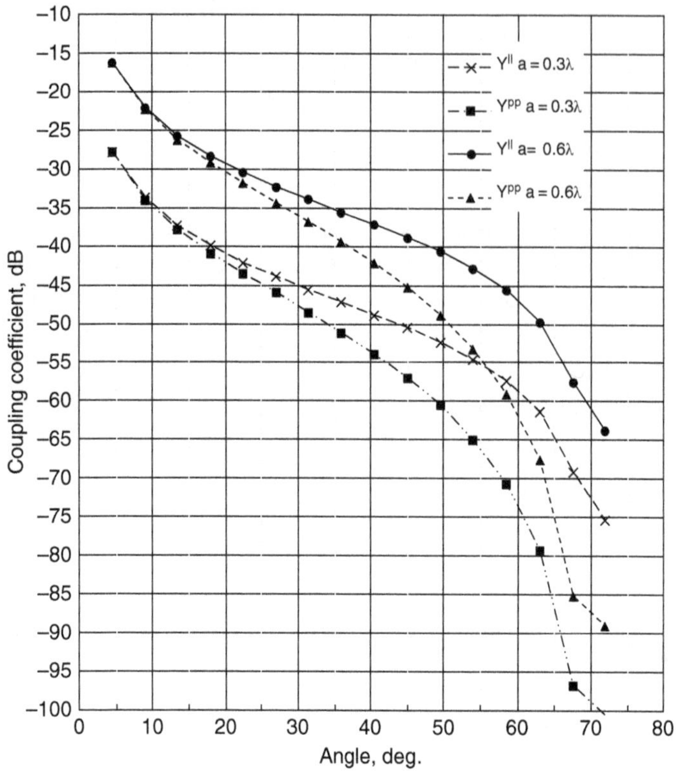

Figure 7.10 Computed E-($Y^{\|}$) and H-(Y^{pp}) plane coupling coefficients of two 0.3λ and 0.6λ radii waveguides terminated in the surface of a conducting sphere of radius 20 λ. Key: –×– 0.3λ $Y^{\|}$; –■– 0.3λ Y^{pp}; –●– 0.6λ $Y^{\|}$; --▲-- 0.6λ Y^{pp}.

and

$$\frac{Z_n''(R_s)}{\eta_o} = \frac{\eta_o}{Z_n'(R_s)}. \tag{7.34b}$$

It is interesting to compare the mutual admittance expressions of the cylinder and sphere, respectively, Eqs. (7.28) and (7.31). For instance, it is seen in the expression for the cylinder that there is an additional cross-coupling term due to the asymmetry of its geometry.

As an example of the results predicted by Eq. (7.31), consider pairs of identical circular waveguides of radius $a = 0.3\lambda$ and 0.6λ in a sphere with a radius of 20 λ. The coupling coefficients have been calculated over most of a 90° sector on the sphere, and the results are shown in Figure 7.10. The decay at large distances in each plane is almost identical and up to about 60° is 0.375 dB/deg. in the E-plane and 0.75 dB/deg. in the H-plane. With spheres of radius 10 λ and 40 λ, the computed results differ little from those given in Figure 7.10 by <1 dB at the lowest levels. After about 60°, the decay rate increases to >1 dB/deg. as coupling depends more on creeping waves on the sphere than the now smaller free-space radiation component.

The mutual coupling between rectangular slots on the surface of a sphere using the series solution has been studied by Inami et al. (1982). The field expressions were obtained as a combination of ray-optical fields and asymptotic approximations to the integrals under the assumption that the radius of the spherical surface is much greater than the wavelength. Numerical results were obtained, and the calculated results for the surface fields and coupling coefficient agreed well with experiment as shown in Figure 7.11. This shows the E-plane coupling as a function of arc

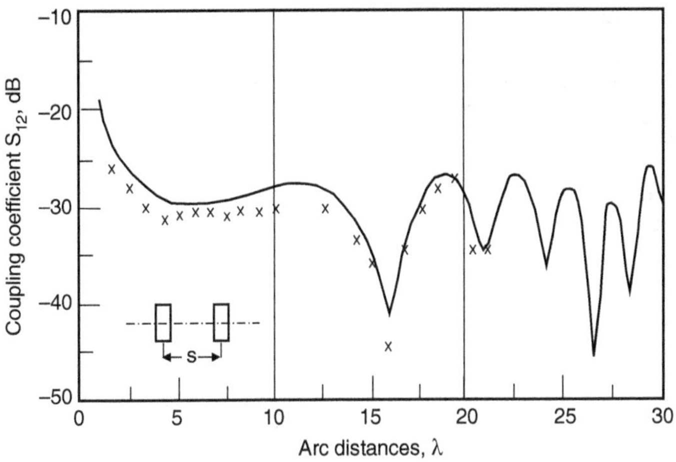

Figure 7.11 Computed E-plane coupling coefficient of two side-by-side slots with dimensions $0.797 \times 0.397\ \lambda$ on the surface of a sphere of radius $a = 12.5\ \lambda$ as a function of separation. Source: Modified from INAMI, K., SAWAYA, K. and MUSHIAKE, Y., ' Mutual coupling between rectangular slot antennas on a concave conducting spherical surface', IEEE Trans. Antennas & Propagat., Vol. AP-30. No. 5, pp. 927-933. © 1982 IEEE. Key: solid line – computed; X – experiment.

distance s on the surface of a sphere of radius $a = 12.5\ \lambda$. The dimensions of the slot antennas are $0.797 \times 0.397\ \lambda$. Not surprisingly, when the arc distance exceeds $s = 20\ \lambda$ or an angle >90° from the source slot, the response is oscillatory due the interference between the direct wave and creeping waves traveling from the opposite direction around the sphere.

7.4.3 Mutual Coupling Between Microstrip Patches on a Cylinder

Suppose a microstrip patch on a substrate is mounted on a large conducting cylinder as illustrated in Figure 7.12. The radius of the cylinder R_o is assumed very much greater than the microstrip dimensions, i.e. $R_o >> w, h, \ell$. The aim is develop a model of coupling using a resonator model for the patch. Suppose patch i is aligned so that the radiating end faces of width w_i are at an angle ω_i as shown in Figure 7.12. The electric field is polarized in the radial direction, and the normal to the apertures is at angle $\omega_i \pm \pi/2$ to the z-axis.

The substrate thickness is h_o and is assumed to be very thin ($h_o \ll \lambda$) and covers the cylinder only about the width of the patches so that the cylinder surface can be assumed to be metallic. The radiating apertures have the same filling as the substrate and appear as two uniform magnetic sources with dimensions $w_i \times h_o$, which are separated by a length ℓ_i. Patch i appears to provide the magnetic currents

$$\boldsymbol{M}_{si} = \hat{n}_{si} \times (\hat{\rho} E_{oi} e^{\mp j\Delta\phi_i}); z = \mp\frac{\ell_i}{2}; -w_i/2 \le \phi_i R_o \le w_i/2, \tag{7.35}$$

where $\Delta\phi = k\sqrt{\varepsilon_r}\,\ell_i$, $E_{oi} = V_i/h_o$ is a constant, V_i is the voltage applied to the patch, and $\hat{n}_{si} = \hat{\phi}\cos\omega_i - \hat{z}\sin\omega_i$ is the unit vector parallel to the normals of the apertures. Although the normal points in a negative direction on one face, the field is also inverted so there no sign change on the amplitude as there is in the phase. The substrate axis bisects the origins of both apertures and is part of a rotated coordinate system turned at an angle ω_s to the cylinder coordinates (R_o, ϕ, z)). This is denoted by the coordinate triad located at the surface of the substrate (R_o, ϕ_s, z_s) and where

$$v_s = \phi_s R_o = -z\ \sin\omega_s + \phi\, R_o \cos\omega_s, z_s = z\ \cos\omega_s + \phi\, R_o \sin\omega_s.$$

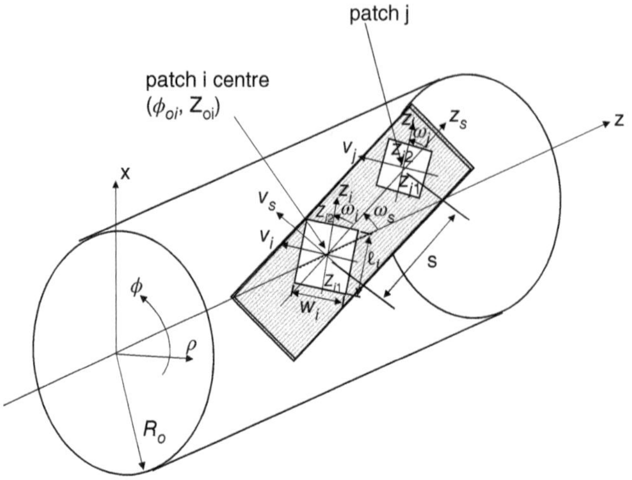

Figure 7.12 Two microstrip patches on a substrate mounted to a cylinder.

Patch i has width w_i and resonant length ℓ_i and is rotated at an angle ω_i. It has coordinates relative to the substrate coordinates as follows:

$$v_i = \phi_i R_o = -z_s \sin \omega_i + v_s \cos \omega_i,\ z_i = z_s \cos \omega_i + v_s \sin \omega_i,$$

and similarly, for patch j

$$v_j = \phi_j R_o = -(z_s - s) \sin \omega_j + v_s \cos \omega_j,\ z_j = (z_s - s) \cos \omega_j + v_s \sin \omega_j,$$

where s is the spacing between the rotation centers of the patches as shown in Figure 7.12. All sources are assumed to radiate in the presence of a bare metallic cylinder. The mutual admittance between the patches can be shown to be

$$\begin{aligned} y_{21} = -\frac{1}{h_o^2} \Bigg\{ & e^{-j(\Delta\phi_j + \Delta\phi_i)} \iint_{S_j} dS_j \iint_{S_i} dS_i \left[\sin \omega_{sj} \sin \omega_{si} G_{\phi\phi}(\phi_{j1}, z_{j1} \mid \phi_{i1}, z_{i1}) \right. \\ & \left. + \cos \omega_{sj} \cos \omega_{si} G_{zz}(\phi_{j1}, z_{j1} \mid \phi_{i1}, z_{i1}) + \sin(\omega_{sj} + \omega_{si}) G_{\phi z}(\phi_{j1}, z_{j1} \mid \phi_{i1}, z_{i1}) \right] \\ & - e^{j(\Delta\phi_j - \Delta\phi_i)} \iint_{S_j} dS_j \iint_{S_i} dS_i \left[\sin \omega_{sj} \sin \omega_{si} G_{\phi\phi}(\phi_{j2}, z_{j2} \mid \phi_{i1}, z_{i1}) \right. \\ & \left. + \cos \omega_{sj} \cos \omega_{si} G_{zz}(\phi_{j2}, z_{sj2} \mid \phi_{i1}, z_{si1}) + \sin(\omega_{sj} + \omega_{si}) G_{\phi z}(\phi_{j2}, z_{j2} \mid \phi_{i1}, z_{i1}) \right] \\ & - e^{-j(\Delta\phi_j - \Delta\phi_i)} \iint_{S_j} dS_j \iint_{S_i} dS_i \left[\sin \omega_{sj} \sin \omega_{si} G_{\phi\phi}(\phi_{j1}, z_{j1} \mid \phi_{i2}, z_{i2}) \right. \\ & \left. + \cos \omega_{sj} \cos \omega_{si} G_{zz}(\phi_{j1}, z_{j1} \mid \phi_{i2}, z_{i2}) + \sin(\omega_{sj} + \omega_{si}) G_{\phi z}(\phi_{j1}, z_{j1} \mid \phi_{i2}, z_{i2}) \right] \\ & - e^{j(\Delta\phi_j + \Delta\phi_i)} \iint_{S_j} dS_j \iint_{S_i} dS_i \left[\sin \omega_{sj} \sin \omega_{si} G_{\phi\phi}(\phi_{j2}, z_{j2} \mid \phi_{i2}, z_{i2}) \right. \\ & \left. + \cos \omega_{sj} \cos \omega_{si} G_{zz}(\phi_{j2}, z_{j2} \mid \phi_{i2}, z_{i2}) + \sin(\omega_{sj} + \omega_{si}) G_{\phi z}(\phi_{j2}, z_{j2} \mid \phi_{i2}, z_{i2}) \right] \Bigg\} \end{aligned} \quad (7.36)$$

where the Green's dyadic is placed in local source (patch) vectors, $\omega_{si} = \omega_i + \omega_s$, $S_i = w_i h_o$, and the subscript jk refers to the patch j edge coordinate on axis z_j of edge k ($k = 1, 2$) in the cylinder coordinates. Thus, for patch i,

$$\phi_{ik} = \left(-\frac{z_{eik}}{R_o} \sin \omega_{si} + \phi_{eik} \cos \omega_{si} \right); z_{ik} = (z_{eik} \cos \omega_{si} - \phi_{eik} R_o \sin \omega_{si})$$

and on patch j

$$\phi_{jk} = \left(-\frac{z_{ejk}}{R_o}\sin\omega_{sj} + \phi_{ejk}\cos\omega_{sj} + s\,sin\omega_j\right); z_{jk} = (z_{ejk}\cos\omega_{sj} - \phi_{ejk}R_o\sin\omega_{sj} + s\,cos\omega_j),$$

where $(R_o, \phi_{ejk}, z_{ejk})$ are the coordinates of radiating edge k of patch j as shown in Figure 7.12; for example when $\omega_j = \pi/2$ and $\omega_s = 0$, $z_{ej1} = s - \ell_j/2$, $z_{ej2} = s + \ell_j/2$ where $s = (z_{oj} - z_{oi})$

When the radius of the cylinder is large, the dyadic components can be replaced by their asymptotic approximations $T_{\phi\phi}$, T_{zz}, and $T_{\phi z}$. Taking this into account, the amplitude is approximately constant at small heights in the substrate above the surface, i.e. $\rho - R_o = h \approx h_o$; the magnitude of the magnetic fields of a point source is assumed approximately constant although its dominant phase factor varies as

$$\exp(-jk\sqrt{(\rho - R_o)^2 + t^2}) \cong \exp(-jkt)\exp\left(-\frac{jk}{2t}(\rho - R_o)^2\right)$$

as $t \ll (\rho - R_o)$. Let $h = (\rho - R_o)$ so that all integrals in the radial direction ρ can be expressed approximately as

$$\exp(-jkt)\int_0^{h_o} dh\exp\left(-\frac{jk}{2t}h^2\right) = \exp(-jkt)\sqrt{\frac{2t}{k}}\mathcal{K}\left(\sqrt{\frac{k}{2t}}h_o\right)$$

where $\mathcal{K}$ is the complex Fresnel integral function (Appendix A). As the argument of this function is small, a two-term small argument approximation can be employed so that

$$\exp(-jkt)\sqrt{\frac{2t}{k}}\mathcal{K}\left(\sqrt{\frac{k}{2t}}h_o\right) \cong \exp(-jkt)h_o\left(1 - j\frac{kh_o^{\,2}}{6t}\right)$$
$$\cong \exp(-jkt)\,h_o$$

With this approximation, the original phase factor is returned to the dyadic so that Eq. (7.36) becomes

$$\begin{aligned}
y_{21} \cong &-\int_{-w_j/2}^{w_j/2} dv_j \int_{-w_i/2}^{w_i/2} dv_i\,\{e^{-j(\Delta\phi_j+\Delta\phi_i)}\,[\sin\omega_{sj}\sin\omega_{si}G_{\phi\phi}(\phi_{j1}, z_{j1} \mid \phi_{i1}, z_{i1})\\
&+\cos\omega_{sj}\cos\omega_{si}G_{zz}(\phi_{j1}, z_{j1} \mid \phi_{i1}, z_{i1}) + \sin(\omega_{sj}+\omega_{si})\,G_{\phi z}(\phi_{j1}, z_{j1} \mid \phi_{i1}, z_{i1})]\\
&+e^{j(\Delta\phi_j-\Delta\phi_i)}\,[\sin\omega_{sj}\sin\omega_{si}G_{\phi\phi}(\phi_{j2}, z_{j2} \mid \phi_{i1}, z_{i1}) + \cos\omega_{sj}\cos\omega_{si}G_{zz}(\phi_{j2}, z_{j2} \mid \phi_{i1}, z_{i1})\\
&+\sin(\omega_{js}+\omega_{si})G_{\phi z}(\phi_{j2}, z_{j2} \mid \phi_{i1}, z_{i1})] + e^{-j(\Delta\phi_j-\Delta\phi_i)}\,[\sin\omega_{sj}\sin\omega_{si}G_{\phi\phi}(\phi_{j1}, z_{j1} \mid \phi_{i2}, z_{i2})\\
&+\cos\omega_{sj}\cos\omega_{si}G_{zz}(\phi_{j2}, z_{j2} \mid \phi_{i1}, z_{i1}) + \sin(\omega_{sj}+\omega_{si})G_{\phi z}(\phi_{j1}, z_{j1} \mid \phi_{i2}, z_{i2})]\\
&+e^{j(\Delta\phi_j+\Delta\phi_i)}\,[\sin\omega_{sj}\sin\omega_{si}G_{\phi\phi}(\phi_{j2}, z_{j2} \mid \phi_{i2}, z_{i2}) + \cos\omega_{sj}\cos\omega_{si}G_{zz}(\phi_{j2}, z_{j2} \mid \phi_{i2}, z_{i2})\\
&+\sin(\omega_{sj}+\omega_{si})G_{\phi z}(\phi_{j2}, z_{j2} \mid \phi_{i2}, z_{i2})]\}
\end{aligned} \tag{7.37}$$

The Green's functions in Eq. (7.37) are now expressed in their series form, Eq. (7.2). Upon reversing the order of integration and integrating over the aperture widths inside the summation over n, it is found that

$$\begin{aligned}
y_{21} \cong &-c_h\,w_iw_j\int_{-\infty}^{\infty} d\xi\sum_{n=-\infty}^{\infty} e^{jn\bar{\phi}_o}\,S\left(\frac{w_i}{2}A_{ci}\right)S\left(\frac{w_j}{2}A_{cj}\right)P(n,\xi)\,[sin\omega_{si}\,sin\omega_{sj}G^c(n,\xi)\\
&+cos\omega_{si}\,cos\omega_{sj}G^a(n,\xi) + sin(\omega_{si}+\omega_{sj})G^x(n,\xi)],
\end{aligned} \tag{7.38}$$

where

$$\begin{aligned}
P(n,\xi) = &\,[e^{-j(\Delta\phi_j+\Delta\phi_i)+j(z_{sj1}B_{cj}+z_{si1}B_{ci})/R_o} + e^{j(\Delta\phi_j-\Delta\phi_i)+j(z_{sj2}B_{cj}+z_{si1}B_{ci})/R_o} + e^{-j(\Delta\phi_j-\Delta\phi_i)+j((z_{sj1}B_{cj}+z_{sj2}B_{ci}))/R_o}\\
&+e^{j(\Delta\phi_j+\Delta\phi_i)+j((z_{sj2}B_{ci}+z_{si2}B_{c2}))/R_o}]
\end{aligned} \tag{7.39}$$

and

$$A_{ci}(n,\xi) = n\cos\omega_{si} + \xi R_o \sin\omega_{si},\ B_{ci}(n,\xi) = n\sin\omega_{si} - \xi R_o \cos\omega_{si},\ \overline{\phi}_o = \phi_{oj} - \phi_{oi},$$

where (ϕ_{oi}, z_{oi}) are patch i center coordinates, the functions $G^c(n, \xi)$, etc. and c_h are defined under Eq. (7.28) while $S(x)$ is the sinc function (see Appendix A).

In this form, the infinite integral in Eq. (7.38) can be integrated numerically. The summation over n can be completed directly without difficulty, although care must be taken with the Hankel function for $|\xi| > 1$. This allows the integral over ξ to be evaluated numerically or asymptotically when the cylinder is large. The coupling coefficient can be calculated from $\boldsymbol{S} = (\boldsymbol{I} - \boldsymbol{y})/(\boldsymbol{I} + \boldsymbol{y})$ in which the self-admittance is approximated by the expression for planar microstrip, such as Eq. (6.7), as the cylinder has a large radius and the patch appears almost flat.

As an example, consider two identical patches on a substrate with a relative permittivity of $\varepsilon_r = 2.33$ and thickness $h_o = 0.0508$ cm, which corresponds to a Rogers Duroid 5887, which are mounted on a cylinder of radius 5.057 cm. A signal of 1 V rms is applied to patch 1 at a frequency of 9 GHz so that $\lambda_g = 2.09$ cm. The patches have dimensions $w = \lambda_g/4 = 0.52$ cm, $\ell = \lambda_g/2 = 1.05$ cm, and $h_o \cong 0.1\lambda_g$ cm. The mutual admittance has been calculated with varying spacings. Two cases are considered: (i) the patches are oriented so that $\omega_i = 0 = \omega_j$ and (ii) they are rotated through $\omega_i = 90^o = \omega_j$. In both cases, the patches are moved along the axis when $\omega_s = 0$, and secondly, they are moved around the circumference when $\omega_s = 90°$. The E- and H-plane coupling in both cases has been calculated at a frequency of 9 GHz, and the results are shown in Figures 7.13 and 7.14. The coupling was calculated as a function of separation in both instances for polarization directions

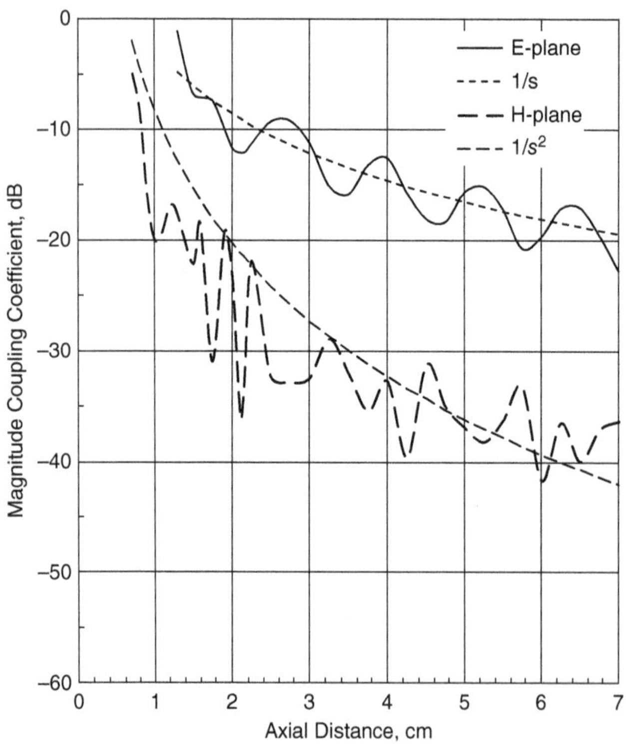

Figure 7.13 Mutual coupling at 9 GHz between patches on surface parallel to the axis $\omega_s = 0$ of a cylinder of radius 5.057 cm. The patches are identical and have dimensions $w_i = 0.52$ cm, $\ell_i = 1.05$ cm and are on a substrate of thickness $h_o = 0.0508$ cm and dielectric constant $\varepsilon_r = 2.33$. Cases shown: E-plane $\omega_i = 0 = \omega_j$; H-plane $\omega_i = 90° = \omega_j$; short dash – decay of $1/s$; longer dash – decay of $1/s^2$.

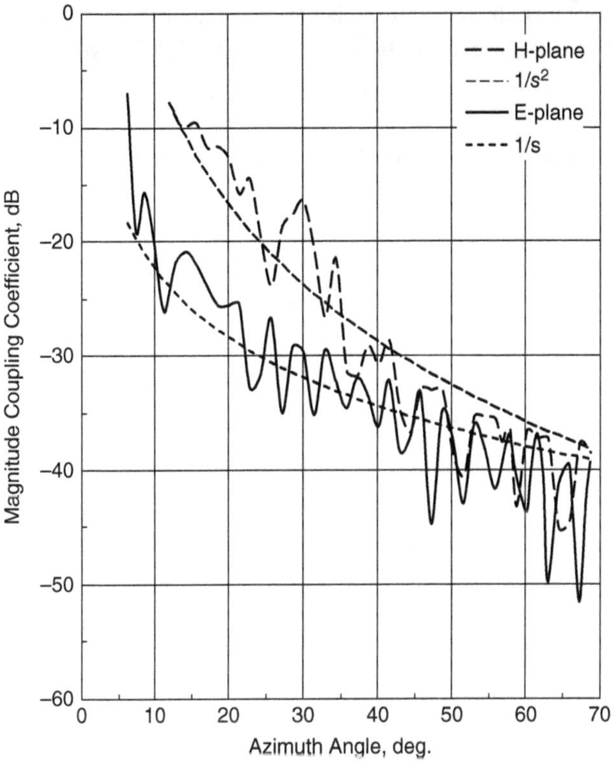

Figure 7.14 Mutual coupling at 9 GHz between patches around the circumference $\omega_s = 90°$ of a cylinder of radius 5.057 cm. The patches are identical and have dimensions $w_i = 0.52$ cm, $\ell_i = 1.05$ cm and are on a substrate of thickness $h_o = 0.0508$ cm and dielectric constant $\varepsilon_r = 2.33$. Cases shown: E-plane $\omega_i = 0 = \omega_j$; H-plane $\omega_i = 90° = \omega_j$; short dash – decay of $1/s$; longer dash –decay of $1/s^2$.

axial and circumferential to the cylinder. Note that the patches are rotated about their axes on the substrate to determine E- and H-plane coupling. When the resonant slots are orthogonal to the axis, the dominant coupling is in the axial direction, and the E plane coupling should decay at the inverse of the distance, while in the H-plane, it decays as the square of the separation as shown in Figure 7.13. Coupling for patches arranged around the circumference are expected to experience slightly lower coupling. However, as Figure 7.14 shows, the coupling in both planes at wide angles decays at a comparable rate as over a plane. With increasing distance s around the circumference the H-plane coupling level, which was initially higher than in the E-plane, decays at a high rate as it becomes increasingly oscillatory and ultimately reaches a similar level to the E-plane coupling.

7.5 Extension of Canonical Solution to Large Convex Surfaces with Slowly Varying Curvature

In Section 7.3, the field on a circular cylinder was represented by creeping waves whose geodesic is a helix. This concept will now be generalized for arbitrary convex surfaces. To commence this, the ray-based vector triad $(\hat{t}, \hat{n}, \hat{b})$ shown in Figure 7.1 is utilized, where $\hat{t}$ is a unit vector tangential to the ray path, $\hat{b}$ is a unit vector transverse to the ray direction, and $\hat{n}$ is the unit normal to the surface.

An arbitrary surface can be defined by two radii of curvature. Let R_t be the radius of curvature in the direction of $\hat{t}$, and R_b be the radius of curvature in the direction of $\hat{b}$. For a circular cylinder,

$$R_t = \frac{R_o}{cos^2\theta} \text{ and } R_b = \frac{R_o}{sin^2\theta}, \tag{7.40}$$

where, as shown in Figure 7.1, θ is the angle from the circumferential direction, and therefore,

$$\frac{1}{R_t} + \frac{1}{R_b} = \frac{1}{R_o}. \tag{7.41}$$

This can be used to express the ray direction angle in terms of the radii of curvature as follows:

$$cos^2\theta = \frac{R_t}{R_o} = \frac{R_b}{R_t + R_b}, sin^2\theta = \frac{R_b}{R_o} = \frac{R_t}{R_t + R_b}, \text{ and } tan^2\theta = \frac{R_b}{R_t}. \tag{7.42}$$

There are two parameters that connect the general surface representation to the canonical solution. These are the size parameter of the asymptotic expansion defined as

$$M = \left(\frac{k_o R_o}{2\,cos^2\theta}\right)^{1/3} = \left(\frac{k_o R_t}{2}\right)^{1/3} \tag{7.43}$$

and the distance parameter, which is the argument of the Fock functions, given by

$$\beta = k_o t\left(\frac{cos^2\theta}{\sqrt{2}\,k_o R_o}\right)^{2/3} = \frac{k_o t}{2M^2}. \tag{7.44}$$

Equation (7.44) is a special result applicable to surfaces with constant curvature R_t. A more general distance parameter has been defined by (Goodrich et al. 1959) as

$$\beta = \int_{P\prime}^{P} ds' \frac{k_o}{2\,[M(s')]^2} \tag{7.45}$$

where s' denotes a point on the surface ray path between P' and P. For general surfaces, Eq. (7.45) can be very time-consuming to compute. If the variation in the surface curvature is not too great, useful results can be obtained from Eq. (7.43). As well, approximate the ray path by straight lines as shown in Figure 7.15 where each section of the surface is modeled with a cylinder having "best-fit" radii of curvature $\overline{R}_t$ and $\overline{R}_b$.

One way of obtaining "best-fit" radii of curvature is via the approach of Lee (1978), who let them equal to the geometric mean of the curvature at the start and end points. Extending this approach to n points $P' = P_1, P_2, \ldots, P_{n-1}, P_n = P$ on the ray path shown in Figure 7.15, the radii of curvature of the surface are defined as

$$\overline{R}_t = [R_t(P_1)\, R_t(P_2) \ldots R_t(P_n)\,]^{1/n} \tag{7.46a}$$

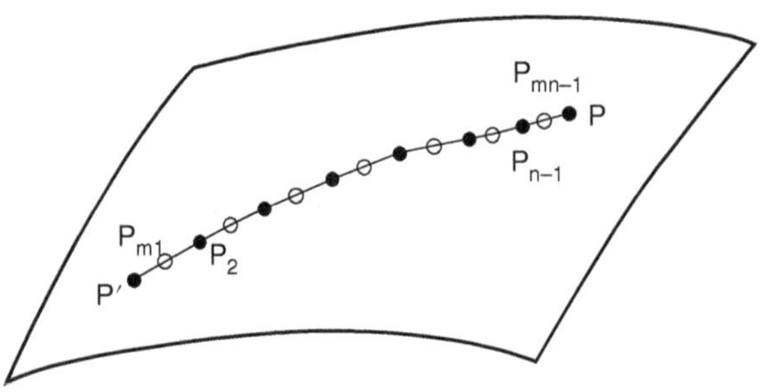

Figure 7.15 Mean path defined by n-1 cylindrical sections on an arbitrary convex surface showing points P_i and segment midpoints P_{mj}.

and

$$\overline{R}_b = [R_b(P_1)\, R_b(P_2) \ldots R_b(P_n)\,]^{1/n} \tag{7.46b}$$

Similarly, a geometric mean size parameter is defined as

$$\overline{M} = [M(P_1)\, M(P_2) \ldots M(P_n)\,]^{1/n} \tag{7.46c}$$

where $M(P_1)$, etc. is given by Eq. (7.43) on each segment. If the path length between P_i, P_{i+1} (i = 1, 2,…,n) is small, the curvature of the surface is almost constant. It can, therefore, be approximated by a cylinder of radius equal to the radius of curvature at the midpoint P_{mj} (j = 1,2,…,n−1). If Δs_i be the path length of the i-th segment, a mean distance parameter, based on Eq. (7.45), can be defined as

$$\overline{\beta} = \frac{k_o}{s} \sum_{i=1}^{n-1} \frac{\Delta s_i}{[M(P_{mi})]^2} \tag{7.46d}$$

An initial part of the analysis in the evaluation of Eqs. (7.46) for $n > 2$, is that the shortest path between P' and P needs to be determined. A way of doing this will be discussed later.

In another approach to model the surface, the radii of curvature at P on a general convex surface are given by Pathak and Wang (1981) could be used. These are

$$\frac{1}{R_t(P)} = \frac{cos^2\alpha}{\rho_{p1}(P)} + \frac{sin^2\alpha}{\rho_{p2}(P)} \tag{7.47a}$$

$$\frac{1}{R_b(P)} = \frac{sin^2\alpha}{\rho_{p1}(P)} + \frac{cos^2\alpha}{\rho_{p2}(P)} \tag{7.47b}$$

where ρ_{p1} and ρ_{p2} are the principal radii of curvature of the surface with $\rho_{p1} \le \rho_{p2}$. The principal radii of curvature are orthogonal at P. In other words, the surface normal planes at the point and in the principal directions are perpendicular to one another, and both are perpendicular to the surface tangent plane at the point. As shown in Figure 7.16, α is the angle between the ray direction $\hat{t}$ and the vector in the direction of ρ_{p1} at P′.

The principal radii of curvature are obtained for a three-dimensional surface by making use of results from differential geometry. A simple approach is to represent a surface parametrically in a functional form such as $f(x, y, z(x, y))$ and to use the formula for curvature, which is the reciprocal of the radius of curvature at a point with co-ordinates x', y'; thus,

$$\begin{aligned} \kappa = \frac{1}{\rho} &= f_{xx}(x',y')x^2 + 2f_{xy}(x',y)xy + f_{yy}(x',y')y^2 \\ &= \begin{bmatrix} x & y \end{bmatrix} \begin{bmatrix} f_{xx}(x',y') & f_{xy}(x',y) \\ f_{xy}(x',y) & f_{yy}(x',y') \end{bmatrix} \begin{bmatrix} x \\ y \end{bmatrix} \\ &= \boldsymbol{X}^T \boldsymbol{H} \boldsymbol{X} \end{aligned} \tag{7.48}$$

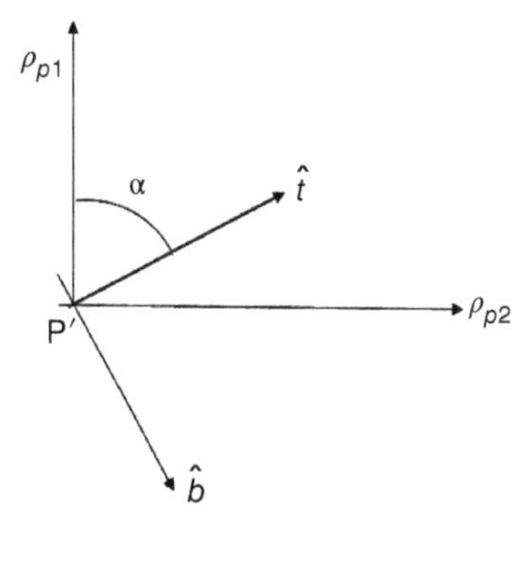

Figure 7.16 Relation of surface ray trajectory to principal radii of curvature ρ_{p1} and ρ_{p2} of an arbitrary surface.

where f_{xx}, etc. are partial derivatives with respect to x, $\boldsymbol{T}$ denotes transpose, $\boldsymbol{H}$ is Hessian matrix at (x', y'), and $\boldsymbol{X}$ is the coordinate vector. The eigenvalues of the Hessian matrix are the principal curvatures of the surface at (x', y'), and the corresponding eigenvectors are related to the principal curvatures (Struik 1961) κ_1 and κ_2. In terms of these curvatures, Eq. (7.48) reduces to

$$\kappa = \kappa_1 \cos^2\alpha + \kappa_2 \sin^2\alpha$$

which is known as Euler's theorem. For example, the principal radii of curvature of an ellipsoid can be found this way. In this case,

$$f(x, y, z(x, y)) = \sqrt{c^2 - \left(\frac{xc}{a}\right)^2 - \left(\frac{yc}{b}\right)^2}$$

so that

$$\boldsymbol{H} = -\left(c^2 - \left(\frac{cx'}{a}\right)^2 - \left(\frac{cy'}{b}\right)^2\right)^{-\frac{3}{2}} \begin{bmatrix} \left(\frac{c}{a}\right)^2\left(c^2 - \left(\frac{cy'}{b}\right)^2\right) & \left(\frac{cx'}{a}\right)\left(\frac{cy'}{b}\right)\left(\frac{c}{a}\right)\left(\frac{c}{b}\right) \\ \left(\frac{cx'}{a}\right)\left(\frac{cy'}{b}\right)\left(\frac{c}{a}\right)\left(\frac{c}{b}\right) & \left(\frac{c}{b}\right)^2\left(c^2 - \left(\frac{cx'}{a}\right)^2\right) \end{bmatrix}.$$

The eigenvalues of this Hessian matrix are

$$\lambda = \frac{-(\alpha + \beta) \pm \sqrt{(\alpha - \beta)^2 + \gamma^2}}{2Z}$$

where $\alpha = \left(\frac{c}{a}\right)^2\left(c^2 - \left(\frac{cy'}{b}\right)^2\right)$, $\beta = \left(\frac{c}{b}\right)^2\left(c^2 - \left(\frac{cx'}{a}\right)^2\right)$, $\gamma = 2\left(\frac{cx'}{a}\right)\left(\frac{cy'}{b}\right)\left(\frac{c}{a}\right)\left(\frac{c}{b}\right)$, and $Z = \left(c^2 - \left(\frac{cx'}{a}\right)^2 - \left(\frac{cy'}{b}\right)^2\right)^{-3/2}$.

In the special case when $y' = 0$, then the two eigenvalues are $\lambda = -\frac{c}{a^2}$ and $\lambda = -\frac{c}{b^2}/\sqrt{1 - (x'/a)^2}$, where the minus sign indicates convexity. Thus, the two principal radii of curvature are $\rho_{p1} = \frac{1}{\kappa_1} = -a^2/c$ and $\rho_{p2} = \frac{1}{\kappa_2} = -b^2\sqrt{1 - (x'/a)^2}/c$.

Another quantity associated with a trajectory on the surface is called the surface ray torsion $\tau(\mathrm{P})$ (Struik 1961). The torsion of a surface ray is the rate of change of the ray's osculating plane, which is the plane through the surface that is tangential to the ray path. In general, this is defined as

$$\tau(\mathrm{P}) = \frac{\sin 2\alpha}{2}\left[\frac{1}{\rho_{p1}(P)} - \frac{1}{\rho_{p2}(P)}\right]. \tag{7.49}$$

For a cylinder, $\tau = \sin 2\theta/2R_o$, and for a sphere, $\tau = 0$. For general surfaces, a normalized geodesic factor relative to that of a cylinder can be defined as

$$\mathrm{T}_o(\mathrm{P}) = \tau(\mathrm{P})R_t(\mathrm{P})/\cot\theta \tag{7.50}$$

Since $R_t(P)$ is R_o for a sphere and $R_o/\sin^2\theta$ for a cylinder, the torsion factor for a sphere is 0 and clearly for a cylinder it is 1.

From geometric optics, it is known that a ray bundle has a divergence factor (DF) owing to the fact that rays within a bundle are not parallel. A DF is, therefore, required to account for the variation of the width of the surface ray cross-section. The DF of a ray bundle is defined by

$$DF = \left(\frac{t\,d\Omega_1}{\rho_c d\Omega_2}\right)^{1/2} \tag{7.51}$$

where, as illustrated in Figure 7.17, $d\Omega_1$ is the angle subtended by the bundle at P', $d\Omega_2$ is the angle subtended by the bundle at P, and ρ_c is the caustic distance of the wavefront at P.

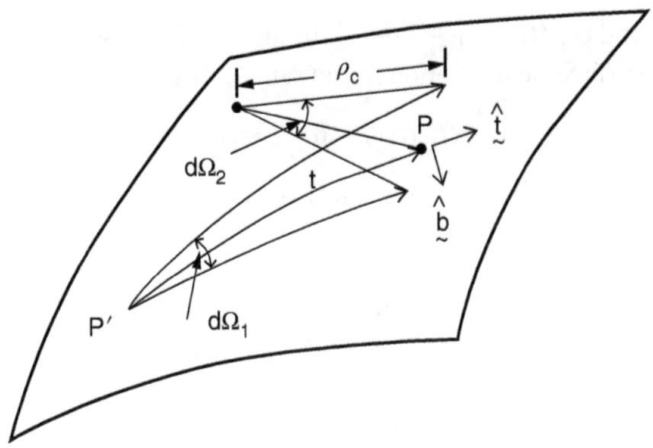

Figure 7.17 Divergence of surface ray bundle.

For a cylinder, $DF = 1$, and for a sphere, $DF = (\theta/sin\theta)^{1/2}$.

Using the parameters defined above, the dyadic components of the asymptotic solution for the cylinder are now translated into the ray-based coordinate system defined in Figure 7.1b where

$$\hat{\phi} = \cos\theta\,\hat{t} + \sin\theta\,\hat{b}$$
$$\hat{z} = \sin\theta\,\hat{t} - \cos\theta\,\hat{b} \tag{7.52}$$

It can be readily verified that

$$\begin{aligned}
T_{tt} &= T_{\phi\phi}cos^2\theta + T_{zz}sin^2\theta + sin\theta cos\theta(T_{\phi z} + T_{z\phi})\\
T_{bb} &= T_{\phi\phi}sin^2\theta + T_{zz}cos^2\theta - sin\theta cos\theta(T_{\phi z} + T_{z\phi})\\
T_{tb} &= sin\theta cos\theta(T_{\phi\phi} - T_{zz}) - cos^2\theta T_{\phi z} + sin^2\theta T_{z\phi}\\
T_{bt} &= sin\theta cos\theta(T_{\phi\phi} - T_{zz}) + sin^2\theta T_{\phi z} - cos^2\theta T_{z\phi}
\end{aligned} \tag{7.53}$$

Substituting the expressions defined in Eq. (7.20) into Eq. (7.53) transforms the asymptotic solution for a cylinder into ray-based dyadic components as follows:

$$\begin{aligned}
T_{tt} &\sim 2G_{ho}(t)q\, v(\beta) + C_{\phi\phi}cos^2\theta + C_{zz}sin^2\theta + 2sin\theta cos\theta\, C_{\phi z}\\
&= G_{ho}(t)q\,(u(\beta) + v(\beta))
\end{aligned} \tag{7.54a}$$

$$\begin{aligned}
T_{bb} &\sim G_{ho}(t)\, v(\beta)(1-q) + C_{\phi\phi}\, sin^2\theta + C_{zz}\, cos^2\theta - 2\, sin\theta\, cos0\, C_{\phi z}\\
&= G_{ho}(t)\, v(\beta)(1-q) + G_{ho}(t)q\,\left[\, tan^2\theta\,(u(\beta) - v(\beta)) + v(\beta)\, l(\theta) + v_1(\beta) l_1(\theta)\right.\\
&\quad \left. + v_2(\beta) l_2(\theta) + v_0(\beta) l_0(\theta)\right]
\end{aligned} \tag{7.54b}$$

$$\begin{aligned}
T_{tb} &\sim sin\theta\, cos\theta\,(C_{\phi\phi} - C_{zz}) - cos2\,\theta\, C_{\phi z}\\
&= G_{ho}(t)q\tan\theta\left[(u(\beta) - v(\beta)) + \frac{1}{3}\,(v(\beta) - v_1(\beta)) + \frac{v_2(\beta)}{120}\, cos^2\theta\right]
\end{aligned} \tag{7.54c}$$

$$T_{bt} = T_{tb}$$

where

$$\begin{aligned}
l(\theta) &= \frac{8}{9}\, sec^2\theta - \frac{49}{72}\\
l_1(\theta) &= \frac{257}{180}\, cos^2\theta - \frac{17}{36}\, cos^4\theta - \frac{43}{45}\, sec^2\theta + \frac{11}{60}\\
l_2(\theta) &= \frac{1}{15}\, sec^2\theta - \frac{1}{24}\\
l_0(\theta) &= \frac{j\,\beta}{5}\, sec^2\theta
\end{aligned} \tag{7.55}$$

and $G_{ho}(t) = -j\frac{kY_o}{2\pi}\frac{\exp(-jkt)}{t}$. In the limiting case of a magnetic dipole on a plane conductor, the exact surface field solution in the same coordinates is reproduced

$$T_{tt}^{(o)} = 2\,G_{ho}(t)\,(1-q)$$
$$T_{bb}^{(o)} = G_{ho}(t)\,[1-q\,(1-q)]$$
$$T_{bt}^{(o)} = 0. \tag{7.56}$$

As the surface will be approximately planar in the vicinity of the source (Boersma and Lee 1978), an extra radial dependence is included in Eq. (7.54) as described earlier so that

$$\underline{\underline{\boldsymbol{T}}}(t,\theta,\beta) \sim v(\beta)\,\underline{\underline{\boldsymbol{T}}}^{(o)}(t,\ \theta) + \underline{\underline{X}}(t,\theta,\beta) \tag{7.57}$$

where

$$X_{tt}(t,\theta) = G_{ho}(t)\,q\,(u(\beta)-v(\beta)) \tag{7.58a}$$

$$X_{bb} = G_{ho}(t)q\,[\,tan^2\theta\,(u(\beta)-v(\beta)) + v(\beta)\,l(\theta) + v_1(\beta)l_1(\theta) + v_2(\beta)l_2(\theta) + v_0(\beta)l_0(\theta)] \tag{7.58b}$$

$$X_{tb} = G_{ho}(t)q\tan\theta\left[(u(\beta)-v(\beta)) + \frac{1}{3}\,(v(\beta)+v_1(\beta)) + \frac{v_2(\beta)}{120}\,cos^2\theta\right] \tag{7.58c}$$

We now recast Eq. (7.57) for the canonical cylinder in a suitable form for general convex surfaces using the quantities defined earlier in this section. All components in Eq. (7.57) in both the field and source directions in the unprimed and primed coordinates undergo a beam divergence or compression relative to the cylinder and, therefore, should be multiplied by the DF, Eq. (7.51). As well, the part of the dyadic component due to the cylinder direction orthogonal to the ray, i.e. X_{bb} and X_{bt}, will undergo ray torsion relative a cylinder so that those components should be multiplied by Eq. (7.50), once in the case of the *bt* component and twice for the *bb* one.

After these heuristic assumptions, an approximate asymptotic solution for a surface of arbitrary curvature is expressed as

$$T_{tt}(t,\beta,\ \theta) \sim DF\,G_{ho}(t)\,q\,[2\,v(\beta)(1-q) + (u(\beta)-v(\beta))] \tag{7.59a}$$

$$T_{bb}(t,\beta,\ \theta) \sim DF\,G_{ho}(t)[\,v(\beta)(1-q\,(1-q)) + q\,T_o^2\,X_{bb}(\beta,\theta)] \tag{7.59b}$$

$$T_{tb}(t,\beta,\ \theta) \sim DF\,G_{ho}(t)q\tan\theta\,T_o\,X_{tb}(\beta,\theta) \tag{7.59c}$$

where the quantities t, β, and θ are taken along the ray trajectory of the arbitrary surface.

As an example, consider a sphere of radius R_o. From Eqs. (7.46) and (7.47), we find $R_t = R_o$ and $R_b = R_o$. Hence, $M = (k_oR_o/2)^{1/3}$, $\beta = M\,t/R_o = M\,\theta$, $T_o = 1$, $R_c = 1$, and $DF = \sqrt{\theta/sin\,\theta}$. Substituting these quantities into Eq. (7.56), we have an approximation to the surface field on a sphere as

$$T_{tt}(t,\theta) \sim \sqrt{\theta/sin\,\theta}\,G_{ho}(t)\,q\,(2\,v(M\,\theta)(1-q) + (u(M\,\theta) - v(M\,\theta))] \tag{7.60a}$$

$$T_{bb}(t,\theta) \sim \sqrt{\theta/sin\,\theta}\,G_{ho}(t)[\,v(M\,\theta)(1-q\,(1-q)\,)\,] \tag{7.60b}$$

$$T_{tb}(t,\theta) \sim 0 \tag{7.60c}$$

This compares with the actual asymptotic solution for the sphere (Lee and Safavi-Naini 1976) of

$$T_{tt}(\mathrm{t},\theta)\,|_{sphere} \sim \sqrt{\theta/sin\,\theta}\,G_{ho}(t)\,q\,[\,v(M\,\theta)\,(1-q) + u(M\,\theta)(1-2q) + q\,(M\,\theta)u'(M\,\theta)] \tag{7.61a}$$

$$T_{bb}(t,\theta)\,|_{sphere} \sim \sqrt{\theta/\sin\theta}\, G_{ho}(t)[\, v(M\,\theta)(1-q) - q^2\, u(M\,\theta) + q\,(M\,\theta)\, v'(M\,\theta)\,] \tag{7.61b}$$

$$T_{bt}(t,\theta)\,|_{sphere} \sim 0 \tag{7.61c}$$

where $\beta u'(\beta) = 3\,(u(\beta) - u_1(\beta))/2$ and $\beta v'(\beta) = (v(\beta) - v_1(\beta))/2$.

Comparing Eqs. (7.60) and (7.61), far from the source ($q \ll 1$), we note that both T_{tt} and T_{bb} in Eq. (7.60) decay more slowly than in Eq. (7.61), and they differ because of the derivative of the Fock functions $u'(M\,\theta)$ and $v'(M\,\theta)$. As a result, it can be shown that $T_{tt}(t,\theta)\,|_{sphere}$ and $T_{bb}(t,\theta)\,|_{sphere}$ decay more slowly with an additional factor proportional to $(-j\beta)^{3/2}$.

Intuitively, one could expect a better solution than Eq. (7.60) would result by blending the two canonical solutions together. To select the appropriate model would require a best-fit approach to the three fundamental surfaces, the plane, the cylinder, and the sphere. One way is to compare the principal radii of curvature of the general surface with those of the cylinder and to choose a suitable blending fraction based on the curvatures. Therefore, define a blending fraction of a "cylinder" for an arbitrary surface as

$$\Lambda_c = \begin{cases} 1 & \text{if } \rho_{p1} \to \infty \\ \sqrt{1 - \left(\frac{\rho_{p2}}{\rho_{p1}}\right)^2} & \text{otherwise} \end{cases}$$
$$= \sqrt{1 - \Lambda_s^2} \tag{7.62}$$

where Λ_s is the blending function for the sphere, ρ_{p1} and ρ_{p2} are principal radii of curvature where $|\rho_{p1}| \geq |\rho_{p2}|$. $\Lambda_c = 1$ and $\Lambda_s = 0$ for a cylinder and for a sphere $\Lambda_c = 0$ and $\Lambda_s = 1$. If there is a large difference in the ratio $|\rho_{p2}/\rho_{p1}|$, Λ_c is close to unity and will mainly consist of the cylinder solution. If the radii are comparable, Λ_c will be small, and the solution will mainly use the sphere solution.

Therefore, a blended approximate asymptotic solution is given by

$$T_{tt}(t,\beta,\theta) \sim DF\, G_{ho}(t)\, q\, \left[2\, v(\beta)(1-q) + \Lambda_c(u(\beta) - v(\beta)) + \Lambda_s\,(\, u(\beta)\,(1-2q) + q\,\beta\, u'(\beta))\right] \tag{7.63a}$$

$$T_{bb}(t,\beta,\theta) \sim DF\, G_{ho}(t)[\, v(\beta)(1 - q\,(1-q)) + q\,(\Lambda_c\, T_o^2 X_{bb}(\beta,\theta) + \Lambda_s(\,\beta\, v'(\beta) - q\, u(\beta)))] \tag{7.63b}$$

$$T_{tb}(t,\beta,\theta) \sim DF\, G_{ho}(t) q \sqrt{\frac{R_b}{R_t}}\, \Lambda_c\, T_o\, X_{tb}(\beta,\theta) \tag{7.63c}$$

In Eqs. (7.63a) to (7.63c) the trigonometric functions in Eqs. (7.60) are expressed in terms of the local radii of curvature so that

$$X_{bb} = G_{ho}(t) q \left[\left(\frac{R_b}{R_t}\right)(u(\beta) - v(\beta)) + v(\beta)\, l(\theta) + v_1(\beta) l_1(\theta) + v_2(\beta) l_2(\theta) + v_0(\beta) l_0(\theta)\right] \tag{7.63d}$$

$$X_{tb} = G_{ho}(t) q \sqrt{\frac{R_b}{R_t}} \left[(u(\beta) - v(\beta)) + \frac{1}{3}\,(v(\beta) + v_1(\beta)) + \frac{v_2(\beta)}{120}\, \frac{R_t}{R_t + R_b}\right] \tag{7.63e}$$

where $\theta = tan^{-1}\sqrt{\frac{R_b}{R_t}}$. As in the canonical equation, $q = \frac{j}{kt}$. A similar blended solution was described by Pathak and Wang (1981). The solution in Eq (7.63) gives the correct asymptotic solution in the limit for each of the quadric surfaces that of the plane, the cylinder, and the sphere.

Pathak and Wang (1981) have also derived an asymptotic solution for the dual source, the electric dipole, that is short and normal to a conducting surface. For an electric dipole on an arbitrary large surface, the surface dyadic components for the magnetic surface field is given by

$$T_{nb}(t,\beta,\theta,h_1,h_2) \sim DF\,G_{ho}(t)\,q\,[2\,V(\beta,h_1,h_2)(1-q)+T_o^2(U(\beta,h_1,h_2)-V(\beta,h_1,h_2))] \tag{7.64a}$$

$$T_{nt}(t,\beta,\theta,h_1,h_2) \sim DF\,G_{ho}(t)q\,T_o\,(U(\beta,h_1,h_2)-V(\beta,h_1,h_2)) \tag{7.64b}$$

$$T_{nn}(t,\beta,\theta,h_1,h_2) \sim 0 \tag{7.64c}$$

where h_1 and h_2 are the height of the source and the observer, respectively, above the surface. The corresponding electric field dyadic is

$$T_{nb}^{e}(t,\beta,\theta,h_1,h_2) \sim 0 \tag{7.64d}$$

$$T_{nt}^{e}(t,\beta,\theta,h_1,h_2) \sim 0 \tag{7.64e}$$

$$T_{nn}^{e}(t,\beta,\theta,h_1,h_2) \sim \eta_o DF\,G_{ho}(t)\left[V(\beta,h_1,h_2)(1-q)+q\left(T_o^2(U(\beta,h_1,h_2)-V(\beta,h_1,h_2))\right.\right. \\ \left.\left.+q(\Lambda_c\,U(\beta,h_1,h_2)+\Lambda_s\,V(\beta,h_1,h_2))\right)\right] \tag{7.64f}$$

where Taylor series approximations have been made for the Airy function integrals for the normal electric field at two small effective distances h_1, h_2 above the cylinder (Pathak and Wang 1981). These approximations are given by

$$U(\beta,h_1,h_2) \cong u(\beta)-\frac{j}{2}(h_1+h_2)\,\frac{u_1(\beta)}{\beta} \tag{7.65a}$$

and

$$V(\beta,h_1,h_2) \cong v(\beta)-\frac{j}{4}(h_1+h_2)\,\frac{v_1(\beta)}{\beta} \tag{7.65b}$$

where $h_1 = \frac{k_o x_1}{M_1}$ is the effective height of source 1 of length x_1 and similarly $h_2 = \frac{k_o x_2}{M_2}$ is the effective height at the observer position at distance x_2 above the surface. The first-order Fock functions in Eq. (7.65) can be expressed in terms of the zero-order functions u and v and their first derivatives are listed in Appendix D. The parameters M_1 and M_2 are the size parameters at the source and observer end of the ray trajectory given by Eq. (7.43). We see that Eqs. (7.65) reduce to the conventional Fock functions when the source and observer are on the cylinder and and, by means of Appendix D, it can be shown that both approximations are finite when $\beta \to 0$.

The approximate asymptotic solution for quadric surfaces, Eq. (7.63), can be used to compute the mutual coupling between aperture antennas or dipoles on arbitrary surfaces as will be shown in Section 7.6. To do this for general surfaces, the surface field has several contributions both direct fields and those creeping along geodesics as illustrated in Figure 7.18. The principal radii of curvature at points on a surface can be obtained numerically by fitting a function about the surface point under consideration and using an incremental form of Eq. (7.48) to determine the eigenvalues of the Hessian matrix **H**. For surfaces that have rotational symmetry, the functional representation is simpler. For example, the power series cone shape, given by $\rho = R_o(z/L)^n$ where $n \le 1$ and R_o is the initial radius, is generated by rotating the curve about the z-axis. This function is often used for nose cones on aircraft where the factor n controls the bluntness of the shape. For values of n above about 0.7, the tip is quite sharp. As n decreases toward zero, the power series cone shape becomes increasingly blunt. Common values of n are the cylinder ($n = 0$), parabola ($n = \frac{1}{2}$), three-quarter ($n = 3/4$), and linear cone ($n = 1$). In each case, the surface is given by

$$\sqrt{x^2+y^2} = R_o(z/L)^n$$

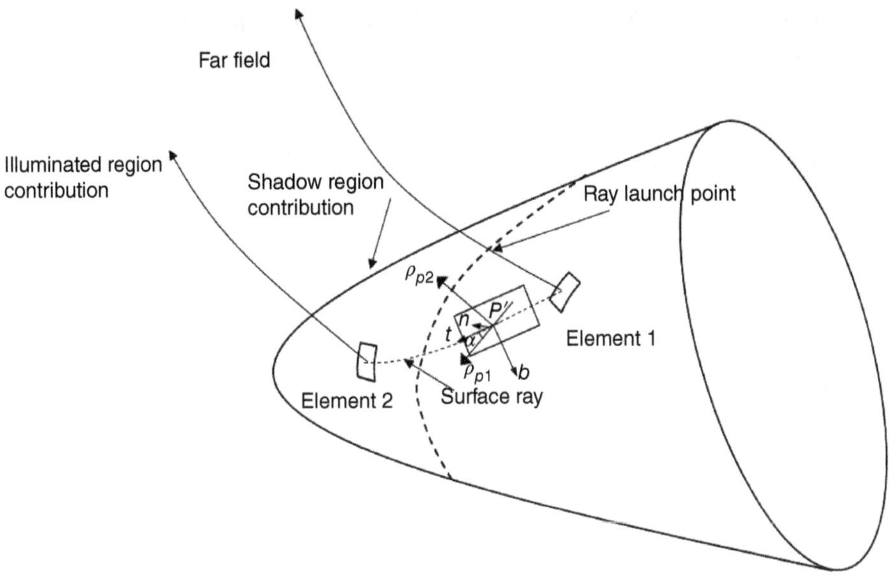

Figure 7.18 Mutual coupling contributions on an arbitrary surface.

so that

$$f(x, y, z(x, y)) = \frac{1}{L}\left(\frac{x^2 + y^2}{R_o^2}\right)^{1/2n}. \tag{7.66}$$

From this equation, Eq. (7.48) enables the principal radii of curvature to be expressed in closed form. Alternatively, the profile $\rho = f(z)$ could be represented by a spline function. Other possibilities are described in the references, e.g. Pathak et al. (1981); Jha and Weisbeck (1995).

Geodesics can be difficult to determine because only the ray end point and its direction are known beforehand. To obtain the geodesic equation, consider a point on the trajectory at point P through which a curve traverses. A geodesic line is a line of shortest distance between two points on the surface. However, more fundamentally, a geodesic is coincident with the principal normal at P and the surface normal ρ_{p2} as illustrated in Figure 7.18. The problem reduces to finding the shortest path along the geodesic that passes through the two points. An incremental length along the geodesic is expressed in rectangular coordinates as

$$ds = \sqrt{dx^2 + dy^2 + dz^2} = d\ell\sqrt{\left(\frac{dx}{d\ell}\right)^2 + \left(\frac{dy}{d\ell}\right)^2 + \left(\frac{dz}{d\ell}\right)^2} \tag{7.67}$$

where $x = x(\ell), y = y(\ell), z = z(\ell)$ are parametric equations over a specified range $\ell = \ell_1$ to ℓ_2. The total length of the curve is

$$S = \int_{\ell_1}^{\ell_2} d\ell L(x(\ell), y(\ell), z(\ell)) \tag{7.68}$$

where

$$L(\ell) = \sqrt{\left(\frac{dx}{d\ell}\right)^2 + \left(\frac{dy}{d\ell}\right)^2 + \left(\frac{dz}{d\ell}\right)^2}. \tag{7.69}$$

The minimum path of S is a standard "calculus of variations" problem to which the solution is known to satisfy the Euler-Lagrange equation in the integrand $L(\ell)$ (Struik 1961). As an example, suppose we wish to find the geodesic on a cylinder. The parametric equations are $x = R\ cos\theta$, $y = R\ sin\theta$, for any z. Equation (7.69) becomes

$$L(\theta) = \sqrt{R^2 + \left(\frac{\partial z}{\partial \theta}\right)^2}$$

The Euler–Lagrange equation in this case is

$$\frac{\partial L}{\partial z} - \frac{\partial}{\partial \theta}\frac{\partial L}{\partial\left(\frac{\partial z}{\partial \theta}\right)} = 0.$$

Any curve on any cylinder is easily found as $\partial L/\partial z = 0$ and

$$\frac{d}{d\theta}\left(\frac{\frac{\partial z}{\partial \theta}}{\sqrt{R^2 + \left(\frac{\partial z}{\partial \theta}\right)^2}}\right) = 0$$

That is

$$\frac{\frac{\partial z}{\partial \theta}}{\sqrt{R^2 + \left(\frac{\partial z}{\partial \theta}\right)^2}} = C_1$$

where C_1 is a constant. Further manipulation gives

$$\frac{\partial z}{\partial \theta} = \frac{C_1\, R}{\sqrt{1 - C_1^2}} = C_2$$

And, therefore, $z = C_2\ \theta + C_3$ in which C_2 and C_3 are constants. As expected, the solution to Eq. (7.68) for the geodesic on a cylinder is a helix.

The Euler–Lagrange equations can be used to find the geodesic on any curved surface. However, more generally, in three dimensions, the solution of the Euler–Lagrange equation presents a mathematical challenge. For general surfaces, a solution is best found numerically using optimization in which the coordinates of a moveable set of points are the variables to be optimized, and the total of the distances between the points is the function to be minimized. Various optimization methods can be used for finding the minimum S such as quasi-Newton or genetic methods. To locate the geodesics of the type shown in Figure 7.18, a double iterative method can be used given the position of one end and the direction of the other. The position of the free end can be allowed to vary along the boundary curve shown in the figure on which the required radiation direction lies on the surface. For each position on the surface, the angle α from it along the trajectory is obtained by finding the direction at which ds as given by Eq. (7.68) from P' is minimum. This involves keeping track of the angle α as the minimum of ds is found.

7.6 Applications of Coupling on Curved Surfaces

7.6.1 Mutual Coupling in a Waveguide Array on a Cylinder

Once the geodesics representing a ray paths are found, it is possible to calculate the mutual coupling between elements. In this section, computed and experimental results are given for a variety of antenna structures. Some results for one antenna have been described by the author (Bird 2016) for an array of rectangular waveguides terminated in a metallic cylinder. The coupling coefficients were calculated for an array of 29 waveguides that is conformal with a cylinder of radius 126.24 mm

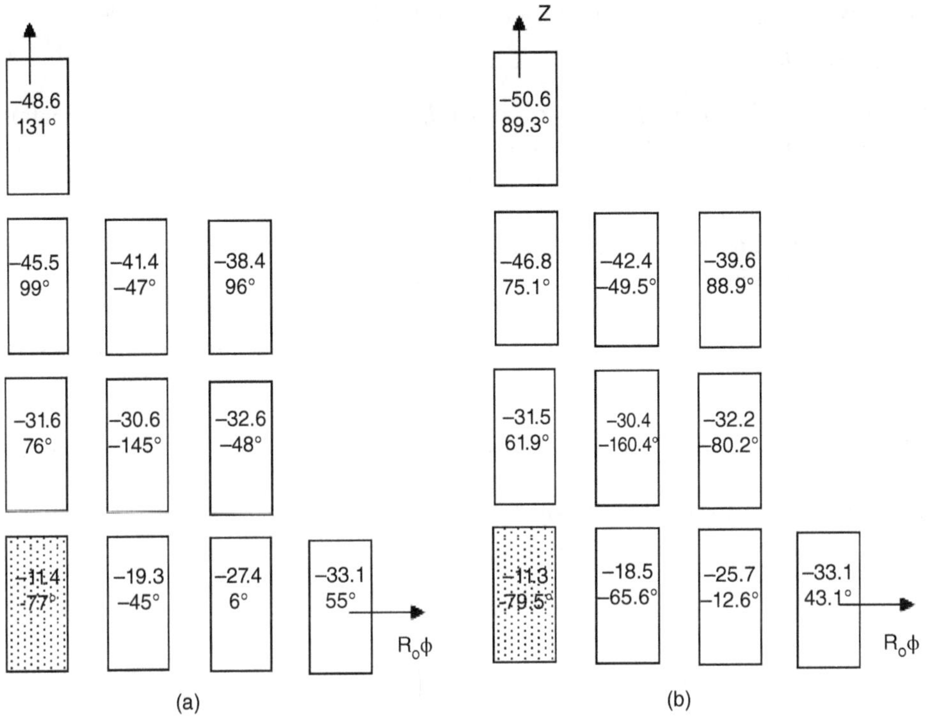

Figure 7.19 Coupling coefficients (magnitude dB, phase deg.) in a quadrant of a 29-element conformal array of WR90 waveguides at a frequency of 9.5 GHz on a cylinder of radius 126.24 mm. The driven element is shown shaded. Element spacing: axial 28.4 mm, azimuth 25.6 mm. (a) Measured (Wills 1983a); (b) finite array analysis with approximate asymptotic solution.

at a frequency of 9.5 GHz. Figure 7.19 shows a comparison of computed and measured coupling coefficients in the first quadrant when the central element was excited, which is shown shaded in the figure. The element spacing is 28.4 mm along the axis and 25.6 mm around the cylinder. A single dominant TE_{10} mode was assumed in each waveguide in the calculation. In this case, the asymptotic formulation in Eq. (7.63) is due to the cylinder solution only as $\Lambda_c = 1$. There is mainly excellent agreement between theory and experiment (Wills 1983a,b), which gives confidence that when extended to more general convex surfaces, the approximate formulation will give satisfactory results.

7.6.2 Mutual Coupling Between Monopoles on a Cylinder

As another example, two short monopoles are shown mounted on a large cylinder in Figure 7.20. The formulation given in Eq. (7.64) can be applied to this situation. Antenna 1 is assumed to be driven with a current $\boldsymbol{J}_{a1}$, and the aim is to find the mutual impedance with a second monopole also normal to the surface. To understand the mechanism of radiation and to help with the coupling analysis, suppose that the monopoles are raised slightly above the cylinder as suggested by Kuehl (1961). When monopole 1 is driven, currents are excited on this monopole, the surface of the cylinder and the second antenna as illustrated in Figure 7.20b. It is assumed that the field has been excited by an electric dipole component $p_{e1x}(x_{01}, y_{01}, z_{01})$ at the base of the monopole 1 on the cylinder where $(x_{01}, y_{01}, z_{01}) = (R_o, 0, 0)$ and $\rho_{e1x}(R_o, 0, 0) = I_{01}(R_o)\ell_1$. At the source element 1,

$$I_{a1}(x') = I_{01}\, i_1(x')$$

and the receiving element $I_{a2}(x') = I_{02}\, i_2(x')$ so that $I_{am} = \hat{\boldsymbol{x}} I_{om}\, i_m(x')$ is the electric current on monopole m (m = 1, 2) with current $i_m(x) = \sin[k(l_m - x + x_{om})]$ where $kl_m \le \pi/2$ (m = 1, 2). The coordinates at the base of the receiving monopole are $(x_{02}, y_{02}, z_{02}) = (R_o, 0, s)$ where s is the separation distance between the monopoles on the cylinder.

The total field impinging on the receiving monopole 2 from the source is found by integrating over all currents due to radiation from monopole 1 as well as the surface field excited on the cylinder. The radiation from one antenna only on a cylinder has been described by Kuehl (1961) and Goldhirsch et al. (1971). In the present situation, of the two monopoles, the mutual impedance is calculated from

$$Z_{21} = \frac{\int_{x_{02}}^{x_{02}+\ell_2} dx_2\, I_{a2}(x_2) \cdot [E_{a1}(x_2) + E_{cyl}(x_2)]}{I_{01} I_{02}}. \tag{7.70}$$

The field radiated directly by the monopole 1 (i.e. no cylinder) is from Eq. (3.21)

$$\boldsymbol{E}_{a1}(r, \theta, \phi) = -\hat{x}\frac{j\,k\,\eta_o}{4\,\pi}\int_{x_{01}}^{x_{01}+\ell_1} dx_1 \left[G_o(|\boldsymbol{r} - \boldsymbol{r}_1|) + \frac{1}{k^2}\frac{\partial^2}{\partial x_1{}^2}G_o(|\boldsymbol{r} - \boldsymbol{r}_1|)\right] I_{a1}(x_1). \tag{7.71}$$

while the contribution from the surface field of the cylinder between the monopoles is

$$E_{cyl}(\boldsymbol{x}) \cong I_{01}\ell_1 \sin(k\ell_1)\mathrm{T}^e_{nn}(t, \beta, \theta_s, x, R_o). \tag{7.72}$$

The monopole lengths are assumed to be small relative to the cylinder radius, i.e. $\ell_1, \ell_2 \ll R_o$. In the diagram shown in Figure 7.20, the monopoles are aligned at an angle $\pi/2 - \theta_s$ to the axis of the cylinder. Therefore, $M(\theta_s) = (k_o R_o/2\, cos^2\theta_s)^{1/3}$, and $\beta = k_o s(cos^2\theta_s/\sqrt{2}\, k_o R_o)^{2/3}$where s is the spacing between the two monopoles on the surface. After the mutual impedance has been calculated

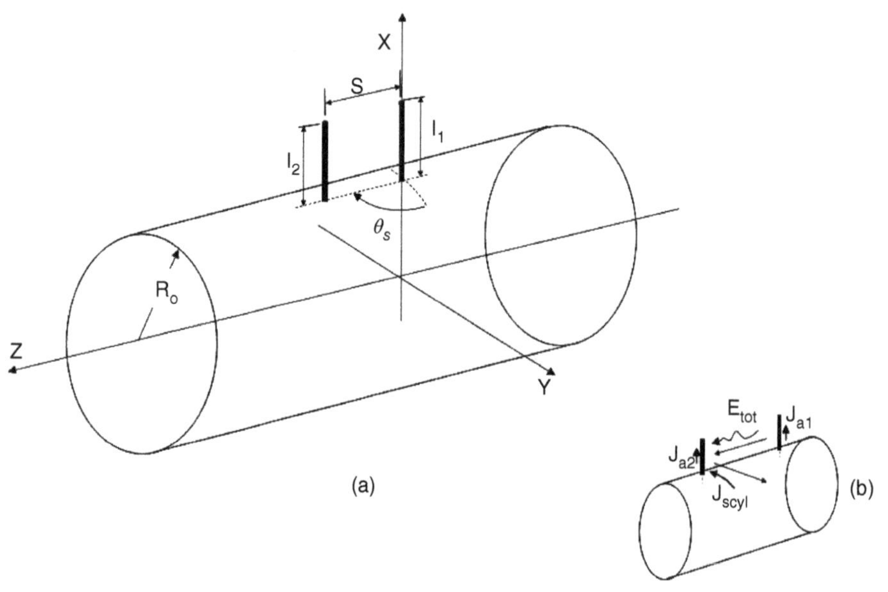

Figure 7.20 Two monopoles normal to the surface of an infinite cylinder of radius R_o. (a) Geometry; (b) radiation from antenna 1 showing induced currents.

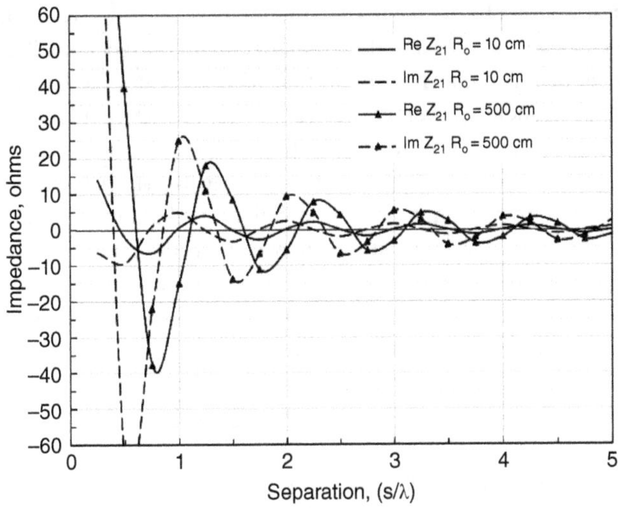

Figure 7.21 Mutual impedance of two quarter-wave monopoles operating at 10 GHz versus separation on a line at an angle $90 - \theta_s = 0.05°$ relative to the axis of a cylinder of radius (a) 10 cm and (b) 500 cm.

from Eq. (7.70), the scattering matrix and hence coupling coefficients can be calculated from Eq. (2.45) by approximating the self-impedance using the impedance of an isolated monopole.

As an example, consider two quarter-wave monopoles aligned at an angle $\theta_s = 89.95°$ operating at 10 GHz on a cylinder of radius 20 cm. The mutual impedance calculated from Eq. (7.70) is shown in Figure 7.21. The impedance and coupling decay with increasing separation at a rate of $1/s$. As the radius of the cylinder is increased, the surface field increases to a higher level effectively improving the image of the monopoles. The mutual impedance on a 500 cm cylinder is also shown in Figure 7.21 and is similar to the mutual impedance of two parallel half-wave dipoles shown in Figure 4.6.

This example shows how a platform, represented here by the cylinder, can affect the performance of antennas, in this case, monopoles, mounted on it. As can be anticipated from Figure 7.21 which only shows the variation in mutual impedance, input impedance and radiation pattern will be modified as well. The topic of coupling of co-sited antennas on platforms is addressed in Chapter 8.

Another approach to solving the problem of a monopole or small antennas located on large objects such as a cylinder, sphere, or an aircraft fuselage is to use hybrid techniques that are the combination of methods such as the method of moments and asymptotic methods typified by the GTD. One such formulation by Thiele and Newhouse (1975) with further results reported by Davidson and Thiele (1980) has proven both efficient and effective. This hybrid technique seeks to combine the moment method with GTD by first representing wire antennas, for instance, using the moment method and then modifying the generalized impedance matrix to account for the effects of nearby conducting bodies via GTD. It differs from other hybrid approaches which use the moment method to extend GTD (e.g. Burnside et al. 1975) by greatly reducing the number of current samples through a diffraction current term, which is based on GTD solutions.

As further introduction to hybrid methods, the Thiele and Newhouse (1975) approach will be considered in more detail. It commences with MoM approach for the orientation and size of all wire segments separately as described in Chapter 4, and then ultimately includes scattering from nearby conducting bodies such as a supporting surface. The next step is to determine the contribution to impedances that come about from the fields either arriving directly from a test element or any field arriving indirectly due to reflection or diffraction from the nearby conducting body, which is much the same as Eq. (7.68). For each pair of wire segments n and m, the direct and indirect contributions are combined to obtain the total impedance matrix term for that wire segment pair. Next, a field source is defined which will generate an incident average field directly on each segment m. For transmitting and receiving wire antenna problems, this source field is nonzero only for the transmitting antenna segments. Finally, the matrix equation $\boldsymbol{I} = \boldsymbol{Z}^{-1}\boldsymbol{V}$ is solved in the usual way (Chapter 4) to obtain the current for every segment of all the wire antennas from which usual antenna parameters are computed such as input impedance and mutual coupling between the antennas.

In assessing the isolation of some systems on a platform from scatter, another parameter, the power coupling factor, is sometimes used instead of S_{21} to assess the interaction between antennas. This is because it is an easier measurement to perform especially when size and access are issues. The power coupling factor for two or more antennas is defined as the ratio of the power developed at the receiving antenna terminals and the input power to the transmitting antenna. Thus, if V_{t1} is the complex voltage applied to the antenna terminals of the transmitting antenna and I_{t1} is the complex current input to these terminals, the input power is $P_{in} = Re(V_{t1}I_{t1}^*)$. The power received at antenna 2 terminals is $P_{out} = C(I_{r2}I_{r2}^*Z_L)$, where I_{r2} is the current at the receiver terminals, C is the coupling efficiency, which is usually close to unity, and Z_L is the load impedance connected across the receiver terminals. Therefore, the power coupling factor is

$$PCF = 10\, log\left(\frac{P_{out}}{P_{in}}\right) = 10\, log\left(\frac{Re\{I_{r2}I_{r2}^*Z_L\}}{Re\{V_{t1}I_{t1}^*\}}\right). \tag{7.73}$$

As an example of the coupling results that can be expected for wire antennas on curved surfaces, the power coupling between two ¼ wavelength monopoles on the circumference of a cylinder of radius 10 wavelengths has been computed (Davidson and Thiele 1980, 1984), as shown in Figure 7.22. The power coupling factor decays at a rate of 20 dB with doubling of spacing until close to an angular separation of 160°. After this angle, any further increase in separation results in commencement of oscillations due to constructive interference with comparable level reverse traveling waves. It was determined (Davidson and Thiele 1980) that the gain of the monopoles was 4.6 dBi.

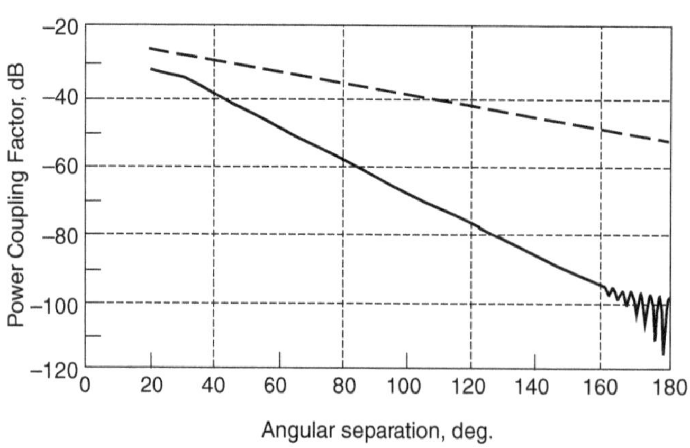

Figure 7.22 Power coupling factor between two quarter-wave monopoles versus the angle around the circumference of a cylinder of radius 10 wavelengths Source: Based on DAVIDSON, S.A. and THIELE, G.A. (1980), 'A hybrid method of moments technique for computing electromagnetic coupling between two monopole antennas on a large cylindrical surface', Technical Report 710816-8. Key: solid line – cylinder; dash line – free space.

By comparison, in free space, the power decays with distance as $20\,log\,(4\pi s)$, where s is the separation distance in wavelengths between the transmitter and receiver, and this is shown in Figure 7.22 for comparison.

7.6.3 Mutual Coupling Between Waveguides on an Ellipsoid

An array of 46 WR-10 waveguides that is conformal with an ellipsoid of revolution (also a prolate spheroid), as shown in Figure 7.23 has been studied by Wills (1986). The array is centered on the z-axis and is configured symmetrically at one end as illustrated. The element shown shaded in Figure 7.23 is driven and the coupling coefficients were calculated using a similar asymptotic expression to Eq. (7.63). The principal radii of curvature at one vertex of the prolate spheroid are identical and equal to a^2/c where a, $b = a$, and c are the three semi-axis lengths where $c > a$. Therefore, $\Lambda_c = 0$. In this example, the operating frequency is 3 GHz, $a = 500$ mm (i.e. five wavelengths), and $a = 1000$ mm as shown in Figure 7.23. In this example, the approximate surface current is dominated by the asymptotic solution for the sphere. Figure 7.24 gives the computed coupling coefficients when the waveguide element surrounded by the dashed rectangle has been excited. The reflection coefficient of this element is shown as $-12.1dB/\_280°$. The distance parameter along a geodesic on the ellipsoidal surface (Jha et al. 1992; Jha and Weisbeck 1995) is given by

$$\beta = \frac{[\sqrt{\pi/c}\,(a^4 - a^2h_{r,m}^2 + c^2\,h_{r,m}^2)]^{2/3}}{a^3}\left\{F\left(\frac{a\cos\ell_2}{\sqrt{a^3 + h_{r,m}^2}}\;\middle|\;\frac{\sqrt{c^2 - a^2}\sqrt{a^2 - h_{r,m}^2}}{ac}\right)\right.$$
$$\left. -F\left(\frac{a\cos\ell_1}{\sqrt{a^3 + h_{r,m}^2}}\;\middle|\;\frac{\sqrt{c^2 - a^2}\sqrt{a^2 - h_{r,m}^2}}{ac}\right)\right\} \tag{7.74}$$

where $F(z\,|\,p)$ is the incomplete elliptic integral function, which is defined in Appendix A, $h_{r,m}$ is the mth-order first geodesic in a right (clockwise) direction, and ℓ_1 is the start of the geodesic and

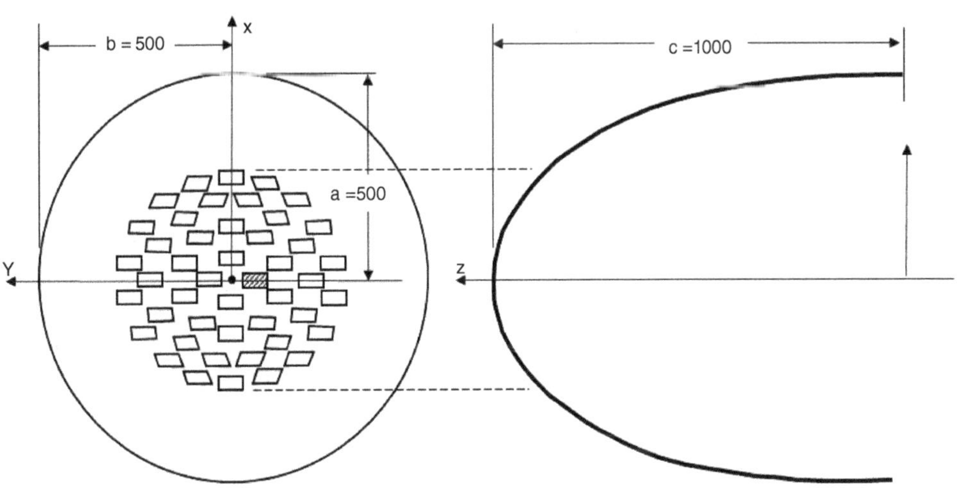

Figure 7.23 Half an ellipsoid with an array of 46 WR-10 waveguides conformal with the surface as shown. The ellipsoid has semi-axes a = b = 500 mm and c = 1000 mm. The shaded element is driven Source: After WILLS, R.W. (1986): 'Mutual coupling between waveguide antenna elements on conformal surfaces', Report, Ref. No. 9996, Plessey Electronic Systems Ltd, Addlestone, Surrey, UK.

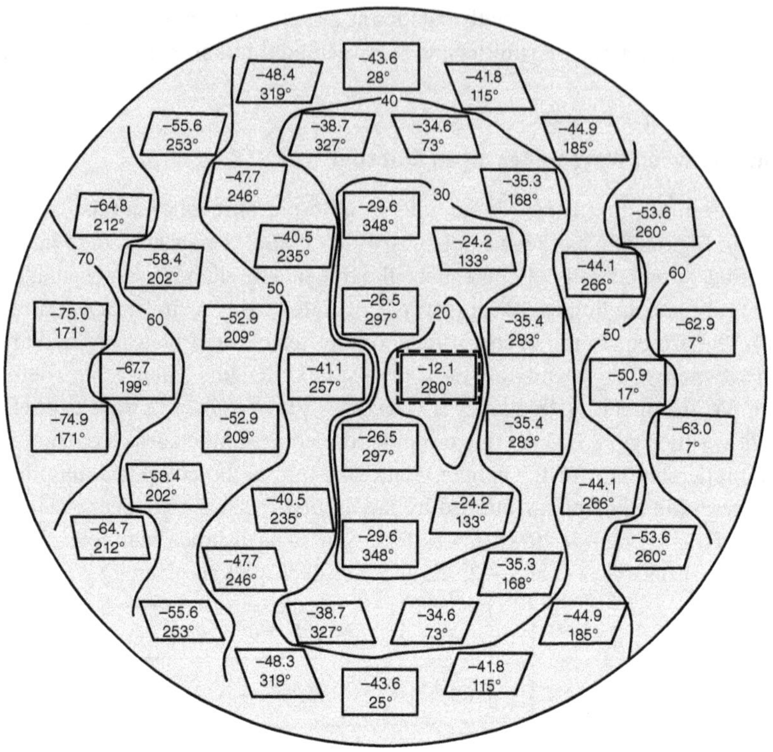

Figure 7.24 Computed coupling coefficients for an array of WG-10 waveguides on the ellipsoid shown in Figure 7.23 when the element with dashed outline has been driven. Operating frequency is 3 GHz Source: After WILLS, R.W. (1986): 'Mutual coupling between waveguide antenna elements on conformal surfaces', Report, Ref. No. 9996, Plessey Electronic Systems Ltd, Addlestone, Surrey, UK.

ℓ_2 is its end point in the coordinates of the geodesic. As the spheroid is large, $m = 1$ only in this case, although there may be a geodesic in the counterclockwise direction. The geodesic constant is usually found iteratively as described by Jha et al. (1992). As expected, the E-plane coupling in the x-direction is strongest and only drops below −40 dB by the fifth row from the source guide, whereas in the H-plane (y-direction), it drops below −40 dB by the second column.

7.7 Conclusion

A general formulation for the analysis of mutual coupling on concave surfaces has been presented in this chapter. This formulation is based on a blend of asymptotic solutions for the electric and magnetic currents on the surface on a cylinder and a sphere. These canonical solutions have been studied separately and results obtained for mutual coupling between apertures on surfaces with small radii. On convex surfaces, the coupling includes the effects of creeping waves on that surface resulting in higher levels of coupling than might be expected if the surface caused blockage only. Some examples of the use of the general formulation have been given for three cases of an array of rectangular waveguides on a circular cylinder, two short monopoles on a circular cylinder, and rectangular apertures in an ellipsoid. If the convex surface is terminated or has edges rather than being smooth, diffraction and reflection from these edges occur which should be considered using techniques such as GTD. An example of this is given in Chapter 8. Concave surfaces can be considered in the same way as described here except that direct radiation effects may need to be included in any ray-based description (Bird 2016).

References

Abramowitz, M. and Stegun, I.A. (1965). *Handbook of Mathematical Functions*. New York: Dover.

Alisan, B. and Ertürk, V.B. (2010). A high frequency based asymptotic solution for surface fields on a source-excited sphere with an impedance surface. *Radio Sci.* 45: 1–14.

Bird, T.S. (1983), 'Mutual coupling analysis of radiation from finite waveguide arrays conforming to cylindrical structures', Research Report, Dept. Elect. & Electron. Eng., James Cook University, North Queensland.

Bird, T.S. (1984). Comparison of asymptotic solutions for the surface field excited by a magnetic dipole on a cylinder. *IEEE Trans. Antennas Propag.* AP-32: 1237–1244.

Bird, T.S. (1985). Accurate asymptotic solution for the surface field due to apertures in a conducting cylinder. *IEEE Trans. Antennas Propag.* AP-33 (10): 1108–1117.

Bird, T.S. (1988). Admittance of rectangular waveguide radiating from a conducting cylinder. *IEEE Trans. Antennas Propag.* AP-36: 1217–1220.

Bird, T.S. (2016). *Fundamentals of Aperture Antennas and Arrays*. Chichester UK: Wiley.

Boersma, J. and Lee, S.W. (1978), 'Surface field due to magnetic dipole on a cylinder: asymptotic expansion of exact solution', University of Illinois at Urbana-Champaign, Dept. of Electrical Engineering, Electromagnetics Laboratory, Report No. 78–17.

Bowman, J.J., Senior, T.B.A., and Uslenghi, P.L.E. (1969). *Electromagnetic and Acoustic Scattering by Simple Shapes*. Amsterdam, the Netherlands, North-Holland..

Burnside, W.D., Yu, C.L., and Marhefka, R.J. (1975). A technique to combine the geometrical theory of diffraction and the moment method. *Trans. Antennas Propag.* 4: 551–558.

Chan, K.K., Felsen, L.B., Hessel, A., and Shmoys, J. (1977). Creeping waves on a perfectly conducting cone. *IEEE Trans. Antennas Propag.* AP-25: 661–670.

Chang, Z.W., Felsen, L.B. and Hessel, A. (1976a), 'Surface ray methods for mutual coupling in conformal arrays on cylinder and conical surfaces', Polytechnic Institute of New York, Final Report, 1976 (prepared under contract N00123-76-C-0236).

Chang, Z.W., Felsen, L.B., Hessel A. and Shmoys, J. (1976b), 'Surface ray methods in the analysis of conformal arrays', AP-S International 'Symposium 1976, Amherst·. Mass. USA, Oct., p. 366–9.

Chew, W.C. (1995). *Waves and Fields in Inhomogeneous Media*. Piscataway, USA: IEEE Press.

Davidson, S.A. and Thiele, G.A. (1980), 'A hybrid method of moments technique for computing electromagnetic coupling between two monopole antennas on a large cylindrical surface', Technical Report 710816-8, ElectroScience Laboratory, Oho State University, Columbus, OH, USA.

Davidson, S.A. and Thiele, G.A. (1984). A hybrid method of moments-GTD technique for computing electromagnetic coupling between two monopole antennas on a large cylindrical surface. *IEEE Trans. Electromagn. Compat.* EMC-26 (2): 90–97.

Duncan, R.H. (1962). Theory of the infinite cylindrical antenna including the feed point singularity in antenna current. *J. Res. Nat. Bur. Std., Part D Radio Propag.* 66D: 181–188.

Felsen, L.B. and Marcuvitz, N. (1973). *Radiation and Scattering of Waves*. NJ, USA: Prentice-Hall.

Fock, V.A. (1965). *Electromagnetic Diffraction and Propagation Problems*. London: Pergamon Press.

Franz, W. (1954). Uber die Green'sche function des zylinders und der kugel. *z. Naturforsch.* 9A: 705–716.

Golden, K.E., Stewart, G.E., and Pridmore-Brown, D.C. (1974). Approximation techniques for the mutual admittance of slot antennas on metallic cones. *IEEE Trans. Antennas Propag.* AP-22: 43–48.

Goldhirsch, J., Kneff, D.L., Doviak, R.J., and Unes, R. (1971). Radiation from a short dipole or monopole near a thick conducting cylinder of resonant length. *IEEE Trans. Antennas Propag.* AP-19: 279–282.

Goodrich, R.F., Kleinman, R.E., Maffett, A.L. et al. (1959). Radiation from slot array and cones. *IRE Trans. Antennas Propag.* AP-7: 213–222.

Hansen, R.C. (ed.), (1981), 'Conformal Antenna Array Design Handbook', Defence Technical Information Centre Publication, No. AD/AllO 091, published by Navair, Sept.

Hasserjian, G. and Ishimaru, A. (1962a). Currents induced on the surface of a conducting circular cylinder by a slot. *J. Res. Nat. Bur. Stand., D. Radio Propag.* 66D: pp. 335–365.

Hasserjian, G. and Ishimaru, A. (1962b). Excitation of a conducting cylindrical surface of large radius of curvature. *IRE Trans. Antennas Propag.* AP-10: 264–273.

Hessel, A., Lin, Y.L., and Shmoys, J. (1979). Mutual admittance between circular apertures on a large conducting sphere. *Radio Sci.* 14: 35–41.

Inami, K., Sawaya, K., and Mushiake, Y. (1982). Mutual coupling between rectangular slot antennas on a concave conducting spherical surface. *IEEE Trans. Antennas Propag.* AP-30 (5): 927–933.

James, G.L. (1976). *Geometrical Theory of Diffraction for Electromagnetic Waves*. London: Peter Peregrinus.

Jha, R. and Weisbeck, W. (1995). The geodesic constant method: a novel approach to analytical surface-ray tracing on convex conducting bodies. *IEEE Antennas Propag. Mag.* 37 (2): 26–38.

Jha, R.M., Edwards, D.J., and Bhakthavathsalam, R. (1992). Surface-ray tracing solution of ellipsoid of revolution for UTD mutual coupling applications. *Electron. Lett* 28 (4): 367–369.

Jones, D.S. (1962). *The Theory of Electromagnetism*. London: Pergamon Press.

Josefson, L. and Persson, P. (2006). *Conformal Array Antenna Theory and Design*. USA: Wiley.

Kouyoumjian, R.G. (1965). Asymptotic high frequency methods. *Proc. IEEE* 53: 864–876.

Kuehl, H.H. (1961). Radiation from a radial electric dipole near a long finite circular cylinder. *IRE Trans. Antennas Propag.* AP-9: 546–553.

Lee, S.W. (1978). Mutual admittance of slots on a cone: solution by ray technique. *IEEE Trans. Antennas Propag.* AP-26: 768–773.

Lee, S.W. and Mittra, R. (1977), 'Mutual admittance between slots on a cylinder or cone', University of Illinois at Urbana-Champaign, Department of Electrical Engineering, Electromagnetics Laboratory, Report No. 77–24.

Lee, S.W. and Mittra, R. (1979). GTD solution of slot admittance on a cone or cylinder. *Proc. IEE* 126: 487–492.

Lee, S.W. and Safavi-Naini, S. (1976), 'Asymptotic solution of surface field due to a magnetic dipole on a cylinder', University of Illinois at Urbana-Champaign, Department of Electrical Engineering, Electromagnetics Laboratory, Report No. 76–11.

Lee, S.W. and Safavi-Naini, S. (1978). Approximate asymptotic solution of surface field due to a magnetic dipole on a cylinder. *IEEE Trans. Antennas Propag.* AP-26: 593–597.

Logan, N.A. and Lee, K.S. (1962). A mathematical model for diffraction by convex surfaces. In: *Electromagnetic Waves* (ed. R.E. Langer), 139–180. Madison, WI: University of Wisconsin Press.

Mittra, R. and Safavi-Naini, S. (1979). Source radiation in the presence of smooth convex bodies. *Radio Sci.* 14: 217–237.

Pappas, C.H. and King, R. (1949). Currents on the surface of an infinite cylinder excited by an axial slot. *Q. Appl. Math.* 7: 175–182.

Pathak, P.H. and Wang, N. (1978), 'An analysis of the mutual coupling between antennas on a smooth convex surface', The Ohio State University Electro-Science Laboratory, Department of Electrical

Engineering, Rep. 784583–7, (prepared under contract N62269–76-C-0554 for Naval Air Development Centre (AD/A 065 591) (NADC-79046-30)).

Pathak, P.H. and Wang, N. (1981). Ray analysis of mutual coupling between antennas on a convex surface. *IEEE Trans. Antennas Propag.* AP-29: 911–922.

Pathak, P.H.P.H., Wang, N., Burnside, W.D., and Kouyoumjian, R. (1981). A uniform GTD solution for the radiation from sources on a convex surface. *IEEE Trans. Antennas Propag.* AP-29: 609–621.

Pearson, L.W. (1986). The electromagnetic edge wave due to a point source of current radiating in the presence of a conducting wedge. *IEEE Trans. Antennas Propag.* AP-34 (9): 1125–1132.

Pistolkors, A.A. (1947). Radiation from longitudinal and transverse slots in a circular cylinder. *J. Tech. Phys.* 17: 365–385.

Pogorzelski, R.J. (1996). On the high-frequency asymptotic evaluation of elemental sources on an anisotropic impedance cylinder. *Radio Sci.* 31: 389–399.

Silver, S. and Saunders, W.K. (1950). The external field produced by a slot on an infinite circular cylinder. *J. Appl. Phys.* 21: 153–158.

Singh, A., Vijay, S., and Baral, R.N. (2017). A review on development and analysis of conformal antennas for aircraft applications. *Int. J. Control Theory Appl* 10 (Part 9): 11–18.

Stewart, G.E. and Golden, K.E. (1971). Mutual admittance for axial rectangular slots in a large conducting cylinder. *IEEE Trans. Antennas Propag.* AP-19: 120–122.

Struik, D.J. (1961). *Differential Geometry*, 2e. Reading, MA: Addison Wesley.

Thiele, G.A. and Newhouse, T.H. (1975). A hybrid technique for combining moment methods with the geometrical theory of diffraction. *IEEE Trans. Antennas Propag.* 23 (1): 62–69.

Togöz, C., Pathak, P.H., and Marhefka, R.J. (2005). An asymptotic solution for the surface magnetic field within the paraxial region of a circular cylinder with an impedance boundary condition. *IEEE Trans. Antennas Propag.* 53 (4): 1435–1443.

Wait, J.R. (1956). Currents excited on a conducting surface of large radius of curvature. *IRE Trans. Antennas Propag.* MTT-4: 143–145.

Wait, J.R. (1959). *Electromagnetic Radiation from Cylindrical Structures*. London: Pergamon Press.

Watson, G.N. (1918). The diffraction of electric waves by the earth. *Proc. Royal Soc. A* 95: 83–99.

Wills, R.W. (1983a), 'Analysis of infinite cylindrical array', unpublished notes.

Wills, R.W. (1983b), 'Analysis of radiation from waveguide arrays on metallic cylinders, including mutual coupling effects', 13th European Microwave Conference, Nürnberg, Germany, pp. 383–388.

Wills, R.W. (1986), 'Mutual coupling between waveguide antenna elements on conformal surfaces', Report, Ref. No. 9996, Plessey Electronic Systems Ltd, Addlestone, Surrey, UK.

8

Mutual Coupling Between Co-Sited Antennas and Antennas on Large Structures

Derek McNamara[1] *and Eqab Almajali*[2]

[1] *School of Electrical Engineering and Computer Science (EECS), University of Ottawa, Ottawa, ON K1N 6N5, Canada*
[2] *Department of Electrical Engineering, University of Sharjah, Sharjah 27272, United Arab Emirates*

8.1 Preliminaries and Assumptions

8.1.1 The Problem at Hand

Earlier chapters have discussed the mutual coupling for several important specific element types used in array antennas. Later chapters continue this and also deal with the implications of mutual coupling in the systems with which such antennas are used. In the present chapter, we deal with ways of evaluating the coupling that exists between disparate antennas that find themselves installed in each other's vicinity (*co-sited*) on vehicular platforms or indeed any platform. In such cases, unwanted coupling occurs between the different antennas; indeed, it is often so undesired that some publications refer to one antenna as the victim and the other as the aggressor.

When an antenna is mounted on a platform, its performance is always influenced by the platform structure (geometry and materials), and one refers to its *installed performance* on the platform. With the goals of the present chapter in mind, we consider two antennas mounted on a platform as indicated somewhat schematically in Figure 8.1. The following can occur:

- The fields surrounding each antenna (in both the near-zone and far-zone regions) are altered by the presence of the platform and the structure of the other antenna (as well as the manner in which the other antenna is terminated).
- The input impedance of each antenna is affected by the platform (albeit principally those parts of the platform in its near vicinity) and the other antenna.
- The coupling between the two antennas is dependent on the platform and the structure and termination of the antennas.

The extent of the effects listed depends on the type of antenna (e.g. high directivity versus low directivity and isolation), its location on the platform, and the geometry of the platform. In this chapter, the determination of the coupling between the two antennas is of interest.

Study of the coupling between platform-mounted antennas is in principle the same as studying coupling between the elements in an array antenna, for instance, as has been discussed in earlier chapters. The circumstances, however, are usually different, for the following reasons:

- Quite often in such co-siting problems, the person responsible for the placement of the antennas on the platform does not have complete knowledge of the internal workings of the antenna(s)

Mutual Coupling Between Antennas, First Edition. Edited by Trevor S. Bird.

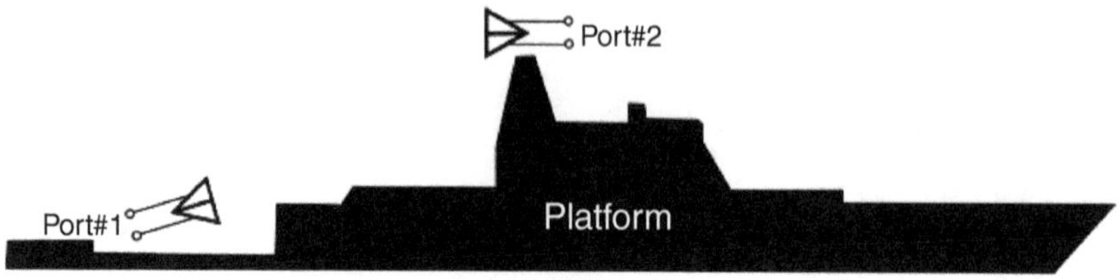

Figure 8.1 Two antennas mounted on a platform.

and so would not have access to as highly customized a model as the antenna designers. This is especially true for complex antennas.

- The details of the platform environment are not usually within the control of the antenna designer and will change from one problem to another. Thus, one needs to use some general-purpose electromagnetic simulation tool rather than some customized electromagnetic model specific to a particular antenna.
- The platform may be electrically very large.
- The coupling between two antennas that are designed to operate at different frequencies is often required. We may want to know the coupling between two antennas at a frequency outside of the nominal operating band of one of the antennas.
- Structures forming part of the platform, not necessarily taken into account during the design of the antenna, may be located in the near-zone of the antenna and might affect its performance.

The points bulleted above should convince us of the multifarious nature of the problem of interest in this chapter. Almost everything available through computational electromagnetics (CEM) can be used to tackle such problems, and texts on CEM can be consulted for such details. The chapter has instead been written in a tutorial tone. The goal is to provide, largely "in words," an organized way in which to think about the modeling of the coupling between platform-mounted co-sited antennas and to be able to interpret information from (existing and future) literature on one or more detailed aspects of the overall antenna placement problem. This allows one to identify the approximations used for, and hence appreciate the limitations of, the route followed in any specific case.

Clarity on a number of subtle ideas from electromagnetic theory greatly facilitates the understanding of the co-sited antenna coupling problem. We will quote quantitative expressions when needed, but the mere mathematical aspects of derivations will not be provided since these can be found discussed in general terms, in one form or another, in many books and journal articles, as well as in other chapters of the present text. Some mathematical expressions are in essence a repetition of material from Chapter 3 but revisited here in the context of the goals of the present chapter. Redundancy has its benefits!

We will always consider just two antennas. All materials are assumed isotropic and linear. This at once implies that they are all reciprocal materials (Lindell 1992). It is possible to arrive at results where reciprocity does not need to apply, but this complicates matters slightly for situations that (at least at present) do not occur too often in antenna placement studies.

It is assumed that the reader has some familiarity with the use, if not the detailed formulations, of one or more of the fundamental methods of CEM (Sheng and Wei 2012). These are the method of moments (MoM), the finite element method (FEM), and the finite-difference time-domain (FDTD) method and its variants such as the finite-integration-technique (FIT). These are full-wave methods because they in essence find, albeit numerically, the solutions to Maxwell's equations for some problem at hand. If the numerical implementation of one of these formulations (that is, CEM models) is faultless, then the results it produces can be considered exact within the physical modeling error inherent to the otherwise rigorous CEM model (Miller 1988).

In antenna placement work, the coverage offered by the far-zone pattern of an installed antenna is usually also under scrutiny (Reddy 2018; MacNamara 2010). We do not specifically discuss this. Normally, any CEM model capable of computing coupling would be able to predict far-zone pattern performance too.

8.1.2 Course Adopted

Sections 8.2 and 8.3 describe antenna operation, and inter-antenna coupling (with a platform present), in the way a full-wave CEM model "thinks" about it. One reason for doing this is that with relatively simple antenna types, and platforms of reasonable electrical size, this rigorous approach is often possible. Secondly, once it is appreciated what the exact approach would be, if only it were computationally feasible, it becomes clear what approximations can be made in the electromagnetic model and yet perform useful coupling computations. Thirdly, with such a rigorous (but not necessarily practical) model in mind, a clearer picture emerges as to precisely what computations are being performed when one uses a commercial code and what assumptions its developers have made. In other words, it becomes possible to make technical judgment calls. Section 8.4 arranges mathematical expressions useful for determining coupling parameters from field solutions, whether the latter are exact or approximate. Section 8.5 serves as a reminder of just enough of the intricacies of the three fundamental full-wave CEM methods to recall why there is a limit to the electrical size of the platform, and the complexity of the antenna structures, that can be modeled as a combined problem in general co-sited antenna coupling problems. It also summarizes how the range of application of CEM modeling has been extended beyond what can be achieved via the full-wave route through the use of hybrid CEM methods. Section 8.6 discusses how and why certain "large" on-platform coupling problems can sometimes nevertheless be dealt with as a single full-wave CEM model (the ideal) and ends outlining the need for incorporating approximations in many practical cases where this is not possible (pragmatic approaches). Section 8.7 reviews reduced complexity models of antennas that make it possible to apply such pragmatic approaches. Section 8.8 discusses such approaches, which are the methods that have made some co-sited antenna coupling problems (almost) routine if existing commercial CEM codes are used. Section 8.9 shows the results of some example computations. Section 8.10 concludes the chapter.

8.2 Full-Wave CEM Modeling View of a Single Antenna

In any CEM model, an impressed source has to be defined. Following Harrington (1961), we consider impressed current densities "to represent sources or known quantities. In this sense, they are independent of the field and are said to cause the field. More generally, the impressed currents are those currents we *view* as sources." Exactly what we select as the impressed source(s) depends on the problem at hand, as will be clarified at various times in this chapter. Figure 8.2a depicts a simplified diagram of an isolated antenna in terms of which we wish to interpret antenna operation. We refer to it as isolated because there are no other material objects present. Quantities $\boldsymbol{J}$ and $\boldsymbol{M}$ are *impressed* electric and magnetic current densities; they are not material objects. CEM models are often set up so that only one or the other type is needed. The rest of the sketch shows the physical structure of the particular antenna, which may be composed of conducting or dielectric material. The rigorous formulations on which full-wave CEM models are based argue as follows: The field generated by the impressed sources if all of the antenna structure were to be removed is the incident field[1] ($\boldsymbol{E}^{inc}$, $\boldsymbol{H}^{inc}$) and exists in all of space. They are not blocked (shadowed) in any way. Due to the

1 In some CEM models, the effect of the impressed source is incorporated by specifying the incident field directly.

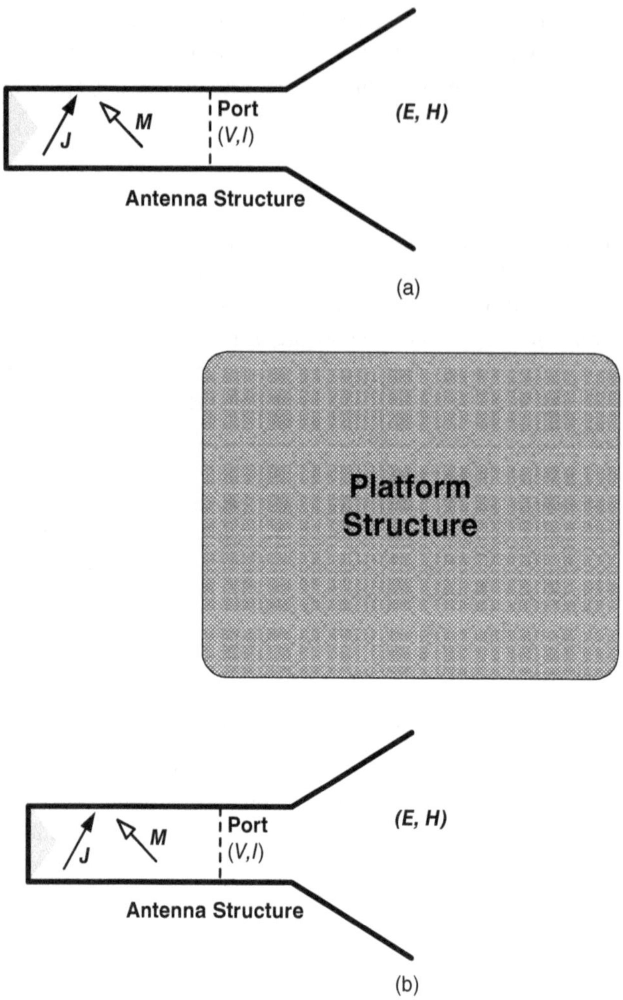

Figure 8.2 Two situations for an antenna. (a) Isolated antenna structure; (b) the same antenna in the presence of a platform structure.

presence of the incident field, there are *induced* conduction and polarization current densities in and on the antenna structure that generate a scattered field ($\boldsymbol{E}^{scat}$, $\boldsymbol{H}^{scat}$) everywhere in space. The difficult part of the CEM modeling is to find the induced currents. Once these have been found, it is a relatively straightforward matter to find the scattered fields, a process we can simply refer to as evaluating an appropriate radiation integral. The superposition of the incident and scattered field gives the total field everywhere. It is this latter total field ($\boldsymbol{E}$, $\boldsymbol{H}$) that is the observed "field of the antenna," which can be observed in the far-zone of the antenna (and hence provide the radiation patterns) or anywhere in the immediate vicinity of the antenna structure itself. If the field is very small at some observation points, it is because the incident and scattered fields are almost canceling each other at that point. We will say that ($\boldsymbol{E}$, $\boldsymbol{H}$) are the fields of the antenna when it is transmitting. It is wholly reasonable to expect that on any antenna a location exists (the port of the antenna) where equivalent circuit voltage (V) and current (I) can be correctly defined in terms of the field quantities there (Elliott 1993; Williams and Alpert 2001). All full-wave CEM antenna models have to somehow incorporate this possibility, as in Eleftheriades and Mosig (1996), for example. At the port, it is then possible, in terms of these circuit quantities, to at once define the input impedance $Z_{in} = R_{in} + jX_{in} = V/I$, the reflection coefficient

$$\Gamma_{in} = \frac{Z_{in} - Z_o}{Z_{in} + Z_o}, \tag{8.1}$$

and the power supplied to the antenna

$$P_{in} = \frac{1}{2}\text{Re}\{VI^*\}. \tag{8.2}$$

Quantity Z_0 is the characteristic impedance of the feedline at which the port is defined. We will refer to these as the port properties of the antenna. Section 8.3 will need to refer to ports being open-circuited or short-circuited; precisely what these mean in physical terms depends on the manner in which voltage and current have been defined in terms of the fields for the specific antenna.

The above full-wave CEM view of the way an antenna operates facilitates an understanding of the influence of other material objects placed in the vicinity of the antenna proper (that is, what we consider to be "the antenna"), as illustrated in Figure 8.2b. The additional material objects may represent the structure of some other unexcited antenna and/or a platform. A complete understanding of the resulting electromagnetic system could only be achieved if these objects were included in the full-wave CEM model. The induced currents on these additional objects would contribute to the total field ($\boldsymbol{E}$, $\boldsymbol{H}$) everywhere and so alter the port and radiation properties of the antenna. We would say that we have determined the radiation pattern of the said antenna in the presence of the platform and/or second unexcited antenna.

8.3 Full-Wave CEM Modeling View of Coupled Antennas in the Presence of a Host Platform

8.3.1 Field Point of View

Consider the situation where we now have two antennas in the presence of a platform, as depicted in Figure 8.3. The antennas are drawn much exaggerated in size relative to the platform, which comprised conducting and dielectric material of arbitrary shape. Each antenna will be modeled by its material structure and impressed sources. A full-wave CEM model will consider the platform plus the antenna structures as a single physical object and be referred to as the "original environment."

Two situations are considered. In situation #a, shown in Figure 8.3a, the impressed source $(\boldsymbol{J}_1, \boldsymbol{M}_1)$ is turned on and generates incident fields $(\boldsymbol{E}_1^{inc}, \boldsymbol{H}_1^{inc})$ everywhere, and $(\boldsymbol{J}_2, \boldsymbol{M}_2)$ is removed. We can also say that in this situation, antenna#1 is transmitting and antenna#2 is receiving. It is assumed that the antenna model is such that, when it is not transmitting (that is, its impressed sources are removed), some of the power arriving at its port can be extracted to some load. Thus, we can also talk of this situation as port#1 being excited and having a load connected to port#2. The total fields everywhere, due to the impressed source in the presence of the physical structures of both antennas, as well as the platform, are ($\boldsymbol{E}_1$, $\boldsymbol{H}_1$). By everywhere, we mean in the port sections of the antennas, at observation points in the near and far-zones, points on the platform and antenna structures, and so on. Voltages and currents can be defined at both ports, in terms of the fields there. These will be denoted by V_{1a} and I_{1a} at port#1, and by V_{2a} and I_{2a} at port#2, in this situation #a. In other words, they are the circuit quantities determined from fields ($\boldsymbol{E}_1$, $\boldsymbol{H}_1$) that are generated when impressed source ($\boldsymbol{J}_1$, $\boldsymbol{M}_1$) is on and ($\boldsymbol{J}_2$, $\boldsymbol{M}_2$) off. We can associate (V_{1a}, I_{1a}) and ($\boldsymbol{E}_1$, $\boldsymbol{H}_1$).

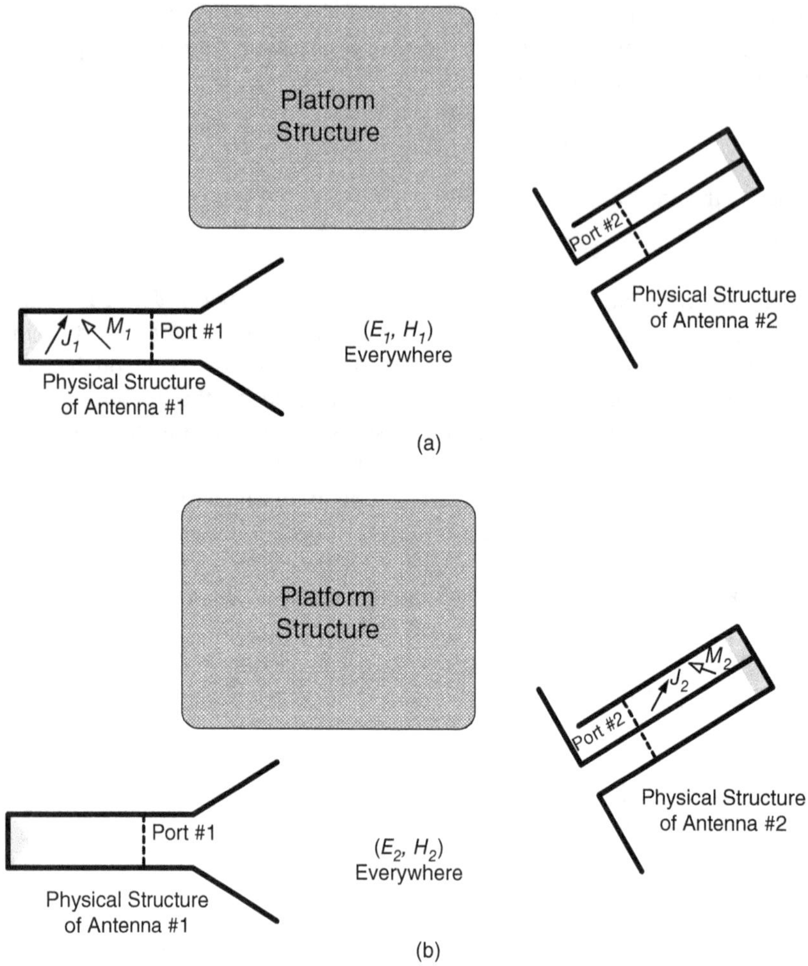

Figure 8.3 Platform and antennas. (a) Situation #a; (b) situation #b.

In situation #b, shown in Figure 8.3b, we keep the structure of the two antennas and the platform unchanged. Nothing is removed. The environment remains the same. However, impressed source $(\boldsymbol{J}_2, \boldsymbol{M}_2)$ is turned on and generates incident fields $(\boldsymbol{E}_2^{inc}, \boldsymbol{H}_2^{inc})$ everywhere and $(\boldsymbol{J}_1, \boldsymbol{M}_1)$ is removed. Antenna#2 is transmitting, and antenna#1 is receiving. The total fields everywhere are now $(\boldsymbol{E}_2, \boldsymbol{H}_2)$.

The circuit quantities at the two ports, once again defined in terms of the fields there, will be called V_{1b} and I_{1b} at port#1, while at port#2, in this situation #b, the corresponding quantities are V_{2b} and I_{2b},. They are the circuit quantities determined from fields $(\boldsymbol{E}_2, \boldsymbol{H}_2)$ that are generated when the impressed source $(\boldsymbol{J}_2, \boldsymbol{M}_2)$ is turned on and $(\boldsymbol{J}_1, \boldsymbol{M}_1)$ off. We can associate (V_{2b}, I_{2b}) and $(\boldsymbol{E}_2, \boldsymbol{H}_2)$. We imagine initially that a full-wave CEM solution of the complete electromagnetic formulation just described can be found.

The two situations are depicted schematically in Figure 8.4, where we view all transmitters and receivers (generators and loads) attached to the ports in the circuit model realm in terms of their Thevenin equivalent circuits. The Norton equivalents could alternatively be used. It will always be

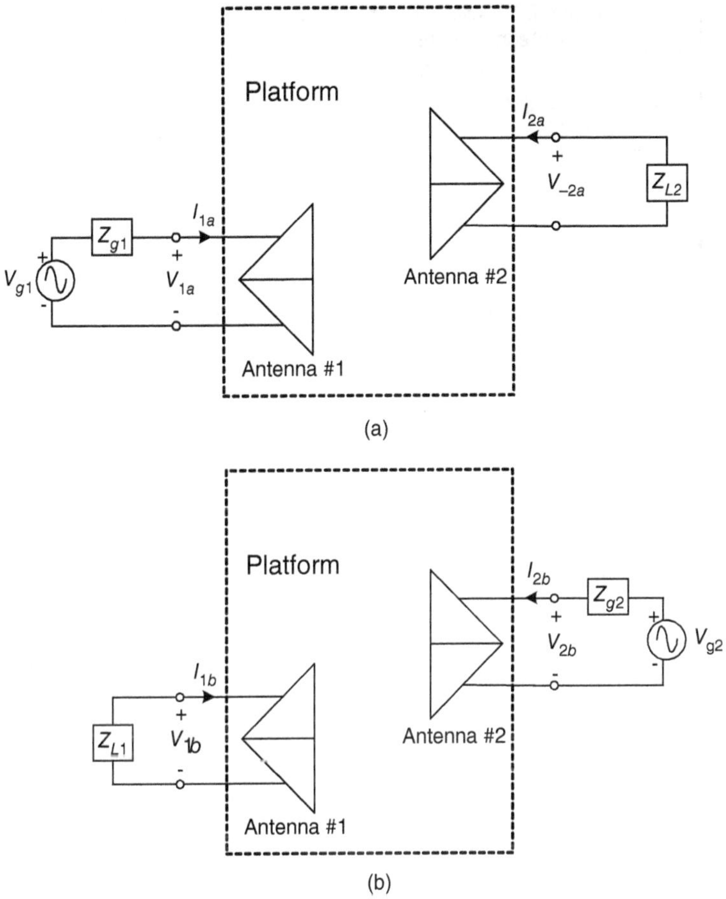

Figure 8.4 Sketch illustrating (a) Situation #a and (b) situation #b.

assumed that $Z_{L1} = Z_{g1}$ and $Z_{L2} = Z_{g2}$. We note that $V_{2a} = -I_{2a}Z_{L2}$ and $V_{1b} = -I_{1b}Z_{L1}$ in each of the two situations, the minus signs arising due to the reference direction assumed for the currents.

8.3.2 Two-Port Network Parameter Point of View

The two antennas and the platform, constituting the electromagnetic environment through which coupling takes place, can be regarded as the two-port network indicated in Figure 8.5, for which the following well-known relations (that define the impedance matrix of the two-port) pertain:

$$V_1 = Z_{11}I_1 + Z_{12}I_2 \tag{8.3}$$

$$V_2 = Z_{21}I_1 + Z_{22}I_2 \tag{8.4}$$

These apply no matter how the two-port voltages and currents arise. The four quantities Z_{ij} define the impedance matrix $[Z]$ of the two-port (Pozar 2005) as described in Section 2.4. The impedance matrix $[Z]$ of the two-port is related (Beatty and Kerns 1964) by simple formulas to its admittance matrix $[Y] = [Z]^{-1}$ and to its scattering matrix $[S]$. The latter requires knowledge of the reference characteristic impedances Z_{01} and Z_{02} at the two ports, in essence that of the access lines on which the ports are located. It is possible to move between one set of parameters and any other.

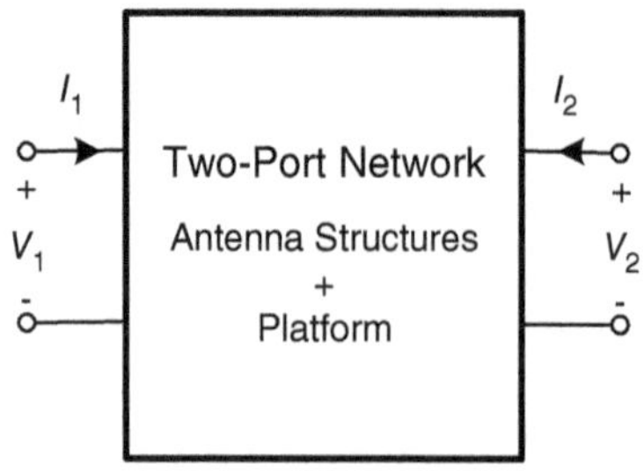

Figure 8.5 Two-port network parameter representation of two-antenna coupling.

From Eqs. (8.3) and (8.4), we recognize that the mutual impedance Z_{21} between the two ports and the self-impedances Z_{11} and Z_{22} can be expressed as[2]

$$Z_{11} = \left.\frac{V_1}{I_1}\right|_{I_2=0} \qquad Z_{21} = \left.\frac{V_2}{I_1}\right|_{I_2=0} \qquad Z_{22} = \left.\frac{V_2}{I_2}\right|_{I_1=0}. \tag{8.5}$$

The open-circuit voltage at port#2 is

$$V_2^{oc} = V_2|_{I_2=0} \tag{8.6}$$

with the superscript "*oc*" added to emphasize it is the open-circuit voltage at port#2. Similarly, the short-circuit current at port#2 is

$$I_2^{sc} = I_2|_{V_2=0} \tag{8.7}$$

with the superscript "*sc*" signifying it is the short-circuit current at port#2, and the mutual admittance between the two ports

$$Y_{21} = \left.\frac{I_2}{V_1}\right|_{V_2=0} \tag{8.8}$$

and so on. If we had access to a full-wave CEM model of the complete original coupling environment, then the field solutions in the antenna excitation regions could in principle be used to find the port voltages and current as secondary parameters from the field solutions. The open-circuit or short-circuit port conditions indicated in Eqs. (8.5) to (8.8) would be effected by setting the values of Z_{L1} or Z_{L2} as needed. Alternatively, the $[S]$ could be found from the field solutions at the ports and any other desired parameters from the conversion formulas. If any changes were made to the physical structure of the platform or the antennas, such as actual changes to their shapes or a change to their relative positions, resimulation of the CEM model would be needed to find the new $[Z]$.

As an illustration of what could be found from the known $[Z]$, observe that the input impedances of the two antennas, with the loads as specified in Section 8.3.1, are

$$Z_1^{in} = \frac{V_{1a}}{I_{1a}} \quad Z_2^{in} = \frac{V_{2b}}{I_{2b}}. \tag{8.9}$$

An equivalent circuit from the vantage point of antenna#1, when it is transmitting, is that shown in Figure 8.6a, and the power supplied to its port is

$$P_{t1} = \frac{1}{2}\text{Re}\{V_{1a}I_{1a}^*\}. \tag{8.10}$$

2 With $Z_{12} = Z_{21}$ due to reciprocity, as will be apparent from Section 8.4.

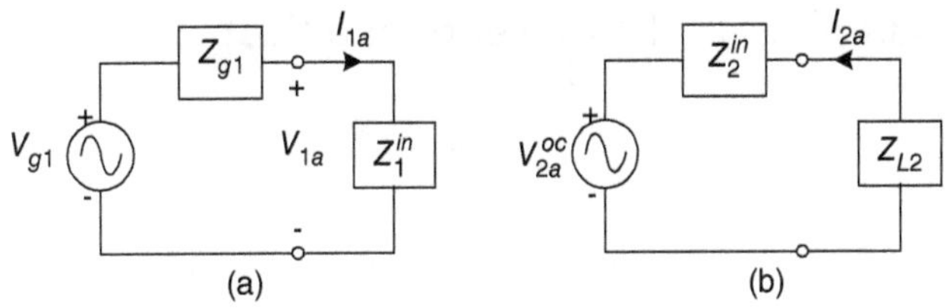

Figure 8.6 Equivalent circuits for transmitting antenna#1 and receiving antenna#2.

A Thevenin equivalent circuit with respect to the terminals of antenna#2, when it is receiving, can be found. The Thevenin voltage is simply the open-circuit voltage in Eq. (8.6), which in present notation is V_{2a}^{oc}, as indicated in Figure 8.6b. The Thevenin impedance is the input impedance observed at port#2 when antenna#2 is transmitting and is

$$Z_2^{in} = \frac{V_{2b}}{I_{2b}} = Z_{22} - \frac{Z_{12}Z_{21}}{Z_{11} + Z_{L1}} \tag{8.11}$$

with

$$I_{2a} = -\frac{V_{2a}^{oc}}{Z_2^{in} + Z_{L2}} \tag{8.12}$$

the power received by antenna#2 becomes

$$P_{r2} = \frac{1}{2}\text{Re}\{V_{2a}(-I_{2a}^*)\} = \frac{1}{2}\frac{|V_{2a}^{oc}|^2}{|Z_2^{in} + Z_{L2}|^2}\text{Re}\{Z_{L2}\}. \tag{8.13}$$

P_{r2} reaches its maximum under conjugate match condition $Z_{L2} = (Z_2^{in})^*$, and so under such matched conditions,

$$P_{r2} = \frac{|V_{2a}^{oc}|^2}{8\ \text{Re}\{Z_{L2}\}}. \tag{8.14}$$

The power coupling between the antennas is P_{r2}/P_{t1}.

At this point, we recognize that, rigorously speaking, the input impedances in Eq. (8.9) are dependent on everything in their environment, namely the platform and antenna structures. In many co-siting situations, Z_1^{in}, for example, will not be significantly influenced by the presence of the platform[3] or the structure of antenna#2, even though its total fields ($\boldsymbol{E}_1$, $\boldsymbol{H}_1$) outside of its excitation region, and hence the level of coupling between the two antennas, are. Thus, in practice, this assumption is often made for Z_1^{in} and Z_2^{in}. Clearly, such an assumption will prove inaccurate if the antennas are extremely close, as is the case with the individual elements in an array antenna, dealt with elsewhere in the present text. If we were to have two antennas in free space, at either end of a communications link, then these antennas are coupled (otherwise, there would not be any received power), yet the presence of the one does not noticeably change the input impedance of the other. In on-platform coupling, the situation is similar, even though the two antennas may not be very distant from each other. We can call this weak coupling. Weak coupling does not imply insignificant coupling.

3 Except possibly those parts of the platform in its immediate vicinity; more on this in Section 8.7.

8.4 Useful Expressions for Coupling in the Presence of a Host Platform

8.4.1 Motivation

It is only rarely possible to solve the complete field problem for the original environment (two antenna structures and a platform) as the single full-wave CEM problem imagined in Section 8.3, which we can refer to as an exact solution, although some remarks in this regard are made in Section 8.6. If exact solutions are not possible, certain expressions, albeit themselves not yet embodying any approximations, based on the reciprocity route and the reaction concept, are useful. These can be used with the exact field solutions to find the secondary parameters. The reason they are next discussed is that, because most often exact solutions are not possible, various approximations have to be made to compute coupling between co-sited antennas, and these expressions facilitate this. It is important to reiterate that $(\boldsymbol{E}_1, \boldsymbol{H}_1)$ and $(\boldsymbol{E}_2, \boldsymbol{H}_2)$ are not necessarily far-zone fields, although they may be approximated as such at times, as discussed in Section 8.8.

8.4.2 Reciprocity and Reaction Theorems Revisited

We recall Figure 8.3 and remind ourselves that $(\boldsymbol{E}_1, \boldsymbol{H}_1)$ and $(\boldsymbol{E}_2, \boldsymbol{H}_2)$ are the fields extant when antenna#1 and antenna#2 are transmitting, respectively, in the original environment. If any parts of the antenna structures or the platform were removed, added to, or shifted around, these fields would change. The general form of the reciprocity theorem as derived in Section 3.2.2, in the context of the two scenarios #a and #b in Eq. (3.6), can be restated as

$$\oiint_S (\boldsymbol{E}_1 \times \boldsymbol{H}_2 - \boldsymbol{E}_2 \times \boldsymbol{H}_1) \cdot \hat{n}\, dS = \iiint_V (\boldsymbol{J}_1 \cdot \boldsymbol{E}_2 - \boldsymbol{M}_1 \cdot \boldsymbol{H}_2)\, dV - \iiint_V (\boldsymbol{J}_2 \cdot \boldsymbol{E}_1 - \boldsymbol{M}_2 \cdot \boldsymbol{H}_1)\, dV \tag{8.15}$$

with S, *any* closed surface whatsoever; V, the volume it encloses; and $\hat{n}$, the outward unit normal to S. Surface S can be chosen so that it encloses none of the impressed sources, such that it encloses all the impressed sources, that it is so formed that it includes the physical cross-section of the port of an antenna, and so on. Judicious selection of the location and shape of S has allowed the reciprocity theorem to be used in many forms and to great effect in electromagnetic engineering, as, for example, in Richmond (1958, 1961), Deschamps (1966a, 1966b), Collin and Zucker (1969), and Monteath (1973). We will harvest solely what is needed for present purposes.

In the quest to obtain a means of computing the coupling parameters between the two antennas, the starting point is that shown in Figure 8.7. Surface S has been selected to be the union of two closed surfaces S_1 and S_2, with surface S_1 enclosing impressed source $(\boldsymbol{J}_1, \boldsymbol{M}_1)$ only, whereas S_2 encloses impressed source $(\boldsymbol{J}_2, \boldsymbol{M}_2)$ only. It is then possible to show in the context of the above scenario that (Richmond 1961) Eq. (3.7) is

$$\oiint_{S_1} (\boldsymbol{E}_1 \times \boldsymbol{H}_2 - \boldsymbol{E}_2 \times \boldsymbol{H}_1) \cdot \hat{n}\, dS = \oiint_{S_2} (\boldsymbol{E}_2 \times \boldsymbol{H}_1 - \boldsymbol{E}_1 \times \boldsymbol{H}_2) \cdot \hat{n}\, dS \tag{8.16a}$$

with $\hat{n}$ the outward unit normal to the respective surfaces, and

$$\iiint_{V_1} (\boldsymbol{J}_1 \cdot \boldsymbol{E}_2 - \boldsymbol{M}_1 \cdot \boldsymbol{H}_2)\, dV = \iiint_{V_2} (\boldsymbol{J}_2 \cdot \boldsymbol{E}_1 - \boldsymbol{M}_2 \cdot \boldsymbol{H}_1)\, dV \tag{8.16b}$$

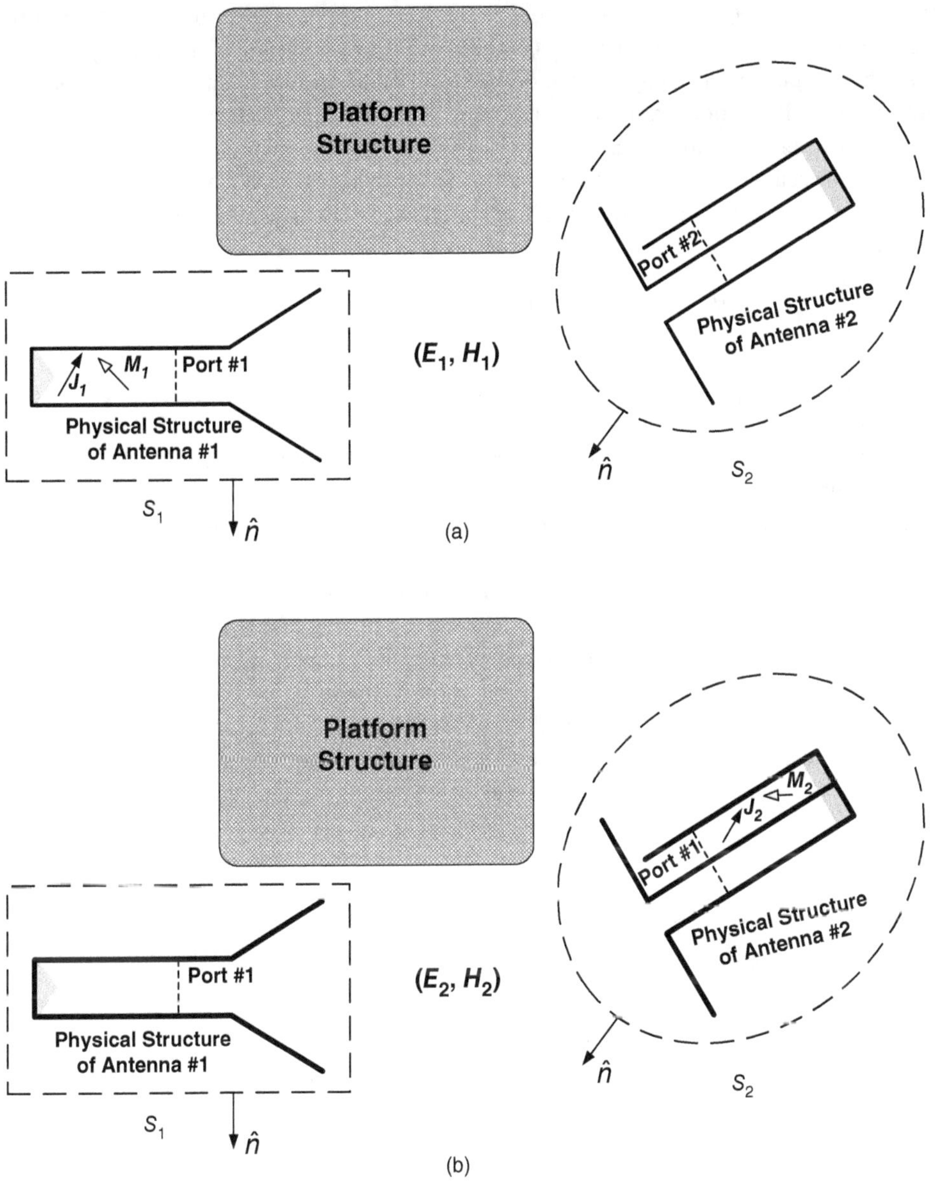

Figure 8.7 Antenna coupling in the presence of a platform. (a) Situation #a; (b) situation #b. Vector $\hat{n}$ is the outward unit normal.

These are "pure field" forms of the reciprocity theorem. Skillful selection of S_1 and S_2, and the fact that deforming them without changing the impressed sources enclosed, reveals (Collin and Zucker 1969) that Eq. (8.16a) implies

$$V_{1a}I_{1b} - V_{1b}I_{1a} = V_{2b}I_{2a} - V_{2a}I_{2b} \tag{8.17}$$

when the ports are located on single-mode transmission lines sections in the antenna model, but a similar result can be found from Eq. (8.16b) if the CEM model incorporates concentrated impressed sources directly at the port location (Harrington 1961; Kong 1986). Expression Eq. (8.17) is a "pure circuit" form of the reciprocity theorem. Textbook authors wisely warn their readers that the absence of complex conjugate operations in the above expressions at once tells us that they do not represent complex power. Indeed, each of the sides of Eqs. (8.16a, b) and (8.17) are referred to as field and circuit reactions, respectively (Rumsey 1954). For an introduction to the concept of reaction, please refer to Section 3.2.2. Rumsey (1963) states that the "definition of reaction makes (it) the same physical entity in field theory and in circuit theory. Hence the difficult problem of connecting the two theories is automatically accomplished." Thus, for instance, we can write (Richmond 1980)

$$V_{2b}I_{2a} - V_{2a}I_{2b} = \oiint_{S_2} (\boldsymbol{E}_2 \times \boldsymbol{H}_1 - \boldsymbol{E}_1 \times \boldsymbol{H}_2) \cdot \hat{n}\, dS \tag{8.18}$$

with a corresponding expression in terms of the voltage and current at port#1. Eq. (8.18) is often referred to as a reaction theorem.[4] The above statements are true as long as S_1 and S_2 enclose only their respective impressed sources. There is no restriction on which portions, if any, of the antenna or platform structures each surface includes within itself. With the usage we have in mind, it will pay to become more specific, and so we will consider S_1 and S_2 to be so chosen that they enclose whatever material structure we consider to be antenna#1 and antenna#2, respectively. Then, the requirement that S_1 and S_2 must each enclose only one of the impressed sources becomes synonymous with saying they must each enclose only one of the antennas. In spite of this decision, considerable flexibility remains as to the shapes selected. For example, Figure 8.8 shows two different options for S_2 that would all yield the same value of the reaction term. If antenna#2 is an aperture antenna, it might be judicious to allow part of S_2 to be coincident with the aperture in order to be able to easily model the antenna without needing detailed design information on it, more on this in Section 8.7. Alternatively, S_2 may be chosen to permit computationally efficient

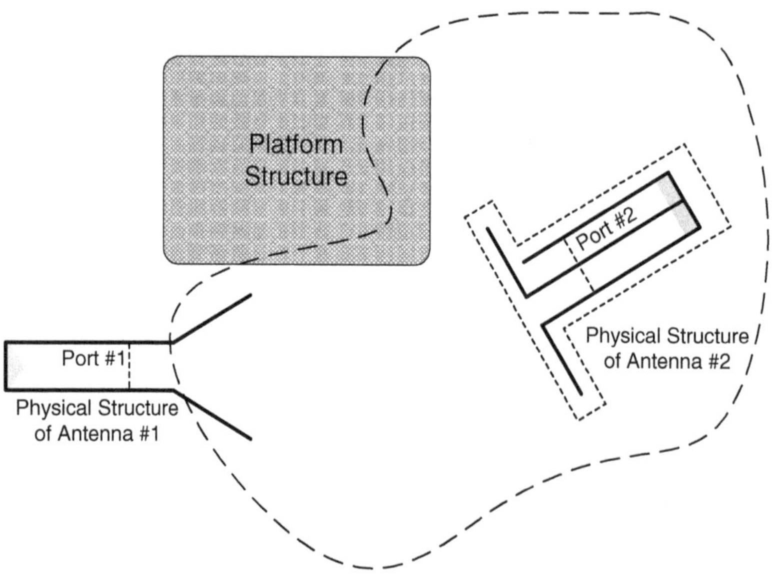

Figure 8.8 Alternative choices for the surface S_2 enclosing antenna#2.

4 This will be the form of interest in this chapter due to the fact that, without loss of generality, we consider antenna#1 the aggressor and antenna#2 the victim.

numerical evaluation of the reaction integral. Indeed, part of S_2 might even be made to coincide with a portion of the platform (e.g. a ground plane) if it is an essential component for the correct operation of antenna#2.

The power radiated by antenna#1 when it transmits is

$$P_{rad1} = \frac{1}{2}\left\{ \oiint_{S_1} \mathrm{Re}\{\boldsymbol{E}_1 \times \boldsymbol{H}_1^*\} \cdot \hat{n}\, dS \right\} \tag{8.19}$$

and similarly, for antenna#2

$$P_{rad2} = \frac{1}{2}\left\{ \oiint_{S_2} \mathrm{Re}\{\boldsymbol{E}_2 \times \boldsymbol{H}_2^*\} \cdot \hat{n}\, dS \right\}. \tag{8.20}$$

In general, $P_{t1} \geq P_{rad1}$, with ratio P_{rad1}/P_{t1} being the radiation efficiency of antenna#1.

8.4.3 Generalized Reaction Theorem

The reaction theorem has been introduced in Section 3.2.3. Suppose now that, unlike the scenario imagined above when enunciating the reciprocity theorem and its outcome, where two different impressed sources were successively considered located in the same original environment, we now allow the environment to change in situation #b. When antenna #1 transmits, the environment is as shown in Figure 8.3a, with fields $(\boldsymbol{E}_1, \boldsymbol{H}_1)$ everywhere, and hence also on S_2, and from which circuit quantities (V_{2a}, I_{2a}) at port#2 can in principle be determined. Suppose that instead of retaining this original environment as in Figure 8.3b, when antenna#2 transmits, it is altered to that as in Figure 8.9, but (and this is important) with that part of the environment within S_2 unchanged. We then have fields $(\boldsymbol{E}'_2, \boldsymbol{H}'_2)$ everywhere in the new environment into which antenna#2 transmits, which is different from the $(\boldsymbol{E}_2, \boldsymbol{H}_2)$ it would have been had the environment not been altered. As outlined in Section 3.2.2, the circuit quantities at port#2 when antenna#2 transmits into the new environment are denoted by (V'_{2b}, I'_{2b}). Richmond (1958, 1961) gave a generalized form of the reaction theorem under such conditions, namely

$$V'_{2b}I_{2a} - V_{2a}I'_{2b} = \oiint_{S_2} (\boldsymbol{E}'_2 \times \boldsymbol{H}_1 - \boldsymbol{E}_1 \times \boldsymbol{H}'_2) \cdot \hat{n}\, dS. \tag{8.21}$$

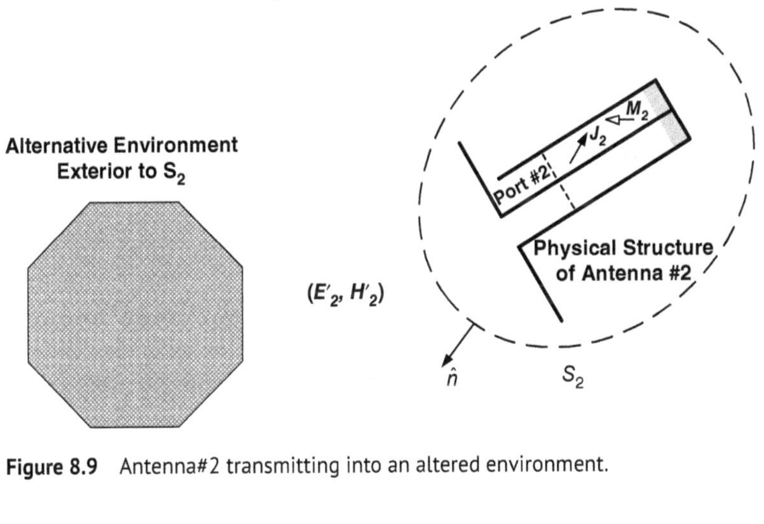

Figure 8.9 Antenna#2 transmitting into an altered environment.

The new environment (Figure 8.7) into which antenna#2 transmits can even be free space; in other words, one where the structure of antenna#1, plus that of the platform, are removed altogether, and nothing else inserted. This is convenient and is often exploited. It must be remembered that the input impedance Z_2^{in} of antenna#2 will not be the same in the altered environment as it is in the actual original one. Obviously, we could swap the roles of the two antennas, and consider the environment into which antenna#1 radiates to be the one that is altered and arrive at corresponding expressions. It should be recalled that the original unaltered environment is considered to be that in which we wish to be able to determine the inter-antenna coupling. The altered environment used in the generalized reaction theorem is merely a means to an end to perhaps more conveniently find various desired quantities, as should be clear from Section 8.4.4.

8.4.4 Expressions for Mutual Impedance and Open Circuit Voltage

To support the present chapter, and to have consistent notation some aspects of Section 3.2.3 is repeated here. Suppose the port of antenna#2 is open-circuited when antenna#1 transmits, then from Figure 8.4a, we have $V_2 = V_{2a}^{oc}$ the open-circuit voltage at port#2, $I_1 = I_{1a}$ is as usual the current into port#1, and $I_2 = I_{2a} = 0$ because port#2 is open-circuited. With $I_{2a} = 0$, the circuit reaction term on the left-hand side of Eq. (8.18) reads $-V_{2a}^{oc} I_{2b}$, and hence

$$V_{2a}^{oc} = -\frac{1}{I_{2b}} \oint\!\!\oint_{S_2} (\boldsymbol{E}_2 \times \boldsymbol{H}_1 - \boldsymbol{E}_1 \times \boldsymbol{H}_2) \cdot \hat{n} dS. \tag{8.22}$$

It in turn follows from Eq. (8.5) that the mutual impedance

$$Z_{21} = \left.\frac{V_2}{I_1}\right|_{I_2=0} = \frac{V_{2a}^{oc}}{I_{1a}} = -\frac{1}{I_{1a} I_{2b}} \oint\!\!\oint_{S_2} (\boldsymbol{E}_2 \times \boldsymbol{H}_1 - \boldsymbol{E}_1 \times \boldsymbol{H}_2) \cdot \hat{n} dS. \tag{8.23}$$

If antenna#2 is terminated in a short circuit when antenna#1 transmits, then $V_2 = V_{2a} = 0$, and $I_2 = I_{2a} = I_{2a}^{sc}$ is the short-circuit current at port#2. Consideration of Eqs. (8.7) and (8.8) results in expression

$$I_{2a}^{sc} = \frac{1}{V_{2b}} \oint\!\!\oint_{S_2} (\boldsymbol{E}_2 \times \boldsymbol{H}_1 - \boldsymbol{E}_1 \times \boldsymbol{H}_2) \cdot \hat{n} dS \tag{8.24}$$

for the short-circuit current at port#2, and the mutual admittance

$$Y_{21} = \left.\frac{I_2}{V_1}\right|_{V_2=0} = \frac{1}{V_{1a} V_{2b}} \oint\!\!\oint_{S_2} (\boldsymbol{E}_2 \times \boldsymbol{H}_1 - \boldsymbol{E}_1 \times \boldsymbol{H}_2) \cdot \hat{n} dS. \tag{8.25}$$

Recall that the fields are those when one antenna transmits and the other is terminated. Fields $(\boldsymbol{E}_1, \boldsymbol{H}_1)$ are those when antenna#1 transmits, with port current I_{2a}, both antenna structures and the platform present (the original environment), and antenna#2 is terminated in an open circuit $Z_{L2} \to \infty$. Fields $(\boldsymbol{E}_2, \boldsymbol{H}_2)$ are those when antenna#2 transmits into the same environment. If we decide to apply the generalized reaction theorem, we can have antenna#2 transmit into any environment we choose. By the substitutions $\boldsymbol{E}_2 \to \boldsymbol{E}_2'$, $\boldsymbol{H}_2 \to \boldsymbol{H}_2'$, $I_{2b} \to I_{2b}'$, and $V_{2b} \to V_{2b}'$ in Eqs. (8.22)–(8.25), the latter expressions will provide the values of the said quantities as they would be in the original unaltered environment even though the fields of antenna#2 were found when it transmits into the altered environment.

8.4.5 Power Coupling

Under the weak coupling assumption (which is valid in many co-sited antenna coupling computations), and the assumption of matched conditions at each antenna, in the sense that $Z_{g1} = Z_{L1} = Z_{01}$ and $Z_{g2} = Z_{L2} = Z_{02}$, Eq. (8.14) becomes

$$P_{r2} = \frac{|V_{2a}^{oc}|^2}{8\, Z_{02}}. \tag{8.26}$$

Also, Eq. (8.10) reads

$$P_{t1} = \frac{1}{2}\mathrm{Re}\{V_{1a}I_{1a}^*\} = \frac{1}{2}|I_{1a}|^2 Z_{01} \tag{8.27}$$

and similarly

$$P_{t2} = \frac{1}{2}\mathrm{Re}\{V_{2b}I_{2b}^*\} = \frac{1}{2}|I_{1b}|^2 Z_{02}. \tag{8.28}$$

If we combine Eqs. (8.26)–(8.28), we obtain

$$\frac{P_{r2}}{P_{t1}} = \frac{\left|\oint\!\!\oint_{S_2} (\boldsymbol{E}_2 \times \boldsymbol{H}_1 - \boldsymbol{E}_1 \times \boldsymbol{H}_2)\cdot \hat{n}\, dS\right|^2}{16\, P_{t1}P_{t2}}. \tag{8.29}$$

Surface S_2 must be that which encloses only the impressed source, and what we consider to be the structure of antenna#2, in order to evaluate P_{rad2} in Eq. (8.20). But we noted in Section 8.4.2 that, as far as the reaction integral in the numerator of Eq. (8.29) is concerned, there is considerable flexibility in the selection of S_2. It must of course include the impressed source of antenna#2 only, and no other impressed sources. It can include any induced sources. Thus, we can replace the surface symbol by S_0, recognize it as any closed surface that separates the two antennas (sometimes called the interaction surface), and write

$$\frac{P_{r2}}{P_{t1}} = \frac{\left|\oint\!\!\oint_{S_0} (\boldsymbol{E}_2 \times \boldsymbol{H}_1 - \boldsymbol{E}_1 \times \boldsymbol{H}_2)\cdot \hat{n}\, dS\right|^2}{16\, P_{t1}P_{t2}}. \tag{8.30}$$

If the antennas are both lossless, we can replace Eqs. (8.27) and (8.28) by Eqs. (8.19) and (8.20), respectively, and then Eq. (8.30) can be written as

$$\frac{P_{r2}}{P_{t1}} = \frac{\frac{1}{4}\left|\oint\!\!\oint_{S_0} (\boldsymbol{E}_2 \times \boldsymbol{H}_1 - \boldsymbol{E}_1 \times \boldsymbol{H}_2)\cdot \hat{n}\, dS\right|^2}{\left\{\oint\!\!\oint_{S_1} \mathrm{Re}\{\boldsymbol{E}_1 \times \boldsymbol{H}_1^*\}\cdot \hat{n}\, dS\right\}\left\{\oint\!\!\oint_{S_2} \mathrm{Re}\{\boldsymbol{E}_2 \times \boldsymbol{H}_2^*\}\cdot \hat{n}\, dS\right\}}. \tag{8.31}$$

Observe that the power coupling between the two antennas can be found solely in terms of the transmitted fields of each antenna, albeit in the presence of each other and that of the platform. No port voltages or currents are required. This result is sometimes referred to as Robieux's theorem (Bird 2016). It can be written in different forms. For instance, because $\boldsymbol{E}_1 \times \boldsymbol{H}_2 = -\boldsymbol{H}_2 \times \boldsymbol{E}_1$, Eq. (8.31) can be expressed as (Hu 1958; Maclean 1986)

$$\frac{P_{r2}}{P_{t1}} = \frac{\frac{1}{4}\left|\oint\!\!\oint_{S_0} (\boldsymbol{E}_2 \times \boldsymbol{H}_1 + \boldsymbol{H}_2 \times \boldsymbol{E}_1)\cdot \hat{n}\, dS\right|^2}{\left\{\oint\!\!\oint_{S_1} \mathrm{Re}\{\boldsymbol{E}_1 \times \boldsymbol{H}_1^*\}\cdot \hat{n}\, dS\right\}\left\{\oint\!\!\oint_{S_2} \mathrm{Re}\{\boldsymbol{E}_2 \times \boldsymbol{H}_2^*\}\cdot \hat{n}\, dS\right\}} \tag{8.32}$$

Some other useful forms of the power coupling theorem are provided in Bird (2016).

8.5 Supplementary Comments on CEM Modeling Methods

The three-fundamental full-wave CEM techniques were named in Section 8.1.1. The full-wave CEM viewpoint for modeling an isolated antenna was outlined in Section 8.2. The same was done in Section 8.3 for a coupling environment (the "original environment" we called it) consisting of two antenna structures in the presence of a platform. If some full-wave CEM model can supply the fields $(\boldsymbol{E}_1, \boldsymbol{H}_1)$ and $(\boldsymbol{E}_2, \boldsymbol{H}_2)$ at the antenna ports, and over the surfaces defined in Section 8.4 in relation to the reaction integrals, for this original environment, we will have everything we need to evaluate coupling accurately. It was hinted that such full-wave CEM modeling of the complete original environment as a single electromagnetic system is very often not possible.

To move forward, CEM developers have supplemented the fundamental full-wave CEM techniques' capabilities through the incorporation of improved algorithms and implementations of some of the internal operations of the various methods, or through their hybridization. A hybrid method is a linkage of two different computational methods; we use the word "linkage" instead of "coupling" because the latter term has a reserved meaning in the present book. If two different full-wave schemes are combined in some way, we could call it a full-wave/full-wave hybrid method (e.g. MM/FEM). If a full-wave method is combined with some asymptotic method (to be defined below), we can call it a full-wave/asymptotic hybrid method and explicitly name the specific methods used in doing so (e.g. MM/PO). There may be more than one way to combine any two methods. We wish to grasp what these enhanced CEM methods are, as they are indispensable in on-platform coupling models. In order to facilitate this, we provide some supplementary comments on CEM methods. These remarks are not meant to replace the more detailed discussions in the references that will be cited. A light reading is intended. We hope the reader will allow us some license. There is not complete unanimity in the CEM literature on terminology; that used here is done without any intentional dogmatism.

First, consider use of the MoM (Volakis and Sertel 2012) for the full-wave CEM models in Sections 8.2 and 8.3. As described in Chapter 3, the MoM formulation is based on some integral equation model of the electromagnetics problem, the unknowns of this equation being a set of equivalent currents (in general, both electric and magnetic) on all the material structures present. This is enabled by the application of the surface equivalence principle (Harrington 1961). The MoM procedure discretizes the integral equation into a matrix equation of the form $[A][X] = [B]$. The unknown column vector $[X]$ contains the coefficients of a set of expansion functions in terms of which the unknown currents are represented. In practice, the surfaces of the material structures are meshed into geometrical elements (e.g. the triangles in Figure 8.10) that define the geometry, and in terms of which expansion functions can be defined. The larger the surface area (in square wavelengths) of the physical structures being meshed, the more mesh elements, and hence unknowns, are required. For situation #a in Section 8.3, the column vector $[B]$ is known from $(\boldsymbol{E}_1^{inc}, \boldsymbol{H}_1^{inc})$. The equivalent currents generate the scattered fields $(\boldsymbol{E}_1^{scat}, \boldsymbol{H}_1^{scat})$. The total fields $(\boldsymbol{E}_1, \boldsymbol{H}_1)$ are the sum of these and the incident fields. Related comments apply to situation #b through a simple change in subscripts. The FEM (Jin 2014), FDTD method (Gedney 2011), and the FIT (Weiland 2003) discretize differential equation models of the electromagnetics problem, the unknowns being the fields $(\boldsymbol{E}_1, \boldsymbol{H}_1)$ or $(\boldsymbol{E}_2, \boldsymbol{H}_2)$ directly in some region of space surrounding the material structures (the scatterers) rather than equivalent current densities from which such fields can be found, and this complete region must be meshed.

All these methods end up in principle having to solve a matrix equation, even if not done explicitly. In full-wave CEM methods, "solving the problem" refers to the solution of this matrix equation. CEM developers are always concerned with the following qualities: numerical accuracy,

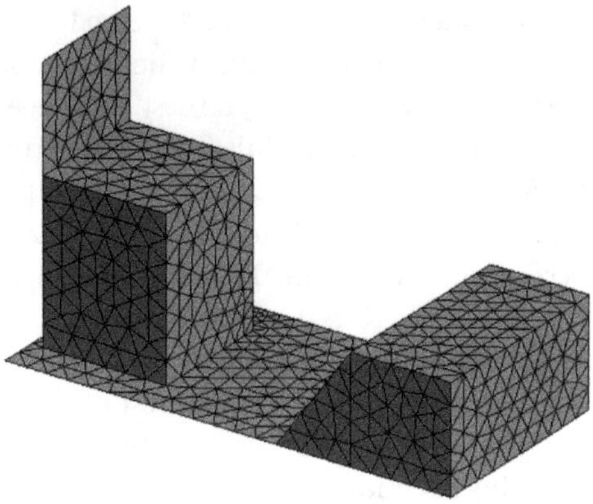

Figure 8.10 Meshed material structure (scattering object). A relatively course mesh is shown for clarity.

convergence rates, scalability, parallelization ability, computation time, and memory consumption. Large matrix equations require use of iterative solvers, and it is the rate at which this converges to an accurate solution that is of concern. If the number of unknowns, and hence $[X]$, is extremely large, the matrix $[A]$ becomes too large in the sense that solution of the matrix equation is not as reliable as required, or takes too much computation time, or requires excessive memory. Some techniques for dealing with this problem are outlined in Section 3.6. There is little sense in thinking one must simply obtain increased computational resources. Electromagnetic engineering, of which on-platform coupling is but one aspect, is insatiable. As more becomes possible, more will be demanded. Ever since the fundamental full-wave CEM methods became established as a technology, it is not surprising that many ingenious ways have been devised to enhance the performance of this technology.

The incorporation of multilevel fast multipole method algorithms (MLFMA) (Song et al. 1997) into MoM implementations over the last 20 years greatly increased the electrical size of a structure that can be handled by the MoM as a single problem. The MLFMA is referred to as an accelerator of the MM, dealing with the manner the matrix $[A]$ is determined and used, and is already incorporated in most commercial CEM codes offering an MoM capability. It is also referred to as the multilevel fast multipole method (MLFMM).

The FEM method can only be applied over a finite region of space and so requires some way to truncate this region (Jin 2014). One rigorous way to do this is to adopt a hybrid FEM/MoM approach, with the FEM applied within some closed surface, and the MoM used to deal with the free space region outside the surface, by placing equivalent electric and magnetic current densities on the truncation boundary surface (Volakis et al. 2006). The internal fields and the surface current densities are found in an electromagnetically consistent manner without approximation. There can be more than one explicit FEM region, as long as each is truncated as indicated above. It is most often called the finite element – boundary integral (FE-BI) method, the term "boundary integral method" being used in many other disciplines for what is most often called the MoM in electromagnetics. This facilitates the accurate full-wave solution of problems involving detailed geometrical and material make-up but which are open to free space.

A more general class of approaches to extend the size and complexity of problem that full-wave CEM methods are able to handle is that of domain decomposition methods (DDMs). The DDM divides the original problem (the domain) into several separate smaller subdomains and is often referred to as a divide and conquer approach by CEM developers. Although the subdomains are rigorously linked, the problem solvers on the subdomains can be done in parallel. A different fundamental full-wave CEM method can be used in each of the subdomains and then would be classified as a full-wave/full-wave hybrid method. The FE-BI method could be considered an example of a DDM where one of the subdomains does not contain any scatterers. DDM frameworks may differ in the type of problem they do best, the range of problems to which they may be applied, and one or more of algorithmic qualities identified earlier in this section. A fairly good overview can be gained through a study of Barka and Caudrillier (2007), Jin and Riley (2009), Wang et al. (2012), Barka (2013), Lu et al. (2019), and Wan et al. (2020), amongst others. It has been said that "as computers become more powerful, the efficiency of algorithms becomes more important, not less" (Trefethen 2011).

We next turn our attention to asymptotic methods. We will think in terms of perfectly conducting (PEC) objects, here and in the rest of this Section 8.5, although penetrable objects can be similarly dealt with. From the mid-1950s onward, asymptotic methods (Kouyoumjian 1965) started receiving attention, in parallel with full-wave CEM methods, in electromagnetic engineering. Both developments were spawned by increasingly advanced applications that necessitated the solution of electromagnetic scattering problems for which analytical solutions cannot found. In the present context, an asymptotic solution to an electromagnetics problem is one that, for a given scattering object size, becomes increasingly accurate as the frequency becomes higher. In other words, the asymptotic modeling approach becomes more accurate as the scattering structures becomes electrically larger, which are also the conditions under which full-wave CEM methods become computationally more burdensome. Hybrid full-wave/asymptotic CEM methods attempt to combine the strong points of each. A simple example applying an asymptotic solution to coupling between antennas on a curved platform was given in Section 7.6.2.

The fields of any finite-sized source becomes almost locally plane (electric and magnetic fields perpendicular to each other and straightforwardly related by the intrinsic impedance of the medium) at observation points sufficiently distant from the source. The smaller the electrical extent of the source, the shorter the distance at which this becomes true. We can, of course, divide any source distribution into smaller parts, with each part's fields looking locally plane relatively close to the source. A full electromagnetic response may be found through superposition. Mathematical expressions for a locally plane field take the form of a reference value multiplied by a so-called spreading factor that accounts for the amplitude decrease and a factor that accounts for the phase change, as the field propagates. An additional term deals with the polarization. We can also impose a direction-dependent pattern factor. Such fields can be interpreted (Deschamps 1972) in a geometrical optics (GO) sense. The field propagation in all directions can be viewed in terms of rays, which fits well with our intuition, with rays undergoing reflection from objects. When using such ray-optic methods to analyze electromagnetic scattering, the field incident from some source is not interpreted in the same sense as stated for a full-wave analysis in Section 8.2. Shadowing of rays by scattering objects occurs. This is what is meant when the textbooks tell us to interpret ray-fields "in a GO sense."

We next reconsider a PEC object and an impressed source, as shown in Figure 8.11a. It may seem an odd choice of object shape but has been done to allow us to think pictorially, as will be seen below. Suppose the intent is to use the MoM to find the scattered, and hence the total, fields everywhere. The surface equivalence principle allows us rigorously to replace this original problem by the equivalent one shown in Figure 8.11b. The equivalent currents exist in free space; invoking them removes the physical object without approximation. An integral equation obtained by imposing the physical condition of zero tangential electric field on surface S_{pec} will have the equivalent surface current density $\boldsymbol{J}_s$ as the unknown quantity to be solved for. As previously stated, the MoM discretizes this integral equation into a matrix equation, and we are able to determine $\boldsymbol{J}_s$, which generates the scattered fields ($\boldsymbol{E}^{scat}$, $\boldsymbol{H}^{scat}$) everywhere. Once we know $\boldsymbol{J}_s$, finding which is the difficult

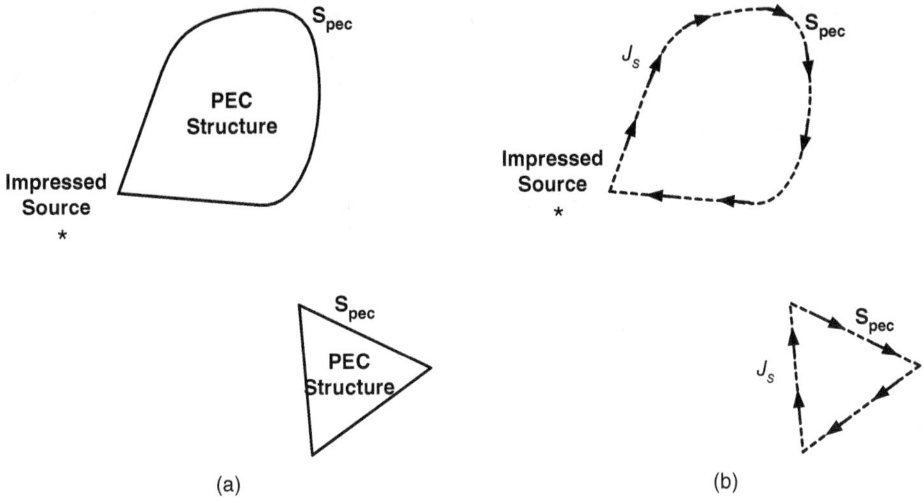

Figure 8.11 (a) Original physical problem; (b) equivalent problem supplied by application of the surface equivalence principle.

part of the task; these scattered fields can be relatively easily found by employing a radiation integral, over S_{pec}, of the known $\boldsymbol{J}_s$. This would be a full-wave CEM approach. We would find that $\boldsymbol{J}_s$ can be nonzero everywhere over S_{pec}, and not only over that portion in line of sight from the location of the impressed source.

Continuing, we return to the physical problem of Figure 8.11a, and consider the sketch in Figure 8.12. If we think of the physical problem strictly in a GO sense, we would say that the front portion of S_{pec}, namely that portion traversed going counterclockwise from Q_1 to Q_2 and from Q_3 to Q_4, is illuminated by the impressed source, but not the rest of it. Exactly which part is illuminated depends on the location of the impressed source with respect to S_{pec}. If we continue to think in a strictly GO sense, we would say that as a result there is a nonzero induced $\boldsymbol{J}_s$ only over the illuminated portion of S_{pec} and none over the remainder. Finally, we would say that the fields are zero in the shadow region of space between the shadow boundaries indicated, in other words behind the PEC object.

The physical optics (PO) procedure now makes the assumption that at each point Q on the finite-sized PEC structure in Figure 8.12, the current density is approximated by the expression

$$\boldsymbol{J}_s^{PO}(Q) = 2\,\hat{n}(Q) \times \boldsymbol{H}^{inc}(Q) \tag{8.33}$$

over the illuminated portion of the object and is assumed zero over the shadowed portion. This current is then used in a radiation integral to find the fields it generates. We will for convenience of later explanation refer to this as the PO current approximation (Smith 1997). One finds that the electrically larger the scatterer is, the more closely the PO current approximates the exact $\boldsymbol{J}_s$. Expressed differently, for a scatterer of a given physical size, this approximation becomes increasingly accurate as the frequency increases. It is an asymptotic method. The form Eq. (8.33) is motivated by the fact that it is the exact analytical solution found when the object in question is an infinite planar conducting ground plane. Observe that there is no matrix equation that has to be solved in the approximation Eq. (8.33). Even for an electrically large object, the PO current Eq. (8.33) near edges on the object (where the object's surface is not smooth) differ appreciably from the exact $\boldsymbol{J}_s$. The accuracy of the approximate current can be improved by adding additional terms to $\boldsymbol{J}_s^{PO}$, using an asymptotic procedure known as the physical theory of diffraction (PTD). The correction terms are

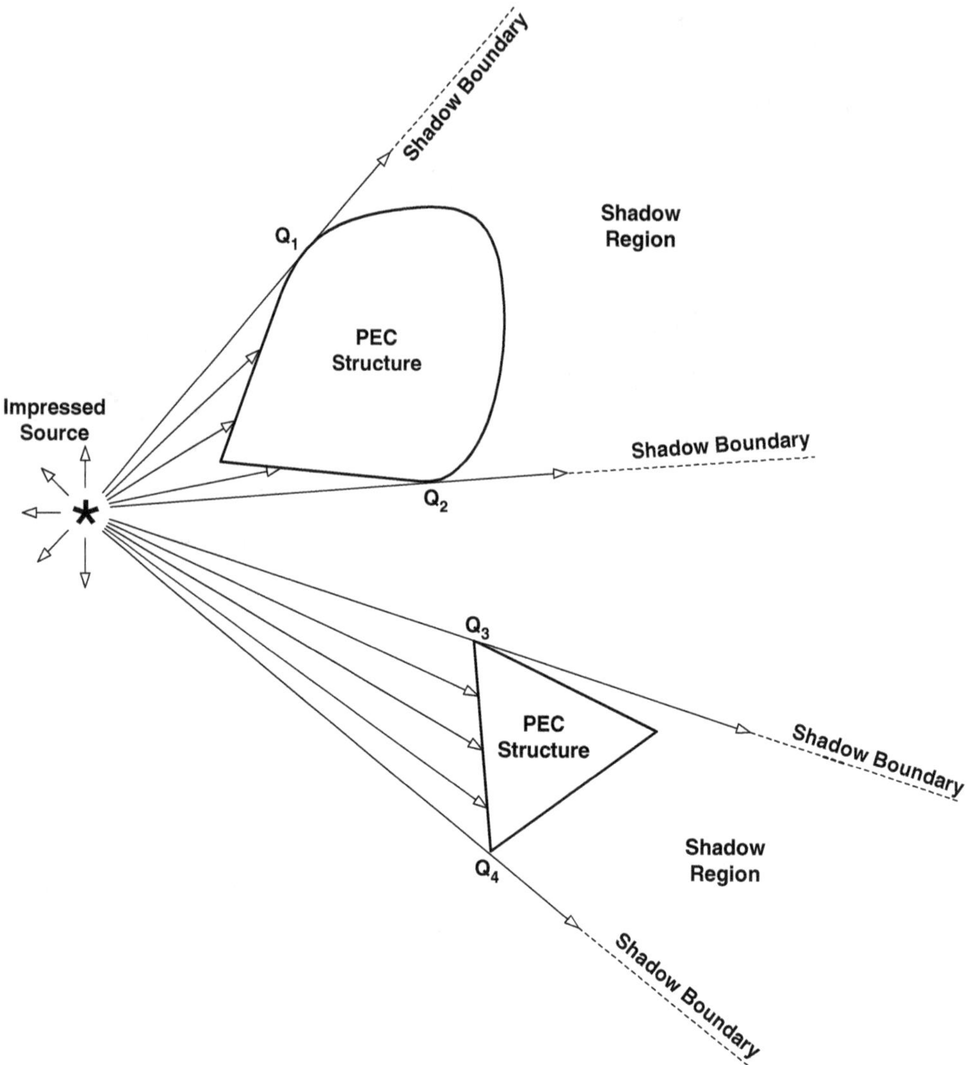

Figure 8.12 Illuminated and shadowed sides of a PEC scatterer when viewed in a GO sense.

inferred from analytically obtained asymptotic solutions of certain geometrically simple, albeit not mathematically simple, problems (Paknys 2016, Chapter 9).

Let us now conduct an imagined experiment. If the above-mentioned exact $\boldsymbol{J}_s$, obtained through use of the MM, were placed in a radiation integral to find the field it generates, then we would discover that, for an electrically large scatterer, and for a given observation point, the total field appears to comprise contributions that appear to be emanate from specific locations on the scatterer. As illustrated in Figure 8.13, these contributions could be classified in terms of whether they emanate from a point on a smooth part of the surface or from an edge, or arrive after appearing to have traveled ("crept") around the surface and then launched toward the observation point. The total field at an observation point could be written as the sum of these direct, reflected, edge-diffracted, or smooth-surface diffraction terms. We would be able to eliminate the obvious error of the classical GO (namely the prediction of identically zero fields in the shadow region) by including these extra

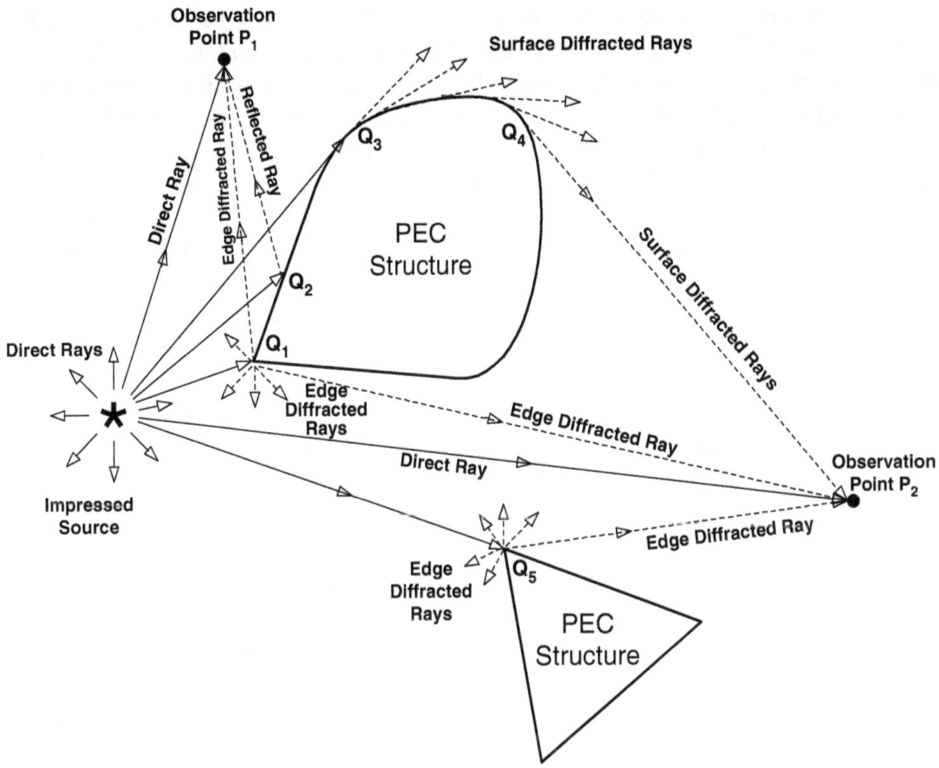

Figure 8.13 GTD view of scattering.

rays, the whole procedure being called the geometrical theory of diffraction (GTD). We would find that the results so obtained improve the larger the electrical size of the object, or, in other words, the higher the frequency for a given sized object. It is an asymptotic approach. As described, the above efforts would not be successful in developing the GTD (or its refinement the uniform GTD, or UTD) as we now know it, as there is much more ingenuity behind the latter. But it encapsulates the idea for initial pedagogical purposes. Indeed, reference to an early analysis in Rusch (1974) greatly assists in grasping what the GTD is about. More complete introductory treatments of the UTD, which encompasses GO, can be found in Paknys (2016) and McNamara et al. (1990), with the references therein pointing to the more advanced seminal works such as Kouyoumjian and Pathak (1974), Pathak et al. (1980, 1981), and Pathak and Wang (1981).

The actual field value (amplitude and phase) and polarization of a reflected ray at some observation point is written in terms of the field impinging on the object at the point of reflection (e.g. Q_2 in Figure 8.13), a reflection coefficient, and the amplitude and phase terms mentioned earlier, dependent on the distance from the reflection point. Like expressions provide the diffracted fields at the observation point, but with diffraction coefficients replacing reflection coefficients, and so on. Such computations are extremely fast. Searching for the location of relevant reflection points and diffraction points (e.g. Q_1 and Q_5 in Figure 8.13) for a given source location and observation point is what uses most of the computer time when UTD methods are implemented in codes that allow analysis of objects of general shape. As with full-wave CEM methods, much work has been done on algorithms that perform such searching, and many commercial codes include a UTD capability. The UTD has been used to model some on-platform coupling problems without the need for hybridization with some other method, but hybridization provides greater generality and flexibility, for example, Theron et al. (2000), Burkholder et al. (2006), and Tzoulis and Eibert(2005).

In the on-platform coupling context, hybrid full-wave/asymptotic methods use some full-wave method over a part of the platform, usually that closest to the antenna(s), and asymptotic methods for determining scattering from, and hence the influence of, the remaining portions. The seminal paper was Ekelman and Thiele (1980). More recent work can be found in Jakobus and Landstorfer (1995), Jakobus (2010), Liu et al. (2014), and Le Lepvrier et al. (2014), as well as many others.

In order to firm up ideas about hybrid full-wave/asymptotic methods, we consider one particular method a little more carefully, namely the MM/PO method. We suppose we have a platform structure, one part (for convenience, call it the full-wave part) of which is meshed for an MoM analysis, and another part (for now, call it the asymptotic part) that is not directly included in the MoM analysis, because it might, for example, make the computational problem intractable. When the MoM is applied to the full-wave part, a typical term A_{mn} of the matrix $[A]$ is obtained (Davidson 2011) by finding the fields $(\boldsymbol{E}_n, \boldsymbol{H}_n)$ due to the n-th expansion function, dot-multiplying it by the m-th expansion function $\boldsymbol{J}_m$ (which will represent an electric current density), and integrating this product over the region D_m occupied by the m-th expansion function. We know from Section 3.4.3 in Chapter 3 that the resulting quantity A_{mn} can in fact be interpreted as the (generalized) mutual impedance or admittance, depending on the integral equation underpinning the particular MoM formulation, between the m-th and n-th expansion functions being used in the MoM. In a hybrid MoM/PO method, we might do the following in finding A_{mn}: We find the fields $(\boldsymbol{E}_n, \boldsymbol{H}_n)$ over D_m as if the asymptotic part of the platform were absent. We also find $(\boldsymbol{E}_n, \boldsymbol{H}_n)$ over the asymptotic part of the platform and find the PO currents there due to them using Eq. (8.33). We then determine the fields, over D_m, due to these PO currents, and add it to the $(\boldsymbol{E}_n, \boldsymbol{H}_n)$ values found previously by ignoring the asymptotic part of the platform. It is this superimposed field that is now dot-multiplied by $\boldsymbol{J}_m$ and integrated over D_m. The term A_{mn} now accounts for the presence of the asymptotic part of the platform, albeit not in a rigorous, but very often satisfyingly accurate, way. The resulting total fields would be the sum of the (scattered) fields due to the resultant MoM currents and the PO currents due to these MoM currents, and the incident field due to the impressed source driving the whole problem. This is the principle. Much detail has been omitted but additional explanation can be found in Jakobus and Landstorfer (1995). The detailed implementation would need to ensure a physically correct interface between the currents on the full-wave part and the asymptotic part of the problem. There are also the usual practical issues of how to implement this hybrid method in a general way so that those parts of the asymptotic portion of the platform illuminated by the n-th expansion function, and those parts that are in shadow, can be automatically and efficiently determined (Xiang and Botha 2016, 2018). The standard PO current need not be used in the MoM/PO implementation, but instead PO currents enhanced using the PTD, or UTD (as will be discussed in another context in the next paragraph). A hybrid method that we could call a MoM/UTD one would use the UTD-determined field reflection and diffraction additions due to $(\boldsymbol{E}_n, \boldsymbol{H}_n)$ impinging on the asymptotic part of the platform. Hybrid methods such as these are sometimes also referred to as DDMs. This is acceptable as long as we appreciate that the linking of the subdomains is not a rigorous one, even if sufficiently accurate for some applications.

The basic (as opposed to hybridized) shooting and bouncing rays (SBRs) asymptotic method takes a slightly different avenue. Many rays are launched from the source (or each part into which a source has been divided, followed by later use of superposition) and used to "paint" an illuminated surface to find the PO currents. The fields due to these PO currents are then in essence converted to ray-optic format, and themselves, as reflected or diffracted rays, allowed to "paint" those parts of the surface lit by each of the many rays. Thus, the SBR extends the standard PO approach by finding an improved approximation $\boldsymbol{J}_s^{SBR}$ to the exact $\boldsymbol{J}_s$, no longer with physically false "dark" shadowed portions. It is this $\boldsymbol{J}_s^{SBR}$ that is used in radiation integrals to find the fields of the antennas in the presence of the platform in order to perform coupling computations. Once again, while the final step in the computation might be rapid, there is much "book-keeping" needed in tracking the various rays. Details can be found in Ling et al. (1989), Kipp (2010), Kipp and Capoglu (2014), Kipp et al. (2015), Weinmann (2009), and Brem and Eibert(2015), and references therein. There are variations in the manner SBR is implemented (Aguilar et al. 2015), and SBR has appeared in hybridized form (Jin et al. 1998).

A review of many hybrid method ideas can be found in Medgyesi-Mitschang and Wang(1991). This would serve as a valuable starting point for the reader interested in the mathematical detail. However, much has been added by CEM researchers, both theoretically and computationally, as well as in the implementation algorithms, over the 30-year period since that publication. Thus, the more recent publications cited in this chapter should be consulted for a complete picture. Insight into the way hybrid frameworks are actually implemented in commercial codes can be obtained from Jakobus (2010).

8.6 Full-Wave CEM Modeling of Coupled Antennas on a Platform – The Ideal

At the risk of repetition, we recall that if exact solutions are forthcoming from some full-wave CEM model of the complete original environment as a single electromagnetic system, then secondary parameters related to coupling could be simply computed from the field solutions and problem needs no further discussion. Such full-wave modeling is possible using general-purpose (commercial) full-wave CEM codes if the antennas in question are relatively simple (e.g. dipoles, slots, or monopoles) and the platform not electrically too large. It may be that the coupling between two relatively closely spaced antennas is most highly influenced by the surrounding structure in their immediate vicinity, and not as much by the more distant parts of the platform. It may then be possible to include only these portions of the platform in the model, and so be able to use a single full-wave CEM model for the entire system, even though this is not feasible were the complete platform to be so included.

If one or both of the antennas whose coupling levels are desired is not simple, the full-wave antenna models will be geometrically and materially complicated. If, in addition, the platform is electrically large, incorporating the antennas and the platform into a single "standard" full-wave CEM model is usually not possible. Customized CEM models (implemented in special purpose rather than general codes) based on DDMs would be needed, and much effort would have to be put into developing the full-wave antenna models themselves if these cannot be transferred from those used in the design of the antennas themselves. Informative examples of such approaches can be found in Wang et al. (2012) and Jin and Riley (2009). What is appealing with domain decomposition is that the most appropriate full-wave CEM method (e.g. FEM) might be used for the antenna model, and another (e.g. MM) for the platform, and then rigorously linked.

Few would doubt that the capabilities of general-purpose codes will continue to be upgraded. However, at the time of writing, the single full-wave CEM model route is most often not possible, due to large memory requirements and long computation times. Naturally, engineers do not simply want to grapple with a coupling problem but want to solve it in order to reach decisions on the appropriateness of a particular location for an antenna on the platform. This, therefore, means that, in practice, approximations need to be made in order to handle realistic problems

(Vukovic et al. 2019). This is the subject of Section 8.8. In order to facilitate the discussion in that section, we first talk about reduced complexity antenna models in Section 8.7. As recognized in Section 8.1.1, the person responsible for the antenna layout on some platform might, for many reasons, not be privy to the details of some antenna design (Alvarez 2014). In such cases, the approaches to be described in Section 8.7 may then be the only way forward.

8.7 Reduced Complexity Antenna Electromagnetic Models

8.7.1 Necessity for Simplified Antenna Models

Reasons why a full-wave CEM model for the complete original environment might not be possible were provided in Section 8.6. In such cases, approximations need to be made, either through simplifications of the individual antenna models or the handling of the platform model. Some ideas on what form the latter take were discussed in Section 8.5. Possible simplified models for an antenna are described here. As will be clarified in Section 8.8, these simplified models must be able to compute the fields of the antenna, when isolated, at any observation point requested in its near- and far-zone. These constitute ($\boldsymbol{E}^{inc}$, $\boldsymbol{H}^{inc}$) when the antenna is placed on the platform and transmits. The simplified model must be chosen and used with care (Mahlström et al. 2018), and calls for numerical experimentation on the changes to the coupling levels computed with slight changes to the details of the simplified model, as well as engineering judgment.

8.7.2 Huygens' Box Model

We suppose that we have access to a detailed full-wave CEM model of an isolated antenna but that it is not feasible (for reasons given in Section 8.6) to incorporate it into an overall full-wave CEM model of the original coupling environment (two antennas plus the platform). We need to replace the antenna by a reduced complexity model. One approach is a direct application of Love's form of the surface equivalence principle (Harrington 1961). The closed surface[5] S_H in Figure 8.14a contains the impressed sources plus the antenna structure; it is assumed to be interior to the surfaces S_1 and S_2 to be used for the coupling evaluation. Surface equivalence says that if electric surface current density $\boldsymbol{J}_s = \hat{n} \times \boldsymbol{H}$ and magnetic surface current density $\boldsymbol{M}_s = -\hat{n} \times \boldsymbol{E}$ are placed on S_H, with $(\boldsymbol{E}, \boldsymbol{H})$ the actual fields on S_H as obtained from the isolated antenna model, and all impressed sources and materials within S_H are removed, then these surface current densities will generate the correct original fields of the isolated antenna at any point external to S_H. This is depicted in 8.14b. The fields $(\boldsymbol{E}, \boldsymbol{H})$ on S_H can be associated via the CEM model with circuit quantities (V_{1a}, I_{1a}) or (V_{2b}, I_{2b}), depending on whether antenna#1 or #2 is being modeled. The surface current densities $\boldsymbol{J}_s$ and $\boldsymbol{M}_s$ are now viewed as impressed sources that provide the fields of the antenna that would be incident on the platform and the other antenna's physical structure. Source/field integrals (usually involving the free space Green's function) must be numerically evaluated to determine the antenna's fields at any observation point. The important point is that such a reduced complexity model allows these fields to be found at both near-field (on the platform and second antenna structure) and far-field observation points. Note also that the physical structure of the antenna that

5 The subscript "H" denotes the fact that such a surface is sometimes referred to as a "Huygens' box." Christiaan Huygens (1629–1695) was a Dutch physicist who expounded the idea that conditions on a given wavefront determine the subsequent wave behavior. The surface equivalence principle is often considered a mathematical formulation of this idea (Smith 1997).

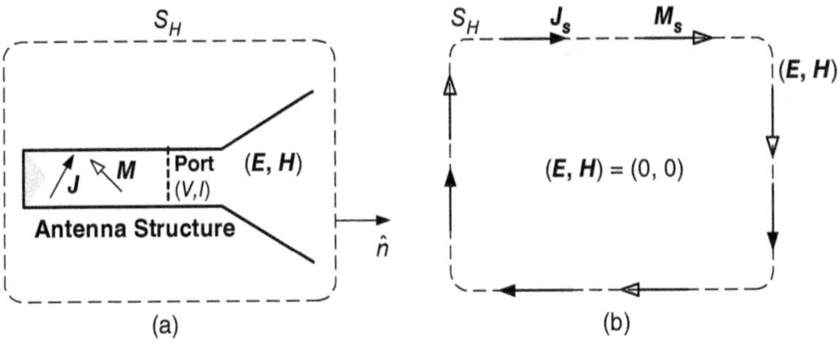

Figure 8.14 Construction of a Huygens' box reduced complexity model for an antenna. (a) The antenna with sources; (b) the model.

has undergone reduced complexity modeling has been removed. Thus, if this were to be done for antenna#1 for instance, adoption of such a model for the antenna is implicitly assuming that the platform and physical structure of antenna#2 have no influence on the port properties of antenna#1 and that physical structure of antenna#1 does not influence the coupling (it has no physical structure on/in which induced currents can occur) apart from its obvious influence on the fields of the isolated antenna.

As noted in Kragalott et al. (2011), although the Huygens' box is in general six-sided (unless it is not rectangular), it could be a five-sided one if the antenna is mounted in front of a large conducting backplane, or simply one-sided (a "planar window") for antennas mounted inside a cavity backing an aperture in such a large conducting plane.

In terms of the two-antenna coupling problem that is foremost in our minds, we observe that if interaction surface S_0 is selected to be S_2, with the latter also the Huygens' surface of the antenna model, then since $\boldsymbol{J}_{s2} = \hat{n} \times \boldsymbol{H}_2$ and magnetic surface current density $\boldsymbol{M}_{s2} = -\hat{n} \times \boldsymbol{E}_2$, then because from a vector identity we have$(\boldsymbol{E}_2 \times \boldsymbol{H}_1) \cdot \hat{n} = \boldsymbol{H}_1 \cdot (\hat{n} \times \boldsymbol{E}_2) = -\boldsymbol{H}_1 \cdot \boldsymbol{M}_{s2}$ and $(\boldsymbol{H}_2 \times \boldsymbol{E}_1) \cdot \hat{n} = -\boldsymbol{E}_1 \cdot \boldsymbol{J}_{s2}$, we can write Eq. (8.31) in the following form:

$$\frac{P_{r2}}{P_{t1}} = \frac{\frac{1}{4}\left|\oint\limits_{S_0}(\boldsymbol{J}_{s2}\cdot\boldsymbol{E}_1 - \boldsymbol{M}_{s2}\cdot\boldsymbol{H}_1)\,dS\right|^2}{\left\{\oint\limits_{S_1}\mathrm{Re}\{\boldsymbol{E}_1\times\boldsymbol{H}_1^*\}\cdot\hat{n}\,dS\right\}\left\{\oint\limits_{S_2}\mathrm{Re}\{\boldsymbol{E}_2\times\boldsymbol{H}_2^*\}\cdot\hat{n}\,dS\right\}} \tag{8.34}$$

In many cases, the assumed detailed full-wave CEM model of an antenna is not in fact available. It is then still possible to construct a Huygens' box (or virtual box) model if near-field measurements of just the tangential electric field $-\hat{n} \times \boldsymbol{E}$ of the antenna are available[6] over some measurement surface S_M shown in Figure 8.15. In practical near-field measurements, $\hat{n} \times \boldsymbol{H}$ is not easily measurable on S_M. The source-reconstruction method (Araque Quijano and Vecchi 2009, 2010a, 2010b) can be used to find the required $\boldsymbol{J}_s$ and $\boldsymbol{M}_s$ on S_H (now also referred to as the reconstruction surface) through numerical solution of a pair of coupled integral equations with the measured tangential electric field as input data. This is an inverse problem and so needs to be handled with great numerical care. One advantage of establishing equivalent models from measurements is that no knowledge of the design details of the measured antenna is needed (Scialacqua et al. 2017). Also, it inherently

6 As is normally the case with standard antenna near-field testing methods (Parini et al. 2014).

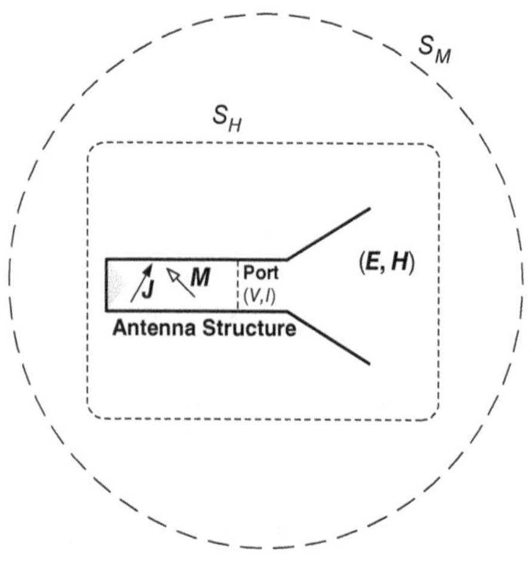

Figure 8.15 Source-reconstruction approach to Huygens' box construction.

incorporates nonideal effects that are difficult to model (Tchorowski and Gupta 2018). It is important to again underscore the fact that a Huygens' box model provides both the near- and far-fields of the antenna so modeled, at any observation point exterior to S_H.

Although the goal is to model the said antenna separated from the platform environment, there might be portions of the platform in the immediate vicinity of the antenna that could considerably modify the near fields of the antenna. Such portions of the platform should be included in establishing the near-field data on S_H in setting up any reduced complexity model, since it could impact the fields used in the coupling evaluation. With this understanding, the resulting reduced complexity model will for convenience still be referred to as that of the isolated antenna. A straightforward example is an antenna that relies for its proper operation on the presence of a planar or curved ground plane that is actually part of the platform. Experimental near-field measurements of the antenna may then be performed with the antenna located on a smaller piece of ground plane of the same shape and the effect of the ground plane's finiteness removed during subsequent construction of the Huygens' box model (Araque Quijano et al. 2011).

8.7.3 Spherical Wave Expansion Models

An alternative to the Huygens' box model is to find the spherical wave expansion coefficients for an antenna. This can be obtained by using the spherical near-field measurement technique to find $-\hat{n} \times \boldsymbol{E}$ of the separated antenna over a measurement sphere S_{meas}, and then compute the spherical wave coefficients (Hansen 1988; Parini et al. 2014). These can then be used in the spherical wave series expansion to determine the field of the antenna at any observation points (near or far) exterior to the surface S_{meas}. If a detailed electromagnetic model of the separated antenna is available, it can be used to find the tangential electric field on S_{meas} and the expansion coefficients determined. In the case of high directivity antennas, the number of significant spherical wave coefficients is large, and then use of an equivalent planar source (see Section 8.7.5) might be the more computationally efficient route to follow.

8.7.4 Infinitesimal Dipole Models

Computed near-fields (both electric and magnetic) on some surface S_M outside the antenna structure have been used (Mikki and Kishk 2007) to find infinitesimal dipole models (IDMs) of antennas. The method uses the supplied near-fields $(\boldsymbol{E}, \boldsymbol{H})$ and an optimization algorithm to determine the complex weights, positions, orientations, and type (electric or magnetic) of a finite number of infinitesimal dipoles within a volume V_{IDM} containing the antenna. It provides closed-form analytical expressions for the antenna fields at any point (near of far) outside the volume V_{IDM}, because the fields of each of the arbitrarily orientated infinitesimal dipoles are known analytically.

8.7.5 Planar Aperture Models

Some antenna types have a distinct physical aperture. A portion of a Huygens' box surface S_H enclosing the antenna, namely that part immediately in front of (but possibly extending slightly beyond) the physical aperture, plays a dominant role in determining the radiated fields (near and far) of the antenna. Examples are horn antennas and reflector antennas. The closed Huygens' box is then replaced by a finite open planar surface on which the equivalent $\boldsymbol{J}_s$ and $\boldsymbol{M}_s$ reside (Bird 2016). In many cases, sufficiently accurate analytical expressions for the tangential fields $-\hat{n} \times \boldsymbol{E}$ and $\hat{n} \times \boldsymbol{H}$, and hence at once $\boldsymbol{J}_s$ and $\boldsymbol{M}_s$, are known over the above-mentioned portion of S_H. These equivalent currents can then be used in a source/field integral (evaluated numerically, or analytically in some cases) to find the antenna fields at any observation point.

There might be instances when a more precise knowledge of the aperture distribution is thought necessary or it is simply not known using physical arguments, as for an antenna with especially low side lobes or one that have a shaped main lobe pattern. In that case, spherical near-field measured data can be back-projected (Janse van Rensburg and Walker 2002) to the equivalent aperture to determine the complex aperture distribution there (Trembinski et al. 2018). Any planar aperture model can, of course, only supply the antenna fields in its forward field of view; for many aperture antennas, this in the predominant path via which coupling to another antenna needs to be quantified anyway. Planar apertures are also used as simplified models for planar array antennas.

Using the spherical wave expansion coefficients of an aperture antenna, obtained as mentioned in Section 8.7.3, it is possible (Serhir et al. 2010) to directly determine the weights of a finite number of infinitesimal dipoles (electric and magnetic) over its extended planar aperture, and tangential to the aperture, to model the antenna fields (near and far).

8.7.6 Point Source Models

This is the simplest of the reduced complexity models. The fields of the antenna at any observation point are taken to be its far-zone fields (as a function of direction) with the usual far-zone distance dependence term e^{-jkr}/r imposed on it. The distance r is measured from a point at which the antenna's phase center is considered to be located, to the observation point. If a particular antenna is being considered for use, and is hence undergoing on-platform coupling studies to be used in decisions as to whether to adopt it or not, all that might be known are its nominal radiation patterns in several cuts. Simple interpolation methods can use this data to build an analytical expression for the far-zone pattern function needed.

8.8 CEM Modeling of Coupled Antennas on a Platform – Pragmatic Approaches

In keeping with what was decided earlier, we continue to think of antenna#1 as the aggressor and antenna#2 as the victim. Whenever the possibility of obtaining solutions for $(\boldsymbol{E}_1, \boldsymbol{H}_1)$ and $(\boldsymbol{E}_2, \boldsymbol{H}_2)$ over the selected interaction surface using a single full-wave CEM model for the entire original environment, as imagined in Section 8.6, is not possible, realistic approximations have to be made. It is not possible to make general statements. There are simply too many possible approximations, and combinations thereof, that have been described in the literature to exhaustively list all of them. This would at any rate not be that useful. Instead, in line with the goals declared in Section 8.1.1, and given the way the chapter has been constructed, we are now able to ponder the modeling of the coupling between platform-mounted co-sited antennas in terms of how the antenna and platform models can be pieced together, in either a rigorous or (more often) approximate way. We are able to distill some procedure we read about into its parts. By now knowing what would be needed for a rigorous treatment, we can better appreciate the impact of the approximations needed for actual work.

A reduced complexity model of an antenna can significantly reduce the computational burden. Thus, a widely used pragmatic approach is to utilize such models for the two antennas. The reduced complexity model for antenna#1 (the transmit antenna) would act as an impressed source (or provide fixed incident fields) for the CEM model describing the platform, which would supply ($\boldsymbol{E}_1$, $\boldsymbol{H}_1$) on the selected interaction surface. If a Huygens' box model is being used for antenna#2, and the interaction surface is chosen to be coincident with the Huygens' surface S_{H2}, where we already know the $\boldsymbol{J}_{s2} = \hat{n} \times \boldsymbol{H}_2$ and $\boldsymbol{M}_{s2} = -\hat{n} \times \boldsymbol{E}_2$, we have all that is needed to evaluate Eq. (8.34) for the power coupling, for instance. If the platform is electrically not too large, a full-wave CEM model could be used to describe the platform. Otherwise, we would use some hybrid CEM model of the platform. The latter is the most widely used approach today. The tools are available commercially and are constantly being improved.

When using approximate methods, it is important to understand what interaction effects are not actually accounted for in the modeling. We can think of the above pragmatic approach as also being domain decomposition, with the two antennas and the platform being three different subdomains. However, unlike the "true" DDM mentioned in Section 8.5, this is an example where the various CEM models for the subdomains are not rigorously linked. Separate CEM models are in effect used to set up each of reduced complexity models for the antennas, as per Section 8.7. The antennas do not form part of the same CEM model as the platform. This at once means that it is assumed that the presence of the platform has no influence on a Huygens' box model, for instance, because the latter model is acting as an impressed source for the platform model, and so we realize that it is impossible to predict any change in the input impedance of the antenna, as already mentioned in Section 8.7.2. We are nevertheless still computing the approximate coupling under the influence of the platform because it alters the near- and far-fields radiated by the given impressed source Huygens' models, and hence over the interaction surface. If full-wave linking between the two antenna subdomains and the platform subdomain were to be used, these Huygens' box currents themselves would be altered by the presence of the platform and the other antenna. Of course, such effects may not be significant in that they might alter the predicted values of the coupling by amounts that are of little consequence to the antenna placement task. Similar comments can be made for any of the reduced complexity antenna model usage.

If antenna#2 might scatter significantly (e.g. reflector antenna), then it should be included as part of the platform when finding ($\boldsymbol{E}_1$, $\boldsymbol{H}_1$) over the interaction surface. In fact, depending on the relative locations of the two antennas, it might even need to be included when establishing the reduced complexity model for antenna#1. Such considerations can only be decided on a case-by-case basis. If there is any doubt, attempts should be made to compute the reduced complexity model with and without the structure of antenna#2 present, even if it is computationally taxing, for decision-making purposes. It can then be confidently ignored in subsequent repeated coupling computations if found to have little effect. As already mentioned in Section 8.6, one

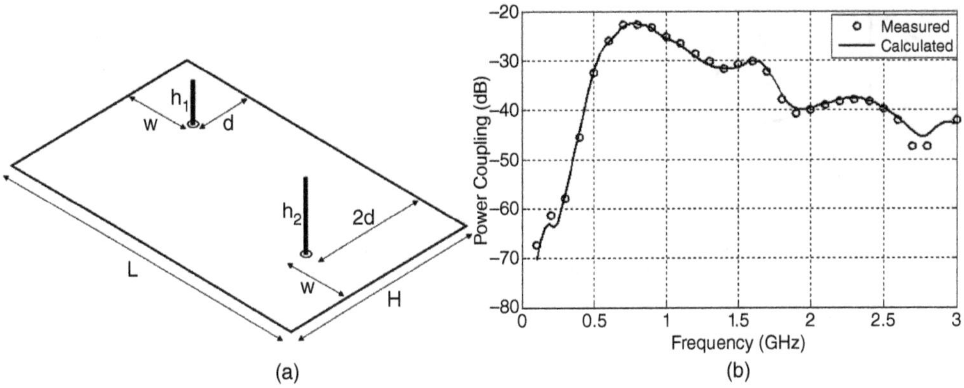

Figure 8.16 Two monopole antennas on a conducting rectangular platform. (a) Geometry; (b) comparison between computed and measured power coupling between the antennas.

should always consider whether only parts of, or the complete platform needs to be modeled for coupling assessment for any two particular antennas. When aperture antennas are involved, careful selection of the interaction surface can be rewarding (Deschênes et al. 2012; Coulombe et al. 2012), as intimated in Section 8.7. Further advice on the use of what we have termed pragmatic approaches can be found in Kragalott et al. (2011) and Vukovic et al. (2019).

8.9 Co-Sited Antenna Coupling Computation Examples

Consider the case of two monopoles on the simple planar rectangular conducting ground plane platform shown in Figure 8.16a. The monopole lengths h_1 and h_2 are selected for center-frequency operation at 1 and 0.5 GHz, respectively. The actual dimensions of the two antennas and the platform are the same as those in Georgakopoulos et al. (1999). The computed power coupling shown Figure 8.16b are for an MoM model of each of the antennas and the platform, and the measured values taken from Georgakopoulos et al. (1999). Similar results are shown in Wang et al. (2012, Figure 5), where a domain decomposition approach was used, with each monopole modeled using the FE-BI method, the ground plane using an MoM approach, and the three subdomains full-wave linked. Coupling levels are low, as expected, but the agreement between computation and measurement is good.

As a second example, we once again take two monopoles whose lengths h_1 and h_2 are selected for center-frequency operation at 1 and 0.5 GHz but now mounted on the conducting platform shown in Figure 8.17. The base of the platform is 20λ by 8λ at 1 GHz, and the other features are sketch in the correct proportions. The platform geometry and electrical size, and the less than complex antennas, were selected to enable the comparison of computed coupling results given below. It might, for instance, be that portion of some larger platform that is expected to most strongly influence the coupling between the two antennas shown. This is not a large platform, and yet the MoM model of the antenna-platform-antenna as a single system requires roughly 117 000 unknowns. It is easy to understand how such exact CEM modeling of much larger real-world platforms might not be feasible. There are several curves in Figure 8.18, some applying with the platform present, and others with it absent. The dashed curve is that for an exact MoM model of the complete antenna-platform-antenna environment, whereas the solid curve is that using a SBR model. The SBR results are computed in a fraction of the computation time taken by the exact MoM model.

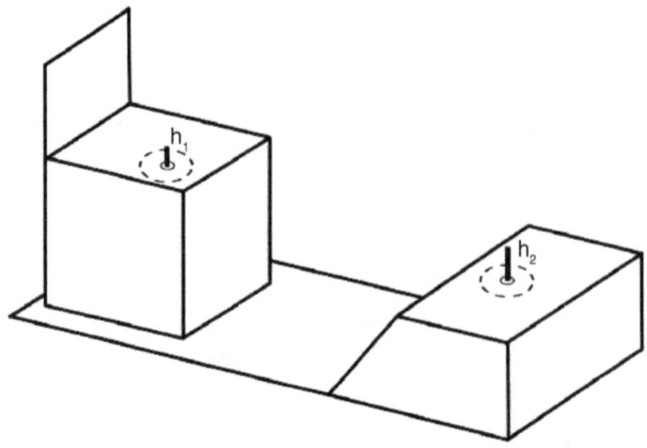

Figure 8.17 Two monopole antennas on a contrived conducting platform.

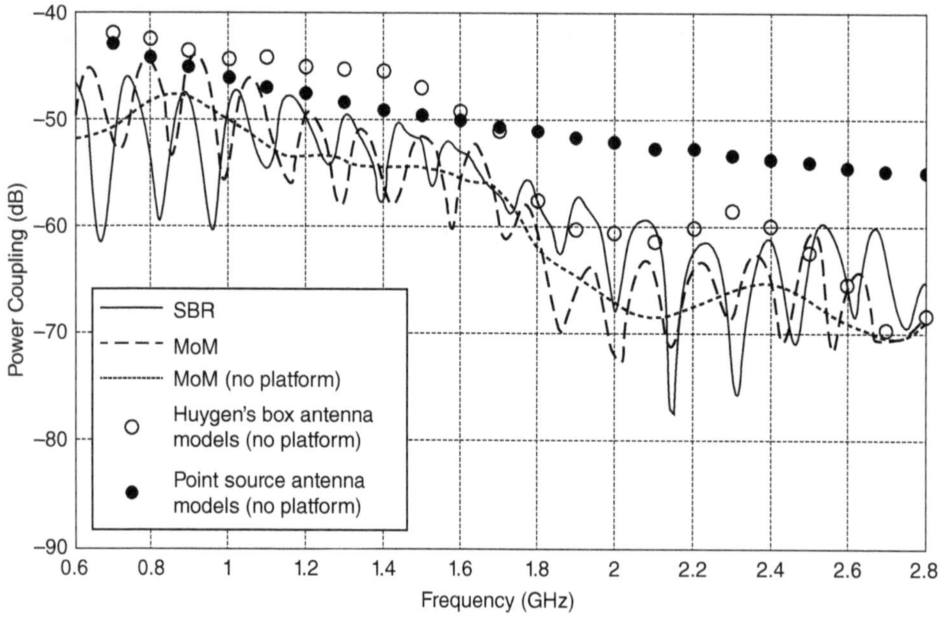

Figure 8.18 Coupling between the monopoles shown in Figure 8.17 computed using the methods identified in the inset. Key: SBR - shooting and bouncing rays, MoM – method of moments.

The remaining results are with the platform absent, but for the same two monopoles, each now on its own finite-sized circular conducting ground planes of diameter 2λ at 1 GHz (suggested by the dashed circles in Figure 8.17). In this case, the Huygens' box approach uses reduced-complexity models obtained from a full-wave model of each monopole in isolation, and then utilizes these in evaluating the coupling integral. A different Huygens' box model was established for the antennas at each frequency. Observe that the Huygens' box antenna model result assumes matched ports and so tends to show increased coupling, whereas the exact MoM model accounts for mismatch. Even the point source models of the two antennas allow reasonable coupling estimates to be obtained. The predictions using the approximate approaches are not identical to that obtained from the exact model but would still be useful in antenna placement studies. Especially so when an exact model is not possible. It should be emphasized that we are not using this to demonstrate typical results in any way, or to imply that results without the platform will be related to those with it present, as they seem to be here. The diverse nature of platforms is one reason installed performance modeling is considered necessary. All we wish to show in Figure 8.18 are the differences in computed results between an exact model and ones that utilize different levels of approximation.

One obviously wants the computed coupling levels to be trustworthy, and so the question of accuracy arises. The issues are no different from any other type of problem undergoing CEM modeling. If some new platform differs markedly from one previously modeled, it is not possible to precisely transfer error estimates from a previous one to the new scenario (Mahlström et al. 2018). The question is often a difficult one, calling for experience and engineering judgment. It is always appropriate to consider some given model as a "numerical laboratory," with small alterations made to the model details and any resulting changes to the predicted coupling results assessed. For instance, the type of reduced-order antenna model, and small changes in its location (e.g. for a Huygens' box model) in relation to the platform model, should be carefully considered (Mahlström et al. 2018). Unless the complete antenna-platform-antenna environment has been modeled rigorously, it is also wise, if possible, to compare results that have been obtained using approaches that utilize different approximations. Their similarity will at once provide some confidence on the robustness of the approximations made.

Some of the computed coupling results shown have been obtained using the commercial software Altair FEKO (Altair Engineering Inc. 2020) and Ansys SAVANT (Ansys Inc. 2020). There are other similarly capable commercial products, including Ansys HFSS (Ansys Inc. 2020), CST Studio Suite (Dassault Systèmes 2020), WiPL-D Suite (WIPL-D 2020), and EMPIRE (IMST GmbH 2020). These should be directly consulted to establish their respective capabilities for on-platform coupling computation. The content covered in this chapter should provide the reader with an ability to evaluate what is needed, understand how a procedure may be pieced together and understood, or picked apart and understood, depending on one's role in the antenna placement study being conducted.

8.10 Concluding Remarks

CEM techniques develop rather rapidly not only because of astounding improvements in the capabilities of computers but also because of new hybridizations and improved implementations. It is safe to expect this to continue because it has been so for more than 40 years. However, any more specific pronouncements of a general nature made right now could be obsolete within a few years (or sooner). Indeed, it has been recommended by at least one Nobel Laureate (Medawar 1984), if one is asked to make pronouncements about future developments in any area, that he had literally one word of advice: *Don't*.

References

Aguilar, A.G., Van Tonder, J. and Jakobus, U. (2015): 'Overview of recent advances in the electromagnetic field solver FEKO', 9th European Conf. Antennas Propagat (EuCAP).

Altair Engineering Inc., 2020 www.altair.com, Accessed April 2020.

Alvarez, J. (2014): 'Simplified antenna models for the prediction of installed performance on large platforms', IEEE Conf. Antenna Measurements and Applications.

Ansys Inc., 2020 www.ansys.com, Accessed April 2020.

Araque Quijano, J.L. and Vecchi, G. (2009). Improved-accuracy source reconstruction on arbitrary 3-D surfaces. *IEEE Antennas Wireless Propag. Lett.* 8: 1046–1049.

Araque Quijano, J.L. and Vecchi, G. (2010a). Field and source equivalence in source reconstruction on 3D surfaces. *Prog. Electromagn. Res* 105: 67–100.

Araque Quijano, J.L. and Vecchi, G. (2010b). Near- and very near-field accuracy in 3-D source reconstruction. *IEEE Antennas Wireless Propag. Lett.* 9: 634–637.

Araque Quijano, J.L., Scialacqua, L., Zackrisson, J. et al. (2011). Suppression of undesired radiated fields based on equivalent currents reconstruction from measured data. *IEEE Antennas Wireless Propag. Lett.* 10: 314–317.

Barka, A. (2013). Integration of antennas onboard vehicles and diffraction by large and complex structures with multiple-domain multiple-method techniques. *Proc. IEEE* 101: 280–297.

Barka, A. and Caudrillier, P. (2007). Domain decomposition method based on generalized scattering matrix for installed performance of antennas on aircraft. *IEEE Trans. Antennas Propag.* 55: 1833–1842.

Beatty, R.W. and Kerns, D.M. (1964). Relationships between different kinds of network parameters, not assuming reciprocity or equality of the waveguide or transmission line characteristic impedances. *Proc. IEEE* 52: 84.

Bird, T.S. (2016). *Fundamentals of Aperture Antennas and Arrays*. Wiley.

Brem, R. and Eibert, T.F. (2015). A shooting and bouncing ray (SBR) modelling framework involving dielectrics and perfect conductors. *IEEE Trans. Antennas Propag.* 63: 3599–3609.

Burkholder, R.J., Pathak, P.H., Sertel, K. et al. (2006). A hybrid framework for antenna/platform analysis. *ACES J.* 21: 177–195.

Collin, R.E. and Zucker, F.J. (eds.) (1969). *Antenna Theory*. McGraw-Hill Book Company.

Coulombe, M., Deschênes, P., Pinchuk, A., and Paknys, R. (2012): 'Electromagnetic environmental effects (E3) computational analysis of a Navy frigate', IEEE International Symposium on Electromagnetic Compatibility, pp. 261-266.

Dassault Systèmes, 2020 www.3ds.com, Accessed April 2020.

Davidson, D.B. (2011). *Computational Electromagnetics for RF and Microwave Engineering*, 2e. Cambridge University Press.

Deschamps, G.A. (1966a): 'I. Le principe de réciprocité en électromagnétisme', Revue du Centre d'Etudes Théoriques de la Détection et des Communications (CETHEDEC), pp. 71-90.

Deschamps, G.A. (1966b): 'II. Application du principe de réciprocité aux antennes et aux guides d'ondes', Centre d'Etudes Théoriques de la Détection et des Communications (CETHEDEC), pp. 91-101.

Deschamps, G.A. (1972). Ray techniques in electromagnetics. *Proc. IEEE* 72: 1022–1035.

Deschênes, P., Coulombe, M., Paknys, R., and Pinchuk, A.R. (2012). Near-field coupling method for a complex navy ship environment. *IEEE Antennas Propag. Symp. Dig.*: 256–260.

Ekelman, E. and Thiele, G. (1980). A hybrid technique for combining the moment method treatment of wire antennas with the GTD for curved surfaces. *IEEE Trans. Antennas Propag.* 28: 831–839.

Eleftheriades, G.V. and Mosig, J.R. (1996). On the network characterization of planar passive circuits using the method of moments. *IEEE Trans. Microwave Theory Tech.* 44: 438–445.

Elliott, R.S. (1993). *An Introduction to Guided Waves and Microwave Circuits*. Prentice-Hall International, Inc.

Gedney, S.D. (2011). *Introduction to the Finite-Difference Time-Domain (FDTD) Method for Electromagnetics*. Morgan & Claypool Publishers.

Georgakopoulos, S.V., Balanis, C.A., and Birtcher, C.R. (1999). Cosite interference between wire antennas on helicopter structures and rotor modulation effects : FDTD versus measurements. *IEEE Trans. Electromagn. Compat.* 41: 221–233.

Hansen, J.E. (1988). *Spherical Near-Field Antenna Measurements*. Peter Peregrinus Ltd.

Harrington, R.F. (1961). *Time-Harmonic Electromagnetic Fields*. McGraw-Hill.

Hu, M. (1958): 'Near-zone power transmission formulas', 1958 IRE National Convention Record, pp. 128-138.

IMST GmbH, 2020 www.imst.com, Accessed April 2020.

Jakobus, U. (2010): 'Overview of hybrid methods in FEKO: Theory and applications', International Conf. Electromagnetics in Advanced Applications (ICEAA), pp. 434-437.

Jakobus, U. and Landstorfer, F.M. (1995). Improvement of the PO-MOM hybrid method by accounting for effects of perfectly conducting wedges. *IEEE Trans. Antennas Propag.* 43: 1123–1129.

Jansevan Rensburg, D.J. and Walker, C. (2002): 'Implementation of back projection on a spherical near-field range', Antenna Measurement Techniques Association (AMTA) Symposium.

Jin, J. (2014). *The Finite Element Method in Electromagnetics*. Wiley.

Jin, J. and Riley, D.J. (2009). *Finite Element Analysis of Antennas and Arrays*', Chapter11. Wiley.

Jin, J., Ling, F., Carolan, S.T. et al. (1998). A hybrid SBR/MoM technique for analysis of scattering from small protrusions on a large conducting body. *IEEE Trans. Antennas Propag.* 46: 1349–1357.

Kipp, R.A. (2010): 'Curved surface scattering geometry in the shooting-and-bouncing rays method', IEEE Antennas & Propagation Symposium Digest.

Kipp, R.A. and Capoglu, I. (2014). Extending the shooting-and-bouncing rays method with uniform theory of diffraction for installed antennas. *IEEE Antennas Propag. Symp. Dig.*: 2198–2199.

Kipp, R.A., Canta, S.M., Courtney, T.A. and Setterdahl, D.L.,(2015): 'Extending the shooting-and-bouncing rays method with creeping waves for radar signature prediction', IEEE Radar Conf., pp. 704-707.

Kong, J.A. (1986). *Electromagnetic Wave Theory*', Sect.5.5. Wiley.

Kouyoumjian, R.G. (1965): 'Asymptotic high-frequency methods', *Proc.IEEE* 53: 864–876.

Kouyoumjian, R.G. and Pathak, P.H. (1974). A uniform geometrical theory of diffraction for an edge in a perfectly conducting surface. *IEEE Trans. Antennas Propag.* 29: 1448–1461.

Kragalott, M., Kluskens, M.S., Zolnick, D.A. et al. (2011). A toolset independent hybrid method for calculatingantenna coupling. *IEEE Trans. Antennas Propag.* 59: 443–451.

Le Lepvrier, B., Loison, R., Gillard, R. et al. (2014). A new hybrid method for the analysis of surrounded antennas mounted on large platforms. *IEEE Trans. Antennas Propag.* 62: 2388–2397.

Lindell, I.V. (1992). *Methods for Electromagnetic Field Analysis*, Sect.3.6. Oxford University Press.

Ling, H., Chou, R., and Lee, S.W. (1989). Shooting and bouncing rays : calculating the RCS of an arbitrarily shaped cavity. *IEEE Trans. Antennas Propag.* 37: 194–205.

Liu, Z., Wang, X., and Wang, C. (2014). Installed performance modelling of complex antenna array mounted on extremely large-scale platform using fast MoM-PO hybrid framework. *IEEE Trans. Antennas Propag.* 62: 3852–3858.

Lu, J., Chen, Y., Li, D., and Lee, J.-F. (2019). An embedded domain decomposition method for electromagnetic modeling and design. *IEEE Trans. Antennas Propag.* 67: 309–323.

Maclean, T.S.M. (1986). *Principles of Antennas: Wire and Aperture*', Sect.14.3. Cambridge University Press.

Macnamara, T. (2010). *Introduction to Antenna Placement and Installation*. Hoboken, NJ, USA: Wiley.

Mahlström, J., Holter, H., and Jonsson, B.L.G. (2018). On the accuracy of equivalent antenna representations – a case study. *IEEE Trans. Antennas Propag.* 66: 3251–3260.

McNamara, D.A., Pistorius, C.W.I., and Malherbe, J.A.G. (1990). *Introduction to the Uniform Theory of Diffraction*. Artech House Inc.

Medawar, P.B. (1984). *Pluto's Republic*. Oxford University Press.

Medgyesi-Mitschang, L.N. and Wang, D. (1991). Hybrid methods in computational electromagnetics: a review. *Comput. Phys. Commun.* 68: 76–94.

Mikki, S.M. and Kishk, A.A. (2007). Theory and applications of infinitesimal dipole models for computational electromagnetics. *IEEE Trans. Antennas Propag.* 55: 1325–1337.

Miller, E.K. (1988). A selective survey and computational electromagnetics. *IEEE Trans. Antennas Propag.* 36: 1281–1305.

Monteath, G.D. (1973). *Applications of the Electromagnetic Reciprocity Principle*. Pergamon Press.

Paknys, R. (2016). *Applied Frequency-Domain Electromagnetics*. Wiley.

Parini, C., Gregson, S., Mccormick, J. and JanseVan Rensburg, D.J. (2014): '*Principles Theory and Practice of Modern Antenna Range Measurements*', IET Electromagnetic Wave Series, Peter Peregrinus, London, UK.

Pathak, P.H. and Wang, N. (1981). Ray analysis of mutual coupling between antennas on a convex surface. *IEEE Trans. Antennas Propag.* 29: 911–922.

Pathak, P.H., Burnside, W.D., and Marhefka, R.J. (1980). A uniform GTD analysis of the diffraction of electromagnetic waves by a smooth convex surface. *IEEE Trans. Antennas Propag.* 28: 631–642.

Pathak, P.H., Wang, N., Burnside, W.D., and Koujoumjian, R.G. (1981). A uniform GTD solution for the radiation from sources on a convex surface. *IEEE Trans. Antennas Propag.* 29: 609–621.

Pozar, D.M. (2005). *Microwave Engineering*, 3e. Wiley.

Reddy, C.J. (2018). Analysis of airborne antennas using hybrid computational techniques. *IEEE Antennas Propag. Symp. Dig.*: 1229–1230.

Richmond, J.H. (1958): 'On the theory of scattering by dielectric and metal objects', Ohio State University Research Foundation, Columbus, Ohio, USA, Rept. 786-3.

Richmond, J.H. (1961). A reaction theorem and its application to antenna impedance calculations. *IRE Trans. Antennas Propag.* 9: 515–520.

Richmond, J.H. (1980): 'The basic theory of harmonic fields, antennas, and scattering', Unpublished Notes, Department of Electrical Engineering, Ohio State University, Columbus, Ohio, USA.

Rumsey, V.H. (1954). Reaction concept in electromagnetic theory. *Physiol. Rev.* 94: 1483–1491.

Rumsey, V.H. (1963). A short way of solving advanced problems in electromagnetic fields and other linear systems. *IEEE Trans. Antennas Propag.* 11: 73–86.

Rusch, W.V.T. (1974). Physical optics diffraction coefficients for a paraboloid. *Electron. Lett* 10: 358–360.

Scialacqua, L., Foged, L.J., Mioc, F.andSaccardi, F. (2017): 'Link between measurement and simulation applied to antenna scattering and placement problems', 11th European Conf. Antennas Propagat (EuCAP), pp. 2952-2955.

Serhir, M., Geffrin, J., Litman, A., and Besnier, P. (2010). Aperture antenna modelling by a finite number of elemental dipoles from spherical wave measurements. *IEEE Trans. Antennas Propag.* 58: 1260–1268.

Sheng, X. and Wei, S. (2012). *Essentials of Computational Electromagnetics*. Wiley.

Smith, G.S. (1997). *An Introduction to Classical Electromagnetic Radiation*, Sect.4.1. Cambridge University Press.

Song, J., Lu, C.-C., and Chew, W.C. (1997). Multilevel fast multipole algorithm for electromagnetic scattering from large complex objects. *IEEE Antennas Propag.* 45: 1488–1493.

Tchorowski, L.and Gupta, I. (2018): 'Antenna modeling on complex platforms through constrained equivalent aperture distributions', Antenna Measurement Techniques Association (AMTA) Symposium, pp. 77-82.

Theron, I.P., Davidson, D.B., and Jakobus, U. (2000). Extensions to the hybrid method of moments/uniform GTD formulation for sources located close to a smooth surface. *IEEE Trans. Antennas Propag.* 48: 940–945.

Trefethen, L.N. (2011). *Trefthen's Index Cards – Forty Years of Notes about People, Words and Mathematics*. World Scientific Publishers Pte. Ltd.

Trembinski, R., McNamara, D.A. and Janse van Rensburg, D.J., (2018): 'Utilizing spherical near-field measurement data in antenna on-platform performance studies', 18th International Symposium on Antenna Technology and Applied Electromagnetics (ANTEM).

Tzoulis, A. and Eibert, T.F. (2005). A hybrid FEBI-MLFMM-UTD method for numerical solutions of electromagnetic problems including arbitrarily shaped and electrically large objects. *IEEE Trans. Antennas Propag.* 53: 3358–3366.

Volakis, J. and Sertel, K. (2012). *Integral Equation Methods for Electromagnetics*. SciTech Publishing Inc.

Volakis, J.L., Sertel, K., and Usner, B.C. (2006). *Frequency Domain Hybrid Finite Element Methods for Electromagnetics*. Morgan Claypool Publishers.

Vukovic, A., Sewell, P., and Benson, T.M. (2019). Holistic appraisal of modelling installed antennas for aerospace application. *IEEE Trans. Antennas Propag.* 67: 1396–1409.

Wan, T., Tang, B., and Li, M. (2020). An iteration-free domain decomposition method for the fast finite element analysis of electromagnetics problems. *IEEE Trans. Antennas Propag.* 68: 400–410.

Wang, X., Peng, Z., Lim, K., and Lee, J.-F. (2012). Multisolver domain decomposition method for modeling EMC effects of multiple antennas on a large platform. *IEEE Trans. Electromagn. Compat.* 54: 375–388.

Weiland T. (2003). Finite integration method and discrete electromagnetism. In: Monk, P., Carstensen, C., Funken, S., Hackbusch, W. and Hoppe, R.H.W. (eds.), *Computational Electromagnetics*. Springer. 183–198.

Weinmann, F. (2009). UTD shooting-and-bouncing extension to a PO/PTD ray tracing algorithm. *ACES J.* 24: 281–293.

Williams, D.F. and Alpert, B.K. (2001). Causality and waveguide circuit theory. *IEEE Trans. Microwave Theory Tech.* 49: 615–623.

WIPL-D, 2020, http://www.wipl-d.com, Accessed April 2020.

Xiang, D.P. and Botha, M.M. (2016). Efficient shadowing determination at grazing incidence, for mesh-based physical optics scattering analysis. *Electron. Lett* 52: 1893–1894.

Xiang, D.P. and Botha, M.M. (2018). MLFMM-based, fast multiple-reflection physical optics for large scale electromagnetic scattering analysis. *J. Comput. Phys.* 368: 69–91.

9

Mutual Coupling and Multiple-Input Multiple-Output (MIMO) Communications

Karl F. Warnick

Department of Electrical and Computer Engineering, Brigham Young University, Provo, UT 84602, USA

9.1 Introduction

Multiple-input multiple-output (MIMO) wireless communication systems use antenna arrays at the transmitter, receiver, or both to increase the achievable data rate in a given bandwidth relative to single-antenna systems in a multipath channel propagation environment. MIMO techniques have their origins in simple antenna diversity schemes that use multiple antennas to improve link performance and have been in use for many decades. The field entered a period of intense research and development in the 1990s, and MIMO techniques are now implemented in commercially available products in a variety of wireless communications applications.

The key measure of the performance of a MIMO system with given antenna arrays and propagation environment is the information-theoretic channel capacity, or the maximum data rate in bits per Hertz of available bandwidth at which information can be transmitted with perfect error correction (Shannon and Weaver 1998). The channel capacity provides an estimate of the data rate that can be achieved in a practical communication system.

A rich multipath environment can theoretically support MIMO channels with capacities much higher than a single-antenna-based system. Whereas higher signal-to-noise ratio (SNR) improves channel capacity only logarithmically, multiple parallel channels increase the achievable information rate multiplicatively. From a physical perspective, a MIMO system transmits different symbols simultaneously along different combinations of propagation paths so that multiple symbols can be received and decoded per use of the channel. The radiation pattern and other characteristics of the antenna elements interact with the propagation environment to influence the performance of a MIMO system.

The improvements in the channel capacity offered by MIMO techniques come at the cost of increased system complexity in the signal processing hardware and software and the need to measure information about the propagation channel and transfer it between nodes in a network. Given the difficulty of implementing a MIMO solution relative to a single-antenna-based transceiver, many research groups have studied in great detail the performance gains available with MIMO systems as a function of the propagation environment and the number and placement of elements in the receive and transmit arrays.

The performance of a MIMO system that uses antenna arrays on the transmit or receive side is influenced by mutual coupling. Since many wireless systems are space-constrained, it is advantageous to use tightly packed and hence highly coupled antenna arrays. One could place a

Mutual Coupling Between Antennas, First Edition. Edited by Trevor S. Bird.

large number of small antennas in a given volume, yet one would not expect unlimited growth in the channel capacity with the number of array elements. Coupling between the elements and resulting overlap in the information available from the array elements limits the improvement in capacity of a multiantenna system over a single-antenna receiver, or the diversity gain of the system.

Consequently, the impact of mutual coupling on MIMO channel capacity has been a topic of considerable research interest over the years. Studies of coupling effects in MIMO systems have ranged from phenomenological studies of channel capacity as a function of array element spacing to deeply theoretical estimates of and bounds on the intrinsic channel capacity of free space or another propagation environment with a given size constraint on the transmit and receive arrays. Mutual coupling considerations are critical to theoretical bounds on channel capacity as well as the performance of practical MIMO systems with space-constrained antennas.

Mutual coupling influences the performance of MIMO systems in four main ways:

- Mutual coupling influences the element patterns at the transmit and receive arrays.
- Mutual coupling affects transmit power due to impedance mismatches at the transmitter.
- Mutual coupling can reduce SNR due to noise mismatches at the receiver.
- Mutual coupling reduces diversity gain through increased signal correlation at the receiving array.

The first area, element pattern perturbation, is typically accounted for automatically as a MIMO system sounds the propagation channel and computes optimal signaling phase and amplitude coefficients for the transmit and receive arrays. The second and third issues related to impedance mismatches are generalizations of the impedance-related effects that can impact transmit power and SNR in single-antenna communication links to multiport antenna systems. The final area, the impact of mutual coupling on signal correlation, is fundamental to the information-theoretic nature of a MIMO system.

By analyzing each of these effects, we will explore the critical connections between mutual coupling and the behavior and effectiveness of MIMO systems. The active impedance concept will be discussed in connection with MIMO systems. We will derive the relationship between S-parameters and envelope correlation in a multipath environment, explore its physical meaning, and consider its ramifications and limitations in terms of the performance of antenna arrays for MIMO communication. We will explore the rich and deep theoretical implications of a uniform distribution of multipath arrival angles. Finally, we will cover methods for mitigating the effects of coupling on MIMO systems.

9.2 Previous Work on Mutual Coupling and MIMO

One of the earliest connections that have been explored between mutual coupling and MIMO communications is the relationship between coupling in the receiving antenna array and the signal correlation of the array outputs when receiving signals in a multipath environment. Studies of coupling effects in multichannel wireless networks date back to Jakes (1974). The effectiveness of a MIMO system is related to the degree of correlation between the output signals at receiving array antenna elements. If the signals are uncorrelated, then the possibility exists to transmit more information to the receiving array than would be possible with a single receiving antenna. Tightly packed antenna elements are highly coupled, and consequently, the received signals in a MIMO system are highly correlated. Further processing cannot recover independent information from the signals received by adjacent array elements, and the diversity gain achieved by the MIMO system is low.

In a rich multipath propagation environment with many obstacles and reflecting objects, scattered signals arrive uniformly from all directions at the receiving array. This is referred to as a Rayleigh propagation environment due to the relevance of the Rayleigh probability distribution to the analysis. For a two-dimensional environment in which scattered waves arrive uniformly in azimuth angle, the corresponding analytical treatment is referred to as *Jakes's model.*

One consequence of the uniform arrival angle distribution is the connection between MIMO system performance and array mutual coupling first published by Vaughan and Andersen (1987, Eq. 14). The correlation coefficient of signals at a pair of antennas in a rich multipath environment is proportional to the mutual resistance of the antennas. This deceptively simple result shows that the open-circuit signal correlation coefficient at a pair of antennas is equal to the normalized mutual resistance. It is possible that the simplicity of this formula led to it largely being overlooked in the literature for over a decade.

The connection between mutual coupling and signal correlation was generalized and extended in the early 2000s by Salonen and Vainikainen (2002) and Blanch et al. (2003). The envelope correlation of signals arriving at a pair of antennas in a rich multipath environment is related to the S-parameters by

$$\rho_{12} = \frac{\left|S_{11}^* S_{12} + S_{21}^* S_{22}\right|^2}{\left(1 - |S_{11}|^2 - |S_{21}|^2\right)\left(1 - |S_{22}|^2 - |S_{12}|^2\right)} \tag{9.1}$$

Expressed in terms of S-parameters, the simple relationship of Vaughan and Anderson takes on a more sophisticated appearance. The intellectual content is the same, but the increased complexity of the formula perhaps attracted more researchers to take note of the result (though the increased attention was certainly also facilitated by improvements in technology in the 2000s that enabled more capable MIMO test beds and implementations). In any case, researchers around the world proceeded to explore energetically practical ramifications of the result for MIMO systems.

The most obvious implication of Eq. (9.1) is that if a pair of antennas are uncoupled ($S_{12} = S_{21} = 0$), the envelope correlation in a rich multipath environment is zero. This remarkable result allows effective arrays for MIMO systems to be designed simply by minimizing the mutual coupling between elements, thus separating the antenna array design from the complexities of the receiver electronics, signal processing methodology, hardware implementation, and all other aspects of the MIMO system.

The envelope correlation formula, Eq. (9.1), has stimulated a vast amount of research in antennas for MIMO. The number of papers on techniques for reducing coupling and small arrays with high isolation between elements is in the thousands. Some of this work is surveyed by Malviya et al. (2017) and Chen et al. (2018). Ground plane adjustments (Chiu et al. 2007) and parasitic elements (Li et al. 2011) can be used to reduce coupling. Mutual coupling can in principle be eliminated with decoupling networks (Dossche et al. 2005; Dossche et al. 2006). This result is of theoretical interest, but the complexity of the required networks has prevented decoupling from having any practical impact on applications of MIMO.

Network theory is a key tool in studying the behavior of MIMO systems with coupled antenna arrays (Svantesson and Ranheim 2001; Janaswamy 2002; Wallace and Jensen 2004a; Waldschmidt et al. 2004; Li and Nie 2004; Wallace and Jensen 2004b). Network theory has been used to study the impact of mutual coupling on the channel capacity of a MIMO system, including detailed models of the receiver electronics (Morris and Jensen 2005; Wallace and Jensen 2002; Getu and Janaswamy

2005). We will use in this chapter the network theoretical framework to identify and analyze the impact of mutual coupling on the performance of a MIMO system.

Experimental results for the performance of MIMO systems with coupled arrays have been available for many years (Wallace et al. 2003). The impact of mutual coupling can be assessed by characterizing channel capacity within a given propagation environment with arrays with different degrees of coupling. To avoid challenges associated with implementing a MIMO test bed and identifying a suitably well-controlled propagation environment, reverberation chamber measurements can be used to assess coupling effects for MIMO arrays (Kildal and Rosengren 2004).

Measured results for the performance of MIMO systems with coupled arrays are, at least on the surface, contradictory and difficult to reconcile. Much of the work in the literature deals with the reduction in channel capacity caused by mutual coupling (Browne et al. 2006). Other papers report that coupling can enhance channel capacity (Jungnickel et al. 2003). Mutual coupling perturbs the embedded element patterns (EEPs), which can increase the diversity of antenna patterns, but mutual coupling also increases correlation between received signals at nearby array elements, which decreases the MIMO channel capacity. The influence of mutual coupling on both element radiation patterns and signal correlation accounts for some of the conflicting observations in the literature (Fletcher et al. 2003). In the limiting case of closely packed, highly coupled elements, the net effect of mutual coupling is to reduce channel capacity, which, as will be seen, is related to conservation of energy for receiving arrays.

9.3 Basics of MIMO Communications

For a general wireless MIMO system, the transmitter and receiver consist of antenna arrays. We will denote the number of transmit antennas as N and the number of receive antennas as M. For the purposes of antenna analysis, we will model the signal as a time-harmonic or single-frequency waveform. The signals will be characterized in terms of the phasor or complex baseband representation of generator voltages that excite the transmit array antenna elements. The generators are equivalent sources for the power amplifiers that drive each array element. For a wideband MIMO system, we will assume that signal processing is done in narrow sub-bands so that within each sub-band, the phasor representation can be used with negligible beam squint or other frequency-dependent effects.

We will arrange the phasor voltages at each driving amplifier into the column vector

$$\boldsymbol{v}_{\mathrm{s}} = \begin{bmatrix} v_{\mathrm{s},1}(n) \\ v_{\mathrm{s},2}(n) \\ \vdots \\ v_{\mathrm{s},N}(n) \end{bmatrix} \tag{9.2}$$

Each element of this vector is a phasor representing the symbol transmitted at the nth symbol period, or the signal waveform in the complex analytic representation. At the receiver, we define

$$\boldsymbol{v}_{\mathrm{r}} = \begin{bmatrix} v_{\mathrm{r},1}(n) \\ v_{\mathrm{r},2}(n) \\ \vdots \\ v_{\mathrm{r},M}(n) \end{bmatrix} \tag{9.3}$$

to be a vector of voltages at the outputs of the receiving antenna elements.

In terms of these quantities, we can represent the propagation channel in matrix form

$$\boldsymbol{v}_{\mathrm{r}}(n) = \boldsymbol{H}(n)\boldsymbol{v}_{\mathrm{s}}(n) + \boldsymbol{v}_{\mathrm{n}}(n) \tag{9.4}$$

where the channel matrix $\boldsymbol{H}$ relates the signals at each transmit array element to the signals at the receiving array outputs. The random vector representing system noise is denoted by $\boldsymbol{v}_{\mathrm{n}}$. In a wireless transceiver, the system noise consists of external noise from blackbody objects and other radiation sources in the propagation environment around the receiver, interference from other transmitters, noise due to ohmic losses in the receiving antenna elements, and receiver noise introduced by amplifiers and other electronic components after the receive array element ports.

For a fixed channel, we can view the channel model as a deterministic matrix with specified values, or for a variable channel, we can model the elements of $\boldsymbol{H}$ as random processes. The transmitted symbols $\boldsymbol{v}_{\mathrm{s}}$ can also be modeled as random processes, since the details of a specific data stream sent across the channel is not important in designing the communication systems. In the following, we will not consider dispersion and time-delay effects. Dispersion and time delay nearly always occur in multipath propagation environments, but we will assume that these effects can be dealt with and removed using signal processing.

9.3.1 MIMO Channel Capacity

To understand the effect of coupling on the performance of a MIMO system, we will review the information-theoretic capacity of a MIMO channel. An information source at the transmitter produces a vector of symbols (complex voltages at each transmit element input) with correlation matrix

$$\boldsymbol{R}_{\mathrm{s}} = \mathrm{E}\left[\boldsymbol{v}_{\mathrm{s}}\boldsymbol{v}_{\mathrm{s}}^{\dagger}\right] \tag{9.5}$$

where $\mathrm{E}[\cdot]$ is the time expectation. The noise at the output of the receiving array has correlation matrix

$$\boldsymbol{R}_{\mathrm{n}} = \mathrm{E}\left[\boldsymbol{v}_{\mathrm{n}}\boldsymbol{v}_{\mathrm{n}}^{\dagger}\right] \tag{9.6}$$

The information-theoretic capacity of a MIMO channel in bits/sec is (Telatar, 1999)

$$C = \max_{\boldsymbol{R}_{\mathrm{s}}} \log_2 \det\left(\boldsymbol{I}_M + \boldsymbol{R}_{\mathrm{n}}^{-1}\boldsymbol{H}\boldsymbol{R}_{\mathrm{s}}\boldsymbol{H}^{\dagger}\right) \tag{9.7}$$

This is the "log-det" time average or ergodic capacity bound for a MIMO channel. If the channel matrix $\boldsymbol{H}$ is variable and modeled stochastically, the capacity is the expectation of this expression over the channel matrix distribution. Because it includes the product of the received signal covariance matrix and the inverse of the noise covariance matrix, the second term inside the determinant is similar to an SNR.

9.3.2 Eigenchannels and the Water-Filling Solution

In the case of uncoupled noise ($\boldsymbol{R}_{\mathrm{n}} = \sigma_{\mathrm{n}}^2\boldsymbol{I}$), we use the singular value decomposition (SVD) $\boldsymbol{H} = \boldsymbol{U}\boldsymbol{S}\boldsymbol{V}^{\dagger}$ of the channel matrix in the definition Eq. (9.4) of the channel matrix to find that

$$\underbrace{\boldsymbol{U}^{\dagger}\boldsymbol{v}_{\mathrm{r}}}_{\widetilde{\boldsymbol{v}}_{\mathrm{r}}} = \boldsymbol{S}\underbrace{\boldsymbol{V}^{\dagger}\boldsymbol{v}_{\mathrm{s}}}_{\widetilde{\boldsymbol{v}}_{\mathrm{s}}} + \underbrace{\boldsymbol{U}^{\dagger}\boldsymbol{v}_{\mathrm{n}}}_{\widetilde{\boldsymbol{v}}_{\mathrm{n}}} \tag{9.8}$$

This derivation relies on the unitary property of $\boldsymbol{U}$ and $\boldsymbol{V}$. This leads to the diagonalized channel relationship

$$\widetilde{\boldsymbol{v}}_{\mathrm{r}} = \boldsymbol{S}\widetilde{\boldsymbol{v}}_{\mathrm{s}} + \widetilde{\boldsymbol{v}}_{\mathrm{n}} \tag{9.9}$$

Motivated by this construction, a MIMO communication system can be implemented by using the columns of $\boldsymbol{V}$ as transmitting array beamformer weights, scaled by symbols selected from a digital modulation scheme constellation such as quadrature phase shift keying (QPSK). The signals at the receiving array are decoded spatially using the columns of $\boldsymbol{U}$ as receive beamformers. Each beam output is sent to a digital demodulator, and the independent data streams transmitted along each eigenchannel are recovered.

One of the effects of mutual coupling on a MIMO system is coupled noise at the receive array, which leads to a nondiagonal noise correlation matrix $\boldsymbol{R}_\mathrm{n}$. In the case of correlated noise, the eigenchannel derivation must be modified by including the noise correlation matrix with the channel matrix when calculating the SVD. The capacity formula can be rearranged to

$$C = \max_{\boldsymbol{R}_\mathrm{s}} \log_2 \det \left[\boldsymbol{I}_M + \boldsymbol{R}_\mathrm{n}^{-1/2}\boldsymbol{H}\boldsymbol{R}_\mathrm{s}\left(\boldsymbol{R}_\mathrm{n}^{-1/2}\boldsymbol{H}\right)^\dagger\right]. \tag{9.10}$$

The singular value decomposition is applied to the noise prewhitened channel matrix $\boldsymbol{R}_\mathrm{n}^{-1/2}\boldsymbol{H}$. Once the channel matrix has been diagonalized, the MIMO channel is transformed into an equivalent set of uncoupled, parallel channels. The squared singular values are the SNRs for each of the parallel eigenchannels.

Once the channel matrix has been diagonalized, the water-filling solution is used to allocate transmitter power to the eigenchannels. This is a classical solution to the problem of optimally signaling over multiple parallel channels with different SNRs. The singular values from the SVD govern the allocation of power so that eigenchannels with higher SNR receive more transmit power (i.e., the singular vector is scaled by a larger value before combining with other singular vectors to form the overall excitations at the transmit array input ports). Eigenchannels with very low SNRs may receive no transmit power and are unused in the MIMO system. The water-filling solution presumes a fixed amount of power available at the transmitter, which corresponds to the compression points of the power amplifiers driving the array, the power available from batteries in a handheld device, regulatory limits on transmit power, or the power allocated to the transmitter in a dynamic spectrum sharing system. As the power accepted by the transmit array is influenced by mutual coupling, the transmit power constraint is relevant to the topic of this chapter and is considered further in Section 9.4.8.

Many techniques have been developed for conveniently combining the spatial modulation from the channel matrix SVD with the temporal modulation into space-time codes. Another large body of work has been devoted to methods for measuring the channel matrix using channel sounding, transmitting channel information from one node to another in a network, and a variety of other practical aspects of MIMO systems. Our concern here is the effects of mutual coupling on MIMO systems, so we turn our attention to the eigenchannels themselves that arise from diagonalizing the channel matrix.

9.3.3 Eigenchannels in MIMO Systems and Beamforming Arrays

The overall performance of a MIMO system and the impact of mutual coupling can be understood either through the overall capacity in terms of the signal and noise correlation matrices in the log-det formula, Eq. (9.10), or through the calculation of individual antenna parameters for each set of array beam former coefficients corresponding to an eigenchannel before combining into the overall array excitations. Both the correlation matrix formulation and the eigenchannel or "beam-by-beam" perspective will be considered in the following sections.

Based on Eq. (9.8), the columns of $\boldsymbol{V}$ and the columns of $\boldsymbol{U}$ from the singular value decomposition of the channel matrix are treated as beamformer weights at the transmitter and receiver, respectively, in a MIMO spatial coding scheme. There is a close relationship between MIMO systems and beamforming arrays. Later in this chapter, we will analyze the performance of a MIMO system on an eigenchannel basis. Each eigenchannel consists of a set of transmit array excitations, referred to below as a vector of antenna port input currents $\boldsymbol{i}$, and a set of receive array combining coefficients, denoted below by $\boldsymbol{w}$. Each eigenchannel is influenced differently by mutual coupling, represents a different impedance mismatch factor, and realized radiation efficiency at the transmitter, and at the receiver a different SNR, receiving efficiency, gain, directivity, antenna efficiency, noise matching efficiency, and other antenna parameters than other eigenchannels. The overall channel capacity in Eq. (9.10) arises from a weighted combination of the antenna parameters and noise characteristics associated with each eigenchannel or set of transmit and receive beam former coefficients.

9.3.4 Reference Planes and the Intrinsic Channel Matrix

The channel matrix can be defined with respect to different reference planes at the transmitter or receiver. At the transmit side, the transmit symbol vector $\boldsymbol{v}_\mathrm{s}$ could be defined as the outputs from digital signal processing hardware before analog signal handling (digital-to-analog conversion, mixing and up conversion in frequency, filtering, and amplification). The transmit signals could be defined instead as input voltages to the final driving power amplifiers connected to the transmitting array elements. Similarly, the output values could be taken at the receive array element terminals, at the loading amplifier output ports, or after digitization in a digital receiver implementation. Here, we view the channel matrix with the reference planes at the transmitter before the amplifiers connected to the transmit array elements and at the outputs of the amplifiers and other analog circuits connected to the receiving array elements.

To highlight how the channel matrix depends on mutual coupling between array elements, we will factor the channel matrix into the product of a matrix that includes multipath effects associated with the propagation environment and transformations that depend on the mutual impedance matrices of the transmit and receive arrays. The intrinsic part of the propagation channel matrix has its reference planes as close as possible to the transmit and receive array antenna elements. It relates the input currents at the transmitting array element ports to the open-circuit voltages induced by the signal and noise waves at the receiving array element ports (Warnick et al. 2018). This portion of the channel matrix depends on coupling between elements through the embedded element radiation patterns of the transmit and receive arrays, but it is independent of the impedance matrices of the transmit and receive arrays. The intrinsic part of the channel matrix includes all multipath effects and other properties of the environment around the transmit and receive arrays.

To identify the intrinsic part of the channel matrix, we can model a combined transceiver system as a multiport network with N ports at the transmit side and M ports at the receiver. The mutual impedance matrix of the combined transmit and receiver system can be written in block form as

$$\begin{bmatrix} \boldsymbol{v}_\mathrm{s} \\ \boldsymbol{v}_\mathrm{r} \end{bmatrix} = \begin{bmatrix} \boldsymbol{Z}_\mathrm{t} & \boldsymbol{Z}_\mathrm{tr} \\ \boldsymbol{Z}_\mathrm{rt} & \boldsymbol{Z}_\mathrm{r} \end{bmatrix} \begin{bmatrix} \boldsymbol{i}_\mathrm{s} \\ \boldsymbol{i}_\mathrm{r} \end{bmatrix} \tag{9.11}$$

where $\boldsymbol{Z}_\mathrm{t}$ is the mutual impedance matrix of the transmitting array and $\boldsymbol{Z}_\mathrm{r}$ is the mutual impedance matrix of the receiving array. The propagation channel is represented by $\boldsymbol{Z}_\mathrm{rt}$. This matrix relates the vector of input currents $\boldsymbol{i}_\mathrm{t}$ into the ports of the transmitting array to the open-circuit voltages $\boldsymbol{v}_\mathrm{r}$ induced at the ports of the receiving array and is the intrinsic part of the channel matrix. It includes the effects of the propagation environment, multipath, and all structures and materials around the antennas.

In terms of the mutual impedances, the channel matrix defined with reference planes at the transmitter amplifier inputs and at the receiver electronics chain outputs is

$$\boldsymbol{H} = \boldsymbol{Q}_{\mathrm{r}}\boldsymbol{Z}_{\mathrm{rt}}\boldsymbol{Q}_{\mathrm{t}} \tag{9.12}$$

where $\boldsymbol{Q}_{\mathrm{r}}$ and $\boldsymbol{Q}_{\mathrm{t}}$ are transformations from the reference planes at which the transmit symbols $\boldsymbol{v}_{\mathrm{s}}$ and the receiver outputs $\boldsymbol{v}_{\mathrm{r}}$ are defined to the transmit array input currents and receive array element port equivalent to open-circuit voltages.

If the details of the transmitter and receiver electronics are unimportant for a given analysis, we could move the channel input and output reference planes to the array element ports. This would place the reference planes in the definition of the channel matrix as close as possible to the array antenna elements. We could then redefine $\boldsymbol{v}_{\mathrm{s}}$ and $\boldsymbol{v}_{\mathrm{r}}$ in Eq. (9.4) to be the transmit element input currents and the receiver element open-circuit equivalent source voltages, respectively. The channel matrix would then be represented directly as $\boldsymbol{H} = \boldsymbol{Z}_{\mathrm{rt}}$. We will see that mutual coupling effects at the transmitting array and receiving array are reflected in the transformations $\boldsymbol{Q}_{\mathrm{t}}$ and $\boldsymbol{Q}_{\mathrm{r}}$. These transformations are derived for typical network models for the transmitter and receiver in the following sections.

9.4 Mutual Coupling and MIMO Transmitting Arrays

For an N-element transmitting array, the relationship between voltages at the element ports and currents flowing into the port referenced to one selected side of a two-conductor transmission line is

$$\boldsymbol{v} = \boldsymbol{Z}_{\mathrm{t}}\boldsymbol{i} \tag{9.13}$$

where $\boldsymbol{v}$ is an N-element column vectors of phasor voltages across the element ports and $\boldsymbol{i}$ is a vector of the element port input currents. For conciseness, we have dropped the subscripts on the element port current and voltage vectors.

The impedance matrix of the array has a real part and an imaginary part. We assume that the real part can be divided into loss due to conductive and dielectric structures in the antenna elements and power that is radiated into the far field. The impedance matrix can be represented in the form

$$\boldsymbol{Z}_{\mathrm{t}} = \boldsymbol{R}_{\mathrm{rad}} + \boldsymbol{R}_{\mathrm{loss}} + j\boldsymbol{X} \tag{9.14}$$

We will see later that the radiative part $\boldsymbol{R}_{\mathrm{rad}}$ of the impedance matrix can be found from the embedded element radiation patterns.

In either the S-matrix or Z-matrix formulation, the off-diagonal matrix elements represent mutual coupling between array elements. $Z_{\mathrm{t},mn}$ is equal to the voltage induced at the open-circuited terminals of the mth element when the nth element is excited by an input current of 1 A. Mutual coupling also perturbs the diagonal matrix elements so that the embedded element self-impedances $Z_{\mathrm{t},nn}$ are different from the isolated element input impedance Z_{in} that would be measured for one array element in the absence of other elements.

For the block diagram shown in Figure 9.1, we can arrange the generator impedances into a diagonal matrix and model the generators as an N-port network with impedance matrix $\boldsymbol{Z}_{\mathrm{g}} = Z_{\mathrm{g}}\boldsymbol{I}$ and N open-circuit phasor voltages given by the vector $\boldsymbol{v}_{\mathrm{s}}$. In most cases, the coupling between the driving

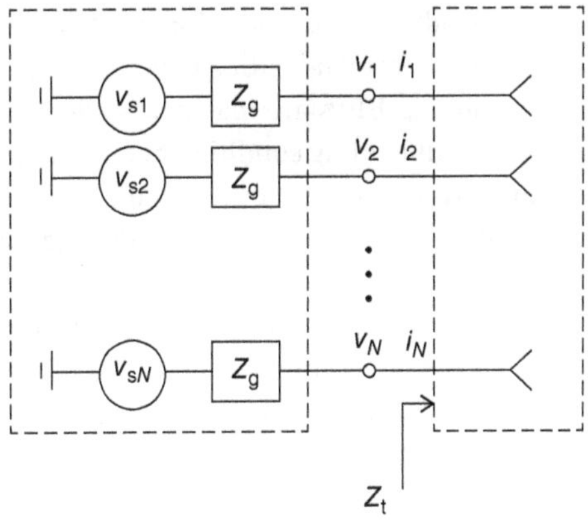

Figure 9.1 Array antenna transmitter block diagram. The transmitted signals for a MIMO system are typically generated digitally, converted to analog voltages, and presented to the array element ports through power amplifiers. The driving network is modeled by Thévenin equivalent sources with phasor voltages $v_{s1}, v_{s2}, \ldots, v_{sN}$ and generator impedance $\boldsymbol{Z}_g$. If the array is mutually coupled, the impedance matrix $\boldsymbol{Z}_t$ is nondiagonal. The array impedance matrix interacts with the impedances of the components directly connected to the array ports to determine the power transfer into the array.

amplifiers can be ignored, but if the coupling in the source network is significant, then $\boldsymbol{Z}_g$ is nondiagonal. For the source network, we must include a generator voltage term to the voltage/current relationship so that the network relationship for the array is

$$\boldsymbol{v} = -\boldsymbol{Z}_g\boldsymbol{i} + \boldsymbol{v}_s \tag{9.15}$$

The negative sign arises because i is defined to be the current flowing into the array ports. Combining Eqs. (9.13) and (9.15) leads to

$$\boldsymbol{i} = \underbrace{(\boldsymbol{Z}_g + \boldsymbol{Z}_t)^{-1}}_{\boldsymbol{Q}_t}\boldsymbol{v}_s \tag{9.16}$$

The transformation $\boldsymbol{Q}_t$ includes the effects of array mutual coupling on the power transfer from the driving sources to the transmitting array.

If the array is uncoupled, the input current at one element port is the generator voltage divided by the sum of the generator impedance and element port self-impedance. If each element self-impedance is well matched to the connected element, the power transfer to the transmitting array is close to the available power from the driving sources. If the array is coupled and $\boldsymbol{Z}_t$ is nondiagonal, then the power transfer depends on the relative amplitudes and phases of the generator voltages. For some sets of excitations, power transfer can be good, and for others, it may be poor. Impedance matching for coupled arrays is therefore dependent on the array excitations.

In beamforming arrays, a dramatic reduction in transferred power for some scan angles is the phenomenon of scan blindness. MIMO arrays do not scan in a traditional sense, but each set of eigenchannel excitation coefficients has a different impedance mismatch factor and realized radiation efficiency. An expression for the impedance mismatch factor for a mutually coupled array is given in Section 9.4.9.

9.4.1 Radiated Electric Field and Embedded Element Patterns

We have considered the network properties of the array at its input ports. Mutual coupling also affects the radiation patterns of elements in the array. We will characterize the radiation characteristics of an antenna array using the EEPs. The open-circuit loaded EEP $\mathbf{E}_n^t(\boldsymbol{r})$ is the radiated electric field with the nth element excited by a current I_0 and with all other elements open circuited so that their input currents are zero.

When an element in the array is excited, the field radiated by that element scatters from other elements into the far field. Because the scattering properties of neighboring antenna elements depend on the loads at their ports, EEPs depend on the loading condition at the ports of the nondriven elements. Common loading conditions for the nondriven elements when computing, measuring, or simulating EEPs are matched loads (e.g. 50 Ω), open circuited, or short circuited. We use the open-circuit loading condition here, as some of the relationships for antenna arrays become mathematically simpler with this loading condition. In particular, the total field radiated by the array is a simple linear combination of the open-circuit loaded EEPs multiplied by the antenna element input currents.[1]

The EEPs could be computed with different loading conditions for the nondriven elements. When EEPs are simulated using a numerical model, it is common to compute the EEPs with matched loads on the nondriven elements. If the EEPs are found with a different loading condition than open circuit, matrix transformations can be used to find the EEPs with open-circuit loads (Warnick et al. 2018). When the loading condition is varied or multiple loading conditions are in use, we might use a superscript on the EEP to denote the load on the nondriven elements so that the open-circuit loaded EEPs would be $\mathbf{E}_n^{\mathrm{t,oc}}(\boldsymbol{r})$.

In view of the open-circuit loading condition used in defining the EEPs, there is a simple relationship between the input currents to the array elements and the total radiated electric field. In terms of the input currents at each element, the total electric field radiated by the transmitting array is

$$\mathbf{E}^{\mathrm{t}}(\boldsymbol{r}) = \frac{1}{I_0}\sum_{n=1}^{N} i_{\mathrm{A},n}\mathbf{E}_n^{\mathrm{t}}(\boldsymbol{r}) \tag{9.17}$$

In a propagation environment, we can think of the EEPs as the fields radiated by the array when in the environment. Or, we could compute the EEPs with the array in free space and then use a scattering model to propagate the EEPs through the environment to find the incident field at the receiving array. In the latter case, we distinguish between the primary EEPs with the array in free space and the secondary EEPs including scattering effects from objects in the propagation environment.

The EEPs from the transmit array radiate through the propagation environment to the receiving array, inducing open-circuit voltages at the receive element terminals that are reflected in the intrinsic channel matrix $\boldsymbol{Z}_{21}$. The nth EEP $\mathbf{E}_n^{\mathrm{t}}$ from the transmit array when propagated through the scattering environment induces open-circuit voltage $Z_{\mathrm{rt},\,mn}I_0$ at the mth receive array element. The intrinsic channel matrix is therefore completely determined by the EEPs together with the propagation environment.

9.4.2 Pattern Overlap Matrix, Conservation of Energy, and Mutual Coupling

The power radiated by the transmitting array is

$$P_{\mathrm{t}} = \frac{1}{|I_0|^2}\boldsymbol{i}^{\dagger}\boldsymbol{A}_{\mathrm{t}}\boldsymbol{i} \tag{9.18}$$

---

1 We choose to parameterize array excitations with the element port input currents. This choice motivates the use of open-circuit loaded EEPs in our treatment. If the port voltages were used as the primary quantities for parameterizing the array excitations, then the short-circuit loaded EEPs would be most useful. If forward waves into the ports were used, then matched loads on the nondriven elements would be the best choice for the EEP loading condition in terms of the simplicity of expressions for the total radiated field and radiated power. In these three cases, the impedance matrix, admittance matrix, and S-parameter matrix formulations, respectively, are most convenient in the network analysis of the array.

where $\boldsymbol{i}$ is a vector of the input currents to each of the N transmit array elements. The matrix $\boldsymbol{A}_\mathrm{t}$ has elements given by the overlap integral

$$A_{\mathrm{t},mn} = \frac{1}{2\eta} \int \mathbf{E}^\mathrm{t}_m(\boldsymbol{r}) \cdot \mathbf{E}^{\mathrm{t}*}_n(\boldsymbol{r})\, r^2 d\Omega \tag{9.19}$$

The resulting matrix $\boldsymbol{A}_\mathrm{t}$ is referred to as the array element pattern overlap matrix.

By conservation of energy, the power input to the array is equal to the sum of the power dissipated in the array and radiated into the far field. Conservation of energy for the array can be expressed in terms of the mutual impedance matrix and pattern overlap matrix as (Warnick et al. 2018)

$$\underbrace{\frac{1}{4}\boldsymbol{i}^\dagger\left(\boldsymbol{Z}_\mathrm{t} + \boldsymbol{Z}_\mathrm{t}^\dagger\right)\boldsymbol{i}}_{\text{Input power}} = \underbrace{\frac{1}{2}\boldsymbol{i}^\dagger \boldsymbol{R}_\mathrm{loss}\boldsymbol{i}}_{\text{Dissipated power}} + \underbrace{\frac{1}{|I_0|^2}\boldsymbol{i}^\dagger \boldsymbol{A}_\mathrm{t}\boldsymbol{i}}_{\text{Radiated power}} \tag{9.20}$$

where $\boldsymbol{R}_\mathrm{loss}$ is the part of the impedance matrix due to ohmic and dielectric losses in the antenna array. If the array is reciprocal (so that $\boldsymbol{Z}_\mathrm{t} = \boldsymbol{Z}_\mathrm{t}^T$), then the impedance matrix is

$$\boldsymbol{Z}_\mathrm{t} = \frac{2}{|I_0|^2}\boldsymbol{A}_\mathrm{t} + \boldsymbol{R}_\mathrm{loss} + j\boldsymbol{X} \tag{9.21}$$

This result shows that the radiative part $\boldsymbol{R}_\mathrm{rad}$ of the impedance matrix in Eq. (9.14) is given by the pattern overlap matrix. If the antenna array is reciprocal and lossless ($\boldsymbol{R}_\mathrm{loss} = 0$), then this relationship simplifies to

$$\mathrm{Re}\left[\boldsymbol{Z}_\mathrm{t}\right] = \frac{2}{|I_0|^2}\boldsymbol{A}_\mathrm{t} \tag{9.22}$$

for the mutual resistance matrix. This relationship, which is a simple statement of conservation of energy for a lossless, reciprocal array, can be used to find the real part of the array mutual impedance matrix from the open-circuit loaded EEPs. We will see below that this relationship has a fundamental connection to the influence of mutual coupling on signal correlation in a MIMO system.

9.4.3 Gain and Directivity in the Overlap Matrix Formulation

In a MIMO system, the singular vector used to generate transmit array excitation coefficients for each eigenchannel has all of the common antenna parameters associated with it, including gain and directivity. The gain and directivity associated with the array excitations are not directly important for a MIMO channel. In a multipath environment, the transmitted power is usually radiated in multiple directions in such a way that maximal power arrives at the receiver, which means that gain or directivity in a single direction is insufficient to predict channel capacity (the relative phases of the radiated signal along different multipath directions are important as well, and the overall phase of a transmit array beam pattern lobe is not reflected in the directivity or gain). The expression for directivity in terms of the pattern overlap matrix, however, does provide a convenient way to understand superdirectivity effects in MIMO systems, so we develop the expression here.

To express the directivity of an array in terms of the EEPs and overlap matrix, we define the signal response matrix $\boldsymbol{B}$ with elements given by

$$B_{mn}(\boldsymbol{r}) = \frac{1}{2\eta}\mathbf{E}_m(\boldsymbol{r}) \cdot \mathbf{E}_n^*(\boldsymbol{r}) \tag{9.23}$$

where $\boldsymbol{r}$ is a point in the far field of the array. In terms of the element driving currents $\boldsymbol{i}$, the directivity is (Warnick et al. 2018)

$$D(\Omega) = \frac{4\pi r^2 \boldsymbol{i}^\dagger \boldsymbol{B}(\boldsymbol{r})\boldsymbol{i}}{\boldsymbol{i}^\dagger \boldsymbol{A}\boldsymbol{i}} \tag{9.24}$$

where Ω is the spherical angle of $\boldsymbol{r}$. We omit the subscript on $\boldsymbol{A}$ because this treatment applies to both transmissing and receiving arrays. The gain is

$$G(\Omega) = \frac{4\pi r^2 \boldsymbol{i}^\dagger \boldsymbol{B}(\boldsymbol{r})\, \boldsymbol{i}}{\frac{1}{2}|I_0|^2 \boldsymbol{i}^\dagger\, Re\, [\boldsymbol{Z}_\mathrm{A}]\, \boldsymbol{i}} \tag{9.25}$$

The denominator of Eq. (9.24) includes only radiated power, whereas the denominator of Eq. (9.25) includes both radiated power and power dissipated in the array.

9.4.4 Overlap Matrix for Isotropic Radiators

For an array of isotropic radiators, the pattern overlap matrix can be evaluated in closed form. The elements of the overlap matrix are

$$A_{mn} = P_\mathrm{el} \frac{\sin\left(kr_{mn}\right)}{kr_{mn}} \tag{9.26}$$

where $k = 2\pi f / c$ is the electromagnetic wavenumber, P_el is the power radiated by one element with input current I_0, and r_{mn} is the distance between the mth and nth elements. In view of Eq. (9.22), this expression gives a rough idea of how mutual coupling between small antenna elements behaves as a function of the element separation distance. Qualitatively, this is a prototypical form for the way mutual coupling depends on separation distance for many types of antennas.

9.4.5 Mutual Coupling for Closely Spaced Elements, Superdirectivity, and Q-Factor Bounds

From Eq. (9.26), the pattern overlap matrix for an array of closely spaced isotropic radiators (i.e. $kr_{mn} \to 0$) is

$$\boldsymbol{A} \simeq P_\mathrm{el} \begin{bmatrix} 1 & \cdots & 1 \\ \vdots & \ddots & \vdots \\ 1 & \cdots & 1 \end{bmatrix} \tag{9.27}$$

This is a rank one matrix and has only one nonzero eigenvalue. For other types of elements, such as dipoles, if the element polarizations are aligned, the pattern overlap matrix is also approximately a rank one matrix. If the polarizations of the elements are not aligned, the rank of the matrix can be higher than one, but regardless of the polarizations of the elements, for a tightly packed array, most of the eigenvalues of the overlap matrix are close to zero.

The directivity of an array is influenced by the spectrum of the pattern overlap matrix and how the beam former weight vector relates to its eigenvectors. If the driving currents are equal, then the directivity of a closely spaced array is usually modest in value. In this case, the vector of driving currents $\boldsymbol{i}$ is close to the eigenvector of the matrix, Eq. (9.27), with a nonzero eigenvalue. If the vector of driving currents $\boldsymbol{i}$ is orthogonal to this eigenvector and lies in the subspace for which the eigenvalues of the overlap matrix are close to zero, then the radiated power $\boldsymbol{i}^\dagger \boldsymbol{A} \boldsymbol{i}$ becomes small. In this case, the directivity of the array is large. This is superdirectivity.[2]

If the array is lossy, then the loss portion of the impedance matrix Eq. (9.21) enters into the denominator of the gain Eq. (9.25). In this case, the input power to the array is appreciable, even though the radiated power is close to zero. Ohmic or dielectric loss in the antenna array regularizes

2 Superdirectivity is "the condition that occurs when the antenna directivity is significantly higher than that with the array or aperture uniformly excited," IEEE_Std_145-2013.

the real part of the mutual impedance matrix and eliminates the small eigenvalues, and the excitation currents may produce superdirectivity, but the gain is modest. With a lossy antenna, supergain is harder to achieve than superdirectivity.

In studies of tightly packed MIMO arrays, when the array elements are modeled as lossless, the existence of superdirectivity current solutions can confound results and lead to erroneous predictions of channel capacity. The quality factor bound defined in Uzsoky and Solymar (1956) and Shamonina and Solymar (2015) is a way to eliminate driving current solutions from an analysis of a MIMO array which encroach too far into the small-eigenvalue subspace of the overlap matrix and lead to unwanted superdirective solutions. Another way to regularize the behavior of the array and avoid superdirective currents for tightly packed arrays with many elements is to include loss in the array model. In the absence of antenna losses or without superdirectivity bounds, the capacity of a MIMO channel comprising an array of very small elements is in principle unlimited. Limits on the superdirectivity of current solutions have been used in obtaining antenna-independent bounds on the capacity of a given propagation environment (Jensen and Wallace 2005).

9.4.6 EEPs, Mutual Coupling, and Minimum Scattering Antennas

The same interaction effects that cause nonzero mutual impedances between element ports also perturb the EEPs. Because the field scattered by the nondriven elements and perturbations to the field caused by the presence of the nondriven elements are included in the radiated field $\mathbf{E}_n^{\mathrm{t}}$, interactions between elements impact the fields radiated by elements in the array.

In some cases, the interactions between elements can have little or no effect on the EEPs. If the loading condition for the nondriven elements is a matched load, the load absorbs energy, so intuitively one might expect that the EEPs with matched loads on the nondriven elements are more strongly affected by interactions between elements than a purely reactive or open-circuit load that does not absorb energy. For a half-wave dipole antenna, the nondriven elements with open-circuit loads are equivalent to pairs of quarter-wave dipole antennas. The matched loaded half-wave dipole and short-circuited dipole are resonant structures, which generally have a higher scattering cross-section than nonresonant structures like a quarter-wave dipole.

This line of thinking can be formalized using the concept of the minimum scattering antenna (MSA). An MSA with a purely reactive load does not scatter an incident field (Kahn and Kurss 1965). For an array of MSAs, the embedded element radiation patterns are the same as the isolated antenna pattern for one element alone. With MSAs, the EEPs defined with nondriven element ports open circuited are not affected by mutual coupling or interactions between elements. In this case, there is a complete separation between mutual coupling effects and the array radiation pattern, because all mutual coupling effects are contained in $\boldsymbol{Q}_{\mathrm{t}}$. For MSAs, the intrinsic channel matrix $\boldsymbol{Z}_{21}$ is independent of interactions between array elements.

For non-MSA elements, the open-circuit loaded embedded element radiation patterns are in general different from the isolated antenna pattern of the array element. In this case, mutual coupling effects are reflected in the transformation $\boldsymbol{Q}_{\mathrm{t}}$. In addition, the intrinsic part of the channel matrix $\boldsymbol{Z}_{\mathrm{rt}}$ is influenced by coupling between elements though the dependence of the embedded element radiation patterns on the presence of other elements in the array.

9.4.7 Mutual Coupling and Interactions Between Elements

In the IEEE standard for antenna terms, mutual coupling is defined to be "interactions of an antenna with one or more other antennas." Mutual coupling effects are defined as "the changes in the radiation pattern and impedance of an antenna due to mutual coupling" (*IEEE Standard Definitions of Terms for Antennas*). To understand the effects of mutual coupling between elements on the behavior of antenna arrays, it is useful to consider the effects of coupling on the radiation patterns and impedances separately.

For an array antenna, this means separating the effects of coupling on the mutual impedance matrix from the influence of coupling on the EEPs. If $Z_{mn} \neq 0$ for $m \neq n$, then at the given frequency, elements m and n are mutually coupled in the network port impedance sense. If the nth EEP $\mathbf{E}_n^{\mathrm{t}}$ is different from the isolated element pattern $\mathbf{E}_{\mathrm{el}}^{\mathrm{t}}$ without other elements present, we say that the nth element EEP is affected by interactions with the other elements. The two effects can be summarized as follows:

- Network port coupling: Nonzero voltage at the terminals of one element when another element is excited ($Z_{mn} \neq 0$, $m \neq n$).
- EEP perturbation: The open-circuit loaded EEP for an element is different from its isolated pattern due to scattering from other elements in the array.

Port coupling and pattern perturbation are certainly related. Port coupling is caused by radiated fields from a driven element that induces a voltage at the terminals of another element. The same radiated fields also scatter from other elements, causing perturbations to the EEP of the driven element. Even though the two effects are connected, the logical distinction between network port mutual coupling and perturbations to the EEPs can be valuable when trying to understand mutual coupling. Two extreme cases can be used to illustrate this idea: a decoupled array and an array of MSAs.

Any array can theoretically be decoupled by adding a $2N$-port decoupling network. An antenna array with a multiport decoupling network has a diagonal mutual impedance matrix. This array is not mutually coupled in the port sense. However, by Eq. (9.22), the EEPs must necessarily be orthogonal so that the EEP overlap matrix is diagonal. Relative to the isolated element patterns, which are generally not orthogonal, the orthogonal EEPs of an array with decoupling network are typically highly distorted relative to the isolated element patterns. The perturbation to the EEPs arises in part because the decoupling network distributes an excitation at a single port to multiple elements in the array. Interestingly, the elimination of port coupling requires that the EEPs of the decoupled array be strongly perturbed relative to the isolated patterns of the individual elements without the decoupling network. For a decoupled array, the ports are not coupled, but the EEPs are strongly perturbed.

The converse of the decoupled array is an array of MSAs. The open-circuit loaded EEPs are identical to the isolated element pattern, but the ports are coupled. Surprisingly, even though the elements of an array of MSAS do not interact in terms of the open-circuit loaded EEPs, an array of MSAs is in general mutually coupled. For an array of MSAs, neighboring open-circuited elements do not scatter incident fields, which means that the EEPs are identical to the isolated element patterns, and there are no interaction effects in the open-circuit loaded EEPs. Even so, the array EEP overlap matrix is nearly always nondiagonal. It follows from Eq. (9.22) that the real part of the mutual impedance matrix is also nondiagonal. Therefore, an array of MSAs is mutually coupled, but the elements do not interact in the sense that the open-circuit loaded EEPs are unperturbed by the presence of other array elements.

A simple example is an array of infinitesimal dipoles. Scattering by infinitesimal dipoles is negligible, so the EEPs are unperturbed in the array. It is easy to show by computing the pattern overlap matrix that the array element ports must be mutually coupled (except for certain element spacings for which the overlap integrals are identically zero). This can be understood physically in the receiving case. No matter how small the dipole, its isolated effective area has a finite size. If the dipoles are closely spaced, the effective area of the array cannot be much larger than the physical extent of the array. Consequently, the embedded effective areas of the dipoles must be smaller than the isolated effective area of one dipole. This can be visualized using the Poynting streamlines technique (Diao and Warnick 2016).

The reduction in effective area of the embedded dipoles is caused by port coupling and is mathematically associated with the off-diagonal elements of the mutual impedance matrix. From a network theory perspective, the available power from the dipoles is restricted by the off-diagonal elements of the impedance matrix, so the array must necessarily be mutually coupled based on conservation of energy. If a decoupling network were added to the array of infinitesimal dipoles, the array would not be mutually coupled, but the EEPs would be very different from the radiation pattern for a single dipole.

9.4.8 Transmitter Power Constraint

As discussed in Section 9.3.2, in order to obtain a bounded channel capacity, the transmit power must be constrained. From a practical perspective, transmit power is limited by spectrum access regulations, limited battery life in a portable device, and other system requirements.

The power accepted by the transmitting array is influenced by mutual coupling. The time average power accepted by the transmitter is the time expectation

$$P_{\mathrm{in}} = \frac{1}{2}\mathrm{E}\left[\mathrm{Re}\left\{\boldsymbol{v}_{\mathrm{s}}^{\dagger}\boldsymbol{Q}_{\mathrm{t}}^{\dagger}\boldsymbol{Z}_{\mathrm{t}}\boldsymbol{Q}_{\mathrm{t}}\boldsymbol{v}_{\mathrm{s}}\right\}\right] \tag{9.28}$$

of the instantaneous input power. The symbol vector $\boldsymbol{v}_{\mathrm{s}}$ represents the excitations delivered to the transmit array by driving amplifiers. If the array losses are significant, the real part of the transmit array impedance matrix can be replaced by the radiation part of the impedance matrix, which is computed from the element pattern overlap matrix, and the transmitted power $P_{\mathrm{t}} = P_{\mathrm{in}} - P_{\mathrm{loss}}$ can be expressed using the overlap matrix according to Eq. (9.18).

If the transmit array is uncoupled and the elements are identical and resonant (so that the input reactance is zero), then $Re\{\boldsymbol{Z}_{\mathrm{t}}\} = R_{\mathrm{rad}}\boldsymbol{I}$. If the elements are conjugate matched to the generators and loads, then $\boldsymbol{Q}_1 = \boldsymbol{Q}_2 = \frac{1}{2}\boldsymbol{I}$. In this case, the transmitted power becomes

$$P_{\mathrm{t}} = \frac{1}{8}R_{\mathrm{rad}}\mathrm{E}\left[\boldsymbol{v}_{\mathrm{s}}^{\dagger}\boldsymbol{v}_{\mathrm{s}}\right] = \frac{1}{8}R_{\mathrm{rad}}\ \mathrm{tr}\,\boldsymbol{R}_{\mathrm{s}} \tag{9.29}$$

where tr denotes the matrix trace. In the uncoupled case, we can express the power constraint in the form

$$\max_{\boldsymbol{R}_{\mathrm{s}}}\ \mathrm{tr}\,\boldsymbol{R}_{\mathrm{s}} \leq P_{\mathrm{t}} \tag{9.30}$$

where we have lumped the scale factors into P_{t}. This is a trace constraint on the signal correlation matrix. If the array elements are mutually coupled, the trace constraint is only an approximation to the actual transmitted power. Differences between MIMO systems performance with the trace constraint and more complex constraints on actual power for a coupled array have been considered in the literature (Wallace and Jensen 2004a).

9.4.9 Impedance Matching at the Transmitter

The impact of mutual coupling on the transmitter is primarily through its influence on impedance matching and power transfer from driving amplifiers to the transmitting array. The power transferred to a mutually coupled array depends on the relative amplitudes and phases of the driving currents. Here, we develop an expression for the ratio of power accepted by the array to the power available from the driving amplifiers, or the impedance mismatch factor.

We assume for a moment that the driving signals input to the transmit array are time-harmonic. The available power from matched power amplifiers with generator voltages v_s is

$$P_a = \frac{\boldsymbol{v}_s^\dagger \boldsymbol{v}_s}{8Z_0} \tag{9.31}$$

The input power into a reciprocal array is

$$P_{in} = \frac{1}{2}\boldsymbol{i}^\dagger \mathrm{Re}\left[\boldsymbol{Z}_t\right] \boldsymbol{i} \tag{9.32}$$

Using Eq. (9.16), the ratio of the input power to the available source power or the impedance mismatch factor is

$$\frac{P_{in}}{P_a} = \frac{4Z_0}{\boldsymbol{v}_s^\dagger \boldsymbol{v}_s}\boldsymbol{v}_s^\dagger \left[\left(Z_0\boldsymbol{I} + \boldsymbol{Z}_t\right)^{\dagger -1} \mathrm{Re}\left[\boldsymbol{Z}_t\right] \left(Z_0\boldsymbol{I} + \boldsymbol{Z}_t\right)^{-1}\right] \boldsymbol{v}_s \tag{9.33}$$

If the transmitting array is matched and uncoupled, then $\boldsymbol{Z}_t = Z_0\boldsymbol{I}$, and the impedance mismatch factor is unity for all vectors of driving voltages $\boldsymbol{v}_s$. If the array is mutually coupled, then the impedance mismatch factor varies between zero and unity, depending on the relative amplitudes and phases of the driving voltages.

In the case of a lossy array, the radiation efficiency $\eta_{rad} = P_t/P_{in}$ is less than unity. The product of the impedance mismatch factor of the array with transmit symbol vector $\boldsymbol{v}_s$ and the radiation efficiency is the realized radiation efficiency. The power absorbed by antenna losses can be quantified using the loss portion of the mutual impedance matrix given in Eq. (9.21). Measuring or modeling loss for an array antenna is challenging. More details on antenna loss modeling and measurement can be found in Warnick et al. (2018).

For a coupled array, the impedance mismatch factor and realized radiation efficiency depend on the relative phases and amplitudes of the driving voltages supplied to the array. In a MIMO system, the driving voltages $\boldsymbol{v}_s$ consist of a combination of right singular vectors of the channel matrix. The singular vectors are multiplied by symbols from the temporal modulation scheme and scaled in amplitude to allocate more transmit power to the eigenchannels with higher SNRs than others as dictated by the water-filling solution. As the transmit symbols change in phase and amplitude over time, for a MIMO system, an expectation over time must be included in Eqs. (9.31)–(9.33).

Each singular vector acts as a beam former weight vector, steering the energy in the temporal symbol applied to that singular vector in a particular direction or set of directions in the multipath propagation environment. Multiple singular vectors weighted by temporal symbols are summed digitally before sending to the transmitter driver amplifiers as the generator voltage vector $\boldsymbol{v}_s$.

Because each singular vector has elements with different phases and amplitudes, each singular vector has a different impedance mismatch factor, realized radiation efficiency, and realized gain. The overall efficiency of power transfer changes over time as the channel matrix and the relative weighting of the eigenchannels evolve. To properly weigh the singular vector, the power matching must be accounted for in the space-time coding. In a conventional MIMO system, this happens when predetermined training symbols are transmitted to the receiver in order to sound the channel and measure the channel matrix $\boldsymbol{H}$, including mutual coupling effects.

If impedance matching is poor for one or more sets of spatial symbols, then power is reflected from the transmitting array and dissipated in the driving amplifiers. In a power-limited system, such as a battery powered device, the SNR at the receiver is reduced, and hence, the channel capacity is smaller. This reduces the realized radiation efficiency of the transmitter. With a highly coupled array, the effect is more pronounced. Remedies for the variability in realized radiation efficiency from one singular vector to another can be dealt with by using decoupling techniques to reduce coupling in the array.

As we have seen, at the transmit side of a MIMO system, mutual coupling primarily affects power transfer. Interaction effects also impact the EEPs, but this is usually not a major factor in overall system performance. At the receive side, mutual coupling affects signal correlation, signal power transfer, and receiver noise. The analysis is richer and more complicated.

9.5 Mutual Coupling and MIMO Receiving Arrays

In a MIMO system, the fields radiated by the transmitter, along with external noise and interference, induce voltages terminals of the receiving array antenna elements. Each eigenchannel exploited in the water-filling solution for a MIMO transmitter can be represented as a classical beamforming array, as shown in Figure 9.2. With digital signal processing, the sampled outputs of the array receiver chains are effectively replicated and provided to multiple beamforming networks with different sets of beam former coefficients $\boldsymbol{w}$. There are various ways to implement MIMO processing, particularly when combining the spatial processing with temporal modulation in a space–time coding scheme, but the effects of the antenna array on the system can be understood through the simple beamforming architecture of Figure 9.2.

9.5.1 Receive Array Signal and Noise Model

Our goal in this section is to model a MIMO receiving array through the signal and noise correlation matrices that appear in the log-det capacity formula Eq. (9.10). The signal correlation matrix

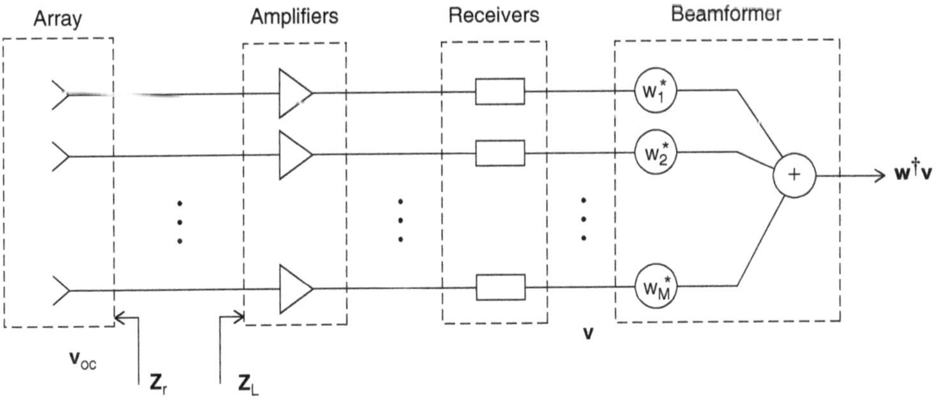

Figure 9.2 Receiver system block diagram for one MIMO eigenchannel. The receive array has mutual impedance matrix $\boldsymbol{Z}_r$ and is loaded by amplifiers or other circuit components with impedance matrix $\boldsymbol{Z}_L$. Open-circuit voltages induced by signal and noise at the receive array element ports are arranged into a column vector $\boldsymbol{v}_{oc}$. After loading by the receiver chains and digital sampling, the phasor voltages before beamforming are denoted by $\boldsymbol{v}$. The beamformer coefficients are taken from the left singular vector of the channel matrix corresponding to one of the eigenchannels in the MIMO signaling water-filling solution. The scalar output $\boldsymbol{w}^\dagger \boldsymbol{v}$ of the eigenbeam is demodulated to recover information transmitted on the eigenchannel. The beamformer block is replicated in digital signal processing for each eigenchannel used in the water-filling solution for optimal MIMO channel capacity.

is a combination of the correlation matrices associated with multipaths in the propagation environment. The noise response consists of external thermal or ambient noise, noise due to losses in the array antenna, and receiver noise. This leads to the receiver output voltage correlation matrix model

$$\boldsymbol{R}_v = \boldsymbol{R}_{\text{sig}} + \underbrace{\boldsymbol{R}_{\text{ext}} + \boldsymbol{R}_{\text{loss}} + \boldsymbol{R}_{\text{rec}}}_{\boldsymbol{R}_{\text{n}}} \tag{9.34}$$

Interference from other transmitters may also be included as an additional contribution to the noise model. Each term in the receiver output voltage correlation matrix is influenced or determined in part by the coupling between array elements. We will see that the signal and noise response of the array both have deep theoretical relationships to array mutual coupling.

The signal correlation matrix in a multipath environment is a combination of the response of the array to plane waves arriving at the receiver over all of the multipath components arriving at the array. The array response to a plane wave is developed in Section 9.5.2. The rank one plane wave correlation matrix is summed over the multipath arrival angles to obtain the signal correlation matrix. As the number of multipaths increase and the propagation environment becomes richer, the signal correlation matrix approaches the array mutual resistance matrix. The history of this important and surprising result was discussed in Section 9.2, and we expand on the discussion below in Section 9.5.5.

The receive array noise response has an equally intriguing connection with array mutual coupling. We will show in Section 9.5.4 that in an isotropic external thermal noise environment, with the array at the same physical temperature as the brightness temperature of the environment, the external and loss noise terms combine and, by Twiss's theorem (Twiss 1955), are together proportional to the array mutual resistance matrix.

9.5.2 Receive Array Thévenin Equivalent Network

We parameterize the received signals in terms of the induced open-circuit voltages at the terminals of the receiving antennas. Together with the array mutual impedance matrix, this provides a network Thévenin equivalent source for the receiving array. The network equivalent source will allow us to use network theory to determine the loaded outputs of the receiver chains attached to the receive array elements.

The impedance matrix associated with the receiving array is $\boldsymbol{Z}_{\text{r}}$. If the array is reciprocal, then the impedance matrix for the receiving network is the same as the impedance matrix with the array operated in transmit mode (de Hoop 1975).

To determine the open-circuit voltages induced at the ports by an incident wave, we use a relationship based on electromagnetic reciprocity (see Sec. 3.2.2). For a single incident plane wave of the form

$$\mathbf{E}^{\text{inc}} = E^{\text{inc}} e^{-j\hat{k}^{\text{inc}}\cdot\boldsymbol{r}}\hat{p} \tag{9.35}$$

with amplitude E^{inc} and polarization $\hat{p}$, the open-circuit voltage induced at the terminals of the nth array element is (de Hoop 1975; de Hoop and de Jong 1974; Davis and Licul 2005; Kildal 2015; Warnick et al. 2018)

$$v_{\text{oc},n} = \frac{1}{I_0}\frac{4\pi jre^{jkr}}{\omega\mu}E^{\text{inc}}\hat{p}\cdot\mathbf{E}_n^{\text{r}}\left(-r\hat{k}^{\text{inc}}\right) \tag{9.36}$$

where $\mathbf{E}_n^{\text{r}}$ is the embedded element radiation field pattern of the nth receive array element. Paralleling the definition of the EEPs for the transmit array, $\mathbf{E}_n^{\text{r}}$ is the electric field radiated by the

nth element of the receive array when excited with the input current I_0 and with all other elements open circuited. The point $\boldsymbol{r}$ at which the EEP is evaluated lies in the direction from which the plane wave arrives, and r is the distance from the receive array to $\boldsymbol{r}$, so that $\boldsymbol{r} = -r\hat{k}^{\text{inc}}$.

Arranging the open-circuit voltages at each element port into a column vector leads to

$$\boldsymbol{v}_{\text{oc}} = c_1 E^{\text{inc}} \boldsymbol{E}_p \tag{9.37}$$

The scale factor is

$$c_1 = \frac{4\pi j r e^{jkr}}{\omega\mu I_0} \tag{9.38}$$

The components of the EEPs in the direction of the incident field polarization $\hat{p}$ are arranged into the vector

$$\boldsymbol{E}_p = \left[\hat{p}\cdot\overline{E}_1^{\text{r}}\,\hat{p}\cdot\overline{E}_2^{\text{r}}\cdots\hat{p}\cdot\overline{E}_M^{\text{r}}\right]^T \tag{9.39}$$

where M is the number of receive array elements. This matrix notation is convenient in joining the electromagnetic properties of the array antenna with the network model and correlation matrix notation used in array signal processing to parameterize the signal and noise response of the array. With the open-circuit voltages $\boldsymbol{v}_{\text{oc}}$ and the receive array mutual impedance matrix $\boldsymbol{Z}_{\text{r}}$, the receiving array can be modeled as a multiport Thévenin equivalent source.

In free space without multipath scattering, the incident field in Eq. (9.35), a plane wave approximation to the field radiated by the transmitter in Eq. (9.17), is evaluated at the location of the receiving array. In a multipath propagation environment, the transmitted field is scattered by objects and structures in the propagation environment to produce the incident field $\mathbf{E}^{\text{inc}}$ at the receiver (or the secondary field as termed in Section 9.4.1 evaluated at the receiver location). In this case, Eq. (9.37) is summed over each plane wave component arriving at the receiver to produce the total open-circuit voltage at the receive array elements. For continuum incident fields such as blackbody radiation, Eq. (9.37) can be integrated over angle of arrival $\hat{k}^{\text{inc}}$ to obtain the second-order statistics or the correlation matrix of the noise waveforms at the receive array (Warnick et al. 2018).

9.5.3 Loaded Receive Array Output Voltages

To develop an electrical model for the array, we need to relate the open-circuit voltages $\boldsymbol{v}_{\text{oc}}$ to the voltages $\boldsymbol{v}$ at the receiver outputs before the beamformer coefficients are applied. The element ports are terminated with loads represented by the impedance matrix $\boldsymbol{Z}_{\text{L}}$. For an active array, the loads at the element terminals represent the input impedances of low-noise amplifiers (LNAs). If the amplifiers and receiver chains are identical and uncoupled, then $\boldsymbol{Z}_{\text{L}} = Z_L\boldsymbol{I}$.

We will arrange the input currents at the receive array element ports into the column vector $\boldsymbol{i}$ and the voltages at the ports similarly into the vector $\boldsymbol{v}$. Using the network relationships,

$$\boldsymbol{v} = \boldsymbol{Z}_{\text{r}}\boldsymbol{i} + \boldsymbol{v}_{\text{oc}} \tag{9.40}$$

and

$$\boldsymbol{v} = -\boldsymbol{Z}_{\text{L}}\boldsymbol{i} \tag{9.41}$$

we find that

$$\boldsymbol{v} = \boldsymbol{Z}_{\text{L}}\left(\boldsymbol{Z}_{\text{L}} + \boldsymbol{Z}_{\text{r}}\right)^{-1}\boldsymbol{v}_{\text{oc}} \tag{9.42}$$

This is a network generalization of the voltage divider expression from circuit theory and is similar in form to Eq. (9.16) for a transmitting array.

The loaded receiver output voltages after amplification and complex baseband demodulation before beamforming are arranged into the M element column vector $\boldsymbol{v}$. The receiver output voltages are related to the open-circuit voltages by a matrix $\boldsymbol{Q}_\mathrm{r}$ defined such that

$$\boldsymbol{v} = \boldsymbol{Q}_\mathrm{r}\boldsymbol{v}_\mathrm{oc} \tag{9.43}$$

If the amplifiers and receivers are identical and uncoupled with complex voltage gain g, then in view of Eq. (9.42), the transformation matrix is

$$\boldsymbol{Q}_\mathrm{r} = g\boldsymbol{Z}_\mathrm{L}\left(\boldsymbol{Z}_\mathrm{L} + \boldsymbol{Z}_\mathrm{r}\right)^{-1} \tag{9.44}$$

In terms of S-parameters, this becomes $\boldsymbol{Q}_\mathrm{r} = g\frac{1}{2}\left(\boldsymbol{I}+\boldsymbol{S}_\mathrm{L}\right)\left(\boldsymbol{I}-\boldsymbol{S}_\mathrm{r}\boldsymbol{S}_\mathrm{L}\right)^{-1}\left(\boldsymbol{I}-\boldsymbol{S}_\mathrm{r}\right)$. This matrix represents the relationship between the open-circuit voltages at the array element terminal reference plane and the array output voltages after the signal chains connected to each array elements.

9.5.4 External Noise and Loss Noise

Referring to the signal and noise model in Eq. (9.34), noise at the receiving array outputs consists of three contributions: external noise from sources in the environment around the array, loss noise from conductive and dielectric materials in the array, and receiver noise from electronic components after the array element ports. The external noise and loss noise will be considered in this section. Receiver noise is treated in Section 9.5.7.

The correlation matrix of the noise voltages induced by external sources at the array outputs depends on the spatial distribution of the noise or interference arriving at the receiving array. Under the simplifying assumption that the noise arrives with equal amplitude from all directions and with no correlation between noise waves arriving from different angles, the noise correlation matrix is (Warnick et al. 2018)

$$\boldsymbol{R}_{ext,\mathrm{iso}} = \frac{1}{|I_0|^2}16k_\mathrm{B}T_\mathrm{iso}B\boldsymbol{Q}_\mathrm{r}\boldsymbol{A}_\mathrm{r}\boldsymbol{Q}_\mathrm{r}^\dagger \tag{9.45}$$

where $\boldsymbol{A}_\mathrm{r}$ is the pattern overlap matrix defined in Eq. (9.19) but evaluated for the receiving array. T_iso is the equivalent brightness temperature of the external noise environment in Kelvin. This interesting relationship highlights a link between the noise received by the array and the transmitting properties of the receiving array (the EEPs in the overlap matrix $\boldsymbol{A}_\mathrm{r}$). The transmitting properties of the receiving array are not to be confused with the transmitting array in the MIMO system, of which Eq. (9.45) is independent.

The noise contributed by losses in the receiving array antenna elements has the correlation matrix

$$\boldsymbol{R}_\mathrm{loss} = 8k_\mathrm{B}T_\mathrm{ph}B\boldsymbol{Q}_\mathrm{r}\boldsymbol{R}_\mathrm{r,loss}\boldsymbol{Q}_\mathrm{r}^\dagger \tag{9.46}$$

where T_ph is the physical temperature of the array in Kelvin. In the case of thermal equilibrium between the array and the external environment, in view of Eq. (9.20), the isotropic external noise together with the loss noise combines to the thermal noise correlation matrix

$$\boldsymbol{R}_\mathrm{t,iso} = 8k_\mathrm{B}T_\mathrm{iso}B\boldsymbol{Q}_\mathrm{r}\, Re\left[\boldsymbol{Z}_\mathrm{r}\right]\boldsymbol{Q}_\mathrm{r}^\dagger \tag{9.47}$$

This expression is Twiss's theorem with minor modifications for the receiving array (Twiss 1955).

9.5.5 Signal Correlation Matrix

As pointed out in Section 9.2, one of the critical effects of coupling on a MIMO system is through the correlation of signals at receiving array elements in a multipath environment. For use in analyzing this effect, we will derive an expression for the signal correlation matrix at the array output ports in terms of the distribution of waves arriving at the receiver in a multipath propagation environment.

For a single incident plane wave, the signal correlation matrix is the rank one matrix

$$\boldsymbol{R}_{\text{sig}} = \text{E}\left[\boldsymbol{v}_{\text{sig}}\boldsymbol{v}_{\text{sig}}^{\dagger}\right] = \boldsymbol{v}_{\text{sig}}\boldsymbol{v}_{\text{sig}}^{\dagger} \tag{9.48}$$

where we have assumed that the amplitude of the incident wave is constant and the time expectation has no effect. We can use Eqs. (9.37) and (9.43) to express the signal correlation matrix in terms of the EEPs as

$$\boldsymbol{R}_{\text{sig}} = \boldsymbol{Q}_{\text{r}}\underbrace{c_2 S_{\text{sig}}\boldsymbol{E}_p(\boldsymbol{r})\boldsymbol{E}_p^{\dagger}(\boldsymbol{r})}_{\boldsymbol{R}_{\text{sig,oc}}}\boldsymbol{Q}_{\text{r}}^{\dagger} \tag{9.49}$$

where $c_2 = 2\eta|c_1|^2$, S_{sig} is the incident power density of the signal at the location of the array, and r is a point such that $\hat{r}$ is in the direction from which the plane wave arrives. In this expression, we have identified the correlation matrix $\boldsymbol{R}_{\text{sig, oc}}$ of the element port open-circuit voltages induced by the plane wave.

In a multipath environment, we must add terms of the form of Eq. (9.49) weighted by the amplitudes of all arriving waves at the receiver. If the multipath environment is characterized statistically, then we can use the fundamental theorem of expectation to find the signal correlation matrix in the multipath environment using

$$\boldsymbol{R}_{\text{sig,multipath}} = \int d\hat{\mathfrak{p}} \int d\Omega\, \boldsymbol{Q}_{\text{r}} c_2 S_{\text{sig}}\left(\Omega, \hat{\mathfrak{p}}\right) \boldsymbol{E}_p(r)\boldsymbol{E}_p^{\dagger}(\boldsymbol{r})\,\boldsymbol{Q}_{\text{r}}^{\dagger} \tag{9.50}$$

where $S_{\text{sig}}\left(\Omega, \hat{\mathfrak{p}}\right)$ is the distribution of the power in the incident wave as a function of arrival angle Ω and polarization $\hat{\mathfrak{p}}$. The integrals are over the direction of arrival Ω, or the spherical angle of $\boldsymbol{r}$, and the polarization of the incident field.

9.5.6 Signal Correlation in a Rich Multipath Environment

In a rich multipath environment, the distribution of incident power over arrival angle and polarization in Eq. (9.50) is uniform, and the power distribution in Eq. (9.50) is constant. In this case, the integrals in Eq. (9.50) are equivalent to the pattern overlap integral in Eq. (9.19) so that

$$\boldsymbol{R}_{\text{sig,uniform}} \sim \boldsymbol{Q}_{\text{r}}\boldsymbol{A}_{\text{r}}\boldsymbol{Q}_{\text{r}}^{\dagger} \tag{9.51}$$

The correlation matrix of the open-circuit voltages at the array element ports due to signals arriving in a rich multipath environment is

$$\boldsymbol{R}_{\text{sig,uniform,oc}} \sim \boldsymbol{A}_{\text{r}} \tag{9.52}$$

In terms of S-parameters, and in the case of a lossless array, the overlap matrix can be related to the S-parameter matrix according to

$$\boldsymbol{A}_{\text{r}} = \frac{|I_0|^2}{2} Z_0\left(\boldsymbol{I} - \boldsymbol{S}_{\text{r}}^{\dagger}\right)^{-1}\left(\boldsymbol{I} - \boldsymbol{S}_{\text{r}}^{\dagger}\boldsymbol{S}_{\text{r}}\right)\left(\boldsymbol{I} - \boldsymbol{S}_{\text{r}}\right)^{-1} \tag{9.53}$$

This expression is equivalent for a two-element array to Eq. (9.1) for the correlation between signals received by array elements in a rich multipath environment.

These results highlight interesting connections between array antenna theory, multiport noise theory, and array signal processing. The signal correlation Eq. (9.52) for a lossless array in a rich multipath environment, the external isotropic thermal noise correlation matrix Eq. (9.47), and the array mutual resistance matrix are each related to the pattern overlap matrix in Eq. (9.22).

The connection between the isotropic noise correlation matrix $\boldsymbol{R}_{\mathrm{t,\,iso}}$ and the rich multipath environment signal correlation matrix $\boldsymbol{R}_{\mathrm{sig,\,uniform}}$ can be understood by appreciating that in the case of a thermal environment, noise waves arrive with equal probability from all angles at the array, just as in a rich multipath environment reflected signal waves arrives with equal probability from all directions. This explains the proportionality between $\boldsymbol{R}_{\mathrm{t,\,iso}}$ and $\boldsymbol{R}_{\mathrm{sig,\,uniform}}$.

The relationship between the overlap matrix and the mutual resistance matrix can be understood from the denominator of the directivity of the array in transmit mode, Eq. (9.24). The radiated power is obtained by integrating the EEPs to accumulate the power transmitted by the array over all angles. Since the pattern overlap matrix obtained in this way controls radiated power, it is equivalent up to a scale factor to the radiation part of the mutual resistance matrix.

For a given array, the isotropic noise response, the receive mode signal correlation in a rich multipath environment, and the total radiated power when the array is operated in transmit mode all involve in some sense an integral over a full sphere. Since this integration over a sphere is common to all three formulas, in the lossless case, the resulting matrices share a common factor – the array pattern overlap integral matrix.

From Eq. (9.52), strong mutual coupling in an array causes received signals in a rich multipath environment to be highly correlated. This reduces the diversity gain and channel capacity that can be achieved by a MIMO system. As we saw above, mutual coupling can reduce the power radiated by the transmitter. We will see in Section 9.5.7 that coupling can increase receiver noise and reduce SNR, further impacting the channel capacity of a MIMO system.

9.5.7 Mutual Coupling, Noise Matching, and Equivalent Receiver Noise

From the classical theory of noisy networks, the noise properties of a two-port device can be characterized with two real parameters and one complex parameter. Various formulations for the noise parameters are available (Engberg and Larsen 1995). Low-noise transistor device specifications typically include the minimum noise temperature T_{min}, optimum source impedance $\boldsymbol{Z}_{\mathrm{opt}}$ or equivalently the optimal source reflection coefficient Γ_{opt}, and the noise resistance R_{n}. Forward and reverse noise wave parameters are used in S-parameter treatments of two-port noise theory. Op amps are often specified by the RMS voltage noise spectral density v_n (V/$\sqrt{Hz}$) and the RMS current noise spectral density i_n (A/$\sqrt{Hz}$). For full noise characterization, the voltage and current noise spectral densities must be augmented by the correlation admittance parameter, which specifies the degree of correlation of the current and voltage noise waveforms.

In the impedance matrix formulation used in this chapter, the voltage and current noise spectral density parameters are most convenient. The correlation matrix associated with noise due to front-end amplifiers in the receiver system is (Warnick et al. 2018)

$$\boldsymbol{R}_{\mathrm{rec}} = 2B\boldsymbol{Q}_{\mathrm{r}}\left(\boldsymbol{V}^2_{\mathrm{n,R}} + \boldsymbol{Z}_{\mathrm{r}}\boldsymbol{Y}_{\mathrm{c}}\boldsymbol{V}^2_{\mathrm{n,R}} + \boldsymbol{V}^2_{\mathrm{n,R}}\boldsymbol{Y}^{\dagger}_{\mathrm{c}}\boldsymbol{Z}^{\dagger}_{\mathrm{r}} + \boldsymbol{Z}_{\mathrm{A}}\boldsymbol{I}^2_{\mathrm{n,R}}\boldsymbol{Z}^{\dagger}_{\mathrm{r}}\right)\boldsymbol{Q}^{\dagger}_{\mathrm{r}} \tag{9.54}$$

where $\boldsymbol{V}^2_{\mathrm{n,R}}$, $\boldsymbol{I}^2_{\mathrm{n,R}}$, and $\boldsymbol{Y}_{\mathrm{c}}$ are diagonal matrices with the squared RMS noise voltage spectral density v_n^2, squared current spectral density i_n^2, and the correlation admittance Y_{c} for each amplifier as the diagonal elements. In this expression, we assume that noise from components after the first amplifiers in the receiver is negligible, but the contribution from other components can be added if necessary.

The receiver noise correlation matrix depends on the amplifier noise parameters and on the source mutual impedance matrix $\boldsymbol{Z}_{\mathrm{r}}$ connected to the amplifier inputs. With the correlation matrix, the receiver noise realized at the output of the beamformer associated with one MIMO eigenchannel can be readily computed. It is not obvious from Eq. (9.54) that receiver noise at the beam former output is minimized for a particular value of the source impedances connected to the amplifier inputs. Moving to the active impedance and active reflection coefficient formulation makes this clear.

9.5.7.1 Active Impedances for Receiving Arrays

Mutual coupling effects on receiver noise are typically understood in terms of active impedances or active reflection coefficients (Warnick et al. 2009). The receiver noise correlation matrix in Eq. (9.54) is useful in developing a model for a MIMO system that includes the effects of mutual coupling on the performance of the system, but it is not very helpful in understanding mutual coupling effects from an intuitive point of view. For a mutually coupled array, the active impedances in a MIMO system as well as other types of array receivers determine the receiver noise contribution for each eigenchannel in a relatively simple way.

The active reflection coefficients associated with each array element port for a MIMO eigenchannel can be found from the array impedance matrix or S-parameters together with beam former combining coefficients. The active impedance looking into the mth element port is

$$Z_{\mathrm{act},m} = \frac{1}{w^*_{\mathrm{oc},m}} \sum_{n=1}^{N} Z_{\mathrm{A},mn} w^*_{\mathrm{oc},n} \tag{9.55}$$

using the open-circuit referenced beam former weights $\boldsymbol{w}_{\mathrm{oc}}$, and the active admittance is the inverse of the active impedance at each element port. The equivalent open-circuit voltage referenced beam former weight vector is

$$\boldsymbol{w}_{\mathrm{oc}} = \boldsymbol{Q}^{\dagger} \boldsymbol{w} \tag{9.56}$$

The equivalent open-circuit voltage beam former weights act on the array element open circuit voltages to produce the combined output voltage

$$v_{\mathrm{out}} = \boldsymbol{w}^{\dagger} \boldsymbol{v} = \boldsymbol{w}^{\dagger}_{\mathrm{oc}} \boldsymbol{v}_{\mathrm{oc}} \tag{9.57}$$

The corresponding active reflection coefficients are

$$\Gamma_{\mathrm{act},m} = \frac{1}{w^*_{\mathrm{f},m}} \sum_{n=1}^{N} S_{\mathrm{A},mn} w^*_{\mathrm{f},n} \tag{9.58}$$

The beam former weights referred to forward wave amplitudes at the amplifier input ports are

$$\boldsymbol{w}_{\mathrm{f}} = \boldsymbol{G}^{\dagger} \boldsymbol{w} \tag{9.59}$$

where the matrix $\boldsymbol{G}$ is similar to $\boldsymbol{Q}$ as defined in Eq. (9.44) but transforms forward wave amplitudes at the array when connected to matched loads to loaded receiver output voltages. If the receivers are uncoupled so that $\boldsymbol{Q} = gZ_{\mathrm{L}}(Z_{\mathrm{L}}\boldsymbol{I} + \boldsymbol{Z}_{\mathrm{r}})^{-1}$, it can be shown that $\boldsymbol{G} = g\sqrt{Z_0}\left(\boldsymbol{I} + \boldsymbol{S}_{\mathrm{L}}\right)\left(\boldsymbol{I} - \boldsymbol{S}_{\mathrm{r}}\boldsymbol{S}_{\mathrm{L}}\right)^{-1}$, where $\boldsymbol{S}_{\mathrm{L}}$ is the S-parameter matrix corresponding to the diagonal load impedance matrix $\boldsymbol{Z}_{\mathrm{L}} = Z_{\mathrm{L}}\boldsymbol{I}$. The active impedances and active reflection coefficients are related by the standard transmission line reflection coefficient formula.

The impedances and reflection coefficients in Eqs. (9.55) and (9.58) are referred to as active because they depend on beam former combining coefficients and are not fixed properties of the array and receiver electronics. This complicates the design of high-performance array systems where mismatch must be carefully controlled. In a single-antenna system, the antenna can be designed to a standard impedance such as 50 Ω and the amplifiers designed with a Z_{opt} of the same value. In an active receiving array, the impedance matching conditions between array elements and active electronics cannot be fully known until the array is embedded in a full system and the beam combining coefficients computed.

9.5.7.2 Equivalent Receiver Noise Temperature and Active Impedance Matching

Mutual coupling affects SNR at an array receiver through active impedances at the array element ports. To understand the effect of coupling on receiver noise, it is helpful to consider briefly the case of a single port receiver system without coupling effects.

By convention, in microwave noise theory, system noise powers are referred to an equivalent noise temperature at the input to a system. This convention is used because minimizing equivalent noise temperature or noise figure maximizes SNR at the system output. The equivalent system noise temperature for a receiver is a combination of the external, loss, and receiver noise contributions.

$$T_{\text{sys}} = T_{\text{ext}} + T_{\text{loss}} + T_{\text{rec}} \tag{9.60}$$

Equivalent system noise temperatures include the effect of impedance mismatch between the antenna and connected active electronics. A poor match between a receiving antenna and an amplifier means that a larger equivalent noise source must be placed at the input to the system to yield the noise power level at the system output. Equivalent noise temperatures also include the interaction between the source impedance and the physical mechanisms within the active device in the attached amplifier. The amplitude of the noise added by the amplifier depends on the impedance of the source attached to the input. Therefore, the equivalent noise temperature includes both the reduction in signal coupling from the antenna to the amplifier and the increase in noise added by the amplifier caused by impedance matches.

In a multiport receiver system, impedance matching between the source (the array antenna) and the devices attached to the element ports (typically LNAs) also includes both signal coupling effects and noise added by the amplifiers. For a given set of combining coefficients (see Figure 9.2) associated with one MIMO eigenchannel, the equivalent system noise temperature is

$$T_{\text{sys}} = T_{\text{iso}} \frac{\boldsymbol{w}^{\dagger} \boldsymbol{R}_{\text{n}} \boldsymbol{w}}{\boldsymbol{w}^{\dagger} \boldsymbol{R}_{\text{t,iso}} \boldsymbol{w}} \tag{9.61}$$

where the thermal noise correlation matrix with an isotropic thermal environment at temperature T_{iso} is given in Eq. (9.47). This represents the equivalent system noise temperature associated with one eigenchannel in a MIMO system. If there is a poor match between the front-end amplifiers and the active impedances looking into array element ports, the isotropic noise that couples from the array to the receiver outputs is reduced by the impedance mismatch factor, which makes the denominator of Eq. (9.62) small, and the system noise temperature becomes large. Impedance mismatch also causes the receiver noise temperature T_{rec} to increase, further raising the system noise temperature.

The effect of impedance mismatch on receiver noise can be seen more explicitly by developing an expression for the receiver noise temperature in terms of active impedances or active reflection coefficients. The equivalent receiver noise temperature is

$$T_{\text{rec}} = T_{\text{iso}} \frac{\boldsymbol{w}^{\dagger} \boldsymbol{R}_{\text{rec}} \boldsymbol{w}}{\boldsymbol{w}^{\dagger} \boldsymbol{R}_{\text{t,iso}} \boldsymbol{w}} \tag{9.62}$$

The receiver noise temperature can be expressed in terms of active reflection coefficients as (Warnick et al. 2009, 2018)

$$T_{\text{rec}} = \frac{\sum_{m=1}^{M} \left|w_{\text{f},m}\right|^2 \left(1 - \left|\Gamma_{\text{act},m}\right|^2\right) T_m}{\sum_{m=1}^{M} \left|w_{\text{f},m}\right|^2 \left(1 - \left|\Gamma_{\text{act},m}\right|^2\right)} \tag{9.63}$$

where

$$T_m = T_{\min,m} + \frac{4R_{\text{n},m} T_0 \left|\Gamma_{\text{act},m} - \Gamma_{\text{opt},m}\right|^2}{Z_0 \left|1 + \Gamma_{\text{opt},m}\right|^2 \mid \left(1 - \left|\Gamma_{\text{act},m}\right|^2\right)} \tag{9.64}$$

We have neglected in this expression noise contributed by components after the first LNAs connected to each array element. The expression Eq. (9.64) is identical to the classical formulas for noise contributed by a two-port network connected to a source with impedance corresponding to the reflection coefficient $\Gamma_{\text{act},\,m}$, where m indexes the array element. Eq. (9.63) shows how the noise contributions from each amplifier are combined across the array to produce the total receiver noise contribution for the eigenchannel beam combining coefficients in the vector $\boldsymbol{w}$.

The critical term in these expressions is $|\Gamma_{\text{act},\,m} - \Gamma_{\text{opt},\,m}|^2$. If the amplifiers are noise matched to the antenna element active impedances, then this term is zero, and the receiver noise temperature reduces to the minimum value $T_{\min}$ (assuming that all LNAs have the same minimum noise temperature parameter). If the amplifiers are not noise matched to the element ports, then this term is nonzero, and the receiver noise temperature is larger. From Eq. (9.55) or (9.58), if the array is mutually coupled, the active impedances and active reflection coefficients depend on the beam former coefficients.

The beam weight dependence of equivalent system noise for coupled array receivers is the key difference between impedance matching in a single-antenna system and matching in a multiantenna system. In a single-antenna receiver (or an uncoupled multiantenna system), the antenna and front-end amplifiers can be designed to a common impedance such as 50 Ω, and the matching design process is relatively straightforward. With a coupled array, the impedances that each amplifier must be matched to change as the beam combining coefficients in $\boldsymbol{w}$ are changed, meaning that matching is different for each eigenchannel and depends on the array and the propagation environment.

9.5.7.3 Noise Matching Efficiency

Noise has particular effect on MIMO systems and particular through coupling. A similar background is required for studying mutual coupling impacts on noise in arrays is seen in Chapter 12. For each eigenchannel or set of beam combining coefficients $\boldsymbol{w}$, we can parameterize the active impedance match between the antenna array and the connected electronics in the receiver chains using the noise matching efficiency. Noise matching efficiency is the ratio of the noise power contributed by receiver electronics at the output of a formed beam, with receiver impedance matched to the array elements for minimum receiver noise temperature, to the actual receiver electronics noise power at the formed beam output, per unit bandwidth and at a specified frequency (IEEE 145-2013 - IEEE Standard for Definitions of Terms for Antennas).

Assuming that receiver chains connected to the receive array elements are identical and that each has the minimum equivalent noise temperature $T_{\min}$, the noise matching efficiency can be computed using

$$\eta_{\text{n}} = \frac{T_{\min}}{T_{\text{rec}}} \tag{9.65}$$

If each amplifier and receiver chain is noise matched to the active impedance at its connected array element port, then $T_{rec} = T_{min}$ and the noise matching efficiency is unity. In the case of imperfect matching, the receiver noise temperature is greater than T_{min} and the noise matching efficiency is less than one.

For transmitters, impedance matching between a driving amplifier and an antenna is included in the realized gain. For receivers, both signal and noise effects must be considered, so the treatment is more complicated. The noise matching efficiency replaces the realized gain for active array receivers as the key figure of merit for impedance matching. The noise matching efficiency includes the impact of both signal coupling from the array elements to receiver electronics and the increased noise added by front-end amplifiers and receiver electronics if the impedance looking from the amplifier port into the connected source is different from the optimal source impedance parameter Z_{opt} of the amplifier.

Unlike noise matching efficiency, all other antenna parameters, such as antenna efficiency, aperture efficiency, radiation, and receiving efficiencies, gain, directivity, and spillover efficiency for receiving antennas are related to the available power at the antenna ports and are not influenced by impedance mismatch between the array elements and connected receiver electronics. For receiving antenna systems, no other key antenna parameter depends on impedance matching between the antenna and connected components. By convention in the IEEE standard for antenna terms (*IEEE Standard Definitions of Terms of Antennas*), all impedance mismatch effects in an active array receiver system are included in the noise matching efficiency parameter.

Finally, we provide a few thoughts on the practical importance of this analysis of receiver noise in a mutually coupled receiver system. For all types of receivers, including MIMO communications, high SNR is desirable. For high-sensitivity array applications, careful analysis and optimization of the system to reduce receiver noise can be critical. For other applications, external noise and interference dominate the noise environment, and complicated active impedance matching effects can be ignored. In practical MIMO implementations, a modest effort to noise match the array elements on an isolated basis (i.e. without considering mutual coupling effects) to the terminating amplifiers generally suffices to achieve adequate performance.

9.6 Conclusion

In this chapter, we have developed a framework for understanding and computing the effects of mutual coupling on MIMO systems. Mutual coupling impacts MIMO systems in four areas:

- Transmit and receive EEPs
- Power coupling into the transmitter
- Receiver noise
- Signal correlation at the receiving array

In the case of a highly coupled array, the impedance mismatch factor at the transmitting array can be a strong function of the array excitations and can lead to poor power coupling into the array. At the receiving array, strong coupling leads to poor active impedance matching and a decrease in SNR at the receiver output. Most importantly, a highly coupled array generally has high correlation between the received signals at the array elements and fails to achieve the antenna diversity that is at the heart of the increased capacity that MIMO systems are expected to realize. The analytical framework developed in this chapter can be used to model fully all coupling effects from end to end in a MIMO system, predict the channel capacity of the system including mutual coupling, and

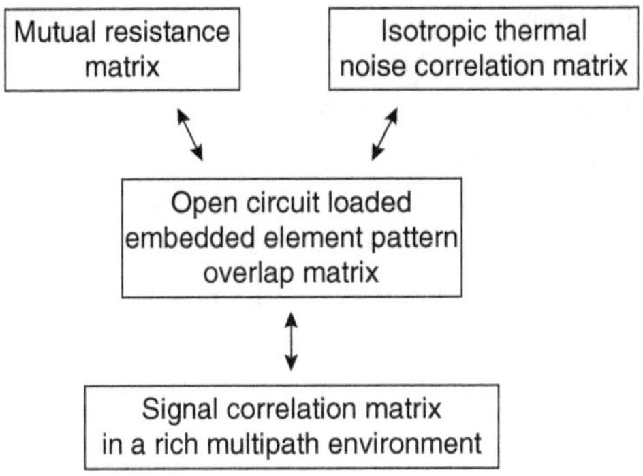

Figure 9.3 Relationships between the receive array mutual resistance matrix Re[$\boldsymbol{Z}_r$], isotropic thermal noise correlation matrix $\boldsymbol{R}_{t,\,iso}$, the open-circuit loaded EEP overlap matrix $\boldsymbol{A}$, and the signal correlation matrix $\boldsymbol{R}_s$ in a rich multipath environment. Neglecting antenna losses, these matrices are each proportional to one another and can be viewed as identical.

optimize the channel capacity achieved by tuning the array or other components in the system to mitigate the effects of coupling.

Of great theoretical interest is the connection between the array EEP overlap matrix, the real part of the mutual impedance matrix, isotropic noise correlation matrix, and the signal correlation matrix in a uniformly distributed multipath environment diagrammed in Figure 9.3. This set of theorems and relationships combine results from antenna theory and array mutual coupling, array signal processing, microwave noise theory, and MIMO communications in a fundamental way. Physically speaking, these quantities are connected because they each involve waves either arriving at the array or radiating from the array over a full sphere.

Modeling and understanding coupling effects on MIMO system performance has proved valuable for many applications. Mutual coupling in MIMO array antennas can have significant impact in some cases, but in others, these effects may be minor when the MIMO system is considered as a whole. When mutual coupling is modest, its effects can usually be ignored in the capacity analysis and design of the MIMO system. In an operational MIMO system with channel sounding, mutual coupling effects on the transmit and receive array element patterns are automatically measured along with the propagation channel and incorporated into the optimal solution for maximum channel capacity. As communications technologies transition from existing low-N, GHz-frequency systems to massive MIMO and 5G systems at higher frequency bands with large numbers of tightly packed elements, mutual coupling considerations are likely to become more important in the design and performance of multiantenna communication systems.

References

Blanch, S., Romeu, J., and Corbella, I. (2003). Exact representation of antenna system diversity performance from input parameter description. *Electron. Lett.* 39 (9): 705–707.

Browne, D.W. et al. (2006). Experiments with compact antenna arrays for MIMO radio communications. *IEEE Trans. Antennas Propag.* 54 (11): 3239–3250.

Chen, X., Zhang, S., and Li, Q. (2018). A review of mutual coupling in MIMO systems. *IEEE Access* 6: 24706–24719.

Chiu, C.-Y., Cheng, C.-H., Murch, R.D. et al. (2007). Reduction of mutual coupling between closely-packed antenna elements. *IEEE Trans. Antennas Propag.* 55 (6): 1732–1738.

Davis, W.A. and Licul, S. (2005). *Antennas: An Introduction to Ultra Wideband Communication Systems* (ed. J. Reed). Englewood Cliffs, NJ: Prentice Hall.

De Hoop, A.T. (1975). The N-port receiving antenna and its equivalent electrical network. *Philips Res. Rep.* 30: 302–315.

De Hoop, A.T. and De Jong, G. (1974). Power reciprocity in antenna theory. *Proc. IEEE* 121 (10): 1051–1056.

Diao, J. and Warnick, K.F. (2016). Poynting streamlines, effective area shape, and the design of superdirective antennas. *IEEE Trans. Antennas Propag.* 65 (2): 861–866.

Dossche, S., Blanch, S., and Romeu, J. (2005). Three different ways to decorrelate two closely spaced monopoles for MIMO applications. In: *Proceedings of IEEE/ACES International Conference on Wireless Communications and Applied Computational Electromagnetics*, 849–852. Piscataway, NJ: IEEE Press.

Dossche, S. et al. (2006). Decoupling of a two-element switched dual band path antenna for optimum MIMO capacity. In: *Proceedings of IEEE Antennas and Propagation Society International Symposium*. Piscataway, NJ: IEEE Press.

Engberg, J. and Larsen, T. (1995). *Noise Theory of Linear and Nonlinear Circuits*. New York: Wiley.

Fletcher, P.N., Dean, M., and Nix, A.R. (2003). Mutual coupling in multi-element array antennas and its influence on MIMO channel capacity. *Electron. Lett.* 39 (4): 342–344.

Getu, B.N. and Janaswamy, R. (2005). The effect of mutual coupling on the capacity of the MIMO cube. *IEEE Antennas Wireless Propag. Lett.* 4: 240–244. IEEE Standard Definitions of Terms for Antennas, IEEE Std 145-2013.

Jakes, W.C. (1974). *Mobile Wireless Communications*. IEEE Press, Piscataway, NJ.

Janaswamy, R. (2002). Effect of element mutual coupling on the capacity of fixed length linear arrays. *IEEE Antennas Wireless Propag. Lett.* 1: 157–160.

Jensen, M.A. and Wallace, J.W. (2005). Antenna-Independent Capacity Bound of Electromagnetic Channels. In: *Proc. IEEE Antennas and Propagation Society International Symposium*. Piscataway, NJ: IEEE Press.

Jungnickel, V., Pohl, V., and Von Helmolt, C. (2003). Capacity of MIMO systems with closely spaced antennas. *IEEE Commun. Lett.* 7 (8): 361–363.

Kahn, W. and Kurss, H. (1965). Minimum-scattering antennas. *IEEE Trans. Antennas Propag.* 13 (5): 671–675.

Kildal, P.-S. (2015). *Foundations of Antenna Engineering: A Unified Approach for Line-of-Sight and Multipath*. Gothenburg, Sweden: Kildal Antenn AB.

Kildal, P.-S. and Rosengren, K. (2004). Correlation and capacity of MIMO systems and mutual coupling, radiation efficiency, and diversity gain of their antennas: simulations and measurements in a reverberation chamber. *IEEE Commun. Mag.* 42 (12): 104–112.

Li, X. and Nie, Z.-P. (2004). Mutual coupling effects on the performance of MIMO wireless channels. *IEEE Antennas Wireless Propag. Lett.* 3: 344–347.

Li, Z. et al. (2011). Reducing mutual coupling of MIMO antennas with parasitic elements for mobile terminals. *IEEE Trans. Antennas Propag.* 60 (2): 473–481.

Malviya, L., Panigrahi, R.K., and Kartikeyan, M. (2017). MIMO antennas with diversity and mutual coupling reduction techniques: a review. *Int. J. Microwave Wireless Technol.* 9 (8): 1763–1780.

Morris, M.L. and Jensen, M.A. (2005). Network model for MIMO systems with coupled antennas and noisy amplifiers. *IEEE Trans. Antennas Propag.* 53: 545–552.

Salonen, I. and Vainikainen, P. (2002). Estimation of signal correlation in antenna arrays. *Proc. JINA* 2: 383–386.

Shamonina, E. and Solymar, L. (2015). Maximum directivity of arbitrary dipole arrays. *IET Microwaves Antennas Propag.* 9 (2): 101–107.

Shannon, C.E. and Weaver, W. (1998). *The Mathematical Theory of Communication*. Champaign, IL: University of Illinois Press.

Svantesson, T. and Ranheim, A. (2001). Mutual coupling effects on the capacity of multielement antenna systems. *ICASSP* 4: 2485–2488.

Telatar, E. (1999). Capacity of multi-antenna Gaussian channels. *Trans. Emerg. Telecommun. Technol.* 10 (6): 585–595.

Twiss, R.Q. (1955). Nyquist's and Thevenin's theorems generalized for nonreciprocal linear networks. *J. Appl. Phys.* 26 (5): 599–602.

Uzsoky, M. and Solymar, L. (1956). Theory of super-directive linear arrays. *Acta Physica Academiae Scientiarum Hungaricae* 6 (2): 185–205.

Vaughan, R.G. and Andersen, J.B. (1987). Antenna diversity in mobile communications. *IEEE Trans. Veh. Technol.* 36 (4): 149–172.

Waldschmidt, C., Schulteis, S., and Wiesbeck, W. (2004). Complete RF system model for analysis of compact MIMO arrays. *IEEE Trans. Veh. Technol.* 53 (3): 579–586.

Wallace, J. W. and M. A. Jensen (2002): 'The capacity of MIMO wireless systems with mutual coupling', IEEE Vehicular Technology Conference.

Wallace, J.W. and Jensen, M.A. (2004a). Mutual coupling in MIMO wireless systems: a rigorous network theory analysis. *IEEE Trans. Wireless Commun.* 3 (4): 1317–1325.

Wallace, J.W. and Jensen, M.A. (2004b). Termination-dependent diversity performance of coupled antennas: network theory analysis. *IEEE Trans. Antennas Propag.* 52 (1): 98–105.

Wallace, J.W. et al. (2003). Experimental characterization of the MIMO wireless channel: data acquisition and analysis. *IEEE Trans. Wireless Commun.* 2 (2): 335–343.

Warnick, K.F. et al. (2009). Minimizing the noise penalty due to mutual coupling for a receiving array. *IEEE Trans. Antennas Propag.* 57 (6): 1634–1644.

Warnick, K.F. et al. (2018). *Phased Arrays for Radio Astronomy, Remote Sensing, and Satellite Communications*. Cambridge University Press, Cambridge.

10

Mutual Coupling in Beamforming and Interferometric Antennas

Hoi Shun Antony Lui[1] *and Trevor S. Bird*[1,2]

[1]*Global Big Data Technology Centre, University of Technology Sydney, Ultimo, NSW 2007, Australia*
[2]*Antengenuity, Eastwood, NSW 2122, Australia*

10.1 Introduction

An obvious way to overcome the impact of mutual coupling in arrays to be used for beamforming or direction finding is to design for this effect in the first instance. However, that has not been done in the past, and only afterward when the array performance is not as expected, corrections and adjustments are made. The effect of mutual coupling in adaptive arrays was first described by Gupta and Ksienski (1983) and also Yirnin et al. (1985). They showed its effect to be significant even for spacings greater than half a wavelength. For direction finding antennas, Jenkins (1991) recommended that their accuracy could be considerably improved when mutual coupling was an integral part of the design. For shaped beam antennas using arrays, Bird (1982), Bird and Poulton (1985), and Poulton (1986b) described techniques for including it in their design. All these and other applications involving beam shaping or steering are significantly influenced by mutual coupling. Often, when this effect is included, the overall performance can be considerably improved.

The implementation of mutual coupling in the two main applications of beam shaping and direction-of-arrival (DOA) estimation is described in this chapter. In beam shaping, two power synthesis approaches are described which are based on computer optimization of a constrained performance index. In DOA estimation, one of the most important applications in array signal processing; we consider the estimation of the azimuth angle using the matrix pencil method (MPM; Sarkar et al. 2003) together with uniform linear arrays (ULAs) with different apertures and different number of array elements. DOA estimation results using ideal arrays are first presented, which provide us some ideas on how the algorithm performs when there is no mutual coupling. Then, monopole, dipole, and helical antenna elements are used. The received signals are determined using a full-wave moment method solution in which mutual coupling effects are incorporated. DOA estimation examples using the coupled received signals and different compensated signals show that accurate DOA estimation can be achieved when mutual coupling effects are properly compensated. In particular, our results show that increasing the number of array elements within a small array aperture enhances DOA estimation accuracies. Such findings are observed in both small aperture ideal array and compact monopole arrays when mutual coupling effects are compensated.

Mutual Coupling Between Antennas, First Edition. Edited by Trevor S. Bird.

10.2 The Array Manifold

Consider an incoming plane wave $s(t)$ incident on an $N+1$ $(0 \le n \le N)$ element ULA from an angle (θ, ϕ) relative to the axis of the array as shown in Figure 10.1 (Liberti and Rappaport 1999). Assume for the meantime that all the array elements are noiseless isotropic antennas, which have uniform gain in all directions. The separation between adjacent elements is Δ, and there is no mutual coupling between the array elements. The signal received at the antenna element n on the array is

$$u_n(t) = s(t)e^{-j\beta n\Delta \cos\phi \sin\theta}, \tag{10.1}$$

where $\beta = 2\pi/\lambda$ is the phase propagation factor, and λ is the wavelength of the incoming signal. The signal $z(t)$ at the array output is

$$z(t) = \sum_{n=0}^{N-1} w_n u_n(t) = s(t)\sum_{n=0}^{N-1} w_n e^{-j\beta n\Delta \cos\phi \sin\theta} = s(t)f(\theta, \phi) \tag{10.2}$$

The term $f(\theta, \phi)$ is well known as the array factor. When working with arrays, it is convenient to use vector notation. Define the weight vector as

$$\mathbf{w} = [w_0 \quad w_1 \quad \cdots \quad w_{N-1}]^H \tag{10.3}$$

and the steering vector as

$$\mathbf{a} = \begin{bmatrix} 1 & a_1(\theta, \phi) & \cdots & a_{N-1}(\theta, \phi) \end{bmatrix}^T \tag{10.4}$$

where the superscript H represents the Hermitian transpose, which is a transposition combined with complex conjugate. The superscript T represents a transpose. For the case of a $N+1$ element ULA, the elements of the steering vector are given by

$$a_n(\theta, \phi) = e^{-j\beta n\Delta \cos\phi \sin\theta}. \tag{10.5}$$

The array factor, $f(\theta, \phi)$, can now be written as

$$f(\theta, \phi) = \mathbf{w}^H \mathbf{a}(\theta, \phi). \tag{10.6}$$

The steering vector essentially describes the phase of the signal available at each antenna element relative to the phase of the signal at reference element (element 0). A set of steering vectors either measured or calculated, over all values of θ and ϕ, is called the array manifold.

Equations (10.4) and (10.5) give the expressions for the ideal array manifold where the antenna elements are ideal point isotropic sensors with uniform gain in all directions and no coupling between them.

10.3 Direction-of-Arrival Algorithms

DOA algorithms fall within the context of array signal processing. In general, these algorithms can be grouped into three categories: conventional methods, subspace methods, and maximum likelihood techniques (Liberti and Rappaport 1999; Krim and Viberg 1996; Godara 1997; Paulraj et al. 1993). Conventional methods are based on classical beamforming techniques where the resolution depends on the array aperture and the number of array elements. To achieve high resolution, a large number of elements are usually required. Subspace methods are high-resolution suboptimal techniques which exploit the eigen-structure of the input data matrix. The main advantage of subspace methods over the conventional beamforming method is that the resolution is not limited by

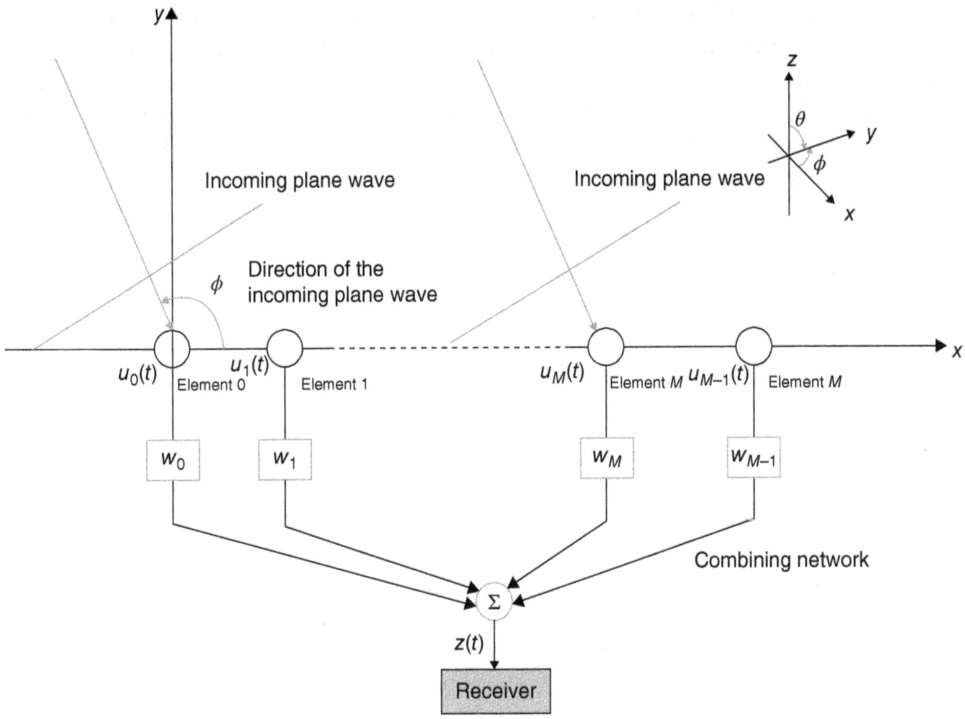

Figure 10.1 A uniform plane wave impinging on an M-element uniform linear array along the x-axis.

the array aperture, and subspace methods are therefore also known as super resolution techniques (Madurasinghe 2003). Maximum likelihood techniques are optimal techniques which perform well even under low signal-to-noise ratio (SNR) but are computationally intensive.

Throughout the last three decades, subspace methods have achieved great success due to their low complexity and reasonable performance. Schmidt (1986) studied the signal properties of a sensor array and introduced the Multiple Signal Classification (MUSIC) algorithm. The MUSIC algorithm exploits the geometrical structure of the signal. Performance evaluations of the MUSIC algorithm have been well studied in the 1980s to mid-1990s (Friedlander and Weiss 1991; Swindlehurst and Kailath 1992). Another well-known subspace method is Estimation of Signal Parameters via Rotational Invariance Techniques (ESPRIT) (Paulraj et al. 1986; Roy and Kailath 1989), which exploits the array structure such that the specific knowledge of the array manifold is not required. Compared to MUSIC, it is computationally more efficient without the need of searching peaks throughout the entire range of angles. A review of subspace methods from the statistical perspective can be found in Paulraj et al. (1993). In addition, there is another subspace method, known as the MPM, which was first applied to damped exponential extractions from transient electromagnetic signals (Sarkar and Pereira 1995; Hua and Sarkar 1990; Lui and Shuley 2004, 2008) and later to DOA estimations (Ouibrahim et al. 1988; Sarkar et al. 2003, 2008; Yilmazer et al. 2006, 2007; Lau et al. 2004; Hwang et al. 2006; Hua 1992). One significant advantage of MPM over MUSIC and ESPRIT is that DOA estimation can be done using a single snapshot instead of the covariance matrix that requires a large number of snapshots (Lau et al. 2004; Sarkar et al. 2003). The incoming directions can be determined directly without the need for searching among all the directions as in MUSIC.

Instead of comparing DOA estimation performance of different algorithms, the focus of this section is to evaluate the impact of mutual coupling effect to DOA estimation performance for antenna arrays with different apertures and different number of elements. Out of all the aforementioned DOA estimation algorithms, MPM will be used as our platform for this purpose. Details of other DOA estimation algorithms can be found in Liberti and Rappaport (1999), Krim and Viberg (1996), Godara (1997), and Paulraj et al. (1993).

10.3.1 Matrix Pencil Method for Direction of Arrival Estimation

MPM was originally used for estimating damped exponentials ($e^{\sigma \pm j2\pi f}$, $\sigma < 0$) in transient electromagnetic signatures from definitive targets, where these damped exponentials correspond to the complex natural resonance ($\sigma \pm j2\pi f$) of the target (Sarkar and Pereira 1995). Lau et al. (2004) and Sarkar et al. (2003, 2008) proposed using it for DOA estimations – estimation of undamped exponentials ($e^{j\phi}$). In common with the well-known MUSIC and ESPRIT, MPM is a subspace method in which the resolution of the algorithm is not highly dependent on the array aperture as compared to the conventional beamforming techniques (Liberti and Rappaport 1999). Both MUSIC and ESPIRT require a large number of samples to construct the covariance matrix, which needs a fair amount of memory for storing the received signals before the estimation commences. As for MPM, however, only a single snapshot of the signal is required, and estimations can be performed without forming the covariance matrix. Even for coherent signals, spatial smoothing is not required (Sarkar et al. 2003). These nice features provide another degree of flexibility for real-time applications.

Consider a space where there are $P+1$ signals coming from different directions. If we have a ULA of omnidirectional antenna elements located along the x-axis and the separation between any two of them is Δ, the voltage $x(n)$ induced in each of the n antenna elements can be written as (Lau et al. 2004)

$$x(n) = \sum_{p=0}^{P} A_p \exp\left(j\gamma_p + \frac{j2\pi\Delta n \cos\phi_p}{\lambda}\right), 0 \le n \le N \tag{10.7}$$

where A_p, γ_p, and ϕ_p are the amplitude, phase, and the incoming direction for each of the $P+1$ plane waves incident on the ULA. Figure 10.2 illustrates the DOA estimation problem of an eight-element linear array with three incoming signals from 20°, 35°, and 100°. Equation (10.7) can also be written as

$$x(n) = \sum_{p=0}^{P} c_p y_p^n, \tag{10.8a}$$

where

$$c_p = A_p \exp(j\gamma_p), \tag{10.8b}$$

$$y_p = \exp\left[\frac{(j2\pi \cos\phi_p)}{\lambda}\right] \tag{10.8c}$$

and λ is the wavelength of the operation frequency. In matrix form, this can be written as

$$\begin{bmatrix} x(0) \\ x(1) \\ \vdots \\ x(n) \end{bmatrix} = \begin{bmatrix} 1 & 1 & \cdots & 1 \\ y_0 & y_1 & \cdots & y_p \\ \vdots & \vdots & \ddots & \vdots \\ y_0^N & y_1^N & \cdots & y_p^N \end{bmatrix} \begin{bmatrix} c_0 \\ c_1 \\ \vdots \\ c_p \end{bmatrix} \tag{10.8d}$$

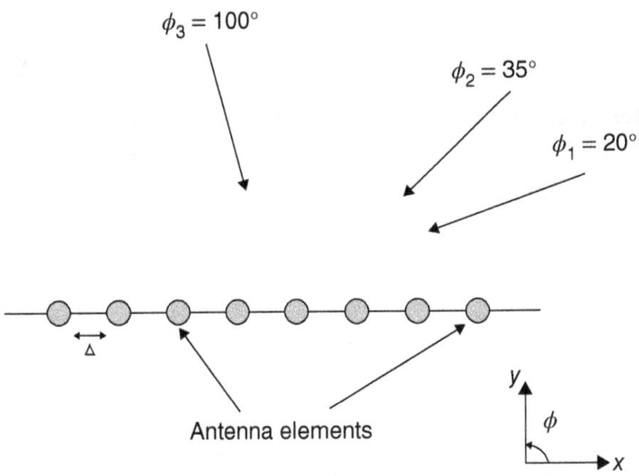

Figure 10.2 The top view of the DOA estimation problems. The ULA is aligned with the x-axis, and the incoming signals are coming from the azimuth angles of $\phi_1 = 20°$, $\phi_2 = 35°$ and $\phi_3 = 100°$. The antenna separation is denoted as Δ. (Lui and Hui 2010b).

or

$$[G] = [Y][C]. \tag{10.8e}$$

Next, we define a $L' \times (N - L' + 2)$ matrix $[X]$ using $x(k)$,

$$[X] = \begin{bmatrix} x(0) & x(1) & \cdots & x(N-L'+1) \\ x(1) & x(2) & \cdots & x(N-L'+2) \\ \vdots & \vdots & \ddots & \vdots \\ x(L'-1) & x(L') & \cdots & x(N) \end{bmatrix} \tag{10.9}$$

where L' is the pencil parameter which must satisfy the criteria of $(N+1-P) \geq L' \geq (P+2)$. Applying a singular value decomposition of $[X]$

$$[X] = [U]\left[\sum\right][V]^H, \tag{10.10}$$

where $[U]$ is the $L' \times L$ unitary matrix whose columns are the eigenvalues of $[X][X]^H$, $[V]$ is the $(N-L'+2)\times(N-L'+2)$ unitary matrix whose columns are the eigenvalues of $[X][X]^H$, and $[\sum]$ is the $L' \times (N-L'+2)$ diagonal matrix with singular values of $[X]$ located along the main diagonal in descending order. If the data is not contaminated by noise, the first $P+1$ singular values are nonzero, and hence, P can be determined. If the data is noise contaminated, the parameter P is estimated by observing the ratio of the various singular values to the largest one as defined by Sarkar and Pereira (1995)

$$\frac{\sigma_c}{\sigma_{max}} \approx 10^{-w}, \tag{10.11}$$

where w is the number of accurate significant decimal digits of the data $x(n)$. The singular values for which the ratio in Eq. (10.11) is below 10^{-w} are essentially noise singular values, and they should be discarded. Once P is determined, the submatrix of $[U]$ with the first $P+1$ columns can be formed and denoted as $[U]_s$. To estimate the DOA, the following problem is solved:

$$[U_1]^+[U_2] - \lambda'[I] = 0 \tag{10.12a}$$

Here, $[U_1] = [U]_s$ with the last row deleted, $[U_2] = [U]_s$ with the first row deleted, and $[I]$ is the $(P+1)\times(P+1)$ identity matrix. The superscript + denotes the Moore–Penrose pseudoinverse of a matrix (Golub and Van Loan, 1996) defined as

$$[U_1]^+ = \{[U_1]^H[U_1]\}^{-1}[U_1]^H. \tag{10.12b}$$

The superscripts H and -1 denote the Hermitian conjugate (i.e. conjugate transpose) and inverse of the matrix, respectively. The λ_p's in Eq. (10.12a) are in the same form as those as the y_ps in Eq. (10.3). Once Eq. (10.12a) is solved, the DOA can be obtained by

$$\phi_p = \cos^{-1}\left(\frac{\ln(\lambda_p')}{j2\pi\Delta}\right) \tag{10.13}$$

It is well known that if the $\cos^{-1}$ argument in Eq. (10.13) is greater than 1, the resultant angle will become complex, which is considered as failure estimation. The resultant complex angle can be used as a sanity check which indicates the algorithm fails to resolve the signals in that particular snapshot. The amplitude and phase of the incoming signal can also be obtained by solving Eq. (10.8e) in a least-square sense using the Moore–Penrose pseudoinverse method (Golub and Van Loan 1996)

$$[C] = [Y^+][G]. \tag{10.14}$$

10.4 Maximum Gain Design for Single and Multiple Beams

Before examining the impact of mutual coupling on more complicated directional beam arrays, the conditions for fixed maximum gain designs for single and multiple beam arrays are first obtained in the presence of mutual coupling. This is followed by the design of a single fixed shaped beam under a variety of practical constraints. Consider an array of N apertures of arbitrary antennas. Under these conditions, the gain function of the antenna is

$$G(\theta, \phi, \boldsymbol{a}) = \frac{2\pi r^2}{\eta_o}\frac{\mathbf{E}\cdot\mathbf{E}^*}{P_{in}} \tag{10.15}$$

where $\mathbf{E}$ is the electric field radiated by the array at a far-field distance of radius r, and P_{in} the power input to the array. Compared with the gain of a uniformly illuminated aperture with the same dimensions, G_o , the array has an efficiency

$$\eta_i(\theta, \phi, \boldsymbol{a}) = \frac{2\pi r^2}{G_o\eta_o}\frac{\mathbf{E}\cdot\mathbf{E}^*}{P_{in}} = C_o\frac{\left(\sum_{k=1}^{N}A_k e_k\right)\left(\sum_{k=1}^{N}A_k e_k\right)^*}{P_{in}} = \frac{QQ^*}{P_{in}} \tag{10.16}$$

where C_o is a constant, $\mathbf{e_k}$ is the radiated electric field at accessible port k and A_k is its amplitude, which can be in the presence of mutual coupling as will be shown, and $Q = \sqrt{C_o}\sum_{k=1}^{N}A_k e_k$. It is assumed that the total radiated power $P_r = P_{in}$ and therefore

$$P_r = \boldsymbol{a}_I^T\boldsymbol{D}\boldsymbol{a}_I$$

where

$$\boldsymbol{D} = \frac{1}{2}\left\{\left[(\boldsymbol{I}-\boldsymbol{S}_{22}\boldsymbol{S}^{(0)})^{-1}\boldsymbol{S}_{21}\right]^H\left[(\boldsymbol{I}+\boldsymbol{S}^{(0)})^H(\boldsymbol{I}-\boldsymbol{S}^{(0)}) + (\boldsymbol{I}-\boldsymbol{S}^{(0)})^H(\boldsymbol{I}+\boldsymbol{S}^{(0)})\right](\boldsymbol{I}-\boldsymbol{S}_{22}\boldsymbol{S}^{(0)})^{-1}\boldsymbol{S}_{21}]\right\}$$

$\boldsymbol{S}^{(0)}$ is the scattering matrix of the array, $\boldsymbol{S}_{21}$], and $\boldsymbol{S}_{22}$ are the submatrices of the scattering matrix of the network feeding the array (see Sec. 2.4.3).

Equation 10.16 gives the antenna gain relative to a uniformly illuminated aperture and is suitable as a basis for a synthesis procedure. In addition, it is convenient to express the efficiency in matrix notation. Let

$$Q = \mathbf{u}^T\mathbf{v}$$

where $\mathbf{v}^T = [a_1\, a_2 \ldots a_{JN}]$ is a subset of the excitation coefficients corresponding to accessible ports, T denotes the matrix transpose operation,

$$\mathbf{u}^T = \mathbf{t}^T\mathbf{W} \tag{10.17}$$

with

$$\mathbf{t}^T = [t_1\; t_2\; \ldots\; t_{MN}] \tag{10.18}$$

where t_k is proportional to the electric field radiated by element k, and

$$w_{ij} \begin{cases} 1 + S_{ij};\, i = 1, 2, \ldots, JN \\ S_{ij};\, \text{otherwise} \end{cases}$$

In this notation, the power radiated by the array is

$$P_r = \mathbf{v}^T\, \mathbf{D}\, \mathbf{v}$$

where $\mathbf{D} = \mathbf{P} - \mathbf{S}^\mathbf{T}\mathbf{P}\mathbf{S}$ is a JN x JN positive definite diagonal matrix given by

$$P = \begin{bmatrix} p_1 & 0 & \ldots & 0 \\ 0 & p_2 & 0 & 0 \\ \vdots & & \ddots & \vdots \\ 0 & 0 & \ldots & p_{MN} \end{bmatrix}$$

where p_i the mode power at the accessible ports i and [S] is the scattering matrix of the array and associated feed network.

Equation (10.16) can now be re-expressed as (Collin and Zucker, 1969).

$$\eta(\theta, \varphi, \mathbf{v}) = \frac{(\mathbf{v}^\mathbf{T}\mathbf{u})^*(\mathbf{u}^\mathbf{T}\mathbf{v})}{(\mathbf{v}^\mathbf{T})^*\mathbf{D}\,\mathbf{v}}$$

and further refined to

$$\eta = \frac{\boldsymbol{x}^H(\boldsymbol{\alpha}^H\boldsymbol{\alpha})\boldsymbol{x}}{\boldsymbol{x}^H\boldsymbol{x}} \tag{10.19}$$

where $\boldsymbol{x} = \mathbf{D}^{1/2}\,\mathbf{v}$, $\boldsymbol{\alpha} = \mathbf{u}^T\mathbf{D}^{1/2}$. The maximum value of Eq. (10.19) is the maximum eigenvalue of $(\boldsymbol{\alpha}^H\boldsymbol{\alpha} - \lambda \boldsymbol{I})\boldsymbol{x} = \mathbf{0}$. Since $\boldsymbol{\alpha}^H\boldsymbol{\alpha}$ is a rank one matrix, the maximum eigenvalue is simply $\boldsymbol{\alpha}^H\boldsymbol{\alpha}$, and the associated eigenvector is $\mathbf{x} = \boldsymbol{\alpha}$. Therefore, the beam efficiency is maximized when

$$\boldsymbol{v} = \boldsymbol{D}^{-1}\boldsymbol{W}^H\boldsymbol{t}^*. \tag{10.20}$$

When there is no coupling or reflection at the ports, i.e. the scattering matrix $\mathbf{S} = 0$, $\mathbf{W} = \mathbf{I}$, and $\boldsymbol{v} = \boldsymbol{D}^{-1}\boldsymbol{t}^*$. Further, as $\mathbf{D}$ is diagonal, the i-th accessible port is proportional to the electric field radiated by element i. Mutual coupling causes the excitation coefficient to depend on all far-field vectors through the matrix $\mathbf{D}$. When the power radiated at each port is normalized to unity, $\mathbf{D} = \mathbf{I} = \mathbf{W}$, and Eq. (10.20) predicts that, for maximum gain, the excitation coefficient of a particular mode should equal the complex conjugate of its own far-field vector.

The formulation above can be extended to include more beams in other directions. Equation (10.19) will be different depending on the beam direction. Let $\boldsymbol{u}_1$ in Eq. (10.17) be due to beam 1 and $\boldsymbol{u}_2$ be the corresponding vector for beam 2. Two efficiency functions in the form of Eq. (10.19) can be defined for both beams in terms of a common desired excitation $\mathbf{v}$, as follows:

$$\eta_i(\theta, \varphi, \boldsymbol{v}) = \frac{Q_i^{\,*}Q_i}{(\boldsymbol{v}^T)^*\boldsymbol{D}\boldsymbol{v}} = \frac{(\boldsymbol{v}^T\boldsymbol{u}_i)^*(\boldsymbol{u}_i^{\,T}\boldsymbol{v})}{(\boldsymbol{v}^T)^*\boldsymbol{D}\boldsymbol{v}} \tag{10.21}$$

where the subscript i indicates quantities relating to beam i. To meet the common objective of achieving maximum efficiency for both beams, a single performance index can be defined as

$$\xi = \eta_1(\theta_1, \varphi_1, \mathbf{v}) + |\boldsymbol{\mu}|^2 \eta_2(\theta_1, \varphi_1, \mathbf{v}) \tag{10.22}$$

where μ is an arbitrary complex constant. Once again, intermediate vectors are chosen in this case $\mathbf{x} = \mathbf{D}^{1/2}\mathbf{v}$, and $\boldsymbol{\alpha}_i = \boldsymbol{u}_i\boldsymbol{D}^{-1/2}$. Substituting these intermediate vectors into Eq. (10.22) results in

$$\xi = \frac{\mathbf{x}^{\dagger}(\boldsymbol{\alpha}_1^H \boldsymbol{\alpha}_1 + |\mu|^2 \boldsymbol{\alpha}_2^H \boldsymbol{\alpha}_2)\boldsymbol{x}}{\boldsymbol{x}^H \boldsymbol{x}} \tag{10.23}$$

As with the case of a single efficiency, Eq. (10.23) is a ratio of quadratic forms. The maximum is readily found when

$$(\boldsymbol{\alpha}_1^H \boldsymbol{\alpha}_1 + |\mu|^2 \boldsymbol{\alpha}_2^H \boldsymbol{\alpha}_2 - \lambda \boldsymbol{I})\boldsymbol{x} = 0. \tag{10.24}$$

This corresponds to the eigenvalue λ that is closest to 1. To find this, define $\boldsymbol{\tau} = \mu\, \boldsymbol{\alpha}_2$. Resolve $\boldsymbol{\alpha}$ into components parallel and orthogonal to $\boldsymbol{\tau}$. Thus, $\boldsymbol{\alpha}_1 = \boldsymbol{\beta}_1 + \boldsymbol{\beta}_2$ where $\boldsymbol{\beta}_1 = k\boldsymbol{\tau}$ and $\boldsymbol{\beta}_2^H \boldsymbol{\tau} = \mathbf{0}$. From this, it follows that $k = (\boldsymbol{\beta}_1^H \boldsymbol{\tau})/(\boldsymbol{\tau}^H \boldsymbol{\tau})$. In addition, the $\mathbf{x}$ vector can be expressed as a linear combination of $\boldsymbol{\beta}_1$ and $\boldsymbol{\beta}_2$; thus, $\mathbf{x}=\boldsymbol{c}\boldsymbol{\beta}_1 + \boldsymbol{d}\boldsymbol{\beta}_2$. Making the above substitutions in Eq. (10.24) results in

$$\boldsymbol{\tau}\, c[(\boldsymbol{\alpha}_1^H\, \boldsymbol{\tau})(1+k^2) + (\boldsymbol{\alpha}^H \boldsymbol{\beta}_2)k - \lambda k] + \boldsymbol{\beta}_2 \frac{d}{k}[(\boldsymbol{\alpha}_1^H\, \boldsymbol{\tau})(1+k^2) + (\boldsymbol{\alpha}^H \boldsymbol{\beta}_2)k - \lambda k] = 0 \tag{10.25}$$

Clearly, Eq. (10.25) is satisfied when

$$\lambda = (\boldsymbol{\alpha}_1^H\, \boldsymbol{\beta}_1)\left(1 + \frac{1}{k^2}\right) + (\boldsymbol{\alpha}_1^H\, \boldsymbol{\beta}_2) \tag{10.26}$$

The associated eigenvector is

$$\boldsymbol{x} = \boldsymbol{c}\boldsymbol{\beta}_1 + \boldsymbol{d}\boldsymbol{\beta}_2 \tag{10.27}$$

where

$$\mathrm{c} = \left(1 + \frac{1}{k^2}\right) \text{ and } \mathrm{d} = 1.$$

From Eq. (10.27), and $Q_i = \boldsymbol{\alpha}_i^H \boldsymbol{x}$, the maximum efficiency of beam i is

$$\eta_i(\theta, \varphi, \mathbf{v}) = \frac{Q_i{}^* Q_i}{x^H\, x} \tag{10.28}$$

where

$$Q_i = \boldsymbol{\alpha}_i^H\, \boldsymbol{x} = c(\boldsymbol{\alpha}_i^H \boldsymbol{\beta}_1) + (\boldsymbol{\alpha}_i^H \boldsymbol{\beta}_2).$$

In the special case when $\mu = 0$ (i.e. only one beam), $\boldsymbol{\tau} = 0$, $\boldsymbol{\beta}_1 = 0$, and $\boldsymbol{\alpha}_1 = \boldsymbol{\beta}_2$. Thus, $Q_1 = \boldsymbol{\alpha}_1^H\, \boldsymbol{\alpha}_1$, which is the single beam result.

The previous approach has also been applied to cross-polarization (Bird 2016). In that case, $\mathbf{u_c}$ is a vector corresponding to currents excited due to copolar illumination, and $\mathbf{u}_x$ is the corresponding vector due to cross-polar illumination. The beam efficiency is defined for each polarization as follows:

$$\eta_c(\theta, \phi, \mathbf{v}) = \frac{Q_c{}^* Q_c}{(\mathbf{v}^{\mathrm{T}})^* \mathbf{D}\, \mathbf{v}} = \frac{(\mathbf{v}^{\mathrm{T}} \mathbf{u}_c)^* (\mathbf{u}_c{}^{\mathrm{T}}\, \mathbf{v})}{(\mathbf{v}^{\mathrm{T}})^* \mathbf{D}\, \mathbf{v}} \tag{10.29a}$$

$$\eta_x(\theta, \phi, \mathbf{v}) = \frac{Q_x{}^* Q_x}{(\mathbf{v}^{\mathrm{T}})^* \mathbf{D}\, \mathbf{v}} = \frac{(\mathbf{v}^{\mathrm{T}} \mathbf{u}_x)^* (\mathbf{u}_x{}^{\mathrm{T}}\, \mathbf{v})}{(\mathbf{v}^{\mathrm{T}})^* \mathbf{D}\, \mathbf{v}} \tag{10.29b}$$

where "c" and "x" indicate copolar and cross-polar illumination, respectively. To meet the common objective of maximum efficiency and minimum cross-polarization, a single performance index is defined as

$$\xi = (1 - \eta_c) + |\mu^2|\eta_x \tag{10.30}$$

By introducing intermediate vectors as in the previous case, Eq. (10.30) can be re-expressed as

$$\xi = \frac{\boldsymbol{x}^H(\boldsymbol{I} - \boldsymbol{\alpha}^H\boldsymbol{\alpha} + |\mu|^2\boldsymbol{\beta}^H\boldsymbol{\beta})\boldsymbol{x}}{\boldsymbol{x}^H\boldsymbol{x}} \tag{10.31}$$

As with the previous cases, the minimum of Eq. (10.31) is readily found from

$$(|\mu|^2\boldsymbol{\beta}^H\boldsymbol{\beta} - \boldsymbol{\alpha}^H\boldsymbol{\alpha} - \lambda\boldsymbol{I})\boldsymbol{x} = 0 \tag{10.32}$$

where the solution is the smallest eigenvalue λ. Once again, define $\boldsymbol{\tau} = \mu\boldsymbol{\beta}$, and $\boldsymbol{\alpha}$ is resolved into components parallel and orthogonal to $\boldsymbol{\tau}$. Thus, $\boldsymbol{\alpha} = \boldsymbol{\alpha}_1 + \boldsymbol{\alpha}_2$ where $\boldsymbol{\alpha}_1 = k\boldsymbol{\tau}$ and $\alpha_2^H\boldsymbol{\tau} = 0$. In addition, let $\mathbf{x} = c\boldsymbol{\alpha}_1 + d\boldsymbol{\alpha}_2$. Making the above substitutions in Eq. (10.32) results in

$$\boldsymbol{\tau}[c(k^2 - 1)\boldsymbol{\alpha}^H\boldsymbol{\tau} - kd\boldsymbol{\alpha}^H\boldsymbol{\alpha}_2] - \boldsymbol{\alpha}_2[ck\boldsymbol{\alpha}^H\boldsymbol{\tau} + d(\boldsymbol{\alpha}^H\boldsymbol{\alpha}_2)] = \lambda(ck\boldsymbol{\tau} + d\boldsymbol{\alpha}_2).$$

By equating the coefficients, the homogeneous equations in c and d are as follows:

$$c[\lambda k + (k^2 - 1)\boldsymbol{\alpha}^H\boldsymbol{\alpha}] + d(k\boldsymbol{\alpha}^H\boldsymbol{\alpha}_2) = 0 \tag{10.33a}$$

$$c(k\boldsymbol{\alpha}^H\boldsymbol{\tau}) + d(\lambda + \boldsymbol{\alpha}^H\boldsymbol{\alpha}_2) = 0. \tag{10.33b}$$

For a nontrivial solution of Eqs. (10.33), the determinant of the matrix relating the coefficients c and d must be zero. This leads to the following characteristic equation in λ:

$$\lambda^2 + \lambda\,[(\boldsymbol{\alpha}^H\boldsymbol{\alpha}) - (\boldsymbol{\tau}^H\boldsymbol{\tau})] + [(\boldsymbol{\alpha}^H\boldsymbol{\tau})^2 - (\boldsymbol{\alpha}^H\boldsymbol{\alpha})(\boldsymbol{\tau}^H\boldsymbol{\tau})] = 0 \tag{10.34}$$

The solutions of Eq. (10.34) are

$$\lambda = \frac{-[(\boldsymbol{\alpha}^H\boldsymbol{\alpha}) - (\boldsymbol{\tau}^H\boldsymbol{\tau})] \pm \boldsymbol{\Delta}}{2} \tag{10.35}$$

where the discriminant is

$$\Delta = |\boldsymbol{\alpha} + \boldsymbol{\tau}||\boldsymbol{\alpha} - \boldsymbol{\tau}|.$$

The smallest eigenvalue is

$$\lambda = -\Delta\cos^2(\theta/2) \tag{10.36}$$

where

$$\cos\theta = \left[\frac{(\boldsymbol{\alpha}^H\boldsymbol{\alpha}) - (\boldsymbol{\tau}^H\boldsymbol{\tau})}{\boldsymbol{\Delta}}\right].$$

Substituting this eigenvalue into Eqs. (10.33) results in $k(c - d) + \nu d = 0$ where

$$\nu = \frac{1}{2}\left[\frac{(\boldsymbol{\alpha}^H\boldsymbol{\alpha}) + (\boldsymbol{\tau}^H\boldsymbol{\tau}) - \boldsymbol{\Delta}}{(\boldsymbol{\alpha}^H\boldsymbol{\tau})}\right]$$

Finally, $\mathbf{x} = (c\boldsymbol{\alpha}_1 + d\boldsymbol{\alpha}_2) = d(\boldsymbol{\alpha} - \mu\nu\boldsymbol{\beta})$. Without loss of generality, the multiplier can be set to unity, i.e. $d = 1$. Therefore, in Eqs. (10.29), we have

$$Q_c = \boldsymbol{\alpha}^H\boldsymbol{x} = \boldsymbol{\alpha}^H\boldsymbol{\alpha} - \mu\nu\,\boldsymbol{\alpha}^H\boldsymbol{\beta} \tag{10.37a}$$

and

$$Q_x = \boldsymbol{\beta}^H \mathbf{x} = \boldsymbol{\beta}^H \boldsymbol{\alpha} - \mu \nu \boldsymbol{\beta}^H \boldsymbol{\beta}. \tag{10.37b}$$

Zero cross-polarization on axis is achieved when $\eta_x(0, 0, \mathbf{v}) = 0$. It follows from Eq. (10.37b) that $Q_x = 0$ when

$$\mu \nu = \boldsymbol{\beta}^H \boldsymbol{\alpha} / \boldsymbol{\beta}^H \boldsymbol{\beta}. \tag{10.38}$$

The copolar beam efficiency with zero cross-polarization (condition Eq. (10.38)) is

$$\eta_c(\theta, \phi, \mathbf{v}) = \boldsymbol{\alpha}^H \boldsymbol{\alpha} - \frac{(\boldsymbol{\beta}^H \boldsymbol{\alpha})}{(\boldsymbol{\beta}^H \boldsymbol{\beta})} (\boldsymbol{\alpha}^H \boldsymbol{\beta}) \tag{10.39}$$

and the corresponding eigenvector is

$$\mathbf{x} = \boldsymbol{\alpha} - \frac{(\boldsymbol{\beta}^H \boldsymbol{\alpha})}{(\boldsymbol{\beta}^H \boldsymbol{\beta})} \boldsymbol{\beta} \tag{10.40}$$

This result is not unexpected because when the cross-polar component is small, the difference between the single beam excitation is the projection of the cross-polar contribution onto the copolar component (i.e. tangent of the angle between the components).

The design of the array excitation to form shaped patterns is an important part of antenna array design. The two common approaches are field and power synthesis. The latter is often preferred as it is simpler and normally gives stable convergence to an optimum. Two power pattern synthesis methods that include mutual coupling are described in the following subsections.

10.4.1 Penalty Function Optimization of Array Parameters

The design of the array excitation to form shaped patterns or to control sidelobes is an important part of antenna array design. The two common approaches are field and power synthesis. As the latter normally gives stable convergence, two power pattern synthesis methods in the presence of mutual coupling are described by Bird (2016). The first approach is based on the minimization of a performance index that is based on power either radiated in specific directions or reflected at the driving ports. For the radiated power, the copolarized contribution is maximized across many stations or observing points in a selected coverage area and is minimized outside the coverage or main beam. This is illustrated in Figure 10.3 as a cut through the shaped beam. The pattern is minimized below the specification G_U across the beam and sidelobes. As well, in some applications, the beam should not fall below a specified level G_L across the same region. The specified limits will be different for copolar and cross-polar radiation patterns.

In order to specify pattern constraints, consider a cut through the far-field. At a given angle at position P in the radiation pattern, the following limits and weight parameters are defined as follows:

c_{UP} = the maximum copolar efficiency level
c_{LP} = the minimum copolar efficiency level
x_{UP} = the maximum cross-polar efficiency level
w_{UP} = the weighting factor on copolar maximum
w_{LP} = the weighting factor on copolar minimum
w_{xP} = the weighting factor on cross-pol maximum

In the same way, limits can be placed on the input match, gain, and element dimensions could also be included through constraints and weight parameters. Some of these quantities may have frequency-dependent constraints such as the input reflection coefficient of the feeders. For example, at frequency k, where the reflection coefficient is Γ_k, specify

ρ_{Uk} = the maximum feed reflection level
w_{rk} = the weighting factor on reflection coefficient.

When limits are placed on the geometry or aperture dimensions, there would need to be constraints on size and location of the components to ensure a physically possible solution.

All constraints can be included in a single performance index as will be outlined. If the co- and cross-polar efficiencies of the array are optimized over the field of view, respectively, η_{cp} and η_{xp}, these can be specified via Eqs. (10.29). As a result, a number of positive discrete differences can be formed at each frequency k ($k = 1, 2, \ldots, N_f$) as follows:

$$\Delta_k = \begin{cases} -w_{Lk}(\eta_{ck} - c_{Lk}); & k = P; P = 1, 2, \ldots, N \\ w_{Uk}(\eta_{ck} - c_{Uk}); & k = N + P; P = 1, 2, \ldots, N \\ w_{xk}(\eta_{xk} - x_{Uk}); & k = 2N + P; P = 1, 2, \ldots, N \\ w_{rk}(\Gamma_k - \rho_{Uk}); & k = 3N + P; P = 1, 2, \ldots, N_f \end{cases} \tag{10.41}$$

where N is the number of sample points in the radiation pattern, and N_f is the number of frequency points. The objective of the optimization is to find the vector of coefficients for which $\Delta_k \leq 0$ for all $k = 1, 2, \ldots, 3N + N_f$. A single performance index that incorporates all constraints is the least p-th index (Bandler and Charalambous 1972)

$$I = H\left[\sum_{k\in\kappa}\left(\frac{\Delta_k}{H}\right)^p\right]^{1/p}; \mathrm{k} = 1, 2, \ldots, N_f \tag{10.42}$$

where $H = \max(\Delta_k)$, $p = \mathrm{sgn}(\Delta_k)\, q$, and κ is the set of specifications

$$\kappa = \begin{cases} \text{select all } \Delta_k & \text{if } H \leq 0 \\ \text{select only positive } \Delta_k & \text{if } H > 0. \end{cases}$$

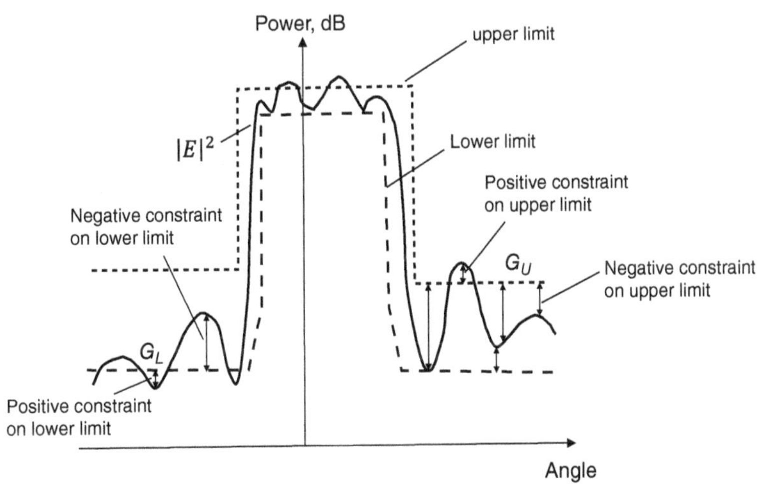

Figure 10.3 Radiation pattern envelopes for array pattern synthesis.

Any even-number integer index q can be chosen in p although $q = 2, 4, 10$, and 100 has been found most useful in antenna designs. Eq. (10.42) can be minimized with most standard optimization methods (Walsh 1977). One such method that has proved reliable in several antenna applications is based on gradient search and numerically calculated differentials (Fletcher 1972). Techniques such as the genetic algorithm (Goldberg 1989) and particle swarm optimization (Kennedy and Eberhart 1995) can be very effective at the commencement of the search for the minimum.

Many examples of the use of optimization of a penalty function in the design of shaped beams for geostationary satellites can be found in the literature (e.g. Bird 2016; Bird and Sroka 1992) and in Section 10.4.3 where results are compared with the method of successive projections, which is the topic of the next section.

10.4.2 Method of Successive Projections

A second approach for creating a shaped beam is the method of successive projections. This is a general iterative technique that can be used to determine a common intersection point among a number of conflicting requirements or sets. The technique is used in image processing and in two-dimensional filter design (Abo-Taleb and Fahmy 1984). In this approach, at each far-field point in the coverage region, there is a set $\boldsymbol{Q}$ of possible excitation vectors that satisfies an appropriate constraint equation. As a result, any point in the intersection of the several sets satisfies all constraints and represents a solution to the problem (Gubin et al. 1967). In relation to antenna arrays, the technique has been modified by Poulton (1986a,b) to determine the array excitation coefficients as will now be described. The aim is to achieve a shaped beam that has its directivity constrained between lower and upper bounds, respectively, G_L and G_U. The approach is illustrated in Figure 10.4.

The first step is to specify a collection of sets of $\{X\}$, which consist of complex valued functions on a rectangular region containing the projected aperture, a weighting function w constrained to have a unit magnitude (i.e. $|w| = 1$), and a weight with continuous phase, e.g. $w = \exp(jp)\,; p \in \{P\}$. The next step (1) (see Figure 10.4) is to provide an initial excitation vector $\boldsymbol{x}_o$ for the array excitation coefficients. These could be uniform or approximate assuming the absence of mutual coupling. The radiation patterns are then computed (2) using the space of complex weights for the radiation patterns of the elements. These may be in closed form or obtained by Fourier transforms. In this way, a vector $\boldsymbol{\Phi}_i$ is determined, which is a vector of the elementary field components, to obtain the far-field radiation at points $i = 1, 2, \cdots, M$ in a number of specified far-field directions (θ_i, ϕ_i). Next step (3) is to find the point greatest error, bring this error to zero by projecting the excitation vector onto the set of available excitations by an application of the successive projection algorithm. This is equivalent to minimizing the distance in excitation space between $\boldsymbol{x}_o$ and the boundary of all available excitations $\boldsymbol{Q}$. The formula to do this is simple to compute and, therefore, the process can

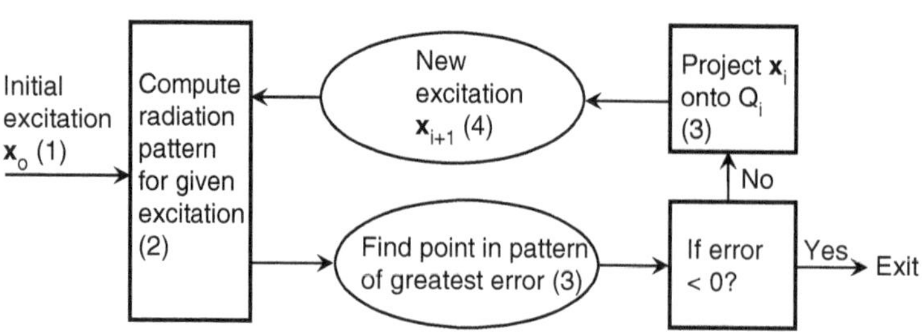

Figure 10.4 Method of successive projections.

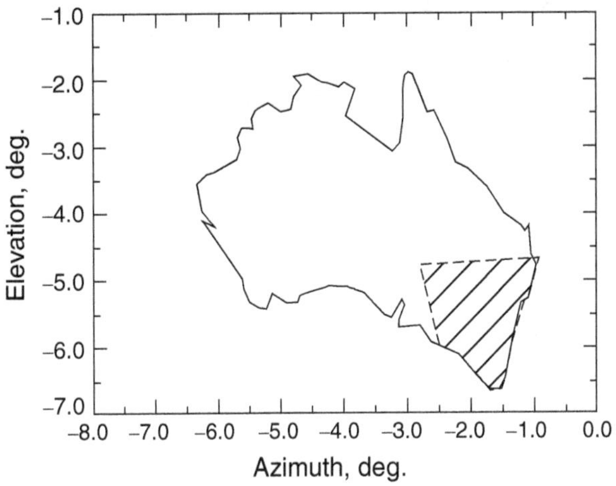

Figure 10.5 Coverage region of shaped beam on SE Australia.

be very fast. A sequence of iterations $\{\mathbf{x}^n\}$ is generated in the following form:

$$\boldsymbol{x}^{n+1} = (y_1/|y_1|, y_2/|y_2|, \ldots, y_N/|y_N|) \tag{10.43}$$

$$\boldsymbol{y} = \boldsymbol{x}^n + r\left(\frac{\sqrt{G_i}}{|F_i^n|} - 1\right)\left(\frac{F_i^n}{\boldsymbol{\Phi}_i \cdot \boldsymbol{\Phi}_i^*}\right)\boldsymbol{\Phi}_i^*, \quad i = 1, \ldots, N \tag{10.44}$$

where $N = 4P_xP_y$, $r \sim 4$ is a relaxation factor, $F_i^n = \boldsymbol{x}^n \cdot \boldsymbol{\Phi}_i$ where $n = 1, 2, \cdots$ is the iteration number, and

$$G_i = \begin{cases} G_L \text{ if } |F_i|^2 \le G_L \\ G_U \text{ if } |F_i|^2 \ge G_U \end{cases}.$$

Step (4) is repeated from (2) until convergence is achieved. This process can continue by tightening the upper or lower bounds, until they can be tightened no further.

At this point, no constraint has been placed on the total input power. This is not necessary if the pattern is specified over the entire visible region. However, for pattern design over a limited angular range, power must be constrained, either by use of a second Lagrange multiplier or more simply by expressing the constraint as an additional set. In the latter case, no change to the structure of the algorithm is required, illustrating the flexibility of the method. Virtually any constraint that can be represented as an inequality may be treated in this way.

10.4.3 Comparison of Penalty Functions and Successive Projections

As an illustration of both methods described in the previous section, results are described for a shaped beam to cover the south-east corner of Australia, Figure 10.5, as seen from a geostationary satellite located at 160°E (Bird and Poulton 1985). The aim was to achieve gain in excess of 36 dBi over this region with a reduction of at least 8 dB within a half a beamwidth of the edge of coverage. Additionally, a cross-polar isolation of 40 dB was required within the region. For this application, the conventional approach is a reflector antenna with an array feed. A single-offset reflector with an aperture of 83.375λ was chosen along with a feed cluster of 18 circular horns of diameter 1.14λ, and to give good beam control, each horn was excited in two orthogonal TE_{11} modes as shown in Figure 10.6.

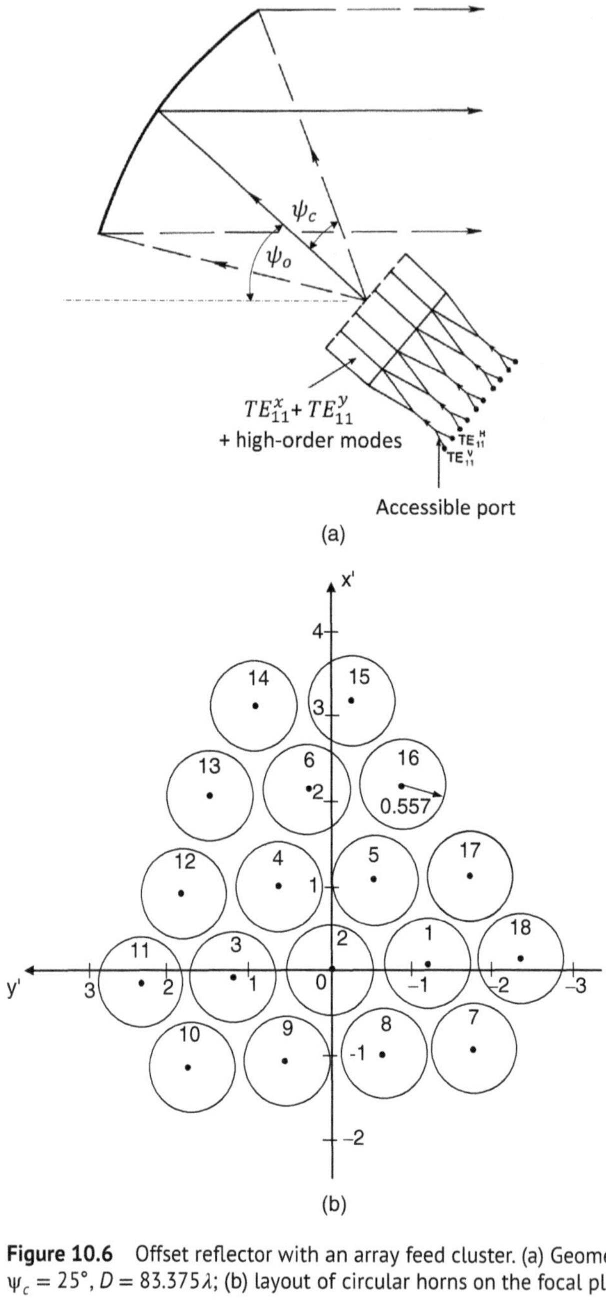

Figure 10.6 Offset reflector with an array feed cluster. (a) Geometry offset angle $\psi_o = 40°$, half-cone angle $\psi_c = 25°$, $D = 83.375\lambda$; (b) layout of circular horns on the focal plane.

After an initial setup of the mutual coupling matrices and radiation patterns both of which are common to both the penalty function and successive projections methods, the methods were applied to determine the shaped beam for SE Australia. All horns were excited in both polarized modes namely TE^x_{11} (copolar) and TE^y_{11} (x-polar). Remarkably, the two methods produced almost identical results, although successive projections proved to be about five times faster for this problem. The amplitude and phase of the excitations from both methods are compared in Figure 10.7. The co- and cross-polar radiation patterns obtained from the results of both excitations

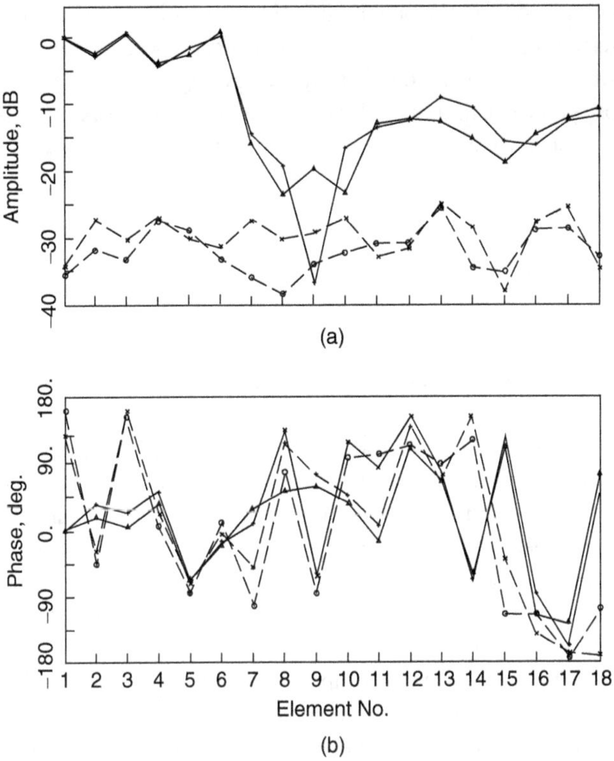

Figure 10.7 Array excitations obtained from two synthesis methods for the SE Australia coverage region for copolar and x-polar modes. (a) Amplitude; (b) phase. Key: Magnitude: solid curve; phase: dash curve; penalty function: +, copolar; x, cross-polar; successive projections: △, copolar; o, x-polar.

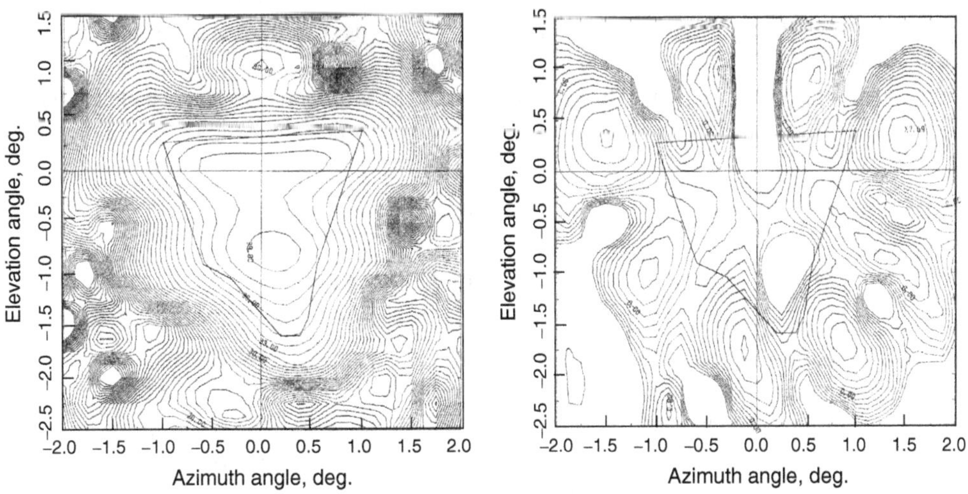

Figure 10.8 Shaped beam for SE Australia coverage with excitations obtained by penalty function. Levels are at 1 dB intervals (a) copolar levels above 0 dBi; (b) cross-polar levels above −6 dBi.

are surprisingly similar. The patterns obtained from the penalty functions excitation are shown in Figures 10.8a and b. The copolar pattern has a peak gain of 39.4 dBi with a good beam shape both inside and outside the coverage region. Almost all stations within the coverage exceed the minimum gain requirement as is the cross-polar isolation, which is −40 dB below the peak. This is especially noteworthy since the offset reflector with a single feed horn has a cross-polar level of about −26 dB below the peak.

10.5 Direction-of-Arrival Estimation

Traditionally, DOA estimation performance of different algorithms has mainly been performed using a standard array with half-wavelength interelement separation. With the increasing demand of portable mobile devices as well as the evolution of Internet-of-Things, there has been a growing demand for small aperture arrays with an element separation of a factor of a half wavelength (Volmer et al. 2008; Yong and Thompson 2004, Browne et al. 2006). In devices equipped with such a function as DOA estimation, the array aperture is substantially reduced compared to standard array with half-wavelength separation. In ideal situation, although the performance of subspace signal processing method is not significantly dependent on the array aperture, there is still a limit such that the array aperture is so small such that the algorithm fails to perform. For instance, due to physical constraints, if we want to design an array with an aperture of 0.1λ with five array elements, does the array still be able to resolve signals coming from 50° and 80° for the application with known algorithms? If the answer is no, perhaps we should consider a larger array aperture, or increase the operational frequency such that the aperture becomes electronically larger, adopt or develop alternative signal processing solutions. At the same time, the interelement separation of these compact arrays embedded inside portable devices can be at least four to five times smaller than the half-wavelength element separation in standard arrays. Subsequently, electromagnetic interactions among all the antenna elements becomes significant, which severely deteriorates the DOA performance. Therefore, it is important to identify strategies to reduce or suppress these undesirable effects.

To answer these questions, we will first evaluate the DOA estimation performance under an ideal situation. The incoming signals are captured using ideal point detectors and the directions of the signals are resolved using MPM. In particular, we consider the two cases:

(a) the impact of the array aperture to DOA estimation performance with a fixed number of elements, and
(b) the impact of the number of elements to DOA estimation performance within in a fixed aperture.

This study is performed using a Monte-Carlo approach where the incoming signals are corrupted by Additive Gaussian White Noise (AGWN). Details are shown in Figure 10.9, path (i).

In the second part, we will then incorporate mutual coupling into the model. Instead of ideal point scatters, these two cases ((a) and (b)) will be studied using arrays with dipole, monopole, and normal-mode helical antenna (NMHA) elements. The DOA problem together with the arrays are simulated using full-wave moment method solver FEKO (2021) such that the electromagnetic interactions between all the elements are accurately considered. Each antenna is terminated with a 50 Ω load. To realize how mutual coupling could deteriorate DOA performance, as shown in Figure 10.9, path (ii), the terminal voltages obtained from FEKO, which have taken mutual coupling effect into accounts, will be directly fed to MPM. To compensate for mutual coupling, decoupling methods such as conventional mutual impedance (Figure 10.9, path (iii)) and receiving mutual impedance (Figure 10.9, path (iv)) will be applied to demonstrate the superiority of mutual coupling compensation. Details about these compensation methods can be found in Chapter 11.

10.5.1 No Coupling Situation

To study the impact of mutual coupling effect in direction finding application, it is valuable to examine the situation first where there is no coupling to provide a future comparison. The following sections will describe several numerical case studies of direction finding performance under ideal arrays where the array apertures and number of elements are varied. In these studies, the incoming signals are detected using ideal arrays where we assume there is no coupling between the antennas. The practical situations where signals are detected by antenna elements where mutual coupling is taken into account, will be studied in section 10.5.2.

10.5.1.1 Cramer-Rao Lower Bound

The DOA estimation problem is essentially an estimation of undamped exponentials ($e^{j\phi}$) based on measured signal samples. The Cramer-Rao Lower Bound (CRLB), which gives the lower bound on the variance of any unbiased estimator, has been well studied for the estimation of a single damped/undamped exponential (e.g. Steedly and Moses 1993; Stoica and Nehoral 1989; Yao and Pandit 1995; Yilmazer et al. 2007; Hua and Sarkar 1989, 1990; Djermoune and Tomczak 2009). Given the incoming single frequency signal with a wavelength of λ, the CRLB of the estimation of a single-mode undamped exponential can be written as a function of the array aperture (physical length of the array, L), number of array elements (N), SNR (η), and the incoming signal direction

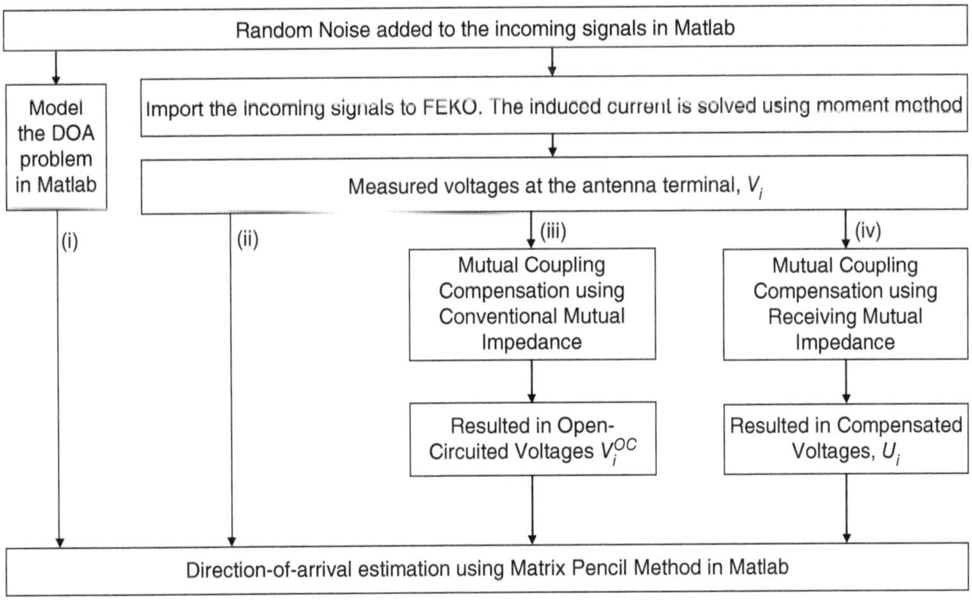

Figure 10.9 Monte-Carlo simulation of arrays with different configurations. (i) Using ideal point detectors (no mutual coupling). In (ii)–(iv), dipole, monopole, or helical antenna elements are used, and the direction-finding problem is solved using moment method in FEKO such that mutual coupling is considered. In (ii), the terminal voltages obtained from FEKO are directly used for direction finding. In (iii) and (iv), the terminal voltages are compensated using conventional and receiving mutual impedance, respectively.

(ϕ_i), given by (Kay 1993)

$$\text{var}(\hat{\phi}_i) \geq \frac{12}{(2\pi)^2 N\eta \frac{N+1}{N-1}\left(\frac{L}{\lambda}\right)^2 \sin^2\phi_i}. \tag{10.45}$$

From Eq. (10.45), it is apparent that the variance of the estimated angle can be enhanced by increasing the number of array elements (N), or alternatively increasing the array aperture (L) Kay (1993). This is the same as the case presented in Figure 10.2, $0 < \phi_i \leq 90°$ where $\phi_i = 90°$ corresponds to broadside direction and $\phi_i = 0°$ corresponds to end-fire direction.

For cases where there are multiple modes of undamped exponentials, the CRLB of an unbiased estimator (Steedly and Moses 1993; Stoica and Nehoral 1989) as well as the variance of the estimation using MPM (Hua and Sarkar 1990) are usually given in the forms of matrix expressions. The impact of the variations of array aperture on the estimation accuracy, however, cannot be easily observed from these matrix expressions. Also, perturbation due to mutual coupling has not been taken into account in Eq. (10.45). To better understand situations with multiple signals coming from different directions, as well as a more practical situation where mutual coupling is taken into account, Monte-Carlo simulation using uniform arrays with different array apertures, different number of array elements, and different antenna elements results will be presented in the next two subsections, which shall provide us with better insights.

10.5.1.2 Four-Element Linear Arrays with Different Apertures (Two Incoming Signals)

Monte-Carlo simulations of DOA estimations using four-element ULAs with different array apertures are first studied. The interelement separation is reduced from $\Delta = 0.5\lambda$ down to 0.05λ, with a step size of 0.05λ. This configuration corresponds to an array aperture (defined as the physical length of the array) of 1.5λ down to 0.15λ. The case with $\Delta = 0.5\lambda$ is the "standard" interelement separation used in those conventional arrays. With the same DOA estimation problem, we aim to investigate the impact of array aperture reduction to the DOA estimation accuracies.

Two equal-power and equal-phase coherent sources come from the azimuth angles of $\phi_1 = 5°$ and $\phi_2 = 6°$, which are close to the end-fire direction ($\phi = 90°$ corresponds to the broadside direction, as shown in Figure 10.2). The signals (both magnitude and phase) are contaminated with AGWN at an SNR of 5 dB, as illustrated in Figure 10.9, path (i). To evaluate the performance of the DOA estimations, the average bias, $\bar{b}$, is introduced and used as a metric throughout Section 10.5.1. It is given by

$$\bar{b} = \sum_{i=1}^{Q} \frac{b_i}{Q}, \tag{10.46}$$

where $b_i = |\hat{\phi}_i - \phi_i|$ is the absolute value of the bias of the estimated direction ($\hat{\phi}_i$ being the mean of the estimated directions. $\hat{\phi}_i$ is determined from 10 000 independent Monte-Carlo simulations), and Q is the number of incoming signals which is 2 in this case. For a successful DOA estimation, all the signals should be accurately estimated. A small value of average bias is obtained only if all the incoming signals are accurately estimated. At each value of interelement separation (Δ) shown in the figures in this subsection, the average bias is obtained by 10 000 independent Monte-Carlo simulations.

The average bias and biases for each signal are shown in Figure 10.10. It is found that the average biases decrease as the interelement separation increases. This indicates that the estimated directions are more accurate when the interelement separation is large. However, when $\Delta = 0.05\lambda$, the average bias is 3° when the SNR levels of the two incoming signals are both 5 dB. This means MPM

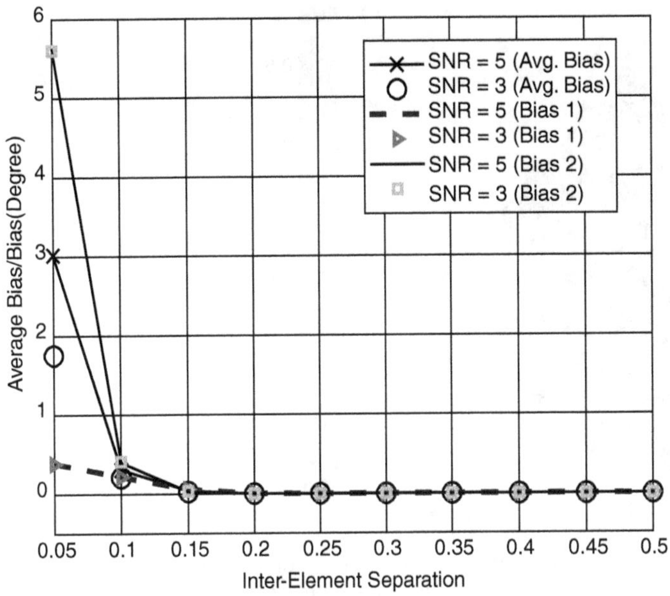

Figure 10.10 Summary of DOA estimations using MPM for a four-element ULA with interelement separation varies from 0.05λ to 0.5λ. Two equal-power incoming signals are from the directions of $\phi_1 = 5°$ and $\phi_2 = 6°$ with an SNR of 5 dB and 3 dB, respectively. The graph shows the average bias of the estimated directions and the bias of the estimation of signals 1 and 2. Source: LUI, H.S. and HUI, H. T. (2013): 'Direction-of-arrival estimation of closely spaced emitters using compact arrays', International Journal of Antennas and Propagation, Volume 2013, Article ID 104848, 9 pages. Licensed under CC BY 3.0.

fails to resolve the two incoming signals as the difference between the two signals is only 1°. The biases of the estimated directions of the two signals are also included. It is interesting to see that for the case of $\Delta = 0.05\lambda$, the bias of ϕ_1 is about 0.4° (i.e. the estimated direction is about 5.4°), which is close to the mean of the two incoming signals (5° and 6°). The bias of ϕ_2 is around 6° which is far from the actual incoming direction, which further confirms that MPM fails to resolve the two incoming signals under such an array configuration. To investigate if the performance varies at different SNR levels, the same DOA problem is repeated except that the SNR level is 3 dB. The results show that there are no significant differences.

Next, we consider a series of DOA estimation problems with the two signals coming from the directions with signal 1 coming from $\phi_1 = 10°$ to 90°, with a step size of $\Delta\phi_1 = 5°$, while signal 2 is only 1° different from signal 1, i.e. $\phi_2 = \phi_1 + 1°$. Therefore, the angular range of 90°–180°, is simply the mirror image of that of 0°–90°. The same DOA estimation calculation procedure is repeated for all cases independently, and the results are shown in Figure 10.11. Similarly, the average biases decrease as Δ increases for the same incoming signals. The results show that the average bias is below 0.001° when the array element separation is greater than or equal to 0.2λ. Furthermore, we also note from the figure that as the signals are coming from a direction closer to the broadside direction of 90°, the accuracy of estimation increases quite significantly even though the element separation remains the same. For instance, when the element separation is 0.05λ with incoming signals of $\phi_1 = 10°$ and $\phi_2 = 11°$, an average bias of around 0.7° is resulted. When the incoming signals become $\phi_1 = 55°$ and $\phi_2 = 56°$, the average bias is less than 0.0001° (with less than 0.01%

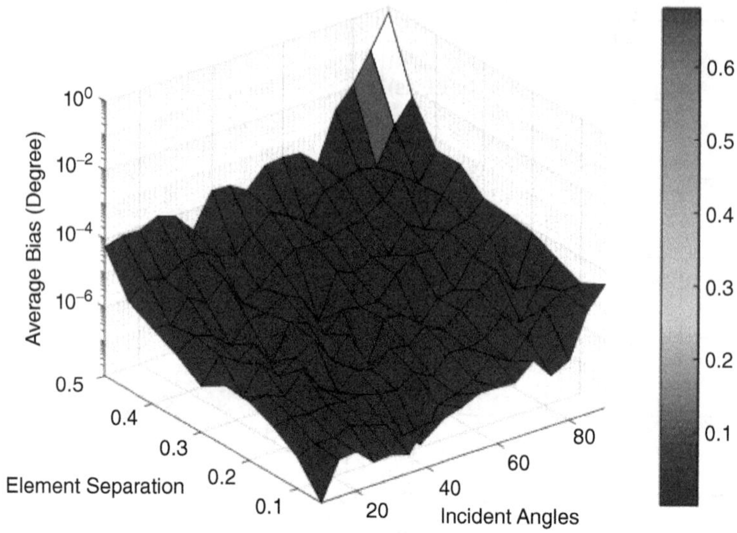

Figure 10.11 Summary of DOA estimations using MPM for a four-element ULA with interelement separation varies from 0.05λ to 0.5λ. Two equal-power incoming signals are from the directions of $\phi_1 = 10°$ to 90° and $\phi_2 = \phi_1 + 1°$ with an SNR of 5 dB. The step size of $\Delta\phi_1 = 5°$ is considered. The figure shows the average biases of the estimated directions. Source: LUI, H.S. and HUI, H. T. (2013): 'Direction-of-arrival estimation of closely spaced emitters using compact arrays', International Journal of Antennas and Propagation, Volume 2013, Article ID 104848, 9 pages. Licensed under CC BY 3.0.

error). This is also in line with the observation from the CRLB expression in Eq. (10.45) that signals coming from the broadside direction will result in a smaller variance under the same array configuration, although we now have two signals instead of one.

10.5.1.3 Fixed Aperture Uniform Linear Arrays with Different Numbers of Elements (Two Incoming Signals)

In this example, DOA estimations using linear arrays with apertures within the range of 0.25λ–1λ are considered. Two incoming signals remain to be the same with $\phi_1 = 10°$ to 90° and $\phi_2 = \phi_1 + 1°$, but the number of array elements is now varying from 4 to 20. Notice that under a fixed array aperture, the interelement separation has to be changed with the change in the number of elements in the array. For instance, the interelement separation for a four-element array with 0.5λ aperture is $0.5/(4-1) = 0.1667\lambda$, while the interelement separation becomes $0.5/(20-1) = 0.0263\lambda$ when the aperture is filled with 20 array elements.

The array apertures of 0.25λ and 0.5λ are first considered, and the results are shown in Figures 10.12a,b. When the incoming signals are close to the end-fire directions (between 5° and 15°), the average biases decrease when the number of elements increases from 4 to 20. As the incoming signals are moving toward the broadside direction (between 20° and 90°), the average biases decrease rapidly with an average bias less than 0.0001° regardless of the number of elements.

Now, the same estimation procedure is repeated with an array aperture of 0.75λ. When the aperture is increased from 0.5λ to 0.75λ, as shown in Figure 10.12c, the average biases have been significantly reduced (less than 0.0022° for all cases). The maximum average bias is reduced from 0.0156° (0.5λ aperture) to 0.0022° (0.75λ aperture), which is nearly a factor of 7. This indicates a significant improvement of the DOA estimations compared with the two cases of aperture size of 0.25λ and 0.5λ. When the array aperture is further increased to 1λ, as shown in Figure 10.12d, it is not apparent if increasing the number of elements further enhances the DOA accuracy under this

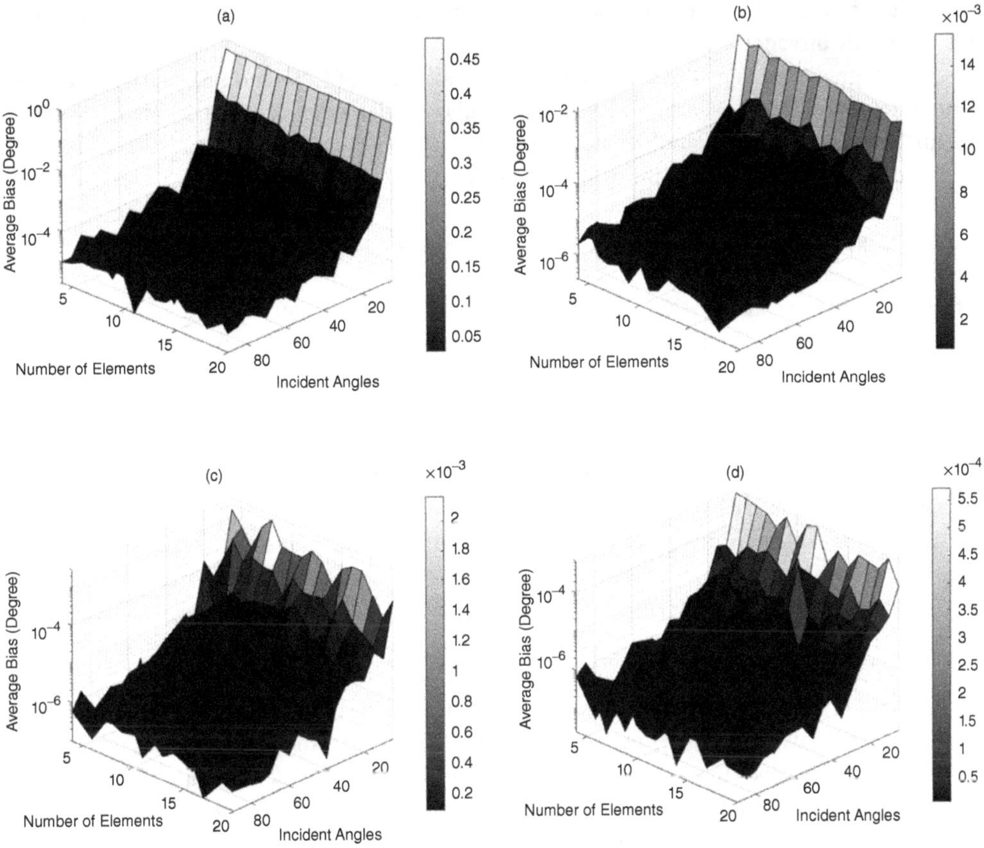

Figure 10.12 Summary of DOA estimations using MPM for a four-element ULA with the number of array elements varies from 4 to 20. Two equal-power incoming signals are from the directions of $\phi_1 = 10°$ to $90°$ and $\phi_2 = \phi_1 + 1°$ with an SNR of 5 dB. The step size of $\Delta\phi_1 = 5°$ is considered. The figures show the average biases of the estimated direction. The apertures of the arrays are (a) 0.25λ, (b) 0.5λ, (c) 0.75λ, and (d) 1λ. Source: LUI, H.S. and HUI, H. T. (2013): 'Direction-of-arrival estimation of closely spaced emitters using compact arrays', International Journal of Antennas and Propagation, Volume 2013, Article ID 104848, 9 pages. Licensed under CC BY 3.0.

aperture. On the other hand, it is worth noting that as the array aperture increases from 0.25λ to 1λ, the average biases for all cases (different number of elements and incoming signals) decrease. This indicates that the accuracy of DOA estimation depends more on the array aperture than on the number of elements when the aperture is sufficiently large. It is also worth noting that in our first example using a four-element linear array to resolve two closely-spaced emitters (5° and 6°), an array aperture $\geq 0.6\lambda$ ($\Delta \geq 0.2\lambda$) is required which is in line with the findings here.

The above results show that for a fixed small array aperture (especially for the cases of 0.25λ and 0.5λ), the accuracies of DOA estimation can be improved by increasing the number of elements when the incoming signals are close to the end-fire directions. When the incoming signals are moving toward to the broadside direction, increasing the number of elements under a fixed aperture do not necessarily improve the accuracies of the estimation. When the array aperture is larger than 0.5λ, increasing the number of array elements does not significantly enhance the accuracies of the estimation. Comparing between signals coming from the broadside direction and the end-fire directions, under the same array configuration, better accuracies are expected for signals coming from broadside direction.

While not presented here, the results on the array apertures of 1.25λ and 1.5λ indicate that there is some improvement of DOA accuracy as the number of elements increases. These results show that further increase in the number of elements for large aperture arrays (>1λ) only yields minor improvement to the accuracy of the estimated angles.

10.5.1.4 Fixed Aperture Uniform Linear Arrays with Different Number of Elements (Three Incoming Signals)

In this example, we consider DOA estimation problems of three closely-spaced, equal-power, and equal-phase incoming signals with an SNR of 5 dB coming from the directions from $\phi_1 = 40°$ to $90°$ with $\Delta\phi_1 = 5°$, $\phi_2 = \phi_1 + 1°$, and $\phi_3 = \phi_1 + 2°$. Figure 10.13a shows the average bias for the case with a 0.75λ aperture. The number of array elements varies from 6 to 20 (the minimum number of sensors required for resolving three incoming signals using MPM is six). It can be clearly seen that the average biases decrease with the number of the elements. Compared to the earlier case with two signals of $\phi_1 = 45°$, $\phi_2 = 46°$ with the array aperture of 0.25λ (Figure 10.12a) and the case of $\phi_1 = 5°$, $\phi_2 = 6°$ with the array aperture of 0.75λ (Figure 10.12c) with six array elements (0.0001° and 0.001° respectively), the average biases presented in Figure 10.13a (6°) are much higher when we have the third incoming signals. This is because resolving three incoming signals using MPM and resolving two incoming signals are different mathematical problems with different outcomes. As expected, when the incoming signals are moving toward the broadside directions, the average biases reduce to 2° for the case of six antenna elements. The average biases are also reduced as the number of elements increases. However, even with 20 elements, the average biases are between 0.5° and 3°. These bias values are still considered as high as the actual separations between the three sources is only 1°, and thus, it cannot be concluded that the sources are clearly resolved.

The array aperture is now increased to 1λ and 1.25λ, and the corresponding results are shown in Figures 10.13b,c, respectively. For the same DOA problem, the average biases decrease as the number of array element increases. At an array aperture of 1λ with 20 array elements, the average biases are at a level less than 0.1° for DOA when the incoming signals are more than 75°. For the DOA problem with incoming signals ($\phi_1 = 45°$, $\phi_2 = 46°$, $\phi_3 = 47°$) with 20 array elements, the average bias is 0.628°. However, when the array aperture is further increased to 1.25λ with 20 array elements, the average bias reduces to smaller than 0.1° (as shown in Figure 10.13c). The average biases for the same DOA estimation problem but with an array aperture of 1.5λ are shown in Figure 10.13d. As can be seen, when the signal sources are closer to the broadside direction, the average biases are smaller. However, an increasing number of array elements do not necessarily increase the accuracy of DOA estimation in this aperture. On the other hand, when the signal sources are closer to the end-fire direction, the accuracy of DOA estimation increases substantially when the number of array elements is increased (see Figure 10.13d).

10.5.2 Perturbation Due to Mutual Coupling

In the previous section, DOA estimation performance of MPM was evaluated under different array configurations using point detectors together with closely placed emitters with only one-degree difference of incoming directions. These simulation results provide us with further insights on the performance of MPM in an ideal situation. In this section, mutual coupling is included and the consequences are studied. Compared to the ideal situation where the measured signal is purely based on the incoming signals, the reradiated field from other array elements that will also form a part in the measured signals. The measured signals are now "corrupted" by the coupled field. If we use the "corrupted" signals for DOA estimation (Figure 10.9, path (ii)), which will be demonstrated in Section 10.5.2.1(a) through a set of examples of eight-element arrays with different interelement separations and different antenna elements, the estimation performance will be significantly deteriorated. Results in Section 10.5.2.1(b) and (c) will demonstrate the impact of mutual coupling compensation using conventional and receiving mutual impedance (Figure 10.9, paths (iii) and (iv)). Also, the effect of array aperture to the DOA estimation performance is also demonstrated through these examples. Then, in Section 10.5.2.2, examples with different array apertures with four and eight array elements will be used to illustrate how DOA performance varies with the number of elements. Also, it is worth noting that the incoming signals results presented in this section is at least 15° differences (except the last one with 5° differences from broadside direction) rather than 1° difference in earlier sections. We are not aiming to replicate the setting as the ideal case simply because it is computationally extensive. At the same time, it is practically challenging to resolve signals with only 1° separation when mutual coupling and other array perturbations are

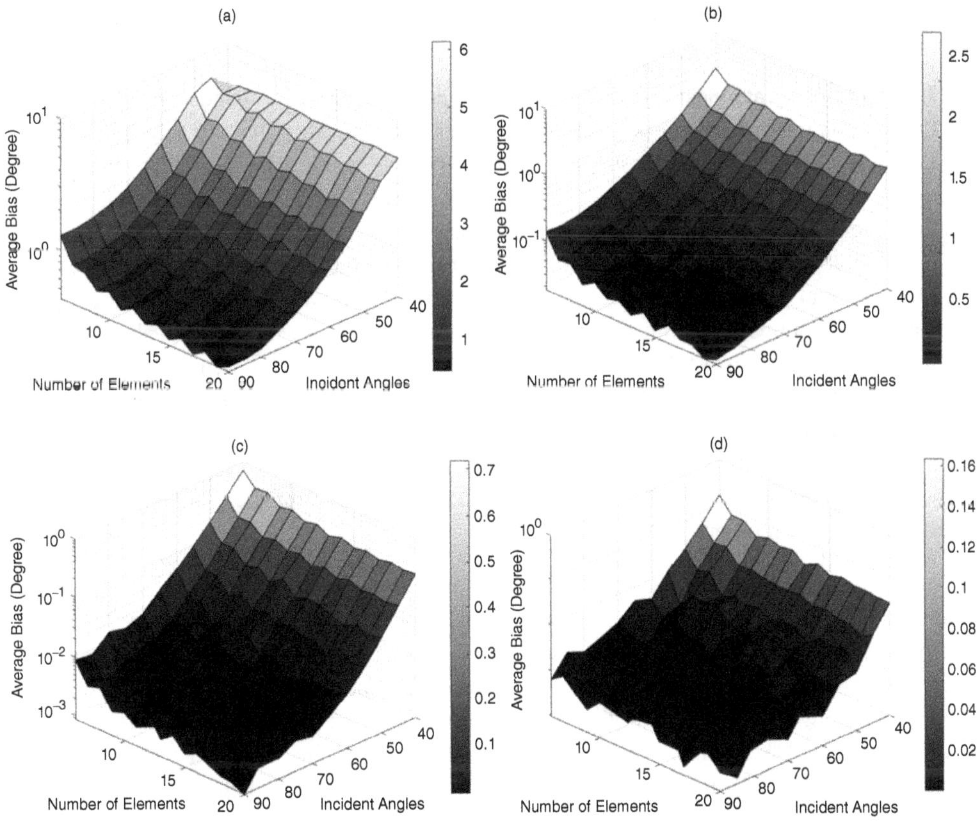

Figure 10.13 Summary of DOA estimations using MPM for a four-element ULA with the number of array elements varies from 6 to 20. Three equal-power incoming signals are from the directions of $\phi_1 = 40°$ to $90°$, $\phi_2 = \phi_1 + 1°$, and $\phi_3 = \phi_1 + 2°$ with an SNR of 5 dB. The step size of $\Delta\phi_1 = 5°$ is considered. The figures show the average biases of the estimated directions with array aperture of (a) 0.75λ, (b)1.00λ, (c) 1.25λ, and (d) 1.50λ. Source: LUI, H.S. and HUI, H. T. (2013): 'Direction-of-arrival estimation of closely spaced emitters using compact arrays', International Journal of Antennas and Propagation, Volume 2013, Article ID 104848, 9 pages. Licensed under CC BY 3.0.

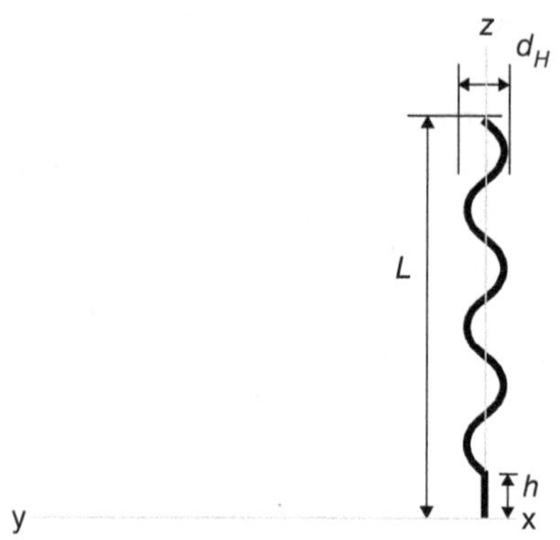

Figure 10.14 The NMHA with $L = 25$ mm, $d_H = 2.6$ mm, $h = 3$ mm, $L = 25$ mm, and $d_w = 0.6$ mm (Lui and Hui 2010b).

considered in practice. Therefore, more realistic DOA estimation problems are selected to illustrate how mutual coupling affects the estimation accuracies, as well as how conventional and receiving mutual impedance methods (RMIMs) can effectively compensate for these undesirable effects.

10.5.2.1 Eight-Element Linear Arrays with Different Apertures (Three Incoming Signals)

The DOA problem of three equal-power coherent signals coming from the azimuth angles of $\phi_1 = 20°$, $\phi_2 = 35°$, and $\phi_3 = 100°$, shown in Figure 10.2, is considered. These three signals come from an elevation angle of $\theta = 90°$ contaminated with AGWN with a SNR of 10 dB. The incoming signals are measured using ULAs with eight omnidirectional antenna elements. The layout configuration of the ULA and the incoming signals are shown in Figure 10.2. Three different types of omnidirectional elements – dipoles, monopoles, and NMHA – are studied independently. The dipole and the monopole antennas are with lengths of 0.48λ and 0.24λ, respectively. The design of the right-hand circular NMHA is shown in Figure 10.14, with the dimension of $L = 25$ mm, $d_H = 2.6$ mm, $h = 3$ mm, and all the NMHAs are three turns. The wire radii of all the three types of antennas are the same and equal to 0.3 mm (i.e. diameter: $d_w = 0.6$ mm). The geometry of the arrays and the DOA problem are modeled using the moment method solver FEKO (2021). Using FEKO, the current distribution and the induced terminal voltages of each antenna element are calculated using the moment method.

(a) DOA Estimation Under the Effect of Mutual Coupling

DOA estimation using the terminal voltages is first presented. An eight-element dipole ULA with an interelement separation of $\Delta = 0.5\lambda$ is considered. To perform Monte-Carlo simulation, shown in Figure 10.9, path (ii), AGWN is generated randomly with SNR = 10 dB and added to each incoming signal. The "noisy" incoming signals are then imported to FEKO. The current distribution and the terminal voltages under the excitation of the "noisy" incoming signals are then calculated using moment methods. The terminal voltages are then imported to MPM to estimate the incoming directions. This process is repeated 1000 times independently.

The mean, bias, standard deviation, Root-Mean Square Error (RMSE) of the estimated angles, and the number of failures (complex angle) are tabulated in Table 10.1. Notice that the results are calculated without taking the failure estimation (complex angle, Eq. (10.8c)) into consideration.

The mean of the estimated angle is found to be $\hat{\phi}_1 = 27.0621°$, $\hat{\phi}_2 = 42.0673°$, and $\hat{\phi}_3 = 104.3980°$, respectively, and the bias of the estimated angles is in the range 4°–7°. The standard

Table 10.1 DOA estimation using a ULA with eight dipole antenna elements shown in Figure 10.2. The results are obtained using the voltages measured at the antenna terminals without any compensation.

	Direction	Mean	Bias	Standard dev	RMSE	Failure
$\Delta = 0.5\lambda$	20°	27.0621°	7.0621°	5.6563°	9.0462°	0
	35°	42.0673°	7.0673°	17.8727°	19.2109°	0
	100°	104.3980°	4.3980°	15.5139°	16.1178°	0
$\Delta = 0.3\lambda$	20°	25.2140°	5.2140°	3.7863°	6.4426°	2
	35°	99.5810°	64.5810°	3.7409°	64.6889°	693
	100°	100.5187°	0.5187°	0.8303°	0.9785°	306
$\Delta = 0.15\lambda$	20°	34.7225°	14.7225°	5.3657°	15.6261°	979
	35°	77.6171°	42.6171°	12.3615°	44.3720°	20
	100°	168.0939°	68.0939°	5.3877°	68.3055°	830

Source: Based on LUI, H.S., YU, Y. and HUI, H. T. (2010d): 'Effect of mutual coupling on the performance of direction-of-arrival estimation of compact array', IEEE International Symposium on Antennas and Propagation, pp. 1-4, Toronto, Ontario, Canada.

deviation and the RMSE are at the levels 10°–20° which indicates that the results are not accurate. Out of the 1000 simulations, there is no failure estimation.

The interelement separation of the antenna element is now reduced to $\Delta = 0.3\lambda$. The same procedures are repeated, and the results for DOA estimation are shown in Table 10.1. The mean of the estimated angles of $\hat{\phi}_1 = 25.2140°$, $\hat{\phi}_2 = 99.5810°$, and $\hat{\phi}_3 = 100.5187°$ are found. The estimated angle of $\hat{\phi}_1 = 25.2140°$ is roughly the average of the incoming signals of $\phi_1 = 20°$ and $\phi_2 = 35°$, which indicates that MPM fails to resolve the signals in this instance. The direction of $\phi_3 = 100°$ is appeared as $\hat{\phi}_2 = 99.5810°$ and $\hat{\phi}_3 = 100.5187°$, respectively. Notice that there is a large number of failures (complex angles) resulted (306 and 693 failure results for directions 2 and 3 out of 1000 simulations, i.e. 30.6% and 69.3% failure). To probe further, the interelement separation of the elements is further reduced to $\Delta = 0.15\lambda$, and the estimated angles are tabulated in Table 10.1. The estimated angle of $\hat{\phi}_1 = 34.7225°$, $\hat{\phi}_2 = 77.6171°$, and $\hat{\phi}_3 = 168.0939°$ are totally different from the actual incoming directions, and the numbers of failures for the three signals are 979, 20, and 830, respectively.

Now, the entire procedure is repeated with the dipole elements in the ULA replaced by monopole and NMHA elements. The results are tabulated in Tables 10.2 and 10.3. Similar observations are found. At $\Delta = 0.5\lambda$ separation, the ULA with monopole antenna gives the most accurate solution, and the bias of the estimated angle is within the range of 4° (Table 10.2) as compared to dipole (7°, Table 10.1) and helical antenna (11°, Table 10.3). As the element separation is further reduced to $\Delta = 0.3\lambda$ and $\Delta = 0.15\lambda$, none of the ULAs is able to resolve the incoming signals without any compensation. The results indicate that MPM is capable of resolving the incoming signals when the element separation is large ($\Delta = 0.5\lambda$). Among the three types of antenna, the monopole antenna gives the best performance of revolving the incoming signals with the bias less than 4° for $\Delta = 0.5\lambda$ (Table 10.2).

The possible reasons of the failures can either be the reduction of the aperture or the existence of the mutual coupling. As previously evaluated, the subspace method is not significantly dependent on the array aperture unless the aperture is smaller than 0.75λ for three incoming signals. Here, our smallest aperture is 1.05λ ($0.15\lambda \times (8-1)$) such that MPM should be able to resolve the direction. As will be demonstrated in the following examples, it will be clear that the failures of detection are mainly due to the existence of the antenna mutual coupling effect rather than the reduction of the array aperture.

(b) Compensation Using Conventional Mutual Impedance

Instead of using the measured voltage from the antenna terminals directly, the voltages are first decoupled using conventional mutual impedance (Gupta and Ksienski 1983). The results for the three types of antenna arrays with 0.5λ element separation based on 1000 independent simulations are shown in Table 10.4. As compared to the results in Tables 10.1–10.3 for the case of $\Delta = 0.5\lambda$, the bias of the estimated angles is now within the range of ~1°. The standard deviations and RMSE become smaller. DOA estimations with better accuracies indicate that the conventional mutual impedance provide a certain level of mutual coupling characterization. While not presented here, simulations of the DOA estimations problem using ULAs with $\Delta = 0.3\lambda$ and $\Delta = 0.15\lambda$ have also been done but the MPM failed to resolve the signals. The results indicated that decoupling using conventional mutual impedance cannot effectively compensate the mutual coupling effect when the element separation is smaller than 0.5λ.

In another study presented in section 11.4.3.2, DOA estimation of two incoming signals detected using three eight-element monopole arrays with inter-element separations of 0.5λ, 0.25λ and 0.05λ using MUSIC algorithm is considered. The corresponding MUSIC spectrums are shown in Figure 11.6. As shown in Figure 11.6(d), when the inter-element separation is 0.05λ, the result show that MUSIC algorithm fails to resolve the incoming signals when conventional mutual impedance is used for compensation.

(c) Compensation Using Receiving Mutual Impedance

DOA estimation using the decoupled voltages based on the RMIM is now considered (Hui 2003, 2004a,b; Hui et al. 2006; Lui et al. 2009; Lui and Hui 2010a,b,c; Yu et al. 2011). Compared to the conventional mutual impedance in which the mutual impedances are determined when the array is in transmitting mode, the receiving mutual impedances are determined when the array is in receiving mode such that the mutual coupling behavior for a receiving mode array is better characterized (Lui et al. 2009). In the context of RMIM, there are three different ways to

Table 10.2 DOA estimation using a ULA with eight monopole antenna elements shown in Figure 10.2. The results are obtained using the voltages measured at the antenna terminals without any compensation.

	Direction	Mean	Bias	Standard dev	RMSE	Failure
$\Delta = 0.5\lambda$	20°	21.5583°	1.5583°	5.8677°	6.0683°	0
	35°	38.4740°	3.4740°	15.3053°	15.6872°	0
	100°	103.5860°	3.5860°	14.8124°	15.2331°	0
$\Delta = 0.3\lambda$	20°	26.5789°	6.5789°	3.9215°	7.6580°	32
	35°	87.1652°	52.1652°	24.6567°	57.6931°	5
	100°	103.8674°	3.8674°	11.5115°	12.1269°	677
$\Delta = 0.15\lambda$	20°	12.7295°	7.2705°	4.8668°	8.7471°	273
	35°	85.1422°	50.1422°	9.9984°	51.1283°	1
	100°	166.7431°	66.7431°	5.2349°	66.9477°	555

Source: Based on LUI, H.S., YU, Y. and HUI, H. T. (2010d): 'Effect of mutual coupling on the performance of direction-of-arrival estimation of compact array', IEEE International Symposium on Antennas and Propagation, pp. 1-4, Toronto, Ontario, Canada.

Table 10.3 DOA estimation using a ULA with eight NMHA elements shown in Figure 10.2. The results are obtained using the voltages measured at the antenna terminals without any compensation.

	Direction	Mean	Bias	Standard dev	RMSE	Failure
$\Delta = 0.5\lambda$	20°	20.7130°	0.7130°	7.6949°	7.7241°	0
	35°	46.0530°	11.0530°	25.7596°	28.0189°	0
	100°	110.4468°	10.4468°	22.7821°	25.0528°	0
$\Delta = 0.3\lambda$	20°	27.3914°	7.3914°	3.1763°	8.0443°	3
	35°	97.5892°	62.5892°	8.0523°	63.1046°	3
	100°	140.0130°	40.0130°	22.2108°	45.7510°	592
$\Delta = 0.15\lambda$	20°	28.0054°	8.0054°	3.4016°	8.6975°	0
	35°	85.9473°	50.9473°	7.6631°	51.5198°	0
	100°	168.8891°	68.8891°	4.4975°	69.0355°	457

Source: Based on LUI, H.S., YU, Y. and HUI, H. T. (2010d): 'Effect of mutual coupling on the performance of direction-of-arrival estimation of compact array', IEEE International Symposium on Antennas and Propagation, pp. 1-4, Toronto, Ontario, Canada.

Table 10.4 DOA estimation using a ULA with eight antenna elements with the element separation of $\Delta = 0.5\lambda$. The results are obtained using the voltages decoupled via conventional mutual impedance.

	Direction	Mean	Bias	Standard dev	RMSE	Failure
Dipole	20°	19.8914°	0.1086°	3.1319°	3.1322°	0
	35°	36.1299°	1.1299°	7.0849°	7.1709°	0
	100°	100.8440°	0.8440°	7.2640°	7.3093°	0
Monopole	20°	19.8296°	0.1704°	2.5119°	2.5164°	0
	35°	35.3918°	0.3918°	5.4661°	5.4774°	0
	100°	100.4485°	0.4485°	4.9696°	4.9874°	0
NMHA	20°	19.3197°	0.6803°	3.0567°	3.1300°	0
	35°	35.6740°	0.6740°	8.2172°	8.2407°	0
	100°	101.0097°	1.0097°	7.8320°	7.8929°	0

determine the receiving mutual impedance. In this section, we consider two of them – namely methods 2 and 3. In method 2 (Hui 2004b; Hui et al. 2006), the receiving mutual impedances are determined with only two elements considered in the geometry. For instance, if we want to determine the receiving mutual impedance between elements 2 and 6, only these two elements are considered, and all other array elements are removed in the geometry, shown in Figure 10.15a. In other words, reradiating field from all other elements is not considered in this approach. In method 3 (Lui and Hui 2010a,b,c), however, $Z_t^{2,6}$ is determined with all other elements that exist such that mutual coupling from all these elements will be taken into account in the resultant $Z_t^{2,6}$, as shown in Figure 10.15b. Strictly speaking, method 2 is an approximated solution, while method 3 is a more accurate solution. Examples of the receiving mutual impedance determined using methods 2 and 3 can be found in Table 11.1. Further details about these two determination methods, as welll as the RMIM can also be found in Chapter 11.

We first consider the receiving mutual impedance determined using method 2 for decoupling. The simulation is set up based on Figure 10.9 (path (iv)), and the results based on 1000 independent simulations with $\Delta = 0.5\lambda$ and $\Delta = 0.3\lambda$ are shown in Tables 10.5 and 10.6, respectively. Comparing the decoupling using conventional mutual impedance for the case of $\Delta = 0.5\lambda$ (Table 10.4), the bias, standard deviation, and RMSE are all below 0.5° when the receiving mutual impedance is utilized for decoupling (Table 10.5). The three types of arrays with the element separation of $\Delta = 0.5\lambda$ (Table 10.5), all show a similar level of accuracy in terms of mean, bias, standard deviation, and RMSE. Comparison between the three types of arrays, arrays with monopole and helical antenna elements, resulted in a lower level of bias, standard deviation, and RMSE. This demonstrates that the RMIM provides a better solution for mutual coupling characterization and compensation than the conventional mutual impedance for all the three types of antenna elements. The results for $\Delta = 0.15\lambda$ are also included in Table 10.7. For this case, only the incoming direction of $\phi_3 = 100°$ is able to be resolved. The first estimated direction of $\hat{\phi}_1 \approx 26°$ roughly corresponds to the average of the incoming signals at $\phi_1 = 20°$ and $\phi_2 = 35°$. However, the estimated angle of $\hat{\phi}_2 \approx 70°$ does not provide any correct information

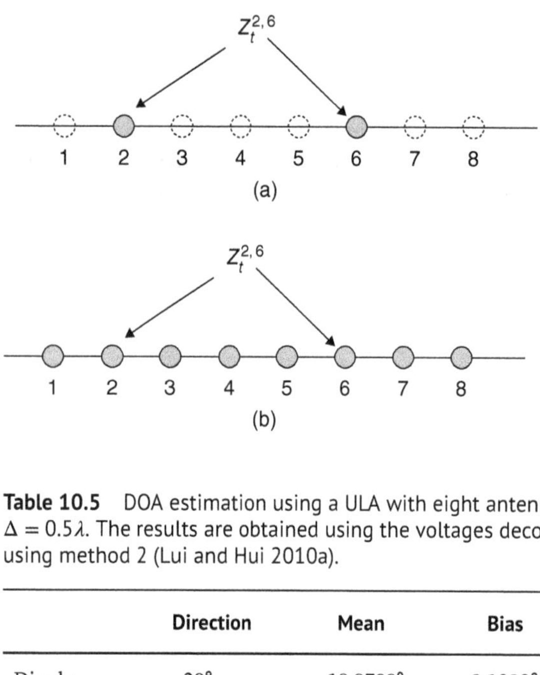

Figure 10.15 Determination of the receiving mutual impedance $Z_t^{2,6}$ using the (a) method 2 and (b) method 3. In method 2, only elements 2 and 6 are considered, and the other elements are removed and the electromagnetic problem changes. In method 3, the mutual impedance is determined with the existence of all the elements and the electromagnetic problem remains the same.

Table 10.5 DOA estimation using a ULA with eight antenna elements with the element separation of $\Delta = 0.5\lambda$. The results are obtained using the voltages decoupled via receiving mutual impedance calculated using method 2 (Lui and Hui 2010a).

	Direction	Mean	Bias	Standard dev	RMSE	Failure
Dipole	20°	19.8708°	0.1292°	0.0059°	0.1293°	0
	35°	34.9603°	0.0397°	0.0177°	0.0435°	0
	100°	100.0074°	0.0074°	$8.6712° \times 10^{-6}$	0.0074°	0
Monopole	20°	19.8742°	0.1258°	0.0035°	0.1259°	0
	35°	34.9585°	0.0415°	0.0066°	0.0420°	0
	100°	100.0073°	0.0073°	$7.3923° \times 10^{-6}$	0.0073°	0
NMHA	20°	19.8794°	0.1206°	0.0058°	0.1207°	0
	35°	34.9481°	0.0519°	0.0057°	0.0522°	0
	100°	100.0071°	0.0071°	$1.6703° \times 10^{-5}$	0.0071°	0

of the actual incoming signal directions, which means that incoming signal directions are not properly resolved.

Next, we consider using the receiving mutual impedance determined using method 3 for decoupling. The corresponding DOA estimation results for all the three types of arrays with element separations of $\Delta = 0.5\lambda$, $\Delta = 0.3\lambda$, and $\Delta = 0.15\lambda$ are tabulated in Tables 10.8 to 10.10, respectively. The bias of the estimated angle is below 0.1° for $\Delta = 0.5\lambda$ and that for $\Delta = 0.3\lambda$ is around 0.2°. More importantly, for the array with $\Delta = 0.15\lambda$, all the three incoming signals can be clearly resolved, and the bias of the estimated angles are below 0.3° for monopole and NMHA and 0.7° for dipole antenna, respectively.

Comparing between the three types of antenna, the results indicate that they all have a similar level of accuracies for $\Delta = 0.5\lambda$ as shown in Table 10.8, which is in line with our previous observation in Table 10.5 for receiving mutual impedance determined using method 2. As the

Table 10.6 DOA estimation using MPM using a ULA with eight antenna elements with the element separation of $\Delta = 0.3\lambda$. The results are obtained using the voltages decoupled via receiving mutual impedance calculated using method 2 (Lui and Hui 2010a).

	Direction	Mean	Bias	Standard dev	RMSE	Failure
Dipole	20°	19.0827°	0.9173°	0.4843°	1.0372°	0
	35°	34.3502°	0.6498°	0.3469°	0.7365°	0
	100°	99.9957°	0.0043°	0.0055°	0.0070°	0
Monopole	20°	19.3181°	0.6819°	0.3882°	0.7846°	0
	35°	34.5166°	0.4834°	0.2255°	0.5334°	0
	100°	99.9988°	0.0012°	0.0047°	0.0049°	0
NMHA	20°	19.6199°	0.3801°	0.0938°	0.3915°	0
	35°	34.6402°	0.3598°	0.6327°	0.7276°	0
	100°	100.0012°	0.0012°	0.0102°	0.0102°	0

Table 10.7 DOA estimation using MPM using a ULA with eight antenna elements with the element separation of $\Delta = 0.15\lambda$. The results are obtained using the voltages decoupled via receiving mutual impedance calculated using method 2 (Lui and Hui 2010a,b; Lui et al. 2010d).

	Direction	Mean	Bias	Standard dev	RMSE	Failure
Dipole	20	26.8544°	6.8544°	2.0403°	7.1514°	0
	35	75.0806°	40.0806°	11.8023°	41.7805°	0
	100	99.7631°	0.2369°	3.3735°	3.3801°	3
Monopole	20	26.5502°	6.5471°	1.8963°	6.8189°	1
	35	72.3251°	37.3251°	10.0761°	38.6599°	1
	100	99.4895°	0.5105°	3.9179°	3.9491°	1
NMHA	20	26.8300°	6.8300°	1.9839°	7.1121°	0
	35	79.8916°	44.8916°	8.9505°	45.7743°	0
	100	99.3991°	0.6008°	3.4001°	3.4511°	3

Table 10.8 DOA estimation using MPM using a ULA with eight antenna elements with the element separation of $\Delta = 0.5\lambda$. The results are obtained using the voltages decoupled via receiving mutual impedance calculated using method 3 (Lui and Hui 2010a).

	Direction	Mean	Bias	Standard dev	RMSE	Failure
Dipole	20°	19.9070°	0.0930°	0.0215°	0.0954°	0
	35°	34.9622°	0.0378°	0.0290°	0.0477°	0
	100°	100.0067°	0.0067°	2.1285×10^{-4}°	0.067°	0
Monopole	20°	19.9022°	0.0978°	0.0086°	0.0982°	0
	35°	34.9565°	0.0435°	0.0078°	0.0442°	0
	100°	100.0068°	0.0068°	1.6790×10^{-4}°	0.0068°	0
NMHA	20°	19.8911°	0.1089°	0.0017°	0.1089°	0
	35°	34.9507°	0.0493°	0.0122°	0.0508°	0
	100°	100.0068°	0.0068°	7.7219×10^{-5}°	0.0068°	0

antenna element separation becomes closer as shown in Tables 10.9 and 10.10, the results again indicate that ULAs with monopole and helical antenna elements outperform the ULA with dipole antenna elements with lower level of bias, standard deviation, and RMSE. Furthermore, it is found that there are three failure angles for the dipole array at $\Delta = 0.15\lambda$ separation, but there is no failure angle for monopole and helical arrays. The results indicate that monopole and helical arrays slightly outperform the dipole array for DOA estimation when the receiving mutual impedances are determined using method 3.

Through these examples, it is demonstrated that the DOA estimation performance is significantly degraded when the array elements are closely spaced such that mutual coupling dominates. The undesirable mutual coupling effect can be decoupled using the conventional mutual impedances when the element separation is $\Delta = 0.5\lambda$. As the interelement separation becomes smaller, conventional mutual impedances fail to compensate for mutual coupling, while the receiving mutual impedance (method 2) still managed to compensate for the mutual coupling effect at $\Delta = 0.3\lambda$. When the separation is further reduced to $\Delta = 0.15\lambda$, however, only the receiving mutual impedance determined with all elements taken into account (method 3) is capable of compensating for the undesirable mutual coupling effect.

In the another DOA estimation study presented in section 11.4.3.2, the result show that the MUSIC algorithm able to resolves two incoming signals using a eight-element monople array with interelement separation is 0.05λ. when the antenna terminal signals are compensated using receiving mutual impedance (method 3).

10.5.2.2 Fixed Array Aperture with Different Numbers of Elements (Two Incoming Signals)

DOA estimation performance of eight-element arrays with monopole, dipole, and helical antenna elements were studied in Section 10.5.2.1. Our results showed that the existence of mutual coupling deteriorated the estimation accuracies. The accuracies of the estimation can be significantly improved when the coupling effect is compensated using the RMIM. To probe further, we evaluate how the number of array elements within a fixed array aperture will impact the DOA estimation accuracies when mutual coupling effect is compensated using RMIM (method 3). Here, the simulation is performed using FEKO following the setup in Figure 10.9 (path (iv)) such that mutual coupling is accurately modelled. The array aperture size is varied from 0.25λ to 1.50λ with a step

Table 10.9 DOA estimation using MPM using a ULA with eight antenna elements with the element separation of $\Delta = 0.3\lambda$. The results are obtained using the voltages decoupled via receiving mutual impedance calculated using method 3 (Lui and Hui 2010a,b).

	Direction	Mean	Bias	Standard dev	RMSE	Failure
Dipole	20°	19.7697°	0.2303°	0.0570°	0.2372°	0
	35°	34.8191°	0.1809°	0.1025°	0.2079°	0
	100°	100.0060°	0.0060°	0.0016°	0.0062°	0
Monopole	20°	19.7742°	0.2258°	0.0401°	0.2294°	0
	35°	34.8612°	0.1388°	0.0638°	0.1527°	0
	100°	100.0063°	0.0063°	0.0011°	0.0064°	0
NMHA	20°	19.8486°	0.1514°	0.0210°	0.1529°	0
	35°	34.8913°	0.1087°	0.1068°	0.1524°	0
	100°	100.0068°	0.0068°	0.0015°	0.0069°	0

Table 10.10 DOA estimation using MPM using a ULA with eight antenna elements with the element separation of $\Delta = 0.15\lambda$. The results are obtained using the voltages decoupled via receiving mutual impedance calculated using method 3 (Lui and Hui 2010a,b; Lui et al. 2010d).

	Direction	Mean	Bias	Standard dev	RMSE	Failure
Dipole	20°	20.6475°	0.6475°	0.4631°	0.7959°	2
	35°	35.4054°	0.4054°	0.3032°	0.5062°	1
	100°	100.0082°	0.0082°	0.0060°	0.0102°	0
Monopole	20°	20.2686°	0.2686°	0.2357°	0.3573°	0
	35°	35.2034°	0.2034°	0.4572°	0.5002°	0
	100°	100.0083°	0.0083°	0.0092°	0.0124°	0
NMHA	20°	20.1861°	0.1861°	0.4111°	0.4511°	0
	35°	35.1614°	0.1614°	0.4504°	0.4783°	0
	100°	100.0078°	0.0078°	0.0078°	0.0110°	0

size of 0.25λ, i.e. 0.25λ, 0.50λ, 0.75λ, 1.00λ, 1.25λ, and 1.50λ. Four-element and eight-element uniform arrays are used. We aim to investigate if the DOA estimation performance can be improved if we increase the number of array elements within a small array aperture and at the same time as when mutual coupling is properly compensated. This is in parallel with our earlier findings in Sections 10.5.1.3 and 10.5.1.4 where ideal point detectors are used.

The corresponding estimation results of two incoming signals with azimuth angles of $\phi_1 = 30°$ and $\phi_2 = 57°$ using four-element arrays with different aperture are shown in Figure 10.16a and the first two rows of Table 10.11. As shown in Figure 10.16a (curves (i) $\phi_1 = 30°$ and (ii) $\phi_2 = 57°$), the four-element array fails to accurately estimate the incoming signals with mean and standard deviation values ($\mu \pm \sigma$) of 23.9° ± 14.66° and 52.5° ± 6.41°, as well as 26.4° ± 6.65° and 55.6° ± 4.76° when the array apertures are 0.25λ and 0.5λ, respectively. Out of the 1000 estimations, the number of failure estimations (a complex angle, Eq. (10.13)) are 406 (40.6%) and 51 (5.1%), respectively, for

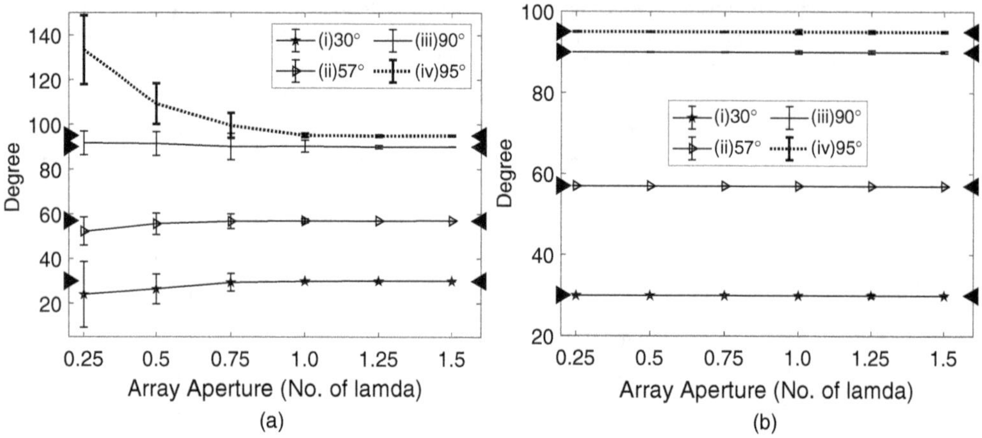

Figure 10.16 Error bars (mean ± standard deviation) of the two DOA estimation examples of the (a) four-element and (b) eight-element uniform linear array with different array apertures using MPM with 1000 independent trials. Example 1: The two signals are coming from (i) $\phi_1 = 30°$ and (ii) $\phi_2 = 57°$. Example 2: The two signals are coming from (iii) $\phi_1 = 90°$ and (iv) $\phi_2 = 95°$. Source: LUI, H.S. and HUI, H. T., 'Direction-of-arrival estimation: measurement using compact arrays under the influence of mutual coupling', IEEE Antennas and Propagation Magazine, Vol.57, No. 6, pp. 62-68. © 2015 IEEE.

Table 10.11 Number of failure estimations out of 1000 estimation trials using the four-element ULA.

		Array aperture					
	Example	**0.25λ**	**0.50λ**	**0.75λ**	**1.00λ**	**1.25λ**	**1.50λ**
1	$\phi_1 = 30°$	406	59	13	2	0	0
	$\phi_2 = 57°$	51	8	5	0	0	0
2	$\phi_1 = 90°$	12	6	6	2	0	0
	$\phi_2 = 95°$	231	32	14	0	0	0

Source: LUI, H.S. and HUI, H. T., 'Direction-of-arrival estimation: measurement using compact arrays under the influence of mutual coupling', IEEE Antennas and Propagation Magazine, Vol.57, No. 6, pp. 62-68. © 2015 IEEE

each direction when a four-element array with an aperture of 0.25λ is used. This reveals that the four-element array fails to resolve the signals accurately at such a small array aperture. As the array aperture increases to 0.75λ and above, the mean values approach to the actual incoming directions of 30° and 57°, respectively. The standard deviation as well as the number of failure estimations drop significantly. Next, we consider the cases for eight-element arrays. The results in Figure 10.16b (curves (i) $\phi_1 = 30°$ and (ii) $\phi_2 = 57°$) show that the mean values are within ±0.2° from the actual incoming directions, and the standard deviations are less than 0.5° in all cases. There is no failure estimation. This shows that the incoming signals can be resolved when eight antenna elements are used regardless of the array aperture.

The results reveal that that when the array aperture is less than 0.75λ, increasing the number of elements from four to eight (under a fixed array aperture) will enhance the accuracy of DOA estimation. When the array aperture further increases (>0.75λ), both arrays are able to accurately resolve the signals. This is only possible when mutual coupling is properly characterized and decoupled using receiving mutual impedance determined using method 3.

In the second example, we consider the case where the two incoming signals are closely located at $\phi_1 = 90°$ and $\phi_2 = 95°$. The elevation angle and SNR remain unchanged. The corresponding DOA estimation results are shown in Figure 10.16a,b (curves (iii) $\phi_1 = 90°$ and (iv) $\phi_2 = 95°$) and Table 10.11 (last two rows). For the four-element arrays with 0.25λ aperture, the means and standard deviations ($\mu \pm \sigma$) of the estimation are $91.63° \pm 5.31°$ and $133.38° \pm 15.38°$, respectively, for each incoming signal direction. The difference between the two sources is only 5°, and the result of $91.63° \pm 5.31°$ is not acceptable for ϕ_1. For ϕ_2, the mean value of 133.38° with a standard deviation of 15.38° away from the actual source, which is again not acceptable. As the array aperture increases to 0.5λ, the mean and standard deviation values of the two signals become $91.34° \pm 5.26°$ and $109.23° \pm 9.16°$. The signals are not properly resolved in these two cases. When the array aperture further increases to 1λ, the results are now $90.24° \pm 2.56°$ and $95.12° \pm 0.78°$. The mean values are now within ±0.25° from the actual incoming directions, and the standard deviations are within ±3° which are much accurate. The corresponding numbers of failure estimation shown in Table 10.11 are also relatively large when the array aperture is 0.25λ and the numbers decrease in a similar manner as it is in the first example when the aperture increases. When eight antenna elements are used, the mean values of the estimation are within ±0.1°, and the standard deviations are within ±0.5° for all array apertures, shown in Figure 10.16b. Same as the first example, there is no failure estimation when the number of antenna elements is increased from four to eight.

In summary, the results show that when the mutual coupling effect is properly compensated, increasing the number of array elements within a small array aperture (<0.75λ) gives a higher accuracy in DOA estimation. This is consistent with our findings in Sections 10.5.1.3 and 10.5.1.4 when ideal point detectors are used.

10.6 Conclusion

Different approaches of incorporating and handling the undesirable mutual coupling effects in beam shaping and direction finding, which are the two main applications for transmitting and receiving arrays, are discussed in detail and exemplified through various examples. For beam shaping applications, the excitation of single and multiple beams was described that includes mutual coupling. This showed that to maximize the gain, the excitation was related to the conjugate of the element far-fields in the chosen direction. With no coupling these excitations are independent but with coupling all element far-fields contribute to every element excitation. Two power pattern synthesis methods that included mutual coupling were described. These were a penalty function method and the method of successive projections. Both techniques were applied to a practical shaped beam synthesis problem and were shown to give remarkably similar results although successive projections proved to be about five times faster in this example.

The performance DOA estimation using linear arrays with different interelement separation and array apertures is quantitatively evaluated through a Monte-Carlo approach with full-wave moment method simulations. First, we evaluate the DOA estimation performance under an ideal situation using point detectors such that the impact of interelement separation and array aperture to the accuracies of the estimation becomes apparent. Our findings show that when the array aperture is small (less than 0.5λ for two and less than 0.75λ for three closely placed signals), the four-element array fails to resolve the signal coming from nearly end-fire directions. The DOA estimation performance can be enhanced by increasing the number of array elements within a small aperture. The findings provide us with the performance and limitations of MPM for DOA estimation.

In practice, when actual antenna elements are used, the mutual coupling effect between them will naturally form a part in the received signals that degrades the estimation performance. Eight-element arrays with element separations of 0.5λ, 0.3λ, and 0.15λ using dipole, monopole, and helical antenna elements were studied. The DOA problem together with the antenna elements are solved using a moment method solution such that the mutual coupling is accurately modeled. For the smallest element separation of 0.15λ, the corresponding array aperture is 1.05λ (0.15 × (8–1)) which is large enough to resolve three incoming signals based on our earlier findings. Without mutual coupling compensation, the array fails to resolve the three signals of 20°, 35°, and 100° accurately even with 0.5λ interelement separation. Using the receiving-mutual-impedance method (RMIM), the incoming signals can be accurately resolved with less than 0.3° for arrays with monopole and helical elements and 0.7° for arrays with dipole element. These results illustrate that mutual coupling does deteriorate the DOA estimation performance, although it can be handled through the RMIM.

Lastly, the impact of number of array elements within a fixed aperture with monopole antenna elements was studied. Four- and eight-element monopole arrays with apertures ranging from 0.25λ to 1.50λ were considered. For the four-element array, our results show that accurate estimation can only be obtained when the array aperture is above 1λ. When the number of array element is increased from four to eight, however, the incoming signals can be accurately estimated even at 0.25λ aperture. This finding is in line with our earlier findings using ideal point detectors where increasing the number of elements within a small aperture does enhance the DOA estimation performance. Once again, the success relies heavily on effective mutual coupling compensation using the RMIM.

References

Abo-Taleb, A. and Fahmy, M. (1984). Design of FIR two-dimensional digital filters by successive projections. *IEEE Trans. Circuits Syst.* 31(9): 801–805.

Bandler, J.W. and Charalambous (1972). Practical least p-th optimization of networks. *IEEE Trans. Microwave Theory Tech.* MTT-20: 834–840.

Bird, T.S. (1982). Contoured-beam synthesis for array-fed reflector antennas by field correlation. *IEE Proc. H. Microwaves Antennas Propag.* 129: 293–298.

Bird, T.S. (2016). *Fundamentals of Aperture Antennas and Arrays.* Chichester UK: Wiley.

Bird, T.S. and Poulton, G.T. (1985): 'Envelope-constrained pattern synthesis for satellite antennas', Convention Digest, IREECON, Melbourne, Aust., pp. 567-570.

Bird, T.S. and Sroka, C. (1992). Design of the Ku-band antennas for the Galaxy HS601C satellites. *J. Electr. Electron. Eng. Aust.* 12: 267–273.

Browne, D.M., Manteghi, M., Fritz, M.P., and Ramit-Sami, Y. (2006). Experiments with compact antenna arrays for MIMO radio communications. *IEEE Trans. Antennas Propag.* 54 (11): 3239–3250.

Collin, R.E. and Zucker, F.J. (eds.) (1969). *Antenna Theory, Part 1.* New York, USA: McGraw-Hill.

Djermoune, E.H. and Tomczak, M. (2009). Perturbation analysis of subspace-based methods in estimating a damped complex exponential. *IEEE Trans. Signal Process.* 57 (11): 4558–4563.

FEKO (2021), Altair Engineering Inc., www.altair.com/feko/, accessed May 2021.

Fletcher, R. (1972): 'Fortran subroutines for minimization by quasi-Newton methods', A.E.R.E., Harwell, UK, AERE Report 7125.

Friedlander, B. and Weiss, A.J. (1991). Direction finding in the presence of mutual coupling. *IEEE Trans. Antennas Propag.* 39 (3): 273–284.

Godara, L.C. (1997). Application of antenna arrays to mobile communications, Part II: Beam forming and direction-of-arrival considerations. *Proc. IEEE* 85 (8): 1195–1245.

Goldberg, D. (1989). *Genetic Algorithms in Search, Optimization and Machine Learning*. Addison-Wesley Longman Publishing Co. Inc.

Golub, G.H. and Van Loan, C.F. (1996). *Matrix Computations*, 3e. The John Hopkins University Press.

Gubin, L.G., Polyak, B.T., and Raik, E.V. (1967). The method of projections for finding the common point of convex sets. *USSR Comput. Math. Math. Phys.* 7 (6): 1–24.

Gupta, I.J. and Ksienski, A.A. (1983). Effect of mutual coupling on the performance of adaptive arrays. *IEEE Trans. on Antennas & Propag* 31 (5): 785–791.

Hua, Y. (1992). Estimating two-dimensional frequencies by matrix enhancement and matrix pencil. *IEEE Trans. Signal Processing* 40 (9): 2267–2280.

Hua, Y. and Sarkar, T.K. (1989). Generalized pencil-of-function method for extracting poles of an EM system from its transient response. *IEEE Trans. Antennas Propag.* 37 (2): 229–234.

Hua, Y. and Sarkar, T.K. (1990). Matrix pencil method for estimating parameters of exponentially damped/undamped sinusoids in noise. *IEEE Trans. Acoust. Speech Signal Process.* 38 (5): 814–824.

Hui, H.T. (2003). Improved compensation for the mutual coupling effect in a dipole array for direction finding. *IEEE Trans. Antennas Propag.* 51 (9): 2498–2503.

Hui, H.T. (2004a). A new definition of mutual impedance for application in dipole receiving antenna arrays. *IEEE Antennas Wireless Propag. Lett.* 3: 364–367.

Hui, H.T. (2004b). A practical approach to compensate for the mutual coupling effect in an adaptive dipole array. *IEEE Trans. Antennas Propag.* 52 (5): 1262–1269.

Hui, H.T., Low, H.P., Zhang, T.T., and Lu, Y.L. (2006). Receiving mutual impedance between two normal-mode helical antennas (NMHAs). *IEEE Antennas Propag. Mag.* 48 (4): 92–96.

Hwang, S., Burintramart, S., Sarkar, T.K., and Best, S.R. (2006). Direction of arrival (DOA) estimation using electrically small tuned dipole antennas. *IEEE Trans. Antenna Propag.* 54 (11): 3292–3301.

Jenkins, H.H. (1991). 'Small-aperture radio direction-finding', Artech House Inc., Norwood, MA USA.

Kay, S.M. (1993). *Fundamentals of Statistical Signal Processing: Estimation Theory*. Prentice Hall.

Kennedy, J. and Eberhart, R. (1995): 'Particle swarm optimization', Proc. IEEE Conference on Neural Networks, Nov/Dec, pp. 1942–1948.

Krim, H. and Viberg, M. (1996). Two decades of array signal processing research: the parametric approach. *IEEE Signal Process. Mag.* 13 (4): 67–94.

Lau, C.K.E., Adve, R.S., and Sarkar, T.K. (2004). Minimum norm mutual coupling compensation with applications in direction of arrival estimation. *IEEE Trans. Antennas Propag.* 52 (8): 2034–2040.

Liberti, J.C. and Rappaport, T.S. (1999). *Smart Antennas for Wireless Communications, IS-95 and Third Generation CDMA Applications*. New Jersey, Prentice Hall: Upper Salle River.

Lui, H.S. and Hui, H.T. (2010a). Effective mutual coupling compensation for direction-of-arrival estimations using a new, accurate determination method for the receiving mutual impedance. *J. Electromagnet. Waves Appl.* 24 (2/3): 271–281.

Lui, H.S. and Hui, H.T. (2010b). Mutual coupling compensation for direction-of-arrival estimations using the receiving-mutual-impedance method. *Int. J. Antennas Propag.*, Special Issue on "Mutual Coupling of Antenna Array" 2010, Article ID: 373061: 7.

Lui, H.S. and Hui, H.T. (2010c). Improved mutual coupling compensation of compact array. *IET Microwave Antennas Propag.* 4 (10): 1506–1516.

Lui, H.S. and Hui, H.T. (2013). Direction-of-arrival estimation of closely spaced emitters using compact arrays. *Int. J. Antennas Propag.* 2013, 104848: 9.

Lui, H.S. and Hui, H.T. (2015). Direction-of-arrival estimation: measurement using compact arrays under the influence of mutual coupling. *IEEE Antennas Propag. Mag.* 57 (6): 62–68.

Lui, H.S. and Shuley, N. (2004). Consequence of incorrect sampling procedures in resonance based radar target identification. *Electron. Lett.* 40 (8): 507–508.

Lui, H.S. and Shuley, N.V.Z. (2008). Sampling procedures in resonance based radar target identification. *IEEE Trans. Antennas Propag.* 56 (5): 1487–1491.

Lui, H.S., Hui, H.T., and Leong, M.S. (2009, 2009). A note on the mutual coupling problems in transmitting and receiving antenna array. *IEEE Antennas Propag. Mag.* 51 (5): 171–176.

Lui, H.S., Yu, Y., and Hui, H.T. (2010d). Effect of mutual coupling on the performance of direction-of-arrival estimation of compact array. *IEEE Int. Symp. Antennas Propag.*: 1–4, Toronto, Ontario, Canada.

Madurasinghe, D. (2003). Super resolution technique for source power and bearing estimation. *Electron. Lett.* 39 (17): 1232–1233.

Ouibrahim, H., Weiner, D.D., and Sarkar, T.K. (1988). A generalized approach to direction finding. *IEEE Trans. Acoust. Speech Signal Process.* 36 (4): 610–612.

Paulraj, A., Roy, R., and Kailath, T. (1986). A subspace rotation approach to signal parameter estimation. *Proc. IEEE* 74 (7): 1044–1046.

Paulraj, A., Ottersten, B., Roy, R. et al. (1993). Subspace methods for directions-of-arrival estimation. In: *Handbook of Statistics*, vol. 10, Chapter 16, 693–739. Netherlands: Elsevier Science.

Poulton, G.T. (1986a). Power pattern synthesis using the method of successive projections. *IEEE Antenna Propagat. Symp.*, 8-13 June, Philadelphia, PA, USA: 667–668.

Poulton, G.T. (1986b). Antenna power pattern synthesis using method of successive projections. *Electron. Lett.* 22 (20): 1042–1043.

Roy, R. and Kailath, T. (1989). ESPRIT – Estimation of signal parameters via rotational invariance techniques. *IEEE Trans. Acoust. Speech Signal Process.* 37 (7): 984–995.

Sarkar, T.K. and Pereira, O. (1995). Using the matrix pencil method to estimate the parameters of a sum of complex exponentials. *IEEE Antennas Propag. Mag.* 37 (1): 48–55.

Sarkar, T.K., Wicks, M.C., Salazar-Palma, M., and Bonneau, R.J. (2003). *Smart Antennas*. Wiley.

Sarkar, T.K., Salazar-Palma, M., and Molole, E.L. (2008). *Physics of Multiantenna Systems and Broadband Processing*. Wiley.

Schmidt, R.O. (1986). Multiple emitter location and signal parameter estimation. *IEEE Trans. Antennas Propag.* 34 (3): 276–280.

Steedly, W.M. and Moses, R.L. (1993). The Cramer-Rao bound for pole and amplitude coefficient estimates of damped exponential signals in noise. *IEEE Trans. Signal Process.* 41 (3): 1305–1318.

Stoica, P. and Nehoral, A. (1989). MUSIC, maximum likelihood, and Cramer-Rao bound. *IEEE Trans Acoust. Speech Signal Process.* 37 (5): 720–741.

Swindlehurst, A. and Kailath, T. (1992). A performance analysis of subspace method in the presence of model errors – Part 1: The MUSIC algorithm. *IEEE Trans. Signal Process.* 40 (7): 1758–1774.

Volmer, C., Weber, J., Stephan, R. et al. (2008). An eigen-analysis of compact antenna arrays and its application of port decoupling. *IEEE Trans. Antennas Propag.* 56 (2): 390–370.

Walsh, G.R. (1977). *Methods of Optimization*. London, UK: Wiley.

Yao, Y.X. and Pandit, S.M. (1995). Cramer-Rao lower bounds for a damped sinusoidal process. *IEEE Trans. Signal Process.* 43 (4): 878–885.

Yilmazar, N., Koh, J., and Sarkar, T.K. (2006). Utilization of a unitary transform for efficient computation in the matrix pencil method to find the direction of arrival finding. *IEEE Trans. Antenna Propag.* 54 (1): 175–181.

Yilmazcr, N., Burintramart, S., and Sarkar, T.K. (2007). Performance analysis of direct data domain approach and ESPRIT method for DOA estimation. *Proc. IEEE Antenna Propag. Symp.*: 5295–5298, Honolulu, Hawaii, USA.

Yirnin, Z., Hirasawa, K., and Fujimoto, K. (1985). Performance of power inversion adaptive array with the effect of mutual coupling. In: *Proc. ISAP'85*, Kyoto, Japan, 20–22 Aug, 803–806.

Yong, S.K. and Thompson, J.S. (2004). Effect of channel conditions and antenna parameters on the performance analysis of compact antenna arrays. *IEE Proc. – Commun.* 151 (4): 394–400.

Yu, Y., Lui, H.S., Niow, C.H. et al. (2011). Improved DOA estimation using measured receiving mutual impedances for mutual coupling compensation: an experimental study. *IEEE Trans. Wireless Comm.* 10 (7): 2228–2233.

11

Techniques for Minimizing Mutual Coupling Effects in Arrays

Hoi Shun Antony Lui[1] *and Trevor S. Bird*[1,2]

[1] *Global Big Data Technology Centre, University of Technology Sydney, Ultimo, NSW 2007, Australia*
[2] *Antengenuity, Eastwood, NSW 2122, Australia*

11.1 Introduction

The algorithms used mostly for designing arrays are based on two major assumptions: (i) the radiated field is a vectorial sum of the field radiated from individual elements, which is a consequence of the linearity of Maxwell's equations, and (ii) the received signals at each array element is purely based on the incoming signals. The former corresponds to the situation where the array elements are used as a transmitter with signals fed from the antenna terminals, while the latter corresponds to the situation where the array elements serve as a receiver that aims to spatially sample the incoming wave. In practice, the array elements interact with each other through mutual coupling. These interactions form a part in the total radiated field and received signals measured in the antenna terminals, which are usually not considered in the signal model in which mutual coupling is neglected and regarded as an undesirable effect.

In transmitting applications, the effect of mutual coupling is essentially higher-order interactions between antenna elements, which form a part in the total radiated field. In most transmitting applications, each antenna element can be fed with a different signal amplitude and phase such that the radiated beam is steered in a desired direction. Ideally, the total radiated field is the sum of the radiated field from the individual array elements. With the existence of mutual coupling, higher-order interactions between the antenna elements will naturally be incorporated such that the radiated beam will not be same as the ideal pattern. In receiving applications, the incoming electromagnetic field induces electromagnetic currents on the antenna elements and thus the terminal voltages. In most applications, it is desired that the received signal is purely dependent on the incoming signal. In practice, the receiving antenna element also serves as a scatterer, and the reradiated field will also be detected by other antenna elements. In other words, the received signal is a sum of the incoming signals and the reradiated field.

As has been described in earlier chapters, mutual coupling in antenna arrays has been well known for years. With different excitation sources, studies have also found that the mutual coupling behavior is different when the same array is operating in transmitting and receiving modes (Allen and Diamond 1966; Balanis 1997; Daniel 1974; Lui et al. 2009). In this chapter, we first revisit the mutual coupling behavior of transmitting and receiving arrays qualitatively through the mutual coupling path, followed by a full-wave moment method (MoM) analysis that provides insights into how the mutual coupling effect of transmitting and receiving arrays can

Mutual Coupling Between Antennas, First Edition. Edited by Trevor S. Bird.

be modeled and quantified. Mutual coupling compensation methods, including the conventional mutual impedance method, full-wave method, receiving-mutual-impedance method (RMIM), and calibration methods, are also discussed.

11.2 Mutual Coupling in Transmitting and Receiving Arrays

11.2.1 The Mutual Coupling Path

In general, mutual coupling behavior of an antenna array in the transmitting and receiving modes is not identical (Allen and Diamond 1966; Balanis 1997; Daniel 1974; Lui et al. 2009). To gain some physical insights into the physical problem, the coupling paths of the antenna elements *m* and *n* of an array in the transmitting and receiving modes are shown in Figure 11.1a,b. In the transmitting mode, shown in Figure 11.1a, both antenna elements are connected to an active source with an internal impedance $\underline{Z}_s$. Consider the case where the source at antenna element *n* is active and the source at antenna element *m* is off. The power generated from the active source first travels to antenna element *n* (labeled as 0 in Figure 11.1) and then radiates to free-space (labeled as 1). Part of the radiated energy is received by antenna element *m* (labeled as 2), and subsequently, a current is induced on it. Now, antenna element *m* acts as a scatterer and radiates part of the energy to free-space (labeled as 3). The remaining energy is dissipated in the passive load of antenna element *m* (labeled as 4). It is possible that some of the radiated energy (3) is received by antenna element *n* (labeled as 5). This field received by antenna element *n* (5) will alter the current distribution of the antenna and thus the radiated field (labeled as 6). Some of this radiated field will also be received by antenna element *m*, and the entire process will continue indefinitely.

In this instance, only antenna element n is excited and the total radiated field originates from the active source in antenna element n. The total radiated field from antenna n is thus given by the sum of the initial radiated field (1), the reradiated field on antenna element *n* due to the coupled field from antenna element *m* (6), as well as the reradiated field in antenna element *m* (3). The latter two components, (6) and (3), are reradiated field due to mutual coupling, which are not usually taken into account in ideal arrays. Certainly, if there are more than one antenna surrounding antenna element *n*, the reradiated field on these antenna elements and antenna element *n* is going to contribute to the total radiated field. Finally, if antenna *m* and other antenna elements are excited, the total radiated field will be the sum of the radiated field of the individual antenna elements, together with the higher-order reradiated field from all the elements.

In the receiving mode, shown in Figure 11.1b, the two antenna elements of the array are now terminated with a load impedance Z_L. In the following discussion, we focus on the signal received by antenna element *m*. The incident wavefront coming from free-space (0) strikes on antenna element *m*, and a current is subsequently induced. The induced current on antenna element *m* reradiates part of the energy to free-space (2), while the rest of the energy travels toward the load (1). Part of the radiated energy (2) is received by antenna element *n* (3). This radiated field induces current on antenna element *n*, and the induced current reradiates. Part of the reradiated field in antenna n is detected by antenna element *m* (4). At the same time, if the terminal of antenna element *m* is a mismatch, some of the received energy (1) will also be reflected (5) and radiated to free-space (6). Similar to the radiated field (2), part of this reradiated field (6) will also be detected by antenna element *n*. Antenna element *n* will scatter some of this reradiated field, and part of this reradiated field will be detected by antenna element *m*. This process continues indefinitely until all the energy is dissipated. In summary, the received signal in antenna element *m* is the sum of the incoming signal (0), reradiated field from antenna element *n* (4), and the signals received by antenna element *m* due higher-order interaction between the two antennas. When there is more than one antenna

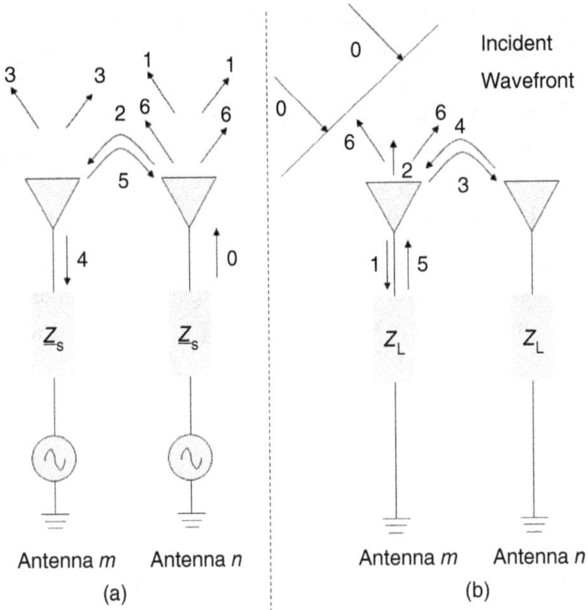

Figure 11.1 The coupling paths for the antenna elements *m* and *n* for an antenna array in (a) transmitting modeand (b) receiving mode. Source: From LUI, H.S., HUI, H.T. and LEONG, M.S. (2009): 'A note on the mutual coupling problems in transmitting and receiving antenna array, 'IEEE Antennas and Propagations Magazine', Vol. 51, No.5, pp. 171-176, Oct. 2009. © 2009 IEEE.

element nearby, due to mutual coupling, the reradiated field from these antennas and the higher interactions between all the elements will form a part to the received signal at antenna element *m*.

11.2.2 Moment Method Analysis

The differences of the mutual coupling phenomena between transmitting and receiving antenna arrays can be best described using a full-wave analysis. According to the standard MoM analysis of dipole arrays that has been described in Chapter 4 and commencing from Eq. (4.15) for a two-element dipole array with each antenna element discretized into $M+1$ segments that requires M basis functions, the situation can be expressed as (Harrington 1993)

$$\boldsymbol{Z}_{MoM}\boldsymbol{I}_{MoM} = \boldsymbol{V}_{MoM} \tag{11.1}$$

where $\boldsymbol{Z}_{MoM}$ is a $2M \times 2M$ impedance matrix that models the interactions of the segments between the two dipoles, $\boldsymbol{V}_{MoM}$ is $2M \times 1$ excitation voltage vector that describes the excitation for the individual segment, and $\mathbf{I}_{MoM}$ is the induced current on the individual segment. For the case of a two-dipole array, Eq. (11.1) can be explicitly written as

$$\begin{bmatrix} \boldsymbol{Z}_{11} & \boldsymbol{Z}_{12} \\ \boldsymbol{Z}_{21} & \boldsymbol{Z}_{22} \end{bmatrix} \begin{bmatrix} \boldsymbol{I}_1 \\ \boldsymbol{I}_2 \end{bmatrix} = \begin{bmatrix} \boldsymbol{V}_1 \\ \boldsymbol{V}_2 \end{bmatrix} \tag{11.2}$$

where the matrix blocks are given by

$$\boldsymbol{Z}_{\alpha\beta} = \begin{bmatrix} z_{11}^{\alpha\beta} & z_{12}^{\alpha\beta} & \cdots & z_{1M}^{\alpha\beta} \\ z_{21}^{\alpha\beta} & z_{22}^{\alpha\beta} & \cdots & z_{2M}^{\alpha\beta} \\ \vdots & \vdots & \ddots & \vdots \\ z_{M1}^{\alpha\beta} & z_{M1}^{\alpha\beta} & \cdots & z_{MM}^{\alpha\beta} \end{bmatrix}, \boldsymbol{I}_\alpha = \begin{bmatrix} I_1^\alpha \\ I_2^\alpha \\ \vdots \\ I_M^\alpha \end{bmatrix} \text{ and } \boldsymbol{V}_\alpha = \begin{bmatrix} V_1^\alpha \\ V_2^\alpha \\ \vdots \\ V_M^\alpha \end{bmatrix}$$

respectively with superscripts α and β ranging from 1 to 2. The elements in the $\mathbf{Z}_{\alpha\beta}$ matrix, $z_{ab}^{\alpha\beta}$, correspond to the impedance between the segment a of antenna α and segment b of antenna β. The elements I_m^α and V_m^α of the $\boldsymbol{I}_\alpha$ and $\boldsymbol{V}_\alpha$ matrices, respectively, are the induced current and the excitation voltage to the corresponding m^{th} segment of the antenna α, where $m = 1, 2, \ldots, M$. The mutual coupling effect represented by Eq. (11.2) is due to the off-diagonal matrix blocks, i.e. $\mathbf{Z}_{\alpha\beta}$ where $\alpha \neq \beta$.

In transmitting applications, the excitation vectors on the right-hand side of Eq. (11.2) can be written as (Harrington 1993) $\boldsymbol{V}_1^{TX} = \begin{bmatrix} 0 & \cdots & \underline{V}_{s1} & \cdots & 0 \end{bmatrix}^T$, $\boldsymbol{V}_2^{TX} = \begin{bmatrix} 0 & \cdots & \underline{V}_{s2} & \cdots & 0 \end{bmatrix}^T$, where $\underline{V}_{s1}$ and $\underline{V}_{s2}$ are the complex amplitudes of the voltage sources, and the superscript T corresponds to the transpose of the matrix. When the excitation vectors are substituted into Eq. (11.1), the induced current on the antenna array is given by

$$\boldsymbol{I}_{MoM}^{TX} = \mathbf{Y}_{MoM}\boldsymbol{V}_{MoM}^{TX} \quad or \quad \begin{bmatrix} \boldsymbol{I}_1^{TX} \\ \boldsymbol{I}_2^{TX} \end{bmatrix} = \begin{bmatrix} \mathbf{Z}_{11} & \mathbf{Z}_{12} \\ \mathbf{Z}_{21} & \mathbf{Z}_{22} \end{bmatrix}^{-1} \begin{bmatrix} \boldsymbol{V}_1^{TX} \\ \boldsymbol{V}_2^{TX} \end{bmatrix} \tag{11.3}$$

where the superscript -1 corresponds to the matrix inverse operator.

Instead of having voltage sources, in receiving applications, the antenna array is illuminated by external electromagnetic sources (e.g. incident plane wave). Consider an incident plane wave $\boldsymbol{E} = \hat{\mathbf{u}}\underline{E}_0 e^{-j\mathbf{k}\cdot\mathbf{r}_n}$, where $\hat{\mathbf{u}}$ is the unit vector that specifies the polarization of the incident wave, $\underline{E}_0$ is the complex amplitude of the electric field of the incident wave, $\mathbf{k}$ is a wavenumber vector pointing in direction of travel of the wave, and $\mathbf{r}_n$ is the radius vector to a point n on the antenna. The excitation vector can be written as $\begin{bmatrix} \boldsymbol{E}_i(1)\cdot\boldsymbol{\Delta l}_1 & \boldsymbol{E}_i(2)\cdot\boldsymbol{\Delta l}_2 & \cdots & \boldsymbol{E}_i(N)\cdot\boldsymbol{\Delta l}_N \end{bmatrix}^T$. Similarly, the induced current can be given by

$$\mathbf{I}_{MoM}^{RX} = \mathbf{Y}_{MoM}\mathbf{V}_{MoM}^{RX} \quad or \quad \begin{bmatrix} \mathbf{I}_1^{RX} \\ \mathbf{I}_2^{RX} \end{bmatrix} = \begin{bmatrix} \mathbf{Z}_{11} & \mathbf{Z}_{12} \\ \mathbf{Z}_{21} & \mathbf{Z}_{22} \end{bmatrix}^{-1} \begin{bmatrix} \mathbf{V}_1^{RX} \\ \mathbf{V}_2^{RX} \end{bmatrix}. \tag{11.4}$$

From Eqs. (11.3) and (11.4), it is clear that the induced currents on the antenna for transmitting and receiving arrays, $\boldsymbol{I}_{MoM}^{TX}$ and $\boldsymbol{I}_{MoM}^{RX}$, are different due to different excitation vectors $\boldsymbol{V}_{MoM}^{TX}$ and $\boldsymbol{V}_{MoM}^{RX}$. With $\boldsymbol{V}_1^{TX} = \begin{bmatrix} 0 & \cdots & V_{s1} & \cdots & 0 \end{bmatrix}^T$ and $\boldsymbol{V}_2^{TX} = \begin{bmatrix} 0 & \cdots & V_{s2} & \cdots & 0 \end{bmatrix}^T$, where the ith element corresponds to the feed, Eq. (11.3) can now be explicitly written as

$$\boldsymbol{I}_{MoM}^{TX} = \begin{bmatrix} Y_{11} & \cdots & Y_{1i} & \cdots & Y_{1(M+i)} & \cdots & Y_{1(2M)} \\ Y_{21} & \cdots & Y_{2i} & \cdots & Y_{2(M+i)} & \cdots & Y_{2(2M)} \\ Y_{31} & \cdots & Y_{3i} & \cdots & Y_{3(M+i)} & \cdots & Y_{3(2M)} \\ \vdots & \ddots & \vdots & \ddots & \vdots & \ddots & \vdots \\ Y_{(2M-2)1} & \cdots & Y_{(2M-2)i} & \cdots & Y_{(2M-2)(M+i)} & \cdots & Y_{(2M-2)(2M)} \\ Y_{(2M-1)1} & \cdots & Y_{(2M-1)i} & \cdots & Y_{(2M-1)(M+i)} & \cdots & Y_{(2M-1)(2M)} \\ Y_{(2M)1} & \cdots & Y_{(2M)i} & \cdots & Y_{(2M)(M+i)} & \cdots & Y_{(2M)(2M)} \end{bmatrix} \begin{bmatrix} 0 \\ \vdots \\ V_{s1} \\ \vdots \\ V_{s2} \\ \vdots \\ 0 \end{bmatrix}$$

or alternatively

$$\boldsymbol{I}_{MoM}^{TX} = V_{s1} \begin{bmatrix} Y_{1i} \\ Y_{2i} \\ Y_{3i} \\ \vdots \\ Y_{(2M-2)i} \\ Y_{(2M-1)i} \\ Y_{(2M)i} \end{bmatrix} + V_{s2} \begin{bmatrix} Y_{1(M+i)} \\ Y_{2(M+i)} \\ Y_{3(M+i)} \\ \vdots \\ Y_{(2M-2)(M+i)} \\ Y_{(2M-1)(M+i)} \\ Y_{(2M)(M+i)} \end{bmatrix}. \tag{11.5}$$

It is apparent that only two columns (column i and column $M+i$) of $\mathbf{Y}_{MoM}$ are used when determining the induced current $\boldsymbol{I}_{MoM}^{TX}$ in transmitting application. With plane wave incidence where all the elements in $\boldsymbol{V}_{MoM}^{RX}$ are nonzero in general, the induced current distribution $\boldsymbol{I}_{MoM}^{RX}$ is a weighted sum of all the column elements in $\boldsymbol{Y}_{MoM}$. Therefore, in receiving applications, the computation is more complicated than transmit.

In short, the current distributions obtained from moment method in both transmitting and receiving arrays are given by two sources: the excitation and coupled field from other antenna elements. In transmitting application, the current distribution of the antenna elements and subsequently the coupled field which will be seen later, can be given by Eq. (11.5) which quantitatively reveals that the current distribution is a weighted sum of the columns of the admittance matrix $\mathbf{Y}_{MoM}$ that corresponds to the antenna terminals. In a similar manner, the current distribution of the antenna elements and the coupled field of a receiving array is a weighted sum of *all* columns from the admittance matrix, which is more complicated in nature.

Once the current distribution ($\boldsymbol{I}_{MoM}^{TX}$ or $\boldsymbol{I}_{MoM}^{RX}$) is known, the mutual coupling effect, or more precisely the coupled field, can be numerically determined based on the elements in $\boldsymbol{Z}_{\alpha\beta}$ (with $\alpha \neq \beta$) and the current distribution through Eq. (11.2), i.e. $\boldsymbol{V}_1 = \mathbf{Z}_{11}\boldsymbol{I}_1 + \boldsymbol{Z}_{12}\boldsymbol{I}_2$ and $\boldsymbol{V}_2 = \mathbf{Z}_{21}\boldsymbol{I}_1 + \boldsymbol{Z}_{22}\boldsymbol{I}_2$, where $\boldsymbol{Z}_{12}\boldsymbol{I}_2$ and $\boldsymbol{Z}_{21}\boldsymbol{I}_1$ are the coupled voltages that appear at terminals 1 and 2, respectively. Notice that as $\boldsymbol{Z}_{12}\boldsymbol{I}_2$ and $\boldsymbol{Z}_{21}\boldsymbol{I}_1$ are matrix products, the entire current distribution and elements in the submatrices $\boldsymbol{Z}_{\alpha\beta}$ are required. In other words, the coupled voltages are determined based on the geometry and physical attributes of the array (e.g. dielectric properties, size, etc.) that appears in $\boldsymbol{Z}_{MoM}$ and the current distrubutions. As we will see in later sections, the coupled voltages, which are the voltages due to coupled field at the antenna terminals, can also be modeled as a product of the conventional (Gupta and Ksienski 1983) or receiving mutual impedance (Hui 2003) and the current at the antenna terminals, which is a product of two complex numbers, rather than a matrix product of the $\mathbf{Z}_{\alpha\beta}$ matrix and the entire current distribution. This is simply because the $\mathbf{Z}_{MoM}$ matrix and the entire current distribution are not practically available in most cases, while the antenna terminal current is measurable. Compared with the above full-wave analysis, mutual impedance is a relatively simple, practical solution to characterize mutual coupling effects. The mutual impedance is an approximation to the products of the current distribution and $\mathbf{Z}_{\alpha\beta}$.

As we have already seen in both Eqs. (11.4) and (11.5), the current distributions for transmitting and receiving situations are not identical. Also, the coupled voltages, are based on the current distributions multiplied by the elements in the $\mathbf{Z}_{\alpha\beta}$ (with $\alpha \neq \beta$) matrix matrices (i.e. $\mathrm{Z}_{12}\mathrm{I}_2$ and $\mathrm{Z}_{21}\mathrm{I}_1$), are not the same in transmitting and receiving applications even though the $\boldsymbol{Z}_{\alpha\beta}$ matrices remain to be unchanged. The conventional (Gupta and Ksienski 1983) and receiving mutual impedances (Hui 2003), denoted by $Z_{\alpha\beta}$, and $\mathrm{Z}_t^{\alpha\beta}$, mathematically relate the total current at antenna terminal β and the coupled voltage at antenna terminal α due to the current at antenna terminal β in transmitting and receiving modes, respectively, can be used to determine the antenna terminal coupled voltages by multiplying the terminal current with the corresponding mutual impedance. This is essentially a multiplication operation of two complex numbers rather than two matrices. In other words, mutual impedance allows us to determine the coupled terminal voltages by knowing only the measurable terminal current rather than the entire current distribution, which is usually not measurable in practice.

To probe further, as we should be able to obtain the same coupled terminal voltages using both approaches (i.e. multiplication of the moment method matrix and current distributions vs multiplication of mutual impedance and terminal current), the mutual impedances ($Z_{\alpha\beta}$ and $\mathrm{Z}_t^{\alpha\beta}$ respectively) have therefore taken into account of the effects due to the elements in the $Z_{\alpha\beta}$ matrix and current distributions in transmitting and receiving modes respectively. In general, the mutual impedance obtained in transmitting and receiving modes should not be identical (Hui 2003). Further details about conventional and receiving mutual impedance are given in sections 11.4.1 and 11.4.3.

11.3 Typical Methods for Minimizing Mutual Coupling

A significant concern is that if mutual coupling is neglected, the expected performance can depart significantly from that achieved in practice. Apart from taking the effect into account, what can be done about it? There are several possible approaches:

- Increase antenna element spacing (not always practical) or randomize the location of elements. This can have a bonus of reducing the array side lobes.
- Reduce the near-field in the direction of adjacent elements by increasing the field taper (not always practical without compromising performance).
- Isolate elements with electromagnetic "fences."
- Compensate for mutual coupling through signal processing methods.
- Take account of mutual coupling at the outset, as in a normal coupled circuit, using one of the methods outlined above.

Some of the most common techniques are illustrated in Figure 11.2.

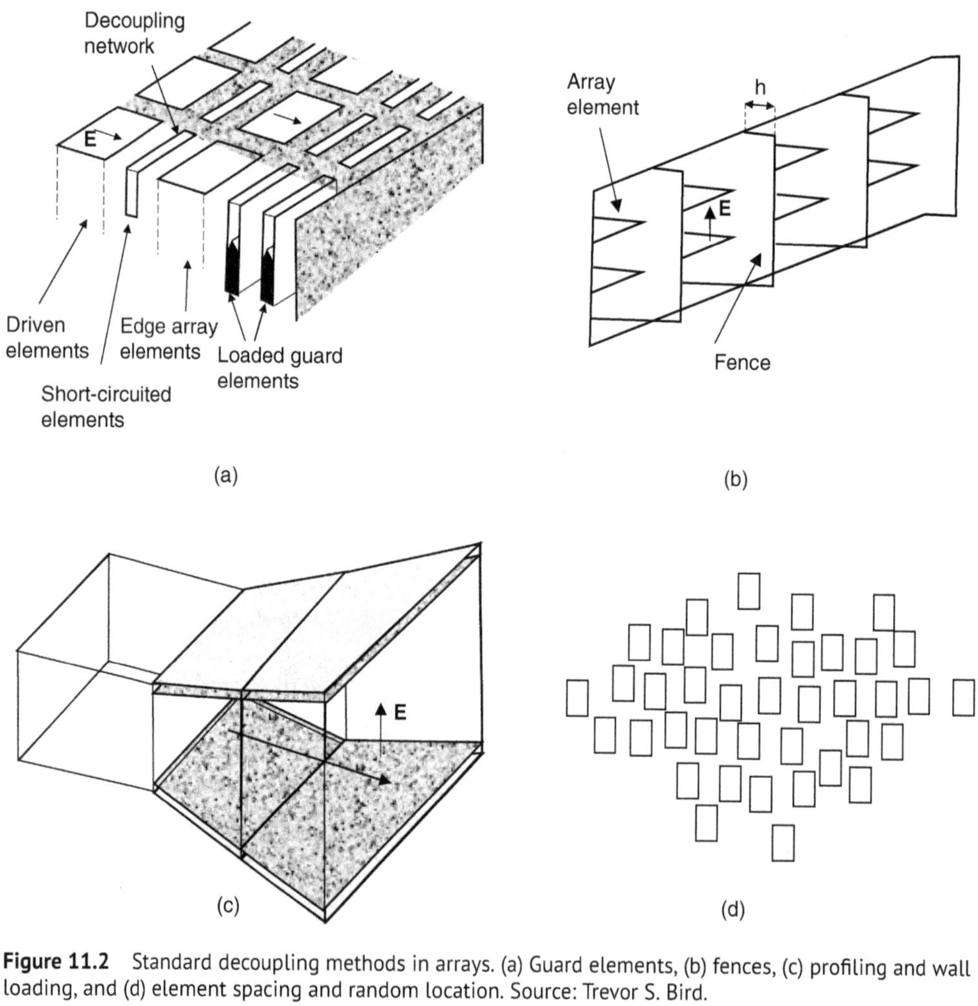

Figure 11.2 Standard decoupling methods in arrays. (a) Guard elements, (b) fences, (c) profiling and wall loading, and (d) element spacing and random location. Source: Trevor S. Bird.

11.3.1 Aperture Field Taper

In some situations, the aperture field can be designed to give a low or matching taper that leads to a lower coupling in one plane. Electromagnetic horns can have aperture field distributions synthesized by means of high-order modes using inserts such as dielectric and corrugations that produce little radiated power in the direction of neighboring array elements (see Chapter 5). For example, the aperture field distribution of rectangular horns, which are excited mainly in the TE_{10} mode, is almost uniform in the E-plane. This implies the coupling will be strong along the E-plane of such horns. Corrugations or dielectric loading on the walls parallel to the H-plane increase the field taper in the E-plane and reduce the field at the edge of the aperture (Figure 11.2c). The same effect is created in circular horns by exciting the correct proportions of TE_{12} and TM_{12} modes, and this results in almost equal E- and H-plane patterns and low cross-polarization (Bird 2016). As the coupling to neighboring apertures is low, the radiation properties of the elements of array are only slightly impacted in the array environment.

In most antennas, the mutual coupling level decreases on average with increasing element separation although the actual level passes through a series of maxima and minima. The location of these minima depends on the geometry of the element, and the source field taper as well. On some occasions, judicious choice of geometry can be made where the coupling is a minimum, and as a result, the array elements interact very little. An example of this is a circular aperture. The first minimum occurs when $ka \cong 3.6$, where a is the aperture radius. At this condition, the field taper in both principal planes is the same and cross-polarization is also low. This unique condition was used for the Parkes multibeam feed design (Bird 2016).

11.3.2 Electromagnetic Fences

The use of metallic guards or fences between elements or at the edges of large arrays (see Figure 11.2b) is sometimes used to reduce element coupling. The fence is usually perpendicular to the ground plane or impedance surface located between elements. For example, in arrays of dipoles, fitting them in circular cups can lower the coupling using the same mechanism described in Section 11.3.1. The purpose of the cup, and other fences, is to reduce the radiation in the direction of the other elements. The height of the fence is important, and its effect is direction dependent. For example, Mailloux (1971) showed that if a fence is placed between two parallel-plate waveguides, which couple in the E-plane, coupling initially increases with the height of the fence up to a height h (in Figure 11.2b) of about 0.25λ above the level of the element, and decreases thereafter, but not monotonically, with fence height depending on the plate spacing. The minimum coupling occurs at heights of $0.4 - 0.5\lambda$, and thereafter, it increases again until near a height of 0.75λ when it is minimum again. By contrast, for coupling in the H-plane, with a fence placed half-way between the elements, the coupling decreases monotonically with fence height up to about 0.6λ before the coupling increases again. Reactively loaded elements placed between adjacent elements can also be used to reduce coupling in waveguide arrays (Edelberg and Oliner 1960; Hockham 1974). Such indirectly excited elements provide extra degrees of freedom for radiation pattern control.

11.3.3 Other Approaches to Compensation

In some arrays, it is possible to compensate for the effect of mutual coupling through appropriate design of the beamforming network. Compensation is possible also by connecting a network between the input ports and the antenna ports. Anderson and Ramussen (1976) describe necessary conditions for such a lossless network to achieve complete decoupling and descattering. There are three main planes for the compensation. These are (i) in the array aperture plane, (ii) the inputs to the array elements which are the usual input plane, and (iii) a new input plane. Techniques for all three methods will be described later in this chapter in Section 11.4.

Large arrays with many coupled elements may not be practical owing to complexity, cost, and weight, and mutual coupling can degrade performance if not included in the design. To achieve the same array size for a given beam width, a thinned array is a possibility (Unz 1962; Ishimaru and Chen 1965). Mutual coupling effects are also reduced in a thinned array (Agrawal and Lo 1972) or an array where the elements have been randomized (Figure 11.2d). The average spacing in a thinned array is a determining parameter for mutual coupling effects. Mutual coupling has negligible effect on the side lobe level if the average spacing is $>2.5\lambda$, and the main-beam fluctuations is small if the average spacing is $>5\lambda$. For smaller average spacing, both these effects are significant. A major issue with randomizing and average spacing is their selection. One approach that was used by Agrawal and Lo (1972) was to examine the probability that the pattern function crossed a specified side lobe level. The resulting integral equation gave in the limit a combined parameter $\alpha = N/SNR$, where N is the number of elements in the array length, and SNR is the sidelobe ratio in dB. Further details can be found in the references (Agrawal and Lo 1972; Hansen 1998). For two-dimensional arrays, computer optimization techniques are often used for thinning (Haupt 1994) or randomizing. One approach used in thinning implemented design constraints on the number of elements, the aperture, and the minimum element spacing, with a genetic algorithm based on chromosome element resetting (Chen et al. 2007). Compared with the synthesis method of thinned arrays, this approach or similar could exploit more of the degrees of freedom of the elements, such as rotation, to control the characteristics of sparse arrays including mutual coupling. Other techniques are based on sparse placement in rings, in fans, or in more general arrangements (e.g. Angeletti et al. 2014).

Most array signal processing applications are developed based on the ideal manifold. In practice, when antenna arrays are employed to either transmit or receive the signals, electromagnetic interactions between the array elements will become part of the signal. These effects, as previously discussed, are functions of antenna element separation, the choice of antenna element, the number of antenna elements, operating frequency, and polarization, which are usually not taken into account in the array manifold (i.e. the basis of how the signal processing algorithms are developed) or part of the signal processing algorithms.

The objective of the mutual coupling compensation technique aims to recover the ideal array manifold by knowing the mutual coupling effect such that the signal processing algorithm can be properly operated without being interfered by mutual coupling.

11.4 Techniques for Practical Mutual Coupling Compensation

11.4.1 Conventional Mutual Impedance Method

Gupta and Ksienski (1983) studied the mutual coupling effect by considering an N-element antenna array and the external incoming signal as an $N+1$ terminal network. As shown in Figure 11.3, the array is driven by an external source with open-circuit voltage V_g and internal impedance Z_g. Consider an array with N antenna elements, and their corresponding terminal voltages can be written as

$$V_1 = I_1 Z_{11} + \cdots + I_j Z_{1j} + \cdots + I_N Z_{1N} + I_s Z_{1s}$$

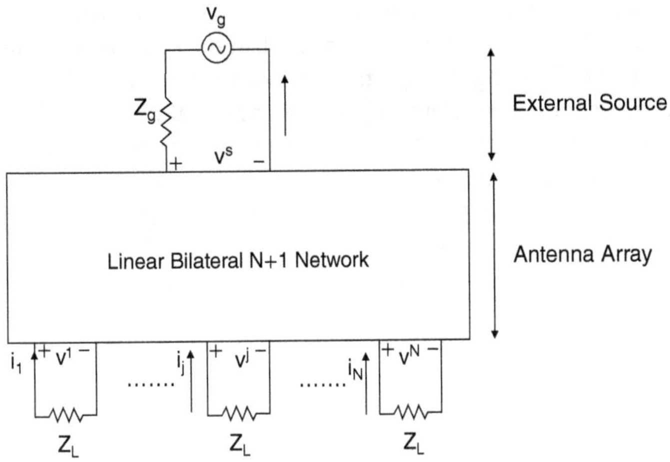

Figure 11.3 Antenna array as a $N+1$ port network. Source: after Gupta and Ksienski, 1983.

$$\vdots$$
$$V_j = I_1 Z_{j1} + \cdots + I_j Z_{jj} + \cdots + I_N Z_{jN} + I_s Z_{js}$$
$$\vdots$$
$$V_N = I_1 Z_{N1} + \cdots + I_j Z_{Nj} + \cdots + I_N Z_{NN} + I_s Z_{Ns} \tag{11.6}$$

where Z_{ij} corresponds to the mutual impedance between antenna elements i and j, and Z_{jj} corresponds to the self-impedance of antenna element j. The terminal current for antenna element j, I_j, can be related to the load impedance Z_L and terminal voltage V_j by

$$I_j = -\frac{V_j}{Z_L}, \quad j = 1, 2, \ldots, N. \tag{11.7}$$

Now, if all the antenna elements are in an open-circuit condition, there is no current induced in the antenna terminal such that

$$I_j = 0, \quad j = 1, 2, \ldots, N, \tag{11.8}$$

the antenna terminal voltage V_j, from Eq. (11.6), is now solely dependent on the source, $I_s Z_{js}$, such that

$$V_j^{OC} = I_s Z_{js}, \quad j = 1, 2, \ldots, N. \tag{11.9}$$

Together with Eqs. (11.7–11.9), Eq. (11.6) can be rewritten as

$$\begin{bmatrix} 1+\frac{Z_{11}}{Z_L} & \frac{Z_{12}}{Z_L} & \cdots & \frac{Z_{1N}}{Z_L} \\ \frac{Z_{21}}{Z_L} & 1+\frac{Z_{22}}{Z_L} & \cdots & \frac{Z_{2N}}{Z_L} \\ \vdots & \vdots & \ddots & \vdots \\ \frac{Z_{N1}}{Z_L} & \frac{Z_{1N2}}{Z_L} & \cdots & 1+\frac{Z_{NN}}{Z_L} \end{bmatrix} \begin{bmatrix} V_1 \\ V_2 \\ \vdots \\ V_N \end{bmatrix} = \begin{bmatrix} V_1^{OC} \\ V_2^{OC} \\ \vdots \\ V_N^{OC} \end{bmatrix} \tag{11.10}$$

or

$$\mathbf{Z}_0 \mathbf{V} = \mathbf{V}^{OC} \tag{11.11}$$

Equation (11.10) describes the antenna terminal voltages in a receiving array as a sum of the incoming signal and the coupled signal from other array elements. When the antenna elements become open-circuit, there is no current flowing through the antenna terminal such that the open-circuit terminal voltages, V^{OC}, are solely dependent on the incoming signals.

This formulation provides a mathematical model that relates the coupled terminal voltages $\mathbf{V}$ with a "signal-only" voltages $\mathbf{V}^{OC}$. In the same publication, Gupta and Ksienski (1983) applied the model to evaluate the impact of mutual coupling on the signal-to-interference-plus-noise ratio (SINR) of linear adaptive arrays. Yeh et al. (1989) modified the well-known Multiple Signal Classification (MUSIC) algorithm (Schmidt 1986) for direction-of-arrival (DOA) estimation by including the $\mathbf{Z}_0$ matrix to model the mutual coupling effect. This resulted in better estimation accuracies. Instead of using the coupled terminal voltage $\mathbf{V}$, this formulation inspired other researchers who have considered using the open-circuit voltages $\mathbf{V}^{OC}$ as the "decoupled" signal for direction finding and other applications (Adve and Sarkar 2000; Lau et al. 2004), which has a significant impact to the upcoming development in mutual coupling compensation.

Mutual coupling compensation using the conventional mutual impedance method requires the knowledge of the $\mathbf{Z}_0$ matrix. To simplify our discussion, instead of an N-element array, let us consider the case for a two-element dipole array. The terminal voltage V_js and terminal currents in Eq. (11.6) can now be written as

$$V_1 = I_1 Z_{11} + I_2 Z_{12}, \tag{11.12a}$$

$$V_2 = I_1 Z_{21} + I_2 Z_{22} \tag{11.12b}$$

where $Z_{11} = \left.\frac{V_1}{I_1}\right|_{I_2=0}$ is the self-impedance of port 1 with port 2 being open-circuited, $Z_{12} = \left.\frac{V_1}{I_2}\right|_{I_1=0}$ is the mutual impedance of port 1 due to a current at port 2 with port 1 being open-circuited, $Z_{21} = \left.\frac{V_2}{I_1}\right|_{I_2=0}$ is the mutual impedance of port 2 due to a current at port 1 with port 2 being open-circuited, and $Z_{22} = \left.\frac{V_2}{I_2}\right|_{I_1=0}$ is the self-impedance of port 2 due to a current at port 2 with port 1 being open-circuited.

The self-impedance of each antenna element is essentially the input impedance of the antenna element when each is radiating in an unbounded medium. This can be determined by evaluating the self-impedance of the element individually, or alternatively exciting the corresponding antenna element and then measuring the corresponding terminal voltage and current while having all other array elements in open-circuited condition. This is under the assumption that the open-circuited elements do not radiate as they are basically minimum scattering antennas (see Sec. 3.5.5). The mutual impedance, Z_{12}, is the ratio of the induced voltage at the terminal of antenna 1 under open-circuited condition and the terminal current at antenna element 2 when antenna 2 is radiating. Similarly, the mutual impedance Z_{21} can thus be the ratio of the terminal voltage at antenna 2 under open-circuit condition and the terminal current at antenna terminal 1 when antenna 1 is exciting.

Furthermore, Eq. (11.12) can also be expressed as

$$Z_{1d} = \frac{V_1}{I_1} = Z_{11} + \frac{I_2}{I_1} Z_{12}. \tag{11.13a}$$

$$Z_{2d} = \frac{V_2}{I_2} = Z_{22} + Z_{21} \frac{I_1}{I_2} \tag{11.13b}$$

Z_{1d} and Z_{2d} are known as the driving point impedance of antenna elements 1 and 2, respectively. The driving point impedance depends on the self-impedance of the antenna element, the mutual impedance, and the ratio of the terminal current. When we attempt to match the antenna in the array, it is the driving point impedance that must be matched. Therefore, it is apparent that mutual coupling effect does play a significant role in the performance of antenna array.

Before we finish the discussion on conventional mutual impedance, let us review how the mutual impedance is calculated for dipole antennas (for more details see Chapter 4). Consider the case of a two-element dipole array in transmitting mode with voltage sources V_1 and V_2. The antenna elements are made of perfect electric conductor (PEC) and are vertically oriented along the z-axis. The z component of the electric field on the surface of antenna element 2 $E_2(z)$ consists of the contribution of electric field that originates from the currents $I_1(z)$ and $I_2(z)$

$$E_2(z) = E_{21}(z) + E_{22}(z) \tag{11.14}$$

where $E_{21}(z)$ is the electric field on the surface of antenna element 2 due to current $I_1(z)$ and $E_{22}(z)$ is the electric field due to $I_2(z)$. To satisfy the boundary condition, $E_2(z)$ must be zero on the surface of the antenna element except at the terminal where the voltage V_2 is applied such that $E_2(z) = -E_0$ at the terminal and zero elsewhere. Multiplying both sides of Eq. (11.14) by $I_2(z)$ and integrating over the entire antenna, we obtain

$$\int_{\ell} \left[E_{21}I_2(z) + E_{22}I_2(z)\right] dz = -V_2 I_2(0) \tag{11.15}$$

Thus, Eq. (11.12b) can be written as

$$V_2 = I_1(0) Z_{21} + I_2(0) Z_{22} \tag{11.16}$$

where the mutual impedance Z_{21} and the self-impedance Z_{22} of antenna element 2 are expressed

$$Z_{21} = -\frac{1}{I_1(0) I_2(0)} \int_{\ell} E_{21} I_2(z)\, dz \tag{11.17}$$

$$Z_{22} = -\frac{1}{I_2^2(0)} \int_{\ell} E_{22} I_2(z)\, dz \tag{11.18}$$

Similarly, Eq. (11.12a) can also be written as

$$V_1 = I_1(0) Z_{11} + I_2(0) Z_{12} \tag{11.19}$$

where the corresponding mutual and self-impedances can be given by

$$Z_{12} = -\frac{1}{I_1(0) I_2(0)} \int_{\ell} E_{12} I_1(z)\, dz \tag{11.20}$$

$$Z_{11} = -\frac{1}{I_1^2(0)} \int_{\ell} E_{11} I_1(z)\, dz \tag{11.21}$$

11.4.2 Full-Wave Method

The conventional mutual impedance method provides a relatively simple and practical solution to model and compensate for the mutual coupling effect for receiving applications. The open-circuit voltages in Eq. (11.9), which are determined when all antenna terminals are under open-circuited condition, assume that the measured voltages are purely dependent on the incoming signal and the open-circuited antenna elements do not scatter. Strictly speaking, this assumption may only be valid for small antennas. In reality, a current is induced on open-circuited antenna elements. These open-circuited elements serve as a secondary source and reradiate. The coupled field from open-circuited antennas could be insignificant when the antenna elements are electrically small or the antenna elements are widely spaced (with at least half-wavelength separation). Furthermore, conventional mutual impedance is determined between a pair of antennas. In arrays where there are more than two elements, the existence of other elements will also serve as a scatterer, and the reradiated field from these elements will alter the current distribution and thus the mutual impedance. The impact of how the scattering effect impact on the mutual coupling and adaptive nulling applications is discussed in Hui (2002).

In this section, we look into another mutual coupling compensation method, namely the full-wave method. The moment method analysis summarized in Section 11.2.2 is based on the entire current distribution of the antenna elements as described in the references (Harrington 1993; Adve and Sarkar 2000; Lau et al. 2004)

$$\mathbf{I}_{MoM}^{RX} = \mathbf{Y}_{MoM}\mathbf{V}_{MoM}^{RX} \quad or \quad \begin{bmatrix} \mathbf{I}_1^{RX} \\ \mathbf{I}_2^{RX} \end{bmatrix} = \begin{bmatrix} \mathbf{Z}_{11} & \mathbf{Z}_{12} \\ \mathbf{Z}_{21} & \mathbf{Z}_{22} \end{bmatrix}^{-1} \begin{bmatrix} \mathbf{V}_1^{RX} \\ \mathbf{V}_2^{RX} \end{bmatrix}. \tag{11.22}$$

Instead here, we adopt a relatively simple model where the load impedance is not considered. The measured voltage at the antenna terminals, $\mathbf{V}_{mea}$, can be given by

$$\boldsymbol{V}_{mea} = \boldsymbol{Z}_L\boldsymbol{I}_{Port} = \boldsymbol{Z}_L\boldsymbol{Y}_{Port}\boldsymbol{V}_{MoM}^{RX} = \boldsymbol{C}\boldsymbol{V}_{MoM}^{RX}, \tag{11.23}$$

or equivalently

$$[\boldsymbol{V}_{mea}]_{N\times 1} = [\boldsymbol{C}]_{N\times NM}[\boldsymbol{V}_{MoM}^{RX}]_{NM\times 1} \tag{11.24}$$

For the case of an N-element array with each antenna element modelled using M basis functions, $\mathbf{V}_{mea}$ and $\mathbf{I}_{port}$ are $N\times 1$ array that corresponds to the port, and $\mathbf{Z}_L$ is an $N\times N$ diagonal matrix with load impedance Z_L at its entries. $\mathbf{Y}_{port}$ is a $N\times NM$ submatrix of $\mathbf{Y}_{MoM}$ with the N rows corresponding to the ports of the antenna elements, and $\mathbf{V}_{MoM}^{RX}$ is an $NM\times 1$ excitation vector corresponding to the incoming signals directly from the incident fields and is free from mutual coupling. The objective of the full-wave method is to determine $\mathbf{V}_{MoM}^{RX}$, or at least the elements of $\mathbf{V}_{MoM}^{RX}$ that correspond to the antenna terminals, from the measured terminal voltages $\mathbf{V}_{mea}$. Out of all the elements inside the matrix $\mathbf{V}_{MoM}^{RX}$, the elements that correspond to the antenna elements are the ones that we needed.

For dipole arrays with elements oriented along the z axis (e.g. Figure 4.13) and in situations where the incoming signals (including the jammers) are incident from approximately the same elevation angle θ, the elements in the $\mathbf{V}_{MoM}^{RX}$ can be related to each other by $V_{i+1} = (e^{jk\Delta z\cos\theta})V_i$ such that all elements can be related to the each other. As an example, for the case of $N = 2$ and $M = 3$, Eq. (11.24) can be written as follows (Adve and Sarkar 2000):

$$\begin{bmatrix} V_{meas1} \\ V_{meas2} \end{bmatrix} = \begin{bmatrix} Z_L & 0 \\ 0 & Z_L \end{bmatrix} \begin{bmatrix} Y_{21} & Y_{22} & Y_{23} & Y_{24} & Y_{25} & Y_{26} \\ Y_{51} & Y_{52} & Y_{53} & Y_{54} & Y_{55} & Y_{56} \end{bmatrix} \begin{bmatrix} V_{MoM}^1 \\ V_{MoM}^2 \\ V_{MoM}^3 \\ V_{MoM}^4 \\ V_{MoM}^5 \\ V_{MoM}^6 \end{bmatrix}, \tag{11.25}$$

where segments 2 and 5 correspond to the terminals of antennas 1 and 2, respectively. Using $V_{MoM}^{i+1} = \left(e^{jk\Delta z\cos\theta}\right) V_{MoM}^i$ and let $\alpha = e^{jk\Delta z\cos\theta}$,

$$V_{MoM}^1 = e^{-jk\Delta z\cos\theta}V_{MoM}^2 = \alpha^{-1}V_{MoM}^2 \tag{11.26}$$

$$V_{MoM}^3 = e^{jk\Delta z\cos\theta}V_{MoM}^2 = \alpha V_{MoM}^2 \tag{11.27}$$

$$V_{MoM}^4 = e^{-jk\Delta z\cos\theta}V_{MoM}^5 = \alpha^{-1}V_{MoM}^5 \tag{11.28}$$

$$V_{MoM}^6 = e^{jk\Delta z\cos\theta}V_{MoM}^5 = \alpha V_{MoM}^5 \tag{11.29}$$

Equation (11.25) can now be written as

$$\begin{bmatrix} V_{meas1} \\ V_{meas2} \end{bmatrix} = \begin{bmatrix} Z_L & 0 \\ 0 & Z_L \end{bmatrix} \begin{bmatrix} \alpha^{-1}Y_{21} + Y_{22} + \alpha Y_{23} & \alpha^{-1}Y_{24} + Y_{25} + \alpha Y_{26} \\ \alpha^{-1}Y_{51} + Y_{52} + \alpha Y_{53} & \alpha^{-1}Y_{54} + Y_{55} + \alpha^{-1}Y_{56} \end{bmatrix} \begin{bmatrix} V_{MoM}^2 \\ V_{MoM}^5 \end{bmatrix}, \tag{11.30}$$

or

$$\begin{bmatrix} V_{meas1} \\ V_{meas2} \end{bmatrix} = \mathbf{B} \begin{bmatrix} V_{MoM}^2 \\ V_{MoM}^5 \end{bmatrix} \tag{11.31}$$

Given the knowledge of the elevation angle θ, $\mathbf{Y}_{MoM}$(or $\mathbf{Z}_{MoM}$) and the load impedance Z_L, $\begin{bmatrix} V_{MoM}^2 & V_{MoM}^5 \end{bmatrix}^T$ can thus be solved.

In situations where the signals are coming from different elevation angles or the information of the elevation angles is not available, or the array is made up of more complicated antenna element, Eq. (11.31) is not applicable. Instead, the voltage vector $\mathbf{V}_{MoM}^{RX}$ in Eq. (11.24) can be solved using a minimum norm formulation (Lau et al. 2004)

$$\mathbf{V}_{MoM}^{RX} = \mathbf{C}^H (\mathbf{C}\mathbf{C}^H)^{-1} \mathbf{V}_{mea} \tag{11.32}$$

where $\mathbf{C}^H$ is the conjugate transpose (Hermitian) of matrix $\mathbf{C}$. Entries of $\mathbf{V}_{MoM}^{RX}$ corresponding to the antenna terminals can be used for further signal processing.

Through a full-wave formulation, the mutual coupling effects of all the antenna elements have been considered simultaneously instead of two antennas at a time. The mutual coupling behavior of all the elements in receiving modes is accurately characterized using the full-wave method. More importantly, the retrieved excitation voltage vector is free from mutual coupling effects, which is more superior than the open-circuited voltages.

11.4.3 Receiving-Mutual-Impedance Method

With the knowledge of the impedance matrix $\mathbf{Z}_{MoM}$, the full-wave method provides an accurate solution to retrieve $\mathbf{V}_{MoM}^{RX}$ from the measured terminal voltages. The elements in the $\mathbf{V}_{MoM}^{RX}$ matrix that correspond to the antenna terminals are essentially incoming signals from the external sources without any perturbation from mutual coupling. The voltages we obtained from $\mathbf{V}_{MoM}^{RX}$ are theoretically better than the open-circuited voltages that are perturbed by reradiated field from other antenna elements. However, the knowledge of the impedance or admittance matrix is not usually available, which limits the practicality of the full-wave method.

Hui (2003, 2004a, 2016) adopted the flexibility and idea of the mutual impedance and introduced the RMIM. As opposed to the conventional mutual impedance, which is determined based on an open-circuit assumption, and with the antenna array is in a "transmitting mode", the receiving mutual impedance is calculated under the conditions that (i) the antenna elements are terminated with a known impedance Z_L and (ii) the antenna elements are in the receiving mode under an external plane wave excitation.

As previously discussed, the mutual coupling path between transmitting and receiving arrays is different. The outcome from Eqs. (11.3) and (11.4) in Section 11.2 quantitatively indicates that the current distribution of receiving array requires all the elements from the admittance matrix $\mathbf{Y}_{MoM}$ while that of transmitting array requires only the part of the same admittance matrix that corresponds to the antenna terminals. As a result, the overall mutual coupling effect in receiving array in general is more complicated than the coupling effect in transmitting array. With proper load terminations, together with an external plane wave excitation, the receiving-mutual-impedance method (RMIM), which was mentioned in Chapter 10, is a "receiving-mode defined" method which characterizes the mutual coupling behavior as shown in Figure 11.1b.

Consider a receiving array with N elements each of which is modeled using M basis functions. In this instance, the MoM solution in Eq. (11.1) can be expressed as

$$\begin{bmatrix} \mathbf{Z}^{11} & \mathbf{Z}^{12} & \cdots & \mathbf{Z}^{1N} \\ \mathbf{Z}^{21} & \mathbf{Z}^{22} & \cdots & \mathbf{Z}^{2N} \\ \vdots & \vdots & \ddots & \vdots \\ \mathbf{Z}^{N1} & \mathbf{Z}^{N2} & \cdots & \mathbf{Z}^{NN} \end{bmatrix} \begin{bmatrix} \mathbf{I}^{1} \\ \mathbf{I}^{2} \\ \vdots \\ \mathbf{I}^{N} \end{bmatrix} = \begin{bmatrix} \mathbf{V}^{1} \\ \mathbf{V}^{2} \\ \vdots \\ \mathbf{V}^{N} \end{bmatrix} \tag{11.33}$$

where the matrix blocks are

$\mathbf{Z}^{\alpha\beta} = \begin{bmatrix} z_{11}^{\alpha\beta} - Z_L & z_{12}^{\alpha\beta} & \cdots & z_{1M}^{\alpha\beta} \\ z_{21}^{\alpha\beta} & z_{22}^{\alpha\beta} & \cdots & z_{2M}^{\alpha\beta} \\ \vdots & \vdots & \ddots & \vdots \\ z_{M1}^{\alpha\beta} & z_{M1}^{\alpha\beta} & \cdots & z_{MM}^{\alpha\beta} \end{bmatrix}$, $\mathbf{I}^{\alpha} = \begin{bmatrix} I_1^{\alpha} \\ I_2^{\alpha} \\ \vdots \\ I_M^{\alpha} \end{bmatrix}$, and $\mathbf{V}^{\alpha} = \begin{bmatrix} V_1^{\alpha} \\ V_2^{\alpha} \\ \vdots \\ V_M^{\alpha} \end{bmatrix}$, respectively. The kth row of Eq. (11.33) is given by

$$\mathbf{Z}^{kk}\mathbf{I}^{k} = -\mathbf{Z}^{k1}\mathbf{I}^{1} - \cdots - \mathbf{Z}^{k(k-1)}\mathbf{I}^{k-1} - \mathbf{Z}^{k(k+1)}\mathbf{I}^{k+1} - \cdots - \mathbf{Z}^{kN}\mathbf{I}^{N} + \mathbf{V}^{k}. \tag{11.34}$$

The left-hand side of Eq. (11.34) $\mathbf{Z}^{kk}\mathbf{I}^{k}$ is exactly the system of equations resulting from a standard moment method analysis of the kth antenna element as if it was standing alone without the existence of all other antenna elements. The right-hand side of Eq. (11.34) can be regarded as two independent sources. The term $\mathbf{V}^{k}$ corresponds to the voltage induced on the antenna element due to incident field from the external sources. The remaining terms $-\mathbf{Z}^{k1}\mathbf{I}^{1} - \cdots - \mathbf{Z}^{k(k-1)}\mathbf{I}^{k-1} - \mathbf{Z}^{k(k+1)}\mathbf{I}^{k+1} - \cdots - \mathbf{Z}^{kN}\mathbf{I}^{N}$ correspond to the voltages due to the current distributions on other array elements, i.e. the mutual coupling effect. According to the principle of superposition, the overall effect on the kth antenna due to these two sources is independent. Mathematically, this means that the field generated by any one of these two excitation sources satisfies the boundary condition alone, and thus, they can exist independent of the field generated by the other source. Therefore, if the exact current distributions and the impedance matrices of all the antenna elements are known, we are able to determine the following: $-\mathbf{Z}^{k1}\mathbf{I}^{1} - \cdots - \mathbf{Z}^{k(k-1)}\mathbf{I}^{k-1} - \mathbf{Z}^{k(k+1)}\mathbf{I}^{k+1} - \cdots - \mathbf{Z}^{kN}\mathbf{I}^{N}$ and $\mathbf{Z}^{kk}\mathbf{I}^{k}$. The voltage due to the incident field $\mathbf{V}^{k}$ can then be determined from Eq. (11.34).

The problem is now to determine the current distributions of the antennas. The current distribution of any antenna element α, i.e. all the matrix elements inside the array $\mathbf{I}^{\alpha}$, can only be obtained through numerical modeling. In practical applications, only the voltages or currents at the antenna terminal are measurable. To solve this problem, we adopt a "single mode" approximation, i.e. we treat the antenna elements in the array as consisting of one segment only. Furthermore, we use the terminal voltages and currents to describe the antenna array as an N-port network. Using the "single mode" approximation, Eq. (11.33) can now be written as

$$\begin{bmatrix} Z_t^{11} - Z_L & Z_t^{12} & \cdots & Z_t^{1N} \\ Z_t^{21} & Z_t^{22} - Z_L & \cdots & Z_t^{2N} \\ \vdots & \vdots & \ddots & \vdots \\ Z_t^{N1} & Z_t^{N2} & \cdots & Z_t^{NN} - Z_L \end{bmatrix} \begin{bmatrix} I_t^{1} \\ I_t^{2} \\ \vdots \\ I_t^{N} \end{bmatrix} = \begin{bmatrix} V_t^{1} \\ V_t^{2} \\ \vdots \\ V_t^{N} \end{bmatrix}, \tag{11.35}$$

where the subscript t is used to signify that these equations refer to the current and voltages at the antenna terminal. The term $Z_t^{\alpha\beta}$ $(\alpha \neq \beta)$ corresponds to the "modified" mutual impedance between antenna elements α and β due to terminal current I_t^{β}. In parallel with Eq. (11.34), the induced voltage in antenna terminal k can be written as

$$\left(Z_t^{kk} - Z_L\right) I_t^k = -Z_t^{k1} I_t^1 - \cdots - Z_t^{k(k-1)} I_t^{k-1} - Z_t^{k(k+1)} I_t^{k+1} - \cdots - Z_t^{kN} I_t^N + V_t^k \tag{11.36}$$

In Eq. (11.36), the measurable quantities are the terminal current, i.e. I_t^α for $\alpha = 1, 2, \ldots, N$. We then remove the voltage dropped across the self-impedance $Z_t^{kk} I_t^k$ of the kth antenna from the left-hand side of Eq. (11.33), and removing that part of the incident voltage that cause this voltage drop on the self-impedance from the right-hand side of Eq. (11.36), we obtain

$$-Z_L I_t^k = -Z_t^{k1} I_t^1 - \cdots - Z_t^{k(k-1)} I_t^{k-1} - Z_t^{k(k+1)} I_t^{k+1} - \cdots - Z_t^{kN} I_t^N + U_t^k \tag{11.37}$$

U_t^k is the terminal voltage of antenna element k due to the incoming signal alone. The antenna terminal current I_t^k can be given by

$$I_t^k = -\frac{V_t^k}{Z_L}, \tag{11.38}$$

where V_t^k is the terminal voltage of antenna element k. Together with Eq. (11.38), Eq. (11.37) can thus be written as

$$V_t^k = Z_t^{k1} \frac{V_t^1}{Z_L} - \cdots - Z_t^{k(k-1)} \frac{V_t^{k-1}}{Z_L} - Z_t^{k(k+1)} \frac{V_t^{k+1}}{Z_L} - \cdots - Z_t^{kN} \frac{V_t^N}{Z_L} + U_t^k \tag{11.39}$$

Rearranging Eq. (11.39), the relationship between U_t^k and V_t^k can be given by

$$\begin{bmatrix} 1 & -\frac{Z_t^{1,2}}{Z_L} & \cdots & -\frac{Z_t^{1,N-1}}{Z_L} & -\frac{Z_t^{1,N}}{Z_L} \\ -\frac{Z_t^{2,1}}{Z_L} & 1 & \cdots & -\frac{Z_t^{2,N-1}}{Z_L} & -\frac{Z_t^{2,N}}{Z_L} \\ \vdots & \vdots & \ddots & \vdots & \vdots \\ -\frac{Z_t^{N-1,1}}{Z_L} & -\frac{Z_t^{N-1,2}}{Z_L} & \cdots & 1 & -\frac{Z_t^{N-1,N}}{Z_L} \\ -\frac{Z_t^{N,1}}{Z_L} & -\frac{Z_t^{N,2}}{Z_L} & \cdots & -\frac{Z_t^{N,N-1}}{Z_L} & 1 \end{bmatrix}_{N\times N} \begin{bmatrix} V_t^1 \\ V_t^2 \\ \vdots \\ V_t^{N-1} \\ V_t^N \end{bmatrix}_{N\times 1} = \begin{bmatrix} U_t^1 \\ U_t^2 \\ \vdots \\ U_t^{N-1} \\ U_t^N \end{bmatrix}_{N\times 1}. \tag{11.40}$$

Given the antenna terminal voltages V_t^k ($k = 1,2,\ldots,$N) from the array in a receiving application, the antenna terminal voltages due to the incoming signals alone U_t^k can be obtained through Eq. (11.40) once the "modified" mutual impedance $Z_t^{\alpha\beta}$ ($\alpha \neq \beta$) and the load impedance Z_L are known.

From Eq. (11.35), it can be seen that the terminal voltages across antenna element k, V_t^k, can be expressed as a sum of products, each of which is a multiplication of the terminal current, I_t^β, on an antenna element $\beta(\neq k)$ with a quantity known as the "modified" mutual impedance, $Z_t^{k\beta}$. It is called "modified" simply because this is different from the conventional mutual impedance (commonly denoted as $Z_{k\beta}$, as discussed in Section 11.3.1). This "modified" mutual impedance $Z_t^{k\beta}$ measures only the coupled voltage across the terminal load of the kth antenna element, while the conventional mutual impedance $Z_{k\beta}$ measures the open-circuit voltage on the kth antenna. The open-circuit voltage is the voltage dropped across the total impedance (i.e. the terminal load plus the self-impedance) of the antenna.

Another difference is that the modified mutual impedance $Z_t^{k\beta}$ refers to the situation where the antenna array is in the receiving mode so that the current distributions and terminal currents of all the array elements are induced by incoming signal(s). The conventional mutual impedance $Z_{k\beta}$, on the other hand, refers to the situation where antenna element k is in a passive receiving mode while the antenna element β is in an active transmitting mode. Although this latter difference is not fundamental, it does result in two different current distributions on antenna element β under these two different situations, as discussed in Section 11.2.2. The two different current distributions on antenna element β thus induce different coupled voltages on the antenna element k. As a result, the modified mutual impedance $Z_t^{k\beta}$ is different from the conventional mutual impedance $Z_{k\beta}$ such that a new name is given to this mutual impedance – "receiving mutual impedance." The receiving mutual impedance is defined as the ratio of the coupled voltage dropped across the terminal load of the antenna element k (due to the reradiation of the current distribution on the antenna element β) to the terminal current on the antenna element β, while the antenna array is in the passive receiving mode and excited by an external electromagnetic plane wave. As the definition of the receiving mutual impedance requires an external plane wave source the result is that the value of the receiving mutual impedance is in general dependent on the direction (incident angle) of the plane wave.

Equation (11.37) can also be written as

$$V_t^k = -Z_L I_t^k = -Z_L \left(I_{t,c}^k + I_{t,s}^k \right) = W_t^k + U_t^k \tag{11.41}$$

where $W_t^k = -Z_t^{k1} I_t^1 - \cdots - Z_t^{k(k-1)} I_t^{k-1} - Z_t^{k(k+1)} I_t^{k+1} - \cdots - Z_t^{kN} I_t^N$ corresponds to the coupled voltages from other antenna elements. The terminal current at antenna element k is now written in two parts $I_{t,c}^k$ and $I_{t,s}^k$, which, respectively, correspond to the induced current due to coupled field from other antenna elements and the induced current due to the incoming sources. The subscripts C and S correspond to coupled and source, respectively. Therefore, we can write

$$\begin{cases} -Z_L I_{t,c}^k = W_t^k \\ -Z_L I_{t,s}^k = U_t^k \end{cases} \tag{11.42}$$

It can be clearly seen that the terminal voltage V_t^k in an array is a sum of the voltages from two sources: the signal due to incoming wave from external sources U_t^k and the signal due to reradiated field from other antenna elements W_t^k.

11.4.3.1 Determination of the Receiving Mutual Impedance

(a) Method 1: Theoretical Method Based on Current Distribution

Let us first consider the case of a two-element array. According to Eq. (11.41), the terminal voltages of each antenna element is given by

$$\begin{cases} V_t^1 = U_t^1 + W_t^1 = U_t^1 - Z_t^{12} I_t^2 \\ V_t^2 = U_t^2 + W_t^2 = U_t^2 - Z_t^{21} I_t^1 \end{cases} \tag{11.43}$$

W_t^1 and W_t^2 correspond to the portion of the terminal voltage due to coupling effect from other antenna elements. By definition, the receiving mutual impedance is the ratio of the coupled voltage dropped across the terminal load of the second antenna (say antenna 2) due to the reradiated field from the first antenna (say antenna 1), i.e. W_t^2, to the terminal current in the first antenna I_t^1, whereas both antennas are in passive receiving operation. To obtain the reradiate field from the first antenna under receiving mode W_t^2, we can first excite the first antenna using the receiving-mode current distribution. For a simple antenna, such as dipole antenna, this can be found in the literature. The corresponding terminal voltages induced in the second antenna W_t^2 and terminal current in the first antenna I_t^1 can be determined through moment method or other numerical methods. The receiving mutual impedance, Z_t^{21}, can thus be determined by

Hui (2003) and Hui (2004b)

$$Z_t^{21} = -\frac{W_t^2}{I_t^1} \tag{11.44}$$

The receiving mutual impedance Z_t^{12} can also be obtained in a similar manner.

The success of this method relies on the knowledge of the receiving-mode current distribution of the antenna elements and exciting the antenna using the receiving-mode current distribution in the computation domain. In situations where the current distribution of the antenna element in receiving mode is not analytically available, the current distribution can be numerically obtained by exciting the element using external plane wave source. Exciting an antenna element using a given current distribution can be done using in-house moment method code or a commercial electromagnetic solver. Since some solvers do not allow us to excite the antenna using current distribution, this limits the practically of this method.

It is worth noting that receiving-mode current distribution in general is a function of incident angle. This implies that the receiving mutual impedance is also a function of incident angle which fundamentally distinguishes it from the conventional mutual impedance. In arrays where there are more than two antenna elements, it can be considered as different pairs of two-elements arrays with different interelement separation, and the corresponding receiving mutual impedance can also be obtained in the same manner.

(b) Method 2: Practical Method Based on Antenna Terminal Voltages

In situations where electromagnetic solver does not allow us to excite the antenna using current distribution, or in practical situations where it is difficult to measure the current distribution and feed the antenna using a specific current distribution, an alternative solution to determine the receiving mutual impedance which does not require the current distribution was introduced (Hui 2004a).

The measurable terminal voltages, terminal currents, and the corresponding receiving mutual impedances of a two-element antenna array under external plane wave excitation are related by Eq. (11.43). Through numerical modeling, the terminal voltages V_t^1 and V_t^2 can be obtained when the array is illuminated under plane wave excitation. The corresponding terminal currents, I_t^1 and I_t^2, are related to the terminal voltages through the load impedance Z_L, Eq. (11.38). The uncoupled voltages U_t^1 and U_t^2, which correspond to the terminal voltages induced when

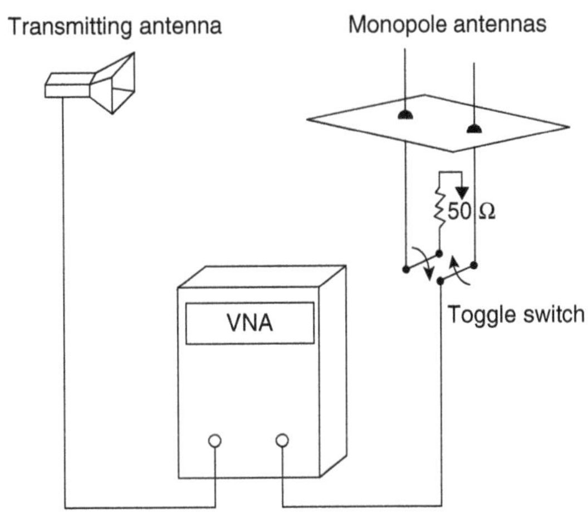

Figure 11.4 The measurement of the receiving mutual impedance inside an anechoic chamber using a vector network analyzer. Source: From HUI, H.T. (2016), 'Concept and applications of receiving mutual impedance' In: Z. Chen, D. Liu, H. Nakano, X. Qing, T. Zwick eds, Handbook of antenna technologies, pp. 235–283, Springer, Singapore. © 2016 Springer Nature.

the antenna element is isolated (without the presence of other antenna elements) and illuminated under plane wave excitation, can also be obtained numerically through simulations. With V_t^1, V_t^2, U_t^1, and U_t^2 obtained through simulation as well as I_t^1 and I_t^2 obtained through Eq. (11.38), the receiving mutual impedances Z_t^{12} and Z_t^{21} can be given by

$$Z_t^{12} = -\frac{V_t^1 - U_t^1}{I_t^2} \quad \text{and} \quad Z_t^{21} = -\frac{V_t^2 - U_t^2}{I_t^1} \tag{11.45}$$

The above calculation can also be performed experimentally with a vector network analyzer (VNA). Figure 11.4 shows the corresponding experimental setting. Port 1 of the VNA is connected to the transmitting antenna, while Port 2 is connected to one of the monopole antennas. According to Hui (2004a), the measured scattering parameters, for instance, S_{21}, can be converted to voltage as

$$S_{21} = \frac{\beta}{\alpha} \tag{11.46}$$

where α is the square root of the power emitted by the transmitting horn antenna inside the anechoic chamber, $\beta = V/\sqrt{Z_0}$ is the square root of the power received by a monopole antenna element, and Z_0 is the system impedance (50 Ω). Both α and β are complex values with magnitudes and phase. To determine the receiving mutual impedance, the detailed measurement procedures are given as follows (Hui 2004a):

(i) Measure S_{21} at the terminal of monopole 1 with monopole 2 terminated with a load impedance Z_0. Denote it as S_{21_1}.
(ii) Measure S_{21} at the terminal of monopole 2 with monopole 1 terminated with a load impedance Z_0. Denote it as S_{21_2}.
(iii) Measure S_{21} at the terminal of monopole 1 with monopole 2 removed from the array. Denote it as S'_{21_1}.
(iv) Measure S_{21} at the terminal of monopole 2 with monopole 4 removed from the array. Denote it as S'_{21_2}.

Notice that the positions of the antenna must remain the same since any change in their positions will alter the phase and magnitude of S21. The above values of S_{21} can be converted to voltages using Eq. (11.46)

$$V_t^1 = S_{21\_1}\alpha\sqrt{Z_0} \tag{11.47}$$

$$V_t^2 = S_{21\_2}\alpha\sqrt{Z_0} \tag{11.48}$$

$$U_t^1 = S'_{21\_1}\alpha\sqrt{Z_0} \tag{11.49}$$

$$U_t^2 = S'_{21\_2}\alpha\sqrt{Z_0} \tag{11.50}$$

Using Eq. (11.45), the receiving mutual impedance can be given by Hui (2004a) and Hui (2016):

$$Z_t^{12} = -\frac{V_t^1 - U_t^1}{I_t^2} = -\frac{S_{21\_1} - S'_{21\_1}}{(-S_{21\_2}/Z_0)} = \frac{S_{21\_1} - S'_{21\_1}}{S_{21\_2}} Z_0 \tag{11.51}$$

$$Z_t^{21} = -\frac{V_t^2 - U_t^2}{I_t^1} = -\frac{S_{21\_2} - S'_{21\_2}}{(-S_{21\_1}/Z_0)} = \frac{S_{21\_2} - S'_{21\_2}}{S_{21\_1}} Z_0 \tag{11.52}$$

Therefore, the receiving mutual impedance can be obtained practically from scattering parameters.

(c) Method 3: Practical Method Taking all Array Elements into Consideration

The previous methods determine the receiving mutual impedance with only two antenna elements at a time, which provides an accurate solution for a two-element array. When there are more than two elements, the previous two methods, however, do not consider the mutual coupling effects of the other antennas. For an N-element array, the relationship between U_t^k and V_t^k can be given by Eq. (11.40). Consider the first row of Eq. (11.40), i.e.

$$V_t^1 - Z_t^{12}\frac{V_t^2}{Z_L} - \cdots - Z_t^{1k}\frac{V_t^k}{Z_L} - \cdots - Z_t^{1N}\frac{V_t^N}{Z_L} = U_t^1 \tag{11.53}$$

It is not difficult to understand that due to different positions of the antenna elements and the unequal mutual coupling effect, the terminal voltages (especially their phases) in Eq. (11.53) are dependent on the incoming directions (θ, ϕ). For arrays with monopole and dipole antenna elements oriented along the z direction, however, their radiation pattern is omnidirectional in the azimuth plane with the elevation angle $\theta = 90°$ ($\theta = 0°$ corresponds to the z axis) such that the current distributions of the antenna elements due to the incoming signals are independent of the azimuth angles ϕ, as long as the signals are coming from the elevation plane at $\theta = 90°$. Consequently, the receiving mutual impedances, which can be determined based on the receiving-mode current distribution according Eq. (11.44), are independent of azimuth angles ϕ at $\theta = 90°$. With these observations, the receiving mutual impedances can be determined by taking multiple measurements from N-1 different azimuth angles ϕ at $\theta = 90°$, which results in

$$\begin{cases} V_t^{1,\phi_1} - Z_t^{12}\frac{V_t^{2,\phi_1}}{Z_L} - \cdots - Z_t^{1k}\frac{V_t^{k,\phi_1}}{Z_L} - \cdots - Z_t^{1N}\frac{V_t^{N,\phi_1}}{Z_L} = U_t^{1,\phi_1} \\ V_t^{1,\phi_2} - Z_t^{12}\frac{V_t^{2,\phi_2}}{Z_L} - \cdots - Z_t^{1k}\frac{V_t^{k,\phi_2}}{Z_L} - \cdots - Z_t^{1N}\frac{V_t^{N,\phi_2}}{Z_L} = U_t^{1,\phi_2} \\ \vdots \\ V_t^{1,\phi_{N-1}} - Z_t^{12}\frac{V_t^{2,\phi_{N-1}}}{Z_L} - \cdots - Z_t^{1k}\frac{V_t^{k,\phi_{N-1}}}{Z_L} - \cdots - Z_t^{1N}\frac{V_t^{N,\phi_{N-1}}}{Z_L} = U_t^{1,\phi_{N-1}} \end{cases}. \tag{11.54}$$

The system of equations consist of N-1 unknowns, i.e. the receiving mutual impedances Z_t^{12}, Z_t^{13},…,Z_t^{1N}. With N-1 equations obtained from N-1 independent measurements, the receiving mutual impedances can thus be determined by solving Eq. (11.54) (Lui and Hui 2010b). The N-1 independent measurements can be chosen to be equally spaced within the angular range of 0–π. The same procedure is repeated for the second and all other rows in Eq. (11.40), and the receiving mutual impedances Z_t^{k1},…,$Z_t^{k(k-1)}$,$Z_t^{k(k+1)}$…,Z_t^{kN} are obtained.

11.4.3.2 Comparison Between Different Mutual Impedances and Direction-Finding Applications

The RMIM was first applied to direction finding applications using uniform linear arrays (Hui 2004a,b) and circular arrays Zhang et al. (2008). This decoupling method has also been applied to magnetic resonance imaging (Hui 2005; Hui et al. 2006), multiple-input multiple-output (MIMO) communication (Lu et al. 2008, 2009). In this section, discussions are limited to direction finding applications using linear arrays. Details about other applications can be found in the aforementioned references.

To illustrate the differences between the conventional mutual impedance and the receiving mutual impedance, consider a two-element quarter-wavelength monopole antenna array with half-wavelength interelement separation (Lui et al. 2009). The array is designed to operate at the frequency of 2.4 GHz. The corresponding results for the conventional and receiving mutual impedance as a function of frequencies are shown in Figures 11.5a and 11.5b, respectively. It is

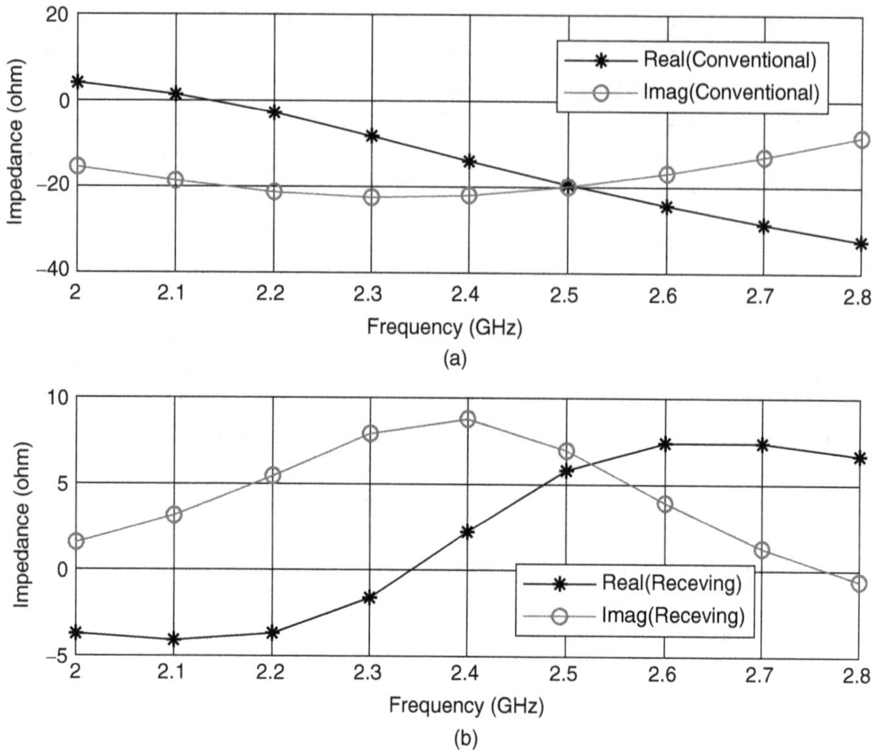

Figure 11.5 The mutual impedances of two quarter-wavelength monopole antennas (designed for an operating frequency of 2.4 GHz) with half-wavelength interelement separation as a function of frequency: (a) conventional mutual impedance and (b) receiving mutual impedance. Source: Lui et al. (2009).

apparent that the two mutual impedance are different – simply because they characterized the mutual coupling behavior when the array is in transmitting and receiving mode, respectively, as mentioned in earlier sections. It is also apparent that, in general, the mutual impedances vary as a function of frequencies in both cases.

Next, we compare the receiving mutual impedance determined using the two practical methods mentioned in Section 11.4.3.1, i.e. methods 2 and 3 (Lui and Hui 2010b). An eight-element uniform linear array with monopole elements is considered. The monopole elements are placed on a PEC ground plate, and the calculation is performed using moment method solver FEKO (2021.). The length of each monopole is $\ell = 0.24\lambda$ with $\ell/a = 100$, where a is the radius of the wire. Each monopole is terminated with a load impedance of 50 Ω. The receiving mutual impedances $Z_t^{1,j}$ determined using methods 2 and 3 are tabulated in Table 11.1. Here, three different interelement separations of $d = 0.5\lambda$, 0.25λ, and 0.05λ are considered.

The receiving mutual impedances obtained using method 2 are based on two antenna elements at a time while those obtained using method 3 are based on all the array elements. In this example, it is clear that the receiving mutual impedances obtained using the two methods are not significantly different when $d = 0.5\lambda$ with less than 0.1 Ω. This is because the reradiated field from the other six antenna elements is not significant when they are located apart from each other. When the interelement separation goes down to $d = 0.05\lambda$, the reradiated field from the other elements becomes significant. Comparing the receiving mutual impedances $Z_t^{1,3} Z_t^{1,4} Z_t^{1,5} Z_t^{1,6}$ and $Z_t^{1,7}$ (column 6 and 7), the differences of the numerical values can go beyond 10 Ω.

Table 11.1 Comparison of the receiving mutual impedances calculated using methods 2 and 3 for three eight-element arrays with 0.24λ monopole elements. The inter-element separations are (a) 0.5λ, (b) 0.25λ and (c) 0.05λ. The unit for each value in the table is ohm (Ω).

	(a) $d = 0.5\lambda$		(b) $d = 0.25\lambda$		(c) $d = 0.05\lambda$	
(Ω)	Method 2	Method 3	Method 2	Method 3	Method 2	Method 3
$Z_t^{1,2}$	3.8746 + j8.6013	3.8801 + j8.5828	−11.6361 + j8.5475	−11.6889 + j8.6014	−21.0952–j5.9569	−25.1223 + j6.4038
$Z_t^{1,3}$	−1.3148–j5.1149	−1.3111–j5.0993	3.8746 + j8.6013	4.0434 + j8.5955	−19.6363–j1.5685	−8.8547 + j0.9401
$Z_t^{1,4}$	0.6597 + j3.5504	0.6596 + j3.5382	6.4790–j2.1096	6.2243–j2.1050	−17.5055 + j2.5477	−36.4891–j1.1149
$Z_t^{1,5}$	−0.4054–j2.7039	−0.4072 + j2.6941	−1.3148–j.1149	−1.0735–j5.0845	−14.7868 + j5.9668	3.2473 + j10.2154
$Z_t^{1,6}$	0.2816–j2.1792	0.2834 + j2.1711	−4.1987 + j0.9009	−4.3391 + j0.8941	−11.6361 + j8.5475	−23.1459 + j6.6162
$Z_t^{1,7}$	−0.2116–j1.8232	−0.2130–j1.8170	0.6597 + j3.5504	0.7334 + j3.5275	−8.2414 + j10.2293	−4.5729 + j11.2133
$Z_t^{1,8}$	0.1679 + j1.5665	0.1693 + j1.5626	3.0712–j0.5075	3.0375–j0.4937	−4.8038 + j11.0072	−5.5475 + j11.0662

The interelement separations are a 0.5λ, b 0.25λ, and c 0.05λ. The unit for each value in the table is ohm (Ω).
Source: Taken from Lui and Hui (2010b).

To demonstrate the impact of different mutual impedances that are used for decoupling, a direction finding example with two signals coming from $\phi_1 = 30°$, $\theta_1 = 90°$ and $\phi_2 = 75°$, $\theta_2 = 90°$ is presented. The incoming signals are captured using the above three arrays. In each of the three cases, the corresponding DOA estimation problem is simulated using FEKO (2021) such that all the electromagnetic interactions between the antenna elements are considered using MoM. The incoming direction is resolved using the well-known MUSIC algorithm (Schmidt 1986). The corresponding spatial spectrums are shown in Figures 11.6a to 11.6d. In each subfigure, we considered five different ways to process the incoming signals: (i) measured (without compensation), (ii) conventional (using conventional mutual impedance), (iii) receiving (method 2), (iv) receiving (method 3), and (v) ideal. When the interelement separation is 0.5λ, shown in Figure 11.6a, the incoming signals can be clearly resolved for all cases. A zoom-in plot of Figure 11.6a is included in Figure 11.6b which clearly shows that the MUSIC spectrum obtained using conventional mutual impedance (dash-dot), receiving mutual impedance (Method 3, dot) and ideal voltage (dash) are clearly distinguishable. When the interelement separation becomes 0.25λ, as shown in Figure 11.5c, the measured signal (without compensation) fails to resolve. Using the conventional mutual impedance,

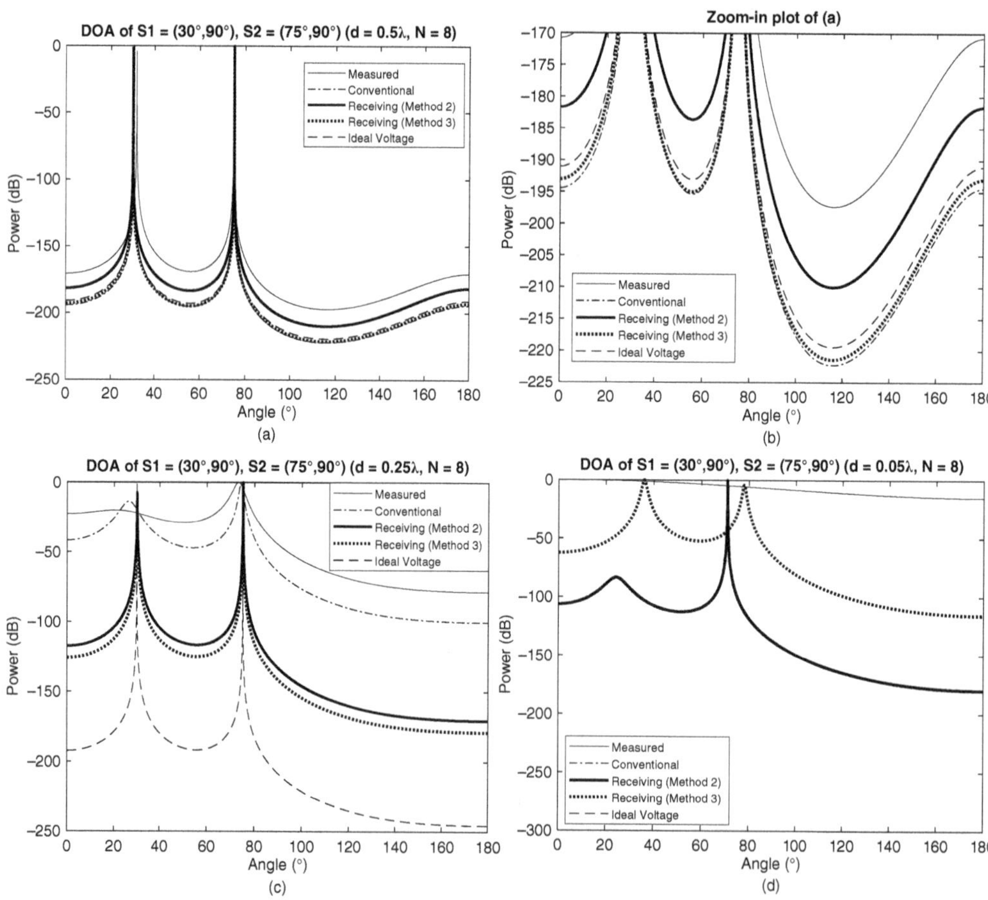

Figure 11.6 Spatial spectrum of the MUSIC algorithm for direction finding of two coherent signals coming from $\phi_1 = 30°$, $\theta_1 = 90°$ and $\phi_2 = 75°$, $\theta_2 = 90°$. The incoming signals are captured by three eight-element monopole arrays with interelement separation of (a) 0.5λ, (b) 0.5λ (Zoom in of (a), (c) 0.25λ, and (d) 0.05λ.

we can clearly see the two peaks, but the first signal at 30° is deviated from the actual incoming directions. Using the receiving mutual impedances (both methods 2 and 3), the directions are accurately resolved with sharper peaks. When the interelement separation becomes 0.05λ, only the receiving mutual impedance method determined using method 3 allows us to resolve the incoming directions. Although the two peaks are still visible using method 2 (solid), the peak at 30° is not very sharp. This result indicates the superior decoupling capabilities of receiving mutual impedance for directional finding applications. More results about DOA estimation using MUSIC together with RMIM for arrays with less than 0.25λ interelement separation can be found in Lui and Hui (2010a). Furthermore, the DOA estimation results using the "ideal voltage" (dash) is also included for comparison. These ideal voltages are found by having the two incoming signals with only one antenna element at a time in FEKO (2021) such that the obtained voltage is purely based on the incoming signals without any mutual coupling. As clearly shown in Figures 11.6, the incoming signals are resolved in all cases, and the resultant MUSIC spectra are nearly independent of the array size and the null between the two peaks is nearly 200 dB below.

In addition to the simulation results presented here and Lui and Hui (2010a), the RMIM has also been applied on experimentally obtained results for direction finding applications using linear arrays. Details can be found in Lui et al. (2010) and Yu et al. (2011).

11.4.4 Calibration Method

Another well-known method for mutual coupling compensation is the calibration method. There have been a number of variations to the calibration methods found in the open literature (e.g. Friedlander and Weiss 1991; Pierre and Kaveh 1991; Su et al. 2000; Su and Ling 2001; Dandekar et al. 2000, 2002). Here, the focus will be on the method described by Su et al. (2000), Su and Ling (2001), and Dandekar et al. (2000, 2002).

The main objective of this method is to determine the "true" array manifold, $\mathbf{A}_{true}$. The true array manifold is essentially the manifold in the physical antenna array, which has taken into account the mutual coupling, magnitude gain, and phase of the antenna element. In general, $\mathbf{A}_{true}$ is a function of the incoming signal direction. The true array manifold $\mathbf{A}_{true}$ and the theoretical array manifold, $\mathbf{A}_{theo}$, can be given by Su et al. (2000) as

$$\mathbf{A}_{true} = \mathbf{C}\mathbf{A}_{theo} \tag{11.55}$$

where $\mathbf{C}$ is known as the distortion matrix. The theoretical array manifold $\mathbf{A}_{theo}$ is the ideal array manifold that is widely used in array signal processing. For the case where the array and the incoming signal(s) are located on the x–y plane, the theoretical array manifold can be expressed as

$$\mathbf{A}_{theo} = \left[a_{theo,1}(\phi), a_{theo,2}(\phi), \dots, a_{theo,M}(\phi)\right]^T,$$

where

$$a_{theo,i}(\phi) = e^{j2\pi(x_i\cos\phi + y_i\sin\phi)/\lambda} \text{ and } i = 1, 2, \dots, N. \tag{11.56}$$

N is the total number of element, (x_i, y_i) are the coordinates of the ith array element, ϕ is the azimuth angle of the incoming signal, and λ is the wavelength of the operation frequency. For simple antennas such as dipoles whose current distributions have the same shape but different amplitudes, $\mathbf{C}$ can be determined from $\mathbf{A}_{theo}$ together with $\mathbf{A}_{true}$ computed at a few observation angles (under plane wave excitation) using moment method. Mathematically, $\mathbf{C}$ can be given by

$$\mathbf{C} = \mathbf{A}_{true}\mathbf{A}_{theo}^H\left(\mathbf{A}_{theo}\mathbf{A}_{theo}^H\right)^{-1} \tag{11.57}$$

where the superscripts H and -1 are the complex conjugate transpose and the inverse of the corresponding matrix, respectively. Once $\mathbf{C}$ is known, $\mathbf{A}_{true}$ at all different angles can be interpolated based on Eq. (11.55). Instead of "interpolation," $\mathbf{A}_{true}$ under plane wave excitation from all the azimuth angles ($0° < \theta < 360°$) can also be directly computed using MoM. The values of $\mathbf{A}_{true}$ as a function of θ are stored in a "look-up" table.

Simulation and measurement results show that both "interpolation" and direct "look-up" table are capable of compensating for mutual coupling effects (Su et al. 2000; Su and Ling 2001; Dandekar et al. 2000; Dandekar et al. 2002). The interpolation approach can be easily applied, and it provides a reasonably good mutual coupling compensation solution for arrays with simple antenna elements. The look-up table approach provides a more accurate solution for mutual coupling compensation but with the price being paid on large memory storage of the "look-up" table. Notice that, for both methods, $\mathbf{A}_{true}$ is determined when the array is in the "receiving mode" which characterizes the mutual coupling effect as described in Figure 11.1b. Upon its introduction, the calibration methods have been widely adopted in the antenna community.

11.4.5 Compensation Through Beamforming Network

A further way to address the changes in performance due to mutual coupling is to add a network between the excitation and the antenna array with the aim of producing an array excitation that is very similar or identical as possible as the desired radiation (Steyskal and Herd 1990). In many instances, the network can be considered as a cancellation or inverse network. The idea is to introduce such a network between the inputs and the antenna as shown in Figure 11.7 to compensate for mutual coupling. In receive, the array excitations are chosen to give the desired output complex excitations $\boldsymbol{c}$.

The electric field radiated by an array with this excitation is

$$\boldsymbol{E}(r,\theta,\varphi) = \boldsymbol{e}(r,\theta,\varphi)\,\boldsymbol{c} \tag{11.58}$$

where $\boldsymbol{e}$ is a row vector of electric field vectors of M radiation modes, and $\boldsymbol{c}$ is a column vector of the desired excitation such as for a Chebyshev distribution. Suppose the actual radiated field is

$$\boldsymbol{E}(r,\theta,\varphi) = \boldsymbol{e}(r,\theta,\varphi)\,\boldsymbol{W}\boldsymbol{a}_I \tag{11.59}$$

where $\boldsymbol{W}$ is an M × N matrix, and $\boldsymbol{a}_I$ is a column vector of N inputs. It can be shown that when a network with scattering matrix $\boldsymbol{S}^{(c)}$ is placed ahead of the antenna array represented by scattering

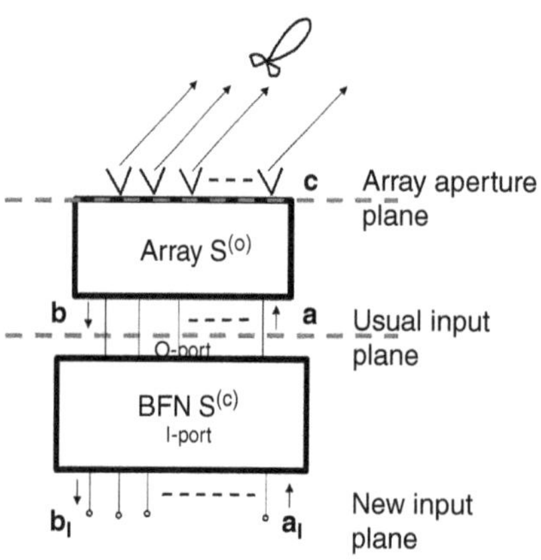

Figure 11.7 Model of array mutual coupling compensation with an external network inserted at the input of the array.

matrix $S^{(0)}$ as illustrated in Figure 11.7

$$W = (I + S^{(0)}) \left(I + S_{OO}^{(c)} S^{(0)} \right)^{-1} S_{OI}^{(c)} \tag{11.60}$$

where

$$S^{(c)} = \begin{bmatrix} S_{II}^{(c)} & S_{IO}^{(c)} \\ S_{OI}^{(c)} & S_{OO}^{(c)} \end{bmatrix}.$$

If no network is attached to the apertures, then $W = (I + S^{(0)})$. Thus, when the actual excitation is applied, it should be

$$a_I = W^{-1} c = S_{OI}^{(c)^{-1}} \left(I + S_{OO}^{(c)} S^{(o)} \right) \left(I + S^{(o)} \right)^{-1} c \tag{11.61}$$

If $S^{(o)}$ is known and fixed, then $S^{(c)}$ needs to be tuned to give the desired output.

As example, consider an eight-element linear array of X-band rectangular waveguides that are mounted in a ground plane. Each element is a column of eight rectangular waveguides and driven with a fixed 1: 8 power divider. The array axis was parallel to the E-plane, and in this plane, the spacing was d = 1.35 cm or 0.517 λ at 12.4 GHz (Steyskal and Herd 1990). The aim was to achieve a 30 dB Chebyshev pattern at a 0° scan angle from boresight. This pattern is shown in Figure 11.8 as a dashed curve. Radiation pattern results are given in Figure 11.8 without (a) and with coupling compensation (b).

It is seen in Figure 11.8 that the beamforming network has indeed removed most of the effects of mutual coupling as was predicted. The compensation technique gives a 10 dB reduction in the side lobe level, resulting in a pattern that is close to the theoretical prediction. Steyskal and Herd (1990) showed that the technique is most beneficial with a digital beamforming, where W^{-1} was realized by Fourier decompensation of the measured array element patterns.

11.4.6 Compensation in the Aperture

Another compensation technique is to introduce parasitic elements into the array and allow the mutual coupling between them and the excited elements to create currents on the parasitic elements as was suggested by Deng et al. (2018). A circuit representation of this is shown in Figure 11.9.

Suppose the currents and voltages at the input to the network are related as follows:

$$\underline{I} = \underline{\underline{Y}}\,\underline{V} \tag{11.62}$$

where $\underline{\underline{Y}} = [Y]$ is the array admittance matrix. Equation (11.62) may be partitioned as follows:

$$\begin{bmatrix} \underline{I}^{(1)} \\ \underline{I}^{(2)} \end{bmatrix} = \begin{bmatrix} \underline{\underline{Y}}^{(11)} & \underline{\underline{Y}}^{(12)} \\ \underline{\underline{Y}}^{(21)} & \underline{\underline{Y}}^{(22)} \end{bmatrix} \begin{bmatrix} \underline{V}^{(1)} \\ \underline{V}^{(2)} \end{bmatrix} \tag{11.63}$$

where in Eq. (11.63), the superscript (1) refers to primary array elements and superscript (2) indicates parasitic elements. Suppose the latter elements are loaded with admittances $\underline{\underline{Y_L}}$ such that

$$\underline{I}^{(2)} = -\underline{\underline{Y_L}}\,\underline{V}^{(2)} \tag{11.64}$$

resulting from the second row of Eq. 11.63 in

$$\underline{V}^{(2)} = -\left(\underline{\underline{Y}}^{(22)} + \underline{\underline{Y_L}} \right)^{-1} \underline{\underline{Y}}^{(21)} \underline{V}^{(1)}. \tag{11.65}$$

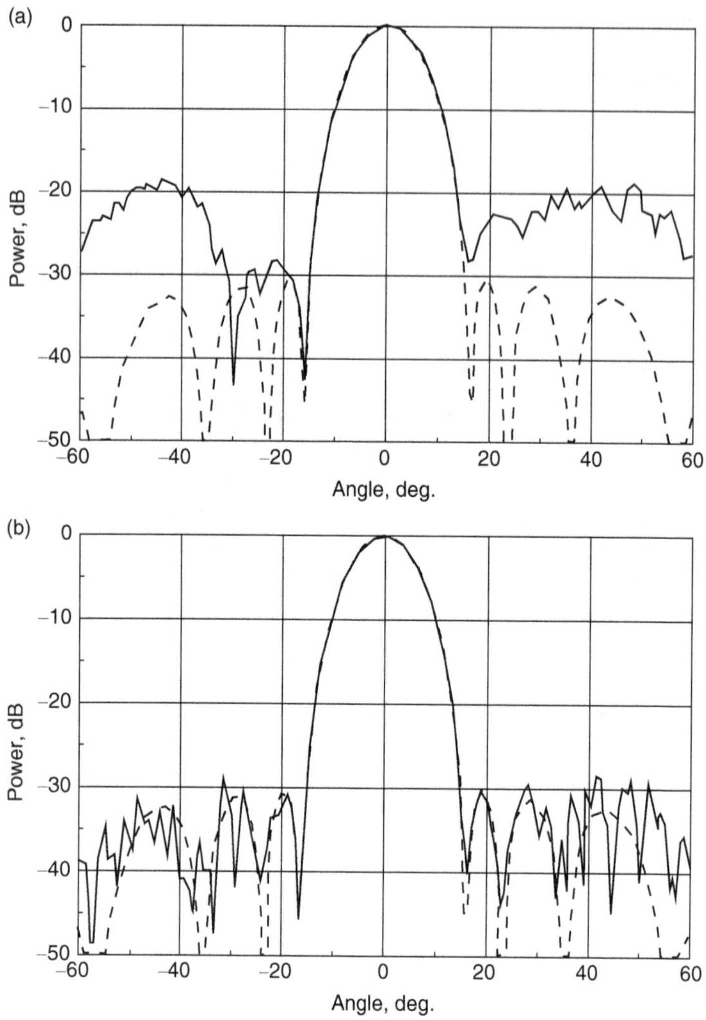

Figure 11.8 30 dB Chebyshev power pattern obtained with an eight-element waveguide array (a) without and (b) with beamforming network compensation at 0° scan angle at 12.4 GHz. Key: full curve – experiment; dashed curve – ideal without mutual coupling.

Substituting this into the first line of the partitioned matrix equation Eq. (11.63) gives

$$\underline{I}^{(1)} = \left[\underline{\underline{Y}}^{(11)} - \underline{\underline{Y}}^{(12)}\left(\underline{\underline{Y}}^{(22)} + \underline{\underline{Y}}_L\right)^{-1}\underline{\underline{Y}}^{(21)}\right]\underline{V}^{(1)} \tag{11.66}$$

Now, let

$$\underline{\underline{Y}}^{(11)} = \underline{\underline{Y}}_I^{(11)} + \underline{\underline{Y}}_C^{(11)}$$

where $\underline{\underline{Y}}_I^{(11)}$ is a diagonal matrix of self-admittances, and $\underline{\underline{Y}}_C^{(11)}$ is an off-diagonal matrix giving cross-coupling between elements. If the primary elements of the array are completely decoupled by the parasitic elements, then

$$\underline{I}^{(1)} = \underline{\underline{Y}}_I^{(11)}\underline{V}^{(1)}$$

For this to occur, the parasitic elements and loads should cancel as follows:

$$\underline{\underline{Y}}_C^{(11)} - \underline{\underline{Y}}^{(12)}\left(\underline{\underline{Y}}^{(22)} + \underline{\underline{Y}}_L\right)^{-1}\underline{\underline{Y}}^{(21)} = 0 \tag{11.67}$$

If the parasitic elements are fixed, the only variable that is free for adjustment is the load. In that case,

$$\underline{\underline{Y}}_L = \underline{\underline{Y}}^{(21)}\left(\underline{\underline{Y}}_C^{(11)}\right)^{-1}\underline{\underline{Y}}^{(12)} - \underline{\underline{Y}}^{(22)} \tag{11.68}$$

A dual expression to Eq. (11.68) can also be readily found by inspection for impedances. Another form of Eq. (11.66) given in terms of scattering waves and associated matrices of the network elements shown in Figure 11.9 can be shown to be

$$\underline{b}^{(1)} = \left[\underline{\underline{S}}^{(11)} + \underline{\underline{S}}^{(12)}\left(\underline{\underline{S}}_L - \underline{\underline{S}}^{(22)}\right)^{-1}\underline{\underline{S}}^{(21)}\right]\underline{a}^{(1)} \tag{11.69}$$

As a result, zero input reflection at the array occurs when

$$\underline{\underline{S}}^{(11)} + \underline{\underline{S}}^{(12)}\left(\underline{\underline{S}}_L - \underline{\underline{S}}^{(22)}\right)^{-1}\underline{\underline{S}}^{(21)} = 0. \tag{11.70}$$

The results summarized in Eqs. (11.69) and (11.70) is similar to the discussion on achieving a reflectionless array in Section 2.4.3.

Using the impedance matrix form of Eq. (11.68) for the case of a three-element dipole array consisting of a loaded dummy element (no. 3) which is located between two driven elements (nos. 1 and 2), results in a load impedance of

$$Z_L = z_{31}z_{12}z_{23} + z_{32}z_{21}z_{13} - z_{33} \tag{11.71}$$

In general,

$$\underline{V}^{(1)} = \left[\underline{\underline{Z}}^{(11)} - \underline{\underline{Z}}^{(12)}\left(\underline{\underline{Z}}^{(22)} + \underline{\underline{Z}}_L\right)^{-1}\underline{\underline{Z}}^{(21)}\right]\underline{I}^{(1)}$$

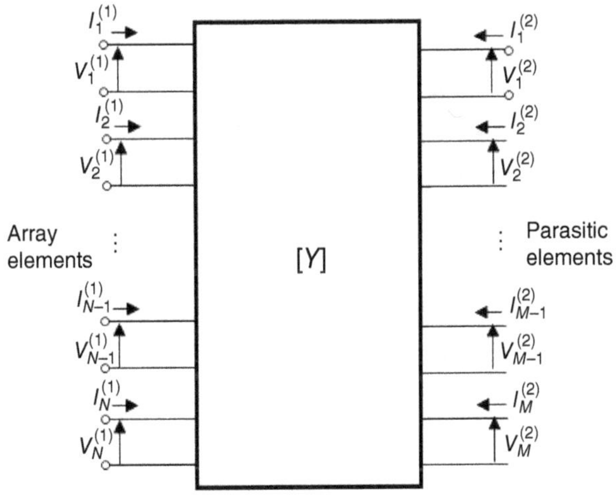

Figure 11.9 Admittance matrix representation of compensation in the aperture.

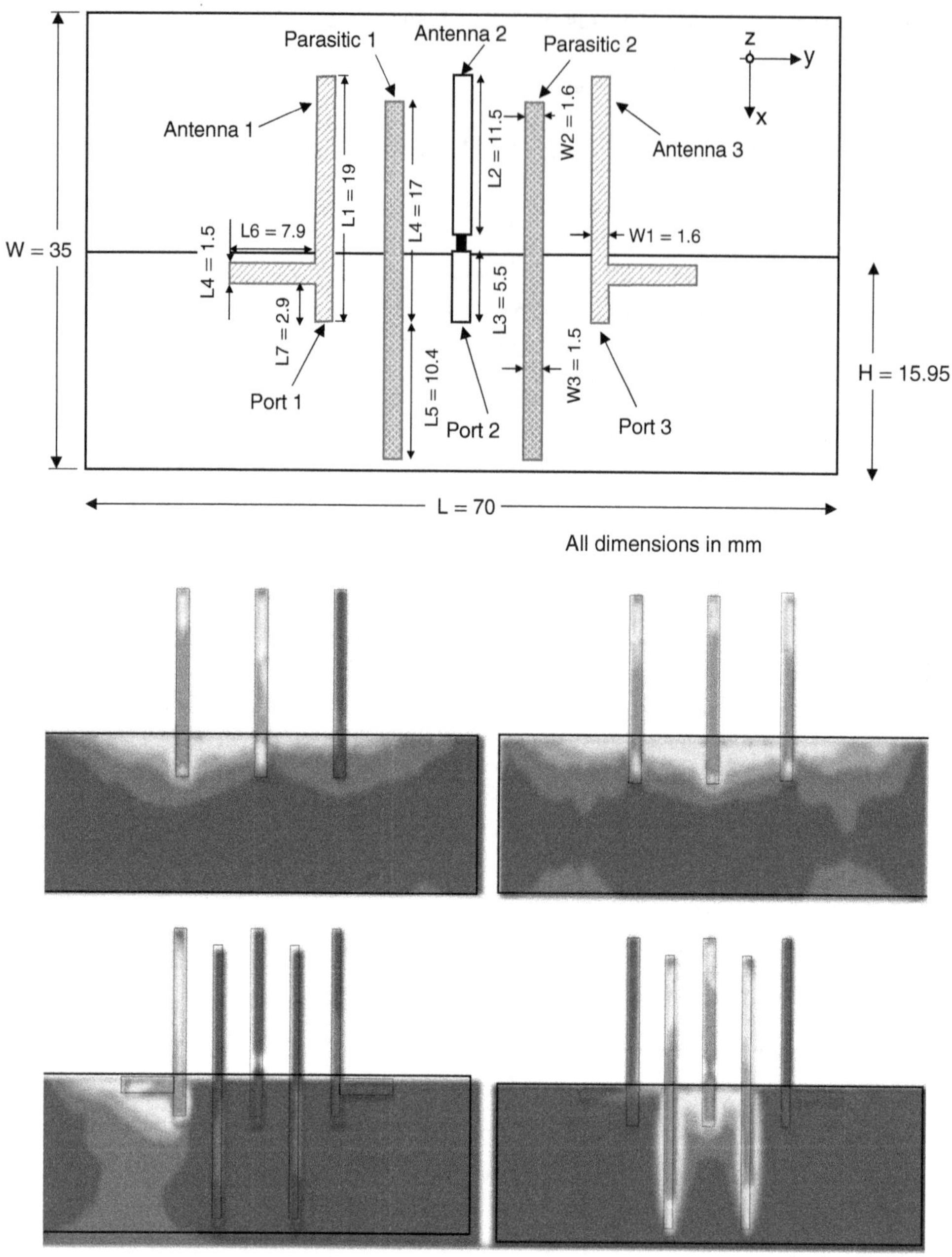

Figure 11.10 Decoupling of three monopoles by two parasitic elements. (a) Configuration with parasitic elements and (b) currents on dipoles for cases with driven elements only (top) and with two parasitic elements (below). Source: From DENG, J-Y, LI, J-Y and GU, L-X. (2018): 'Decoupling of a three-port MIMO antenna with different impedances using reactively loaded dummy elements', IEEE. Antennas Wireless Propagat. Lett, Vol. 17, No. 3, pp. 430–433. © 2018 IEEE. Color code from maximum to minimum signal levels: red, orange, yellow, green, light blue, dark blue. Source: © 2018 IEEE. Reprinted, with permission from Deng et al. (2018).

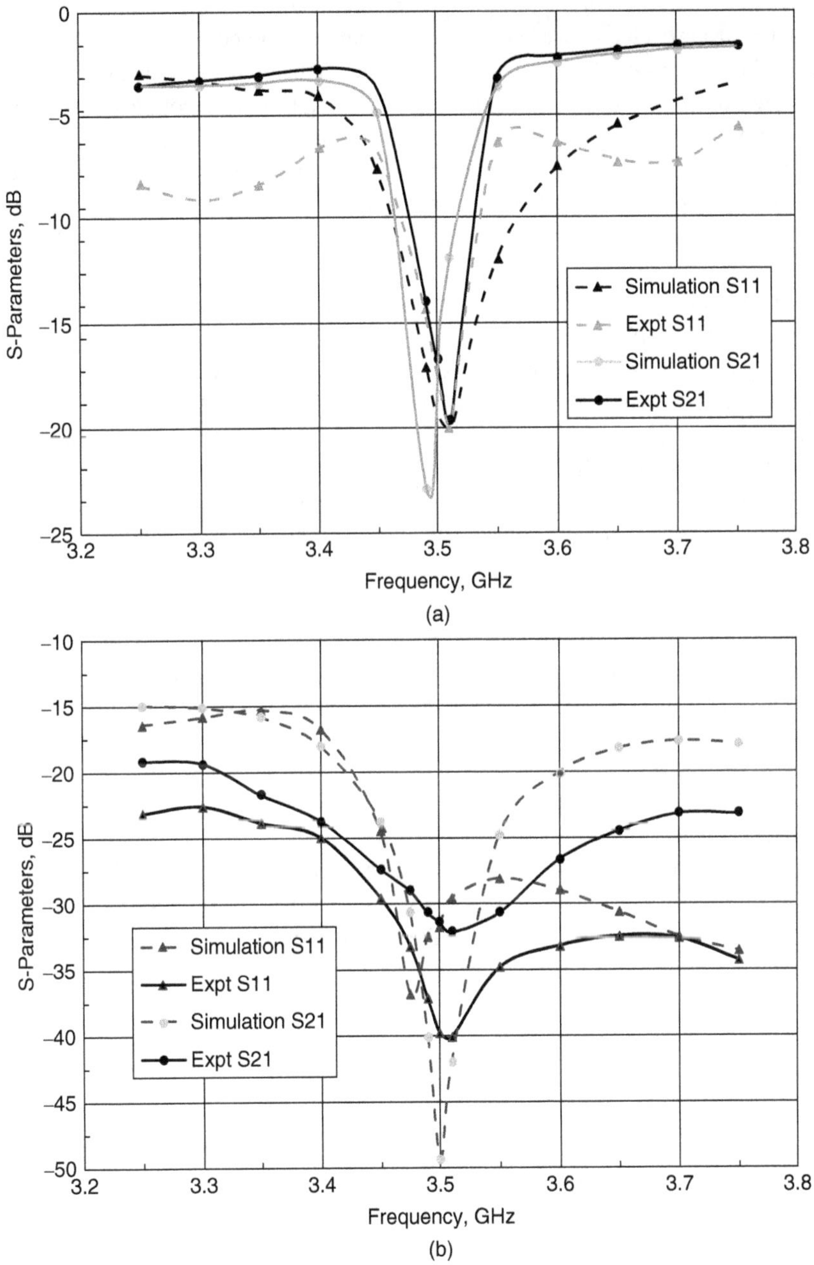

Figure 11.11 Simulated and measured results of three MIMO antennas shown in Figure 11.1 (a) without parasitic elements and (b) with loaded parasitic elements. Source: After Deng et al. (2018). Key: solid line – experiment; dashed – simulation; dot – S_{12}; triangle S_{11}.

Therefore, if $\underline{\underline{Z_L}}$ is very large,

$$\underline{V}^{(1)} \to \underline{\underline{Z}}^{(11)}\,\underline{I}^{(1)}$$

showing in that case the primary elements are very close to being decoupled.

As an example, consider the decoupling in a MIMO antenna (Deng et al. 2018) as illustrated in Figure 11.10. This shows reactively loaded parasitic elements being used to decouple a three-port MIMO antenna consisting of monopole antennas. The design of the antenna has been for operation at 3.5 GHz. When port 1 is excited, the current from the radiating element #1 that couples to elements #2 and #3 is largely cut off by the reactively loaded parasitic elements. The monopole antennas are printed on FR4 substrate ($\varepsilon_r = 4.4$, $\tan\delta = 0.02$) having a dimension of 70 mm (L) × 35 mm (W) × 0.8 mm (H). The distance between the adjacent active elements is only about 0.15λ at the center frequency, and therefore, the isolation of the antenna system without the parasitic elements is low and is about 6.3 and 13.6 dB, respectively. To improve the MIMO antenna isolation, two dummy elements were placed along the axis of symmetry of the two outside elements as shown in Figure 11.10a. Together, the two parasitic elements and the three unmatched radiating elements form a five-port array. Once the parasitic elements were placed in position, their dimensions were adjusted using Agilent ADS to achieve the best isolation. Optimal loads given by Eqs. (11.68) were included, resulting in a revised three-port network as shown in Figure 11.10a. The reactive loads of the parasitic elements can be realized with either lumped or distributed elements, although for convenience of tuning, two open-circuited transmission lines were used in the present design.

To illustrate the result of this decoupling, the simulated current distributions for the coupled and decoupled antenna cases are shown at the center frequency in Figure 11.10b. When port 1 is excited, the current from the radiating element 1 coupling to elements 2 and 3 is largely blocked by the reactively loaded parasitic elements. The same mechanism occurs when port 2 is excited. Decoupling of the radiating elements affects all aspects of matching characteristics. Therefore, a matching network was added to each port of the antenna element as shown in Figure 11.10b. Since the three radiating elements are well isolated from each other after decoupling, each element can be independently matched. In this example, two open-circuited stubs are used for elements 1 and 3, while a lossless capacitor is used for antenna 2 with a dimension of 1.6 mm × 0.8 mm due to space constraints. The simulation and measured results are given in Figures 11.11a and 11.11b. It can be seen that the isolation of the decoupled antenna at the center frequency has been improved by about 25 dB compared to the original antenna (no parasitic elements).

11.5 Concluding Remarks

This chapter has discussed various techniques for minimizing mutual coupling in arrays. These have included as conventional mutual impedance method, the full-wave method, receiving mutual impedance method, and beamforming network method. The details have been given for each method, and their use frequently depends on the application. For instance, another method that uses additional parasitic elements in the array can complement the original approach when there is sufficient space available on the array surface, whereas the beamforming network method may be prohibitive due to cost as an additional network has to designed and included at the input of the array. The selection may depend also on whether the array is primarily transmitting or receiving where a different approach for reducing coupling must be used. The desire to achieve the array performance at or near the uncoupled case means that techniques for minimizing mutual coupling will always be in demand.

References

Adve, R.S. and Sarkar, T.K. (2000). Compensation for the effects of mutual coupling on direct data domain adaptive algorithm. *IEEE Trans. Antennas Propagat.* 48 (1): 86–94.

Agrawal, V.D. and Lo, Y.T. (1972). Mutual coupling in phased arrays of randomly spaced antennas. *IEEE Trans. Antennas & Propag.* 20 (3): 288–295.

Allen, J. L. and Diamond, B. L. (1966): 'Mutual coupling in array antennas', Technical Report EDS-66-443, Lincoln Lab., MIT, Oct. 4, 1966.

Anderson, J.B. and Ramussen, H.H. (1976). Decoupling and descattering networks for antennas. *IEEE Trans. Antennas & Propag.* 24 (6): 841–846.

Angeletti, P., Toso, G. and Ruggerini, G. (2014): 'Sparse array antennas with optimized elements positions and dimensions', Proc. 8th European Conference on Antennas and Propagation (EuCAP 2014), pp. 3142–3145.

Balanis, C.A. (1997). *Antenna Theory and Design*, 2e. New York: Wiley.

Bird, T.S. (2016). *Fundamentals of Aperture Antennas and Arrays*. Chichester, UK: Wiley.

Chen, K., Yun, X., He, Z., and Han, C. (2007). Synthesis of sparse planar arrays using modified real genetic algorithm. *IEEE Trans. Antennas & Propag.* 55 (4): 1067–1073.

Dandekar, K.R., Ling, H., and Xu, G. (2000). Effect of mutual coupling on direction finding in smart antenna applications. *Electron. Lett.* 36 (22): 1889–1891.

Dandekar, K.R., Ling, H., and Xu, G. (2002). Experimental study of mutual coupling compensation in smart antenna applications. *IEEE Trans. on Wireless Commun.* 1 (3): 480–487.

Daniel, J. (1974). Mutual coupling between antennas for emission or reception – application to passive and active dipole. *IEEE Trans. Antennas Propagat.* 22: 347–349.

Deng, J.-Y., Li, J.-Y., and Gu, L.-X. (2018). Decoupling of a three-port MIMO antenna with different impedances using reactively loaded dummy elements. *IEEE Antennas Wireless Propagat. Lett* 17 (3): 430–433.

Edelberg, S. and Oliner, A.A. (1960). Mutual coupling effects in large arrays II: compensation effects. *IRE Trans. Antennas Propagat.* AP-8: 360–367.

FEKO, Altair Engineering Inc., (2021). www.altair.com/feko/, Accessed May 2021

Friedlander, B. and Weiss, A.J. (1991). Direction finding in the presence of mutual coupling. *IEEE Trans. Antennas Propagat.* 39 (3): 273–284.

Gupta, I.J. and Ksienski, A.A. (1983). Effect of mutual coupling on the performance of adaptive arrays. *IEEE Trans. Antennas Propagat.* 31 (5): 785–791.

Hansen, R.C. (1998). *Phased Array Antennas*. New York, USA: Wiley.

Harrington, R.F. (1993). *Field Computation by Moment Methods*. New York: IEEE Press.

Haupt, R.L. (1994). Thinned arrays using genetic algorithms. *IEEE Trans. on Antennas & Propag.* 42 (7): 993–999.

Hockham, G.A. (1974). TEM waveguide coupling in the presence of short-circuited slot. *Electron. Lett.* 10: 62–63.

Hui, H.T. (2002). Reducing the mutual coupling effect in adaptive nulling using a re-defined mutual impedance. *IEEE Microw. Wirel. Compon. Lett.* 12 (5): 178–180.

Hui, H.T. (2003). Improved compensation for the mutual coupling effect in a dipole array for direction finding. *IEEE Trans. on Antennas & Propag.* 51 (9): 2498–2503.

Hui, H.T. (2004a). A new definition of mutual impedance for application in dipole receiving antenna arrays. *IEEE Antennas & Wireless Propag. Lett.* 3: 364–367.

Hui, H.T. (2004b). A practical approach to compensate for the mutual coupling effect in an adaptive dipole array. *IEEE Trans. on Antennas & Propag.* 52 (5): 1262–1269.

Hui, H.T. (2005). An effective compensation method for the mutual coupling effect in phased arrays for magnetic resonance imaging. *IEEE Trans. Antennas Propag.* 53 (10): 3576–3583.

Hui, H.T. (2016). Concept and applications of receiving mutual impedance. In: *Handbook of Antenna Technologies* (eds. Z. Chen, D. Liu, H. Nakano, et al.), 235–283. Singapore: Springer.

Hui, H.T., Li, B.K., and Crozier, S. (2006). A new decoupling method for quadrature coils in magnetic resonance imaging. *IEEE Trans. Biomed. Engineering.* 53 (10): 2114–2116.

Ishimaru, A.A. and Chen, Y.-S. (1965). Thinning and broadbanding antenna arrays by unequal spacings. *IEEE Trans. On Antennas & Propagat.* 13 (1): 34–42.

Lau, C.K.E., Adve, R.S., and Sarkar, T.K. (2004). Minimum norm mutual coupling compensation with applications in direction of arrival estimation. *IEEE Trans. on Antennas & Propag.* 52 (8): 2034–2040.

Lu, S., Hui, H.T., and Bialkowski, M.E. (2008). Optimizing MIMO channel capacities under the influence of antenna mutual coupling. *IEEE Antennas Wireless Propag. Lett.* 7: 287–290.

Lu, S., Hui, H.T., and Bialkowski, M.E. (2009). Performance analysis of MIMO-OFDM systems under the influence of antenna mutual coupling effect. *IET Microw., Antennas, Propag.* 3 (2): 288–295.

Lui, H.S. and Hui, H.T. (2010a). Improved mutual coupling compensation of compact array. *IET Microw. Antennas and Propagation* 4 (10): 1506–1516.

Lui, H.S. and Hui, H.T. (2010b): 'Mutual coupling compensation of compact antenna array for direction-of-arrivals estimations', 4th International Conference on Signal Processing and Communication Systems, pp. 1–10, Gold Coast, Australia.

Lui, H.S., Hui, H.T., and Leong, M.S. (2009). A note on the mutual coupling problems in transmitting and receiving antenna array. *IEEE Antennas and Propag. Magaz.* 51 (5): 171–176.

Lui, H.S., Hui, H.T. and Leong, M.S. (2010): "Experimental study of mutual coupling compensation in direction finding using a compact antenna array", Asia Pacific Symposium of Electromagnetic Compatibility, pp. 830–833, Beijing, China.

Mailloux, R.J. (1971). Reduction of mutual coupling using perfectly conducting fences. *IEEE Trans. Antennas Propagat.* AP-19: 166–173.

Pierre, J. and Kaveh, M. (1991): "Experimental performance of calibration and direction-finding algorithms," IEEE International Conference on Acoustics, Speech, and Signal Processing, Toronto, Ont., Canada, vol. 2, pp. 1365–1368.

Schmidt, R.O. (1986). Multiple emitter location and signal parameter estimation. *IEEE Trans. Antennas Propag.* 34 (3): 276–280.

Steyskal, H. and Herd, J.S. (1990). Mutual coupling compensation in small array antennas. *IEEE Trans. Antennas & Propag.* 38 (12): 1971–1975.

Su, T. and Ling, H. (2001). On modeling mutual coupling in antenna arrays using the coupling matrix. *Microw. Opt. Technol. Lett.* 28 (4): 231–237.

Su, T., Dandekar, K., and Ling, H. (2000). Simulation of mutual coupling effect in circular arrays for direction-finding applications. *Microw. Opt. Technol. Lett.* 26 (5): 331–336.

Unz, H. (1962). Nonuniform arrays with spacings larger than one wavelength. *IRE Trans. on Antennas & Propag.* AP-10: 647–648.

Yeh, C.C., Leou, M.L., and Ucci, D.R. (1989). Bearing estimations with mutual coupling present. *IEEE Trans. on Antennas & Propag.* 37 (10): 1332–1335.

Yu, Y., Lui, H.S., Niow, C.H. et al. (2011). Improved DOA estimation using measured receiving mutual impedances formutual coupling compensation - an experimental study. *IEEE Trans. Wireless Communications* 10 (7): 2228–2233.

Zhang, T.T., Lu, Y.L., and Hui, H.T. (2008). Compensation for the mutual coupling effect in uniform circular arrays for 2D DOA estimations using a new searching algorithm. *IEEE Trans. Aero. and Elect. Sys.* 44 (3): 1215–1221.

12

Noise Performance in the Presence of Mutual Coupling

Christophe Craeye[1], Jean Cavillot[1] and Eloy de Lera Acedo[2]

[1] *Louvain School of Engineering, Institute of Information and Communication Technologies, Electronics and Applied Mathematics (ICTEAM), Université Catholique de Louvain, Ottignies-Louvain-la-Neuve, Belgium*
[2] *Cavendish Astrophysics, Cavendish Laboratory, Physics Department, University of Cambridge, Cambridge, UK*

12.1 Generalities About Noise in Receiving Arrays

Besides noise stemming from the environment, which may sometimes be regarded as the signal of interest, noise can be generated by the front-end amplifiers or by lossy media around or in the antennas, radiating as black bodies. That radiation follows Planck's law (Planck 1914), which will be exploited here in its low-frequency asymptotic version (valid below about 300 GHz). In this chapter, special attention will be given to noise that leaves one element (or a given low-noise amplifier [LNA] that terminates the element) and couples back toward the receiver via the other elements. A general formulation will also be given for noise from the environment impinging on coupled arrays. Before delving into the details of the effects of mutual coupling between antennas, it is worth reminding how noise can be taken into account for isolated elements and for arrays, assuming first no mutual coupling between elements, which we may labeled as a quite "crude" approximation.

Firstly, in highly sensitive array systems, noise essentially originates from the leading amplifiers. This impacts the sensitivity in similar ways for both analog and digital scanning. It is worth reminding how the signal-to-noise ratio (SNR) improves as the array size increases in the less intuitive case of digital beamforming. The point is that signal voltages from a strongly localized source are correlated at the level of loads terminating the antennas. By contrast, noise "signals" originating from the different amplifiers can be assumed to be uncorrelated. Let us consider a source located in the intermediate-to-far-field. To first order, beamforming essentially compensates for phase differences, which means that the total signal magnitude, after phase correction and addition, will be proportional to the number N of antennas. Therefore, when power is computed, the result is proportional to N^2, which at first glance seems to defeat power conservation; indeed, that result amounts to power increasing faster than the surface covered by the array. Of course, in the case of digital beamforming, this power is an arbitrary number affected by multiplicative constants which sometimes become difficult to trace down. The physical meaning of such a number is restored when computing the SNR. Indeed, noise voltages being uncorrelated, after beamforming, the noise power is simply proportional to N (all cross-terms involving different amplifiers being zero) such that the SNR increases as N^2/N, i.e. linearly with the number of antennas, or with the array surface in the case of regular or very dense arrays. For large arrays, mutual coupling will not fundamentally change this scaling rule, except that the SNR will become dependent on scan angle, as will be explained in Section 12.2.

Mutual Coupling Between Antennas, First Edition. Edited by Trevor S. Bird.

Secondly, noise from the lossy antennas themselves can be computed based on thermodynamic reciprocity. Considering first a single antenna, one may start from the noise model of a resistor whose value R is equal to the real part of the load Z_L terminating the receiving antenna. By virtue of the Johnson–Nyquist noise theory, the corresponding noise generator has a series voltage with a variance equal to $4KTBR$, where $K = 1.380649 \times 10^{-23} J \cdot K^{-1}$ is Boltzmann's constant T is temperature in Kelvin, and B is bandwidth in Hz. From there, the noise power radiated in both near and far-fields can be obtained; this involves a factor that stands for matching of the antenna impedance with respect to the load Z_L. A fraction $1-\varepsilon$ of that power is absorbed by a nearby lossy (and hence noisy) scatterer. Thermodynamic equilibrium imposes that the fraction of noise stemming from the resistor R and absorbed by the scatterer is equal to the one transferred from the scatterer to the resistor. This result can be easily extended to lossy antennas. In this case, the "scatterer" is the antenna itself, and ε simply corresponds to the antenna efficiency (Collin and Zucker 1969). This result can also be extended to arrays. Given that the noise originates from a region located in the near-field of the array, a certain correlation occurs between corresponding noise voltages on the antenna loads. That correlation also contributes to a dependence of SNR on scan angle. Sections 12.2 and 12.3, respectively, explain how noise correlation, coming either from LNAs or from lossy materials, may be obtained numerically. An extension is also presented regarding noise from the far-field environment (Section 12.4). Appendix G is a reminder of a standard result regarding reciprocity, and Appendix E reviews the four-parameter noise model of an amplifier.

The examples provided in the following sections concern the SKALA antenna (de Lera Acedo et al. 2015), devoted to astronomic observation. The shape of the antennas is provided in Figure 12.1, along with – in shaded scale where dark is strongest – current distributions corresponding to different Macro Basis Functions used to solve for current distributions (Bui Van et al. 2018). Regarding the analysis of mutual coupling, the main quantity of interest corresponds to the currents on all antennas when one of them is excited, while the other elements are passively terminated. This yields the array impedance matrix, as well as the embedded element patterns. As a result of mutual coupling, these embedded patterns can vary significantly, as is illustrated in Figures 12.2 and 12.2b, obtained for the two elements that are indicated in, Figure 12.2c which shows an array comprising 256 elements operating at 110 MHz.

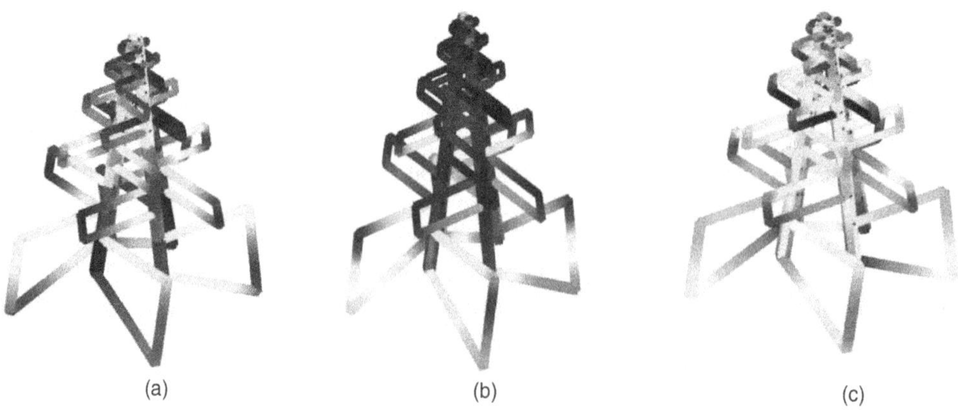

(a) (b) (c)

Figure 12.1 Three current distributions resulting from different numbers of Macro Basis Functions on SKALA antenna at 110 MHz. (a) 4, (b) 12, and (c) 19 MBFs.

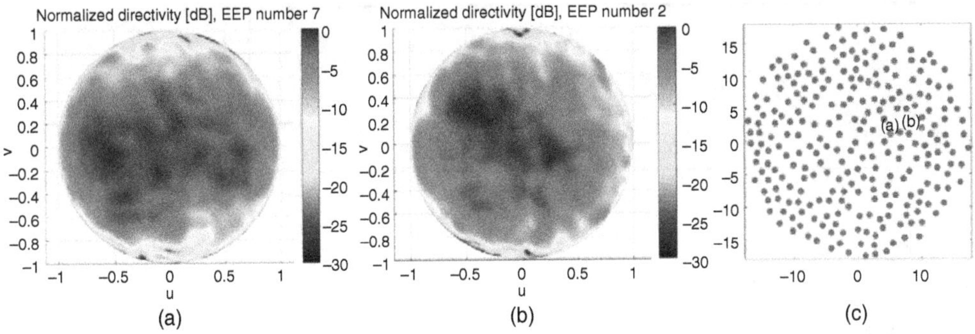

Figure 12.2 (a) and (b) Two embedded element patterns appearing in a 256-element array (c) of SKALA elements at 110 MHz.

12.2 Coupling of Noise Originating from LNAs

Many representations exist for noise generated by LNAs, one of the most popular models being based on voltage and current noise parameters, subject of IRE standards on noise measurements and summarized in Haus et al. (1960). Power-wave representations and many others have also been proposed. Engberg and Larsen (1995) established the equivalence between all those representations. In the present explanation, an impedance matrix representation is used for the noiseless equivalent amplifier, while equivalent noise sources are represented by mutually correlated series voltage and parallel-current sources. Appendix E provides a reminder regarding these parameters. The noise voltage, current, and related correlation are described by parameters R_n, G_n, and Y_γ^*, respectively. In the paragraphs below, a second index may appear to identify a given amplifier in the array.

The noise sources associated with the different amplifiers are not correlated with each other. Therefore, one will simply sum the corresponding noise powers, as obtained at the output of the beamformer. When doing so, one should bear in mind that the noise from a given amplifier travels through virtually all elements, and the impact of this multipath propagation on total noise after beamforming needs to be estimated. To ease the derivation, the signal and noise currents will be taken as corresponding to those appearing across the input impedances of the different LNAs. The underlying assumption is that the components that come after the LNA either do not add significant noise, or that their noise figures are the same for all elements and hence uniformly affect the final SNR. Given the input impedances of the amplifiers $Z_{L,i}$ on one side and the array impedance matrix $\boldsymbol{Z}$ on the other side, it is quite easy to determine the noise voltages across each load, as resulting from the (correlated) current and noise voltages due to one given amplifier. Figure 12.3 shows a general circuit representation of the problem, including noise sources and loads on which the (correlated) signal and noise voltages will be evaluated. If $\boldsymbol{Z}_L$ is the diagonal matrix comprising the loads (denoted as $Z_{L,i}$, for element i), and $\boldsymbol{Y}_a = (\boldsymbol{Z} + \boldsymbol{Z}_L)^{-1}$, then the noise voltage contribution of amplifier m at the output of the beamformer with weights $\boldsymbol{w}$ reads (Craeye et al. 2004)

$$c_m = r_m \, v_{n,m} + s_m \, i_{n,m} \tag{12.1}$$

with $\boldsymbol{r} = \boldsymbol{w}^T \boldsymbol{Z}_L \, \boldsymbol{Y}_a G_v$ and $\boldsymbol{s} = \boldsymbol{r} \, \boldsymbol{Z}$, with G_v the voltage gain common to all amplifiers, while $v_{n,m}$ and $i_{n,m}$ correspond to the voltage and current noise sources that characterize amplifier m. The column vector $\boldsymbol{w}$ contains the beamforming weights. The above expression is a minor extension of a result found in Craeye et al. (2004) to amplifiers with different input impedances. The total noise

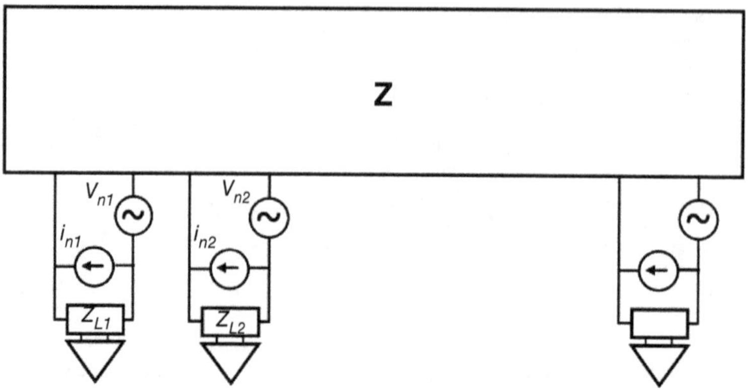

Figure 12.3 Terminated coupled array with equivalent noise sources of amplifiers.

power at the output of the beamformer then reads

$$P_n = 4\,KTB\,\Re\,\{Z_{L,o}^{-1}\}\sum_{m=1}^{N}(|r_m|^2\,R_{n,m} + |s_m|^2\,G_{n,m} + 2\,\Re\,\{r_m s_m^*\,Y_{\gamma,m}^*\}\,R_{n,m}) \tag{12.2}$$

The factor $\Re\,\{Z_{L,o}^{-1}\}$ that may be viewed as a dummy factor is also present when signals are treated. It may be interpreted as the admittance at the output of the beamformer and is kept only to preserve dimensional consistency.

An alternative representation makes use of noise power waves and of the active reflection coefficient; which may be viewed as the S-parameter approach, in contrast with the above Z-parameters method. In this perspective, the interpretation of the outcome is very simple: at the output of the beamformer, the obtained noise contribution is just equal to what it would be if the impedance seen by each antenna were just equal to the active impedance. This conditional formulation indicates the non-entirely-physical aspect of that representation at the level of antenna ports: it only makes sense from the perspective of the output of the beamformer. The active impedance Z_m^a is the ratio between voltage and current observed at a given antenna port m when all elements are active, with the excitation law corresponding to the desired beamforming law, described by the vector of weights $\boldsymbol{w}$. This explains why the total noise perceived at the output of the beamformer is dependent on scan angle. Let us denote by a_m and b_m the downward- and upward-going noise waves of amplifier m (Engberg and Larsen 1995). Then, the available noise power contributed by amplifier m is (Weem and Popovic 2001)

$$P_{n,m} = \left| w_m a_m + G_v \sum_{i=1}^{N} w_m S_{i,m}\, b_m \right|^2 = |\,w_m|^2\,|\,a_m + G_v\,\Gamma_m^a\,b_m|^2 \tag{12.3}$$

where Γ_m^a is the active reflection coefficient (Galindo and Wu 1968), equal to $(Z_m^a - Z_{L,m}^*)/(Z_m^a + Z_{L,m})$. Also, using a power-wave approach, Ivashina et al. (2008) provide an equivalent single-channel receiver model (single antenna and two-port amplifier) for the SNR of a finite array. This leads to a compact and convenient way to obtain the sensitivity of an arbitrary coupled array.

For the special case of regular arrays, it is interesting to link the noise power obtained for finite and infinite arrays. Indeed, supposing that the beamforming law just corresponds to a linear phase shift, one can see that the noise power per element in an infinite array is given by

$$P_n = 4\,KTB\,\Re\,\{Z_L^{-1}\}(|r|^2\,R_n + |s|^2\,G_n + 2\,\Re\,\{r\,s^*\,Y_\gamma^*\}\,R_n) \tag{12.4}$$

where $r = Z_L/(Z_L + Z^a)\, G_v$ and $s = Z^a\, r$ and Z^a is the (angle-dependent) active impedance of the antenna in the array. With the power-wave formulation, one has

$$P_n = |a + b\ G_v \Gamma^a|^2 \tag{12.5}$$

with Γ^a a scan-dependent quantity, which can be either larger or smaller than in the case where noise coupling is not included. Coming back to the formulation based on voltage and current sources, the noise power may also be written as $\Re\,\{Z_L^{-1}\}\,(\boldsymbol{w}^T\,\boldsymbol{Z}_L\,\boldsymbol{R}_r^i\,\boldsymbol{Z}_L^*\boldsymbol{w}^*)\,|G_v|^2$, where $\boldsymbol{R}_r^i$ is the receiver noise current correlation matrix, and $\boldsymbol{w}$ is a vector containing the beamforming weights. Based on the above derivations, one may write the correlation matrix between currents as

$$R_r^i = 4\,KTB \sum_{m=1}^{N} (G_m^{vv}\, R_{n,m} + G_m^{ii}\, G_{n,m} + 2\,\Re\,\{G_m^{vi}\, Y_\gamma^*\}\, R_{n,m}) \tag{12.6}$$

where the $\boldsymbol{G}$ matrices are built in the following way: $\boldsymbol{G}_m^{vv} = Y_a^{cm}\, Y_a^{rm,*}$, where the superscript "cm" means "column m of the matrix" and "rm" means "transposed of column m." $\boldsymbol{G}_m^{ii} = (Y_a Z)^{cm}\,(Y_a Z)^{rm,*}$ and, $\boldsymbol{G}_m^{vi} = Y_a^{cm}\,(Y_a Z)^{rm,*}$. That result can be proven as follows. Column m of matrix $\boldsymbol{Y}_a$, when multiplied by the noise voltage $v_{n,\,m}$, produces the corresponding port currents for each amplifier. Likewise, column m of matrix $\boldsymbol{P} = \boldsymbol{Y}_a\boldsymbol{Z}$, when multiplied by the noise current $i_{n,\,m}$ produces the corresponding port currents for each amplifier. The matrix that provides correlations between all pairs of noise currents is then given by the following expectation:

$$R_{n,m} = \langle (Y^{cm} v_{n,m} + P^{cm} i_{n,m})(Y^{rm,*}\, v_{n,m}^* + P^{rm,*}\, i_{n,m}^*) \rangle$$

In the above expression, for each of the four terms, the correlations between noise currents and/or voltages appear (see Appendix E), leading to Eq. (12.6).

A formulation based on power waves has been previously developed by Warnick and Jensen (2007). Beyond the S-parameter formulation for noise in the multiport case, the authors also derive an important result regarding the introduction, if possible, of a matching network between the array and the receivers: for optimal noise performance, the matching network must decouple the array and present isolated, individually noise-matched ports to the amplifier inputs. As mentioned above, another representation of receiving arrays goes through the definition of an equivalent single-port antenna and amplifier, as explained by Ivashina et al. (2008). The definitions proposed by Warnick et al. (2010) entirely encompass the models proposed by Warnick and Jeffs (2008) and Ivashina et al. (2008), which – in passing – also proves their full consistency. An interesting alternative representation, applied to the analysis and design of focal-plane arrays, is provided by Roshi et al. (2019), who make use of Lange parameters (Lange 1967), deriving directly from the noise parameters proposed as a standard by Haus et al. (1960). The Lange parameters are also used by Hay (2010) for the LNA noise model. In Roshi et al. (2019), the SNR is described with a single correlation matrix (as opposed to different matrices in numerator and denominator). The use of a single matrix is made possible by normalizing the weights by the square root of the noise correlation matrix. This representation facilitates the computation of optimal weights.

For the array introduced in Section 12.1, the equivalent noise temperature is computed versus scan angle at 110 GHz. The amplifiers are assumed identical for all elements, with the following example parameters: $R_{n,\,m}$=50 Ohm, $G_{n,\,m}$ = 0.17 S, and Y_γ^* = 5.6 + 9i mS. The result is shown in Figure 12.4 in direction-cosine coordinates (i.e. half unit sphere seen from top). The physical temperature is assumed to be 290 K, and the value is given per channel (256 elements here). In this case, the noise temperature is higher by about 50% near grazing incidences.

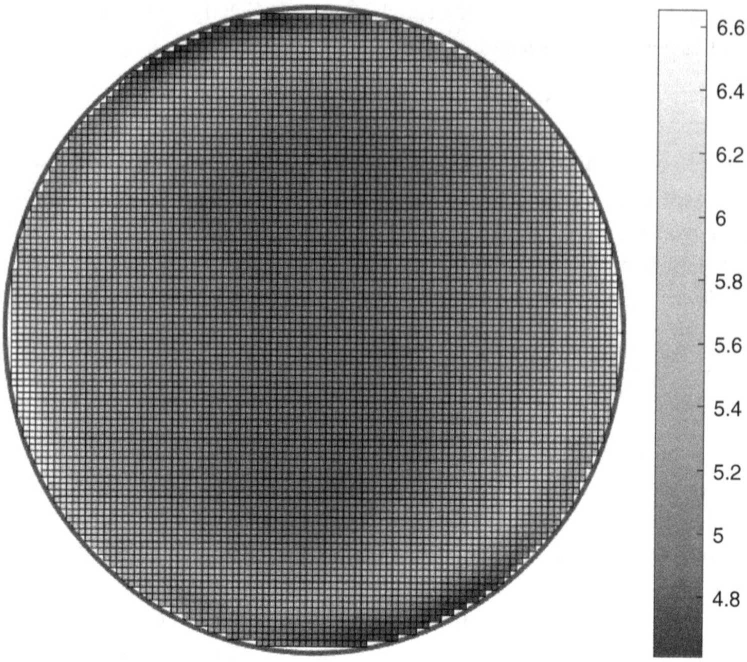

Figure 12.4 Noise temperature (Kelvin) per channel for the SKA1-LOW array at 110 MHz. Noise parameters of the amplifiers are $R_{n,m}$=50 Ohm, $G_{n,m}$ = 0.17 S, and Y_{γ}^{*} = 5.6 + 9i mS.

12.3 Coupling of Noise Originating from Lossy Antenna Arrays

Consider a random noise field within a given frequency band emitted from a point-like source located in the very far-field of an array. In that case, a priori, nothing distinguishes the "noise" from an actual signal. In those circumstances, at any frequency, the signal obtained at a given antenna can be causally related to the one received at another antenna, considering their respective radiation patterns and relative positions.[1] Hence, those two "noise signals," received by two antennas, are perfectly correlated. This is no longer the case when the noise source has a certain spatial extent and/or when it is not located in the far-field (Wolf 1986). In that case, the signals received at different antennas are partially correlated. Roughly speaking, this has a double origin: (i) the signals emitted by different parts of the thermal noise source are not correlated, and (ii) those sources exhibit different transmittances toward any given antenna. That is also true when the source of noise is the antenna itself because it comprises lossy materials, i.e. when the antenna array has a limited radiation efficiency. The goal of this section, hence, consists of studying the correlation matrix $\boldsymbol{R}_l$ of noise signals received at different antenna loads, as a result of the array itself not having 100% efficiency.

Two different perspectives can be taken to model noise from coupled antennas. A quite general one results from the Twiss theorem (1955) and essentially views the array as a lossy N-port circuit. From this perspective, the loss includes Joule effects on the antennas themselves as well as radiation loss (Warnick and Jeffs 2008). The noise correlation at those ports can then be delineated from the scattering parameters (S-matrix) or, equivalently, from the impedance matrix of the N-port. According to Twiss (1955), the correlation matrix between open-circuit noise voltages in an N-port

1 In case a common reference point for the phase is taken for all patterns, then relative positions do not need to be considered.

system is simply related to the real part of the array impedance matrix:

$$\boldsymbol{R}_{oc} = 8\, KTB\, \Re\, \{\boldsymbol{Z}\} \tag{12.7}$$

From there, the correlation matrix of noise currents on the loads of the terminated array reads

$$\mathbf{R}_t = 8\, \mathrm{KTB}\, \mathrm{Y_A} \Re\, \{\mathrm{Z}\}\, \mathrm{Y_A^H}. \tag{12.8}$$

To isolate the contribution from losses on the antennas themselves, it becomes necessary to remove the contribution from the environment (Warnick and Jeffs 2008). This is done by considering an isotropic environment at the physical temperature of the array, which contributes a correlation matrix $\boldsymbol{R}_i$. Its expression is provided in Section 12.4.

Another approach is similar to the formalism used in the field of thermographic interferometry (Mamouni et al. 1991). It extends the link between efficiency and noise contribution, recalled in Section 12.1, to antennas with multiple ports. In this case, only noise from the antennas themselves is considered. First, the noise "signals" generated by the different micro-volumes of the object are added up. Given the strongly local aspect of thermal emission, from a mathematical perspective, "micro-volumes" can be replaced with "infinitesimal volumes," which allows direct integration over the lossy body responsible for thermal radiation. As is the case for the isolated antenna, thermodynamic reciprocity is exploited to calculate the noise power dynamically exchanged between the lossy object and the loads terminating the antennas. The correlation between noise voltages at the different ports is obtained from the following two principles: (i) noise voltages contributed at two different ports by a given infinitesimal volume are fully correlated, and (ii) their ratio can be computed based on electromagnetic reciprocity. The second point is expressed as the ratio between electrical currents induced in the infinitesimal volume when the ports of interests are excited with unit sources. The Thevenin representation of such sources has a unit voltage and a series impedance equal to the antenna termination impedance on receive. From there, the correlation of thermal noise currents observed at two ports can be written as a scalar product over the volume of the lossy object (e.g. the whole receiving array) of induced currents when the two ports are active. For the case of antennas made of good conductors (Craeye 2005), integration over the surface is sufficient and the correlation between currents reads

$$R_l^i(i,j) = 8\, KTB \iint R_s \boldsymbol{J}_i . \boldsymbol{J}_j^*\, dS \tag{12.9}$$

where R_s is the surface resistance and $\boldsymbol{J}_i$ is the surface current on the antenna when port i is excited with a Thevenin source, while other elements are passively terminated. Those currents are normalized with respect to the source voltages (their units are thus Ohm^{-1} m^{-1}). Similarly to what is done with amplifier noise, the noise power at the output of the beamformer is computed as $\Re\,\{Z_L^{-1}\}/2\,(\boldsymbol{w}^T \boldsymbol{Z}_L\, \boldsymbol{R}_l^i\, \boldsymbol{Z}_L^* \boldsymbol{w}^*)$, where the presence of the beamforming weights leads to a dependence on direction of observation.

12.4 Coupling of Noise Originating from the Far-Field Environment

The noise current correlation matrix has three additive components, $\boldsymbol{R}_n = \boldsymbol{R}_e + \boldsymbol{R}_l + \boldsymbol{R}_r$, as follows: originating from the exterior environment (see below), losses on the antennas (see Section 12.3), and active components of the receiver (see Section 12.2). The external noise contribution can be computed as $\boldsymbol{R}_e = 8\, KT_0 B\, \boldsymbol{B}^s/\eta$, where $\boldsymbol{B}^s$ contains the sky-weighted[2] pattern coupling factors:

$$\mathrm{B^s(i,j)} = \iint \mathbf{F}_i(\Omega).\mathbf{F}_j^*(\Omega)\, \mathrm{T}(\Omega)/\mathrm{T_0}\, d\Omega \tag{12.10}$$

2 Other environments than the sky can be included.

In the new definitions proposed by Warnick et al. (2010) the external isotropic noise correlation matrix, $\boldsymbol{R}_i$, plays a key role. It is defined in the same way as $\boldsymbol{R}_e$, except that the noise temperature of the environment is assumed isotropic and corresponds to the physical temperature of the array.[3] In Eq. (12.10), $\boldsymbol{F}_i(\Omega)$ corresponds to the embedded element pattern (with Thevenin source) of element *i*. A first use that can be made of this quantity is obtaining $\boldsymbol{R}_l$ as the difference between $\boldsymbol{R}_t$ (see Section 12.3), obtained from the Twiss theorem (1955), and $\boldsymbol{R}_i$ (Warnick and Jeffs 2008), as already outlined in Section 12.3. It is important to notice that the correlation matrices provided above correspond to correlations between port currents. The correlations between port voltages will be denoted by a superscript v, i.e. $\boldsymbol{R}_n^v = \boldsymbol{Z}_L \boldsymbol{R}_n^i \boldsymbol{Z}_L^*$ (no superscript will assume correlation between currents).

The above-mentioned subtraction technique to obtain $\boldsymbol{R}_l$ may be ill-posed for high-efficiency antennas because it may entail the difference between two almost equal quantities, each affected by significant uncertainties. Therefore, if current distributions over the whole array can be determined for excitation at individual elements, it may be more accurate to determine $\boldsymbol{R}_l$ directly from scalar products between such currents, as explained in Section 12.3. Craeye (2012) provides an explicit link between this approach and that of Warnick and Jeffs (2008) for the case of antennas made of good conductors. Beyond this, the isotropic noise correlation $\boldsymbol{R}_i$ has appeared as a cornerstone of a new set of definitions that extend standard efficiency and receiver temperature definitions to general receiving arrays, as detailed by Warnick et al. (2010). For instance, radiation efficiency is given by $\boldsymbol{w}^T \boldsymbol{R}_i^{v,o} \boldsymbol{w}^* / \boldsymbol{w}^T \boldsymbol{R}_i^v, \boldsymbol{w}^*$, where the "o" superscript indicates noiseless antennas and receiver electronics. The (beam-equivalent) receiver noise temperature is given by $T_r = T_o \boldsymbol{w}^T \boldsymbol{R}_r^v \boldsymbol{w}^* / \boldsymbol{w}^T \boldsymbol{R}_t^v \boldsymbol{w}^*$, with $\boldsymbol{R}_t^v$ obtained in isotropic environment and beamformer weights referred to forward-wave amplitudes.

The external noise contribution results from the far-field environment or from the sky, in the form of diffuse radio-frequency interference, low-frequency galactic noise, cosmic microwave background, or radiation from specific celestial bodies (sun, other galaxies, etc.). Globally, noise will be associated with the thermal sources that are considered outside the scope of ongoing observations. As will be shown below, when stemming from the far-field region, such noise contribution can be quite simply connected with the directivity of the antenna array.

As mentioned in Section 12.1, it is difficult to determine the effective area of a phased array, given the active components it comprises, and the weighting factors applied in analog or digital form. It is hence safer to talk in terms of SNR, both being affected by the beamforming operation. With a given vector of weights $\boldsymbol{w}$, the SNR can be written as the ratio between two quadratic forms involving signal and noise voltage correlation matrices (Warnick and Jeffs 2008). Then, we have

$$\mathrm{SNR} = \frac{\mathrm{w^T\, R_S^V\, w^*}}{\mathrm{w^T\, R_n^V\, w^*}} \tag{12.11}$$

Denoting by $E_0 \hat{e}_0$ the complex incoming electric field at the origin of coordinates ($\hat{e}_0$ being the unit polarization vector), based on reciprocity (see Appendix G), the currents[4] signal correlation matrix is given by

$$\mathbf{R_s} = |\mathrm{E_0}|^2\, \mathbf{S_0}\, (4\pi/\mathrm{k}\eta)^2 \tag{12.12}$$

---

3 In our notation, given the use of embedded patterns $\boldsymbol{F}_i(\Omega)$, $\boldsymbol{R}_i$ refers to the correlation matrix between port currents of the terminated array, in an isotropic environment.

4 Here, identical loads (input impedances of amplifiers) are assumed. With voltage-gain amplification, different loads can be introduced by multiplying voltage V_i by $Z_{L,i}/Z_{L,o}$ with $Z_{L,o}$, a reference impedance.

in which k is the wavenumber, η is the free-space impedance, and matrix $\boldsymbol{S}_o$ is expressed with the help of the embedded element patterns $\boldsymbol{F}_n(\Omega)$:

$$S_o(i,j)(\Omega) = \mathbf{F}_i(\Omega_o).\hat{e}_o\, \mathbf{F}_j^*(\Omega_o).\hat{e}_o^* \tag{12.13}$$

It is also interesting to link the $A_{\text{eff}}/T_{\text{sys}}$ ratio (effective area to system temperature) with the array pattern directivity. Let us first consider an isolated antenna with embedded pattern $\boldsymbol{F}(\Omega)$ and load impedance Z_L. If the incident field is a plane wave with amplitude E_o and polarization $\hat{e}_o$, the signal power absorbed by the load is (see Appendix G)

$$S = \left| E_o \frac{4\pi}{jk\eta_o} \mathbf{F}(\Omega_o).\hat{e}_o \right|^2 \frac{\Re\{Z_L\}}{2} = \frac{A_{\text{eff}}\,|E_o|^2}{2\eta} \tag{12.14}$$

where $\boldsymbol{F}(\Omega_o)$ is the embedded element pattern with a Thevenin source, in the direction of the incoming signal. The noise power is obtained by expressing the dynamic equilibrium of power between the (noisy) load (r.m.s. noise voltage of 4 $KTB\,\Re\{Z_L\}$) and from the environment at temperature T. That equilibrium (Appendix F) is expressed as

$$N = \iint \frac{|\mathbf{F}(\Omega)|^2}{\eta_o} 4\,\text{KTB}\,\Re\{Z_L\}\, d\Omega = T_{\text{sys}}\text{KB} \tag{12.15}$$

From Eqs. (12.14) and (12.15), after a few simple algebraic transformations, the following ratio is readily obtained

$$\frac{A_{\text{eff}}}{T_{\text{sys}}} = \frac{\lambda^2}{4\pi} \frac{D^s}{T_o} \varepsilon_p \tag{12.16}$$

where ε_p is the polarization matching factor and

$$D^s = \frac{4\pi\ \|\mathbf{F}(\Omega_o)\|^2}{\iint \|\mathbf{F}(\Omega)\|^2\, T(\Omega)/T_o\, d\Omega} \tag{12.17}$$

is a weighted directivity factor, equal to the traditional directivity when the sky temperature is homogeneous and equal to T_o.

As proven in Appendix F, the noise current received at different loads can be obtained from two orthogonal equivalent currents $I_s\,dl$ distributed over a sphere of radius R in the far-field, with magnitudes defined by $\eta_o\ |I_s|^2\ dl^2 = 8\ KTB\ R^2 d\Omega$. As also shown in Appendix F, that result leads to the following noise current correlation matrix $\boldsymbol{R}_e = 8\ KT_oB\ \boldsymbol{B}^s/\eta_o$. The signal correlation matrix is given by Eqs. (12.12) and (12.13). The $A_{\text{eff}}/T_{\text{sys}}$ factor can be written quite explicitly based on port signal and noise currents ($I_{s,i}$ and $I_{n,i}$, respectively), as the two factors below.

$$\frac{A_{\text{eff}}}{T_{\text{sys}}} = \frac{\left(\sum_i w_i I_{s,i}\right)\left(\sum_j w_j^* I_{s,j}\right)^*}{|E_o|^2/(2\eta_o)} \left(\frac{1}{KB}\sum_{i,j} w_i \langle I_{n,i} I_{n,j}^* \rangle w_j^*\right)^{-1} \tag{12.18}$$

The signal currents are then rewritten using the basic reciprocity result recalled in Appendix G and definition (12.14), while the correlation between noise currents is obtained from Appendix F and definition (12.10). Then, establishing a link between $A_{\text{eff}}/T_{\text{sys}}$ and SNR from Eqs. (12.14) and (12.15), we arrive at

$$\frac{A_{\text{eff}}}{T_{\text{sys}}} = \frac{\lambda^2}{4\pi} \frac{4\pi\, w^T S_o w^*}{w^T B^s w^*} \frac{1}{T_o} \tag{12.19}$$

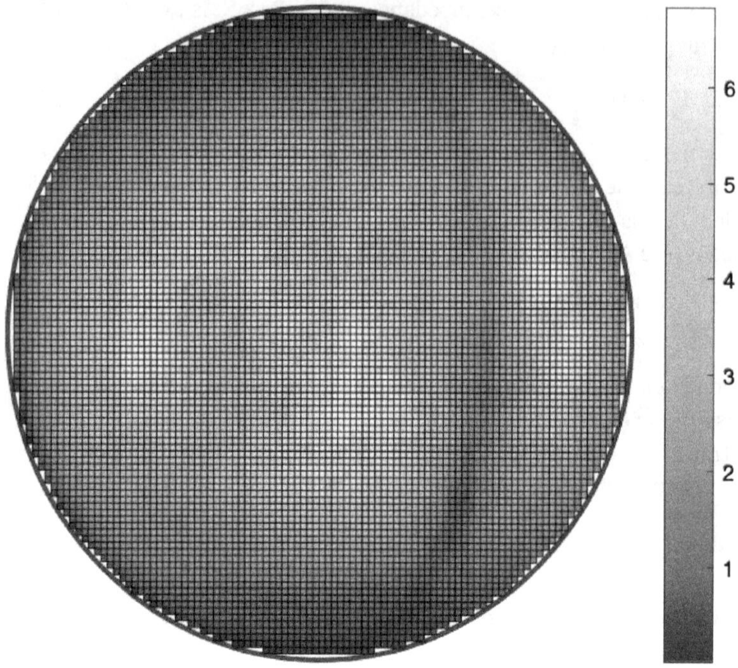

Figure 12.5 Estimate of $A_{eff}f/T_{sys}$, in m^2/K, at the MRO site in Western Australia, at midnight 01/01/2020, for the SKA1-LOW array at 110 MHz.

Let us denote by $f_{c,i} = \boldsymbol{F}_i.\hat{e}_o$ the component of the pattern that is co-polar to the incident field and by $f_{x,i} = \boldsymbol{F}_j.\hat{e}_x$ the cross-polar part. Then, the quadratic forms that appear in the numerator and in the denominator can be factorized as

$$\frac{A_{eff}}{T_{sys}} = \frac{\lambda^2}{4\pi}\frac{1}{T_o}\frac{4\pi\,|\,f_c^{o,T}w|^2}{\iint(|\,f_c^T w|^2 + |\,f_x^T w|^2)T/T_o\,d\Omega} \tag{12.20}$$

which smoothly extends Eq. (12.16), a basic result of antenna theory, to antenna arrays in an environment at a nonhomogeneous temperature. Indeed, for $T = T_0$, the fraction in the right-hand side corresponds to the co-polar directivity of the array pattern. An example has been computed for the array referred to in Section 12.1 at 110 MHz. The result is shown in Figure 12.5[5] in direction-cosine coordinates. The lower-value arc on the right side results from the higher noise temperature arising from the Milky Way. Sky temperature is obtained from Remazeilles et al. (2015), with isotropic spectral index[6] (SI) equal to 2.5.

12.5 Conclusion

High-sensitivity arrays have a noise behavior that strongly depends on the performance of the front-end amplifiers. In densely packed arrays, the traditional noise budget based on noise figure of the amplifier is no longer complete, since noise sent toward the antenna can couple back via neighboring elements. At the output of beamformer, this amounts to direction-dependent noise.

5 Orientation of the dipoles is NS and EW. Location: lat. = −26.704 08, long. = 116.670 23.
6 Sky brightness is proportional to f^{SI}, with the frequency f.

That conclusion also holds for thermal noise originating from lossy materials composing the antennas. At low frequencies, galactic noise can also strongly impact the noise budget; in that case, we established the link with the co-polar directivity of the array. In all cases, the accurate computation of the noise budget increasingly relies on the ability to efficiently and precisely evaluate quantities such as the array impedance matrix and embedded element patterns.

References

Bui Van, H., Abraham, J., Arts, M. et al. (2018). Fast and accurate simulation technique for large irregular arrays. *IEEE Trans. Antennas Propag.* 66: 1805–1817.

Collin, R.E. and Zucker, F.J. (1969). The receiving antenna', Chap. 4. In: *Antenna Theory Part 1*. New York: McGraw Hill.

Craeye, C. (2005). Including spatial correlation of thermal noise in the noise model of high-sensitivity arrays. *IEEE Trans. Antennas Propag.* 53: 3845–3848.

Craeye, C. (2012). On the equivalence of two formulations for noise correlation in arrays made of imperfect conductors. *IEEE Trans. Antennas Propag.* 60: 3531–3533.

Craeye, C., Parvais, B., and Dardenne, X. (2004). MOM simulation of signal-to-noise patterns in infinite and finite receiving antenna arrays. *IEEE Trans. Antennas Propag.* 52: 3245–3256.

Engberg, J. and Larsen, T. (1995). *Noise Theory of Linear and Nonlinear Circuits*. Chichester, UK: Wiley.

Galindo, V. and Wu, C.P. (1968). On the asymptotic decay of coupling for infinite phased arrays. *Proc. IEEE* 56: 1872–1880.

Haus, H.A., Atkinson, W.R., Branch, G.M. et al. (1960). Representation of noise in linear two ports. *Proc. IRE* 48: 69–74.

Hay, S.G. (2010). Maximum-sensitivity matching of connected-array antennas subject to Lange noise constants. *Int. J. Microw. Opt. Tech.* 5: 375–383.

Ivashina, M.V., Maaskant, R., and Woestenbrug, B. (2008). Equivalent system representation to model the beam sensitivity of receiving antenna arrays. *IEEE Antennas Wireless Propag. Lett.* 7: 733–737.

Lange, J. (1967). Noise characterization of linear two-ports in terms of invariant parameters. *IEEE J. Solid-State Phys.* 2: 37–40.

de Lera Acedo, E., Razavi-Ghods, Troop, N. et al. (2015). SKALA, a log-periodic array antenna for the SKA-low instrument: design, simulations, tests and system considerations. *Exp. Astron.* 39: 567–594.

Mamouni, A., Leroy, Y., Bocquet, B. et al. (1991, 1991). Computation of near-field microwave radiometric signals: definition and experimental verification. *IEEE Trans. Microw. Theory Technol.* 39: 124–132.

Planck, M. (1914). *The Theory of Heat Radiation*(trans. Masius, M.). USA: P. Balkiston's Son & Co.

Remazeilles, M., Dickinson, C., Banday, A.J. et al. (2015). An improved source-subtracted and destriped 408 MHz all-sky map. *Mon. Not. R. Astron. Soc.* 451: 4311.

Roshi, D.A., Shillue, W., and Fisher, J.R. (2019). Model for a noise-matched phased array feed. *IEEE Trans. Antennas Propag.* 67: 3011–3021.

Twiss, R.Q. (1955). Nyquist's and Thevenin's theorems generalized for nonreciprocal linear networks. *J. Appl. Phys.* 26: 599–602.

Warnick, K.F. and Jeffs, B.D. (2008). Efficiencies and system temperature for a beamforming array. *IEEE Antennas Propag. Lett.* 8: 565–568.

Warnick, K.F. and Jensen, M.A. (2007). Optimal noise matching for mutually coupled arrays. *IEEE Trans. Antennas Propag.* 55: 1726–1731.

Warnick, K.F., Ivashina, M.V., Maaskant, R., and Woestenburg, B. (2010). Unified definitions of efficiencies and system noise temperature for receiving antenna arrays. *IEEE Trans. Antennas Propag.* 58: 2121–2125.

Weem, J.P. and Popovic, Z. (2001). A method for determining noise coupling in a phased array antenna. In: *Proc. IEEE MTT-Int. Microwave Symp*, vol. 1, 271–274. Phoenix, USA.

Wolf, E. (1986). Coherent-mode propagation in spatially band-limited wave fields. *JOSA A* 3: 1920–1924.

13

Methods for Analyzing Mutual Coupling in Large Arrays

Christophe Craeye and Ha Bui Van

Louvain School of Engineering, Institute of Information and Communication Technologies, Electronics and Applied Mathematics (ICTM), Université Catholique de Louvain, Ottignies-Louvain-la-Neuve, Belgium

13.1 Goals of Numerical Mutual Coupling Analysis

Antenna arrays may be viewed in a very general sense, i.e. they may correspond to any structure radiating from several well-defined ports. Those ports correspond to small surfaces with two conductors, where – by virtue of a proper quasi-static approximation – voltages and currents can be defined (Pozar 2005, Section 3.1). Looking from the ports toward the outside, the array may in principle be made of any material configuration. Strictly speaking, the nearby environment of the array should be considered as part of the ("extended") array. This is, for instance, the case for a vehicle or mast that carries an antenna array as discussed in Chapter 8. In general, a pragmatic distinction will need to be made between the (extended) array and its environment. Workable models, for instance, based on ray tracing, will then need to be developed regarding the interaction between the array and its environment. A finite ground plane may typically be considered as part of the extended array, and it is beneficial to include at the design level any already-known features of the near-field environment, such as dimensions and materials of a casing.

In the following, we will assume that, looking toward the generator, the ports are uncoupled. The reader is referred to De Hoop (1975) for the extension to coupled ports, which assume a linear relationship between source and port voltages. In case they are coupled, an equivalent representation of the coupled generator in terms of a set of uncoupled ones can be found (Bui Van et al. 2019). The analysis will be carried out in spectral domain with a $\exp(j\omega t)$ factor suppressed, where ω is radian frequency and t is time. The main characteristics of the finite array correspond to the N-port impedance matrix (or, equivalently, its scattering matrix) and the embedded element patterns (EEPs), i.e. the patterns of each element determined in the presence of the other elements of the (extended) array. Two important comments need to be made here. First, mutual coupling is a phenomenon, which is not entirely described by the N-port impedance or scattering matrix (in this respect, the expression "coupling matrix" is often misleading). That representation is in general not sufficient to account for the effects of mutual coupling on the embedded patterns. Second, the embedded patterns can also be defined in the near-field of the array. Those patterns depend on the three spatial coordinates, i.e. the $\exp(-jkR)/R$ dependence on range (with k the wavenumber and R the radial distance) is simply no longer factorized.

Regarding embedded patterns, one needs to state the excitation conditions under which they are defined. A common representation, quite consistent with the IEEE definition of radiation pattern,

Mutual Coupling Between Antennas, First Edition. Edited by Trevor S. Bird.

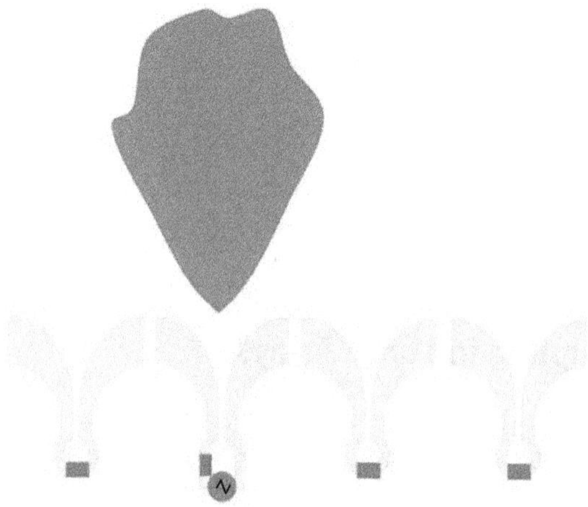

Figure 13.1 View of an embedded element pattern, defined with an equivalent Thevenin representation of the generator connected to the element of interest and with the other elements passively terminated.

corresponds to the open-circuit pattern as described in Warnick and Jeffs (2008) and in Craeye and Gonzalez-Ovejero (2011), i.e. the pattern of a given element when it is excited with a unit current, while the other elements are open-circuited. Its main advantage is that this definition is given independently from the transmitter or receiver. A second representation assumes that the generators (or receivers) are given in terms of uncoupled equivalent Thevenin circuits. In that case, a unit voltage is associated with the port of interest, while the other ones are passive (i.e. they are just terminated with a passive impedance), as sketched in Figure 13.1. While that representation does depend on the generators or receivers, it is quite convenient when relating transmitting and receiving properties of the array. Indeed, for a given incoming field, the current across the load is simply proportional to the scalar product between the electric field of the incoming wave and the (embedded) radiation pattern obtained in transmitting conditions (Collin and Zucker 1969). A reminder about this relation is provided in Appendix G. Finally, reciprocal to the open-circuit pattern representation, short-circuit patterns can also be provided (Kelley 2002). They are less common but, as explained in Section 13.5, they can be exploited to provide an exact extension to some (quite crude) approximation often encountered in the literature.

The challenges from a numerical point of view consist of determining the N-port matrix of large (extended) arrays as well as their embedded patterns, within limited time on affordable computational platforms and with sufficient accuracy. This essentially consists of solving Maxwell's equations (with any of the different excitations referred to in the previous paragraph) and of observing radiated fields. By virtue of the unicity and equivalence principles (Harrington 1961; Rao and Wilton 1990; Van de Water et al. 2005), those fields may be limited to the tangential components of electric or magnetic fields on a closed boundary containing the array, as well as voltages or currents on the nonexcited ports of the array. The fields on the closed boundary can be translated into equivalent currents onto the antenna surfaces, which correspond to physical electrical currents in the case of perfectly conducting antennas. In the case of printed antennas, for instance, fields over the whole space can be computed through convolution of the electrical currents on the metallic printed pattern using the Green's function that accounts for the presence of the dielectric layers (Rana and Alexopoulos 1981).

In other words, the most intensive part of the computations is to obtain the currents (or equivalent quantities) on the surfaces of the antennas when one element is excited at a time. Far-fields, as well as the array impedance matrix, can be computed very rapidly from there. Four categories of methods can be imagined here, depending on whether Maxwell's equations are solved in the time domain or the frequency domain and on whether a differential or an integral formulation is adopted. In this chapter, integral-equation approaches will be preferred because they do not require absorbing boundary conditions when open-domain problems are studied, and frequency-domain approaches will be preferred given the stability issues that still may occur in time-domain integral-equation methods (Li and Weile 2019). Also, given the wide analytical background on which frequency-domain integral equations rely, they allow quite interesting physical interpretation of either the final results or some intermediate quantities. In integral-equation formulations, the unknowns correspond to (equivalent) currents on the interfaces between homogeneous media, and each equation imposes the continuity of tangential fields across those same interfaces (Rao and Wilton 1990). The fields are computed from the current, through convolution with the appropriate Green's function. Those equations are usually discretized and solved using the method of moments (MoM), for which the reader is referred to Chapter 4 (Section 4.4.2).

In the following, the classical MoM will be considered for the analysis of single elements of the array, while special accelerations will be used for studying large arrays. For the analysis of isolated elements, the MoM produces a linear system of equations, denoted here as $\boldsymbol{Z}\,x = b$. In that equation, x contains the unknown coefficients of the currents in terms of the predetermined basis functions, while $-b$ contains the tested incident fields (typically an incoming wave or a voltage "gap" between two conductors). The system matrix $\boldsymbol{Z}$ is often referred to as the "impedance matrix" and contains the reactions between basis (columns) and testing functions (rows). Those reactions correspond to fields radiated by a given basis function and then are multiplied by a testing function and integrated over the latter's domain. As described in Section 3.4.3, most MoM implementations make use of Galerkin testing, i.e. the set of testing functions corresponds to the set of basis functions. Figure 13.2 shows simulation results for printed patch antennas on a dielectric substrate (relative permittivity: 2.2; thickness: 0.381 mm). Here, to compute the impedance matrix, a Green's function specialized to layered media has been used. Simulations have been carried out with an in-house code, as well as with the IE3D commercial software (Mentor Graphics 2005). Figure 13.2a shows the rectangular mesh used to define basis functions on the patch. Figure 13.2b shows the comparisons between input impedances obtained with the two software packages, and Figures 13.2c and 13.2d shows the EEPs obtained with the two codes. One can see that an excellent agreement is achieved. This element will also be used to study the effects of mutual coupling in Sections 13.3 and 13.4 of the chapter.

Three special categories of methods can be exploited for the analysis of large arrays:

1) Methods starting from infinite-array analysis. They apply to large regular arrays and may be combined with corrections to account for truncation and edge effects.
2) Iterative solution techniques, in particular techniques that do not require the storage of the, sometimes, huge system of equations resulting from discretization of the whole finite-array structure, combined with special techniques for fast matrix-vector products.
3) Methods involving a reduction of the number of unknowns, through the definition of Macro Basis Functions (MBFs), or equivalent quantities.

These three approaches will be detailed in Sections 13.2 to 13.4. Section 13.5 summarizes findings regarding the modal representation of currents on a given antenna and proposes an extension to single-mode approximations to describe the transformation of embedded patterns when antennas are placed in dense arrays. Section 13.6 then shows how those fast techniques can be applied to array optimization, with a special focus on the particularly hard problem of antenna placement in nonregular arrays. A brief summary and future research directions are provided in Section 13.7.

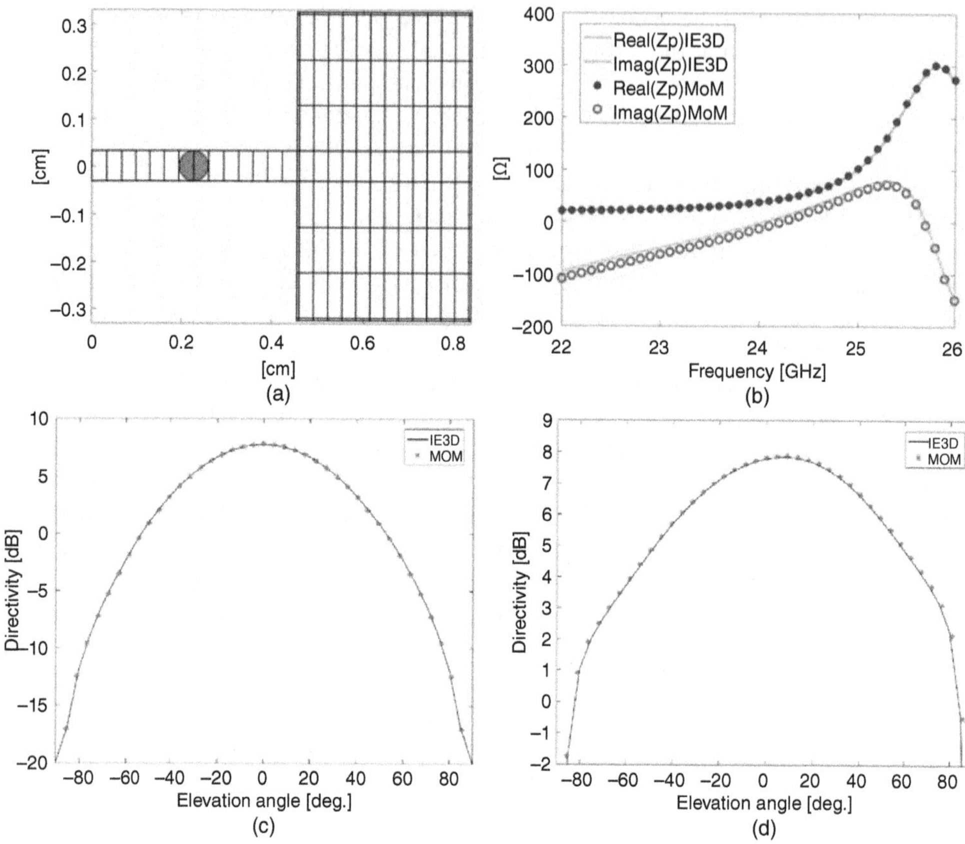

Figure 13.2 Printed patch antenna on a dielectric substrate. (a) Mesh of printed antenna; (b) input impedances obtained with two different software packages; and embedded patterns at 24 GHz, in (c) H-plane and (d) E-plane, respectively.

13.2 Periodic Method of Moments

Large regular arrays are conveniently analyzed assuming a unit-cell approach, as described in Section 3.5.1. Such analysis includes all the effects of mutual coupling, even in tightly coupled arrays, except for truncation of the array. It is precious for initial design of the array and is an important tool to avoid scan blindness (Pozar and Schaubert 1984), a major risk factor in large wideband arrays. According to this analysis approach, which we will name the infinite-array approach (Craeye and Arts 2004), the amplitude of the currents will be assumed the same in consecutive cells, while a linear phase shift is introduced, in order to scan a beam in a given direction. Even if this assumption will be the starting point of this section, we will show that it is not a necessary condition for efficiently obtaining embedded patterns and, to a wide extent, the impedance matrix in large finite arrays.

For the moment, assume a constant amplitude and linear phase shift across the array. In this case, if differential techniques are used (Jin 1993), that condition can be imposed through periodic boundary conditions enforced on the contours of the unit cell. If the MoM is used, the presence of the (infinite number of) other cells is accounted by using periodic Green's functions. Such a Green's function corresponds to the fields due to an infinite series of point sources located in consecutive cells and with the imposed linearly progressing phase difference. The space-domain summation of the Green's function associated with individual sources has very poor convergence, and many alternative formulations have been proposed in the literature, many of which have exponential convergence (Schuchinsky et al. 2006; Valerio et al. 2007; Singh and Singh 1993; Stevanovic and Mosig 2007; Craeye and Capolino 2006; Guerin et al. 2009; Valerio et al. 2016; Florencio et al. 2015; Rogier 2005; Dardenne and Craeye 2008). A reference quantity obtained in this way corresponds to the active impedance Z_a, i.e. the voltage-to-current ratio observed at a given port when the whole array is scanned in the direction of interest. It is important to stress that quantity is dependent on scan angle.

The results obtained with the unit-cell approach may find applications beyond what is usually thought. Rather paradoxically, the excitation does not need to be constant in amplitude and linear in phase for the result to be exploitable. Indeed, it can be proven that the EEP of a given antenna in an infinite array can be obtained through simulations in which the whole array is excited (Craeye and Gonzalez-Ovejero 2011). Suppose the value of such a pattern is looked for in direction Ω. First, an infinite fully excited array scanned in direction Ω is analyzed, with interelement phase-shift $-\psi$. ($-\psi_x$ and $-\psi_y$ for 2D arrays). Then, the current obtained on the unit cell is isolated, and the pattern radiated by that current is computed in direction Ω. A minor drawback of this approach is that one infinite-array solution needs to be computed for every direction Ω, but such computations can be run efficiently (and possibly in parallel) since the computational domain is very small. In a similar way, assuming that active impedances are obtained for all directions Ω, including those in invisible space (i.e. with interelement phase shifts beyond what is needed to scan in a visible direction of space), then the mutual impedance can be obtained through a Fourier transform between every pair of ports. This can be immediately delineated from the Array Scanning Method (ASM) (Munk and Burrell 1979), which reads as

$$I(m) = \frac{1}{2\pi} \int_0^{2\pi} I^{\infty}(\psi)\, e^{-j\,m\,\psi} d\psi \tag{13.1}$$

where $I(m)$ is the current (or any field quantity) on antenna m, while $I^{\infty}(\psi)$ is the infinite-array solution obtained on cell 0 with a phase shift $-\psi$ between consecutive cells. This result is readily extended to two-dimensional arrays. It is important to notice that I can correspond to any vector field quantity, assuming that $I(m)$ and $I^{\infty}(\psi)$ are taken on corresponding points on cell m and cell 0, respectively. If, in the numerical experiment, the antennas are fed with a pure voltage source, then the currents at each port correspond to mutual admittances.

From there, an approximate treatment of finite regular arrays can be obtained. It consists of assuming that, when a given element is excited, it induces on the finite array a current distribution equal to the one it would induce in an infinite array, except for the (obvious) truncation of the array support. Mutual impedances are obtained in the same way such that it simply depends on the difference between indices of antennas in the regular structure (Roscoe and Perrott 1994). This does not fully account for the effects of array truncation. Indeed, when a given element is excited, the currents on neighboring elements may be viewed as a "wave" that starts from the excited element and is reflected or "scattered" by the edges of the periodic structure. This effect can be modeled in two different ways. First, a perturbation approach consists of assuming, when excitation at a given element is treated, that the current on all other elements is the same, except for multiplication by a coefficient proportional to the excitation law (Skrivervik and Mosig 1993). Then, based on that initial assumption, an integral equation is set up for the given element. Another approach consists of studying arrays that are finite in one direction and infinite in the orthogonal direction and to combine the results obtained for arrays made infinite in two orthogonal directions. To a wide

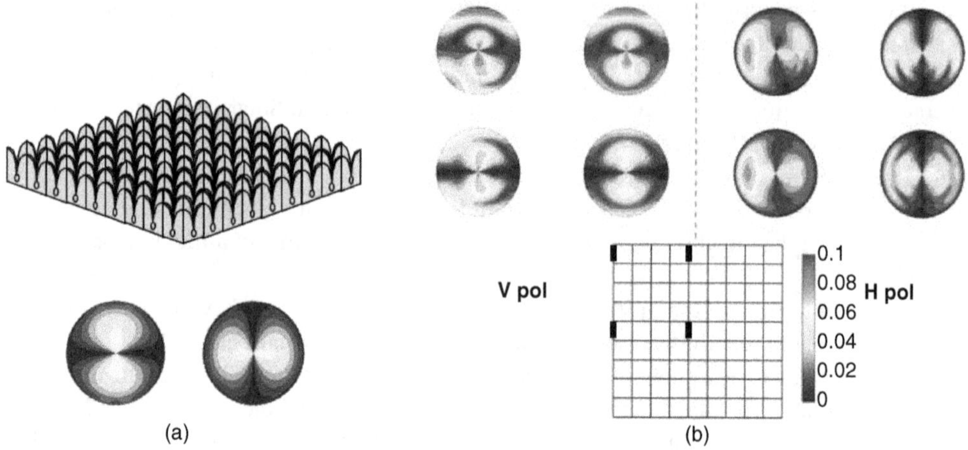

Figure 13.3 Embedded element patterns of dual-polarized array of tapered-slot array. (a) Tapered-slot antennas and infinite-array patterns (V and H polarizations); (b) top view of array grid with period of 0.212 wavelengths. Thick bars on the corners of the top left quadrant indicate the four elements for which patterns are given as follows: top four left patterns: vertical (V) polarization, and top four right patterns: horizontal (H) polarization.

extent, this accounts for scattering by all edges of the array but is limited to rectangular arrays. This technique has been applied in Craeye and Dardenne (2006) to arrays made of electrically connected tapered-slot (or "Vivaldi") antennas. Those arrays are so tightly coupled that there is virtually no relation between the behavior of the antenna taken in isolation and its performance in the array. The embedded patterns for such an array are shown in Figure 13.3 for four elements of the connected array (see bottom-left), for vertical and horizontal polarizations. The patterns are given versus direction-cosine coordinates (i.e. drawn on hemisphere seen from top). It can be seen that the variability among patterns is very significant (see infinite-array patterns, bottom, Figure 13.3b). The finite-by-infinite array approach described in Craeye and Dardenne (2006) allows catching to a wide extent that variability. The same reference also provides a correction for embedded patterns in the array corners.

Finally, another way to include infinite-array calculations into the analysis of finite arrays will be presented in Section 13.4. It makes use of the current distributions obtained according to Eq. (13.1), which are used as MBFs.

13.3 Iterative Solution Techniques

The huge system of equations that needs to be solved when treating large finite arrays may be solved iteratively. Some methods are described in Section 3.6.3. The iterative solution of large linear systems of equations, say $\boldsymbol{Z}\boldsymbol{x} = \boldsymbol{b}$ in matrix form, has been a matter of research on its own, from the 1960s to the 1990s. Initially, several corrective terms were evaluated, but convergence was not necessarily guaranteed. This changed drastically with the appearance of Krylov subspace iteration (Van Der Vorst 2000) and with methods such as Generalized Minimum Residual (GMRES, Saad and Schultz 1986). Other techniques, specific to the solution of Hermitian systems of equations, such as the Conjugate Gradient (CG) method (Saad and Schultz 1986), are quite efficient too, but unfortunately, the problems at hand here are – based on reciprocity – symmetric but not Hermitian. The probably best known basic iterative technique is the Jacobi method, which relatively naively re-estimates the solution by inserting the approximate solution in each consecutive equation and determining one unknown at a time. This is not necessarily converging, but if all correction vectors are kept in memory, associated with an unknown scalar multiplier and reinserted into the original system of equations, a small system of equations is obtained for the multipliers, which can be solved rapidly, in a convergent manner. This is the spirit of the GMRES or Full Orthogonalization Method (FOM) (Saad and Schultz 1986). More formally, in this approach, the solution will lie in the subspace generated by vectors $\boldsymbol{b}$, $\boldsymbol{Zb}$, $\boldsymbol{Z}^2\boldsymbol{b}$, $\boldsymbol{Z}^3\boldsymbol{b}$, etc., which is the Krylov subspace. Doing so, more equations than unknowns are generated. In this case, the problem is solved by imposing that the residue, i.e. $\boldsymbol{b} - \boldsymbol{Z}\,\boldsymbol{x}$, is to be orthogonal to either the Krylov subspace (FOM method) or to that subspace left-multiplied by $\boldsymbol{Z}$ (GMRES method). Each of these two approaches has its own advantages, but which exhibit only very small differences in terms of convergence rate. Such iterations are sometimes "restarted." Indeed, as the iterations go on, the obtained (slowly growing) system of equations may become ill-conditioned. In that case, a new system of equations is set, which corresponds to the difference between the initial one and $\boldsymbol{Z}\,\boldsymbol{x}_n = \boldsymbol{b}_n$, where $\boldsymbol{x}_n$ is the solution obtained thus far. A new iteration is then started to solve that new system of equations to obtain $\boldsymbol{x} - \boldsymbol{x}_n$. It is difficult to predict the performance of the restart procedure, which should rather be regarded as a pragmatic way around ill-conditioning issues. A simple reformulation of Krylov methods is outlined in Craeye et al. (2014).

Such iterative procedures involve many matrix-vector multiplications (typically one or two per iteration), and it is desirable to perform those multiplications without even computing the whole matrix. A relatively "ad hoc" procedure assumes that the matrix, or at least some blocks of the matrix, has a very low rank. Besides, a low-rank representation of the matrix is obtained based on the computation of a limited number of well-selected rows and columns. This is the principle underlying methods such as incomplete QR (Ozdemir and Lee 2004) iterations or Adaptive Cross Approximation methods (ACA) (Bebendorf and Rjasanow 2003). Other methods make a more extensive use of the knowledge of the Green's function. Indeed, the matrix-vector product referred to above may be viewed as a discretized representation of the convolution between currents and Green's function to obtain fields. With this perspective, two main families of methods have been developed. A first one is based on multipole decomposition, while another one makes use of fast Fourier transforms (FFT).

Multipole methods are based on the decomposition (with spherical-wave expansions as an intermediate step) of the Green's function into a continuous spectrum of propagating plane waves (Coifman et al. 1993). The advantages of plane waves are that, in Cartesian coordinates, their expression is separable versus observation and source coordinates. That property allows a separate treatment at both levels, through the definition of radiation patterns of basis and testing functions. In a nutshell, the computational domain is divided into several cells. From each of them, a finite number of plane waves are spawn; they are then translated between all pairs of cells and are then projected on every testing function. A more precise, and quite pedagogical, description can be found in Coifman et al. (1993). If N is the total number of unknowns, the complexity is of order $N^{3/2}$ and can be extended to $N \log N$ if a multilevel approach (with cells containing cells) is developed (Song et al. 1997). The advent of multipole methods allowed a big leap in terms size of problems that could be solved, which happened a little before year 2000, going from thousands to millions of unknowns. More recent implementations now even led to billions (Pan et al. 2012). It is important to notice that multipole decompositions do not apply to the near-field where interactions over distances below half-wavelength need to be treated through a local computation of the MoM impedance matrix. This becomes a problem when the meshing of the structure is very fine compared to the wavelength, which is the case when complex antenna elements are treated. In that case, indeed, the near-field region contains many unknowns (up to a few thousands), and the treatment of near-field interactions can dominate the analysis time. This subject, which may be viewed as one of the currently most important ones in computational electromagnetics, is receiving increasing attention (Bogaert et al. 2008; Xia et al. 2018; Karki 2018). Iterative techniques can also be combined with reduced-order modeling, such as MBFs described in Section 13.4, as done in Pairon et al. (2019) for the analysis of arrays devoted to Synthetic Aperture Radar.

Another method to speed up matrix-vector multiplications makes use of the convolution between currents and Green's function in the (spatial) spectral domain, via the FFT. While the idea may sound as old as the FFT algorithm itself (Cooley and Tukey 1965), its application to high-frequency electromagnetics took a while. The issue to be solved consisted of knowing what to do with the near-field interactions, for which the Green's function is nearly singular. That singularity translates into slow spectral fall-off such that the unavoidable spectral truncation makes the FFT approach inapplicable to very close interactions. This is where the idea of pre-correction comes in (Bleszynski et al. 1996; Yuan et al. 2003). It entails computing those interactions anyway with a finite spectral representation of current and Green's function, then undoing that interaction only for the close interactions and finally replacing those interactions with explicit computations based on the MoM impedance matrix, as is also done with the multipoles approach. In this process, the spectral Green's function is in general obtained through the FFT of the Green's function estimated on a regular grid in spatial domain. An alternative approach, which takes advantage of the analytical knowledge of the Green's function in a spectral domain, is presented in Hubert et al. (2018). With such methods, totally arbitrary arrays, with elements of different sizes, and orientations, in nonregular arrangements can be analyzed efficiently.

An important component of iterative techniques is preconditioning. Convergence can be much faster if an approximate solution is available at each iteration. Suppose one can compute the solution of $\boldsymbol{M}\,\boldsymbol{x} = \mathbf{c}$ much faster than that of $\boldsymbol{Z}\,\boldsymbol{x} = \boldsymbol{b} = \mathbf{b}$, with $\mathbf{M}$ close to $\mathbf{Z}$, then it is better to iteratively solve $\boldsymbol{M}^{-1}\boldsymbol{Z}\,\boldsymbol{x} = \boldsymbol{M}^{-1}\boldsymbol{b}$, for which much faster convergence will in general be observed. This improvement can be linked with the eigenvalues of the $\boldsymbol{I} - \boldsymbol{Z}$ matrix (with I the unit matrix), as explained by Greenbaum (1997). It is important to notice that the solution of this "preconditioned" system of equations does not require the calculation of $\boldsymbol{M}^{-1}$ but just needs a fast way to solve $\boldsymbol{M}\,\boldsymbol{x} = \mathbf{c}$. Preconditioning has become a quite involved subject of research, and efficient preconditioners for electric field integral equations (EFIE) and magnetic field integral equations (MFIE) from DC to high-frequency electromagnetics can be found in Andriulli et al. (2008) and Adrian et al. (2019).

Drawbacks of iterative techniques for array problems are that a new iteration needs to be run for excitation at every element and that the number of iterations is not known a priori. Therefore, alternative techniques, based on reduced-order models, have been developed, as summarized in Section 13.4.

13.4 Macro Basis Functions

Antennas often comprise complex details, especially near the feed point. Therefore, the meshing of the surfaces may need to be much finer than the traditional 10 triangles per wavelength advised for smooth surfaces. This often leads to hundreds or thousands of unknowns per antenna and hence may result in millions of unknowns for very large arrays, which need to be determined for excitation at every element. At the same time, the actual number of degrees of freedom of fields on a small object, compared to the wavelength, is quite small (Bucci and Franceschetti 1989; see also the contribution from Laviada in Craeye et al. 2014). Therefore, it appears reasonable to search for solutions on a specific subspace, i.e. to write currents as a linear combination of predetermined current distributions. The choice of the set of current distributions (or "modes") is still a subject of research: many approaches have been proposed in the literature. In most of them, the current distributions are obtained after illumination of a single element by specific sources surrounding it. Depending on the specific way used to determine those distributions, they are named "Macro Basis Functions" (MBF) (Suter and Mosig 2000), "Synthetic Functions" (SFX) (Matekovits et al. 2007), or "Characteristic Modes" (CBFM) (Prakash and Mittra 2003). It has been shown in Ozdemir et al. (2013) that current distributions obtained from a multiple-scattering process between neighboring antennas can reproduce the Krylov subspace. The MBFs can also be obtained from infinite-array solutions (Craeye and Sarkis 2008). That approach is particularly well suited to the analysis of connected arrays since infinite-array solutions capture very accurately the continuity of currents between consecutive elements. Given that a definitive way to determine the subspace has still not been proposed, this approach is sometimes subject to criticism. However, the authors observed that the accuracy of the solution rapidly improves as the number of MBFs is increased through a multiple-scattering process. In this respect, the use of MBFs (or closely related macro-models) may be viewed as a valuable way to extend the (often deficient) single-mode representation of currents on antennas to an arbitrary number of modes. In practice, order of 25 "modes" per cubic wavelength is found enough (i.e. about 4 MBFs for three-dimensional antennas of size close to half wavelength). Another modal approach (with spherical modal fields instead of induced currents) is described by Rubio et al. (2005), who combined finite-elements at the element level with spherical-decompositions at the array level.

Using the MBF approach, the number of unknowns per antenna typically decreases from about 500 to 15, i.e. a saving of unknowns between 1 and 2 orders of magnitude. Therefore, the reduced system of equations may remain manageable, even for large arrays, and the corresponding matrix can be inverted or undergo any alternative factorization, such as LU factorization. Once such inversion or factorization is available, solutions for excitation at any port (see quantities of interest, described in Section 13.1) are obtained extremely fast. Let us first consider the case of arrays made of identical elements. In that case, it is reasonable to consider the same MBFs on every element. Let us denote by $\boldsymbol{Q}$ the matrix whose columns represent the MBFs in terms of elementary basis functions. Then, the solution for currents on antenna i can be written as $\boldsymbol{x}_i = \boldsymbol{Q}\,\boldsymbol{y}_i$, where $\boldsymbol{y}_i$ is a much smaller vector than $\boldsymbol{y}_i$, which explains why we have a reduced-order representation. Let us denote by $\boldsymbol{Z}_{ij}$ a block of the original MoM impedance matrix, with basis functions on antenna j and testing functions on antenna i, and by $\boldsymbol{b}_i$ the segment of the excitation vector relating to antenna i. Then, the reduced system of equations reads as $\boldsymbol{Z}'\boldsymbol{y} = \boldsymbol{b}'$, with $\boldsymbol{Z}'_{ij} = \boldsymbol{Q}^H \boldsymbol{Z}_{ij}\,\boldsymbol{Q}$ and $\boldsymbol{b}'_i = \boldsymbol{Q}^H \boldsymbol{v}_i$. Matrix $\boldsymbol{Q}$ typically has 1–2 orders of magnitude fewer columns (number of MBFs) than rows (number of elementary basis functions per antenna). This explains why the reduced system of equations has much fewer unknowns y. One may also consider a different set of MBFs on each antenna, which makes the $\boldsymbol{Q}$ matrix dependent on the antenna index. With the use of MBFs, solutions can be obtained very fast for excitation at any element, and from there, the array impedance matrix, as well as all embedded patterns (Craeye et al. 2009). Such a direct solution may no longer be tractable when the array becomes extremely large. In that case, a tessellation approach can be applied (Bui Van et al. 2018): the array is decomposed into hexagonal domains, each of which can be extended with an auxiliary domain, containing the closest half of each neighboring hexagonal domain. The problem is then solved for each extended domain, and the results are kept for the core domains only. This approach neglects coupling with elements beyond the auxiliary domain. For arrays made of connected elements, the MBFs are inspired from infinite-array solutions (ASM-MBF approach; Craeye and Sarkis 2008). Figure 13.4 shows results obtained for an array of transverse-electromagnetic mode (TEM)

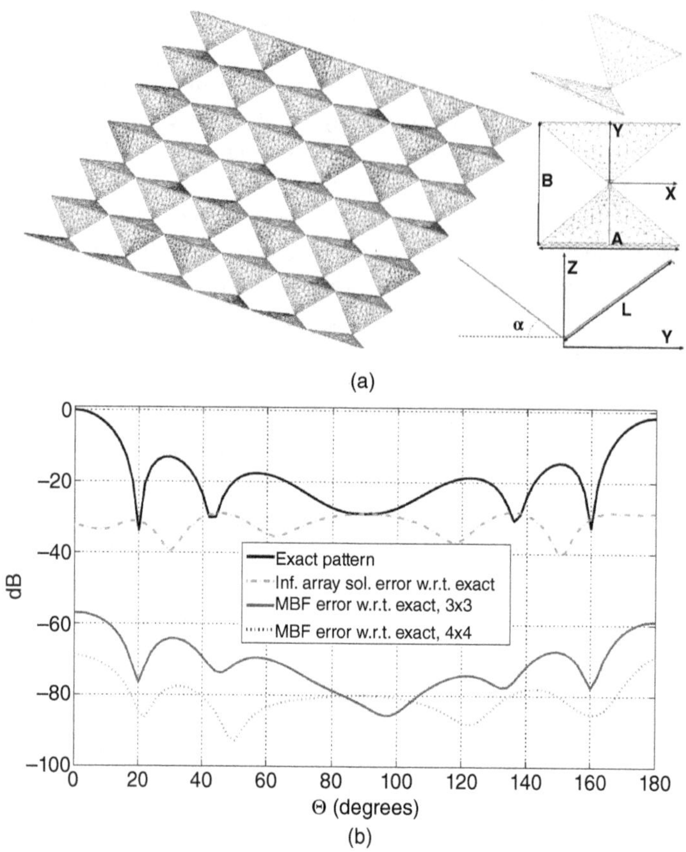

Figure 13.4 Array of TEM horn antennas. (a) Configuration and (b) array patterns computed with the ASM-MBF method, with error levels obtained with different methods: the infinite-array approach (with very high error level) and the ASM-MBF method with 3×3 and 4×3 MBFs. Source: CRAEYE, C. and GONZALEZ-OVEJERO, D. (2011): 'A review on array mutual coupling analysis', Radio Science, 46, pp. 1-25.© 2011. American Geophysical Union.

horn antennas (Craeye and Gonzalez-Ovejero 2011). Figure 13.4a shows the elements geometry, while Figure 13.4b shows the pattern of the array scanned at broadside, while a quite limited number of MBFs is used.

The calculation of the interactions between MBFs may require a huge computation time. To avoid the preliminary calculation of the huge system of equations standing for the whole array, numerous techniques have been proposed in the literature; many of them are inspired from acceleration techniques found in iterative methods (see Section 13.3). Indeed, methods based on multipoles (Craeye 2006) and ACA (Maaskant et al. 2008) have been proposed. Large savings in terms of computation can be achieved, thanks to precomputations based on the a priori knowledge of the radiating element. The interactions between MBFs on different elements can be mapped a priori, versus relative positions between antennas. As such, this technique is not tractable, since the needed sampling is too fine, given affordable time and memory resources. This challenge can be dramatically alleviated through a few transformations applied to the function to be mapped. A first approximation is based on the Friis equation, where the MBF corresponds to the transmitting antenna and the Macro Testing Function acts as the receiving antenna. Hence, only the difference between the exact solution and that far-field approximation needs to be tabulated. The remaining function is rapidly oscillating versus relative positions and is somewhat singular near the origin: extraction of the phase and appropriate change of variables allow mapping based with about 50 samples only (Gonzalez-Ovejero and Craeye 2011). This led to the HARP software, which, for instance, allowed the computation of all the embedded patterns for arrays comprising 256 complex elements (all embedded patterns and N-port matrix) in less than a minute on a desktop computer (Bui Van et al. 2018). For antennas printed on layered media, another way to obtain the interactions table can be delineated from the spectral representation of the Green's function; it is well suited to the analysis of printed antennas. The table for interactions between MBFs is then obtained through an FFT; more precisely, a series of FFTs named Contour-FFT, especially developed to avoid poles in the spectral Green's function (CFFT) (Jha and Craeye 2014). In this way, the tabulation of the interaction between MBFs takes little time, compared to the solution time itself (i.e. the time needed to solve the reduced system of equations). To test the accuracy of this approach, a relatively small array, made of 32 of the elements shown in Figure 13.2a, is analyzed at 24 GHz using this fast method, and the obtained array pattern is compared with the brute-force MoM approach, as shown in Figure 13.5. Figure 13.5a shows the array configuration, while Figures 13.5b and 13.5c show the array pattern when scanned at broadside. The patterns and array impedance matrix are computed with the help of a combination between HARP (for average-medium part) and CFFT techniques (for remaining

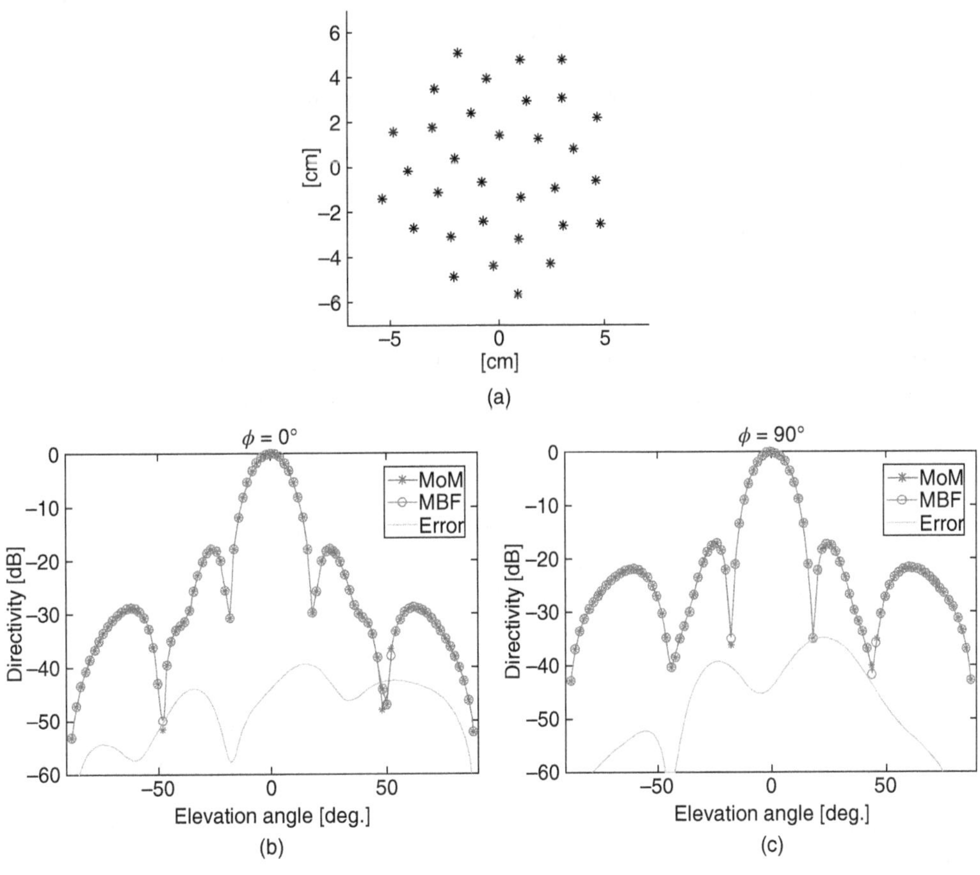

Figure 13.5 Array of 32 patch elements shown in Figure 13.2a. (a) Array configuration, and comparison between array patterns computed by brute-force of using MBFs for (b) E-plane and for (c) H-plane.

layered medium part). In Figure 13.5b,c, on the same scale, the difference between brute-force and MBF + CFFT techniques is displayed. It can be seen that the error level is of the order of −40 dB.

13.5 Pattern Transformations

As has been explained through the different sections above, mutual coupling modifies the antenna's EEPs. This section tries to determine to what extent the array impedance matrix appears in that pattern transformation and what is the minimal set of parameters needed to describe such transformation.

Embedded patterns obtained with different sets of loads terminating the antenna can be related using the array impedance matrix $\boldsymbol{Z}$. We first describe the two sets of loads using two diagonal matrices, $\boldsymbol{Z}_{L,1}$ and $\boldsymbol{Z}_{L,2}$. Then, the corresponding embedded patterns $\boldsymbol{F}_1$ and $\boldsymbol{F}_2$ can be connected as follows (Craeye and Gonzalez-Ovejero, 2011; Warnick and Davidson, 2021):

$$(\boldsymbol{Z} + \boldsymbol{Z}_{L,1})\,\boldsymbol{F}_1 = (\boldsymbol{Z} + \boldsymbol{Z}_{L,2})\boldsymbol{F}_2 \tag{13.2}$$

In that matrix representation, different columns of $\boldsymbol{F}_1$ and $\boldsymbol{F}_2$ correspond to different directions, while each line is associated with a different antenna (i.e. a different embedded pattern). Here, the embedded patterns correspond to a unit-voltage excitation, with a series impedance, while the other elements are passively terminated. That representation is particularly convenient when it comes to obtaining the current through the loads in receiving conditions, as reminded in Appendix G. One should also note that the patterns referred to here can also be defined in the near-field; in that case, for convenience, one simply refers to the electric or magnetic fields, functions of three spatial coordinates.

Open-circuit patterns are defined with a unit current at the element of interest, while other elements are open. Since they correspond to fields per unit current instead of unit voltage, we may expect different units. Their relationship with embedded patterns is easy to prove; it reads

$$\boldsymbol{F}^{o.c.} = (\boldsymbol{Z} + \boldsymbol{Z}_{L,1})\boldsymbol{F}_1 \tag{13.3}$$

Likewise, short-circuit patterns can also be defined (Kelley 2002). They can be simply obtained from Eq. (13.3) by setting $\boldsymbol{Z}_{L,2} = 0$. This, for instance, leads to the following link between open- and short-circuit patterns:

$$\boldsymbol{F}^{s.c.} = \boldsymbol{Y}(\boldsymbol{Z} + \boldsymbol{Z}_{L,1})\boldsymbol{F}_1 \tag{13.4}$$

where $\boldsymbol{Y} = \boldsymbol{Z}^{-1}$ is the array N-port admittance matrix. An approximation often made consists of assuming that open elements do not scatter, which is approximately true for half-wave dipoles. This is equivalent to considering that open-circuit patterns are identical to isolated-element patterns when the latter are also defined using a unit port current

$$\boldsymbol{F}^{is.} \sim (\boldsymbol{Z} + \boldsymbol{Z}_{L,1})\boldsymbol{F}_1 \tag{13.5}$$

If the isolated patterns are defined with a unit voltage and generator impedance Z_L (instead of a unit-current source), the above result is simply divided by scalar $Z_A + Z_L$, where Z_A is the impedance of the isolated antenna. This approximation may leave the impression that, based on the knowledge of the array impedance matrix $\boldsymbol{Z}$, voltages across the loads of a receiving array can be easily corrected for the effects of mutual coupling. Indeed, as seen in Appendix G, the current across the loads corresponds to a scalar product between the incident plane wave and the antenna pattern. A left-multiplication of the column vector of received voltages by $(\boldsymbol{Z} + \boldsymbol{Z}_{L,1})/(Z_A + Z_L)$ would then be sufficient to correct for the effects of mutual coupling, since it would produce the voltage across the load terminating the isolated element. That would imply that all the information regarding mutual coupling is contained in the impedance matrix $\boldsymbol{Z}$. Unfortunately, it can be easily shown that this approximation is valid only if, when the antennas are left open they do not significantly scatter (Craeye and Gonzalez-Ovejero 2011). Huge errors can arise when that condition is not satisfied.

Equivalently, if the current distribution on any element can be described with several "modes," the above approximation corresponds to a single-mode assumption, where no port current amounts to no current at all. It is equivalent to the very popular coupling correction technique proposed by Gupta and Ksienski (1983). The question then arises as to whether that approximation can be extended to the case where the antennas support a finite number of modes. It appears that this may be explained more compactly using short-circuit patterns. In that case, the above approximation reads

$$\boldsymbol{F}^{s.c.} \sim \boldsymbol{Y}\boldsymbol{F}^{is.} \tag{13.6}$$

where $\boldsymbol{Y}$ is the array admittance matrix. If the port currents associated with the single mode are set to 1, then Y_{ij} simply corresponds to the (i, j) entry of the inverse of the reaction matrix between the modes. The (i, j) entry of the reaction matrix between such modes is the field radiated by the mode on antenna j and tested by the mode on antenna i. That reaction matrix is also the MoM matrix of the array when a single (modal) basis function is used per antenna.

Let us now consider several current modes on each antenna, defined without loss of generality with unit port currents. Now, the reaction matrix involves several modes on each antenna; that number of modes can be different from one antenna to another. The inverse of that matrix is then partitioned into blocks corresponding to different antennas, the number of rows of block (i,j) being equal to the number of modes on antenna i and the number of columns of block (i,j) equal to the number of modes on antenna j. That matrix is sketched in Figure 13.6. A rectangular matrix is obtained by summing up the entries of each row within each block (gray columns in Figure 13.6). Let us denote by $\boldsymbol{Y}^e$ the transpose of that matrix; its number of rows is the number of antennas, and its number of columns is the total number of modes on all antennas. Now, let us define a matrix $\boldsymbol{F}^m$ that provides all the patterns of the modes, with as many rows as there are modes (i.e. in general, larger than the number of antennas). All patterns are defined with a common point for the phase,

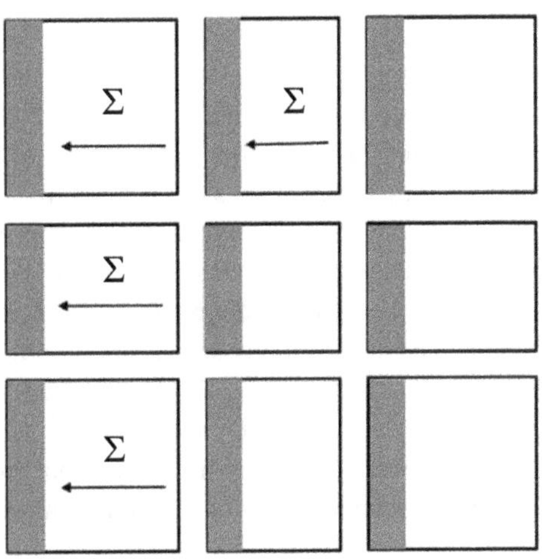

Figure 13.6 Configuration of reactions matrix, partitioned per antenna. If, in each block, the rows are summed up (cf. gray columns), then the $\mathbf{Y}^e$ matrix is obtained. That matrix is used to transform the modal patterns into short-circuit patterns. If, in each block, all elements are summed up (i.e. sum of row segments of $\mathbf{Y}^e$), then the array admittance matrix is obtained.

located near the center of the array. Then, we can write

$$F^{s.c.} = Y^e F^m \tag{13.7}$$

A proof of this matrix equation is provided in Appendix H in linear equation form.

In this case, an equality sign has been used because we assumed that enough (i.e. well-defined) modes are used on each antenna. Those modes can be directly identified with MBFs on the different antennas, provided those MBFs are scaled to produce a unit current at the level of the ports. The rows of Y^e can be divided into segments, the number of entries of segment i being equal to the number of modes on antenna i. If the entries of each segment are summed up, then the N-port admittance matrix $\mathbf{Y}$ is obtained. Equation 13.7 can hence be viewed as an extension of approximation Eq. (13.6), in which matrix Y^e obviously contains more information than Y. This explains why a "decoupling" correction in general cannot be simply based on the knowledge of the array impedance (or admittance) matrix. That formulation also explains what is the number of independent parameters that describe the effect of mutual coupling: it corresponds to the number of antennas multiplied by the total number of currents modes.

Finally, it is interesting to notice that, in discretized form (i.e. with current distributions described in terms of elementary basis functions), this modal approach is equivalent to the use of MBFs (see Section 13.4).

13.6 Optimization

The design of large antenna arrays is often assisted by optimization tools, where antenna positions and/or excitations are optimized to produce certain radiation patterns. Different techniques have been applied in this area to efficiently and effectively synthesize the arrays. Most of them, however, often work on an assumption that the antenna patterns are identical for all elements, thus simplifying the calculation of array pattern to the array factor, which allows a fast evaluation of fitness functions. As the embedded elements patterns are different for each antenna, depending on its position within the array, the synthesis without coupling might not work as expected, e.g. higher sidelobe levels. In that sense, the fast determination of embedded patterns is essential in an optimization loop, especially when tackling the hard problem consisting of optimization of the positions of elements while including mutual coupling. In this respect, a quasi-convex method has been developed by Fuchs et al. (2012), without coupling, and recently extended to the case including coupling, as shown in Bui Van et al. (2016). In that work, the fast simulation techniques are called at each iteration to very quickly update the EEP of each antenna, while it is shifted within the array, along a direction determined by the quasi-convex method described in Fuchs (2014). The updated EEPs are then reintegrated in the next iteration of the optimization loop. The final array will therefore ensure the performance of the array with respect to predefined requirements as well as any physical constraints, such as physical size, while including all the effects of mutual coupling. This approach has been recently applied to arrays made of printed elements in Bui Van et al. (2016): one could observe that the antennas can be shifted relatively far without updating the embedded patterns; indeed, such steps could be of order $\lambda/6$, which means a number of full-wave simulations smaller than initially expected. Interestingly, also, the sidelobe level could sometimes be made lower with coupling than without!

To demonstrate this optimization approach, an example regarding concentric circular arrays made of rectangular microstrip antennas is presented in Figure 13.7. It is a problem similar to

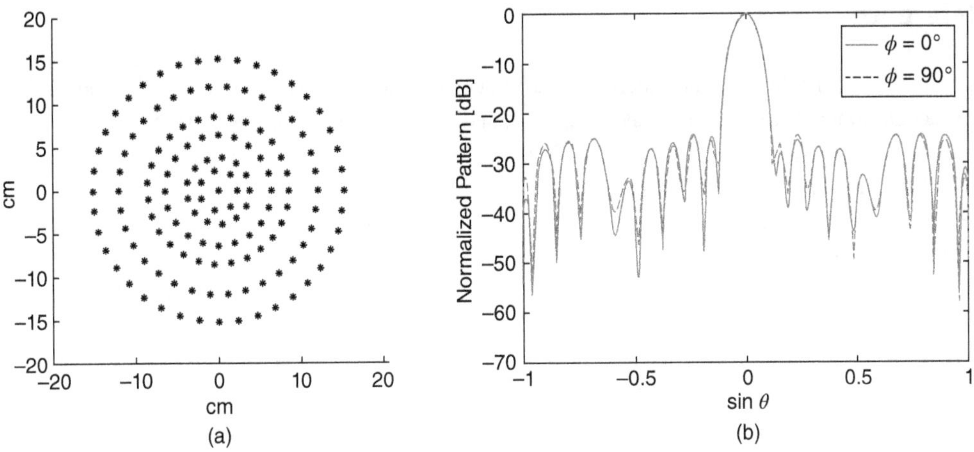

Figure 13.7 Concentric circular array of rectangular microstrip patches. (a) Array configuration, with elements from Figure 13.2a, and (b) optimized array patterns in E- and H-planes.

the one analyzed in Bui Van et al. (2016), where the (more realistic) microstrip antenna from Figure 13.2a is now used instead of the bowtie antenna. The objective is to generate a pencil beam at broadside while minimizing the sidelobe level by optimizing the radii of the circles. Figure 13.7a shows the obtained array configuration, while Figure 13.7b shows cuts in the optimized array pattern in E- and H-planes. The optimization time, including regular refresh of EEPs, is of the order of five minutes on a laptop computer.

13.7 Conclusion

Numerical methods are essential tools for array design, when mutual coupling needs to be included, which appears necessary in most cases. Different categories of methods are available, based on Maxwell's equations in differential or integral forms. The latter present an advantage for the study of large arrays made of identical elements, although clever combinations between integral and differential techniques also facilitate the analysis of arrays made of very complex elements. The definition of MBF (or any type of synthetic modal representations) has proven quite favorable and efficient to the analysis of very large arrays. Several techniques have been developed for the direct interaction between MBFs; they allowed the acceleration by several orders of magnitude of array analysis and optimization techniques. Finally, parallel to the MBF concept, it has been shown how pattern-transformation models can be extended beyond single-mode approximations through the definition of an extended admittance matrix. The number of coefficients involved in that transformation appears to be equal to the number of antennas, multiplied by the total number of modes on the – possibility different – antennas composing the array. Among important future research directions, one should note the coupling between antennas and near-field platforms, such as finite ground plane (Cavillot et al. 2020), vehicles, or buildings on which the arrays are installed, preferably while still benefiting from compact representations for the currents on the elements composing the array.

References

Adrian, S.B., Andriulli, F.P., and Eibert, T.F. (2019). On a refinement-free Calderón multiplicative preconditioner for the Electric Field Integral Equation. *J. Comput. Phys.* 376: 1232–1252.

Andriulli, F., Cools, C., Bagci, H. et al. (2008). A multiplicative Calderón preconditioner for the electric field integral equation. *IEEE Trans. Antennas Propagat.* 56: 2398–2412.

Bebendorf, M. and Rjasanow, S. (2003). Adaptive low-rank approximation of collocation matrices. *Computing* 70: 1–24.

Bleszynski, E., Bleszynski, M., and Jaroszewicz, T. (1996). AIM: adaptive integral method for solving large-scale electromagnetic scattering and radiation problems. *Radio Sci.* 31: 1225–1251.

Bogaert, I., Peeters, J., and Olyslager, F. (2008). A nondirective plane wave MLFMA stable at low frequencies. *IEEE Trans. Antennas Propag.* 56: 3752–3767.

Bucci, O.M. and Franceschetti, G. (1989). On the degrees of freedom of scattered fields. *IEEE Trans. Antennas Propag.* 37: 918–926.

Bui Van, H., Jha, S.N., and Craeye, C. (2016). Fast full-wave synthesis of printed antenna arrays including mutual coupling. *IEEE Trans. Antennas Propag.* 64: 5163–5171.

Bui Van, H., Abraham, J., Arts, M. et al. (2018). Fast and accurate simulation technique for large irregular arrays. *IEEE Trans. Antennas Propag.* 66: 1805–1817.

Bui Van, H., Arts, M., Craeye, C., and de Lera Acedo, E. (2019). On the maximum absorbed power in receiving antenna arrays. ; *IEEE Trans. Antennas Propag.* 67: 1993–1995.

Cavillot, J., Tihon, D., Mesa, F. et al. (2020). Efficient simulation of large irregular arrays on finite ground plane. *IEEE Trans. Antennas Propag.* 68: 2753–2764.

Coifman, R., Rokhlin, V., and Wandzura, S. (1993). The fast multipole method for the wave equation: a pedestrian prescription. *IEEE Antennas Propag. Mag.* 35: 7–12.

Collin, R.E. and Zucker, F.J. (1969). The receiving antenna. In: *Antenna Theory Part 1'*, Chap. 4, 100–103. New York: McGraw Hill.

Cooley, J.W. and Tukey, J.W. (1965). An algorithm for the machine calculation of complex Fourier series. *Math. Comput.* 19: 297–301.

Craeye, C. (2006). A fast impedance and pattern computation scheme for finite antenna arrays. *IEEE Trans. Antennas Propag.* 54: 3030–3034.

Craeye, C. and Arts, M. (2004). On the receiving cross section of an antenna in infinite linear and planar arrays. *Radio Sci.* 39: RS2010.

Craeye, C. and Capolino, F. (2006). Accelerated computation of the free space Green's function of semi-infinite phased arrays of dipoles'. *IEEE Trans. Antennas Propag.* 54: 1037–1040.

Craeye, C. and Dardenne, X. (2006). Element pattern analysis of wide-band arrays with the help of a finite-by-infinite array approach. *IEEE Trans. Antennas Propag.* 54: 519–529.

Craeye, C. and Gonzalez-Ovejero, D. (2011). A review on array mutual coupling analysis. *Radio Sci.* 46: 1–25.

Craeye, C. and Sarkis, R. (2008). Finite array analysis through combination of macro basis functions and array scanning methods. *J. Appl. Comput. Electromagn. Soc.* 23: 999–1007.

Craeye, C., Gilles, T., and Dardenne, X. (2009). Efficient full wave characterisation of arrays of antennas embedded in finite dielectric volumes. *Radio Sci.* 44: 1–12.

Craeye, C., Laviada, J., Maaskant, R., and Mittra, R. (2014). Macro basis function framework for solving Maxwell's equations in surface integral equation form. *FERMAT J.* 3: 1–16.

Dardenne, X. and Craeye, C. (2008). Method of moments simulation of infinitely periodic structures combining metal with connected dielectric objects. *IEEE Trans. Antennas Propag.* 56: 2372–2380.

De Hoop, A.T. (1975). The N-port receiving antenna and its equivalent electrical network. *Philips Res. Rep.* 30: 302–315.

Florencio, R., Boix, R., and Encinar, J. (2015). Fast and accurate MoM analysis of periodic arrays of multilayered stacked rectangular patches with application to the design of reflectarray antennas. *IEEE Trans. Antennas Propag.* 63: 2558–2571.

Fuchs, B. (2014). Application of convex relaxation to array synthesis problems. *IEEE Trans. Antennas Propag.* 62: 634–640.

Fuchs, B., Skrivervik, A., and Mosig, J.R. (2012). Synthesis of uniform amplitude focused beam arrays. *IEEE Antennas Wireless Propag. Lett.* 11: 1178–1181.

Gonzalez-Ovejero, D. and Craeye, C. (2011). Interpolatory macro basis functions analysis of nonperiodic arrays. *IEEE Trans. Antennas Propag.* 61: 3117–3122.

Greenbaum, A. (1997), 'Iterative method for solving linear systems', Frontiers in Applied Mathematics, 17, SIAM, Philadelphia, 213 p.

Guérin, N., Dardenne, X., and Craeye, C. (2009). Accelerated computation of the free space Green's function gradient in infinite phased arrays of dipoles. *IEEE Trans. Antennas Propag.* 57: 3430–3434.

Gupta, I. and Ksienski, A. (1983). Effect of mutual coupling on the performance of adaptive arrays. *IEEE Trans. Antennas Propag.* 31: 785–791.

Harrington, R.F. (1961). *Time-Harmonic Electromagnetic Fields*. New York: McGraw-Hill.

Hubert, S., Jha, S.-N., and Craeye, C. (2018). Analysis of large nonregular printed scatterers using the Contour-FFT. *IEEE Trans. Antennas Propag.* 66: 6115–6127.

Jha, S.N. and Craeye, C. (2014). Contour-FFT based spectral domain MBF analysis of large printed antenna arrays. *IEEE Trans. Antennas Propag.* 62: 5752–5764.

Jin, J.M. (1993). *The Finite Element Method in Electromagnetics*. New York: Wiley.

Karki, S. (2018): 'Design and modeling of circularly polarized conical beam antennas for space missions', Ph. D thesis, Université catholique de Louvain, Dec. 2018.

Kelley, D.F. (2002). Relationships between active element patterns and mutual impedance matrices in phased array antennas. In: *Proc. IEEE Antennas Propag. Soc. Int. Symp*, 524–527. San Antonio, USA: IEEE.

Li, L. and Weile, D.S. (2019). Integral accuracy and the stability of two methods for the solution of time-domain integral equations for scattering from perfect conductors. *IEEE Trans. Antennas Propag.* 67: 4924–4929.

Maaskant, R., Mittra, R., and Tijhuis, A.G. (2008). Fast analysis of large antenna arrays using the characteristic basis function method and the adaptive cross approximation algorithm. *IEEE Trans. Antennas Propag.* 56: 3440–3451.

Matekovits, L., Laza, V., and Vecchi, G. (2007). Analysis of large complex structures with the synthetic-functions approach. *IEEE Trans. Antennas Propag.* 55: 2509–2521.

Mentor Graphics (2005). *IE3D User's Manual, Release 11*. Fremont, CA, USA: Zeland Software, Inc.

Munk, B. and Burrell, G. (1979). Plane-wave expansion for arrays of arbitrarily oriented piecewise linear elements and its application in determining the impedance of a single linear antenna in a lossy half-space. *IEEE Trans. Antennas Propag.* 27 (3): 331–343.

Ozdemir, N. and Lee, J.-F. (2004). A low-rank IE-QR algorithm for matrix compression in volume integral equations. *IEEE Trans. Magn.* 40: 1017–1020.

Ozdemir, N.A., Gonzalez-Ovejero, D., and Craeye, C. (2013). On the relationship between multiple-scattering macro basis functions and Krylov subspace iterative methods. *IEEE Trans. Antennas Propag.* 61: 2088–2098.

Pairon, Th., Karki S., Monni, S., and Craeye, C. (2019): 'Efficient analysis of arrays of compact metallic elements devoted to satellite SAR', Proc. 13th Eur. Conf. Antennas Propag., Krakow, Poland, April 2019.

Pan, X.-M., Pi, W.-C., Yang, M.-L. et al. (2012). Solving problems with over one billion unknowns by the MLFMA. *IEEE Trans. Antennas Propag.* 60: 2571–2574.

Pozar, D.M. (2005). *Microwave Engineering*, 3e. New York: Wiley.

Pozar, D. and Schaubert, D. (1984). Scan blindness in infinite phased arrays of printed dipoles. *IEEE Trans. Antennas Propag.* 32: 602–610.

Prakash, V.V.S. and Mittra, R. (2003). Characteristic basis function method: a new technique for efficient solution of method of moments matrix equations. *Microwave Opt. Technol. Lett.* 36: 95–100.

Rana, I.E. and Alexopoulos, N.G. (1981). Current distribution and input impedance of printed dipoles. *IEEE Trans. Antennas Propag.* 29: 99–105.

Rao, S.M. and Wilton, D.R. (1990). E-field, H-field, and combined field solution for arbitrarily shaped three-dimensional dielectric bodies. *Electromagnetics* 10: 407–421.

Rogier, H. (2005). Derivation of fast converging 2-D periodic Green's function series using PML-based Eigenmodes. In: *Proc. Int. Conf. Electromag. Adv. Appl*, 241–244. Torino, Italy.

Roscoe, A.J. and Perrott, R.A. (1994). Large finite array analysis using infinite array data. *IEEE Trans. Antennas Propag.* 42: 983–992.

Rubio, J., Gonzalez, M.A., and Zapata, J. (2005). Generalized-scattering-matrix analysis of a class of finite arrays of coupled antennas by using 3-D FEM and spherical mode expansion. *IEEE Trans. Antennas Propag.* 53: 1133–1144.

Saad, Y. and Schultz, M.H. (1986). GMRES: a generalized minimal residual algorithm for solving nonsymmetric linear systems. *SIAM J. Sci. Stat. Comput.* 7: 856–869.

Schuchinsky, A., Zelenchuk, D., Lerer, A., and Dickie, R. (2006). Full-wave analysis of layered aperture arrays. *IEEE Trans. Antennas Propagat.* 54: 490–502.

Singh, S. and Singh, R. (1993). On the use of Levin's T-transform in accelerating the summation of series representing the free-space periodic Green's functions. *IEEE Trans. Microwave Theory Tech.* 41: 884–886.

Skrivervik, A.K. and Mosig, J.R. (1993). Analysis of finite phase arrays of microstrip patches. *IEEE Trans. Antennas Propag.* 41: 1105–1114.

Song, J., Lu, C.-C., and Chew, W.-C. (1997). Multilevel fast multipole algorithm for electromagnetic scattering by large complex objects. *IEEE Trans. Antennas Propag.* 45: 1488–1493.

Stevanovic, I. and Mosig, J. (2007). Periodic Green's function for skewed 3-D lattices using the Ewald transformation. *Microw. Optical Techn. Lett.* 49: 1353–1357.

Suter, E. and Mosig, J.R. (2000). A subdomain multilevel approach for the efficient MOM analysis of large planar antennas. *Microwave Opt. Technol. Lett.* 26: 270–277.

Valerio, G., Baccarelli, P., Burghignoli, P., and Galli, A. (2007). Comparative analysis of acceleration techniques for 2-D and 3-D Green's functions in periodic structures along one and two directions. *IEEE Trans. Antennas Propag.* 55: 1630–1643.

Valerio, G., Paulotto, S., Baccarelli, P. et al. (2016). Convergent expressions for periodic potentials in stratified media using asymptotic extractions. *IEEE Trans. Magnetics* 52: 7207804.

Van De Water, A.M., De Hon, B.P., Van Beurden, M.C. et al. (2005). Linear embedding via Green's operators: a modeling technique for finite electromagnetic band-gap structures. *Phys. Rev. E* 72: 056704.

Van Der Vorst, H. (2000). Krylov subspace iteration. *Comput. Sci. Eng.* 2: 32–37.

Warnick, K.F. and Jeffs, B.D. (2008). Efficiencies and system temperature for a beamforming array. *IEEE Antennas Propag. Lett.* 8: 565–568.

Warnick, K.F., Davidson, D.B and Buck D. (2021). Embedded element pattern loading condition transformations for phased array modeling. *IEEE Trans. Antennas Propag*. 69: 1769–1774.

Xia, T., Meng, L.L., Liu, Q.S. et al. (2018). A low-frequency stable broadband multilevel fast multipole algorithm using plane wave multipole hybridization. *IEEE Trans. Antennas Propag.* 66: 6137–6145.

Yuan, N., Yeo, T.S., Nie, X.-C., and Li, L.W. (2003). A fast analysis of scattering and radiation of large microstrip antenna arrays. *IEEE Trans. Antennas Propag.* 51: 2218–2226.

14

Measurement of Mutual Coupling Effects

Alpha O. Bah[1] *and Trevor S. Bird*[1,2]

[1]*Global Big Data Technology Centre, University of Technology Sydney, Ultimo, NSW 2007, Australia*
[2]*Antengenuity, Eastwood, NSW 2122, Australia*

14.1 Introduction

Experimental testing and measurement are vital for verification of the design and assessing whether the antenna is performing as expected, as there may be differences that would make the antenna unacceptable for the intended application. The situation is even more important in arrays and antennas where mutual coupling occurs. There are many possible tests that can be carried out on an antenna system, and this is usually more complex and involved for an array antenna. These include interaction of the antenna with the radio frequency (RF) electronics or mechanical parts that can lead to high-order harmonics such as passive intermodulation. Mutual coupling in the array environment can be due to the radiation between the elements, the reflections in the feed network, or the interaction with the RF electronics or mounting platform. These various coupling mechanisms can have significant impacts on the element input impedances, the arrayed gain, and the shape of the array pattern. However, in this chapter, we cover only effects related to the radiative mutual coupling between the antenna elements and in the frequency domain.

A suite of instrumentation is required for testing, and the most important of these are outlined in Section 14.2. The instrumentation has changed over the years, and both old and new approaches will be discussed. For individual elements, in Section 14.3, we will describe the measurement of mutual coupling when all elements apart from the one being tested have a matched load. Radiation patterns under the same passive conditions will be described as well. In arrays, the mutual coupling when the antenna is scanned is important as is the radiation patterns. These are known as active or dynamic measurements. These aspects are detailed in Section 14.4.

14.2 Instrumentation

The basic measurement setup for testing mutual coupling in the frequency domain is illustrated in Figure 14.1. Similar measurements are possible in the time domain using a reflectometer for reflection measurements and an oscilloscope for transmission, but we will not consider this approach here but refer the reader to the references (e.g. Agilent 2007). In the frequency domain, the basic configuration consists of a source and a vector voltmeter/receiver to measure and compare the complex signals from the source and the inputs of the antenna elements. To prevent false levels being

Mutual Coupling Between Antennas, First Edition. Edited by Trevor S. Bird.

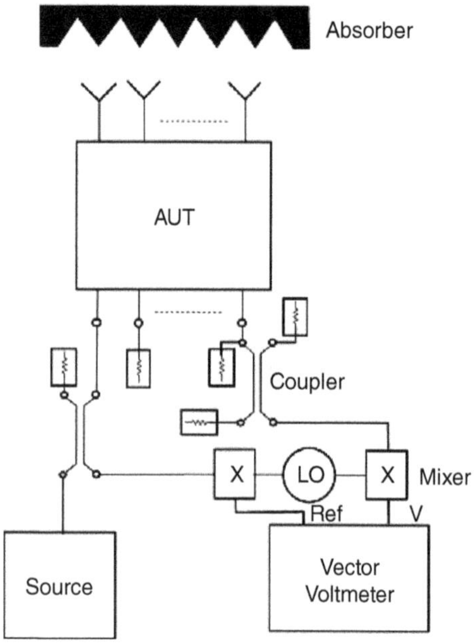

Figure 14.1 Basic setup for measurement of mutual coupling of an antenna under test (AUT). The mixers are required if the operating frequency exceeds the range of the vector voltmeter/receiver.

recorded due to scattering from walls, ceilings, or other nearby objects, a microwave absorbing screen should be placed in front of the array as illustrated in the figure and sufficiently far enough away, which is typically in the Fresnel zone of the antenna under test (AUT).

Depending on the operating frequency of the vector voltmeter, the reference and the coupled voltage can be measured directly or mixed down to a suitable intermediate frequency (IF) for measurement and comparison. Modern network analyzers allow the same type of measurement in a more convenient fashion. The output of either instrument is the relative complex voltage of the transmitted signal to the reference signal.

A vector network analyzer (VNA) measures the amplitude and phase of incident, reflected, and transmitted signals from a device under test (DUT). This may be the full AUT or individual elements of the antenna or components such as an array beamforming network. A basic block diagram of a modern two port VNA is shown in Figure 14.2 (Agilent 2005). There are four general sections to the instrument. The first section is the signal source which provides the stimulus to the system. The signal sources are often synthesized sweepers with either frequency or power level sweeps to provide better measurement performance. The second section comprises the signal separation devices, which together measures the incident reference signal and separates the incident from the reflected signals at the input to the DUT. The signal separation devices are made up of the transfer switch, the splitter, and the directional coupler. The third section of the VNA is the receiver and detector block. The set of receivers within this block is used to determine the changes to the stimulus signal caused by the DUT. The receiver architecture comprises four receivers. There is one main receiver (A or B) and one reference receiver (R1 or R2) for each port. The receiver block downconverts the RF signal to a lower IF signal by mixing it with a local oscillator (LO) signal. The IF signal is then filtered to remove any unwanted harmonics. The filtered signal is digitized using an analog-to-digital converter (ADC), and further digital signal processing is performed to extract the magnitude and phase information. The last and final block of the VNA is the processor and display block which calculates and displays the results in various formats.

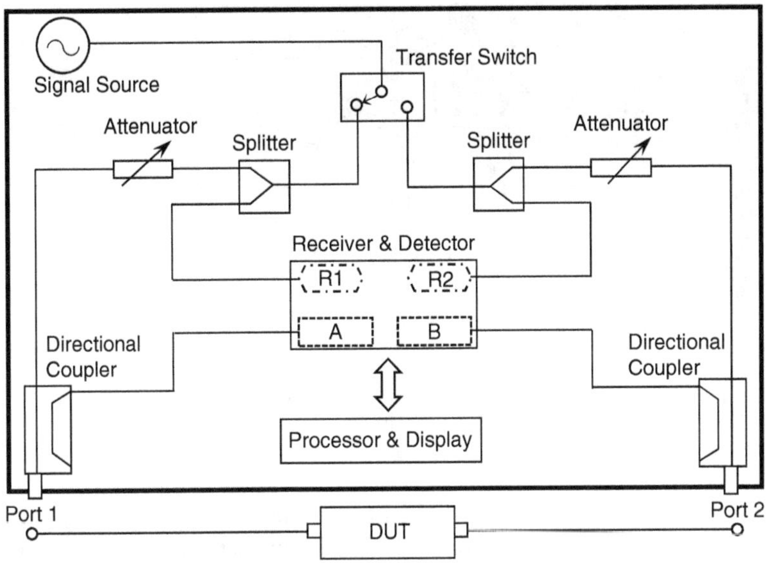

Figure 14.2 Basic setup for measurement of mutual coupling of a device under test (DUT) using a modern two port vector network analyzer. Source: Based on AGILENT TECHNOLOGIES (2005): 'Network Analyzer Basics', online: https://www.keysight.com/upload/cmc_upload/All/BTB_Network_2005-1.pdf.

14.3 Basic Measurement of Static Element Coupling and Radiation

14.3.1 Measurement of Coupling Coefficients

The basic aim of the so-called static coupling or radiation measurement is to measure the complex signal transferring to the DUT through the radiation field when the array is not scanned. This compares with an active measurement when the beam is scanned. The elements of the array scattering matrix $[S]$ can be measured or calculated. If required, the impedance or admittance matrix can be calculated from

$$[Y] = [S + I]^{-1}[I - S] \tag{14.1}$$

where $[I]$ is the unit matrix. As depicted in Figure 14.1, the signal from the source enters the input to the DUT, and part of this signal is used as a reference signal for comparison. The signal exiting from a neighboring DUT is the coupled signal and is measured as a complex transmission coefficient. If the source is connected to antenna m and measurement is taken at the input to antenna n, the result is the complex scattering parameter S_{nm}. If this measurement is repeated for the source connected to the m = 1,…,N DUTs and the coupler connected to n = 1,…,N inputs, the full scattering matrix is obtained. The case when m = n occurs when the signal exiting from the directional coupler is the reflection coefficient at port m, which is found when all other ports are terminated in matched loads (Pozar, 2001).

Using a VNA, the reflection S-parameters of a single port DUT, such as an antenna, can be measured with one of the ports of the VNA. To measure the coupling coefficients of a multiport device such as an antenna array, one port of the VNA is connected to a given port of the antenna array whose coupling coefficients are required with respect to the other ports. The other port of the VNA is then connected to the remaining ports of the antenna array in turn, until all the coupling coefficients are measured.

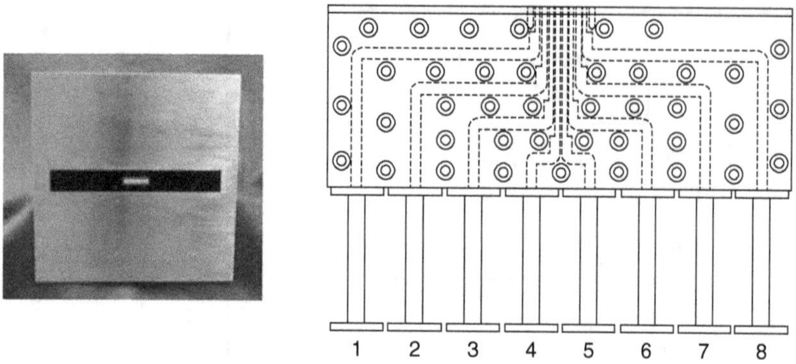

Figure 14.3 10-element array of dielectric loaded square waveguides. (a) Array aperture in a ground plane. The elements at either end are passively excited Source: Trevor S. Bird (b) the feed network to the eight central driven elements.

As an example, the measurement process is described for a 10-element array of vertically polarized elements operating at 59.5 GHz (Bird et al. 2001). The central eight elements are driven while two end-elements are passively excited to control the beam shape of the driven elements. The 10 elements are E-plane stepped horns that are loaded with a dielectric to provide an element spacing of one-half wavelength and are numbered as shown in Figure 14.3. This allows for simple transitions to WR-15 waveguide and gives enough room for an aperture spacing of 2.50 mm. The array was designed using a computer package that is based on mode matching (Bird and Bateman 1992), which uses a sum of rectangular waveguide modes in stepped uniform sections to represent the fields leading to the apertures and also mutual coupling as has been described in Sec. 5.2.3. The eight driven horns are excited by waveguide runs that include transitions from dielectric-filled square waveguide (side length 1.88 mm, ε_r=2.53) to air-filled WR-15 rectangular waveguides, which are the inputs. The software models the 10 apertures as well as the E-plane steps from square dielectric-filled waveguide.

The measurements undertaken included the following:

1. Input reflection coefficients at the eight input ports
2. Mutual coupling coefficients
3. E-plane and H-plane radiation pattern of each element at 59.5 GHz
4. Cross-polarized patterns corresponding to (3)
5. Gain of each element

A VNA was used for the input reflection coefficient and mutual coupling measurements. These measurement procedures are described in the next sub-section. The test-set used had WR15 ports which were calibrated using known standards.

E-plane and H-plane co- and cross-polarized radiation patterns, gain, and relative phase measurements were made in an anechoic chamber with a phase-locked microwave receiver. These radiation measurements were all undertaken at 59.5 GHz and will be described in detail in Section 14.3.2 whilst the gain measurement is described in Section 14.4.4.

14.3.1.1 Input Reflection Coefficient and Insertion Loss

A VNA was used to measure the return loss S_{mm} at the input of each waveguide flange (WR-15 UG385/U) and the signal coupled out of each of the ports S_{mn} $(m \neq n)$ across the frequency range from 59 to 60 GHz when the remaining ports were terminated in matched loads. Reference loads and waveguide lengths were used for calibration. The predicted aperture reflection is shown along the diagonal of Table 14.1.

To obtain the reflection coefficients, conventional S11 measurements were undertaken across the frequency range of 55–65 GHz. Initial measurements showed that the match at ports 3, 6, and 8 were quite poor. To improve the match at these ports, tuning screws were placed at the bend nearest the aperture. As a result, a return loss better than 15 dB was achieved over a 1 GHz bandwidth around the center frequency. The results are shown in Figure 14.4. The insertion loss of each port from input to the aperture via the feed network was estimated by shorting the apertures and measuring the reflection coefficient. The results obtained varied by about 0.5 dB for ports 5 and 8, which have the shortest dielectric waveguide runs in the transitions, to 1.2 dB for ports 4 and 4, which are connected by the longest lengths of dielectric loading.

14.3.1.2 Mutual Coupling Coefficients

Standard VNA transmission measurements (S21) were undertaken to measure the mutual coupling between two ports with all the remaining ports terminated in matched loads. The measured and computed coupling coefficients at 59.5 GHz are compared in Table 14.1. They varied little over the band of interest. The values are typically about 2 dB less than predicted, probably due to losses in the waveguide inputs. The phase departs ±60°, which corresponds to a distance error at 59.5 GHz of 85 μm and is most likely caused by the curved waveguide runs. As can be seen, the elements with the same spacing have similar coupling values. The measurements on the more widely spaced elements, with lower coupling values, are less reliable due to noise in the measurement system.

14.3.2 Measurement of Element Radiation

The measurement of antenna radiation patterns (IEEE Std 149-1979) is adequately covered in several texts (Barlow and Cullen 1950; Hollis et al. 1970; Bird 2016), and only an outline will be given here. There are three main regions of the radiation field that are used to measure the element radiation pattern, and these are illustrated in Figure 14.5. These regions are known as the far-field, the Fresnel or spherical near-field, and the near-field. In a far-field measurement, the distant radiation pattern is measured on a sphere of constant radius. This measurement can be made in transmission or reception as on most occasions the antenna is reciprocal and satisfies the Lorenz reciprocity theorem. The latter type of antenna only is assumed here. The equipment needed in far-field measurement is a rotator in azimuth and/or elevation.

A source antenna is located at a distance greater than or equal to the Rayleigh distance $R = 2\,L^2/\lambda$, where L is the maximum dimension of the AUT, and λ is the shortest wavelength (highest frequency) in the band of interest. The equipment set up is shown in Figure 14.6.

The AUT is placed on a rotator with its phase center as close as possible to the axis or axes of rotation in the vertical and horizontal directions. This can be done with measurement tape and string or more accurately by photogrammetry. The AUT should be located on the rotator according to this estimate of phase center. It is then rotated, and the received signal is recorded with a sensitive receiver. This will give a first estimate of the radiation pattern. If the phase is recorded as the antenna is rotated, this can be used to measure the phase center. The AUT is moved backward and forward along its axial direction on the rotator as depicted in Figure 14.6. Its position should be recorded and note made of this on the antenna. Further adjustment, both forward and backward, will indicate a position for which the phase is most uniform. This is the location of the phase center providing the AUT is in the far-field.

The H-plane co-polar pattern is obtained when the AUT is rotated in a plane perpendicular to the electric field of the source. If the antenna is rotated in the same plane, but with the source

Table 14.1 Measured and computed (shaded) magnitude (dB) and phase (deg.) of the coupling coefficients S_{mn} ($m, n = 1, 2, \ldots 8$) at 59.5 GHz.

	Computed\measured															
Port	**1**		**2**		**3**		**4**		**5**		**6**		**7**		**8**	
1	−31.4	45.5	−17.2	−70.0	−23.8	−141.5	−29.4	−105.2	−33.2	107.0	−36.6	−54.9	−38.2	82.9	−42.7	−102.0
2	−15.5	−114.6	−29.5	79.5	−17.6	−47.5	−24.2	−19.7	−29.2	−168.3	−32.5	*32.3*	−34.3	163.4	−38.3	−16.4
3	−21.9	58.7	−15.3	−112.4	−28.7	82.1	−17.7	−81.1	24.5	127.3	−29.3	−31.8	−31.9	95.7	−36.2	−82.9
4	−26.5	−123.0	−21.6	61.3	−15.2	−112.0	−28.5	82.4	−18.3	166.3	−24.3	2.7	−28.8	125.7	−33.9	−54.8
5	−30.2	57.1	−26.3	−120.3	−21.5	61.8	−15.2	−111.9	−28.5	82.4	−18.2	−142.1	−24.5	−19.3	−30.2	157.9
6	−33.4	−122.0	−30.0	59.8	−26.2	−119.8	−21.5	61.8	−15.2	−112.0	−28.7	82.1	−17.9	−170.6	−24.8	4.1
7	−36.2	58.7	−33.2	−119.7	−30.0	59.8	−26.3	−120.3	−21.6	61.3	−15.3	−112.4	−29.5	79.5	−17.7	134.3
8	−38.9	122.6	−36.2	53.7	−33.4	−122.0	−30.2	57.1	−26.5	−123.0	−21.9	58.7	−15.5	−114.6	−31.4	45.5

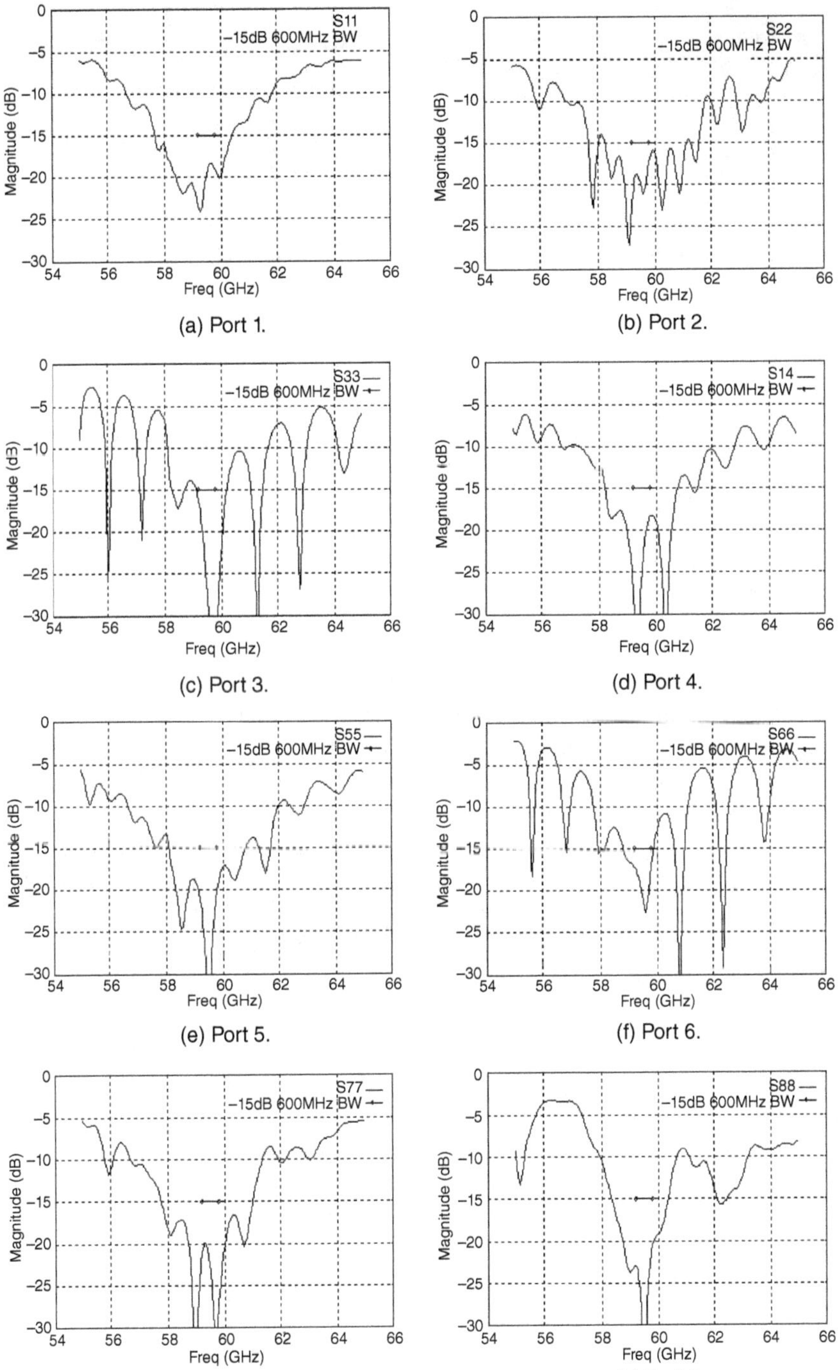

Figure 14.4 Measured reflection coefficient for accessible ports 1–8 of the 10-element array.

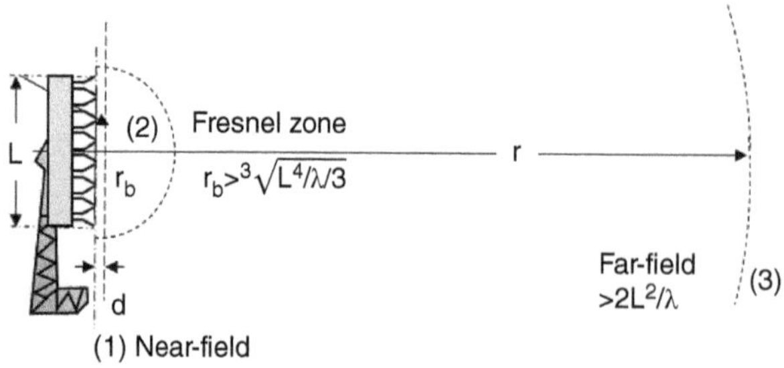

Figure 14.5 Regions for antenna radiation pattern measurement. (1) Near-field, (2) Fresnel zone, and (3) far-field.

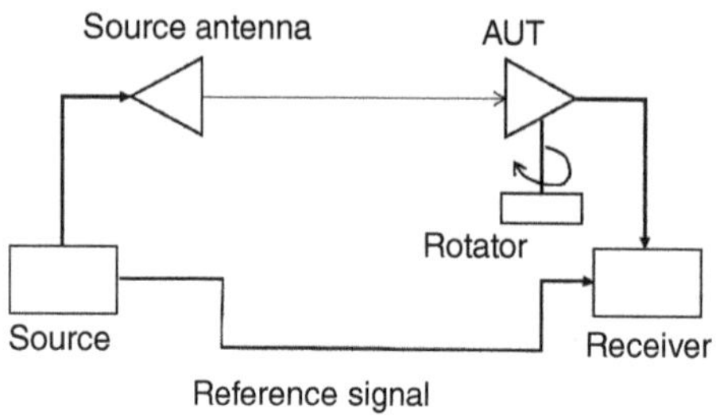

Figure 14.6 Far-field measurement setup.

polarization rotated by 90°, the pattern obtained is the cross-polar pattern in the H-plane. The E-plane pattern is obtained by physically rotating the AUT through 90° as well as the source polarization and making a measurement in the orthogonal plane. The radiation patterns in another plane, such as in the 45°-plane, may be obtained by rotating the AUT through 45° and aligning the polarization in the same plane as the pattern is measured. The cross-polar pattern in the 45°-plane is then found by maintaining the source in the same position as for the co-polar measurement but with the AUT now rotated through −90° relative to the previous location. Thus, the cross-polar pattern is obtained in the plane perpendicular to the plane of parallel to the polarization.

As an example of far-field measurement, the E- and H-plane radiation patterns of the 10-element array at 59.5 GHz were obtained. To measure each element pattern, the array was positioned so that the center face of the element to be measured coincided with the center of azimuthal rotation and the center of polarization rotation for both the element under test and the source horn. As a result, the zero-degree azimuth position of the element under test was directly in line with the source horn. The H-plane patterns are those with the electric field vector perpendicular to the plane of rotation, that is, with the array horizontal. The E-plane patterns are those with the electric field vector parallel to the plane of rotation, that is, with the array vertical. Results are given here for beam 4 near the center of the array and element 1 at the edge, and these are shown in Figures 14.7 and 14.8, respectively. The vertical axes are in decibels relative to an isotropic radiator (dBi).

The E-plane phase patterns have a slight slope. Calculations show that this is commensurate with a slight lateral offset in the mechanical alignment of the antenna of the order of 0.2 mm. A lesser lateral misalignment would explain the slight asymmetry in the H-plane phase patterns.

The cross-polarized patterns were obtained by rotating the source horn through 90° and maintaining the array in the same orientation. The theoretical value for cross-polarization in the principal (E and H) planes is zero, as the antenna is symmetrical in these planes, and so the results are due to asymmetry in the antenna structure, internal or external.

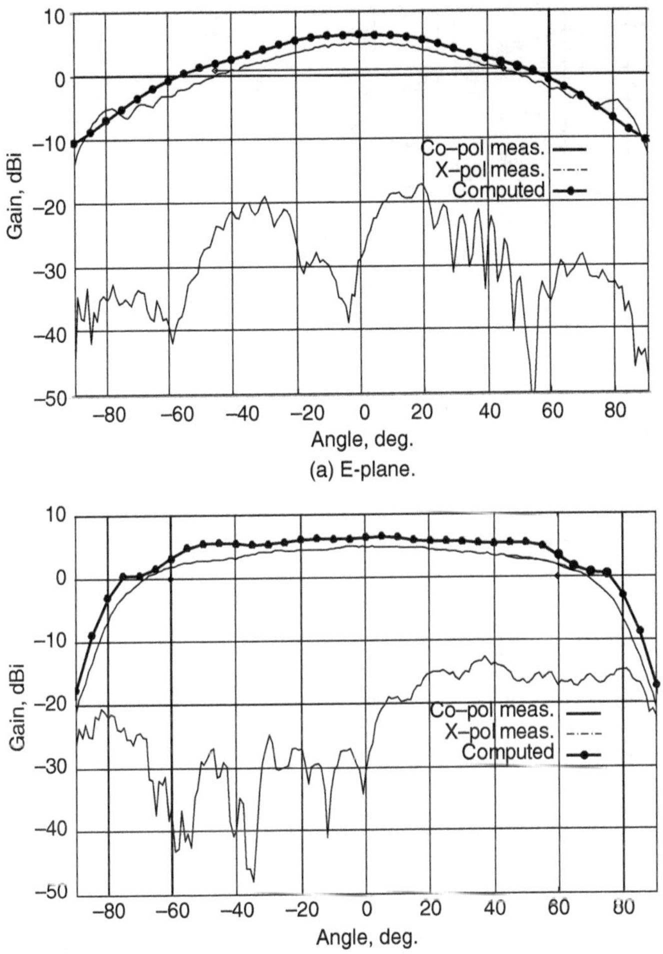

Figure 14.7 Principal plane radiation patterns of beam 4 for the 10-element waveguide array in Figure 14.3 at 59.5 GHz. (a) E-plane and (b) H-plane.

As the far-field region of some arrays can occur several kilometers or more away from the source depending on their size, a very long far-field range may be required. Therefore, alternatives to measuring the far-field of aperture antennas in an outdoor range are required. One particularly powerful technique, which was developed in the early 1960s as the necessary computational power became available, is probe-compensated near-field measurement (NFM) (Bird 2016) (see Figure 14.9). Today, this technique is very widely used in research as well as in industry. The technique requires the near-field of the AUT to be sampled on a specified surface, usually a plane, a cylinder, or a sphere, at about 5–10 wavelengths away, and these data are numerically transformed to obtain the far-field. Some effects of the sampling probe on the measurement can be included in this transformation. Probe-compensated NFM has been shown to be time and cost effective, and results can be obtained that are at least comparable to those obtained with a far-field range or sometimes better. Stray reflections can be controlled by means of microwave absorber, and this, plus the error involved in positioning the probe, usually limits the useful frequency range

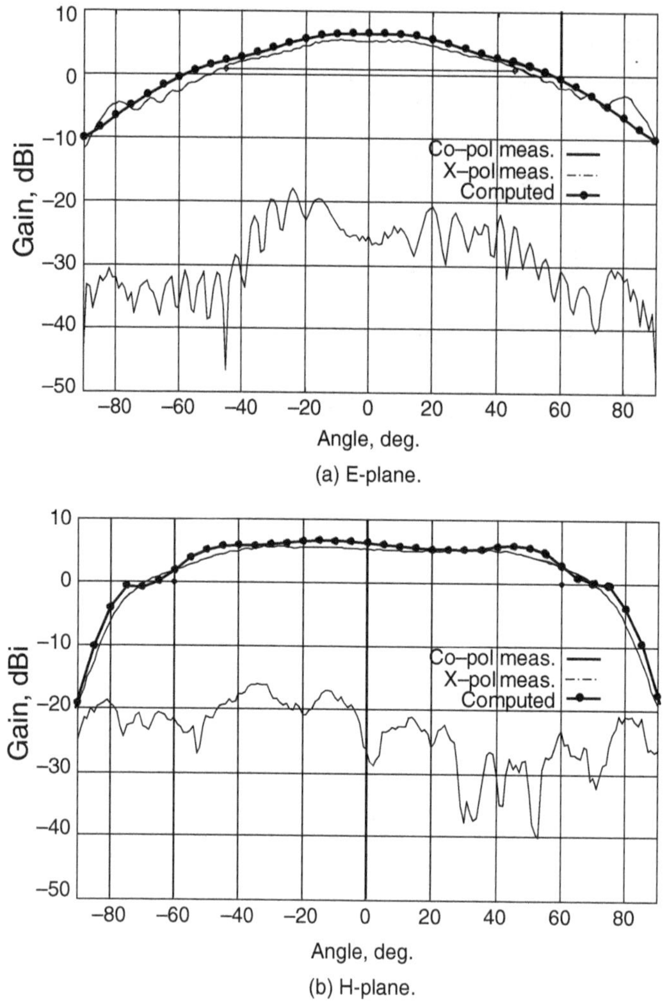

Figure 14.8 Principal plane radiation patterns of beam 1 for the 10-element waveguide array in Figure 14.3 at 59.5 GHz. (a) E-plane and (b) H-plane.

of NFM from low UHF to around 1 THz. As NFM is more complicated than other techniques, it must be automated, and staff must be highly skilled. Also, the probe antenna must be extensively and accurately calibrated at all operating frequencies. The measurement is made under computer control, and the same computer can also be used to transform the near-field data to the desired region in real time, which can be an advantage for assessment and comparison. This may not be possible with electrically large antennas as these measurements can take several hours to complete.

14.4 Measurement of Active Element Coupling and Array Radiation

14.4.1 Measurement of Active Element Patterns

The active element pattern of an antenna array is the radiation pattern of the array when only a single element is excited and all other array elements are terminated in matched loads (Pozar

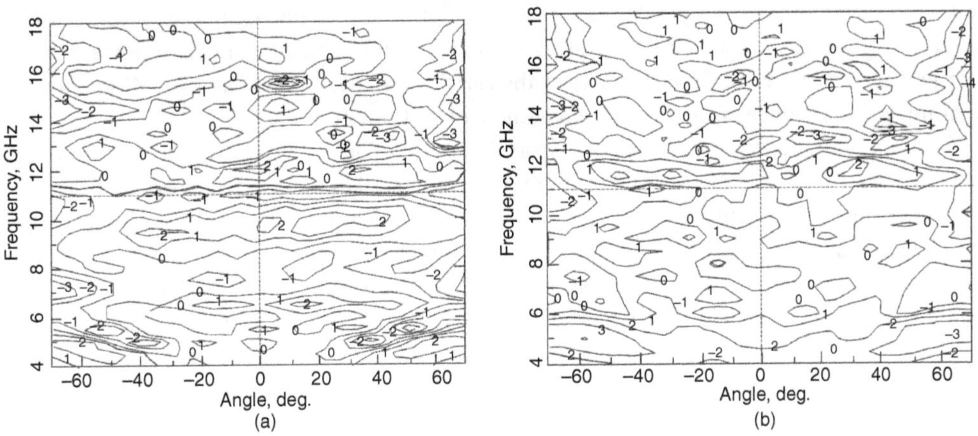

Figure 14.9 Measured normalized embedded pattern in the E-plane. (a) Central element; and (b) edge element. Source: From HOLTER, H. (2007): 'Dual-polarized broadband array antenna with BOR-elements, mechanical design and measurements', IEEE Trans. Antennas Propagat., Vol. 55, no. 2, pp. 305–312. © 2018, IEEE.

1994). The active element pattern g_a, of a given array element, can be approximated from its active reflection coefficient Γ_A (Amitay et al. 1972, pp. 23–25), according to the following equation:

$$g_a(\theta, \phi) = \frac{4\pi A}{\lambda^2}(1 - |\Gamma_A(\theta, \phi)|^2)\cos\theta \tag{14.2}$$

where $A\cos\theta$ is the aperture projected in the scan direction θ which varies as $\cos\theta$, $A = dx\,dy$ is the aperture area of the unit cell, and Γ_A is the active reflection coefficient. The active reflection coefficient is obtained from the measured coupling coefficients (see Section 14.4.2). For an equally spaced triangular grid array,

$$g_a(\theta, \phi) = \frac{2\sqrt{3}\pi s^s}{\lambda^2}(1 - |\Gamma_A(\theta, \phi)|^2)\cos\theta \tag{14.3}$$

where s is the element spacing By plotting the frequency (wavelength) versus scan angle θ in a two-dimensional contour plot, the active element pattern for the range of scan angles can be demonstrated. The patterns shown in Fig. 14.9 (Holter 2007) are calculated from Eq. (14.2) normalized, i.e. with Γ_A set to zero, which results in the "ideal" element pattern. A positive (negative) value in the plots means that the measured element gain is higher (lower) than the gain pattern calculated from Eq. (14.2). The shape of the active element pattern in terms of dips and nulls gives an indication of the shape of the fully excited phased array when scanned to those angles.

The variation of the active element pattern with scan angle and frequency is a particularly useful plot. In Figure 14.9, small variations can be seen between the central and edge elements. Ideally, the pattern should be close to 0 dB on average, which should get better as the frequency increases. For the central element, this is approximately the case. For the edge element, the pattern in Figure 14.7b is asymmetric about 0°, which can be assessed through these types of plots. The active element pattern can also be measured in an anechoic chamber.

14.4.2 Measurement of Array Radiation Patterns

Array radiation patterns can be measured using a variety of techniques as described in Sec. 14.3.2 and the technique chosen depends on the size and geometry of the array, the type of array elements, and the required level of acceptable accuracy. For widely spaced element antenna array, in which the effects of mutual coupling on the element patterns are not significant, and is composed of uniformly spaced elements of similar structure, the classical pattern multiplication (PM) method can be employed to produce fairly accurate assessment of performance.

In cases where the elements are closely-spaced, say a fraction of a wavelength λ, and the effects of mutual coupling on the individual element patterns as well as the full array pattern are significant, other methods are appropriate. These methods depend on the size of the array and whether the elements in the array are similar or different. The element patterns in these cases are usually referred to as the active or embedded or scanned elements patterns (Pozar 1994; Hansen 1998; Holter 2007). The active element pattern shape will depend on the position of the excited element in the array. The terminated elements will be passively excited due to mutual coupling with the excited element. The measured or computed active element patterns can be used to calculate the pattern of the fully excited array.

The analysis of closely spaced arrays can take on three forms (Kelley 1993) depending on the size of the array. For a small closely spaced array of any configuration and any element type, the *unit excitation active element pattern* (UE-AEP) method can be used to accurately compute the array pattern. For a medium-sized closely spaced array composed of either the same kind of elements or a mixture of elements with multiple periodicities, the *hybrid active element pattern* (H-AEP) method can be used to predict the array pattern. The effect of the edge elements is more severely prominent in these types of arrays. For a large or practically infinite closely spaced array composed of either the same kind of elements or a mixture of elements with multiple periodicities, the *average active element pattern* (A-AEP) method can be used to predict the array pattern.

The number of elements used to define the array size as either small, medium, or large is dependent on the level of acceptable accuracy, the availability of computing resources, the allocated time for design and testing, and the available budget assigned to the project. However, as a general rule, arrays with less than 10 elements can be considered small and those with more than 100 elements can be considered large. The methods just mentioned will be analyzed in more detail in the sections to follow.

14.4.2.1 Pattern Multiplication Method

The PM method applies to arrays with widely spaced elements such that the effects of mutual coupling on the element patterns are relatively small. In this case, the isolated element patterns are essentially the same as those in the array environment; hence, it can be factored out of the array pattern expression. Nevertheless, the effect of mutual coupling can significantly impact the element's active input impedance as compared to the isolated element case. For the method to work, the array should be composed of elements that are wide part and very similar in nature. The PM method can be used to obtain the antenna array pattern of widely spaced periodic arrays fairly accurately as long as the element feed currents, or its active input impedances are known. In this case, the measurement of one element near the center of the array will be reasonably representative of the actual element pattern, and this can be used with the array factor to give a good estimate of the overall radiation pattern.

The array radiation pattern, P_A, can be derived from the product of the isolated element radiation pattern, p_{iso}, and the array factor

$$P_A(\theta, \phi) = p_{iso}(\theta, \phi) \cdot AF(\theta, \phi) \tag{14.4}$$

where $p_{iso}(\theta, \phi)$ is the element power pattern either measured as described in Section 14.3.2 or computed; the array factor is

$$AF(\theta, \phi) = \sum_{n=1}^{N} I_n e^{j\Theta_n} \tag{14.5}$$

where I_n is the complex current (complex weight or excitation) induced on the nth radiating element, Θ_n is the observation angle with respect to the nth element, and N is the number of voltage sources applied to the array. The observation phase Θ_n is the scalar product of the magnitude of the wave vector $\boldsymbol{k}$ and the observation position vector $\boldsymbol{r}$

$$\Theta_n = |\boldsymbol{k}|.\boldsymbol{r} = k.\boldsymbol{r} \tag{14.6}$$

where

$$\mathbf{k} = \hat{\boldsymbol{x}}k_x + \hat{\boldsymbol{y}}k_y + \hat{\boldsymbol{z}}k_z \tag{14.7a}$$

$$k = \frac{2\pi}{\lambda} = |\mathbf{k}| = \sqrt{k_x^2 + k_y^2 + k_z^2} \tag{14.7b}$$

wherein $k_x = ksin\theta\ cos\phi$, $k_y = ksin\theta\ sin\phi$, $k_z = kcos\theta$, λ is the wavelength, and (θ, ϕ) are the observation angles $(0° \leq \theta \leq 180°\ and\ 0° \leq \phi \leq 360°)$.

The observation position vector, $\boldsymbol{r}$, is defined as the scalar product of the unit radial vector from the coordinate origin in the (θ, ϕ) direction, $\hat{\boldsymbol{r}}$, and the position vector from the origin to the center of the nth element, $\boldsymbol{r_n}$

i.e.

$$\boldsymbol{r} = \boldsymbol{r_n}.\hat{\boldsymbol{r}} \tag{14.8a}$$

$$\boldsymbol{r_n} = \hat{\boldsymbol{x}}x_n + \hat{\boldsymbol{y}}y_n + \hat{\boldsymbol{z}}z_n \tag{14.8b}$$

and

$$\hat{\boldsymbol{r}} = \hat{\boldsymbol{x}}sin\theta\ cos\phi + \hat{\boldsymbol{y}}sin\theta\ sin\phi + \hat{\boldsymbol{z}}cos\theta \tag{14.8c}$$

allowing Eq. (14.8a) to be expanded as

$$\boldsymbol{r} = x_n sin\theta\ cos\phi + y_n sin\theta\ sin\phi + z_n cos\theta$$

Thus, the observation phase is

$$\Theta_n = k.\boldsymbol{r} = kx_n sin\theta\ cos\phi + ky_n sin\theta\ sin\phi + kz_n cos\theta$$

or equivalently

$$\Theta_n = k_x x_n + k_y y_n + k_z z_n \tag{14.9}$$

The complex array currents are given by

$$I_n = |I_n|e^{j\Theta_{n0}} \tag{14.10}$$

where Θ_{n0} is the phase conjugate of Θ_n that allows the main beam to scan to a given scan angle (θ_0, ϕ_0).

$$\Theta_{n0} = -(kx_n sin\theta_0\ cos\phi_0 + ky_n sin\theta_0 sin\phi_0 + kz_n\ cos\theta_0) \tag{14.11}$$

Finally, the array factor is expressed as

$$AF(\theta, \varphi) = \sum_{n=1}^{N} |I_n| e^{j(\Theta_n + \Theta_{n0})} \tag{14.12}$$

If the isolated element pattern $p_{iso}(\theta, \varphi)$ has been measured and the array factor expression in Eq. (14.12) has been obtained, the antenna array pattern can be easily computed using Eq. (14.4). Clearly, the PM method is not applicable to arrays with different element current distributions such as arrays with electrically large or dissimilar elements. For a uniformly excited array with unity amplitude, when the observation angles θ and φ are equal to the scan angles θ_0 and φ_0, the array factor, $AF(\theta, \varphi) = N$.

14.4.2.2 The Unit Excitation Active Element Pattern Method

When the array is small and the array elements are closely spaced, the unit excitation active element pattern (UE-AEP) method can be employed to accurately estimate the antenna array pattern. This method works equally well for arbitrarily configured arrays or arrays of different element types. The UE-AEP represents the pattern radiated by the entire array when only one element is actively excited with a unit voltage and the remaining elements are terminated in matched loads. This setup is sometimes referred to as a passively terminated array. Terminated elements also radiate due to the parasitic excitation from the active element. The UE-AEP for each element in the array is unique and contains all the effects of radiative mutual coupling due to the surrounding elements and all mismatch losses.

The total current distribution on a fully excited antenna array is the sum of the unit excitations of the individual array elements scaled by their complex excitation voltages, V_n (Kelley 1993). Hence, the far-field pattern radiated by a fully excited antenna array can be expressed as the superposition of the UE-AEPs of the individual elements, p_{ue}^n

$$P_A(\theta, \phi) = \sum_{n=1}^{N} V_n p_{ue}^n(\theta, \phi) \tag{14.13}$$

The patterns obtained using the UE-AEP method have the spatial phase information $e^{j\Theta_n}$ already embedded within them because they are referenced to the origin of the array coordinate system as well as mutual coupling. Using this method in measurement, the array should be mounted with its center facing the source antenna at the beginning of the measurement cycle. Further alignments between the individual array elements and the source antenna are not required for the rest of the measurements. The UE-AEPs of the individual elements could be replaced by the phase-adjusted UE-AEP, which is $p_{pe}^n = p_{ue}^n e^{-j\Theta_n}$. This is referenced to the individual elements. In order to place the center of the phase-adjusted UE-AEP to the array origin, we multiply p_{pe}^n with the spatial phase term $e^{j\Theta_n}$.

14.4.2.3 The Average Active Element Pattern Method

If the array is large, it can be assumed that all the elements will see a uniform neighboring environment with identical radiation patterns. In this scenario, the active element pattern can be approximated by identical elements in the array. The unit-excitation element pattern of the center array element can be selected to represent the average element pattern, $\mathbf{g}_{av}$; thus,

$$\mathbf{E}(\theta, \phi) = \mathbf{g}_{av}(\theta, \phi) \sum_{n=1}^{N} V_n e^{j\Theta_n}. \tag{14.14}$$

In the limit of an array comprised of an infinite number of elements, Eq. (14.14) yields an exact array pattern expression since the phase-adjusted active element patterns of all elements in such an array are identical.

In an infinite array analysis, the following assumptions are made:

- All elements are identical
- Edge effects are insignificant
- It uses uniform excitation amplitude
- The unit cell active element pattern is equal to the average active element pattern.

14.4.2.4 The Hybrid Active Element Pattern Method

Though the average element pattern approach can produce accurate results for large, equally spaced arrays, it usually fails for small arrays in which mutual coupling effects cause significant variation in active input impedances and element patterns across the array. The coupling inherent in large arrays can be used to both calibrate and predict the radiation patterns as shown by Aumann et al. (1989). The *hybrid active element pattern(H-AEP)* method can be used to analyze medium-sized periodic arrays very accurately. In this approach, the array elements are divided into an *interior element* group and an *edge element* group. The edge elements are typically taken to be the first few active elements on each end of a linear array, or the first few rows and columns of active elements on the sides of a planar array. If any passive or "guard" elements are present in the array, they are not included in either group. The effect of the passive elements is included in the active element patterns. If the number of interior elements is N_i, and the number of edge elements is N_e, the total number of active array elements is equal to the number of excitation sources; thus, $N_s = N_i + N_e$. The array pattern is computed using phase-adjusted element patterns as follows:

$$\mathbf{E}(\theta,\phi) = \sum_{n=1}^{N_i} V_n \mathbf{g}_p^n(\theta,\phi)\, e^{j\Theta_n} + \sum_{m=1}^{N_e} V_m \mathbf{g}_p^m(\theta,\phi)\, e^{j\Theta_m} \tag{14.15}$$

where the $\boldsymbol{n}$ subscript refers to interior elements and the $\boldsymbol{m}$ subscript refers to edge elements.

The phase-adjusted element patterns of the interior elements are often similar, which enables Eq. (14.15) to be rewritten as

$$\mathbf{E}(\theta,\phi) = \mathbf{g}_{av}^i(\theta,\phi) \sum_{n=1}^{N_i} V_n\, e^{j\Theta_n} + \sum_{m=1}^{N_e} V_n \mathbf{g}_p^m(\theta,\phi)\, e^{j\Theta_m} \tag{14.16}$$

where $\mathbf{g}_{av}^i$ is the A-AEP of the interior elements.

Equation (14.16) can be used to approximate the pattern of a medium-sized, equally spaced array. Only $N_e + 1$ element patterns must be stored in memory in order to use the hybrid method. This is much less than the $N_i + N_e$ patterns that must be stored in order to use the UE-AEP method.

14.4.3 Measurement of Input Mismatch and Coupling

The input impedance of an active element in the array can be quite different from that of an isolated element as a result of mutual coupling. The active input impedance values also vary depending on the position of the element in the array and the applied excitation for a given scan angle. Therefore, reasonable active impedance values should be obtained via an appropriate impedance matching network for all elements in the array for the desired frequency band and range of scan angles to ensure proper operation.

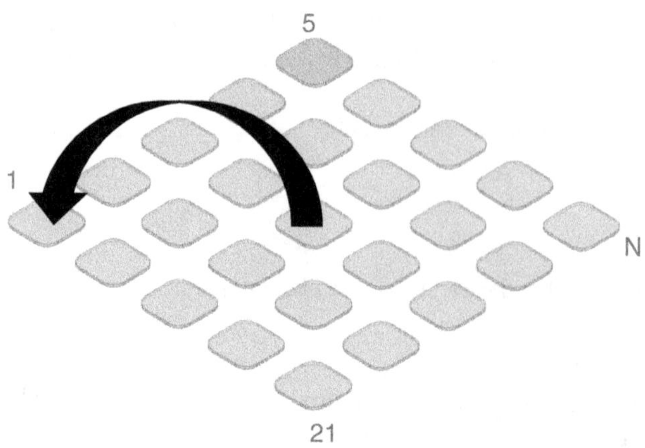

Figure 14.10 Mutual coupling measurement from the center element to all the other elements of the array. When measuring the mutual coupling between any two elements, the other N-2 elements should be terminated with load impedances equal to their input impedance, which is usually 50 Ω.

In the array environment where scanning is required, a more useful concept is the active coupling or scan impedance. These are defined as the coupling or input impedance of an element with all other elements excited as in operation. As a result, the active impedance changes when the excitation is changed due to changed relative amplitudes and phases at the inputs. As in the static case, when the scan impedance is known, the associated scattering matrix can be calculated from Eq. (14.1).

The scattering matrix can be obtained by a variety of ways. One of the earliest techniques is to use a waveguide simulator. In such a simulator, the walls of the unit cell are assumed to be perfectly conducting surface. As a result, the array appears to be an infinite array. There are some symmetry planes that have similar properties to a magnetic conductor and scan angles that allow perfectly conducting walls. As a result, these walls can be a metallic waveguide that is terminated in a section of the array comprising a small number of elements. The most common lattice configurations for large arrays are rectangular and triangular configurations, and these can be readily created with a waveguide simulator. Rectangular waveguide allows principal plane scans. A triangular simulator allows scans along the symmetry planes. Hexagonal array simulators can also be created from combinations of rectangular and triangular waveguides.

14.4.3.1 Mutual Coupling Coefficient Method

One of the ways of obtaining the active or scan reflection coefficient of a given radiating element within an antenna array is by measuring all the mutual coupling coefficients between that element and all the other elements in the array as illustrated in Figure 14.10. During each measurement period, all the other nonactive elements should be terminated with load impedances equal to their input impedance, which is usually 50 Ω. The active reflection coefficient ($\Gamma_a(\theta_o, \phi_o)$) for element i, scanning in the (θ_o, ϕ_o) direction, can then be estimated by means of (Amitay et al. 1972, p. 22)

$$\Gamma_a(\theta_o, \phi_o) = \sum_{n=1}^{N} S_{ni}|a_i|e^{j\Theta_{no}(x_n, y_n)}. \tag{14.17}$$

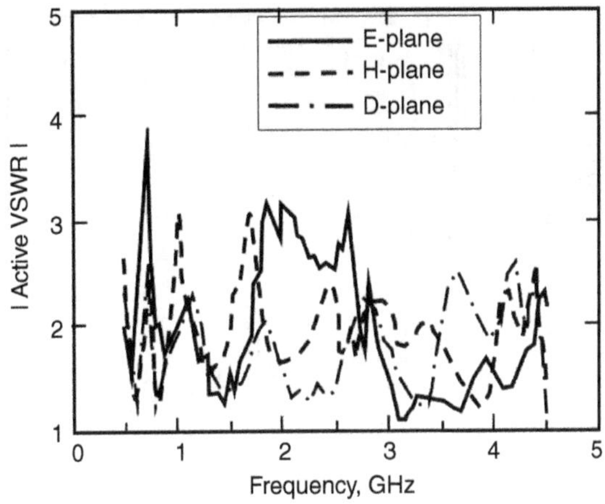

Figure 14.11 The scanning active VSWR of the 10×10 array obtained from the measured coupling coefficients across all three planes. Results shown are for 50° in the E-plane, 55° in the H-plane, and 65° in the D-plane.

Substituting Eq. (14.11) into Eq. (14.17) for the x- and y-directions and knowing that $x_n = xd_x$ and $y_n = yd_y$,

$$\Gamma_a(\theta_o, \phi_o) = \sum_{n=1}^{N} S_{ni} |a_i| e^{-jk(xd_x sin\theta_o cos\phi_o + yd_y sin\theta_o sin\phi_o)} \tag{14.18}$$

where S_{ni} is the measured passive mutual coupling coefficients between the reference radiating element, i, and the nth element in the array, $|a_i|$ is the magnitude of the excitation of the radiating element, N is the total number of elements, k is the free space wave number, x and y are the element numbers, and d_x and d_y are the element spacings along the $\boldsymbol{x}$- and $\boldsymbol{y}$-directions, respectively. To speed up measurement times, a power divider can be used to combine rows or columns in which case the number of measurements will be reduced to the number of rows or columns.

This method is suitable for small- and medium-sized arrays. The problem with this method is that the passive mutual coupling coefficients of all the N radiating elements have to be measured relative to the other elements in the array. The number of measurements to be carried out can quickly get out of hand as the array element numbers increase.

As an example of this approach, the active reflection coefficient of a 10×10 array of tightly coupled elements and wide-angle impedance matching (WAIM) of the type described in Section 6.3.3 was evaluated to assess the overall performance (Bah et al. 2019). The active area of this array is the 6×10 inner elements. The active reflection coefficient of the built array was calculated from the measured coupling coefficients between the element in row 5 and column 5 (element 5, 5) and every other element in the array. All measurements were made with a Keysight Technologies N5225A PNA network analyzer. During each measurement, the 98 nonactive elements were terminated in matched loads. The active reflection coefficient for element i, scanning in the θ_o, ϕ_o direction, can be estimated using Eq. (14.18). The active VSWR was deduced from the active reflection coefficient for broadside and for scanning in the E, H, and D (diagonal) planes. The active VSWR in the three principal planes is shown in Figure 14.11. The broadside active VSWR of the simulated 10×10 finite array with 60 active elements and the measured results obtained from the measured coupling coefficients are displayed in Figure 14.12. There is good general agreement between the two results across the band 0.80 – 4.38 GHz.

14.4.3.2 Directional Coupler Method

A directional coupler can be used to obtain a measurement of each term in Eq. (14.18) directly in two steps. This is illustrated in Figure 14.13 for an array of N elements. The signal coupled

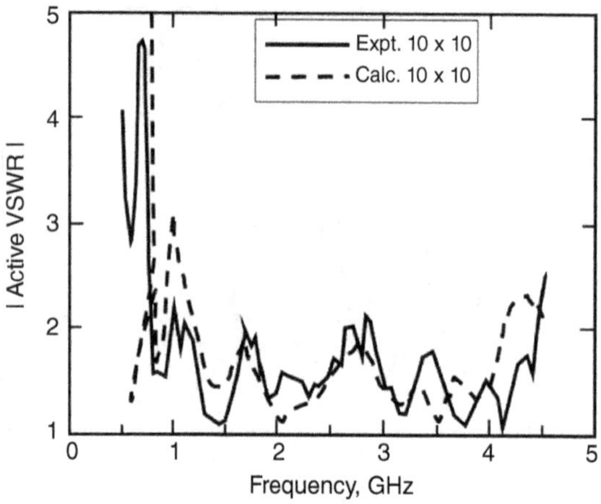

Figure 14.12 The measured broadside active VSWR of a 10×10 array with 60 active elements compared with simulations. The measured results were obtained from the measured coupling coefficients.

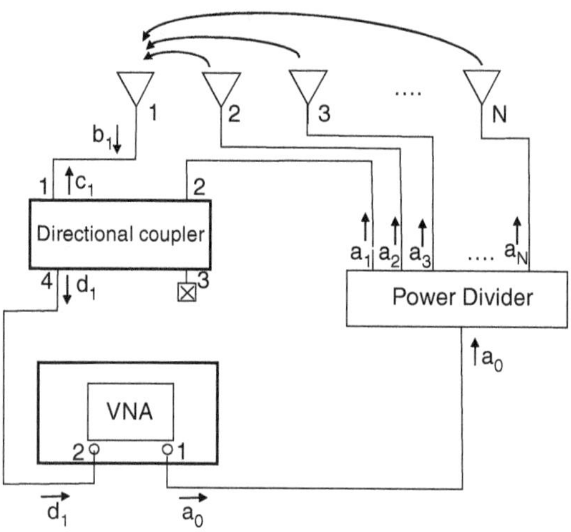

Figure 14.13 First step in the measurement of active coupling coefficient using a directional coupler and vector network analyzer (VNA). In the second step, a load is connected to port 1 of the directional coupler.

to element 1 from all other N elements in the array is measured directly. This is b_1. If the directional coupler has high directivity $d_1 \cong b_1$, and from the VNA, we get a $S_{21}^{(1)} = d_1/a_o$. In the second step, a load is connected to port 1 of the directional coupler, and a second reference transmission $S_{21}^{(2)}$ coefficient is measured with the VNA. The active reflection coefficient of port 1 is

$$S_{11}^a = S_{21}^{(1)}/S_{21}^{(2)}. \tag{14.19}$$

In general, if the excitation conditions of the array elements do not change as the loading is changed, the active coupling coefficient S_{mn}^a can be estimated (Zhang et al. 2017) from the measured active reflection coefficient S_{mm}^a using

$$S_{mn}^a = S_{mm}^a \, a_m/a_n. \tag{14.20}$$

14.4.3.3 Power Divider Method

In this method, an N-port power divider is utilized to measure the active reflection coefficient of the whole array. This method is good for small arrays, but as the element numbers increase, the cost becomes increasingly prohibitive, and the measurement errors increase. It is also not possible to measure the active reflection coefficient of a given element of the array relative to the other elements. A failed element will adversely affect the measurement results.

14.4.4 Measurement of Gain

The gain of static antennas is usually measured by two approaches. The most common method is gain comparison. In this method, the maximum signal is recorded with a reference antenna (the standard) for which gain is accurately known from a previous measurement. The maximum signal of the AUT is recorded under the same conditions such as frequency and similar power level to minimize errors due to nonlinearity of the receiver. From these measurements, the gain of the AUT is calculated as follows:

$$Gain\,(AUT, dB) = Gain\,(\text{Reference, dBi}) - Signal\,(\text{Reference, dBi}) + Signal\,(AUT, dB). \tag{14.21}$$

As an example, suppose a reference horn has a gain of 20 dBi. If the maximum signal received from an AUT is 17.5 dB while for the reference the measured signal is 11.2 dB, the gain of the AUT given by Eq. (14.21) is 26.3 dBi.

Another technique that is used is the "three antenna measurement method." The basis of this method is the Friis transmission formula (Kraus and Carver 1973). The received signal is

$$P_{rx} = P_{tx} G_{tx} G_{rx} \left(\frac{\lambda}{4\pi r}\right)^2 \tag{14.22}$$

where r is the distance between the source antenna and receiver, G_{tx} and G_{rx} are the antenna gains of the transmitting and receiving antennas, and P_{tx} and P_{rx} are the transmitted and received powers in Watts.

The signal of the AUT and two reference antennas is measured as described above for the gain comparison method. However, on this occasion, each of the three antennas is used in turn as the source antenna. All three combinations of measurements are used, and Eq. (14.22) for the three cases results in

$$Gain\,(AUT, dBi) + Gain\,(\text{Ref1, dBi}) = 20\,log_{10}\left(\frac{4\pi r}{\lambda}\right) + 10\,log_{10}\left(\frac{PowerRxAUT}{PowerTx1}\right) \tag{14.23a}$$

$$Gain\,(\text{Ref2}, dBi) + Gain\,(\text{Ref1, dBi}) = 20\,log_{10}\left(\frac{4\pi r}{\lambda}\right) + 10\,log_{10}\left(\frac{PowerRxRef2}{PowerTx1}\right) \tag{14.23b}$$

$$Gain\,(AUT, dBi) + Gain\,(\text{Ref2, dBi}) = 20\,log_{10}\left(\frac{4\pi r}{\lambda}\right) + 10\,log_{10}\left(\frac{PowerRxAUT}{PowerTx2}\right). \tag{14.23c}$$

where *Gain* (*AUT*, *dBi*) is the gain in dBi of the AUT, *PowerRxRefk* is the received power with antenna k as a reference, and PowerTx*k* (k = 1,2) refers to when reference antenna k is used as the source antenna. From the three equations, the three unknowns, the gain of each antenna can be calculated by simultaneous solution of Eq. (14.23).

To continue with the example of the 10-element array shown in Figure 14.3 , the gain comparison method was used to measure the maximum gain of each beam. The gain of each beam was used to scale the radiation pattern of each beam as evidenced by Figures 14.7 and 14.8. As well as determination of peak gain, there was, for the application, a gain target of 0 dBi at ±60° in the horizontal or H-plane and 1 dBi at ±45° in the E-plane. The H-plane target was met for all ports and the E-plane met for ports 1, 2, and 7. The worst case for the other ports was port 3 with −0.15 dBi at +45°. The patterns show that the 1 dBi level was met for ±45°. The gain is lower than expected due to higher losses in the antenna, most likely dominated by the dielectric loss.

The array gain can also be computed from the embedded element patterns (Fenn 2018), as detailed below. The embedded element radiated power density at a distance r is given by

$$P_{de} = \frac{EIRP_e}{4\pi r^2} = \frac{P_{te} g_{te}(\theta, \varphi)}{4\pi r^2} \tag{14.24}$$

where $EIRP_e$ is the effective isotropic radiated power of the embedded element, P_{te} is the transmit power accepted by the embedded element, and g_{te} is the transmit gain of the embedded element including resistive losses. The effective isotropic radiated power is the total power that would be radiated by a hypothetical isotropic antenna to give the same signal strength as the actual source in the direction of the antenna's strongest beam. From Eq. (14.24),

$$g_{te}(\theta, \varphi) = \frac{4\pi r^2 P_{de}}{P_{te}} \tag{14.25}$$

In the far-field, the instantaneous power density of the embedded element can be expressed in terms of its electric field intensity

$$P_{de} = \frac{|E_e(\theta, \varphi)|^2}{\eta_0} \cong \frac{|E_e(\theta, \varphi)|^2}{120\pi} \tag{14.26}$$

Substituting Eq. (14.26) into Eq. (14.25) yields

$$g_{te}(\theta, \varphi) = \frac{r^2 |E_e(\theta, \varphi)|^2}{30\, P_{te}}. \tag{14.27}$$

This embedded element gain pattern includes all mutual coupling and mismatch losses. Rearranging Eq. (14.27) above results in

$$|E_e(\theta, \varphi)| = \frac{\sqrt{30 g_{te}(\theta, \varphi) P_{te}}}{r} \tag{14.28}$$

For the nth array element,

$$|E_{en}(\theta, \phi)| = \frac{\sqrt{30 g_{ten}(\theta, \phi) P_{ten}}}{r} = \frac{\sqrt{30 g_{ten}(\theta, \phi)}}{r} A_n \tag{14.29}$$

where g_{ten} is the embedded gain of the nth element, $\sqrt{P_{ten}} = A_n$, and A_n is the nth array element amplitude or array weighting.

The complex embedded element electric field intensity is defined as

$$E_{en}(\theta, \phi) = |E_{en}(\theta, \phi)| e^{j\Theta_{en}(\theta,\phi)}. \tag{14.30}$$

The array pattern can be obtained by the superposition of the individual embedded element patterns with the required excitations (array amplitude taper and phase) to steer the beam

$$E_{array}(\theta, \phi) = \sum_{n=1}^{N} E_{en}(\theta, \phi) e^{j\Theta_{n0}} \tag{14.31}$$

Substituting Eqs. (14.29) and (14.30) into Eq. (14.31) gives

$$\begin{aligned} E_{array}(\theta, \phi) &= \sum_{n=1}^{N} |E_{en}(\theta, \phi)| e^{j(\Theta_{en}+\Theta_{n0})} \\ &= \sum_{n=1}^{N} \frac{\sqrt{30 g_{ten}(\theta, \phi)}}{r} A_n e^{j(\Theta_{en}+\Theta_{n0})} \end{aligned} \tag{14.32}$$

which has a maximum value

$$[E_{array}(\theta, \phi)]_{max} = \sum_{n=1}^{N} \frac{\sqrt{30 g_{ten}(\theta, \phi)}}{r} A_n \tag{14.33}$$

Modifying the expression for the element gain given in Eq. (14.28) allows the array gain to be expressed as

$$G_{array}(\theta, \phi) = \frac{r^2|E_{array}(\theta, \phi)|^2}{30\, P_{tarray}} \tag{14.34}$$

where

$$P_{tarray} = \sum_{n=1}^{N} A_n^2.$$

Substituting Eq. (14.32) into Eq. (14.34) gives

$$G_{array}(\theta, \phi) = \frac{\left|\sum_{n=1}^{N} \sqrt{g_{ten}(\theta, \phi)} A_n e^{j(\Theta_{en}+\Theta_{n0})}\right|^2}{P_{tarray}} \tag{14.35}$$

Therefore,

$$[G_{array}(\theta, \phi)]_{max} = \frac{\left|\sum_{n=1}^{N} \sqrt{g_{ten}(\theta, \phi)} A_n\right|^2}{\sum_{n=1}^{N} A_n^2} \tag{14.36}$$

The maximum gain for any size array with uniform illumination (i.e. $A_n = 1, n = 1, \ldots N$) is

$$G_{array}(\theta, \phi) = N\left(\sum_{n=1}^{N} \sqrt{g_{ten}(\theta, \phi)}\right)^2 \tag{14.37}$$

For very large-sized arrays, $g_{ten} = g_{te}$. When the illumination is nonuniform, Eq. (14.36) simplifies to

$$G_{array} = g_{te}(\theta, \phi) \frac{\left|\sum_{n=1}^{N} A_n\right|^2}{\sum_{n=1}^{N} A_n^2} \tag{14.38}$$

In case of uniform amplitudes, i.e. $A_n = 1$; n=1, 2, ...N, in a large-sized array, Eq. (14.36) simplifies to

$$G_{array} = g_{te}(\theta, \phi) \frac{1.N^2}{N.1^2} = g_{te}(\theta, \phi) N \tag{14.39}$$

In other words, the array gain is the sum of all the element gain values.

As an example of the type of gain measurements possible that make use of the above equations, the gain of the 10×10 array of elements with a WAIM covering that was described in Section 6.3.3 was determined from Eq. (14.37). The ideal broadside peak gain is obtained when the reflection coefficient is assumed to be zero. The passive reflection coefficient of the central (5, 5) element was measured, from which the broadside peak gain was calculated. This gain was compared to the measured gain obtained by gain comparison with a standard gain horn. The results obtained from the gain comparison method include all mismatch and radiation losses, unlike the reflection coefficient method which only covers mismatch losses. The results of these measurements are shown in Figure 14.14 together with the ideal case.

14.5 Conclusion

Techniques for the test and measurement of mutual coupling in arrays have been provided as a necessary means of confirming the design and fabrication. Two basic sets of approaches are possible

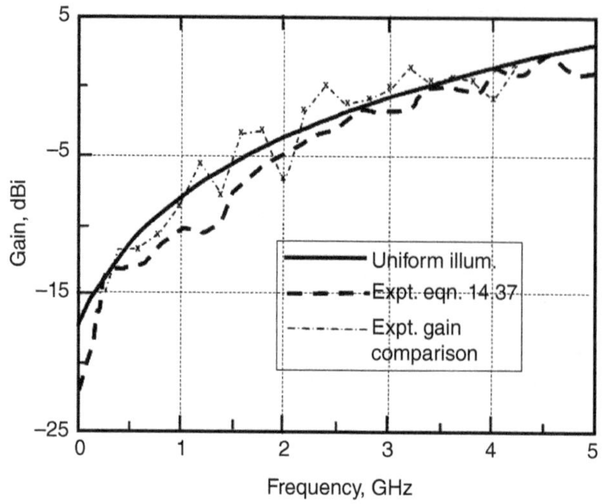

Figure 14.14 Measured gain of 10 × 10 array when element (5, 5) is driven. All other elements are terminated in matched loads.

for arrays that have static beams and phased arrays in which the beams are scanned. Methods using VNAs allow full calibration of the instrumentation and greater accuracy. Radiation methods are possible for the determination of mutual coupling, and these are described as well.

References

Agilent Technologies (2005): 'Network Analyzer Basics', online: https://www.keysight.com/upload/cmc_upload/All/BTB_Network_2005-1.pdf

Agilent Technologies (2007): 'Agilent Time Domain Analysis Using a Network Analyzer', Application Note 1287-12, USA, no. 5989-5723EN.

Amitay, N., Galindo, V., and Wu, C.P. (1972). *Theory and Analysis of :Phased Array Antennas*. New York, USA: Wiley-Interscience.

Aumann, H.M., Fenn, A.J., and Willwerth, F.G. (1989). Phased array antenna calibration and pattern prediction using mutual coupling measurements. *IEEE Trans. Antennas Propag.* 37 (7): 844–850.

Bah, A.O., Qin, P.-Y., Ziolkowski, R.W. et al. (2019). A wideband low-profile tightly coupled antenna array with a very high figure of merit. *IEEE Trans. Antennas & Propagat* 67: 2332–2343.

Barlow, H.M. and Cullen, A.L. (1950). *Microwave Measurements*. London: Constable.

Bird, T.S. (2016). *Fundamentals of Aperture Antennas*', Chap. 10. Chichester, UK: Wiley.

Bird, T.S. and Bateman, D.G. (1992). Software for modelling finite arrays of rectangular waveguides or horns with mutual coupling. In: *IEEE Antennas & Propagation Symposium*, 633–636. Chicago, Illinois.

Bird, T.S., Hayman, D.B., Greene, K.J., Chu, I.W. and Bateman, D.B. (2001): 'A close-packed dielectric-loaded horn array for 60 GHz', 2001 Asia Pacific Microwave Conference, Taipei, Taiwan, 3 - 6 December.

Fenn, A.J. (2018). *Electromagnetics and Antenna Technology*', Chap. 2. Boston, London: Artech House.

Hansen, R.C. (1998). *Phased Array Antennas*. New York, USA: Wiley.

Hollis, J.S., Lyon, T.J., and Clayton, L. (1970). *Microwave Antenna Measurements*. Atlanta, GA: Scientific Atlanta.

Holter, H. (2007). Dual-polarized broadband array antenna with BOR-elements, mechanical design and measurements. *IEEE Trans. Antennas Propag.* 55 (2): 305–312.

IEEE Std 149-1979 (1979). *IEEE Standard Test Procedures for Antennas*. Piscataway, NY: IEEE Inc.

Kelley (1993). Array antenna pattern modeling methods that include mutual coupling effects. *IEEE Trans. Antennas Propag.* 41 (12): 1625–1632.

Kraus, J.D. and Carver, K.R. (1973). *Electromagnetics*, 2e. Tokyo, Japan: McGraw Hill Kogakusha Ltd.

Pozar, D. (1994). The active element pattern. *IEEE Trans. Antennas Propag.* 42 (8): 1176–1178.

Pozar, D. (2001). *Microwave and RF wireless systems*. New York, USA: Wiley.

Zhang, C., Lai, Q., and Gao, C. (2017). Measurement of active S-parameters on array antenna using directional couplers. In: *Proc. Asia Pacific Microwave Symposium*, 1167–1170.

Appendix A

Useful Identities

Trevor S. Bird[1,2]

[1]*Global Big Data Technology Centre, University of Technology Sydney, Ultimo, NSW2007, Australia*
[2]*Antengenuity, Eastwood, NSW2122, Australia*

A.1 Vector Identities

Some useful vector identities and geometrical transformations are summarized below (Harrington, 1961), (Hildebrand, 1962), (Gradshteyn et al., 1994).

A vector quantity is defined in bold lettering e.g. **A**. A unit vector is shown as a standard variable with a caret ("hat") symbol e.g. $\hat{a}$.

$$(\mathbf{A} + \mathbf{B}) \times \mathbf{C} = \mathbf{A} \times \mathbf{C} + \mathbf{B} \times \mathbf{C}$$

$$\mathbf{A} \cdot (\mathbf{B} \times \mathbf{C}) = \mathbf{B} \cdot (\mathbf{C} \times \mathbf{A}) - \mathbf{C} \cdot (\mathbf{A} \times \mathbf{B})$$

$$\mathbf{A} \times (\mathbf{B} \times \mathbf{C}) = (\mathbf{A} \cdot \mathbf{C})\,\mathbf{B} - (\mathbf{A} \cdot \mathbf{B})\,\mathbf{C}$$

$$\nabla \cdot (\mathbf{A} + \mathbf{B}) = \nabla \cdot \mathbf{A} + \nabla \cdot \mathbf{B}$$

$$\nabla \times (\mathbf{A} + \mathbf{B}) = \nabla \times \mathbf{A} + \nabla \times \mathbf{B}$$

$$\nabla \cdot (\phi \boldsymbol{A}) = \phi \nabla \cdot \boldsymbol{A} + \boldsymbol{A} \cdot \nabla \phi$$

$$\nabla \times (\phi \nabla \alpha) = \nabla \phi \times \nabla \alpha$$

$$\nabla \times \nabla \alpha = 0$$

$$\nabla \cdot (\mathbf{A} \times \mathbf{B}) = \mathbf{B} \cdot \nabla \times \mathbf{A} - \mathbf{A} \cdot \nabla \times \mathbf{B}$$

$$\nabla \cdot (\nabla \times \mathbf{A}) = 0$$

$$\nabla \times \nabla \times \mathbf{A} = \nabla (\nabla \cdot \mathbf{A}) - \nabla^2 \mathbf{A}$$

Mutual Coupling Between Antennas, First Edition. Edited by Trevor S. Bird.

Note that

$$
\begin{aligned}
\mathbf{A}\times\mathbf{B} &= \begin{vmatrix} \hat{x} & \hat{y} & z \\ A_x & A_y & A_z \\ B_x & B_y & B_z \end{vmatrix} \\
&= \hat{x}\begin{vmatrix} A_y & A_z \\ B_y & B_z \end{vmatrix} - \hat{y}\begin{vmatrix} A_x & A_z \\ B_x & B_z \end{vmatrix} + \hat{z}\begin{vmatrix} A_x & A_y \\ B_x & B_y \end{vmatrix} \\
&= \hat{x}\left(A_yB_z - A_zB_y\right) - \hat{y}\left(A_xB_z - A_zB_x\right) + \hat{z}\left(A_xB_y - A_yB_x\right)
\end{aligned}
$$

where $|X|$ is the determinant of matrix $\mathbf{X}$.

A.2 Geometric Identities

The three main coordinates systems used here are rectangular (x, y, z), cylindrical (ρ, ϕ, z) and spherical (r, θ, ϕ) coordinates, as shown in Figure A.1. The relationship between the coordinates is

$$x = \rho\cos\phi = r\sin\theta\cos\phi$$

$$y = \rho\sin\phi = r\sin\theta\sin\phi$$

$$z = \rho\cos\theta$$

$$\rho = \sqrt{x^2 + y^2} = r\sin\theta$$

$$\phi = \tan^{-1}\left(\frac{y}{x}\right)$$

$$r = \sqrt{x^2 + y^2 + z^2} = \sqrt{\rho^2 + z^2}$$

$$\phi = \tan^{-1}\left(\frac{\sqrt{x^2 + y^2}}{z}\right) = \tan^{-1}\left(\frac{\rho}{z}\right)$$

Transformations between the coordinate vector components are given by the relationships

$$
\begin{aligned}
A_x &= A_\rho\cos\phi - A_\phi\sin\phi \\
&= A_r\sin\theta\cos\phi + A_\theta\cos\theta\cos\phi - A_\phi\sin\phi \\
A_y &= A_\rho\sin\phi + A_\phi\cos\phi \\
&= A_r\sin\theta\sin\phi + A_\theta\cos\theta\sin\phi + A_\phi\cos\phi \\
A_z &= A_r\cos\theta - A_\theta\sin\theta \\
A_\rho &= A_x\cos\phi + A_y\sin\phi \\
&= A_r\sin\theta + A_\theta\cos\theta \\
A_\phi &= -A_x\sin\phi + A_y\cos\phi \\
A_r &= A_x\sin\theta\cos\phi + A_y\sin\theta\sin\phi + A_z\cos\theta \\
&= A_\rho\sin\theta + A_z\cos\theta \\
A_\theta &= A_x\cos\theta\cos\phi + A_y\cos\theta\sin\phi - A_z\sin\theta \\
&= A_\rho\cos\theta - A_z\sin\theta
\end{aligned}
$$

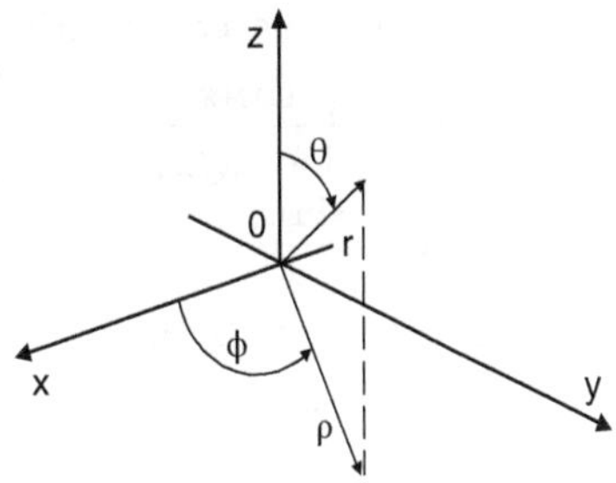

Figure A.1 Coordinate definitions.

The differential surface and volume elements are, respectively,

$$\begin{aligned}\mathbf{dS} &= \hat{x} dy\, dz + \hat{y}\, dx\, dz + \hat{z}\, dx\, dy \\ &= \hat{\rho}\, \rho\, d\phi\, dz + \hat{\phi}\, d\rho\, dz + \hat{z}\, \hat{\rho}\, d\rho\, d\phi \\ &= \hat{r}\, r^2 \sin\theta\, d\theta\, d\phi + \hat{\theta}\, r \sin\theta\, dr\, d\phi + \hat{\phi}\ r\, dr\, d\theta\end{aligned}$$

$$dV = dx\, dy\, dz = \rho\, d\rho\, d\phi\, dz = r^2 \sin\theta\, dr\, d\theta \ \ d\phi.$$

A.3 Transverse Representation of the Electromagnetic Field

In two-dimensional field problems and also in uniform cylindrical structures, it is convenient to use a transverse field representation. Thus, three-dimensional fields are represented as

$$\begin{aligned}\mathbf{A}(x,y,z) &= \mathbf{A}_t(x,y,z) + \hat{z}A_z(x,y,z) \\ &= \mathbf{A}_t(\rho,\phi,z) + \hat{z}A_z(\rho,\phi,z)\end{aligned}$$

The gradient vector is similarly represented as

$$\nabla = \nabla_t + \hat{z}\frac{\partial}{\partial z}$$

where

$$\begin{aligned}\nabla_t &= \hat{x}\frac{\partial}{\partial x} + \hat{y}\frac{\partial}{\partial y} \\ &= \hat{\rho}\frac{\partial}{\partial \rho} + \hat{\phi}\frac{1}{\rho}\frac{\partial}{\partial \phi}\end{aligned}$$

and

$$\begin{aligned}\nabla_t^2 &= \frac{\partial^2}{\partial x^2} + \frac{\partial^2}{\partial y^2} \\ &= \frac{1}{\rho}\frac{\partial}{\partial \rho}\left(\rho\frac{\partial}{\partial \rho}\right) + \frac{1}{\rho^2}\frac{\partial^2}{\partial \phi^2}\end{aligned}$$

A.4 Useful Functions

Some functions that are used throughout are listed below.

The Heaviside unit step function is

$$H(u) = \begin{cases} 0; & u < 0 \\ 1; & \text{otherwise} \end{cases}$$

$$Int\,(x) = \text{Integer value of}\, x \le x$$

$$C\,(x) = \frac{\cos(x)}{1 - (2x/\pi)^2}$$

$$S\,(x) = \frac{\sin(x)}{x}$$

A recursion formula for the Chebyshev polynomial of order n, $T_n(z)$, is as follows:

$$T_n\,(z) = 2zT_{n-1}\,(z) - T_{n-2}\,(z)$$

where $T_0(z) = 1$, $T_1(z) = z$, $T_2(z) = 2z^2 - 1$, etc. Also

$$T_n\,(z) = \begin{cases} \cos\left[n\cos^{-1}(z)\right] & ; |z| < 1 \\ \cosh\left[n\cosh^{-1}(z)\right] & ; |z| > 1 \end{cases}$$

The incomplete elliptic integral function is defined as

$$F\,(z \mid p) = \int_0^z dt \frac{1}{\sqrt{1 - p\sin^2 t}}; 0 < p < 1$$

The sine and cosine integrals are defined as

$$Si\,(z) = \int_0^z dt \frac{\sin(t)}{t};\ Ci\,(z) = -\int_z^\infty dt \frac{\cos(t)}{t}$$

Series expansions of these functions are (Abramowitz and Stegun 1965, p. 231f)

$$Si\,(z) = \sum_{n=0}^{\infty} \frac{(-1)^n z^{2n+1}}{(2n+1)\,(2n+1)!}\ ;\ Ci\,(z) = \gamma + \ln z + \sum_{n=1}^{\infty} \frac{(-1)^n z^{2n}}{(2n)\,(2n)!}$$

where $\gamma = 0.5772156649053$ is Euler's constant.

A.5 Complex Fresnel Integrals

A complex Fresnel integral is here defined by

$$\mathcal{K}\,(z) = \int_0^z \exp\left(-j\xi^2\right) d\xi$$

where z is generally complex also. Note that

$$-\mathcal{K}\,(z) = \mathcal{K}\,(-z) \text{ and } \mathcal{K}\,(\infty) = \frac{\sqrt{\pi}}{2} exp\left(-j\frac{\pi}{4}\right)$$

A variation of the cosine and sine forms of the Fresnel integral function (Abramowitz and Stegun 1965, p. 300) are defined here as

$$C^{(2)}\,(z) = \int_0^z dt\,\cos\left(\frac{\pi}{2}t^2\right) \text{ and } S^{(2)}\,(z) = \int_0^z dt\,\sin\left(\frac{\pi}{2}t^2\right)$$

Thus,

$$C^{(2)}\,(z) = \left(\sqrt{\frac{2}{\pi}}\right) Re\left\{\mathcal{K}\left(\sqrt{\frac{\pi}{2}}z\right)\right\} \text{ and } S^{(2)}\,(z) = -\left(\sqrt{\frac{2}{\pi}}\right) \Im m\left\{\mathcal{K}\left(\sqrt{\frac{\pi}{2}}z\right)\right\}$$

or alternatively

$$\mathcal{K}\,(z) = \sqrt{\frac{\pi}{2}}\left[C^{(2)}\left(\sqrt{\frac{2}{\pi}}z\right) - jS^{(2)}\left(\sqrt{\frac{2}{\pi}}z\right)\right]$$

When $|z| \ll 1$

$$\mathcal{K}(z) \approx z\left(1 - j\frac{z^2}{3}\right)$$

and for large arguments $|z| \gg 1$

$$\mathcal{K}(z) \approx -\frac{1}{2z} exp\left(-j\left(z^2 + \frac{\pi}{2}\right)\right)\left(1 - j\frac{1}{2z^2}\right).$$

A.6 Hypergeometric Function

This function is the solution of the hypergeometric differential equation and it has the property that it converges to other special functions cases or limits to them. It is normally indicated by ${}_2F_1(s, t; u; z)$ (Hildebrand 1962), (Abramowitz & Stegun 1965) and given by

$${}_2F_1(\alpha, \beta; \gamma; z) = \sum_{n=0}^{\infty} \frac{(\alpha)_n(\beta)_n}{n!(\gamma)_n} z^n; \, |z| < 1$$

where $(\ldots)_n$ is Pochhammer's symbol signifying $(\alpha)_0 = 1, (\alpha)_1 = \alpha, (\alpha)_2 = \alpha(\alpha+1), \ldots$. This series representation converges inside the unit circle given by $|z| = 1$. When $z = 1$, the series converges only if $\gamma - \alpha - \beta > 0$ and when $z = -1$, it converges only of if $\gamma - \alpha - \beta + 1 > 0$.

References

Abramowitz, M. and Stegun, I.A. (1965). *Handbook of Mathematical Functions*. New York, USA: Dover Publications, Inc.

Gradshteyn, I.S., Ryzhik, I.M., and Jeffrey, A. (1994). *Table of Integrals, Series, and Products*, 5e. London, UK: Academic Press Inc.

Harrington, R.F. (1961). *Time-Harmonic Electromagnetic Fields*. New York: McGraw-Hill.

Hildebrand, F.B. (1962). *Advanced Calculus for Applications*. Englewood Cliffs, NJ: Prentice Hall.

Appendix B

Bessel and Hankel Functions

Trevor S. Bird[1,2]

[1]*Global Big Data Technology Centre, University of Technology Sydney, Ultimo, NSW 2007, Australia*
[2]*Antengenuity, Eastwood, NSW 2122, Australia*

B.1 Properties

The Bessel equation of integer order n is given by (Watson, 1962)

$$x\frac{d}{dx}\left(x\frac{d\psi}{dx}\right) + \left(x^2 - n^2\right)\psi = 0 \tag{B.1}$$

The solutions are expressed in several forms. Two possible solutions to this equation are called Bessel functions of first and second kinds. These are $J_n(x)$ and $Y_n(x)$. A series expansion of the former type is

$$J_n(x) = \sum_{k=0}^{\infty}\frac{(-1)^k}{k!\,(n+k)!}\left(\frac{x}{2}\right)^{n+2k}$$

For a negative integer order, the expansion shows that $J_{-n}(x) = (-1)^n J_n(x)$ and x real $J_n(-x) = -J_n(x)$. In addition,

$$Y_n(x) = \lim_{\nu\to n}\frac{J_\nu(x)\cos\nu\pi - J_{-\nu}(x)}{\sin\nu\pi}$$

For small arguments,

$$J_n(x) \to \frac{1}{\Gamma(n+1)}\left(\frac{x}{2}\right)^n,\ Y_n(x) \to -\frac{\Gamma(n)}{\pi}\left(\frac{2}{x}\right)^n$$

where for the latter when $n = 0$, $Y_0(x) \to \frac{2}{\pi}\ln(x)$. $\Gamma(n) = (n-1)!$ is the Gamma function of integer argument.

For large arguments (i.e. $|z| >> n$),

$$J_n(x) \sim \sqrt{\frac{2}{\pi x}}\cos\left(x - \frac{1}{2}n\pi - \frac{\pi}{4}\right)$$

$$Y_n(x) \sim \sqrt{\frac{2}{\pi x}}\sin\left(x - \frac{1}{2}n\pi - \frac{\pi}{4}\right)$$

Mutual Coupling Between Antennas, First Edition. Edited by Trevor S. Bird.

On the other hand, for large orders (i.e. $n \to \infty$),

$$J_n(x) \sim \frac{1}{\sqrt{2\pi n}}\left(\frac{xe}{2n}\right)^n$$

$$Y_n(x) \sim -\sqrt{\frac{2}{\pi n}}\left(\frac{xe}{2n}\right)^{-n}$$

and for positive values of n,

$$J_n(n\mathrm{sech}\alpha) \sim \frac{\exp(n\tanh\alpha - \alpha)}{\sqrt{2\pi n \tanh\alpha}}$$

and also

$$J_n'(n\mathrm{sech}\alpha) \sim \sqrt{\frac{\sinh 2\alpha}{4\pi n}}\exp(n\tanh\alpha - \alpha)$$

The ratio of the derivative of J_n with itself for large orders can be expressed as

$$\frac{J_n'(z)}{J_n(z)} \sim \sqrt{\left(\frac{n}{z}\right)^2 - 1}$$

Zeros of the first Bessel functions and its derivative can be approximated by the first three terms of McMahon's expansion. An estimate for the zero of a Bessel function with n fixed and s large order s is

$$j_{n,s} \sim \beta - \frac{(4n^2-1)}{(8\beta)} - \frac{4(4n^2-1)}{3(8\beta)}\frac{(28n^2-31)}{(8\beta)^2}$$

where $\beta = (n/2+s-1/4)\pi$, and similarly for the derivative of the Bessel function,

$$j_{n,s}' \sim \beta' - \frac{(4n^2+3)}{(8\beta')} - \frac{4[7(4n^2)(4n^2)+82(4n^2)-9]}{3(8\beta')(8\beta')^2}$$

where $\beta' = (n/2+s-3/4)\pi$. These formulae give suitable estimates to commence a search for more accurate values.

The Hankel functions of first and second kinds are defined by

$$H_n^{(1)}(x) = J_n(x) + jY_n(x)$$

$$H_n^{(2)}(x) = J_n(x) - jY_n(x)$$

where $H_{-\nu}^{(1)}(x) = e^{j\nu\pi}H_\nu^{(1)}(x)$ and $H_{-\nu}^{(2)}(x) = e^{-j\nu\pi}H_\nu^{(2)}(x)$.

For large arguments, the Hankel functions and their derivative are

$$H_n^{(1,2)}(x) \sim \sqrt{\frac{2}{\pi x}}\exp\left[\pm j\left(x - \frac{1}{2}n\pi - \frac{\pi}{4}\right)\right]$$

$$H_n^{(1,2)\prime}(x) \sim \pm j\sqrt{\frac{2}{\pi x}}\exp\left[\pm j\left(x - \frac{1}{2}n\pi - \frac{\pi}{4}\right)\right]$$

Approximations with greater degree of accuracy than given above are available in the references (e.g. Abramowitz and Stegun 1965).

Recurrence relations for any of the Bessel functions, $Z_n(x)$, which stands for any one of the ordinary Bessel functions or Hankel functions J_n, Y_n, or $H_n^{(1,2)}$, are as follows:

$$Z_n(x) = x\left(Z_{n-1}(x) + Z_{n+1}(x)\right)/2n$$

$$Z_n'(x) = \left(Z_{n-1}(x) - Z_{n+1}(x)\right)/2$$

$$= Z_{n-1}(x) - nZ_n(x)/x$$
$$= -Z_{n+1}(x) + nZ_n(x)/x \tag{B.2}$$

A useful series expansion for $Y_0(x)$ is

$$Y_0(x) = \frac{2}{\pi}\left[ln\left(\frac{x}{2}\right) + \gamma\right] J_0(x) - 2\sum_{k=1}^{\infty}(-1)^k \frac{J_{2k}(x)}{k}$$

where $\gamma = 0.5772156649053$.

The modified Bessel functions of first and second kinds, namely $I_n(x)$ and $K_n(x)$, are solutions of the Bessel equation Eq. (B.1) with an imaginary argument. They are expressed as

$$I_n(x) = \exp(-jn\pi/2)\, J_n(jx);\; -\pi < \arg(x) \le \pi/2$$
$$= \exp(j3n\pi/2)\, J_n(-jx);\; \pi/2 < \arg(x) \le \pi$$
$$K_n(x) = (j\pi/2)\exp(jn\pi/2)\, H_n^{(1)}(jx);\; -\pi < \arg(x) \le \pi/2$$
$$= -(j\pi/2)\exp(-jn\pi/2)\, H_n^{(2)}(jx);\; -\pi/2 < \arg(x) \le \pi$$

with

$$K_0(x) = -\left[\ln\left(\frac{x}{2}\right) + \gamma\right] I_0(x) + 2\sum_{k=1}^{\infty}\frac{J_{2k}(x)}{k}$$

For large arguments, $K_n(x)$ and its derivative are

$$K_n(x) \sim \sqrt{\frac{2}{\pi x}}\exp(-x)$$

$$K_n'(x) \sim -\sqrt{\frac{2}{\pi x}}\exp(-x)$$

Recurrence relations Eq. (B.2) apply also for $I_n(x)$ or $K_n(x)$ with the following changes in the expressions $Z_n(x) \Rightarrow Z_n(x)$, $Z_{n+1}(x) \Rightarrow (-1)Z_{n+1}(x)$ and $Z_{n-1}(x) \Rightarrow Z_{n-1}(x)$.

Some Wronskian relations for the Bessel functions are

$$J_{n+1}(x)\, Y_n(x) - J_n(x)\, Y_{n+1}(x) = 2\pi/x$$

$$J_n'(x)\, Y_n(x) - J_n(x)\, Y_n'(x) = 2\pi/x$$

$$H_{n+1}^{(1)}(x)\, H_n^{(2)}(x) - H_n^{(1)}(x)\, H_{n+1}^{(2)}(x) = -4j/\pi x$$

$$H_n^{(1)\prime}(x)\, H_n^{(2)}(x) - H_n^{(1)}(x)\, H_n^{(2)\prime}(x) = 4j/\pi x$$

$$J_n'(x)\, H_n^{(2)}(x) - J_n(x)\, H_n^{(2)\prime}(x) = 2j/\pi x$$

In coaxial structures, it is convenient to adopt compound Bessel functions of the form

$$Z_n(x,y) = J_n(x) - \frac{J_n'(y)}{Y_n'(y)} Y_n(x)$$

$$\Lambda_n(x,y) = J_n(x) - \frac{J_n(y)}{Y_n(y)} Y_n(x)$$

$$Z_n'(x,y) = J_n'(x) - \frac{J_n'(y)}{Y_n'(y)} Y_n'(x)$$

$$\Lambda_n'(x,y) = J_n'(x) - \frac{J_n(y)}{Y_n(y)} Y_n'(x)$$
$$U_n(x,y) = J_n(x) - \frac{J_n(y)}{H_n^{(2)}(y)} H_n^{(2)}(x)$$
$$V_n(x,y) = J_n(x) - \frac{J_n'(y)}{H_n^{(2)\prime}(y)} H_n^{(2)}(x)$$

where the $'$ refers to the derivative with respect to the first or only argument.

B.2 Series Involving Bessel Functions

Useful series are given below.

$$\exp(jx\cos\theta) = J_0(x) + 2\sum_{k=1}^{\infty} j^k J_k(x)\cos(k\theta)$$
$$\exp(jx\sin\theta) = \sum_{k=-\infty}^{\infty} \exp(jk\theta)\, J_k(x)$$
$$1 = J_0(x) + 2\sum_{k=1}^{\infty} J_{2k}(x)$$
$$\exp(z) = I_0(x) + 2\sum_{k=1}^{\infty} I_{2k}(x)$$
$$1 = I_0(x) + 2\sum_{k=1}^{\infty} (-1)^k I_{2k}(x)$$
$$J_p(az)\, J_q(bzz) = \frac{\left(\frac{1}{2}az\right)^p \left(\frac{1}{2}bz\right)^q}{\Gamma(q+1)} \sum_{r=0}^{\infty} (-1)^r \frac{{}_2F_1\left(-r, -p-r; q+1; (b/a)^2\right)(a\,z/2)^r}{r!\Gamma(p+q+1)}$$

${}_2F_1(s, t; u; z)$ is the hypergeometric function defined in Appendix A.

$$J_\mu(az)J_\nu(bz)J_\lambda(cz) = \left(\frac{1}{2}az\right)^\mu \left(\frac{1}{2}bz\right)^\nu \left(\frac{1}{2}cz\right)^\lambda \sum_{m=0}^{\infty} \left(-\frac{1}{4}a^2z^2\right)^m \sum_{n=0}^{m} \frac{(b/a)^{2n}}{(n)!\Gamma(\nu+n+1)}$$
$$\times \sum_{p=0}^{m-p} \frac{(c/a)^{2p}}{(m-n-p)!\Gamma(\mu+m-n+p+1)(p)!\Gamma(\lambda+p+1)}$$

B.3 Integrals of Bessel Functions

Some useful integrals involving Bessel functions in closed form are (McLachlan 1934; Gradshteyn et al. 1994; Watson 1962)

$$\int_0^{2\pi} \left.\begin{matrix}\cos\\ \sin\end{matrix}\; p\phi \right\} e^{jz\cos(\phi-\phi')}\, d\phi' = 2\pi j^p\, J_p(z) \left\{\begin{matrix}\cos\\ \sin\end{matrix}\; p\phi\right. \tag{B.3}$$

$$\int^z z^p J_{p-1}(z)\, dz = z^p J_p(z)$$

$$\int_0^a \left[J_p(\alpha z)\right]^2 z\, dz = \frac{a^2}{2}\left\{\left[J_p(\alpha a)\right]^2 - J_{p-1}(\alpha a) J_{p+1}(\alpha a)\right\} \tag{B.4}$$

$$\int_0^a J_p(\alpha z)\, J_p(\beta z)\, z\, dz = \frac{a}{\alpha^2-\beta^2}\left(\alpha J_{p+1}(\alpha a) J_p(\beta a) - \beta J_p(\alpha a) J_{p+1}(\beta a)\right)$$
$$= \frac{a}{\alpha^2-\beta^2}\left(\beta J_p(\alpha a) J_{p-1}(\beta a) - \alpha J_{p-1}(\alpha a) J_p(\beta a)\right) \quad \text{(B.5)}$$

$$\int_0^a J_p(\alpha z)\, J_p(\beta z) \exp\left(-j\gamma z^2\right) z dz = \left(\frac{\alpha\beta}{4}a^2\right)^p \frac{a^2}{2\Gamma(p+1)} \sum_{m=0}^{\infty} (-1)^m \left(\frac{\alpha}{2}a\right)^{2m}$$
$$\times \frac{2F_1\left(-m,-m-p;p+1;(\beta/\alpha)^2\right)}{m!\Gamma(p+m+1)}$$
$$\times \sum_{\nu=0}^{\infty} \frac{\left(-j\gamma a^2\right)}{\nu!\,(p+m+\nu+1)} \quad \text{(B.6)}$$

$$\int_0^a dw w^{\alpha-1}\left(1-w^2\right)^{\beta-1/2} \frac{J_p(Aw)\, J_q(Bw)}{\left(w^2-u_m^2\right)^{\lambda}\left(w^2-u_n^2\right)^{\tau}} J_\nu(Cw)$$
$$= \frac{\Gamma(\beta)}{2}\Gamma\left(\frac{A}{2}\right)^p \Gamma\left(\frac{B}{2}\right)^q \sum_{\sigma=0}^{\infty}(-1)^\sigma Q_\sigma(p+\sigma,q+1,A,B)\left\{\sum_{\mu=0}^{\infty}\frac{(-1)^\mu}{\mu!}\left(\frac{C}{2}\right)^{2\mu}\right.$$
$$\times \left[\Gamma\left(\frac{C}{2}\right)^{\nu} \frac{\Gamma(\chi)}{\Gamma(\nu+\mu+1)\,\Gamma\left(\beta+\frac{1}{2}+\chi\right)} g_\nu + j\,sgn(1/2-\beta)\,\Gamma\left(\frac{C}{2}\right)^{1+2(\lambda-\beta-\sigma)-(p+q+\alpha)}\right.$$
$$\left.\times \frac{\Gamma(\theta)}{\Gamma(\nu-\theta+1)\,\Gamma\left(\beta+\frac{1}{2}-\mu\right)} h_\nu\right] \quad \text{(B.7)}$$

where $A, B < C$, $u_m = x_m/A$ and $u_n = x_n/B$ in which $J_p(x_m) = 0$,

$$Q_m(s,t,x,y) = \frac{\left(\frac{x}{2}\right)^{2m} {}_2F_1\left(-m,-s;t;\left(\frac{y}{x}\right)^2\right)}{m!\,\Gamma(s+1)\,\Gamma(t)}, \quad \text{(B.8)}$$

$\chi = (p+q+\nu+\alpha)/2 + (\sigma+\mu)$, $\theta = (p+q+\nu+\alpha)/2 + \left(\sigma-\mu+\beta-\frac{1}{2}-\lambda\right)$, and

$$g_\nu = \begin{cases} 1; \quad \lambda = 0 = \tau \\ \frac{-1}{\left(u_m^2-u_n^2\right)}\left[u_m^{-2}\,{}_1F_2\left(1,\chi;\beta+\frac{1}{2}+\chi;u_m^{-2}\right) - u_n^{-2}\,{}_1F_2\left(1,\chi;\beta+\frac{1}{2}+\chi;u_n^{-2}\right)\right]; \lambda = 1 = \tau \\ u_m^{-2\lambda}\,{}_1F_2\left(\lambda,\chi;\beta+\frac{1}{2}+\chi;u_m^{-2}\right) \quad ;\lambda > 0, \tau = 0 \end{cases}$$

$$h_\nu = \begin{cases} 1; \quad \lambda = 0 = \tau \\ \frac{1}{\left(u_m^2-u_n^2\right)}\left[{}_1F_2\left(1,-\mu;\beta+\frac{1}{2}-\mu;u_m^2\right) - {}_1F_2\left(1,-\mu;\beta+\frac{1}{2}-\mu;u_n^2\right)\right]; \lambda = 1 = \tau \\ {}_1F_2\left(\lambda,-\mu;\beta+\frac{1}{2}-\mu;u_m^2\right) \quad ;\lambda > 0, \tau = 0 \end{cases}.$$

When $\lambda = 0$ and $\tau > 0$ in the g_ν and h_ν functions, the roles of λ and τ reverse in the third case as are those of u_m and u_n.

B.4 Lommel-Type Integrals

Some integrals of this type are Eqs. (B.5) and (B.6). Others include

$$
\begin{aligned}
Q_p(w) &= \int_a^b d\rho\, \rho\, Z_p(k\rho,\ \beta) J_p(w\rho) \\
&= \int_0^b d\rho\, \rho\, Z_p(k\rho,\ \beta) J_p(w\rho) - \int_0^a d\rho\, \rho\, Z_p(k\rho,\ \beta) J_p(w\rho)
\end{aligned}
\tag{B.9}
$$

and

$$O_p(w) = \int_a^b d\rho\, \rho^2\, Z_p(k\,\rho,\ \beta)\, J_p'(w\,\rho)$$

$$Q_p(-w) = (-1)^p Q_p(w)$$

$$O_p(-w) = (-1)^{p+1} O_p(w)$$

$$O_{p+1}(w) = -\frac{(p+1)}{w} Q_{p+1}(w) + \frac{p}{k} Q_p(w) - O_p(w)$$

References

Abramowitz, M. and Stegun, I.A. (1965). *Handbook of Mathematical Functions*. New York: Dover Inc.

Gradshteyn, I.S., Ryzhik, I.M., and Jeffrey, A. (1994). *Table of Integrals, Series, and Products*, 5e. London, UK: Academic Press Inc.

McLachlan, N.W. (1934). *Bessel Functions for Engineers*. London, UK: Oxford University Press.

Watson, G.N. (1962). *A Treatise on the Theory of Bessel Functions*, 2e. Cambridge, UK: Cambridge University Press.

Appendix C

Properties of Hankel Transform Functions

Trevor S. Bird[1,2]

[1]*Global Big Data Technology Centre, University of Technology Sydney, Ultimo, NSW 2007, Australia*
[2]*Antengenuity, Eastwood, NSW 2122, Australia*

Closed form solutions to the integral of products of Bessel functions have been described by several authors (Nicholson 1920; De Hoop 1955; Watson 1962; Luke 1969). In coupling computations involving circular apertures, products of up to three Bessel functions can occur in the mutual admittance expressions (Bird 1996a, 1996b). A general Hankel transform involving triple products of Bessel functions is defined by

$$C^{(1)}_{p,q,\nu}\left(\alpha,\beta;u_m,u_n;s\right)=\int_0^\infty dw\frac{w^3}{\sqrt{1-w^2}}\frac{\left[J_p\left(ka_iw\right)-\alpha J_p\left(kb_iw\right)\right]}{\left(w^2-u_m^2\right)}\times\frac{\left[J_q\left(ka_jw\right)-\beta J_q\left(kb_jw\right)\right]}{\left(w^2-u_n^2\right)}J_\nu\left(ksw\right) \tag{C.1}$$

and a second type involving derivatives of the two compound Bessel functions

$$C^{(2)}_{p,q,\nu}\left(\alpha,\beta;u_m,u_n;s\right)=\int_0^\infty dww\sqrt{1-w^2}\frac{\left[J'_p\left(ka_iw\right)-\alpha J'_p\left(kb_iw\right)\right]}{\left(w^2-u_m^2\right)}\times\frac{\left[J'_q\left(ka_jw\right)-\beta J'_q\left(kb_jw\right)\right]}{\left(w^2-u_n^2\right)}J_\nu\left(ksw\right) \tag{C.2}$$

Both transforms can be expressed in terms of the following general integral:

$$E_{p,q,\nu}\left(\mu,\sigma,\lambda,\tau;x,y,z\right)=\int_0^\infty dw\,w^\mu\left(1-w^2\right)^{\sigma-1}\frac{J_p\left(kxw\right)}{\left(w^2-u_m^2\right)^\lambda}\frac{J_q\left(kyw\right)}{\left(w^2-u_n^2\right)^\tau}J_\nu\left(kzw\right) \tag{C.3}$$

where μ, λ *and* τ are integers, $\sigma=\frac{n}{2}$ where n is an integer, is a multiple of and kx, ky, and kz are normalized lengths. Also, u_m and u_n are the m-th and n-th zeros of J_p and J_q. Expanding $C^{(1)}_{p,q,\nu}$ in terms of Eq. (C.3), it is found that

$$\begin{aligned}C^{(1)}_{p,q,\nu}\left(\alpha,\beta;u_m,u_n;s\right)&=E_{p,q,\nu}\left(3,\frac{1}{2},1,1;a_i,a_j,R_{ij}\right)-\alpha E_{p,q,\nu}\left(3,\frac{1}{2},1,1;b_i,a_j,R_{ij}\right)\\&-\beta E_{p,q,\nu}\left(3,\frac{1}{2},1,1;a_i,b_j,R_{ij}\right)+\alpha\beta\,E_{p,q,\nu}\left(3,\frac{1}{2},1,1;b_i,b_j,R_{ij}\right)\end{aligned} \tag{C.4}$$

Mutual Coupling Between Antennas, First Edition. Edited by Trevor S. Bird.

In addition, since $J'_p(z) = -J_{p+1}(z) + (p/z)J_p(z)$, $C^{(2)}_{p,q,\nu}(\alpha, \beta; u_m, u_n; s)$ can be expanded similarly into 16 terms. In the special case when $\alpha = \beta = 0$, Eq. (C.2) becomes

$$C^{(2)}_{p,q,\nu}(0,0;u_m,u_n;s) = E_{p+1,q+1,\nu}\left(2,\frac{3}{2},1,1;a_i,a_j,R_{ij}\right) - \left(\frac{q}{ka_j}\right)E_{p+1,q,\nu}\left(1,\frac{3}{2},1,1;a_i,a_j,R_{ij}\right) - \left(\frac{p}{ka_i}\right)E_{p,q+1,\nu}\left(1,\frac{3}{2},1,1;a_i,a_j,R_{ij}\right) + \left(\frac{pq}{ka_ika_j}\right)E_{p,q,\nu}\left(0,\frac{3}{2},1,1;a_i,a_j,R_{ij}\right) \quad (C.5)$$

The integral of the type Eq. (C.3) can be evaluated using the series given in Eq. (B.7). However, this representation fails in the vicinity of the poles and where the hypergeometric functions peak on the unit circle or breaks down completely when $u_m, u_n < 1$.

An alternative integral representation, which is better suited to numerical calculation, is (Bird 2004):

$$C^{(1)}_{p,q,\nu}(\alpha,\beta;u_m,u_n;s) = \int_0^1 du\frac{u^3}{\sqrt{1-u^2}}\frac{[J_p(ka_iu) - \alpha J_p(kb_iu)]}{(u^2-u_m^2)} \times \frac{[J_q(k\,u) - \beta J_q(kb_ju)]}{(u^2-u_n^2)}H^{(2)}_\nu(k\,su) - \frac{2j}{\pi}\cos\left[(p+q-\nu)\frac{\pi}{2}\right] \times \int_0^\infty dt\frac{t^3}{\sqrt{1+t^2}}\frac{[I_p(ka_it) - \alpha I_p(kb_it)]}{(t^2+u_m^2)} \times \frac{[I_q(k\,t) - \beta I_q(k\,b_jt)]}{(t^2+u_n^2)}K_\nu(ks\,t) \quad (C.6)$$

$$C^{(2)}_{p,q,\nu}(\alpha,\beta;u_m,u_n;s) = \int_0^1 du\,u\sqrt{1-u^2}\,\frac{[J'_p(k\,a_iu) - \alpha J'_p(k\,b_iu)]}{(u^2-u_m^2)} \times \frac{[J'_q(k\,a_ju) - \beta J'_q(k\,b_ju)]}{(u^2-u_n^2)}H^{(2)}_\nu(ksu) - \frac{2j}{\pi}\cos\left[(p+q-\nu)\frac{\pi}{2}\right] \times \int_0^\infty dt\,t\sqrt{1+t^2}\frac{[I'_p(k\,a_it) - \alpha I'_p(k\,b_it)]}{(t^2+u_m^2)} \times \frac{[I'_q(k_oa_jt) - \beta I'_q(k_ob_jt)]}{(t^2+u_n^2)}K_\nu(ks\,t) \quad (C.7)$$

where $H^{(2)}_p$ is the Hankel function of second kind, I_p and K_p are the modified Bessel functions of order p. The integrals in Eqs. (C.6) and (C.7) have equal and opposite non-integrable singularities at u, t = 0. These singularities are removed by integrating the integrals together in the range 0 to 1. Due to the presence of the modified Bessel functions, both infinite integrals converge rapidly with increasing t.

References

Bird, T.S. (1996a). Cross-coupling between open-ended coaxial radiators. *IEE Proc. Microwaves Antennas Propagat.* 143 (4): 265–271.

Bird, T.S. (1996b). Improved solution for mode coupling in different-sized circular apertures and its application. *IEE Proc. Microwaves Antennas Propagat.* 143 (6): 457–464.

Bird, T.S. (2004). Mutual coupling in arrays of coaxial waveguides and horns. *IEEE Trans. Antennas Propagat.* 52, 3: 821–829.

De Hoop, A.T. (1955). On integrals occurring in the variational formulation of diffraction problems. *Proc. Koninklijke Nederlandse Akademie van Wetenschappen, B* 58: 325–330.

Luke, Y.L. (1969). *The Special Functions and their Approximations*, vol. 2, 422–435. Academic Press.

Nicholson, J.W. (1920). Generalization of a theorem due to Sonine. *Q. J. Pure Appl. Math.* 48: 321–329.

Watson, G.N. (1962). *A Treatise on the Theory of Bessel Functions*, 2e. Cambridge University Press.

Appendix D

Properties of Surface Fock Functions

Trevor S. Bird[1,2]

[1] *Global Big Data Technology Centre, University of Technology Sydney, Ultimo, NSW 2007, Australia*
[2] *Antengenuity, Eastwood, NSW 2122, Australia*

D.1 Definitions

The surface Fock functions occur in the asymptotic representation of the surface currents excited on a circular cylinder and are also used in the description of the currents on general convex surfaces (Bowman et al. 1963; Fock 1965; Bird 1985). They are defined as integrals in the complex τ -plane in terms of the Airy integral $w_2(\tau)$ or its derivative $w_2'(\tau)$. The Airy integral is defined by

$$w_2(\alpha) = \frac{1}{\sqrt{\pi}} \int_{\Gamma_2} d\tau \, exp\left(\alpha\tau - \tau^3/3\right) \tag{D.1}$$

where the contour Γ_2 in the τ-plane is shown in Figure D.1.

A surface Fock function of the n-th kind, order m, is defined by

$$L_m^{(n)}(z) = \int_{\Gamma_1} d\tau \, \tau^n \left(\frac{w_2'(\tau)}{w_2(\tau)}\right)^m exp(-jz\tau) \tag{D.2}$$

where m, n are integers and the contour Γ_1 in the complex plane is shown in Figure D.1. Two families of Fock functions have been defined. These are known as soft and hard surface functions, depending on $m > 0$ or $m < 0$, respectively.

D.2 Soft Surface Functions ($m > 0$)

These are defined as

$$u_n^{(m)}(z) = \frac{\exp\left(j(n+3/2)\pi/2\right) z^{n+3/2}}{2\Gamma(n+3/2)} L_{m+1}^{(n)}(z); \quad m, n = 0, 1, \ldots \tag{D.3}$$

$$u_n(z) \equiv u_n^{(0)}(z)$$

and

$$u(z) \equiv u_0(z) \equiv u_0^{(0)}(z)$$

The first derivative is given by

$$u_n^{(m)\prime}(z) = (n+3/2)\left[u_n^{(m)}(z) - u_{n+1}^{(m)}(z)\right]/z$$

Mutual Coupling Between Antennas, First Edition. Edited by Trevor S. Bird.

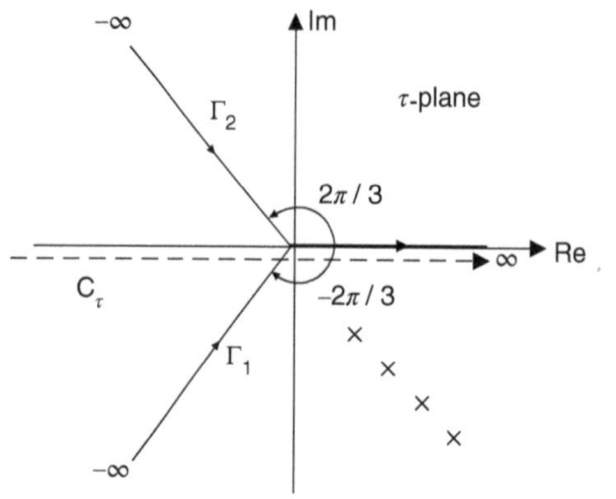

Figure D.1 Contours in the complex τ-plane. Crosses indicate zeros of w_2 or w_2' depending on the integrand of the soft or hard Fock functions.

A residue series to Eq. (D.2) that is sufficiently accurate for $z > 1$. It is found to be

$$u_n(z) = \frac{\pi exp\,(j\,(n+3/2)\,\pi/2)\,z^{n+3/2}}{\Gamma\,(n+3/2)} \sum_{p=1}^{\infty} \left(t_p\right)^n exp\left(-jzt_p\right) \tag{D.4}$$

where t_p (p = 1,2,…, ∞) are the zeros of w_2 and are related to the zeros α_p of the Airy function of first kind, which is defined as

$$Ai\,(x) = \frac{1}{\sqrt{\pi}} \int_{C_\tau} d\tau \cos\left(\alpha\tau + \tau^3/3\right) \tag{D.5}$$

where C_τ is shown in Figure D.1. Relevant series and approximations are given in Abramowitz and Stegun (1965). The Airy function zeros α_p are available in tables (Abramowitz and Stegun 1965) for p = 1 to 10 and

$$t_p = \left|\alpha_p\right| exp\,(-j\pi/3)$$

For $p > 10$, a highly accurate approximation to $|\alpha_p|$ is

$$\left|\alpha_p\right| \approx \chi^{2/3}\left[1 + 5/\left(48\chi^2\right)\right]$$

where $\chi\,(p) = \frac{3\pi}{2}\,(p - 1/4)$.

A small argument approximation to Eq. (D.3) can be found by making the substitution $\tau = (-js)$ in Eq. (D.2). The contour is modified allowing the integral to be expressed as Bromwich integral, and further the Airy function ratio is expressed in inverse powers of $s^{3/2}$ (Hasserjian and Ishimaru 1962). Eq. (D.2). It can be shown that

$$u_n(z) \approx 1 + (-1)^n \sum_{m=1}^{\infty} exp\,(-j3m\pi/4)\,a_m(n)\,z^{3m/2} \tag{D.6}$$

The first five coefficients of Eq. (D.6), $a_m(n)$ (m = 1, 2,…5), are given by

$$a_1(n) = \frac{\pi/4}{\Gamma(1-n)\,\Gamma(n+3/2)},\ a_2(n) = \frac{5\pi/32}{\Gamma(5/2-n)\,\Gamma(n+3/2)},$$

$$a_3(n) = \frac{15\pi/64}{\Gamma(4-n)\,\Gamma(n+3/2)},$$

$$a_4(n) = \frac{1105\pi/2048}{\Gamma(11/2-n)\,\Gamma(n+3/2)},\ \text{and}\ a_5(n) = \frac{1695\pi/1024}{\Gamma(7-n)\,\Gamma(n+3/2)}$$

A list of the first 20 coefficients of the small argument approximation is given in the references (Bird 1985).

D.3 Hard Surface Fock Functions ($m < 0$)

The n-th order hard surface Fock functions are defined as

$$v_n^{(m)}(z) = \frac{\exp(j(n+1/2)\pi/2)\, z^{n+1/2}}{2\Gamma(n+1/2)} L_{-m-1}^{(n)}(z); \quad m, n = 0, 1, \ldots \tag{D.7}$$

The first derivative is

$$v_n^{(m)\prime}(z) = (n+1/2)\left[v_n^{(m)}(z) - v_{n+1}^{(m)}(z)\right]/z$$

It is conventional for $m = 0$ to define

$$v_n(z) \equiv v_n^{(0)}(z)$$

and the zeroth-order function to be given by

$$v(z) \equiv v_0 \equiv v_0^{(0)}(z)$$

As for the soft surface Fock functions, a residue series is available for $z > 1$. In this instance,

$$v_n(z) = \frac{\pi \exp(j(n+1/2)\pi/2)\, z^{n+1/2}}{\Gamma(n+1/2)} \sum_{p=1}^{\infty} \left(t_p'\right)^{n-1} \exp\left(-jzt_p'\right), \tag{D.8}$$

t_p' (p = 1,2,…, ∞) are the zeros of w_2' which are related to the zeros α_p' of the derivative of the Airy function $Ai'(t)$ as follows:

$$t_p' = \left|\alpha_p'\right| \exp(-j\pi/3)$$

α_p' are also tabulated in the references (Abramowitz and Stegun 1965) for p = 1 to 10. For $p > 10$, a highly accurate approximation to $|\alpha_p|$ is

$$\left|\alpha_p'\right| \approx \chi^{2/3}\left[1 - 7/\left(48\chi^2\right)\right]$$

where $\chi(p) = \frac{3\pi}{2}(p - 3/4)$.

A small argument approximation can be found by returning to the definition in Eq. (D.7) and following the approach described for the soft surface function. It can be shown that

$$v_n(z) \approx 1 + (-1)^n \sum_{m=1}^{\infty} \exp(-j3m\pi/4)\, b_m(n)\, z^{3m/2} \tag{D.9}$$

The first five coefficients of this series, $b_m(n)$ (m = 1,2,…,5), are as follows:

$$b_1(n) = \frac{\pi/4}{\Gamma(2-n)\Gamma(n+1/2)}, \quad b_2(n) = \frac{7\pi/32}{\Gamma(7/2-n)\Gamma(n+1/2)},$$

$$b_3(n) = \frac{21\pi/64}{\Gamma(5-n)\Gamma(n+1/2)},$$

$$b_4(n) = \frac{1463\pi/2048}{\Gamma(13/2-n)\Gamma(n+1/2)}, \quad \text{and } b_5(n) = \frac{2121\pi/1024}{\Gamma(8-n)\Gamma(n+1/2)}$$

The first 20 coefficients of the small argument approximation are given in the references (Bird 1985).

D.4 Hard Surface Fock Function of the First Kind

A zero-order hard surface Fock function of the first kind is also required for some asymptotic expansions (Bird, 1985). This function is

$$v_n^{(1)}(z) = \frac{\exp(j\pi/4)\, z^{1/2}}{2\Gamma(1/2)} L_{-2}^{(n)}(z) \tag{D.10}$$

The zeroth function has the residue series

$$\begin{aligned} v_0^{(1)}(z) &= \sqrt{\pi}\exp\left(\frac{j3\pi}{4}\right) z^{\frac{1}{2}} \left(1 + \sum_{p=1}^{\infty} \left(1 + jzt_p'\right) \frac{\exp\left(-jzt_p'\right)}{\left(t_p'\right)^3}\right) \\ &= \sqrt{\pi z}\exp(-j\pi/4) \left(1 - \sum_{p=1}^{\infty} (1+w) \frac{\exp(-w)}{\left|\alpha_p'\right|^3}\right) \end{aligned} \tag{D.11}$$

where $w = z\,|\alpha_p'| \exp(j\pi/6)$.

A small argument approximation is

$$v_0^{(1)}(z) = \sqrt{\pi}\, exp(j\pi/4)\, z^{1/2} \left(1 - \sum_{m=1}^{\infty} c_m(0)(-z)^{3m-j3m/2}\right). \tag{D.12}$$

The first five coefficients are

$$c_1(0) = \frac{1/2}{\Gamma(5/2)}, c_2(0) = \frac{1/2}{\Gamma(4)}, c_3(0) = \frac{49/64}{\Gamma(11/2)}, c_4(0) = \frac{105/64}{\Gamma(7)}, \text{ and}$$

$$c_5(0) = \frac{19019/4096}{\Gamma(17/2)}.$$

A list of the first 10 coefficients is available in Bird (1985). Functions of higher order but still of the first kind are calculated from

$$v_n^{(1)}(z) = \frac{z^2}{(n-1/2)(n-3/2)} \left[(n-1)\, v_{n-2}^{(1)}(z) - (n-3/2)\, v_{n-1}^{(1)}(z)\right].$$

References

Abramowitz, M. and Stegun, I.A. (1965). *Handbook of Mathematical Functions*. New York: Dover Inc.

Bird, T.S. (1985). Accurate asymptotic solution for the surface field due to apertures in a conducting cylinder. *IEEE Trans. Antennas Propagat.* AP-33: 1108–1117.

Bowman, J.J., Senior, T.B.A., and Uslenghi, P.L.E. (1963). *Electromagnetic and Acoustic Scattering by Simple Shapes*. Amsterdam: North-Holland.

Fock, V.A. (1965). *Electromagnetic Diffraction and Propagation Problems*. London: Pergamon Press.

Hasserjian, G. and Ishimaru, A. (1962a). Currents induced on the surface of a conducting circular cylinder by a slot. *J. Res. Nat. Bur. Stand., D. Radio Propag.* 66D, pp. 335–365.

Appendix E

Four Parameter Noise Representation of an Amplifier

Christophe Craeye[1], Jean Cavillot[1] and Eloy de Lera Acedo[2]

[1] *Louvain School of Engineering, Institute of Information and Communication Technologies, Electronics and Applied Mathematics (ICTEAM), Université Catholique de Louvain, Ottignies-Louvain-la-Neuve, Belgium*
[2] *Cavendish Astrophysics, Cavendish Laboratory, Physics Department, University of Cambridge, Cambridge, UK*

The effects of noise on both sides (i.e. for devices connected at input as well as output) of an amplifier can be described by a noiseless amplifier, which is preceded by a series noise voltage and a parallel current as illustrated in Figure E.1, that are correlated with each other. The r.m.s. (root-means square) values of these noise voltages and currents are given by

$$\left\langle v_n\, v_n^* \right\rangle = L\, R_n; \left\langle i_n\, i_n^* \right\rangle = L\, G_n; \quad \left\langle v_n\, i_n^* \right\rangle = L\, R_n\, Y_\gamma^*; \quad \text{with} \quad L = 4\, KTB$$

From the perspective of the amplifier, the parallel noise current can be replaced into a series voltage source $Z_s\, i_n$, with $Z_s = R_s + j\, X_s$, the impedance of the generator (e.g. the antenna connected to the LNA). Then, the total available noise power at the input of the (noiseless) amplifier is

$$P_n = \frac{L}{4\, R_s} \left(R_s + R_n + |Z_s|^2\, G_n + 2\, \Re\, \{Z_s^*\, Y_\gamma^*\}\, R_n \right) \tag{E.1}$$

In general, the noise figure is the ratio between output and input signal-to-noise SNR values. In this configuration, it can be written as the ratio between available powers with and without the noise contribution from the amplifier, i.e.

$$F = \left(R_s + R_n + |Z_s|^2\, G_n + 2\Re\, \{Z_s^*\, Y_\gamma^*\}\;\; R_n \right) / R_s \tag{E.2}$$

The optimal source impedance is then obtained by imposing $\frac{\partial F}{\partial R_s} = 0$ and $\frac{\partial F}{\partial X_s} = 0$. Denoting $Y_\gamma = G_\gamma + j\, B_\gamma$, the following solution is obtained:

$$R_s^2 = R_a^2\, \left(1 - R_a^2\, B_\gamma^2\right) \quad \text{and} \quad X_s = B_\gamma\, R_a^2, \quad \text{with } R_a^2 = R_n / G_n$$

It is easy to verify that $1/\left|Z_{s,\text{opt}}\right|^2 = \left|Y_{s,\text{opt}}\right|^2 = 1/R_a^2$, which upon substitution into Eq. (E.2) yields the optimal noise figure

$$F_{\min} = 1 + 2\, R_n\, \left(Y_\gamma + Y_{s,\text{opt}}\right) = 1 + 2\, \left(R_n G_\gamma + R_{s,\text{opt}} G_n\right). \tag{E.3}$$

The above relations explain why G_n and Y_γ are often not provided; instead of those, $Y_{s,\text{opt}}$ and $F_{\min}$ are given, which have a more directly interpretable impact for the user. The actual noise figure is then (Abrie 1999)

$$F = F_{\min} + \frac{R_n}{G_s} \left|Y_s - Y_{s,\text{opt}}\right|^2 \tag{E.4}$$

where $Y_s = \frac{1}{Z_s} = G_s + j\, X_s$.

Mutual Coupling Between Antennas, First Edition. Edited by Trevor S. Bird.

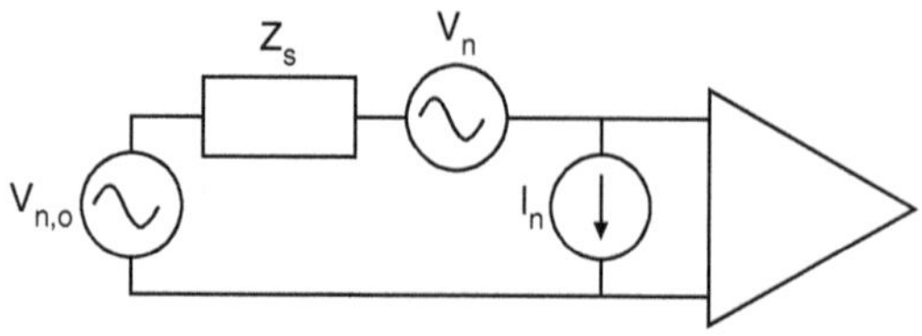

Figure E.1 Classical voltage and current noise representation of an amplifier.

Reference

Abrie, P.L.D. (1999). *Design of Microwave Amplifiers and Oscillators*. Boston.: Artech House.

Appendix F

Equivalent Noise Currents

Christophe Craeye[1]*, Jean Cavillot*[1] *and Eloy de Lera Acedo*[2]

[1] *Louvain School of Engineering, Institute of Information and Communication Technologies, Electronics and Applied Mathematics (ICTEAM), Université Catholique de Louvain, Ottignies-Louvain-la-Neuve, Belgium*
[2] *Cavendish Astrophysics, Cavendish Laboratory, Physics Department, University of Cambridge, Cambridge, UK*

Employing the notation used in Section 12.4, the following relation between received noise current and radiation pattern can be obtained by thermodynamic reciprocity:

$$\frac{\Re\{Z_L\}\left\langle I_{n,i}\, I^*_{n,i}\right\rangle}{2} = \iint \frac{|\boldsymbol{F}_i(\Omega)|^2}{\eta_o} 4\, KTB\, \Re\left\{Z_L\right\}\, d\Omega \tag{F.1}$$

It expresses the equality between noise power absorbed by the load at port i and power absorbed by the environment at infinity through the noise fields radiated by that same load (root-mean-square (r.m.s) value of noise voltage equal to 4 $KTB\, \Re\,\{Z_L\}$) through the i-th radiation pattern $\vec{F}_i\,(\Omega)$. That relation allows us to represent noise collected from the far-field environment as due to a collection of equivalent dipoles. More precisely, the noise stemming from an infinitesimal solid angle $d\Omega$ can also be viewed as stemming from two current sources with amplitudes I_s and length dl with orthogonal polarizations (say "v" and "h" polarizations). The amplitudes of the currents are uncorrelated, but their standard deviations are assumed identical (non-polarized noise). The noise current picked up by the load can then be obtained from the incident field, e.g. $\vec{E}_{t,v} = -jk\eta_o\, I_s\, dl \exp(-j\,k\,R)\,/\,(4\pi\, R)\,\hat{e}_v$ for "v" polarization, based on fields radiated by a dipole. Then, making use of a reciprocity relationship involving the embedded pattern (Appendix G), currents on the load due to sources $I_s\, dl$ can be determined, as well as the corresponding dissipated power. That result is identified to Eq. (F.1) for an infinitesimal solid angle $d\Omega$. Finally, after a few trivial algebraic simplifications, the following expression is obtained for the equivalent noise current sources:

$$\eta_o\, |I_s|^2\, dl^2 = 8\, KTB\, R^2 d\Omega$$

Then, computing the received currents on antennas i and j, multiplying them, and integrating over noise sources in all directions, we obtain

$$\left\langle I_{n,i}\, I^*_{n,j}\right\rangle = \iint \frac{\boldsymbol{F}_i(\Omega)\,.\boldsymbol{F}^*_j(\Omega)}{\eta_o} 8\, KTB\, d\Omega. \tag{F.2}$$

This result, although obtained quite differently, is equivalent to expression (13) given in Warnick and Jeffs (2008).

Mutual Coupling Between Antennas, First Edition. Edited by Trevor S. Bird.

Reference

Warnick, K.F. and Jeffs, B.D. (2008). Efficiencies and system temperature for a beamforming array. *IEEE Antennas Propag. Lett.* 8: 565–568.

Appendix G

Basic Reciprocity Result

Christophe Craeye[1], *Jean Cavillot*[1] *and Eloy de Lera Acedo*[2]

[1] *Louvain School of Engineering, Institute of Information and Communication Technologies, Electronics and Applied Mathematics (ICTEAM), Université Catholique de Louvain, Ottignies-Louvain-la-Neuve, Belgium*
[2] *Cavendish Astrophysics, Cavendish Laboratory, Physics Department, University of Cambridge, Cambridge, UK*

Suppose $\boldsymbol{F}(\Omega)$ is the embedded element pattern for element i, with a unit voltage generator and a series impedance equal to the input impedance of the receiver, while the other elements are passively terminated. When a plane wave with amplitude E_o and polarization $\hat{e}_o$ is received, then the current across the load is

$$I_r = E_o \frac{4\pi}{jk\eta_o} \boldsymbol{F}(\Omega_i) . \hat{e}_o \tag{G.1}$$

where k is the free-space wavenumber and η_o is the free-space wave impedance.

Appendix H

On the Extended Admittance Matrix

Christophe Craeye and Ha Bui Van

Louvain School of Engineering, Institute of Information and Communication Technologies, Electronics and Applied Mathematics (ICTEAM), Université Catholique de Louvain, 1348 Louvain-la-Neuve, Belgium

The aim is to provide details of the format of a matrix equation given in Chapter 13. Let us denote by α_{jq} the coefficients that multiply mode q on antenna j, while element i_0 is excited. In this case, the method of moments system of equations (see Chapter 4) reads

$$\sum_j \sum_q \alpha_{jq} R_{(i,p),(j,q)} = -b_{i,p} \tag{H.1}$$

where the double indexing (i, p) denotes mode p on antenna i and $b_{i,p}$ is the excitation field tested by a mode (i, p).

In case antenna i_o only is excited with a voltage V, $b_{i_o,p} = V$, for all p, given the normalization of all modal currents at the level of the port. Still, in that case, the α coefficients on antenna j are given by $\alpha_j = \boldsymbol{Y}^r_{j,i_o} u\, V$, where $\boldsymbol{Y}^r = \boldsymbol{R}^{-1}$ and u is a column vector with as many unit entries as there are modes defined on antenna i_o. This gives the modal currents on antenna j when a voltage V is impressed on antenna i_0; given the presence of vector u, it amounts to summing up the entries of each row of $\boldsymbol{Y}^r_{j,i_o}$. Then, obtaining the total port current on antenna j when a unit voltage is impressed on antenna i_0 amounts to summing up the entries of α_j, since the modes have a unit port current. In turn, this means summing up all the entries of Y^r_{i,i_o} provides entry (i, j_0) of the N-port array admittance matrix $\boldsymbol{Y}$.

When the patterns associated to each mode are defined, the embedded patterns are obtained by mere superimposition of the modal patterns, through the coefficients α_{jq}, stored in matrix $\boldsymbol{Y}^e$. Indeed, for a given direction of observation, each modal pattern simply needs to be multiplied by the coefficient that multiplies that mode function and the results for all modes are simply summed (see Sec. 2.5).

Mutual Coupling Between Antennas, First Edition. Edited by Trevor S. Bird.

Index

Mutual Coupling Between Antennas, First Edition. Edited by Trevor S. Bird.

b

f

www.ingramcontent.com/pod-product-compliance
Lightning Source LLC
LaVergne TN
LVHW062107200926
848484LV00007B/1615